Hormones in Neuroscience

Hormones in Neuroscience

The Pancreas and Pancreatic Hormones

Edited by

Victor R. Preedy
Department of Clinical Biochemistry, King's College Hospital, London, United Kingdom
Faculty of Life Sciences and Medicine, King's College London, London, United Kingdom
University of Suffolk, Ipswich, United Kingdom

Colin R. Martin
Institute for Health and Wellbeing, University of Suffolk, Ipswich, United Kingdom

Rajkumar Rajendram
Medical Protocol Department, King Abdulaziz Medical City, Ministry of National Guard Heath Affairs, Riyadh, Saudi Arabia

Vinood B. Patel
School of Life Sciences, University of Westminster, London, United Kingdom

ELSEVIER

ACADEMIC PRESS
An imprint of Elsevier

Academic Press is an imprint of Elsevier
125 London Wall, London EC2Y 5AS, United Kingdom
50 Hampshire Street, 5th Floor, Cambridge, MA 02139, United States

ISBN: 978-0-443-30263-3

For information on all Academic Press publications visit our website at https://www.elsevier.com/books-and-journals

Publisher: Megan Ball
Acquisitions Editor: Megan McManus
Editorial Project Manager: Jahnuvi Chauhan
Production Project Manager: Gomathi Sugumar
Cover Designer: Mark Rogers

Typeset by TNQ Tech

Contents

11. Peripheral nerves in diabetic neuropathy: Features and methods

Rajeswari Aghoram, Saranya B. Gomathy and Stalin Viswanathan

12. Linking brain insulin receptors, neural stem cells and synaptic plasticity: Potential implications for neurodegenerative disorders

Sneha Raut, Jaldeep Langhnoja and Prakash Pillai

Section 3
Psychology and behavior as a consequence of changes in the pancreas or the pancreatic endocrine status or the effects of hormones

13. Cognitive links with brain insulin signaling in metabolism: A new narrative

Habib Yaribeygi and Behina Forouzanmehr

Section 4
Treatment regimens modulating pancreatic hormones status and impact on neurological tissues and behavior

Section 5
Nonpancreatic treatment regimens and effects on neuroscientific domains

Section 6
Diseases of the pancreas or pancreas hormones status and impact on the brain and behavior

31. Pancreatic cancer—An overview

Stanisław Surma, Dawid Szkudłapski and Kazimierz Ciechanowski

32. Acute pancreatitis and its impact on mental health

Mikaela Hoard and Zachary L. Smith

Section 7
Diet and nutrition and impact on the pancreas and neuroscientific domains or behavior

33. Diet, diabetes and gene expression in the brain

Jung Han Kim and Lawrence Grover

34. Diet, behavior and diabetes—an overview

Stanisław Surma, Karol Momot, Artur Mamcarz, Kazimierz Ciechanowski and Marcin Barylski

35. Brain insulin signaling and nutritional modulation: the role of microRNAs and the gut–brain axis

Javier Mateu-Fabregat, Nil Novau-Ferré, Laura Panisello-Melo, Jaume Folch and Mònica Bulló

36. Vitamins, analog and effects on insulin signaling in the brain

Camila A.E.F. Cardinali, Yandara A. Martins and Andrea S. Torrão

Disclaimer

Every reasonable effort has been made to ensure that the data and information published are reliable. However, the publisher, editors, and authors cannot take responsibility for all of the materials' validity or any sequelae from their use. Neither the publishers nor the editors endorse any of the substances, treatments, interventions, protocols, or centers mentioned in the chapter contributions.

Contributors

Hana Abbasian, Harvard Medical School, Boston, MA, United States

Amir Abdi, Student Research Committee, School of Medicine, Tehran Medical Sciences, Islamic Azad University, Tehran, Iran

Fatemeh Abediankenari, Immunogenetics Research Center, Mazandaran University of Medical Sciences, Sari, Iran

Rajeswari Aghoram, Department of Neurology, Jawaharlal Institute of Postgraduate Medical Education and Research, Puducherry, India

Lamiaa A. Ahmed, Pharmacology and Toxicology Department, Faculty of Pharmacy, Cairo University, Cairo, Egypt

Kenni Wojujutari Ajele, Faculty of Humanities, North-West University, Mafikeng, South Africa

Khaled F. Al-Massri, Pharmacy and Biotechnology Department, Faculty of Medicine and Health Sciences, University of Palestine, Gaza, Palestine

Alaleh Alizadeh, Faculty of Medicine, Mashhad Branch, Islamic Azad University, Mashhad, Iran

Shanvanth Arnipalli, Department of Human Sciences, College of Human Education and Ecology, The Ohio State University, Columbus, OH, United States

Aryadi Arsyad, Department of Physiology, Faculty of Medicine, Hasanuddin University, Makassar, Indonesia

Marcin Barylski, Department of Internal Medicine and Cardiac Rehabilitation, Medical University of Lodz, Łódź, Poland

Faizan Bashir, School of Medicine, Shiraz University of Medical Sciences, Shiraz, Iran

Mònica Bulló, Department of Biochemistry and Biotechnology, School of Medicine and Life Sciences, University Rovira I Virgili, Reus, Spain

De las Cuevas C., Department of Internal Medicine, Dermatology, and Psychiatry and Instituto Universitario de Neurociencia (IUNE), University of La Laguna, San Cristobal de La Laguna, Spain

Camila A.E.F. Cardinali, Departamento de Fisiologia e Biofísica, Universidade de Sao Paulo, Sao Paulo, Brazil

Ivonne Carosi Arcangeli, Department of Humanities, Philosophy and Education, University of Salerno, Fisciano, Italy

Giovanna Celia, eCampus University, Novedrate, Italy

Xinyue Chang, Eli Lilly and Company, Indianapolis, IN, United States

Kazimierz Ciechanowski, Department of Nephrology, Transplantology and Internal Medicine, Pomeranian Medical University, Szczecin, Poland

Mauro Cozzolino, Department of Humanities, Philosophy and Education, University of Salerno, Fisciano, Italy

Niloofar Deravi, Student Research Committee, School of Medicine, Shahid Beheshti University of Medical Sciences, Tehran, Iran

Tanvi Desai, Symbiosis School of Biological Sciences, Symbiosis International (Deemed University), Pune, Maharashtra, India

Florin Despa, Department of Pharmacology and Nutritional Sciences, College of Medicine, The University of Kentucky, Lexington, KY, United States

Francesco Di Prinzio, Department of Humanities, Philosophy and Education, University of Salerno, Fisciano, Italy

Tanu Dixit, Symbiosis School of Biological Sciences, Faculty of Medical and Health Sciences, Symbiosis International (Deemed University), Pune, Maharashtra, India

Yulia Yusrini Djabir, Laboratory of Clinical Pharmacy, Faculty of Pharmacy, Hasanuddin University, Makassar, Indonesia

Mudra Dwivedi, School of Pharmacy & Technology Management, SVKM'S NMIMS Deemed-to-be University, Shirpur, Maharashtra, India

Soha Farooq, Department of Pharmacy, Shifa Tameer e Millat University, Islamabad, Pakistan

Gerardo J. Félix-Martínez, Department of Electrical Engineering, Universidad Autónoma Metropolitana, México City, México

Jaume Folch, Department of Biochemistry and Biotechnology, School of Medicine and Life Sciences, University Rovira I Virgili, Reus, Spain

Behina Forouzanmehr, Functional Neurosurgery Research Center, Research Institute of Functional Neurosurgery, Shohada Tajrish Comprehensive Neurosurgical Center of Excellence, Shahid Beheshti University of Medical Sciences, Tehran, Iran

Sana Esamaeilzadeh Gharajeh, Faculty of Veterinary Medicine, Islamic Azad University, Tabriz Branch, Tabriz, Iran

Hossein Gharedaghi, School of Medicine, Zanjan University of Medical Science, Zanjan, Iran

Chandraiah Godugu, Department of Biological Sciences (Regulatory Toxicology), National Institute of Pharmaceutical Education & Research (NIPER), Hyderabad, Telangana, India

Saranya B. Gomathy, Department of Neurology, Jawaharlal Institute of Postgraduate Medical Education and Research, Puducherry, India

Lawrence Grover, Department of Biomedical Sciences, Joan C. Edwards School of Medicine, Marshall University, Huntington, WV, United States

Girdhari Lal Gupta, Department of Pharmacology, School of Pharmacy & Technology Management, SVKM'S NMIMS Deemed-to-be University, Shirpur, Maharashtra, India

Shubhangi Gupta, Department of Psychiatry, Seth VC Gandhi & MA Vora Municipal General Hospital, Mumbai, Maharashtra, India

Mikaela Hoard, West Virginia School of Osteopathic Medicine, Lewisburg, WV, United States

Omkar Hulawale, School of Pharmacy & Technology Management, SVKM'S NMIMS Deemed-to-be University, Shirpur, Maharashtra, India

Erhabor Sunday Idemudia, Faculty of Humanities, North-West University, Mafikeng, South Africa

Rosaria Ingrassia, Section of Biotechnologies, Department of Molecular and Translational Medicine, University of Brescia, Brescia, Italy

Angela Marcela Jaramillo-Ospina, Department of Psychiatry, Faculty of Medicine and Health Sciences, McGill University, Ludmer Centre for Neuroinformatics and Mental Health, Douglas Hospital Research Centre, Montreal, QC, Canada

Koji Kamagata, Department of Radiology, Juntendo University Graduate School of Medicine, Tokyo, Japan

Mahsa Kamali, Universal Scientific Education and Research Network (USERN), Tehran, Iran

Smruti Karale, Department of Medicine, GMC Kolhapur, Kolhapur, Maharashtra, India

Aaliyah Khan, Department of Biotechnology and Food Science, Faculty of Applied Sciences, Durban University of Technology, Durban, South Africa

Mohammad Ebrahim Kherad, School of Medicine, Islamic Azad University, Sari, Iran

Amirhossein Khoshnam, Student Research Committee, School of Medicine, Zahedan University of Medical Sciences, Zahedan, Iran

Jung Han Kim, Department of Biomedical Sciences, Joan C. Edwards School of Medicine, Marshall University, Huntington, WV, United States

Takafumi Kitagawa, Department of Radiology, Juntendo University Graduate School of Medicine, Tokyo, Japan

Rupali Kohal, Department of Pharmaceutical Chemistry, ISF College of Pharmacy, Moga, Punjab, India

Zeinab Kolahdouzan, School of Medicine, Kashan University of Medical Sciences, Isfahan, Iran

Navneesh Kumar, Department of Pharmaceutical Chemistry, ISF College of Pharmacy, Moga, Punjab, India

Preety Kumari, Department of Pharmaceutical Chemistry, ISF College of Pharmacy, Moga, Punjab, India

Riya Kumari, Symbiosis School of Biological Sciences, Faculty of Medical and Health Sciences, Symbiosis International (Deemed University), Pune, Maharashtra, India

Jaldeep Langhnoja, Department of Radiation Oncology, University of California, Irvine, Medical Sciences I, Irvine, CA, United States

Natrus Larysa, Department of Modern Medical Diagnostic, and Treatment Technology, Bogomolets National Medical University, Kyiv, Ukraine

Halimat Yusuf Lukman, Department of Biotechnology and Food Science, Faculty of Applied Sciences, Durban University of Technology, Durban, South Africa

Simone Macrì, Centre for Behavioural Sciences and Mental Health, Istituto Superiore di Sanità, Rome, Italy

Thobeka Madide, Department of Human Biology and Integrated Pathology, Medical School, Nelson Mandela University, Gqeberha, South Africa; Division of Pharmacology, Faculty of Pharmacy, Rhodes University, Makhanda, South Africa

Artur Mamcarz, 3rd Department of Internal Medicine and Cardiology, Medical University of Warsaw, Warsaw, Poland

Yandara A. Martins, Departamento de Fisiologia e Biofísica, Universidade de Sao Paulo, Sao Paulo, Brazil

Javier Mateu-Fabregat, Department of Biochemistry and Biotechnology, School of Medicine and Life Sciences, University Rovira I Virgili, Reus, Spain

Koyo Mizuta, Department of Radiology, Juntendo University Graduate School of Medicine, Tokyo, Japan

Karol Momot, Department of Experimental and Clinical Physiology, Medical University of Warsaw, Warsaw, Poland; 3rd Department of Internal Medicine and Cardiology, Medical University of Warsaw, Warsaw, Poland

Naoya Murao, Departments of Endocrinology, Diabetes and Metabolism, School of Medicine, Fujita Health University, Toyoake, Aichi, Japan; Institute for Diabetes and Organoid Technology, Helmholtz Diabetes Center, Helmholtz Zentrum München, Deutsches Forschungszentrum für Gesundheit und Umwelt (GmbH), Neuherberg, Germany

Muhammad Nafees, Department of Pharmacy, University of Faisalabad, Faisalabad, Pakistan; Jinnah Institute of Medical Sciences (JIMS), Faisalabad, Pakistan

Ehsan Noroozi, Department of Internal Medicine, Shahid Labbafi Nezhad Hospital, Shahid Beheshti University of Medical Science, Tehran, Iran

Nil Novau-Ferré, Department of Biochemistry and Biotechnology, School of Medicine and Life Sciences, University Rovira I Virgili, Reus, Spain

Laura Panisello-Melo, Department of Biochemistry and Biotechnology, School of Medicine and Life Sciences, University Rovira I Virgili, Reus, Spain

Vinood B. Patel, School of Life Sciences, University of Westminster, London, United Kingdom

Christina Peter, Department of Biotechnology and Food Science, Faculty of Applied Sciences, Durban University of Technology, Durban, South Africa

Prakash Pillai, Division of Neurobiology, Department of Zoology, Faculty of Science, The Maharaja Sayajirao University of Baroda, Vadodara, Gujarat, India

Edoardo Pisa, Centre for Behavioural Sciences and Mental Health, Istituto Superiore di Sanità, Rome, Italy

Binod Pokharel, Department of Human Sciences, College of Human Education and Ecology, The Ohio State University, Columbus, OH, United States

Samata Pradhan, Department of Biological Sciences (Regulatory Toxicology), National Institute of Pharmaceutical Education & Research (NIPER), Hyderabad, Telangana, India

Victor R. Preedy, Department of Clinical Biochemistry, King's College Hospital, London, United Kingdom; Faculty of Life Sciences and Medicine, King's College London, London, United Kingdom; University of Suffolk, Ipswich, United Kingdom

Fatemeh Naseri Rad, Student Research Committee, Faculty of Medicine, Iran University of Medical Sciences, Tehran, Iran

Mohammad Rahmanian, Gastroenterology and Liver Diseases Research Center, Research Institute for Gastroenterology and Liver Diseases, Shahid Beheshti University of Medical Sciences, Tehran, Iran; Student Research Committee, School of Medicine, Shahid Beheshti University of Medical Sciences, Tehran, Iran

Rajkumar Rajendram, Research Unit, Department of Medical Education, College of Medicine, King Saud bin Abdulaziz University for Health Sciences, Riyadh, Saudi Arabia; Medical Protocol Department, King Abdulaziz Medical City, Ministry of National Guard Health Affairs, Riyadh, Saudi Arabia; Research Offices, King Abdullah International Medical Research Center, Riyadh, Saudi Arabia

Anuradha Rathod, Department of Psychiatry, Government Medical College, Thane, Maharashtra, India

Sneha Raut, Division of Neurobiology, Department of Zoology, Faculty of Science, The Maharaja Sayajirao University of Baroda, Vadodara, Gujarat, India

Selvan Ravindran, Symbiosis School of Biological Sciences, Faculty of Medical and Health Sciences, Symbiosis International (Deemed University), Pune, Maharashtra, India

Saheed Sabiu, Department of Biotechnology and Food Science, Faculty of Applied Sciences, Durban University of Technology, Durban, South Africa

Sayedeh-Fatemeh Sadat-Madani, School of Medicine, Isfahan University of Medical Sciences, Isfahan, Iran

Sweta Sahu, Department of Medicine, JJM Medical College, Davanagere, Karnataka, India

Mansi Sawardekar, Symbiosis School of Biological Sciences, Faculty of Medical and Health Sciences, Symbiosis International (Deemed University), Pune, Maharashtra, India

Yusuke Seino, Departments of Endocrinology, Diabetes and Metabolism, School of Medicine, Fujita Health University, Toyoake, Aichi, Japan

Patrick R. Sellers, Department of Human Sciences, College of Human Education and Ecology, The Ohio State University, Columbus, OH, United States

Ntethelelo Sibiya, Division of Pharmacology, Faculty of Pharmacy, Rhodes University, Makhanda, South Africa

Patricia Pelufo Silveira, Department of Psychiatry, Faculty of Medicine and Health Sciences, McGill University, Ludmer Centre for Neuroinformatics and Mental Health, Douglas Hospital Research Centre, Montreal, QC, Canada

Stefania Sinesi, Department of Humanities, Philosophy and Education, University of Salerno, Fisciano, Italy

Zachary L. Smith, Division of Gastroenterology and Hepatology, Wisconsin Institute for Research in Endoscopy, Medical College of Wisconsin, Milwaukee, WI, United States

Stanisław Surma, Department of Internal Medicine and Clinical Pharmacology, Medical University of Silesia, Katowice, Poland

Atsushi Suzuki, Departments of Endocrinology, Diabetes and Metabolism, School of Medicine, Fujita Health University, Toyoake, Aichi, Japan

Dawid Szkudłapski, Collegium Medicum, WSB University, Dąbrowa Górnicza, Poland

Kaito Takabayashi, Department of Radiology, Juntendo University Graduate School of Medicine, Tokyo, Japan

Andrea S. Torrão, Departamento de Fisiologia e Biofísica, Universidade de Sao Paulo, Sao Paulo, Brazil

Rukeye Tuerxun, Department of Radiology, Juntendo University Graduate School of Medicine, Tokyo, Japan

Sreelakshmi Vaidyanathan, Department of Psychiatry, All India Institute of Medical Sciences, Nagpur, Maharashtra, India

Sant Kumar Verma, Department of Pharmaceutical Chemistry, ISF College of Pharmacy, Moga, Punjab, India

Soumya Verma, Symbiosis School of Biological Sciences, Faculty of Medical and Health Sciences, Symbiosis International (Deemed University), Pune, Maharashtra, India

Stalin Viswanathan, Department of General Medicine, Jawaharlal Institute of Postgraduate Medical Education and Research, Puducherry, India

Jie Xue, Eli Lilly and Company, Indianapolis, IN, United States

Yuichiro Yamada, Yutaka Seino Distinguished Center for Diabetes Research, Kansai Electric Power Medical Research Institute, Kyoto, Japan

Habib Yaribeygi, Research Center of Physiology, Research Institute of Neurosciences, Semnan University of Medical Sciences, Semnan, Iran

Salma Younas, Department of Pharmacy, University of the Punjab, Lahore, Pakistan

Klys Yuliia, Research Institute of Experimental and Clinical Medicine, Bogomolets National Medical University, Kyiv, Ukraine

Osadchuk Yuliia, University Clinic, Bogomolets National Medical University, Kyiv, Ukraine

Amiryousef Zahedi, Department of Internal Medicine, Caucasus Medical Center (Vian Health Holding), Tbilisi, Georgia

Ouliana Ziouzenkova, Department of Human Sciences, College of Human Education and Ecology, The Ohio State University, Columbus, OH, United States

Rui Zou, Department of Radiology, Juntendo University Graduate School of Medicine, Tokyo, Japan

Preface

The neuroendocrine system is a complex biological network of pathways that maintains the homeostatic milieu of neurological tissues. The importance of this system is highlighted by the numerous regulatory processes that modulate these pathways. These include the maintenance of circulating hormone levels, receptor concentrations, post-receptor responses, signaling pathways, and tissue effects. Of course, these simple abstractions belie the many layers of interactions between genes, epigenetic mechanisms, transcription, translation, hormonal turnover, feedback loops, and many other processes, which are interlinked with other hormones. Yet diseases and surgical interventions may induce changes in neuroendocrine profiles that are adaptive and beneficial or maladaptive and harmful, causing further tissue and organ injury. Moreover, there are also connections between other modulators such as development, aging, medications, diet, and nutrition. For example, dietary iodine is a component of thyroid hormones, and iodine deficiency frequently occurs. Maternal iodine deficiency can cause neurological impairment in offspring. Indeed, it has been argued that iodine deficiency is the biggest cause of mental retardation in the world. Some endocrine abnormalities also induce psychological disorders such as anxiety and depression and modulate the expression of emotions and behavior or affect quality-of-life measures. One also needs to consider that some adaptive changes to endocrine status are beneficial. For example, this occurs in trauma and critical care, where the mobilization of metabolic fuels is driven by endocrine stress responses. Thus, it is somewhat erroneous to consider the "neuroendocrine system" as a single pathway between hormones and neurological tissues.

To understand the links between hormones and neurological tissues, all facets of homeostatic regulation must be considered. However, much of the available text that links hormones and neurological tissues is designed for the specialist or does not address the translational context. Moreover, much of the literature fails to address the behavioral consequences of altered endocrine gland pathology or endocrine status. These deficiencies are addressed in the series *Hormones in Neuroscience*. While it embraces hormones and neurological tissues at its core, it also includes causative and modulating influences as well as consequential effects.

Hormones in Neuroscience adopts an illustrative ethos and has the following unique subsections:

- **Abstract** (published online)
- **Other Hormones, Endocrine Tissues, and Pathways**
- **Applications of the Material**
- **Mini-Dictionary of Terms**
- **Summary Points**

Other Hormones, Endocrine Tissues, and Pathways ensure there is holistic coverage within the chapter, while the subsection **Applications of the Material** ensures there is translational and cross-disciplinary information within the chapter. **The Mini-Dictionary of Terms** is particularly useful for those who have expertise in other disciplines or who are novices to the area, and the **Summary Points** encapsulate the entire chapter in succinct sentences, which are also useful for the novice.

In this volume, ***The Pancreas and Pancreatic Hormones***, we have the following sections:

Section 1. Introductory Chapters
Section 2. Focused Areas of Neurological Pathology and Physiology
Section 3. Psychology and Behavior as a Consequence of Changes in Endocrine Status
Section 4. Treatment Regimens Modulating Pancreatic Hormones Status and Impact on Neurological Tissues and Behavior
Section 5. Non-Pancreatic Treatment Regimens and Effects on Neuroscientific Domains
Section 6. Diseases of the Pancreas or Pancreas Hormones Status and Impact on Brain and Behavior
Section 7. Diet and Nutrition and Impact on the Pancreas and Neuroscientific Domains or Behavior
Section 8. Resources

Hormones in Neuroscience: The Pancreas and Pancreatic Hormones is suitable for undergraduates entering university and beyond. Because of the translational and transdisciplinary nature of the book series, it will also be suitable for physicians, academics at the PhD level, researchers, university teachers, lecturers, and professors. Endocrinologists, pathologists, pharmacologists, health care professionals, neuroscientists, research scientists, behavioral scientists, psychologists, and psychiatrists will benefit from this book. Clinical professionals working at the intersection of neurology and endocrinology will also find this useful, as will early-career researchers, including postgraduate research students (e.g., master's by research). Industrial heads or those leading biodiscovery groups in the pharmaceutical sector will also find the books very useful.

The Editors

Professor Victor R. Preedy,
Professor Colin R. Martin,
Dr. Rajkumar Rajendram,
Professor Vinood B. Patel

Section 1

Introductory chapters

Chapter 1

Pancreatic hormones: a focus on amylin—a new narrative

Florin Despa

Department of Pharmacology and Nutritional Sciences, College of Medicine, The University of Kentucky, Lexington, KY, United States

Abbreviations

AD Alzheimer's disease
APP amyloid precursor protein
Aβ amyloid β
CAA cerebral amyloid angiopathy
CaMKII calcium/calmodulin-dependent protein kinase II
CDK5 cyclin-dependent kinase 5
CGRP G protein-coupled receptor
CNS central nervous system
CSF cerebral spinal fluid
ECT electron chain transport
ER endoplasmic reticulum
F6P fructose-6-phosphate
G6P glucose-6-phosphate
GSK3β glycogen synthase kinase-3β
HDAC histone deacetylase
HIF-2α hypoxia-inducible factor 2α
IAPP islet amyloid polypeptide
IL-1β interleukin-1β
IL-6 interleukin-6
NO nitric oxide
PKA protein kinase A
PPARα peroxisome proliferator-activated receptor α
PS1 presenilin 1
RAMP receptor activity-modifying protein
RBC red blood cell
SVCD small vessel cerebrovascular disease
TNF-α tumor necrosis factor α

Introduction

Amylin, also known as islet amyloid polypeptide (IAPP), is a 37 amino acid hormone that is synthesized and co-secreted with insulin by the pancreatic β-cells (Kahn et al., 1990). Its physiological role includes central control of satiety and modulation of energy homeostasis (Hay et al., 2015). The ability of amylin to suppress appetite and slow gastric emptying has led to the development of amylin mimetics (e.g., pramlintide) as novel treatments for overweight, obesity and diabetes (Mullard, 2025). Previous review articles have comprehensively summarized amylin receptor pharmacology and therapeutic potential, detailing receptor biology, the outcomes of receptor modulation, and preclinical as well as clinical evidence, particularly in the context of obesity treatment (Hay et al., 2015; Hay and Pioszak, 2016; Lutz, 2012). However, a balanced perspective must also consider emerging evidence that prediabetes-associated amylin hypersecretion can initiate pathology affecting cardiovascular system (Despa et al., 2012, 2014; Erickson et al., 2013; Ilaiwy et al., 2016; Lambert et al., 2015; Liu et al., 2016, 2018; Popescu et al., 2019; Velmurugan et al., 2023; Verma et al., 2020) and central nervous system (CNS) (Fawver et al., 2014; Jackson et al., 2013; Kotiya et al., 2023; Ly et al., 2017; Ly, Guerreiro et al., 2021; Martinez-Valbuena et al., 2019; Oskarsson et al., 2015; Roostaei, Nazeri et al., 2017; Royall et al., 2020; Schultz et al., 2017, 2018; Verma et al., 2016, 2023; Zhu et al., 2019). This chapter focuses on the pathological aspects of amylin biology to provide an objective framework for evaluating its therapeutic potential, while accounting for the off-target risks associated with elevated amylin secretion in prediabetes.

Amylin dysregulation: pancreatic origins, systemic propagation and organ-specific pathology

Amylin secretion increases in parallel with insulin during the prediabetic stage (Hoppener et al., 2000; Kahn et al., 1990; Westermark et al., 2011). Chronic hypersecretion of amylin is a key driver of islet amylin amyloid deposition (Bhowmick et al., 2022; Westermark et al., 2011; Yong et al., 2021). The resulting amylin aggregation impairs β-cell function, diminishes insulin secretory capacity and

Hormones in Neuroscience. https://doi.org/10.1016/B978-0-443-30263-3.00007-5

accelerates metabolic decline, contributing to pathogenesis of type 2 diabetes (Bhowmick et al., 2022; Westermark et al., 2011; Yong et al., 2021). Individuals with type 1 diabetes exhibit markedly reduced blood amylin levels (Tomita, 2011), reflecting the loss of β-cell mass. However, some newly diagnosed young patients with type 1 diabetes display elevated plasma amylin (Altamirano-Bustamante et al., 2020; Paulsson et al., 2014), suggesting that amylin dysregulation may occur in early type 1 diabetes. Supporting this notion, early islet amylin amyloid deposits have been observed in type 1 diabetes (Beery et al., 2019; Westermark et al., 2017), where amyloid formation may accelerate β-cell loss and contribute to autoimmune progression.

Prediabetic hypersecretion of amylin is likely the cause of accumulation of aggregated amylin in extra-pancreatic tissues. Fig. 1.1 shows examples of aggregated amylin deposits identified in diabetic human pancreatic islets (a), circulating red blood cells (RBCs) (b) and monocytes (c) (Verma et al., 2020, 2023), in the microvasculature (e-g) (Jackson et al., 2013; Ly et al., 2017; Ly, Guerreiro et al., 2021; Oskarsson et al., 2015; Schultz et al., 2017; Verma et al., 2023), and intracellularly in failing hearts (d) (Despa et al., 2014; Liu et al., 2016) and Alzheimer's disease (AD)

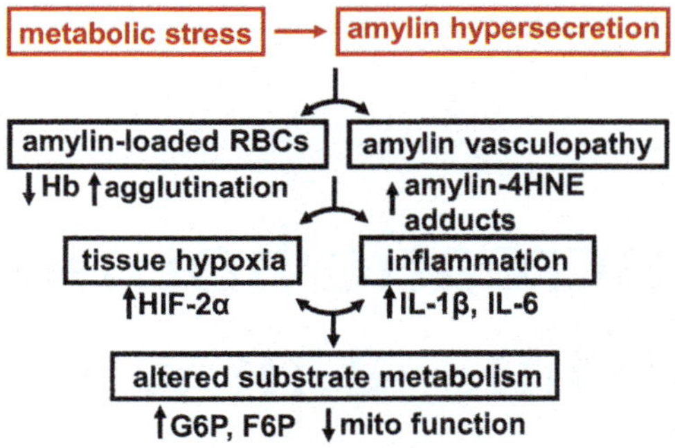

FIGURE 1.2 This figure summarizes pathological effects of amylin hypersecretion identified in "amylin humanized" mice/rats. Metabolic stressed-induced amylin hypersecretion triggers a feed-forward cycle of membrane damage and aggregation of amylin in the microvasculature and circulating RBCs leading to ischemic-hypoxic injury, inflammation and altered substrate metabolism.

brains (h) (Ly, Guerreiro et al., 2021; Schultz et al., 2017; Verma et al., 2016). Fig. 1.2 is a schematic summary of These findings suggest that chronic pancreatic secretion of amyloid-forming amylin during prediabetes promotes pancreatic amyloid pathology that spreads to microcirculation and extra-pancreatic tissues (see schematic summary of these findings in Fig. 1.1i). Because human amylin seeds

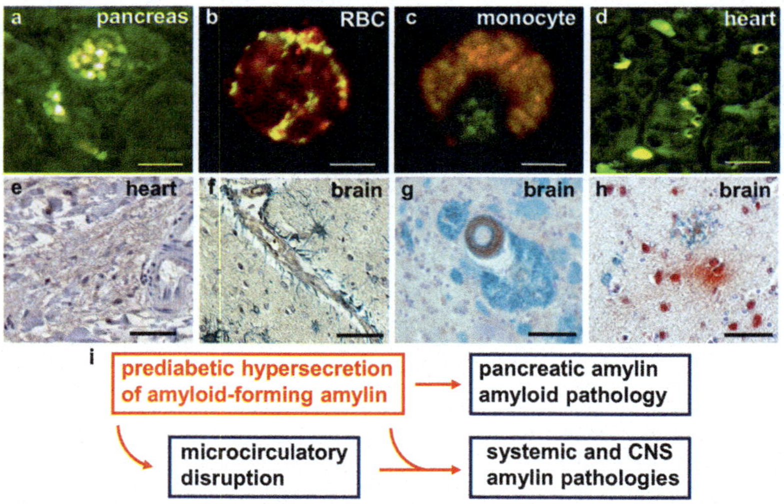

FIGURE 1.1 This figure shows examples of aggregated amylin deposits identified by immunochemical methods in diabetic human pancreatic islets (A), circulating RBCs (B) and monocytes (C), cardiac tissue (D and E) and brain (F–H). *From (A) and (D) published previously in Liu, M., Verma, N., Peng, X., et al. (2016). Hyperamylinemia increases IL-1β synthesis in the heart via peroxidative sarcolemmal injury.* Diabetes, *65(9), 2772–2783. https://doi.org/10.2337/db16-0044. Copyright 2011 by the American Diabetes Association, (B) reprinted from Verma, N., Liu, M., Ly, H., et al. (2020). Diabetic microcirculatory disturbances and pathologic erythropoiesis are provoked by deposition of amyloid-forming amylin in red blood cells and capillaries.* Kidney International, *97(1), 143–155. https://doi.org/10.1016/j.kint.2019.07.028, Copyright (2020), with permission from Elsevier, (C) reprinted from Verma, N., Velmurugan, G.V., Winford, E., et al. (2023). Aβ efflux impairment and inflammation linked to cerebrovascular accumulation of amyloid-forming amylin secreted from pancreas.* Communications Biology, *6(1), 2. Published 2023 Jan 3. https://doi.org/10.1038/s42003-022-04398-2, (open access) with permission from Springer Nature, (E) published previously in Despa, S., Margulies, K.B., Chen, L., et al. (2012). Hyperamylinemia contributes to cardiac dysfunction in obesity and diabetes: A study in humans and rats.* Circulation Research, *110(4), 598–608. https://doi.org/10.1161/CIRCRESAHA.111.258285. Copyright by the American Heart Association, (F) reprinted from Ly, H., Verma, N., Wu, F., et al. (2017). Brain microvascular injury and white matter disease provoked by diabetes-associated hyperamylinemia.* Annals of Neurology, *82(2), 208–222. https://doi.org/10.1002/ana.24992, with permission from WYLEY. (G) and (H) reprinted from Ly, H., Verma, N., Sharma, S., Kotiya, D., Despa, S., Abner, E.L., Nelson, P.T., Jicha, G.A., Wilcock, D.M., Goldstein, L.B., Guerreiro, R., Bras, J., Hanson, A.J., Craft, S., Murray, A.J., Biessels, G.J., Troakes, C., Zetterberg, H., Hardy, J., ... Despa, F. (2021b). The association of circulating amylin with beta-amyloid in familial Alzheimer's disease.* Alzheimer's & Dementia, *7, e12130. (I) Pathological processes identified by analyses of human tissues which suggest that chronic pancreatic secretion of amyloid-forming amylin during prediabetes promotes pancreatic amyloid pathology that spreads to microcirculation and extra-pancreatic tissues.*

amyloid (Westermark et al., 2011), peripheral amylin deposition may expand even as the circulating levels of amyloid-forming amylin declines with β-cell depletion, in persons with full blown type 2 diabetes.

Clinical and molecular correlates of dysregulated amylin and heart failure

Immunochemical analyses of failing hearts from patients with obesity or/and type 2 diabetes showed abundant amylin oligomers and amylin amyloid deposition (Despa et al., 2012, 2014; Liu et al., 2016). Amylin deposits were absent in normal hearts and in failing hearts from lean, non-diabetic individuals (Despa et al., 2012, 2014; Liu et al., 2016). In non-failing hearts of overweight and obese patients, lower molecular weight amylin aggregates were already elevated (Despa et al., 2012; Liu et al., 2016), indicating that cardiac amylin deposition begins early in prediabetes, before the onset of overt diabetes or cardiac failure. The distribution and morphology of amylin aggregates resembled the infiltrative pattern seen in amyloid-related cardiomyopathies (Liao et al., 2001; Palladini et al., 2006). Cardiac amylin accumulation correlated with elevated lipid peroxidation products such as 4-hydroxynonenal (4-HNE) (Liu et al., 2016), a potent apoptotic inducer due to its abilities to form protein adducts and to propagate oxidative stress. Amylin deposits often co-localize with activated macrophages and inflammatory markers interleukin (IL)-1β and IL-6 (Despa et al., 2012; Liu et al., 2016, 2018) resembling atherosclerotic lesions. These results demonstrate that amylin aggregation occurs in the human heart in association with obesity and type 2 diabetes, begins before overt cardiac failure and correlates with pathological remodeling, oxidative stress, and inflammation, supporting a causative role of dysregulated amylin in diabetic cardiomyopathy.

Clinical and molecular correlates of dysregulated amylin and Alzheimer's pathology

Several independent studies report: (1) elevated amylin levels in blood and cerebrospinal fluid (CSF) of individuals with AD-type cognitive impairment (Ly, Guerreiro et al., 2021; Royall et al., 2020; Verma et al., 2023; Zhu et al., 2019); (2) blood amylin as a mediator of age-related effects on dementia severity in large cohorts (Royall et al., 2020); and (3) mixed amylin-amyloid-β (Aβ) deposits in cerebral blood vessels and parenchyma, in both sporadic and familial AD pathologies (Jackson et al., 2013; Kotiya et al., 2023; Ly, Guerreiro et al., 2021; Verma et al., 2023). Other reports describe amylin-Aβ and amylin-tau plaques in both pancreas and brain (Martinez-Valbuena et al., 2019). A genome-wide association study in Alzheimer's Neuroimaging Initiative (ADNI) participants found amylin gene variants associated with CNS amyloidosis measured in CSF (Roostaei, Alzheimer'S Disease Neuroimaging Initiative et al., 2017). Analyses of autopsied brain tissues from the Religious Orders Study and Memory and Aging Project identified an amylin gene-AD pathology interaction associated with cognitive impairment (Roostaei, Alzheimer'S Disease Neuroimaging Initiative et al., 2017).

Immunohistochemical analyses showed that amyloid-forming amylin secreted from the pancreas can co-aggregate with vascular Aβ within arteriole smooth muscle cell layers (Oskarsson et al., 2015; Verma et al., 2023), consistent with cerebral amyloid angiopathy (CAA), and with parenchymal Aβ in mixed plaques and hetero-oligomers (Jackson et al., 2013; Kotiya et al., 2023; Ly, Guerreiro et al., 2021). Brain amylin deposition was confirmed by mass spectrometry (Verma et al., 2016). Triple immunostaining for amylin, Aβ, and α-smooth muscle actin, as well as proximity ligation assays, further demonstrated amylin-Aβ co-deposition in the vasculature (Ly, Guerreiro et al., 2021; Verma et al., 2023). Amylin deposition in brain microvasculature has been associated with tissue alterations, including enlarged perivascular spaces, irregular vascular wall contours, wall disruption, and astrogliosis (Ly et al., 2017; Oskarsson et al., 2015; Schultz et al., 2017; Verma et al., 2023), features consistent with small vessel cerebrovascular disease (SVCD).

Using species variation to define in vivo effects of dysregulated human amylin

Although rodent and human amylin are highly similar, human amylin is substantially more aggregation-prone because it lacks proline residues in the amino acid 20–29 region (Westermark et al., 1990). Therefore, common rodent models of diabetes are limited by the absence of amylin pathology (Butler et al., 2004). This provides a unique opportunity to mechanistically investigate the effects of dysregulated human amylin in vivo. Studies using rodents genetically engineered to secrete pancreatic human amylin at levels comparable to those in people with prediabetes/diabetes show that chronically elevated circulating human amylin exerts widespread in vivo effects, disrupting metabolism and function across multiple organs, including the pancreas, kidneys, heart and brain (Despa et al., 2012; Liu et al., 2016; Ly et al., 2017; Ly, Guerreiro et al., 2021; Srodulski et al., 2014; Verma et al., 2020).

Prediabetic amylin hypersecretion impairs microcirculatory and metabolic function

RBCs, which deliver oxygen to tissues as they pass through capillaries during circulation from arteries to veins, are negatively affected by elevated amylin secretion and diabetes pathogenesis (Verma et al., 2020). The oxygen-carrying capacity of RBCs and the structural integrity of capillaries decline in diabetes, inducing tissue hypoxia and organ dysfunction (Barrett et al., 2017; McMillan et al., 1978; Tonelli et al., 2008). Amylin-loaded RBCs isolated from "amylin-humanized" rats show reduced hemoglobin (Hb) content, increased adherence to vascular endothelial cells and agglutination (Verma et al., 2020). Intravenous infusion of amylin-loaded RBCs into healthy wild-type rats triggered proinflammatory immune responses, activated hypoxia-inducible factors (HIFs) in capillary endothelial cells and induced arginase-nitric oxide (NO) dysregulation (Verma et al., 2020). Inflammation and activated macrophages promote metabolic shifts and alters NO metabolism (El Kasmi and Stenmark, 2015; Koelwyn et al., 2018; Xiong et al., 1998). Thus resulting proinflammatory milieu may further disrupt O_2- and nutrient-sensing pathways within tissues. Therefore, amylin-loaded RBCs may directly impair blood flow, oxygen delivery and metabolic homeostasis. Along this line, non-targeted metabolomics in multiple tissues (brain, heart and liver) from "amylin humanized" rats revealed elevated levels of glycolytic metabolites such as glucose-6-phosphate (G6P) and fructose-6-phosphate in (F6P) (Ilaiwy et al., 2016). Such changes were absent in diabetic rats without amylin hypersecretion, suggesting that high G6P and F6P levels may be a specific signature of increased amylin availability in tissues. These results are consistent with data from skeletal muscle and some cancer cell lines and tumors demonstrating that amylin increases G6P, lowers insulin-stimulated glucose uptake and inhibits glycolysis (Castle, Kuo, Han, et al., 1998, Castle, Kuo, Ivy, 1998; Leighton and Cooper, 1988; Molina et al., 1990; Venkatanarayan et al., 2015; Zierath et al., 1992). Knockout of either component of the amylin receptors abolished the amylin-mediated effect on glycolysis (Venkatanarayan et al., 2015). Furthermore, increased amylin availability in tissues impaired mitochondrial function and induced oxidative stress, lower rates of ATP production and stabilization of hypoxia-inducible factor 2α (HIF-2α) (Lambert et al., 2015; Velmurugan et al., 2023, 2025; Verma et al., 2020, 2021). Activation of hypoxia signaling has a detrimental impact on mitochondrial respiratory capacity, altering electron transport chain (ETC) activity including impaired respiration supported by complex I, and decreased complex I activity and abundance (Cole et al., 2016). Moreover, hypoxia signaling decreases the tissue capacity for fatty acid oxidation via suppression of the fatty acid-activated transcription factor, peroxisome proliferator-activated receptor α (PPARα) and its downstream targets (Cole et al., 2016). In tissues from "amylin-humanized" rats, reduced glycolytic flux, altered mitochondrial metabolism and HIF-2α stabilization (Ilaiwy et al., 2016; Velmurugan et al., 2023; Verma et al., 2020) were accompanied by impaired protein synthesis (Ilaiwy et al., 2016). Thee results are consistent with brain tissue rarefaction reported in rats with chronic pancreatic hypersecretion of human amylin.

Circulating oligomeric amylin causes systemic inflammation and neuroinflammation

Biochemical analyses of heart and brain tissues from amylin-humanized rats revealed accumulation of large amylin oligomers (32–64 kDa), consistent with the cytotoxic species identified in human heart and brain tissues, and in vitro (Despa et al., 2012, 2014; Liu et al., 2016, 2018; Srodulski et al., 2014; Ly et al., 2017, Verma et al., 2016). Vascular accumulation aggregated amylin is associated with elevated oxidative stress markers, including 4-HNE and malondialdehyde (MDA), as well as increased IL-1β expression (amylin vasculopathy) (Verma et al., 2016; Liu et al., 2016). Formation of amylin-lipid aldehyde adducts was also detected (Verma et al., 2016; Liu et al., 2016), suggesting that oxidative modification of amylin may promote vascular cell injury through a feed-forward cycle of membrane damage and aggregation. This finding is noteworthy because clinical trials have shown that reducing inflammation within the NLRP3/IL-1β/IL-6 innate immunity pathway significantly lowers cardiovascular risk (Baylis et al., 2017; Ridker et al., 2017); however, the underlying mechanisms remain incompletely understood.

In the brain, amylin vasculopathy triggers neuroinflammation, evidenced by activation of microglia and macrophages, particularly around amylin-positive vascular lesions (Ly et al., 2017; Srodulski et al., 2014), upregulation of both M1 (pro-inflammatory) and M2 (alternative) microglial markers (Srodulski et al., 2014), and increased levels of pro-inflammatory cytokines (IL-1β TNF-α, IL-6) accompanied by decreased anti-inflammatory cytokines (IL-10) (Srodulski et al., 2014; Verma et al., 2023). This pattern suggests a vascular and parenchymal mechanism in which amylin infiltration along arterioles and capillaries initiates localized inflammation, potentially compromising neuronal and synaptic function. The inflammatory response mirrors the known pathogenic role of amylin in pancreatic β-cell injury (Masters et al., 2010), linking chronically elevated amylin secretion to systemic inflammation and neuroinflammation. Fig. 1.2 is a schematic summary of the mechanistic studies of amylin-related pathological processes using "amylin humanized" mice and rats.

Cerebral amylin vasculopathy impairs Aβ homeostasis

Cerebrovascular accumulation of amyloid-forming amylin secreted from the pancreas impairs brain Aβ clearance (Verma et al., 2023). Three interdependent mechanisms appear to underlie amylin-induced Aβ dysregulation

(Verma et al., 2023): 1, chronically elevated blood levels of amyloid-forming amylin promote its accumulation in circulating monocytes, reflecting systemic inflammation; 2, this systemic amylin deposition leads to cerebrovascular amylin accumulation; and 3, cerebrovascular amylin disrupts LRP1- and P-gp-mediated Aβ transport across the blood-brain barrier and impairs Aβ clearance via interstitial fluid drainage along vascular walls, as shown by amylin-Aβ co-localization in vessel walls and perivascular spaces. Dysregulation of the arginase-NO pathway within vessel walls further reduces cerebral vasodilation and decreases cerebral blood flow (Verma et al., 2023), impairing interstitial fluid drainage and Aβ clearance. Amylin deposition in brain capillaries alters the expression of P-gp and LRP1, which are critical for Aβ transport across the blood-brain barrier (Kang et al., 2000; Storck et al., 2016), suggesting that luminal amylin may downregulate LRP1 and exacerbate cerebrovascular dysfunction. Moreover, infusion of APOE4 protein into "amylin humanized" rats exacerbated amylin-mediated capillary injury and increased brain amylin deposition (Ly et al., 2017), highlighting a potential interaction between circulating amylin and plasma APOE in promoting vascular accumulation. Cerebrovascular amylin accumulation was linked to white matter injury and reduced brain volume (Ly et al., 2017), indicating that microvascular injury and hypoperfusion underlie these structural impairments. Infusion of aggregated human amylin induced microhemorrhages even in the absence of hyperglycemia (Ly et al., 2017), demonstrating that circulating amylin alone can provoke endothelial dysfunction. Furthermore, infusion of aggregated human amylin into amylin knockout rats caused cerebrovascular deposition associated with endothelial cell damage, tight junction degradation, and microhemorrhages with recruitment of perivascular inflammation cells (Ly et al., 2017). In amylin knockout rats, concentric luminal deposition demonstrated that circulating aggregated amylin can directly deposit within the brain microvasculature, narrowing lumens and contributing to hypoperfusion (Ly et al., 2017).

Crossing APP/PS1 rats with human amylin-expressing rats generated "amylin humanized" AD-model rats which show accelerated AD-like pathology (Ly, Guerreiro et al., 2021). In contrast, genetically suppressing amylin secretion protected against cerebral Aβ accumulation (Ly, Guerreiro et al., 2021). In "amylin humanized" APP/PS1 rats, the development of amylin vasculopathy correlated with disrupted CSF-brain Aβ exchange and promoted amylin-Aβ cross-seeding amyloid (Ly, Guerreiro et al., 2021). Intravenous infusion of aggregated human amylin into young APP/PS1 rats induced amylin-Aβ cross-seeding even before endogenous Aβ pathology developed (Ly, Guerreiro et al., 2021), demonstrating that circulating amylin alone can trigger early AD-like changes. These findings collectively show that pancreatic amyloidogenic human amylin contributes to cerebral amyloid accumulation via cross-seeding with Aβ, exacerbating AD pathology. Circulating amylin appears as a modulable factor with potential as a therapeutic target to mitigate AD-type neurodegeneration.

Amylin dysregulation alters intracellular Ca^{2+} and its signaling pathways

In human amylin-expressing rats, myocardial amylin accumulation develops early, prior to overt diabetes, mirroring the preclinical phase of human metabolic disease (Despa et al., 2012, 2014). These oligomers localize to the sarcolemma of cardiac myocytes, where they disrupt Ca^{2+} homeostasis by increasing sarcolemmal Ca^{2+} leak and amplifying Ca^{2+} transients (Despa et al., 2012, 2014; Popescu et al., 2019). This disturbance activates downstream Ca^{2+}-dependent hypertrophic signaling cascades, including the CaMKII-HDAC and calcineurin-NFAT pathways (Despa et al., 2012) resulting in cardiomyocyte hypertrophy, structural remodeling, and diastolic dysfunction (Despa et al., 2012). In contrast, age- and glucose-matched rats, which express non-amyloidogenic rat amylin, exhibit no cardiac amylin deposition, maintain normal myocyte Ca^{2+} cycling, and show preserved cardiac structure and function (Despa et al., 2014; Lambert et al., 2015; Liu et al., 2018; Velmurugan et al., 2023). These comparisons confirm that cardiac injury in "amylin-humanized" rats arises specifically from amylin aggregation, rather than from hyperglycemia or insulin resistance per se.

Cell-based models further support these findings. Amylin exerts a contractile response directly in rat ventricular cardiomyocytes via CGRP1-preferring receptors (Bell and McDermott, 1995). Acute exposure of isolated cardiomyocytes to human amylin rapidly increases sarcolemmal permeability to Ca^{2+} within 1–2 hours, consistent with the formation of small membrane-associated amylin oligomers (Despa et al., 2012). These changes are accompanied by elevated cytosolic Ca^{2+}, activation of CaMKII, and nuclear export of class II HDACs, molecular signatures of hypertrophic signaling (Despa et al., 2012). In contrast, exposure to non-amyloidogenic rat amylin fails to alter Ca^{2+} dynamics or activate these pathways (Despa et al., 2012). Prolonged incubation with human amylin promotes the formation of larger oligomeric species and fibrillar aggregates, suggesting a continuum from early membrane disruption to progressive proteotoxic injury (Despa et al., 2012, 2014; Liu et al., 2018).

The hypothesis of amylin-cAMP-PKA overactivation inducing tau neurodegeneration

From another angle, chronically elevated amylin signaling through its receptor(s) can also impair intracellular Ca^{2+} signaling pathways. Amylin binds to G protein-coupled amylin receptors (calcitonin receptor/RAMP heterodimers) that are expressed in the CNS and peripheral tissues (Hay et al., 2015; Hay and Pioszak, 2016). Amylin receptor activation increases cAMP production (Bower et al., 2018) and stimulates PKA (Singh Alvarado et al., 2024). While transient cAMP-PKA signaling supports metabolism and memory (Singh Alvarado et al., 2024), chronic activation, as occurs in metabolic disorders (Cummings et al., 1996; Yang, 2018), drives pathological signaling. In this context, PKA primes tau for hyperphosphorylation (Jicha et al., 1999) and phosphorylates glycolytic kinases (CHANDEL, 2021; Kriegel et al., 1994; Rider et al., 2004), thereby reducing cerebral glycolysis, two key pathological processes in many dementias (An et al., 2018; Goyal et al., 2023). Mechanistically, PKA further phosphorylates ryanodine receptor 2 (RyR2) at Ser2808, causing ER Ca^{2+} leak (Dridi et al., 2023), which elevates cytosolic Ca^{2+}. Chronic CaMKII activation was documented in brains of human amylin-expressing rats and dementia patients with type 2 diabetes (Erickson et al., 2013). Elevated cytosolic Ca^{2+} activates also GSK3β and CDK5 (Braun and Schulman, 1995; Hudmon and Schulman, 2002; Wen et al., 2008) to promote tau hyperphosphorylation. Although brain amylin signaling is strongly linked to cAMP-PKA activation (Bower et al., 2018; Hay et al., 2015; Hay and Pioszak, 2016) and Ca^{2+} dysregulation (Despa et al., 2012, 2014; Erickson et al., 2013; Popescu et al., 2019; Velmurugan et al., 2023), how these processes converge in vivo remains unclear.

Together, these data suggest a mechanistic link between increased amylin viability in brain tissues and exacerbated amylin receptor signaling leading to cAMP-PKA overactivation and consequent Ca^{2+} dysregulation and neurodegeneration (Fig. 1.3).

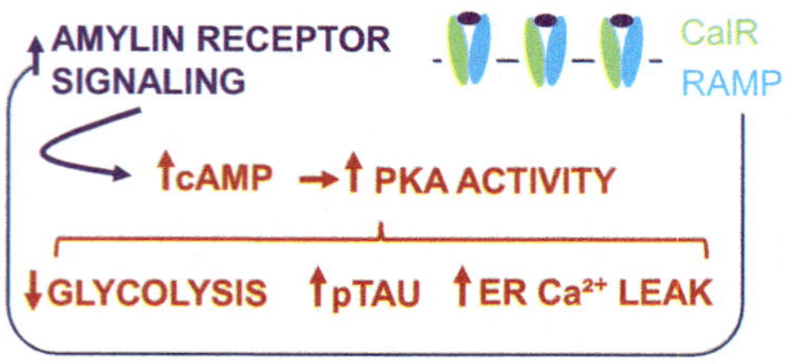

FIGURE 1.3 Proposed mechanistic link between increased amylin viability in brain tissues and exacerbated amylin receptor signaling leading to cAMP-PKA overactivation and consequent Ca^{2+} dysregulation and neurodegeneration.

Amylin behavioral phenotype

Amylin behavioral phenotype is context dependent. In rodents, infusion of amylin into the nucleus accumbens suppresses motor activity and food intake (Baldo and Kelley, 2001), while accumulation in the brain impairs exploration, vestibulomotor coordination, and memory (Ly et al., 2017; Srodulski et al., 2014). Systemic administration of amylin or its analog pramlintide can improve behavior in AD models (Zhu et al., 2015), suggesting potential protective effects. However, genetic manipulation of amylin secretion in AD-model rats has shown opposite outcomes: suppression alleviates behavioral deficits and neuropathology (Ly, Verma et al., 2021), consistent with reports that pancreatic expression of human amylin increases cerebral amyloid burden through diabetogenic effects (Moreno-Gonzalez et al., 2017; Wijesekara et al., 2017). Thus, amylin dysregulation contributes to behavioral deficits by disrupting neural circuits involved in energy balance, stress response and cognition, linking metabolic disturbances to neurobehavioral impairments.

Applications of the material

This chapter reviews evidence from human studies and experimental models supporting a cumulative framework in which compensatory amylin hypersecretion during prediabetes leads to the formation of circulating toxic oligomers that activate immune responses, promote microvascular deposition, and induce hypoxic-ischemic injury, ultimately predisposing to organ failure. It is proposed that early amylin hypersecretion may cause subtle disturbances in energy homeostasis that worsen with metabolic decline, suggesting that amylin-driven microvascular injury can precede hyperglycemia and overt diabetes. This framework identifies several therapeutic targets, including inhibitors of amylin aggregation, blockers of amylin-induced microglial activation and interventions that prevent vascular infiltration, which may help prevent or mitigate diabetes-associated cardiovascular complications and neurological deficits.

Chronic elevation of secreted amylin can also interfere with its signaling through the amylin receptor. Mechanistically, these effects are mediated through amylin receptor activation of cAMP-PKA signaling, influencing intracellular Ca^{2+} homeostasis, energy metabolism, tau phosphorylation and secretase activity, processes that control both amyloid formation and neurodegeneration. These findings underscore the delicate balance between the physiological and pathological actions of amylin in CNS.

Overall, dysregulated amylin secretion in prediabetes and type 2 diabetes drives metabolic dysfunction, microvascular injury and inflammation that compromise both cardiac and cerebral function. In turn, cardiac and neural stress can perturb peripheral amylin dynamics, amplifying systemic dysfunction. Through these reciprocal interactions, amylin emerges as a central mediator linking metabolic stress to organ-specific pathology, contributing to the progression of diabetic heart failure and Alzheimer's-like neurodegeneration. Understanding this bidirectional amylin signaling opens new therapeutic avenues across traditionally distinct disease domains. Integrating the physiological and pathological dimensions of amylin biology will be essential for realizing its therapeutic potential while minimizing risks associated with elevated amylin levels in prediabetes.

Mini-dictionary of terms

- **Circulating amyloid-forming amylin.** Aggregated human amylin that is engulphed by circulating monocytes and deposits on RBCs.
- **Extra-pancreatic amylin amyloid.** Accumulation of amyloid-forming amylin in extra-pancreatic tissues.
- **Amylin-cAMP-PKA activation pathway.** cAMP-PKA activation through amylin receptor signaling.
- **Amylin vasculopathy.** Amylin deposits lining the capillary endothelium associated with peroxidative cell membrane injury and immune responses.
- **Amylin behavioral phenotype.** Neurological deficits induced by chronically dysregulated amylin.

Summary points

- Chronic pancreatic secretion of amyloid-forming amylin during prediabetes promotes pancreatic amyloid pathology that spreads to microcirculation and extra-pancreatic tissues.
- Circulating toxic oligomers promote immune activation, microvascular deposition and hypoxic-ischemic injury, ultimately contributing to end-organ failure.
- Beyond its toxic oligomer effects, increased amylin levels in heart and brain tissues alter intracellular Ca^{2+} and its signaling pathways, impairing cellular metabolism and function.
- Amylin receptor activation increases cAMP production and stimulates PKA, driving metabolic deficits and tau hyperphosphorylation.
- Therapeutic strategies shall target amylin aggregation, receptor signaling and vascular protection.

Statement on the use of AI-assisted technologies

During the preparation of this work, the author used ChatGPT3.5 to correct grammar errors and improve readability. After using this tool, the author reviewed and edited the content as needed and takes full responsibility for the publication's content.

References

Altamirano-Bustamante, N. F., Garrido-Magana, E., Moran, E., Calderon, A., Pasten-Hidalgo, K., Castillo-Rodriguez, R. A., Rojas, G., Lara-Martinez, R., Leyva-Garcia, E., Larralde-Laborde, M., Domiguez, G., Murata, C., Margarita-Vazquez, Y., Payro, R., Barbosa, M., Valderrama, A., Montesinos, H., Dominguez-Camacho, A., Garcia-Olmos, V. H., … Altamirano-Bustamante, M. M. (2020). Protein-conformational diseases in childhood: Naturally-occurring hIAPP amyloid-oligomers and early beta-cell damage in obesity and diabetes. *PLoS One, 15*, e0237667.

An, Y., Varma, V. R., Varma, S., Casanova, R., Dammer, E., Pletnikova, O., Chia, C. W., Egan, J. M., Ferrucci, L., Troncoso, J., Levey, A. I., Lah, J., Seyfried, N. T., Legido-Quigley, C., O'Brien, R., & Thambisetty, M. (2018). Evidence for brain glucose dysregulation in Alzheimer's disease. *Alzheimer's & Dementia, 14*, 318–329.

Baldo, B. A., & Kelley, A. E. (2001). Amylin infusion into rat nucleus accumbens potently depresses motor activity and ingestive behavior. *American Journal of Physiology—Regulatory, Integrative and Comparative Physiology, 281*, R1232–R1242.

Barrett, E. J., Liu, Z., Khamaisi, M., King, G. L., Klein, R., Klein, B. E. K., Hughes, T. M., Craft, S., Freedman, B. I., Bowden, D. W., Vinik, A. I., & Casellini, C. M. (2017). Diabetic microvascular disease: An endocrine Society scientific statement. *Journal of Clinical Endocrinology and Metabolism, 102*, 4343–4410.

Baylis, R. A., Gomez, D., Mallat, Z., Pasterkamp, G., & Owens, G. K. (2017). The CANTOS trial: One important step for clinical cardiology but a giant leap for vascular biology. *Arteriosclerosis, Thrombosis, and Vascular Biology, 37*, e174–e177.

Beery, M. L., Jacobsen, L. M., Atkinson, M. A., Butler, A. E., & Campbell-Thompson, M. (2019). Islet amyloidosis in a child with type 1 diabetes. *Islets, 11*, 44–49.

Bell, D., & Mcdermott, B. J. (1995). Activity of amylin at CGRP1-preferring receptors coupled to positive contractile response in rat ventricular cardiomyocytes. *Regulatory Peptides, 60*, 125–133.

Bhowmick, D. C., Kudaibergenova, Z., Burnett, L., & Jeremic, A. M. (2022). Molecular mechanisms of Amylin turnover, misfolding and toxicity in the pancreas. *Molecules, 27*.

Bower, R. L., Yule, L., Rees, T. A., Deganutti, G., Hendrikse, E. R., Harris, P. W. R., Kowalczyk, R., Ridgway, Z., Wong, A. G., Swierkula, K., Raleigh, D. P., Pioszak, A. A., Brimble, M. A., Reynolds, C. A., Walker, C. S., & Hay, D. L. (2018). Molecular signature for receptor engagement in the metabolic peptide hormone Amylin. *ACS Pharmacology & Translational Science, 1*, 32–49.

Braun, A. P., & Schulman, H. (1995). The multifunctional calcium/calmodulin-dependent protein kinase: From form to function. *Annual Review of Physiology, 57*, 417–445.

Butler, A. E., Jang, J., Gurlo, T., Carty, M. D., Soeller, W. C., & Butler, P. C. (2004). Diabetes due to a progressive defect in beta-cell mass in rats transgenic for human islet amyloid polypeptide (HIP Rat): A new model for type 2 diabetes. *Diabetes, 53*, 1509–1516.

Castle, A. L., Kuo, C. H., Han, D. H., & Ivy, J. L. (1998). Amylin-mediated inhibition of insulin-stimulated glucose transport in skeletal muscle. *American Journal of Physiology, 275*, E531–E536.

Castle, A. L., Kuo, C. H., & Ivy, J. L. (1998). Amylin influences insulin-stimulated glucose metabolism by two independent mechanisms. *American Journal of Physiology, 274*, E6–E12.

Chandel, N. S. (2021). Signaling and metabolism. *Cold Spring Harbor Perspectives in Biology, 13.*

Cole, M. A., Abd Jamil, A. H., Heather, L. C., Murray, A. J., Sutton, E. R., Slingo, M., Sebag-Montefiore, L., Tan, S. C., Aksentijevic, D., Gildea, O. S., Stuckey, D. J., Yeoh, K. K., Carr, C. A., Evans, R. D., Aasum, E., Schofield, C. J., Ratcliffe, P. J., Neubauer, S., Robbins, P. A., & Clarke, K. (2016). On the pivotal role of PPARalpha in adaptation of the heart to hypoxia and why fat in the diet increases hypoxic injury. *The FASEB Journal, 30*, 2684–2697.

Cummings, D. E., Brandon, E. P., Planas, J. V., Motamed, K., Idzerda, R. L., & Mcknight, G. S. (1996). Genetically lean mice result from targeted disruption of the RII beta subunit of protein kinase A. *Nature, 382*, 622–626.

Despa, S., Margulies, K. B., Chen, L., Knowlton, A. A., Havel, P. J., Taegtmeyer, H., Bers, D. M., & Despa, F. (2012). Hyperamylinemia contributes to cardiac dysfunction in obesity and diabetes: A study in humans and rats. *Circulation Research, 110*, 598–608.

Despa, S., Sharma, S., Harris, T. R., Dong, H., Li, N., Chiamvimonvat, N., Taegtmeyer, H., Margulies, K. B., Hammock, B. D., & Despa, F. (2014). Cardioprotection by controlling hyperamylinemia in a "humanized" diabetic rat model. *Journal of the American Heart Association, 3.*

Dridi, H., Liu, Y., Reiken, S., Liu, X., Argyrousi, E. K., Yuan, Q., Miotto, M. C., Sittenfeld, L., Meddar, A., Soni, R. K., Arancio, O., Lacampagne, A., & Marks, A. R. (2023). Heart failure-induced cognitive dysfunction is mediated by intracellular Ca(2+) leak through ryanodine receptor type 2. *Nature Neuroscience, 26*, 1365–1378.

El Kasmi, K. C., & Stenmark, K. R. (2015). Contribution of metabolic reprogramming to macrophage plasticity and function. *Seminars in Immunology, 27*, 267–275.

Erickson, J. R., Pereira, L., Wang, L., Han, G., Ferguson, A., Dao, K., Copeland, R. J., Despa, F., Hart, G. W., Ripplinger, C. M., & Bers, D. M. (2013). Diabetic hyperglycaemia activates CaMKII and arrhythmias by O-linked glycosylation. *Nature, 502*, 372–376.

Fawver, J. N., Ghiwot, Y., Koola, C., Carrera, W., Rodriguez-Rivera, J., Hernandez, C., Dineley, K. T., Kong, Y., Li, J., Jhamandas, J., Perry, G., & Murray, I. V. (2014). Islet amyloid polypeptide (IAPP): A second amyloid in Alzheimer's disease. *Current Alzheimer Research, 11*, 928–940.

Goyal, M. S., Blazey, T., Metcalf, N. V., Mcavoy, M. P., Strain, J. F., Rahmani, M., Durbin, T. J., Xiong, C., Benzinger, T. L., Morris, J. C., Raichle, M. E., & Vlassenko, A. G. (2023). Brain aerobic glycolysis and resilience in Alzheimer disease. *Proceedings of the National Academy of Sciences of the United States of America, 120*. e2212256120.

Hay, D. L., Chen, S., Lutz, T. A., Parkes, D. G., & Roth, J. D. (2015). Amylin: Pharmacology, physiology, and clinical potential. *Pharmacological Reviews, 67*, 564–600.

Hay, D. L., & Pioszak, A. A. (2016). Receptor activity-modifying proteins (RAMPs): New insights and roles. *Annual Review of Pharmacology and Toxicology, 56*, 469–487.

Hoppener, J. W., Ahren, B., & Lips, C. J. (2000). Islet amyloid and type 2 diabetes mellitus. *New England Journal of Medicine, 343*, 411–419.

Hudmon, A., & Schulman, H. (2002). Neuronal CA2+/calmodulin-dependent protein kinase II: The role of structure and autoregulation in cellular function. *Annual Review of Biochemistry, 71*, 473–510.

Ilaiwy, A., Liu, M., Parry, T. L., Bain, J. R., Newgard, C. B., Schisler, J. C., Muehlbauer, M. J., Despa, F., & Willis, M. S. (2016). Human amylin proteotoxicity impairs protein biosynthesis, and alters major cellular signaling pathways in the heart, brain and liver of humanized diabetic rat model in vivo. *Metabolomics, 12.*

Jackson, K., Barisone, G. A., Diaz, E., Jin, L. W., Decarli, C., & Despa, F. (2013). Amylin deposition in the brain: A second amyloid in Alzheimer disease? *Annals of Neurology, 74*, 517–526.

Jicha, G. A., Weaver, C., Lane, E., Vianna, C., Kress, Y., Rockwood, J., & Davies, P. (1999). cAMP-dependent protein kinase phosphorylations on tau in Alzheimer's disease. *Journal of Neuroscience, 19*, 7486–7494.

Kahn, S. E., D'Alessio, D. A., Schwartz, M. W., Fujimoto, W. Y., Ensinck, J. W., Taborsky, G. J., JR., & Porte, D., JR. (1990). Evidence of cosecretion of islet amyloid polypeptide and insulin by beta-cells. *Diabetes, 39*, 634–638.

Kang, D. E., Pietrzik, C. U., Baum, L., Chevallier, N., Merriam, D. E., Kounnas, M. Z., Wagner, S. L., Troncoso, J. C., Kawas, C. H., Katzman, R., & Koo, E. H. (2000). Modulation of amyloid beta-protein clearance and Alzheimer's disease susceptibility by the LDL receptor-related protein pathway. *Journal of Clinical Investigation, 106*, 1159–1166.

Koelwyn, G. J., Corr, E. M., Erbay, E., & Moore, K. J. (2018). Regulation of macrophage immunometabolism in atherosclerosis. *Nature Immunology, 19*, 526–537.

Kotiya, D., Leibold, N., Verma, N., Jicha, G. A., Goldstein, L. B., & Despa, F. (2023). Rapid, scalable assay of amylin-beta amyloid co-aggregation in brain tissue and blood. *Journal of Biological Chemistry, 299*, 104682.

Kriegel, T. M., Rush, J., Vojtek, A. B., Clifton, D., & Fraenkel, D. G. (1994). In vivo phosphorylation site of hexokinase 2 in *Saccharomyces cerevisiae*. *Biochemistry, 33*, 148–152.

Lambert, R., Srodulski, S., Peng, X., Margulies, K. B., Despa, F., & Despa, S. (2015). Intracellular Na+ concentration ([Na+]i) is elevated in diabetic hearts due to enhanced Na+-Glucose cotransport. *Journal of the American Heart Association, 4*, e002183.

Leighton, B., & Cooper, G. J. (1988). Pancreatic amylin and calcitonin gene-related peptide cause resistance to insulin in skeletal muscle in vitro. *Nature, 335*, 632–635.

Liao, R., Jain, M., Teller, P., Connors, L. H., Ngoy, S., Skinner, M., Falk, R. H., & Apstein, C. S. (2001). Infusion of light chains from patients with cardiac amyloidosis causes diastolic dysfunction in isolated mouse hearts. *Circulation, 104*, 1594–1597.

Liu, M., Hoskins, A., Verma, N., Bers, D. M., Despa, S., & Despa, F. (2018). Amylin and diabetic cardiomyopathy—Amylin-Induced sarcolemmal Ca(2+) leak is independent of diabetic remodeling of myocardium. *Biochimica et Biophysica Acta (BBA)—Molecular Basis of Disease, 1864*, 1923–1930.

Liu, M., Verma, N., Peng, X., Srodulski, S., Morris, A., Chow, M., Hersh, L. B., Chen, J., Zhu, H., Netea, M. G., Margulies, K. B., Despa, S., & Despa, F. (2016). Hyperamylinemia increases IL-1beta synthesis in the heart via peroxidative sarcolemmal injury. *Diabetes, 65*, 2772–2783.

Lutz, T. A. (2012). Control of energy homeostasis by amylin. *Cellular and Molecular Life Sciences, 69*, 1947–1965.

Ly, H., Verma, N., Sharma, S., Kotiya, D., Despa, S., Abner, E. L., Nelson, P. T., Jicha, G. A., Wilcock, D. M., & Goldstein, L. B. (2021). The association of circulating amylin with β-amyloid in familial Alzheimer's disease. *Alzheimer's & Dementia: Translational Research & Clinical Interventions, 7*, e12130.

Ly, H., Verma, N., Sharma, S., Kotiya, D., Despa, S., Abner, E. L., Nelson, P. T., Jicha, G. A., Wilcock, D. M., Goldstein, L. B., Guerreiro, R., Bras, J., Hanson, A. J., Craft, S., Murray, A. J., Biessels, G. J., Troakes, C., Zetterberg, H., Hardy, J., … Despa, F. (2021). The association of circulating amylin with beta-amyloid in familial Alzheimer's disease. *Alzheimer's & Dementia, 7*, e12130.

Ly, H., Verma, N., Wu, F., Liu, M., Saatman, K. E., Nelson, P. T., Slevin, J. T., Goldstein, L. B., Biessels, G. J., & Despa, F. (2017). Brain microvascular injury and white matter disease provoked by diabetes-associated hyperamylinemia. *Annals of Neurology, 82*, 208–222.

Martinez-Valbuena, I., Valenti-Azcarate, R., Amat-Villegas, I., Riverol, M., Marcilla, I., De Andrea, C. E., Sánchez-Arias, J. A., Del Mar Carmona-Abellan, M., Marti, G., & Erro, M. E. (2019). Amylin as a potential link between type 2 diabetes and alzheimer disease. *Annals of Neurology, 86*, 539–551.

Masters, S. L., Dunne, A., Subramanian, S. L., Hull, R. L., Tannahill, G. M., Sharp, F. A., Becker, C., Franchi, L., Yoshihara, E., Chen, Z., Mullooly, N., Mielke, L. A., Harris, J., Coll, R. C., Mills, K. H., Mok, K. H., Newsholme, P., Nunez, G., Yodoi, J., … O'Neill, L. A. (2010). Activation of the NLRP3 inflammasome by islet amyloid polypeptide provides a mechanism for enhanced IL-1beta in type 2 diabetes. *Nature Immunology, 11*, 897–904.

Mcmillan, D. E., Utterback, N. G., & La Puma, J. (1978). Reduced erythrocyte deformability in diabetes. *Diabetes, 27*, 895–901.

Molina, J. M., Cooper, G. J., Leighton, B., & Olefsky, J. M. (1990). Induction of insulin resistance in vivo by amylin and calcitonin gene-related peptide. *Diabetes, 39*, 260–265.

Moreno-Gonzalez, I., Edwards III, G., Salvadores, N., Shahnawaz, M., Diaz-Espinoza, R., & Soto, C. (2017). Molecular interaction between type 2 diabetes and Alzheimer's disease through cross-seeding of protein misfolding. *Molecular Psychiatry, 22*, 1327–1334.

Mullard, A. (2025). Amylin takes another shot at the obesity prize. *Nature Reviews Drug Discovery, 24*, 403–406.

Oskarsson, M. E., Paulsson, J. F., Schultz, S. W., Ingelsson, M., Westermark, P., & Westermark, G. T. (2015). In vivo seeding and cross-seeding of localized amyloidosis: A molecular link between type 2 diabetes and Alzheimer disease. *American Journal of Pathology, 185*, 834–846.

Palladini, G., Lavatelli, F., Russo, P., Perlini, S., Perfetti, V., Bosoni, T., Obici, L., Bradwell, A. R., D'Eril, G. M., Fogari, R., Moratti, R., & Merlini, G. (2006). Circulating amyloidogenic free light chains and serum N-terminal natriuretic peptide type B decrease simultaneously in association with improvement of survival in AL. *Blood, 107*, 3854–3858.

Paulsson, J. F., Ludvigsson, J., Carlsson, A., Casas, R., Forsander, G., Ivarsson, S. A., Kockum, I., Lernmark, A., Marcus, C., Lindblad, B., & Westermark, G. T. (2014). High plasma levels of islet amyloid polypeptide in young with new-onset of type 1 diabetes mellitus. *PLoS One, 9*, e93053.

Popescu, I., Yin, G., Velmurugan, S., Erickson, J. R., Despa, F., & Despa, S. (2019). Lower sarcoplasmic reticulum Ca(2+) threshold for triggering afterdepolarizations in diabetic rat hearts. *Heart Rhythm, 16*, 765–772.

Rider, M. H., Bertrand, L., Vertommen, D., Michels, P. A., Rousseau, G. G., & Hue, L. (2004). 6-phosphofructo-2-kinase/fructose-2,6-bisphosphatase: Head-to-head with a bifunctional enzyme that controls glycolysis. *Biochemical Journal, 381*, 561–579.

Ridker, P. M., Everett, B. M., Thuren, T., Macfadyen, J. G., Chang, W. H., Ballantyne, C., Fonseca, F., Nicolau, J., Koenig, W., Anker, S. D., Kastelein, J. J. P., Cornel, J. H., Pais, P., Pella, D., Genest, J., Cifkova, R., Lorenzatti, A., Forster, T., Kobalava, Z., … Group, C. T. (2017). Antiinflammatory therapy with canakinumab for atherosclerotic disease. *New England Journal of Medicine, 377*, 1119–1131.

Roostaei, T., Nazeri, A., Felsky, D., De Jager, P. L., Schneider, J. A., Pollock, B. G., Bennett, D. A., & Voineskos, A. N. (2017). Genome-wide interaction study of brain beta-amyloid burden and cognitive impairment in Alzheimer's disease. *Molecular Psychiatry, 22*, 287–295.

Roostaei, T., Nazeri, A., Felsky, D., De Jager, P. L., Schneider, J. A., Pollock, B. G., Bennett, D. A., Voineskos, A. N., & Alzheimer'S Disease Neuroimaging Initiative. (2017). Genome-wide interaction study of brain beta-amyloid burden and cognitive impairment in Alzheimer's disease. *Molecular Psychiatry, 22*, 287–295.

Royall, D. R., Palmer, R. F., & Alzheimer's Disease Neuroimaging Initiative. (2020). Blood-based protein mediators of senility with replications across biofluids and cohorts. *Brain Communications, 2*, fcz036.

Schultz, N., Byman, E., Fex, M., & Wennstrom, M. (2017). Amylin alters human brain pericyte viability and NG2 expression. *Journal of Cerebral Blood Flow and Metabolism, 37*, 1470–1482.

Schultz, N., Byman, E., Netherlands Brain, B., & Wennstrom, M. (2018). Levels of retinal IAPP are altered in Alzheimer's disease patients and correlate with vascular changes and hippocampal IAPP levels. *Neurobiology of Aging, 69*, 94–101.

Singh Alvarado, J., Lutas, A., Madara, J. C., Isaac, J., Lommer, C., Massengill, C., & Andermann, M. L. (2024). Transient cAMP production drives rapid and sustained spiking in brainstem parabrachial neurons to suppress feeding. *Neuron, 112*, 1416–1425 e5.

Srodulski, S., Sharma, S., Bachstetter, A. B., Brelsfoard, J. M., Pascual, C., Xie, X. S., Saatman, K. E., van Eldik, L. J., & Despa, F. (2014). Neuroinflammation and neurologic deficits in diabetes linked to brain accumulation of amylin. *Molecular Neurodegeneration, 9*, 30.

Storck, S. E., Meister, S., Nahrath, J., Meissner, J. N., Schubert, N., Di Spiezio, A., Baches, S., Vandenbroucke, R. E., Bouter, Y., Prikulis, I., Korth, C., Weggen, S., Heimann, A., Schwaninger, M., Bayer, T. A., & Pietrzik, C. U. (2016). Endothelial LRP1 transports amyloid-beta(1-42) across the blood-brain barrier. *Journal of Clinical Investigation, 126*, 123–136.

Tomita, T. (2011). Islet amyloid polypeptide in pancreatic islets from type 1 diabetic subjects. *Islets, 3*, 166–174.

Tonelli, M., Sacks, F., Arnold, M., Moye, L., Davis, B., Pfeffer, M., For The, C., & Recurrent Events Trial, I. (2008). Relation between red blood cell distribution width and cardiovascular event rate in people with coronary disease. *Circulation, 117*, 163–168.

Velmurugan, S., Liu, T., Chen, K. C., Despa, F., O'Rourke, B., & Despa, S. (2023). Distinct effects of mitochondrial Na(+)/Ca(2+) exchanger inhibition and Ca(2+) uniporter activation on Ca(2+) Sparks and arrhythmogenesis in diabetic rats. *Journal of the American Heart Association, 12*, e029997.

Velmurugan, S., Pandey, V. K., Verma, N., Kotiya, D., Despa, F., & Despa, S. (2025). Cardiac remodelling, recognition memory deficits and accelerated ageing in a rat model of gestational diabetes. *Diabetologia, 68*, 1574–1584.

Venkatanarayan, A., Raulji, P., Norton, W., Chakravarti, D., Coarfa, C., Su, X., Sandur, S. K., Ramirez, M. S., Lee, J., Kingsley, C. V., Sananikone, E. F., Rajapakshe, K., Naff, K., Parker-Thornburg, J., Bankson, J. A., Tsai, K. Y., Gunaratne, P. H., & Flores, E. R. (2015). IAPP-driven metabolic reprogramming induces regression of p53-deficient tumours in vivo. *Nature, 517*, 626–630.

Verma, N., Liu, M., Ly, H., Loria, A., Campbell, K. S., Bush, H., Kern, P. A., Jose, P. A., Taegtmeyer, H., Bers, D. M., Despa, S., Goldstein, L. B., Murray, A. J., & Despa, F. (2020). Diabetic microcirculatory disturbances and pathologic erythropoiesis are provoked by deposition of amyloid-forming amylin in red blood cells and capillaries. *Kidney International, 97*, 143–155.

Verma, N., Ly, H., Liu, M., Chen, J., Zhu, H., Chow, M., Hersh, L. B., & Despa, F. (2016). Intraneuronal Amylin deposition, peroxidative membrane injury and increased IL-1beta synthesis in brains of Alzheimer's disease patients with type-2 diabetes and in diabetic HIP rats. *Journal of Alzheimer's Disease, 53*, 259–272.

Verma, N., Srodulski, S., Velmurugan, S., Hoskins, A., Pandey, V. K., Despa, F., & Despa, S. (2021). Gestational diabetes triggers postpartum cardiac hypertrophy via activation of calcineurin/NFAT signaling. *Scientific Reports, 11*, 20926.

Verma, N., Velmurugan, G. V., Winford, E., Coburn, H., Kotiya, D., Leibold, N., Radulescu, L., Despa, S., Chen, K. C., van Eldik, L. J., Nelson, P. T., Wilcock, D. M., Jicha, G. A., Stowe, A. M., Goldstein, L. B., Powel, D. K., Walton, J. H., Navedo, M. F., Nystoriak, M. A., … Despa, F. (2023). Abeta efflux impairment and inflammation linked to cerebrovascular accumulation of amyloid-forming amylin secreted from pancreas. *Communications Biology, 6*, 2.

Wen, Y., Planel, E., Herman, M., Figueroa, H. Y., Wang, L., Liu, L., Lau, L. F., Yu, W. H., & Duff, K. E. (2008). Interplay between cyclin-dependent kinase 5 and glycogen synthase kinase 3 beta mediated by neuregulin signaling leads to differential effects on tau phosphorylation and amyloid precursor protein processing. *Journal of Neuroscience, 28*, 2624–2632.

Westermark, P., Andersson, A., & Westermark, G. T. (2011). Islet amyloid polypeptide, islet amyloid, and diabetes mellitus. *Physiological Reviews, 91*, 795–826.

Westermark, P., Engstrom, U., Johnson, K. H., Westermark, G. T., & Betsholtz, C. (1990). Islet amyloid polypeptide: Pinpointing amino acid residues linked to amyloid fibril formation. *Proceedings of the National Academy of Sciences of the United States of America, 87*, 5036–5040.

Westermark, G. T., Krogvold, L., Dahl-Jorgensen, K., & Ludvigsson, J. (2017). Islet amyloid in recent-onset type 1 diabetes-the DiViD study. *Upsala Journal of Medical Sciences, 122*, 201–203.

Wijesekara, N., Ahrens, R., Sabale, M., Wu, L., Ha, K., Verdile, G., & Fraser, P. E. (2017). Amyloid-beta and islet amyloid pathologies link Alzheimer's disease and type 2 diabetes in a transgenic model. *The FASEB Journal, 31*, 5409–5418.

Xiong, M., Elson, G., Legarda, D., & Leibovich, S. J. (1998). Production of vascular endothelial growth factor by murine macrophages: Regulation by hypoxia, lactate, and the inducible nitric oxide synthase pathway. *American Journal of Pathology, 153*, 587–598.

Yang, L. (2018). Neuronal cAMP/PKA signaling and energy homeostasis. *Advances in Experimental Medicine & Biology, 1090*, 31–48.

Yong, J., Johnson, J. D., Arvan, P., Han, J., & Kaufman, R. J. (2021). Therapeutic opportunities for pancreatic beta-cell ER stress in diabetes mellitus. *Nature Reviews Endocrinology, 17*, 455–467.

Zhu, H., Tao, Q., Ang, T. F. A., Massaro, J., Gan, Q., Salim, S., Zhu, R. Y., Kolachalama, V. B., Zhang, X., Devine, S., Auerbach, S. H., Decarli, C., Au, R., & Qiu, W. Q. (2019). Association of plasma Amylin concentration with Alzheimer disease and brain structure in older adults. *JAMA Network Open, 2*, e199826.

Zhu, H., Wang, X., Wallack, M., Li, H., Carreras, I., Dedeoglu, A., Hur, J. Y., Zheng, H., Li, H., Fine, R., Mwamburi, M., Sun, X., Kowall, N., Stern, R. A., & Qiu, W. Q. (2015). Intraperitoneal injection of the pancreatic peptide amylin potently reduces behavioral impairment and brain amyloid pathology in murine models of Alzheimer's disease. *Molecular Psychiatry, 20*, 252–262.

Zierath, J. R., Galuska, D., Engstrom, A., Johnson, K. H., Betsholtz, C., Westermark, P., & Wallberg-Henriksson, H. (1992). Human islet amyloid polypeptide at pharmacological levels inhibits insulin and phorbol ester-stimulated glucose transport in in vitro incubated human muscle strips. *Diabetologia, 35*, 26–31.

Chapter 2

Computational modeling of the pancreas and pancreatic function

Gerardo J. Félix-Martínez
Department of Electrical Engineering, Universidad Autónoma Metropolitana, México City, México

Abbreviations

ATP Adenosine triphosphate
CFTR Cystic Fibrosis Transmembrane Conductance Regulator
ER Endoplasmic reticulum
FDA U.S. Food and drug administration
GIRK G-protein coupled inwardly rectifying potassium channels
IP3 Inositol 1,4,5-triphosphate
IVGTT Intravenous glucose tolerance test
KATP ATP-sensitive potassium channels
OGTT Oral glucose tolerance test

Introduction

The pancreas is a dual-function gland with a highly integrated anatomical and physiological organization that is central to digestive and metabolic processes. Its exocrine compartment, composed primarily of acinar and ductal cells, secretes digestive enzymes and bicarbonate-rich fluids into the duodenum, ensuring efficient nutrient breakdown and absorption. Dispersed throughout the exocrine tissue are the islets of Langerhans, the endocrine compartment, which contains distinct populations of hormone-secreting cells: β-cells (insulin), α-cells (glucagon), δ-cells (somatostatin), PP-cells (pancreatic polypeptide), and ε-cells (ghrelin), which coordinate blood glucose regulation. Disruption of the exocrine and endocrine pancreas function underlies several major diseases, including diabetes mellitus, pancreatitis, and pancreatic cancer, all of which have motivated the development of computational models to unravel the complexity of pancreatic physiology and pathology. Excellent reviews of the anatomy and physiology of the endocrine and exocrine pancreas are available elsewhere (see for example, Brass et al., 2010; Dolenšek et al., 2017).

From a physiological perspective, computational models have proven extremely useful for dissecting the complex multiscale processes that govern pancreatic function. On the one hand, models of exocrine secretion have clarified the roles of ion transport, fluid dynamics, and intracellular signaling, offering mechanistic insights that complement experimental studies. In the endocrine pancreas, mathematical representations of ion channel dynamics, intracellular calcium oscillations, and metabolic pathways have been extremely useful in explaining how islet cells, either isolated or collectively, shape hormone secretion. At the tissue and organ levels, models have elucidated the interplay between electrical coupling and paracrine signaling in coordinating the islet function. Overall, mathematical models of pancreatic physiology have provided a framework for studying pathological states, such as diabetes, guiding both experimental design and therapeutic strategies.

This chapter provides a general overview of the computational models used to study pancreas physiology and related diseases. Given the extensive work in the field, representative examples of the different types of models are mentioned, acknowledging that several relevant works are not explicitly described.

Models of systemic glucose regulation

Mathematical modeling of systemic glucose regulation has been pivotal in elucidating the dynamic interplay between glucose and insulin, particularly in the context of diabetes. Early models developed in the 1960s and the 1970s, such as those by Bolie (1961), Ackerman et al. (1965) and Bergman et al. (1979) laid the foundations by employing simplified compartmental and feedback frameworks. These early models were commonly validated against experimental data from oral (Ackerman et al., 1965) and intravenous glucose tolerance tests (Bolie, 1961; Segre et al., 1973) in subjects with normal blood glucose and diabetes,

Hormones in Neuroscience. https://doi.org/10.1016/B978-0-443-30263-3.00018-X

including animal (Bolie, 1961) and human studies (Ackerman et al., 1965; Bergman et al., 1979; Segre et al., 1973). Innovative techniques across these studies included the use of electronic and digital computers for simulation and parameter estimation, compartmental and lumped parameter modeling, continuous and high-frequency sampling, and the development of new experimental protocols, such as the glucose clamp technique and the use of radioactive glucose for tracer studies. These approaches enabled more precise and physiologically meaningful modeling of glucose regulation, advancing the understanding of metabolic disorders, such as diabetes. These models typically used systems of differential equations to represent physiological processes and focused on model-derived parameters to quantitatively characterize key aspects of glucose and insulin dynamics, such as the rates of glucose disappearance, insulin secretion, and system responsiveness. These parameters allowed discrimination between individuals with normal glucose regulation, prediabetes, and diabetes with high accuracy, and provided valuable criteria for diagnosis and monitoring.

A seminal contribution to this field was the Minimal Model by Bergman et al. (1979), shown schematically in Fig. 2.1A, which reduced glucose-insulin physiology to a compact system of equations. It introduced a remote insulin compartment to capture delayed insulin action, enabling the manageable and interpretable quantification of insulin sensitivity and glucose effectiveness. By treating insulin as an input and glucose as an output, the minimal model yielded identifiable parameters that could be estimated from the IVGTT data. This framework led to the creation of sophisticated in silico simulators of the glucose regulation system in health and disease and provided useful clinical indicators to predict the onset of type 2 diabetes and to stratify and understand the differences in diabetes risk between species and human populations. The Minimal Model has been extensively reviewed in Bergman (1989, 2021).

Some models moved beyond traditional OGTT/IVGTT tests by simulating glucose-insulin dynamics under real-life meal conditions. Based on these ideas, Cobelli and Dalla Man extended the minimal model principles to oral and meal challenges by incorporating explicit representations of meal absorption, hepatic and renal glucose handling, and patient-specific parameterization (Cobelli et al., 2014; Man et al., 2007). Their work was calibrated against large human datasets using physiologically constrained parameters, which enabled the reconstruction of glucose appearance. The Dalla Man Meal Model served as the foundation for the UVA/Padova Type 1 Diabetes Simulator, a sophisticated in silico tool designed to model glucose-insulin dynamics in individuals with type 1

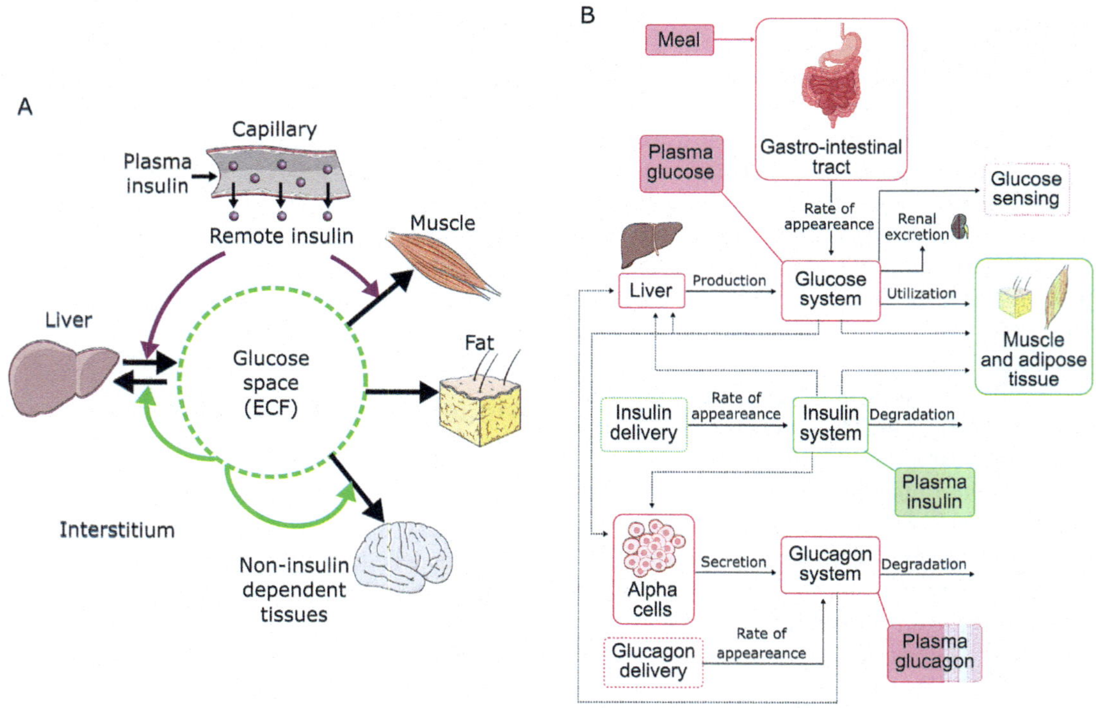

FIGURE 2.1 Representative examples of models of systemic glucose regulation. (A) Scheme of the minimal model by Bergman (1989) developed to reproduce the temporal course of glucose following an IVGTT. (B) Diagram of the UVA/Padova type 1 diabetes simulator, approved by the FDA as a substitute for preclinical trials.

diabetes (Man et al., 2014). This simulator extended the original model by incorporating a diverse virtual population with realistic physiological variability and integrating data from tracer studies and continuous glucose monitoring. A diagram of the UVA/Padova model is shown in Fig. 2.1B. Its predictive capabilities were validated to such an extent that, in 2008, the U.S. Food and Drug Administration (FDA) accepted it as a substitute for animal trials in the preclinical testing of insulin treatment strategies. The UVA/Padova simulator has since become a milestone in the development and testing of artificial pancreas algorithms and closed-loop insulin delivery systems. This exemplifies how systemic models of glucose regulation have transitioned from theoretical constructs to clinically validated tools with a direct translational impact. Details of the UVA/Padova model can be found in the articles by Cobelli and Kovatchev (2023), Man et al. (2014), and Visentin et al. (2018). Similarly, detailed reviews about models of glucose regulation at the systemic level can be found in the works by Ali et al. (2025), Cobelli et al. (2014), Lamlili et al. (2025), Mari et al. (2020) and Palumbo et al. (2013).

Models of islet hormone secretion

Mathematical and computational models of pancreatic hormone secretion, mainly insulin, have progressed from simple descriptions of insulin release based on storage pools, heterogeneous glucose thresholds (Grodsky, 1972; Grodsky & Francisco, 1972), and graded secretory responses (Stamper & Wang, 2013) to sophisticated multiscale model frameworks that integrate morphological aspects and regulatory pathways, such as the triggering and amplifying pathways (Pedersen et al., 2011; Stamper & Wang, 2019). Initial models proposed the two-compartment and threshold distribution hypotheses for insulin secretion based on experiments in perfused pancreas (Grodsky, 1972; Grodsky & Francisco, 1972), schematically described in Fig. 2.2A. According to these hypotheses, insulin is stored in storage (stable) and readily releasable (labile) pools, considering the fast depletion and slow repletion of the labile pool and heterogeneous activation thresholds among its insulin packets (Grodsky, 1972; Grodsky & Francisco, 1972). Based on these foundations, more recent models incorporated granule mobilization, priming, and calcium-triggered fusion with the plasma membrane (Pedersen et al., 2011; Stamper & Wang, 2019). These processes are often represented through kinetic schemes or compartmental approaches, with distinctions made between granule pools based on their readiness, maturation, and calcium sensitivity. As a representative example of these models, the model of insulin secretion proposed by Pedersen et al. (2011) is illustrated in Fig. 2.2B.

While early models favored the heterogeneity hypothesis to explain diverse experimental observations such as the staircase experiment and multiphasic secretion, newer frameworks demonstrate that homogeneous cells or granule pools exhibiting graded responses can also replicate key experimental findings. These results suggested that both mechanisms may contribute to the secretion dynamics.

Subcellular models of insulin secretion were later incorporated into multiscale models (as shown schematically in Fig. 2.3) that focused on modeling insulin secretion from pancreatic β-cells, integrating subcellular mechanisms such as β-cell electrophysiology, calcium dynamics, granule trafficking, and electrical communication in islet architectures to explain biphasic and oscillatory insulin release and its alterations observed in response to glucose protocols and disease states (Pedersen & Cobelli, 2013;

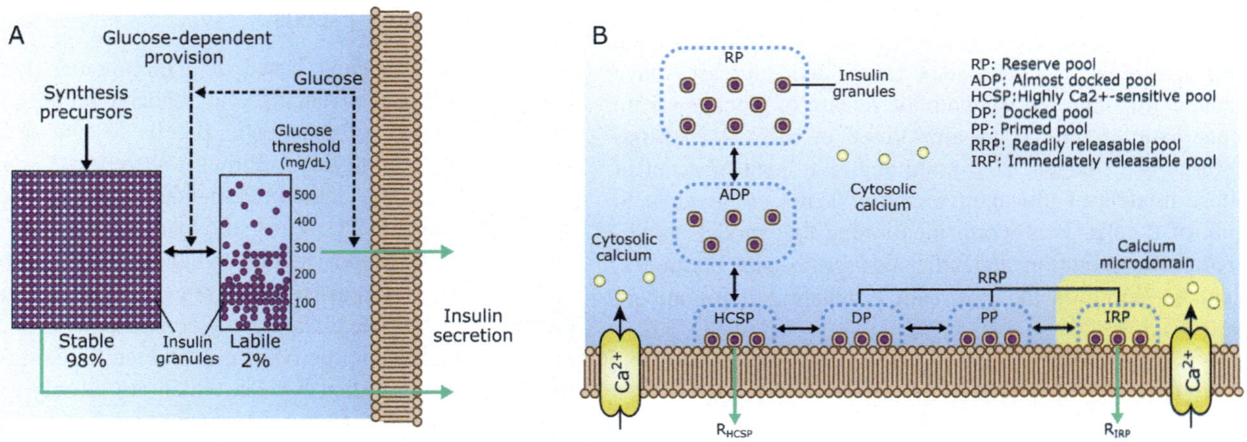

FIGURE 2.2 Evolution of models of insulin secretion. (A) The first models of insulin secretion adopted the idea of different granule pools with heterogeneous secretion thresholds. (B) Recent models of insulin secretion capture granule maturation and calcium signaling. *(A) Adapted from Grodsky, G. M. (1972). A threshold distribution hypothesis for packet storage of insulin and its mathematical modeling.* Journal of Clinical Investigation, *51(8), 2047–2059. https://doi.org/10.1172/jci107011.*

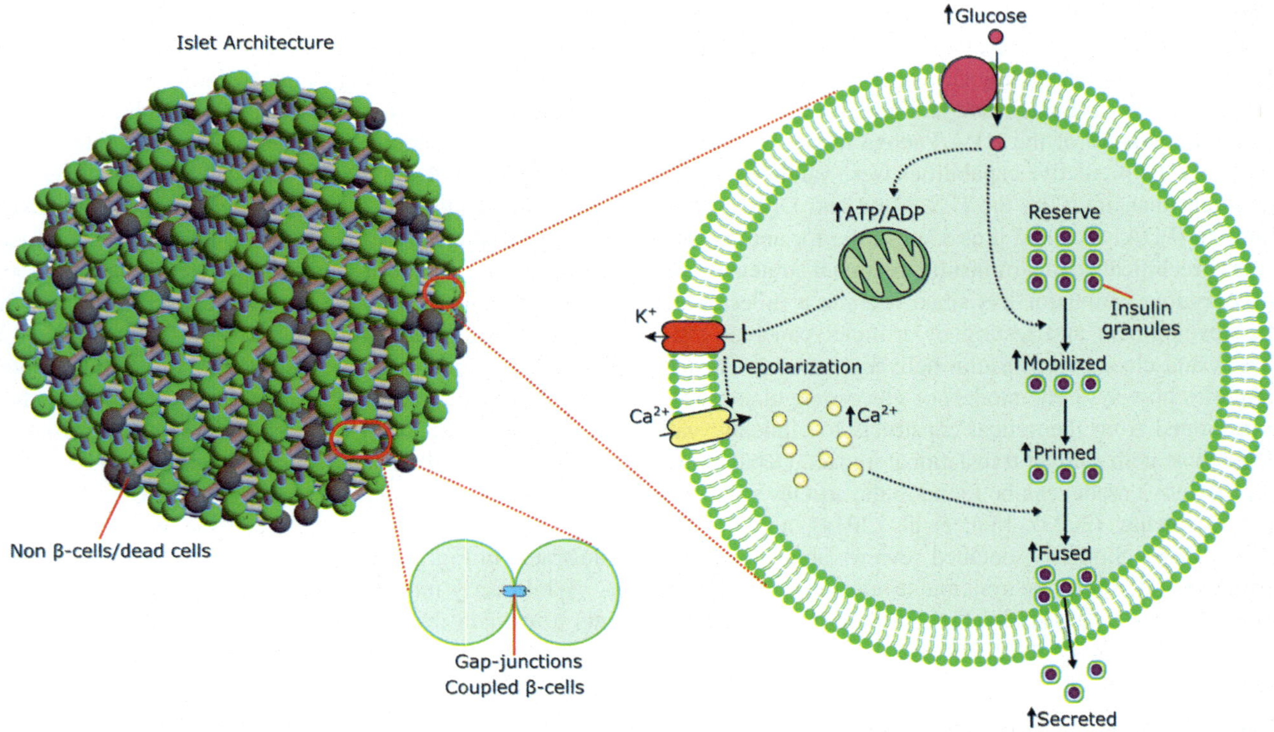

FIGURE 2.3 Multiscale models. More complex models of insulin secretion have integrated models of subcellular mechanisms, islet cell organization, including cell death, electrical activity of β-cells, and intra-islet communication (e.g., gap junctions).

Pedersen et al., 2011; Stamper & Wang, 2019). These models linked cellular and granule-level processes to whole-body clinical minimal models used in IVGTT/OGTT to quantify β-cell function and incretin effects to bridge mechanistic insights with clinical observations. Importantly, these modeling efforts have extended beyond theoretical constructs to serve as predictive tools for understanding disease states, such as diabetes, and interpreting clinical data in terms of the underlying cellular processes.

Many of these ideas, originally devoted to explaining insulin secretion in response to different experimental protocols (e.g., glucose step, ramp, and staircase) have been extrapolated to simulate the release of glucagon from pancreatic α-cells (González-Vélez et al., 2012; Montefusco & Pedersen, 2015), although the level of detail of these models of glucagon secretion is not comparable to that of β-cells. It is worth mentioning that models of somatostatin secretion and other endocrine islet hormones, such as ghrelin or the pancreatic polypeptide hormone are still lacking.

Models of electrical activity of islet cells

The electrical activity of pancreatic α-, β-, and δ-cells is a fundamental mechanism governing the secretion of glucagon, insulin, and somatostatin, respectively. In all cases, the electrophysiological behavior mainly arises from the interplay between ion channels (Fig. 2.4), modulated by metabolic and hormonal signals. Given the difficulty in experimentally isolating every variable involved in vitro, computational modeling has become a valuable tool as a complement to electrophysiology experiments. This section summarizes the key models developed to study the electrical activity of islet cells.

Insulin-secreting β-cells

Mathematical modeling based on experimental data has been relevant in advancing our understanding of the electrical activity of pancreatic β-cells to elucidate the mechanisms underlying stimulus-secretion coupling (Félix-Martinez & Godínez-Fernández, 2014; Pedersen, 2009; Sherman, 1996). Early models, such as the paradigmatic Chay-Keizer model (Chay & Keizer, 1983), adapted the Hodgkin-Huxley framework (Hodgkin & Huxley, 1952) to explain how glucose-induced changes modulate ion channel activity and generate bursting electrical patterns in rodent β-cells, which are closely associated with insulin release. As experimental observations revealed more complex oscillatory behaviors, such as fast, slow, or compound bursting, models evolved to incorporate the interplay between calcium handling by

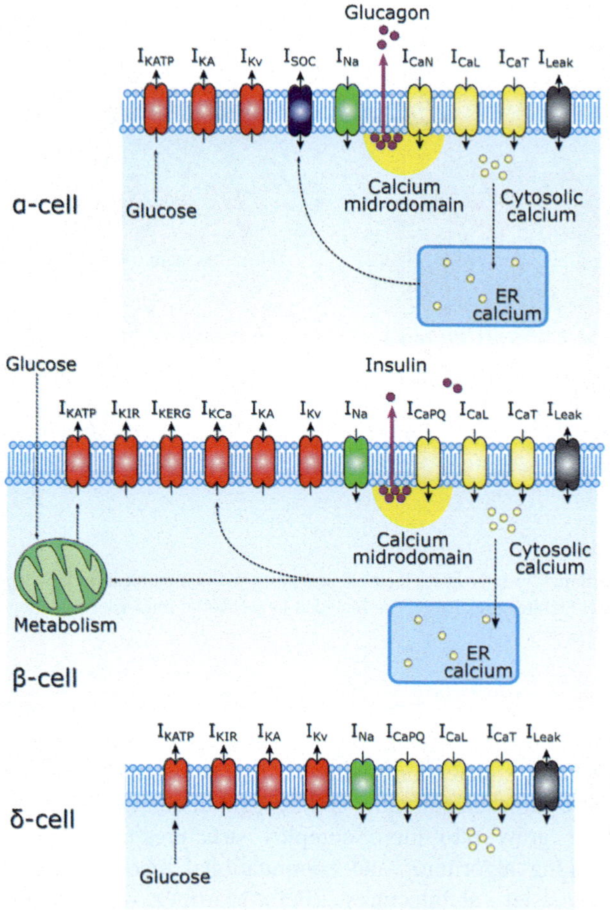

FIGURE 2.4 Models of electrical activity of single islet cells. Diagrams of the models of the electrical activity of human α (top), β (middle) and δ-cells (bottom) highlighting the different mechanisms (e.g., ion channels, endoplasmic reticulum, metabolism, intracellular calcium compartments) and complexity considered so far.

mitochondria and the endoplasmic reticulum and metabolic oscillations (Fig. 2.4A), perfectly exemplified by the Integrate Oscillator (Bertram et al., 2023; McKenna et al., 2016) and Dual Oscillator models (Bertram, Satin, et al., 2007; Bertram, Sherman, et al., 2007). Collectively, these models aimed to explain the origin of the different electrical behaviors observed experimentally and the underlying mechanisms to evaluate the impact of physiological and pathological changes on β-cell function.

Given the substantial interspecies differences, such as the glucose threshold for secretion, kinetics of insulin exocytosis, and repertoire of ionic channels, which limit the translatability of rodent-based models to humans directly, models of human β-cells have been developed (Félix-Martínez & Godínez-Fernández, 2017; Pedersen, 2010; Riz et al., 2015). Despite these advances, human models remain scarce because of the limited availability of human tissues for experimentation.

Glucagon-secreting α-cells

Models of pancreatic α-cells aimed to elucidate how increased glucose inhibits glucagon secretion while stimulating insulin secretion in β-cells, despite both cells sharing essentially the same cellular mechanisms. These models have explored diverse hypotheses, including the direct suppression of glucagon secretion by driving the α-cell into a depolarized, electrically silent state via KATP closure and the consequent inactivation of Na^+ and Ca^{2+} channels (Diderichsen & Göpel, 2006; Watts & Sherman, 2014), the impact of stochastic KATP channel activity and ATP/Ca^{2+} interplay on electrical and Ca^{2+} oscillations (González-Vélez et al., 2020), the role of inward store-operated currents, which at higher glucose levels would be inhibited, thus suppressing the firing of action potentials and therefore the secretion of glucagon (Watts & Sherman, 2014), and the inhibition of glucagon secretion by the inhibitory effect of insulin and somatostatin secreted by nearby β and δ-cells (Briant et al., 2018). Models of α-cells showed that endocrine pancreatic cells, grouped into islets, might rely on external signals and that the study of single cells, either experimentally or theoretically, might not reflect the whole picture of glucose regulation of hormone secretion in vivo. A schematic diagram of a model of a single α-cell is shown in Fig. 2.4B.

Somatostatin-secreting δ-cells

In contrast to α and β-cells, which have been modeled due to the direct effect of glucagon and insulin on glucose regulation, the electrical activity of somatostatin-secreting δ-cells has been modeled mainly for its indirect effect by inhibiting both glucagon and insulin secretion. In fact, models of δ-cells based completely on experimental data are scarce and most of them are adapted versions of models of α-cells (Briant et al., 2017, 2018; Diderichsen & Göpel, 2006; Watts et al., 2016). Only a model of the electrical activity of a single human δ-cell based completely on electrophysiological experiments has been proposed (Félix-Martínez et al., 2020), although such a model still lacks subcellular components already considered in models of α or β-cells (e.g., metabolism and calcium handling), as described graphically in Fig. 2.4C.

Paracrine and electrical communication in pancreatic islets

Eventually, models of the electrical activity of endocrine pancreatic cells added intra-islet communication to explain experimental observations. These models have evolved from minimal models of a few electrically coupled β-cells via gap junctions to increasingly sophisticated multicellular frameworks that integrate both electrical coupling

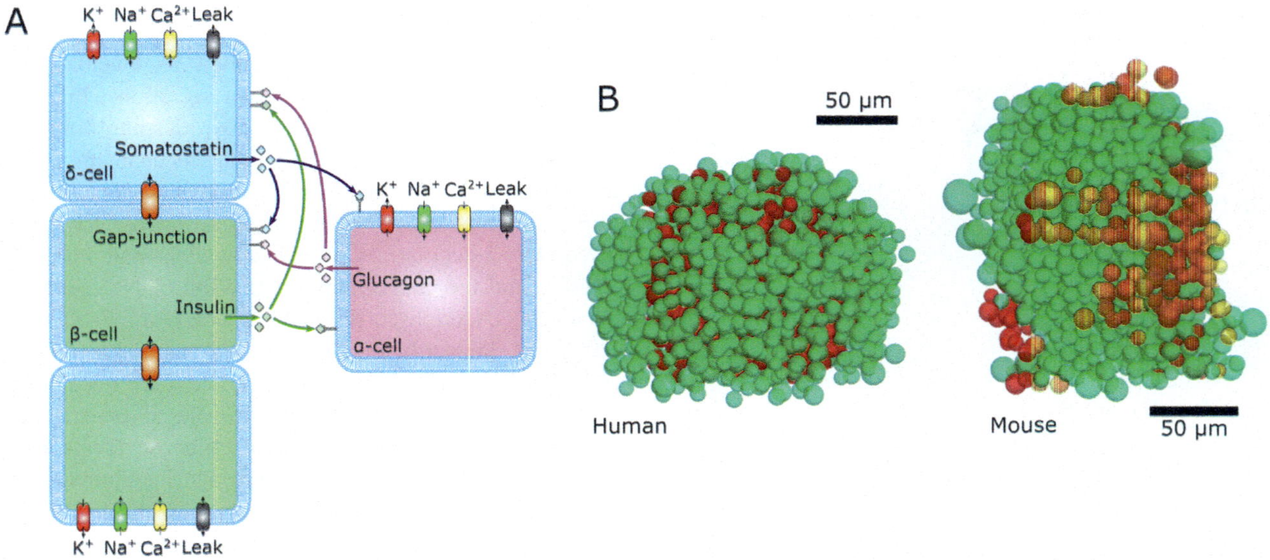

FIGURE 2.5 Multicellular models of paracrine and electrical communication. (A) Inter-islet electrical and paracrine communication have been modeled to explain experimental observation at the islet level. (B) Complex islet architectures have been included in models of intra-islet signaling to study the functional implications of islet cell organization.

and, in recent years, complex paracrine signaling among α-, β-, and δ-cells (Fig. 2.5A). The evolution of multicellular islet models has been recently reviewed by Félix-Martínez and Godínez-Fernández (2023).

Hypotheses tested using models of electrically coupled β-cells included the synchronization of β-cells in early models (Bangham et al., 1986), the diffusion of ions and metabolites through gap-junctions (Tsaneva-Atanasova et al., 2006), the origins of heterogeneous bursting periods in β-cells (i.e., intrinsic vs. emergent behavior due to cell coupling), the role of noise (Jo et al., 2005), and the heterogenous electrophysiological properties of β-cells such as ion channel conductances, metabolism, cell size and gap–junction conductance (Zimliki et al., 2004), the potential for intercellular K^+ diffusion to synchronize islet activity (Stokes & Rinzel, 1993), the emergence and influence of leader or hub cells (Dwulet et al., 2021), the mechanisms of wave propagation within islets (Benninger et al., 2008), the role of paracrine signaling on the regulation of glucagon secretion (Briant et al., 2018) and the functional implications of the islet cytoarchitecture (Nittala et al., 2007). While most of these models were based on experimental data from rodents, some of these ideas have been extrapolated to human islets (Lei et al., 2018).

Accumulating experimental evidence has led to the incorporation of paracrine feedback loops and the recognition of the importance of multicellular organization for synchronized hormone pulses. Advances in imaging, optogenetics, and computational power enabled the development of realistic islet models composed of hundreds or thousands of cells, revealing how these pathways coordinate hormone secretion and maintain glucose homeostasis. These models first adopted relatively simple arrays of cells such as open- or closed-ended linear arrays of cells, squared or hexagonal bidimensional arrays, and three-dimensional cubic and hexagonal configurations that later grew into more complex structures using sphere packing algorithms and computational reconstruction of real islet architectures (Félix-Martínez & Godínez-Fernández, 2023). Examples of reconstructed three-dimensional architectures of mouse and human islets used in multicellular models are shown in Fig. 2.5B.

Representative examples of models of paracrine signaling are the model of Watts et al. (2016), who developed a model of paracrine interactions among α, β, and δ-cells to study the paracrine effects on α-cells, and the model by Briant et al. (2018), where, in addition to paracrine signals, electrical communication between β and δ-cells was also considered. While these models were based on single-cell models of electrical activity, alternative approaches have been used. This is the case of the model of coupled oscillators by Hoang et al. (2016) where the biophysical details were neglected, favoring instead the analysis of how the spatial organization and paracrine/autocrine signaling among α, β, and δ-cells enable synchronous or asynchronous hormone pulses depending on glucose levels, which may be crucial for metabolic regulation. Recent models also account for cell heterogeneity (e.g., leader and follower β-cells) and the dynamic interplay between intrinsic and extrinsic regulatory mechanisms, providing deeper insights into islet physiology and the pathophysiology of diabetes.

Despite these advances, there remain significant gaps. For instance, most models have adopted a simplified representation of paracrine signals and lack important anatomical features, such as islet vasculature and δ-cell projections. In addition, models are often oversimplified by not fully accounting for the diversity and spatial arrangement of islet cells, as technical limitations hinder the ability to map and record the activities of different cell types at high resolution. Future studies should focus on integrating more realistic islet architectures, cellular heterogeneity, spatial dynamics, and more detailed molecular mechanisms to better capture the complexity of intercellular communication within the islets.

Inter-islet synchronization

Synchronization among pancreatic islets has also been investigated using mathematical modeling. Proposed mechanisms for islet synchronization primarily focus on negative feedback loops via glucose-insulin dynamics. According to this hypothesis, insulin secretion lowers blood glucose, a signal sensed by all islets that coordinates their activity. Mathematical models support this idea, indicating that a pancreas-liver feedback, specifically insulin-dependent modulation of hepatic glucose output, could be sufficient to synchronize islet oscillations (Bruce et al., 2022). A different proposal involves a periodic input from intrapancreatic ganglia (modeled as IP3 pulses produced by acetylcholine) as a plausible explanation for the synchronization of a population of islets, coordinating their oscillations so that plasma insulin remains pulsatile, even if only a fraction of the islets is innervated (Adablah et al., 2019). Collectively, these models suggest that both metabolic feedback and neural input may act together to ensure robust and pulsatile insulin secretion.

Models of exocrine pancreas function

The exocrine pancreas has been modeled using a diverse array of mathematical and computational approaches to elucidate its physiological function and pathological mechanisms, focusing on bicarbonate secretion and diseases such as pancreatitis and cystic fibrosis (Dressel & Custer, 2024; Sohma et al., 2000; Whitcomb & Ermentrout, 2004; Yamaguchi et al., 2017). These investigations used mathematical modeling to resolve the physiological paradox of how ductal cells secrete a HCO_3^--rich fluid with a concentration exceeding 140 mM. A primary focus of these studies was to identify the transport mechanisms involved (Fig. 2.6).

Other models involving the exocrine pancreas have focused on gaining a better understanding of the impact of genetic mutations and microenvironmental factors on pancreatic cancer progression and therapy (Chen et al.,

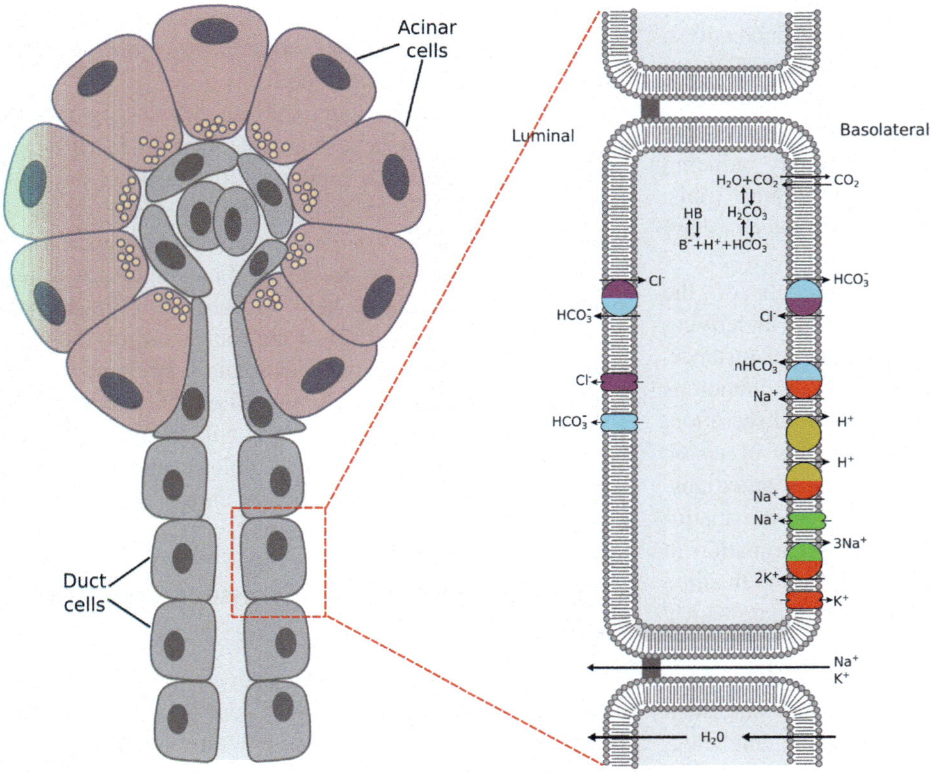

FIGURE 2.6 Models of exocrine pancreas. Models of duct cells have been useful in gaining a better understanding of the transport mechanisms involved in the secretion of HCO3- at high concentrations.

2020; Plaugher & Murrugarra, 2021), and the dynamics of cytokine-mediated inflammation in chronic pancreatitis and cystic fibrosis (Hao et al., 2017). Models of the exocrine pancreas are scarce compared to those of the endocrine pancreas. However, it has been proposed recently that endocrine and exocrine tissues are functionally integrated via a shared, bidirectional vascular network, and that blood supply and signals (e.g., insulin) could affect both compartments rather than being isolated systems, laying the foundation for integrated models that consider both endocrine and exocrine components.

Models of pancreatic disorders and clinical applications

As described throughout this chapter, mathematical modeling, as a complement to experimental work, has emerged as a valuable tool for deciphering the complex pathophysiology of pancreatic disorders, providing a formal framework for integrating experimental data, generating testable hypotheses, and driving clinical translation. The most notable example of this impact is the development and regulatory acceptance of the UVA/Padova Type 1 Diabetes Simulator, which began as an extension of physiological models of glucose-insulin dynamics and evolved into the first FDA-accepted substitute for animal trials in the preclinical testing of insulin therapies and artificial pancreas algorithms. By simulating a virtual population with metabolic variability, including hypoglycemic counterregulation and intraday fluctuations, the simulator provides a robust, cost-effective, and ethical environment for in silico trials. It has been extremely useful in accelerating the development and safety validation of closed-loop control systems, effectively reducing the risk involved in the pathway to human clinical trials, and marking a paradigm shift in how diabetes technology is engineered.

The applications of mathematical models of the exocrine pancreas extend to elucidating the underlying mechanisms of other pancreatic pathologies. For instance, computational models have shown how CFTR mutations severely impair pancreatic bicarbonate and fluid secretion, thus explaining the pathogenic characteristics of cystic fibrosis (i.e., mucus plugging and pancreatic autodigestion). Similarly, models developed to simulate the spatiotemporal dynamics of cytokine-mediated inflammation in chronic pancreatitis have been developed, offering in silico platforms for testing hypothetical anti-inflammatory interventions.

In contrast, islet electrophysiology models have contributed to explaining how cellular mechanisms, such as ion channels, Ca^{2+} handling, metabolism, and electrical and paracrine communication, could fail in diabetes, either impairing insulin secretion, producing hyperglucagonemia, or disrupting islet oscillations. These computational approaches have provided a quantitative approach to study disease progression, ultimately paving the way for more personalized and mechanism-based therapeutic strategies.

Computational models have also been used to simulate the glucose/oxygen dynamics and biphasic insulin release to evaluate the performance of encapsulation devices in preclinical contexts (Félix-Martínez et al., 2024; Komatsu et al., 2017). These models are useful for informing the design and optimization of bioartificial pancreas systems, transplant site selection and oxygen management for transplantation, and shaping strategies to reduce hypoxic graft loss. They also support protocol design for the preservation of islet cultures and early post-transplant phases and in silico testing of specific scenarios to accelerate development and decision-making.

Model limitations and perspective

Despite the significant advances described above, computational models of the pancreas remain limited. For instance, most models rely on simplified assumptions that compromise physiological realism and predictive power. These abstractions often omit regulatory mechanisms while adopting assumptions that limit their identifiability and generalizability across species, populations, and experimental protocols. Similarly, data limitations, particularly the predominance of models based on rodent data with scarce human models, further restrict their applicability and clinical translatability.

Future research should focus on developing mechanistic multiscale models that integrate endocrine and exocrine processes with high-resolution spatial and temporal dynamics. Although this would require access to diverse large datasets to improve parameter estimation and model validation, incorporating artificial intelligence and machine learning could help explore high-dimensional parameter spaces and enhance the robustness of computational models of the pancreas. Ultimately, bridging the gap between experimental observations, computational modeling, and clinical applications will require collaborative efforts across computational biology, physiology, and clinical research to build predictive, interpretable, and usable tools with clinical relevance.

Other hormones, endocrine tissues, and their pathways

The mathematical frameworks and computational methodologies used to model the endocrine and exocrine pancreatic tissue are not exclusive to this area of study, but rather constitute a generalizable toolkit for other biological systems. For instance, the principles of modeling ion transport mechanisms, metabolism, intracellular calcium handling, electrical activity at the cellular level and intracellular signals involved in secretory processes, as well as mechanisms of intercellular communication and the potentiation of secretory responses, are common biological processes. Similarly, multicellular systems based on the same principles are ubiquitous. All these similarities make it possible to extrapolate the modeling structures and algorithms used for modeling the pancreas tissues to simulate a vast array of other systems.

Applications of the material

In this chapter, a general overview of the field of computational modeling of the pancreas is provided, describing how models have been used as a complement to experimental work at different levels of organization, integrating diverse perspectives with a common goal. For experimentalists, this chapter provides a general glimpse of the field of computational/mathematical modeling of the pancreas and could be useful for exploring its use in designing experiments, testing hypotheses, and exploring therapeutic ideas. Clinicians can use this material to lead research projects in which both experimentalists and modelers are involved to improve diagnostic accuracy and design personalized treatments and devices. Hopefully, modelers will benefit from this chapter by becoming aware of some of the gaps, limitations, and emerging trends in the field. Ultimately, the main objective of this chapter is to serve as a bridge connecting experimentalists, modelers, and clinical scientists, promoting collaboration and knowledge exchange to enhance our understanding and to improve patient care.

Mini-dictionary of terms

Computational model. A mathematical representation of a biological system implemented using a computer program.

Compartmental model. A type of mathematical model that divides a system into distinct homogeneous, well-mixed compartments.

Differential equations. Equations that describe the rate of change of a quantity (e.g., glucose) with respect to another variable (e.g., time or distance).

Glucose clamp technique. Experimental method that involves injecting glucose and/or insulin to clamp blood glucose at a predetermined concentration.

In silico simulator. Software implementation of a mathematical or computational model.

Kinetic scheme. A mathematical representation of a dynamic process generally involving reactions, transitions or interactions.

Mathematical model. A formal framework that uses equations, functions, and parameters to describe the behavior of a system.

Multiscale model. Computational framework that integrates processes occurring at different spatial (e.g., molecular, cellular, tissue, organ) and temporal scales (e.g., milliseconds, seconds, minutes, hours).

Lumped parameter model. A simplified mathematical model that aggregates complex processes into a single or small set of parameters.

Tracer study. Experimental technique in which a small amount of labeled glucose is infused into the bloodstream to track glucose production, distribution, and utilization in real time.

Summary points

- Early models of glucose regulation provided quantitative parameters for distinguishing between subjects with normal blood glucose, prediabetes, and diabetes, based on oral or intravenous glucose tolerance tests.
- Models of glucose regulation have been extended to capture real-life meal conditions, eventually resulting in complex simulators approved for preclinical trials.
- Subcellular models of the mechanisms of insulin and glucagon secretion have been explained using computational models validated against diverse experimental protocols.
- The electrical activity of isolated endocrine cells has been modeled, considering ion transport mechanisms, calcium handling, and metabolism.
- Multicellular models have explored intra-islet communication between islet cells, including electrical and paracrine signaling.
- Multiscale models have integrated subcellular and cellular models with increasingly complex representations of islet architectures to capture the functional implications of cell organization in pancreatic islets.
- Potential mechanisms of synchronization between pancreatic islets have been studied using computational modeling.
- Models of the exocrine pancreas have focused on explaining the mechanisms involved in the secretion of high concentrations of bicarbonate and pathological states, such as pancreatitis, cystic fibrosis, and cancer.
- Despite their limitations, models of the pancreas and pancreatic function have been useful as a complement to experimental work in elucidating the mechanisms involved in the development of diverse pathological conditions.
- By integrating detailed models of both the exocrine and endocrine pancreas at different organizational levels, future models could bridge the gap between experiments, theory, and clinical applications.

Declaration of AI and AI-assisted technologies in the writing process

During the preparation of this work, the author used Paperpal to check the grammar and error corrections. Under no circumstances was it used during the writing process. After using this service, the author reviewed and edited the content and take full responsibility for the content of the publication.

References

Ackerman, E., Gatewood, L. C., Rosevear, J. W., & Molnar, G. D. (1965). Model studies of blood-glucose regulation. *Bulletin of Mathematical Biophysics, 27*(Suppl. 1), 21–37. https://doi.org/10.1007/bf02477259

Adablah, J. E., Vinson, R., Roper, M. G., & Bertram, R. (2019). Synchronization of pancreatic islets by periodic or non-periodic muscarinic agonist pulse trains. *PLoS One, 14*(2), e0211832. https://doi.org/10.1371/journal.pone.0211832

Ali, H. A., Daneshkhah, A., Boutayeb, A., Merabet, N., & Mukandavire, Z. (2025). Computational mathematics and modelling for diabetes. *Industrial and Applied Mathematics*, 133–165. https://doi.org/10.1007/978-981-96-1925-2_3

Bangham, J. A., Smith, P. A., & Croghan, P. C. (1986). Biophysics of the pancreatic β-cell. *Advances in Experimental Medicine and Biology, 211*, 265–278. https://doi.org/10.1007/978-1-4684-5314-0_24

Benninger, R. K. P., Zhang, M., Head, W. S., Satin, L. S., & Piston, D. W. (2008). Gap junction coupling and calcium waves in the pancreatic islet. *Biophysical Journal, 95*(11), 5048–5061. https://doi.org/10.1529/biophysj.108.140863

Bergman, R. N. (1989). Toward physiological understanding of glucose tolerance: Minimal-model approach. *Diabetes, 38*(12), 1512–1527. https://doi.org/10.2337/diab.38.12.1512

Bergman, R. N. (2021). Origins and history of the minimal model of glucose regulation. *Frontiers in Endocrinology, 11*, 583016. https://doi.org/10.3389/fendo.2020.583016

Bergman, R. N., Ider, Y. Z., Bowden, C. R., & Cobelli, C. (1979). Quantitative estimation of insulin sensitivity. *American Journal of Physiology - Endocrinology and Metabolism, 236*(6), E667. https://doi.org/10.1152/ajpendo.1979.236.6.e667

Bertram, R., Marinelli, I., Fletcher, P. A., Satin, L. S., & Sherman, A. S. (2023). Deconstructing the integrated oscillator model for pancreatic β -cells. *Mathematical Biosciences, 365*, 109085. https://doi.org/10.1016/j.mbs.2023.109085

Bertram, R., Satin, L. S., Pedersen, M. G., Luciani, D. S., & Sherman, A. (2007a). Interaction of glycolysis and mitochondrial respiration in metabolic oscillations of pancreatic islets. *Biophysical Journal, 92*(5), 1544–1555.

Bertram, R., Sherman, A., & Satin, L. S. (2007b). Metabolic and electrical oscillations: Partners in controlling pulsatile insulin secretion. *The Australian Journal of Pharmacy: Endocrinology and Metabolism, 293*(4), E890–E900. https://doi.org/10.1152/ajpendo.00359.2007

Bolie, V. W. (1961). Coefficients of normal blood glucose regulation. *Journal of Applied Physiology, 16*(5), 783–788. https://doi.org/10.1152/jappl.1961.16.5.783

Brass, B. J., Abelev, Z., Liao, E. P., & Poretsky, L. (2010). *Endocrine pancreas* (pp. 37–55). Springer US.

Briant, L. J. B., Reinbothe, T. M., Spiliotis, I., Miranda, C., Rodriguez, B., & Rorsman, P. (2018). δ-cells and β-cells are electrically coupled and regulate α-cell activity via somatostatin: β-to-δ gap junction coupling. *The Journal of Physiology, 596*(2), 197–215. https://doi.org/10.1113/jp274581

Briant, L. J. B., Zhang, Q., Vergari, E., Kellard, J. A., Rodriguez, B., Ashcroft, F. M., & Rorsman, P. (2017). Functional identification of islet cell types by electrophysiological fingerprinting. *Journal of the Royal Society Interface, 14*(128), 20160999. https://doi.org/10.1098/rsif.2016.0999

Bruce, N., Wei, I.-A., Leng, W., Oh, Y., Chiu, Y.-C., Roper, M. G., & Bertram, R. (2022). Coordination of pancreatic islet rhythmic activity by delayed negative feedback. *American Journal of Physiology - Endocrinology and Metabolism, 323*(6), E492–E502. https://doi.org/10.1152/ajpendo.00123.2022

Chay, T. R., & Keizer, J. (1983). Minimal model for membrane oscillations in the pancreatic beta-cell. *Biophysical Journal, 42*(2), 181–189.

Chen, J., Weihs, D., & Vermolen, F. J. (2020). Computational modeling of therapy on pancreatic cancer in its early stages. *Biomechanics and Modeling in Mechanobiology, 19*(2), 427–444. https://doi.org/10.1007/s10237-019-01219-0

Cobelli, C., & Kovatchev, B. (2023). Developing the UVA/Padova type 1 diabetes simulator: Modeling, validation, refinements, and utility. *Journal of Diabetes Science and Technology, 17*(6), 1493–1505. https://doi.org/10.1177/19322968231195081

Cobelli, C., Man, C. D., Pedersen, M. G., Bertoldo, A., & Toffolo, G. (2014). Advancing our understanding of the glucose system via modeling: A perspective. *IEEE Transactions on Biomedical Engineering, 61*(5), 1577–1592. https://doi.org/10.1109/tbme.2014.2310514

Diderichsen, P. M., & Göpel, S. O. (2006). Modelling the electrical activity of pancreatic α-cells based on experimental data from intact mouse islets. *Journal of Biological Physics, 32*(3–4), 209–229. https://doi.org/10.1007/s10867-006-9013-0

Dolenšek, J., Pohorec, V., Rupnik, M. S., & Stožer, A. (2017). Pancreas physiology. In *Challenges in pancreatic pathology*. https://doi.org/10.5772/65895

Dressel, T. D., & Custer, C. M. (2024). A closed-loop negative feedback model for the pancreas: A new paradigm and pathway to a cure. *Medicine, 103*(28), e38802. https://doi.org/10.1097/md.0000000000038802

Dwulet, J. M., Briggs, J. K., & Benninger, R. K. P. (2021). Small subpopulations of β-cells do not drive islet oscillatory [Ca2+] dynamics via gap junction communication. *PLoS Computational Biology, 17*(5), e1008948. https://doi.org/10.1371/journal.pcbi.1008948

Félix-Martínez, G. J., González-Vélez, V., Godínez-Fernández, J. R., & Gil, A. (2020). Electrophysiological models of the human pancreatic δ-cell: From single channels to the firing of action potentials. *International Journal for Numerical Methods in Biomedical Engineering, 36*(2). https://doi.org/10.1002/cnm.3296

Félix-Martínez, G. J., & Godínez-Fernández, J. R. (2017). Modeling the spatiotemporal distribution of Ca2+ during action potential firing in human pancreatic β-cells. *Biomedical Physics & Engineering Express, 3*(2), 025020. https://doi.org/10.1088/2057-1976/aa669f

Félix-Martínez, G. J., & Godínez-Fernández, J. R. (2023). A primer on modelling pancreatic islets: From models of coupled β-cells to multicellular islet models. *Islets, 15*(1), 2231609. https://doi.org/10.1080/19382014.2023.2231609

Félix-Martínez, G. J., Osorio-Londoño, D., Godínez-Fernández, J. R., & Beard, D. A. (2024). Impact of oxygen and glucose availability on the viability and connectivity of islet cells: A computational study of reconstructed avascular human islets. *PLoS Computational Biology, 20*(8), e1012357. https://doi.org/10.1371/journal.pcbi.1012357

Félix-Martinez, G. J., & Godínez-Fernández, J. R. (2014). Mathematical models of electrical activity of the pancreatic β-cell: A physiological review. *Islets*, e949195. https://doi.org/10.4161/19382014.2014.949195

González-Vélez, V., Dupont, G., Gil, A., González, A., & Quesada, I. (2012). Model for glucagon secretion by pancreatic α-cells. *PLoS One, 7*(3), e32282. https://doi.org/10.1371/journal.pone.0032282

González-Vélez, V., Piron, A., & Dupont, G. (2020). Calcium oscillations in pancreatic α-cells rely on noise and atp-driven changes in membrane electrical activity. *Frontiers in Physiology, 11*, 602844. https://doi.org/10.3389/fphys.2020.602844

Grodsky, G. M. (1972). A threshold distribution hypothesis for packet storage of insulin and its mathematical modeling. *Journal of Clinical Investigation, 51*(8), 2047–2059. https://doi.org/10.1172/jci107011

Grodsky, G. M., & Francisco, S. (1972). A threshold distribution hypothesis for packet storage of insulin: II. Effect of calcium. *Diabetes, 21*(Suppl. 2), 584–593. https://doi.org/10.2337/diab.21.2.s584

Hao, W., Komar, H. M., Hart, P. A., Conwell, D. L., Lesinski, G. B., & Friedman, A. (2017). Mathematical model of chronic pancreatitis. *Proceedings of the National Academy of Sciences, 114*(19), 5011–5016. https://doi.org/10.1073/pnas.1620264114

Hoang, D.-T., Hara, M., Jo, J., & Herrath, M. G. von (2016). Design principles of pancreatic islets: Glucose-dependent coordination of hormone pulses. *PLoS One, 11*(4), e0152446. https://doi.org/10.1371/journal.pone.0152446

Hodgkin, A. L., & Huxley, A. F. (1952). A quantitative description of membrane current and its application to conduction and excitation in nerve. *The Journal of Physiology, 117*(4), 500–544.

Jo, J., Kang, H., Choi, M. Y., & Koh, D.-S. (2005). How noise and coupling induce bursting action potentials in pancreatic β-cells. *Biophysical Journal, 89*(3), 1534–1542. https://doi.org/10.1529/biophysj.104.053181

Komatsu, H., Cook, C., Wang, C.-H., Medrano, L., Lin, H., Kandeel, F., Tai, Y.-C., & Mullen, Y. (2017). Oxygen environment and islet size are the primary limiting factors of isolated pancreatic islet survival. *PLoS One, 12*(8), e0183780. https://doi.org/10.1371/journal.pone.0183780

Lamlili, M. E. N., Boutayeb, W., & Boutayeb, A. (2025). In *Computational mathematics and modelling for diabetes. Industrial and applied mathematics* (pp. 97–131). https://doi.org/10.1007/978-981-96-1925-2_2

Lei, C.-L., Kellard, J. A., Hara, M., Johnson, J. D., Rodriguez, B., & Briant, L. J. B. (2018). Beta-cell hubs maintain Ca 2+ oscillations in human and mouse islet simulations. *Islets, 10*(4), 151–167. https://doi.org/10.1080/19382014.2018.1493316

Man, C. D., Micheletto, F., Lv, D., Breton, M., Kovatchev, B., & Cobelli, C. (2014). The UVA/PADOVA type 1 diabetes simulator. *Journal of Diabetes Science and Technology, 8*(1), 26–34. https://doi.org/10.1177/1932296813514502

Man, C. D., Rizza, R. A., & Cobelli, C. (2007). Meal simulation model of the glucose-insulin system. *IEEE Transactions on Biomedical Engineering, 54*(10), 1740–1749. https://doi.org/10.1109/tbme.2007.893506

Mari, A., Tura, A., Grespan, E., & Bizzotto, R. (2020). Mathematical modeling for the physiological and clinical investigation of glucose homeostasis and diabetes. *Frontiers in Physiology, 11*, 575789. https://doi.org/10.3389/fphys.2020.575789

McKenna, J. P., Ha, J., Merrins, M. J., Satin, L. S., Sherman, A., & Bertram, R. (2016). Ca2+ effects on ATP production and consumption have regulatory roles on oscillatory islet activity. *Biophysical Journal, 110*(3), 733–742. https://doi.org/10.1016/j.bpj.2015.11.3526

Montefusco, F., & Pedersen, M. G. (2015). Mathematical modelling of local calcium and regulated exocytosis during inhibition and stimulation of glucagon secretion from pancreatic alpha-cells. *The Journal of Physiology, 593*(20), 4519–4530. https://doi.org/10.1113/jp270777

Nittala, A., Ghosh, S., Wang, X., & Crampin, E. (2007). Investigating the role of islet cytoarchitecture in its oscillation using a new beta-cell cluster model. *PLoS One, 2*(10), e983. https://doi.org/10.1371/journal.pone.0000983

Palumbo, P., Ditlevsen, S., Bertuzzi, A., & Gaetano, A. D. (2013). Mathematical modeling of the glucose–insulin system: A review. *Mathematical Biosciences, 244*(2), 69–81. https://doi.org/10.1016/j.mbs.2013.05.006

Pedersen, M. G. (2009). Contributions of mathematical modeling of beta cells to the understanding of beta-cell oscillations and insulin secretion. *Journal of Diabetes Science and Technology, 3*(1), 12–20. https://doi.org/10.1177/193229680900300103

Pedersen, M. G. (2010). A biophysical model of electrical activity in human β-cells. *Biophysical Journal, 99*(10), 3200–3207. https://doi.org/10.1016/j.bpj.2010.09.004

Pedersen, M. G., & Cobelli, C. (2013). Multiscale modelling of insulin secretion during an intravenous glucose tolerance test. *Interface Focus, 3*(2), 20120085. https://doi.org/10.1098/rsfs.2012.0085

Pedersen, M. G., Man, C. D., & Cobelli, C. (2011). Multiscale modeling of insulin secretion. *IEEE Transactions on Biomedical Engineering, 58*(10), 3020–3023. https://doi.org/10.1109/tbme.2011.2164918

Plaugher, D., & Murrugarra, D. (2021). Modeling the pancreatic cancer microenvironment in search of control targets. *Bulletin of Mathematical Biology, 83*(11), 115. https://doi.org/10.1007/s11538-021-00937-w

Riz, M., Braun, M., Wu, X., & Pedersen, M. G. (2015). Inwardly rectifying Kir2.1 currents in human β-cells control electrical activity: Characterisation and mathematical modelling. *Biochemical and Biophysical Research Communications, 459*(2), 284–287. https://doi.org/10.1016/j.bbrc.2015.02.099

Segre, G., Turco, G. L., & Vercellone, G. (1973). Modeling blood glucose and insulin kinetics in normal, diabetic and obese subjects. *Diabetes, 22*(2), 94–103. https://doi.org/10.2337/diab.22.2.94

Sherman, A. (1996). Contributions of modeling to understanding stimulus-secretion coupling in pancreatic beta-cells. *American Journal of Physiology - Endocrinology And Metabolism, 271*(2), E362–E372. https://doi.org/10.1152/ajpendo.1996.271.2.e362

Sohma, Y., Gray, M. A., Imai, Y., & Argent, B. E. (2000). HCO3− transport in a mathematical model of the pancreatic ductal epithelium. *Journal of Membrane Biology, 176*(1), 77–100. https://doi.org/10.1007/s00232001077

Stamper, I. J., & Wang, X. (2013). Mathematical modeling of insulin secretion and the role of glucose-dependent mobilization, docking, priming and fusion of insulin granules. *Journal of Theoretical Biology, 318*(C), 210–225. https://doi.org/10.1016/j.jtbi.2012.11.002

Stamper, I. J., & Wang, X. (2019). Integrated multiscale mathematical modeling of insulin secretion reveals the role of islet network integrity for proper oscillatory glucose-dose response. *Journal of Theoretical Biology, 475*, 1–24. https://doi.org/10.1016/j.jtbi.2019.05.007

Stokes, C. L., & Rinzel, J. (1993). Diffusion of extracellular K+ can synchronize bursting oscillations in a model islet of Langerhans. *Biophysical Journal, 65*(2), 597–607. https://doi.org/10.1016/s0006-3495(93)81092-0

Tsaneva-Atanasova, K., Zimliki, C. L., Bertram, R., & Sherman, A. (2006). Diffusion of calcium and metabolites in pancreatic islets: Killing oscillations with a pitchfork. *Biophysical Journal, 90*(10), 3434–3446. https://doi.org/10.1529/biophysj.105.078360

Visentin, R., Campos-Náñez, E., Schiavon, M., Lv, D., Vettoretti, M., Breton, M., Kovatchev, B. P., Man, C. D., & Cobelli, C. (2018). The UVA/Padova type 1 diabetes simulator goes from single meal to single day. *Journal of Diabetes Science and Technology, 12*(2), 273–281. https://doi.org/10.1177/1932296818757747

Watts, M., Ha, J., Kimchi, O., & Sherman, A. (2016). Paracrine regulation of glucagon secretion: The β/α/δ model. *American Journal of Physiology - Endocrinology And Metabolism, 310*(8), E597–E611. https://doi.org/10.1152/ajpendo.00415.2015

Watts, M., & Sherman, A. (2014). Modeling the pancreatic α-cell: Dual mechanisms of glucose suppression of glucagon secretion. *Biophysical Journal, 106*(3), 741–751. https://doi.org/10.1016/j.bpj.2013.11.4504

Whitcomb, D. C., & Ermentrout, G. B. (2004). A mathematical model of the pancreatic duct cell generating high bicarbonate concentrations in pancreatic juice. *Pancreas, 29*(2), e30–e40. https://doi.org/10.1097/00006676-200408000-00016

Yamaguchi, M., Steward, M. C., Smallbone, K., Sohma, Y., Yamamoto, A., Ko, S. B. H., Kondo, T., & Ishiguro, H. (2017). Bicarbonate-rich fluid secretion predicted by a computational model of guinea-pig pancreatic duct epithelium. *The Journal of Physiology, 595*(6), 1947–1972. https://doi.org/10.1113/jp273306

Zimliki, C. L., Mears, D., & Sherman, A. (2004). Three roads to islet bursting: Emergent oscillations in coupled phantom bursters. *Biophysical Journal, 87*(1), 193–206. https://doi.org/10.1529/biophysj.103.038471

Further reading

Kovatchev, B. (2019). A century of diabetes technology: Signals, models, and artificial pancreas control. *Trends in Endocrinology & Metabolism, 30*, 432–444. https://doi.org/10.1016/j.tem.2019.04.008

Metabolic models, in silico trials, and algorithms. *Journal of Diabetes Sciences and Technology, 19*(4), (2025), 895–907. https://doi.org/10.1177/19322968251338300

Zavala, E., et al. (2019). Mathematical modelling of endocrine systems. *Trends in Endocrinology & Metabolism, 30*(4), 244–257. https://doi.org/10.1016/j.tem.2019.01.008

Chapter 3

Pancreatic hormones and the gut-brain axis

Amirhossein Khoshnam[1], Faizan Bashir[2], Fatemeh Abediankenari[3], Mohammad Ebrahim Kherad[4], Mohammad Rahmanian[5] and Niloofar Deravi[5]

[1]Student Research Committee, School of Medicine, Zahedan University of Medical Sciences, Zahedan, Iran; [2]School of Medicine, Shiraz University of Medical Sciences, Shiraz, Iran; [3]Immunogenetics Research Center, Mazandaran University of Medical Sciences, Sari, Iran; [4]School of Medicine, Islamic Azad University, Sari, Iran; [5]Student Research Committee, School of Medicine, Shahid Beheshti University of Medical Sciences, Tehran, Iran

Abbreviations

A-cells Ghrelin-secreting cells
Akt protein kinase B
ARC arcuate nucleus
BBB blood–brain barrier
BDNF brain-derived neurotrophic factor
cAMP cyclic adenosine monophosphate
CNS central nervous system
DCs d-cells (somatostatin-producing cells)
DPP-4 dipeptidyl peptidase-4
DVC dorsal vagal complex
ENS enteric nervous system
GH growth hormone
GIP glucose-dependent insulinotropic polypeptide
GI gastrointestinal
GLP-1RA GLP-1 receptor agonist
GLP-1R GLP-1 receptor
GLP-1 glucagon-like peptide-1
GLP-2 glucagon-like peptide-2
GLUT2 glucose transporter type 2
GPCR G protein-coupled receptor
HPA axis hypothalamic-pituitary-adrenal axis
IL-6 interleukin-6
LPS lipopolysaccharide
mTOR mechanistic target of rapamycin
NEP 24.11 neutral endopeptidase 24.11
NTS nucleus tractus solitarius
PI3K phosphoinositide 3-kinase
PVN paraventricular nucleus
PYY peptide YY
SCFA short-chain fatty acids
SSTR somatostatin receptor
SST somatostatin
T1D type 1 diabetes
T2DM type 2 diabetes mellitus
TNF-α tumor necrosis factor alpha
VIP vasoactive intestinal peptide
VTA ventral tegmental area

Introduction

The pancreas is a retroperitoneal, complex exocrine and endocrine organ that governs metabolic balance via its secretion of critical hormones including insulin and glucagon. Emerging evidence points to a dynamic interaction between these hormones and the gut-brain axis, a two-way network that connects the immunological, endocrine, and neurological systems to the gastrointestinal tract. This connection and association provide new information about metabolic regulation and offers new perspectives on diseases ranging from diabetes to neurovegetative disorders.

Through its dual secretory capabilities, the pancreas exhibits remarkable functional variety. The exocrine component of this organ is responsible for production, secreting, and storing pancreatic enzymes. Digestive enzymes produced by acinar cells include amylases, nucleases, and proteases. The endocrine component of this gland plays a vital role in blood glucose regulation in the body. It is made up of five different types of cells: alpha, beta, gamma, delta, and epsilon cells. The vast majority of these cells are beta cells, which produce amylin and insulin. Alpha cells produce glucagon, while delta cells release somatostatin, gamma cells synthesize pancreatic polypeptide, and epsilon cells generate ghrelin. These hormones play an important role in the body's glucose

Hormones in Neuroscience. https://doi.org/10.1016/B978-0-443-30263-3.00036-1

metabolism; insulin increases blood glucose uptake and conversion into fatty acids via lipogenesis. Glucagon regulates blood glucose levels and its release is triggered by a variety of metabolic stimuli, including changes in blood glucose levels, certain amino acids, and potentially free fats. Additionally, stresses such as sympathetic nervous system activity might cause its release (Valente et al., 2024). The hormone amylin serves multiple functions, including delayed stomach emptying, postprandial glucose control, and meal-induced glucagon secretion (Volčanšek et al., 2025). These hormones work closely together to regulate glucose metabolism, appetite regulation, and nutrient processing.

The gut brain axis is defined as a bidirectional communication or relationship between the gastrointestinal tract and the brain, which happens primarily via the autonomic nervous system and the hypothalamic-pituitary axis. Neural signals from gut to brain are predominantly mediated by vagal and spinal afferent neurons. Multiple mechanisms facilitate this connection for metabolic homeostasis, with major mediators including enteroendocrine cells (EECs), neuropod cells, and enterochromaffin cells (ECs). These specialized gut cells secrete peptides such as glucagon like peptide 1 (GLP-1), Cholecystokinin (CCK), Glucose-dependent insulinotropic polypeptide (GIP), and Peptide YY (PYY), alongside neurotransmitters including serotonin (5-HT). These signaling molecules stimulate vagal afferent neurons either directly or indirectly via the enteric nervous system, passing information to the nucleus tractus solitarius (NTS) and higher brain centers. Gut peptides can also reach systemic circulation and signal directly to the NTS or stimulate splanchnic nerve endings, establishing communication channels to the brain through spinal afferent neurons (Wachsmuth et al., 2022). The gut microbiota, with an estimated 10 trillion of varying microbial organisms, is an important part of this communication network (Cani et al., 2019). Microbial organisms create many metabolites that impact gut-brain communication directly and indirectly. The microbiome directly modulates the expression of nutrient receptors and release of gut peptides from EECs and affects serotonin release from enterochromaffin cells, which can potentially affect enteric nervous system function and vagal afferent activation (Wachsmuth et al., 2022).

The overlap of pancreatic hormones with the gut-brain axis has drawn recent interest in research, with increasing evidence showing the involvement of the brain in glucose regulation and energy homeostasis (Roh et al., 2016). *Obici* et al. showed that insulin operates centrally to regulate hepatic glucose metabolism (Obici et al., 2002). Moreover, glucose-sensing processes in hypothalamic neurons are similar to those for pancreatic β-cells. Brain areas regulating glucose metabolism, i.e., hypothalamic nuclei and brainstem regions, possess glucose-sensing neurons whose excitability changes with fluctuations in glucose levels. These areas are also important for regulating energy balance. GLP-1, released by intestinal L-cells postprandially, acts via its receptors in brain and pancreatic tissues. In the brain, GLP-1 receptors are strategically positioned in major sites such as vagus nerve terminals, the central nucleus of the amygdala, and the hypothalamic regions of paraventricular nucleus (PVN) and arcuate nucleus (ARC), and the caudal brainstem, which points to its involvement in central nervous system signaling (Roh et al., 2016). At the same time, GLP-1 receptors are maximally expressed on pancreatic beta cells, where they activate glucose-stimulated insulin release, thus directly connecting gut signals and pancreatic hormone release (Alavi & Shahmabadi, 2021).

Understanding of this relationship is of particular translational relevance, since gut-brain axis metabolite dysregulation has been noted in conditions such as hyperinsulinemia, obesity, and neurodegenerative disease (Green et al., 2025). The gut-brain axis represents an important signaling network regulating glucose homeostasis, and perturbation is linked with the pathophysiology of type 2 diabetes mellitus. T2DM insulin resistance is related to disturbances in gut-brain axis signaling, pointing to the clinical relevance of these intricate metabolic networks (Bessac et al., 2018).

This chapter examines the complex molecular mechanisms underlying pancreatic hormone integration with gut-brain signaling and aims to provide a comprehensive framework for understanding this multifactorial pathway.

Insulin

Insulin, a 51-amino-acid hormone, is crucial for regulating blood glucose levels, supporting cell growth, and maintaining metabolic functions (Lewis & Brubaker, 2021). Previously thought to be exclusively produced by pancreatic β cells, insulin has now been detected in low concentrations within specific neurons of the central nervous system, according to recent evidence (Csajbók & Tamás, 2016). While both the synthesis and secretion of insulin are regulated by blood glucose levels, each process is triggered at a different glucose concentration threshold (Alarcón et al., 1993; Kaufman et al., 2015). Insulin secretion requires glucose levels to rise above 5 mM, whereas its biosynthesis is stimulated by subtler fluctuations between 2 and 4 mM (Alarcón et al., 1993). After its release, insulin enters systemic circulation and reaches hepatocytes, prompting them to store glucose as glycogen. It also acts on skeletal muscle cells and adipocytes, facilitating glucose uptake and thereby lowering blood glucose levels back to baseline (Vasiljević et al., 2020). Like other protein hormones, insulin activates the tyrosine kinase receptor pathway to promote glucose uptake, stimulate skeletal muscle protein synthesis, and drive both glycogenesis and lipogenesis (Yang et al., 2018). Insulin receptors located on the plasma membrane act as enzymes, catalyzing the transfer of phosphate groups from ATP to tyrosine residues on intracellular proteins, thereby initiating downstream signaling (Yang et al., 2018). Binding of insulin to the α subunits of its receptor induces autophosphorylation of the β subunits, activating the receptor's kinase function. This activation initiates phosphorylation of downstream intracellular proteins that regulate metabolism, promote cell growth, and influence gene expression involved in cell differentiation (Avruch, 1998). The gut–brain axis is a two-way communication system that links the gut and the brain through both hormonal signals and neural pathways (Hales et al., 2020). Multiple mechanisms connect the gut to the brain in maintaining metabolic balance. Traditionally, nutrient intake during a meal triggers gut-derived signals that are transmitted to the brain, providing the central nervous system (CNS) with information about the meal's size and composition (de Lartigue & Diepenbroek, 2016). The hypothalamus serves as the central hub for integrating gut-derived signals and other physiological inputs, coordinating the regulation of appetite, energy expenditure, and glucose homeostasis (Schwartz et al., 2000). Specialized neuroendocrine cells within the intestinal lining play a key role in postprandial gut-to-brain signaling. By sensing the contents of the intestinal lumen, they secrete metabolic hormones—such as GLP-1 and cholecystokinin (CCK)—that help regulate metabolic homeostasis (Gribble & Reimann, 2016).

Insulin is vital for brain function, affecting synaptic plasticity, the release of neurotransmitters, and the survival of neurons (Kleinridders et al., 2014). Some gut microbes, including *Akkermansia muciniphila*, enhance glucose regulation and insulin responsiveness by increasing mucin secretion and reinforcing the integrity of the intestinal barrier (Dang et al., 2020). Kootte et al. showed that fecal microbiota transplantation (FMT) from lean donors to people with metabolic syndrome can enhance insulin sensitivity, underscoring the therapeutic promise of targeting gut microbiota to combat insulin resistance (Kootte et al., 2017). Ferriera et al. demonstrated that insulin exerts its effects in the brain by binding to the insulin receptor (IR) and the insulin-like growth factor-1 receptor (IGF-1R), activating key signaling cascades such as the phosphoinositide 3-kinase (PI3K)/Akt and mitogen-activated protein kinase (MAPK) pathways (Ferreira et al., 2018). In a study Scobar et al. found insulin resistance is linked to elevated production and reduced clearance of amyloid-beta (Aβ), a key feature of Alzheimer's disease pathology (Escobar et al., 2022). Another study by Rodriguez et al. confirmed insulin resistance contributes to tau hyperphosphorylation, another hallmark of AD, by disrupting the regulation of glycogen synthase kinase-3β (GSK-3β) (Rodriguez-Rodriguez et al., 2017) (Fig. 3.1).

Kwoji et al. (2021) and Dias et al. (2022) demonstrated that administering a multi-strain probiotic cocktail can synergistically increase short-chain fatty acid (SCFA) production, enhance gut barrier integrity, and modulate neurotransmitter synthesis more effectively than single-strain probiotics, leading to improvements in insulin sensitivity and cognitive function (Dias et al., 2022; Kwoji et al., 2021). Studies have also proposed that microbiota-derived exosome therapy could be developed to deliver targeted microbial metabolites directly to the brain, thereby improving cognitive function and insulin sensitivity by influencing the gut-brain-metabolic axis (Manzaneque-López et al., 2023) (Fig. 3.2) (Table 3.1).

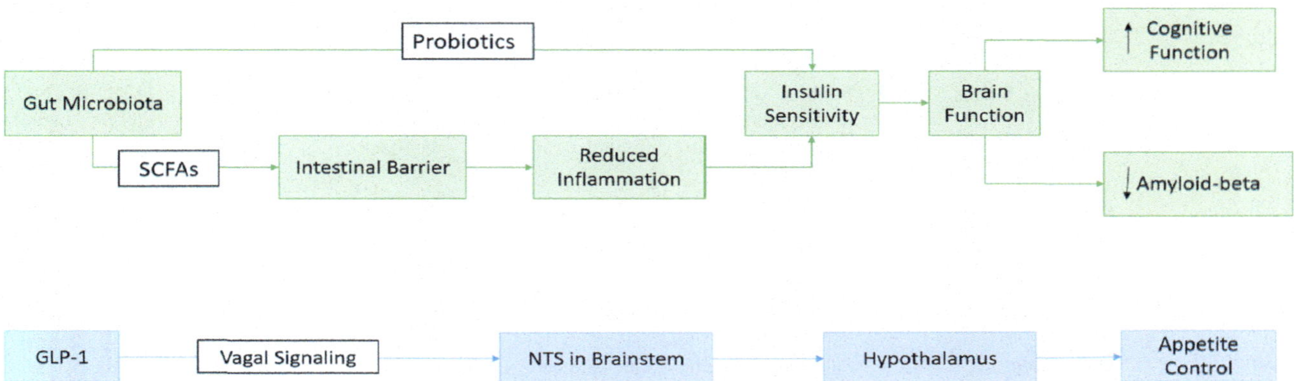

FIGURE 3.1 Gut microbiota influence on brain and metabolic function. This diagram presents how gut microbiota and GLP-1 influence brain function via the gut–brain axis. Microbiota-derived SCFAs enhance the intestinal barrier, reduce inflammation, and increase insulin sensitivity, leading to improved brain function, higher cognition, and reduced amyloid-beta. GLP-1 signals via vagal pathways to the hypothalamus to regulate appetite and metabolic balance.

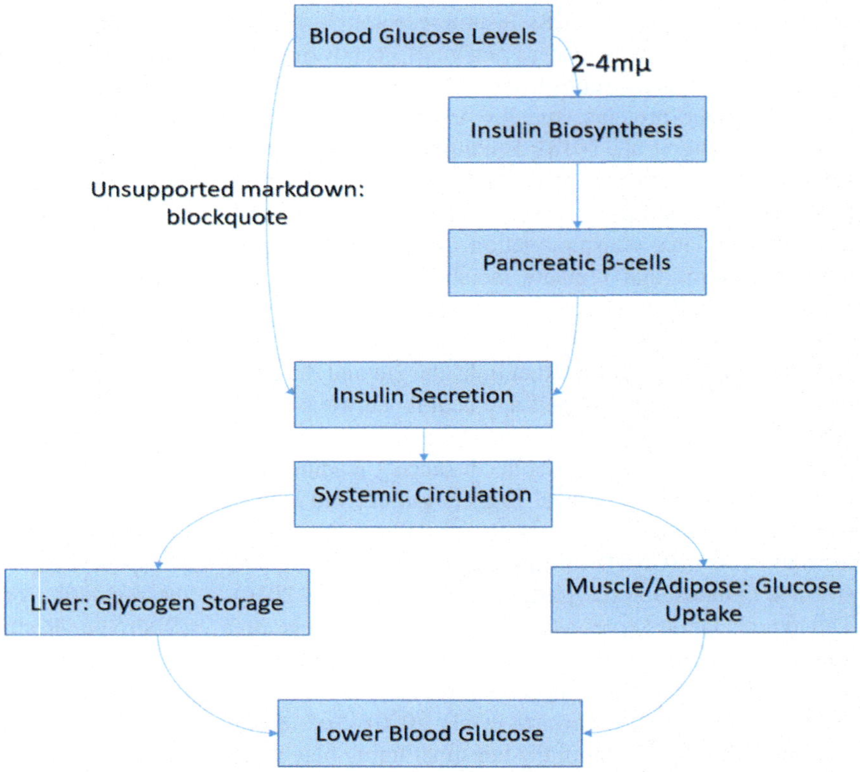

FIGURE 3.2 Insulin biosynthesis and systemic glucose regulation. This figure demonstrates the regulation of insulin synthesis and secretion in response to blood glucose levels. At 2–4 mM glucose, biosynthesis begins, with secretion occurring above 5 mM. Secreted insulin acts on liver (glycogen storage) and muscle/adipose tissue (glucose uptake), reducing systemic glucose and maintaining metabolic balance.

TABLE 3.1 Pancreatic hormones and their functional interactions within the gut–brain axis.

Hormone	Source	Key functions	Gut–brain axis involvement
Insulin	Pancreatic β-cells	• Lowers blood glucose • Promotes lipogenesis and glycogenesis • Supports neuroplasticity and neuron survival	• Signals to hypothalamus for energy regulation • Modulates hepatic glucose output via CNS • Influenced by gut peptides and microbiota
Glucagon	Pancreatic α-cells	• Raises blood glucose • Promotes gluconeogenesis and glycogenolysis • Balances insulin activity	• Modulated by GLP-1 • Participates in nutrient sensing via vagal and spinal pathways
Somatostatin	Pancreatic δ-cells, GI tract	• Inhibits insulin, glucagon, and gastric secretions • Suppresses gut motility and inflammation	• Regulates feeding behavior via SSTR2 in CNS • Controlled via vagal and enteric neural feedback
Ghrelin	Stomach A-cells, pancreas	• Stimulates hunger and food intake • Enhances pancreatic enzyme secretion • Modulates brain inflammation and cognition	• Signals through vagus nerve to the brainstem • Affected by microbiota composition and fasting state
VIP	Enteric neurons, pancreas	• Stimulates insulin and glucagon secretion • Promotes gut motility and vasodilation • Modulates mucosal immunity	• Acts on hypothalamic feeding centers • Maintains gut microbiota via epithelial signaling

Glucagon

Glucagon is a 29-amino-acid single-chain polypeptide with a distinct structural configuration that enables its selective binding to specific receptors (Müller et al., 2017). Glucagon is primarily synthesized in the pancreatic alpha cells, with its production controlled by intricate regulatory pathways that respond to variations in blood glucose levels (Hoosein & Gurd, 1984). This process promotes the release of glucose into the bloodstream, helping to raise blood glucose levels and prevent hypoglycemia. Additionally, glucagon interacts closely with insulin, working in tandem to maintain metabolic balance (Sasaki et al., 1980).

Within the brain and the enteroendocrine L-cells of the intestine, proglucagon undergoes cleavage by the enzyme prohormone convertase 1/3 (PC1/3), resulting in the formation of several peptides, including GLP-1, GLP-2, intervening peptide-2 (IP-2), glicentin, and oxyntomodulin. Oxyntomodulin (OXM) is a gastrointestinal hormone secreted by enteroendocrine cells in the colon, exerting its effects through both GLP-1 receptors and glucagon receptors (Holst, 1997). GLP-1 reduces the release of glucagon from pancreatic alpha cells, leading to a higher insulin-to-glucagon ratio, which in turn enhances the body's sensitivity to insulin (Nauck et al., 1993). Additionally, GLP-1 promotes an increase in the number of pancreatic beta cells by encouraging cell growth, supporting the formation of new islets, and preventing programmed cell death. Due to its strong insulin-stimulating properties, GLP-1 is widely regarded as a leading contender for the role of the proposed "incretin" hormone (Farilla et al., 2003). DPP-4, an enzyme broadly distributed in the endothelial lining of capillaries, swiftly deactivates GLP-1(7-36)-amide within the intestinal vessels, the hepatoportal circulation, and other peripheral tissues. GLP-1 is broken down by DPP-4 through cleavage at the N-terminal end, where the enzyme removes a histidine-alanine pair, producing the inactive form GLP-1(9-36)-amide (Plamboeck et al., 2005).

In the brain, DPP-4 is found to be particularly active in regions such as the hypothalamus, hippocampus, choroid plexus, leptomeninges, and certain circumventricular structures (Matheeussen et al., 2011). Another enzyme, neutral endopeptidase 24.11 (NEP24.11), also contributes to GLP-1(7-36)-amide breakdown and is located in both peripheral tissues and the central nervous system (Plamboeck et al., 2005). NEP24.11 further processes GLP-1(7-36)-amide by cleaving it into smaller fragments, notably GLP-1(28-36)-amide and GLP-1(32-36)-amide. Following its secretion by intestinal L-cells, GLP-1 reaches its peak levels in the extracellular space beneath the intestinal lining, shows moderate concentrations in the portal vein, and appears at significantly lower levels in the general bloodstream (Hupe-Sodmann et al., 1995). Estimates suggest that under 10% of GLP-1 secreted in the intestine remains biologically active after passing through the liver, and only a small fraction circulates to reach distant organs like the brain (Holst & Deacon, 2005).

Within the central nervous system, pre-proglucagon is primarily expressed in the nucleus of the solitary tract (NTS) and other regions of the brainstem (Kreymann et al., 1989; Larsen et al., 1997). Following post-translational modification, proglucagon undergoes a processing pathway similar to that in intestinal L-cells, resulting in the production of GLP-1 along with approximately equal quantities of glicentin and oxyntomodulin (Kreymann et al., 1989; Larsen et al., 1997). GLP-1(7-36)-amide is produced by neurons situated in the caudal portion of the NTS within the brainstem. This region borders the lower part of the fourth ventricle. The NTS is a component of the dorsal vagal complex (DVC), which also consists of the area postrema (AP) and the dorsal motor nucleus of the vagus (DMNX). This anatomical complex plays a crucial role in receiving visceral sensory signals, primarily transmitted by vagal afferents originating from the stomach and duodenum (Grill, 2010). Cells expressing pre-proglucagon are not confined solely to a group of neurons in the caudal section of the NTS; they are also present in a few neurons that project laterally from the NTS, traversing the dorsal reticular zone and reaching into the A1 region of the medulla (Llewellyn-Smith et al., 2011) (Table 3.2).

Studies using in situ hybridization for GLP-1 receptor mRNA, along with binding assays employing radiolabeled GLP-1, have revealed that GLP-1-responsive cells are widely distributed across various brain regions in rodents (Llewellyn-Smith et al., 2011). These investigations have also identified a dense population of neurons expressing GLP-1 receptors in both the hypothalamus and the brainstem—two key regions that regulate autonomic functions and maintain energy balance. Within the hypothalamus, GLP-1 receptors are predominantly located in the arcuate nucleus, paraventricular nucleus, dorsomedial nucleus, and supraoptic nucleus. In the dorsal vagal complex (DVC), the receptors are especially abundant in the nucleus of the solitary tract (NTS) and the area postrema. Additionally, GLP-1 binding sites have been detected in various circumventricular organs, including the subfornical organ and the area postrema (Yamamoto et al., 2003). Given the low concentration of GLP-1 in the bloodstream and its rapid degradation, researchers have questioned whether part of its endocrine action—particularly its influence on pancreatic β-cells and glucose regulation—might be mediated through a non-hormonal route. This has led to the hypothesis that GLP-1 released from intestinal L-cells could impact neuronal activity in the brain through an alternative neural circuit, possibly beginning with sensory mechanisms located in the hepatic portal area (Burcelin et al., 2001).

TABLE 3.2 Major gut–brain signaling elements and their effects on pancreatic and metabolic function.

Signaling element	Source or pathway	Role in gut–Brain communication	Effect on pancreas or metabolism
GLP-1	Intestinal L-cells, NTS neurons	• Signals nutrient presence to brain via vagus and circulation • Activates CNS appetite-regulating regions	• Stimulates insulin secretion Inhibits glucagon • Enhances β-cell survival and proliferation
Vagus nerve	Afferent/efferent autonomic nerve fibers	• Transmits gut-derived signals to brain • Sends CNS commands to pancreas and gut	• Controls insulin and enzyme release • Coordinates satiety and digestive responses
Gut microbiota	Colon and small intestine	• Produces SCFAs and neuroactive compounds • Modulates enteroendocrine and ENS signaling	• Affects insulin sensitivity • Regulates GLP-1, ghrelin, and inflammatory tone
Enteroendocrine cells (EECs)	Intestinal epithelium	• Sense luminal nutrients • Secrete hormones (e.g., GLP-1, CCK, PYY) to activate vagal neurons	• Influence pancreatic hormone secretion • Regulate appetite and energy homeostasis
Area postrema/ NTS	Brainstem (DVC complex)	• Integrates vagal signals from gut • Controls autonomic outputs to pancreas and GI tract	• Mediates ghrelin effects • Controls pancreatic enzyme secretion and appetite regulation

Recent animal studies, including those conducted in our own lab, have provided evidence that the process of glucose sensing triggers GLP-1 release, which then acts on receptors situated on vagal nerve endings in the gut region (Nakagawa et al., 2004). Consequently, metabolic signals are conveyed via the vagus nerve to the NTS in the brainstem, which subsequently forwards the glucose-related information to various hypothalamic nuclei (Washington et al., 2010).

The brain acts as a central hub for processing metabolic information and producing signals that direct energy flow to different tissues, regulating whether energy is utilized immediately or stored for later use. This complex communication pathway is known as the gut-brain-periphery axis. In our initial experiments, we demonstrated that infusing glucose at a low rate directly into the portal vein surprisingly led to a drop in blood glucose levels, a phenomenon referred to as paradoxical hypoglycemia (Burcelin et al., 2003). This finding suggested that glucose sensors in the portal vein were activated, and that the detection of glucose depended on the presence of the GLUT2 transporter (Burcelin et al., 2000). Based on these observations, we proposed that processes akin to those in insulin-secreting β-cells play a key role in triggering the gut-to-brain communication pathway. In this context, GLP-1 acts as a hormone that promotes the transmission of nutritional signals to the brain, whereas somatostatin functions as an inhibitory regulator in this signaling process (Fig. 3.3).

Vasoactive intestinal polypeptide (VIP)

In the early 1970s, a 28-amino acid neuropeptide was identified in pigs, with functions including vasodilation and serving as a gut hormone (Hou et al., 2022). Vasoactive intestinal polypeptide or VIP represents a key mediator in the enteric nervous system (Agibalova et al., 2024). Beyond its presence in the pancreas and other parts of the GI tract, this hormone is also located in the CNS, cardiovascular system, respiratory system, reproductive system, and immune organs (Hou et al., 2022; Semaniakou et al., 2021). Because of its extensive distribution, VIP functions as a neurotransmitter, vasodilator, secretagogue, and immunomodulator (Hou et al., 2022). VIP belongs to the gut-brain peptide family, where its structure is related to hormones such as glucagon and secretin (Semaniakou et al., 2021). The physiologic effects of this hormone are primarily mediated by two receptors, VPAC1 and VPAC2, both of which fall under class B of G-protein-coupled receptors. Since VIP operates and functions in various parts of the body, some of its functions include regulating hormone secretion, smooth muscle relaxation, and promoting gastrointestinal motility (Hou et al., 2022).

As noted previously, pancreatic islets receive innervation from parasympathetic, sympathetic, and sensory nerve fibers. This peptide exists in postganglionic parasympathetic neurons that originate from the dorsal motor nuclei of the vagus nerve (Hou et al., 2022). Within the pancreas, VIP's endocrine roles include stimulating

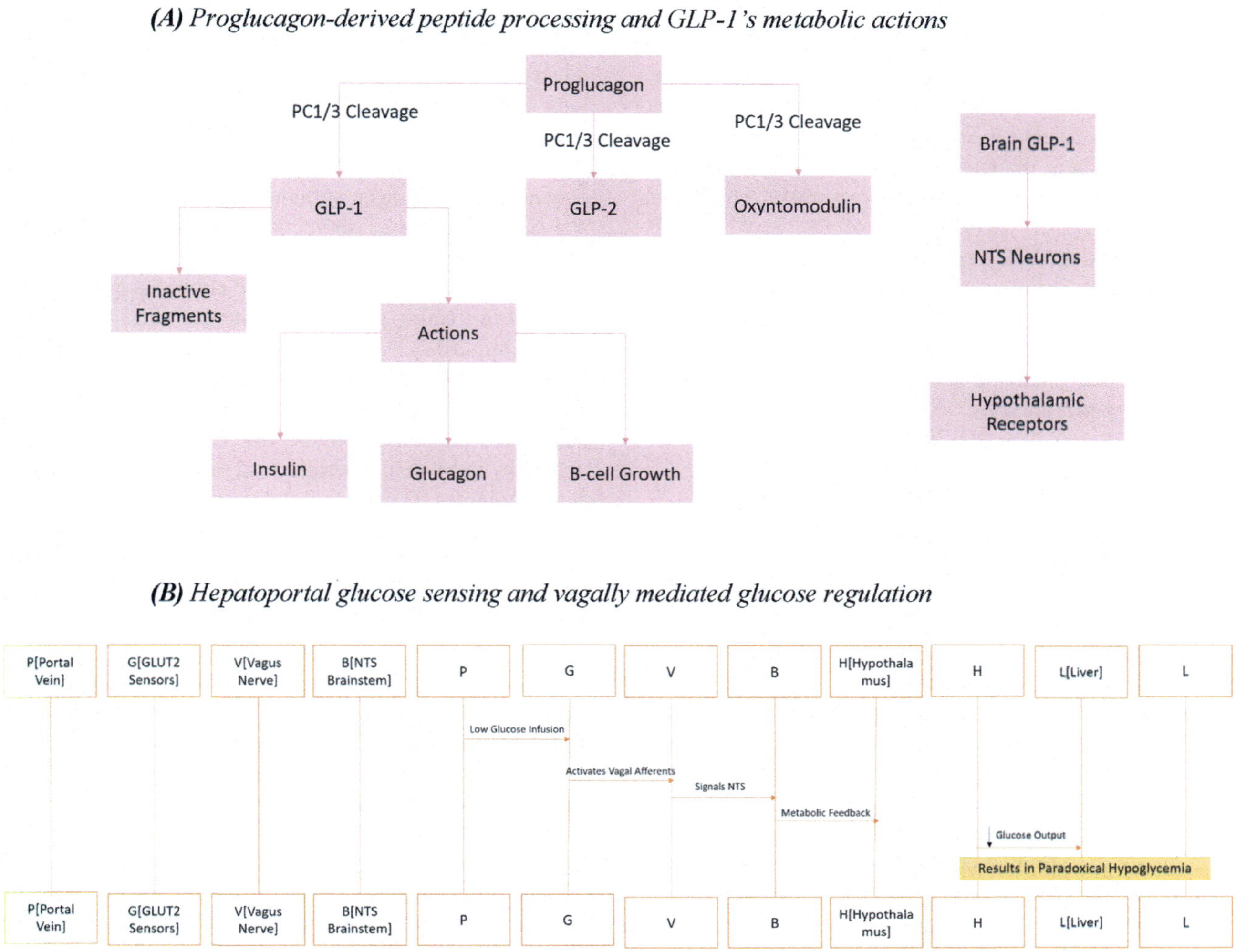

FIGURE 3.3 GLP-1-mediated pathways in peripheral and central glucose regulation. (A) Proglucagon-derived peptide processing and GLP-1's metabolic actions. (B) Hepatoportal glucose sensing and vagally mediated glucose regulation.

glucagon and insulin release, while its additional functions throughout the GI tract encompass mucosal ion transport, gastric acid production, blood flow regulation, gut motility, sphincter relaxation, and mucosal inflammatory immune responses (Hou et al., 2022; Vu et al., 2015).

VIP has multiple key roles in the gut-brain axis. It has been noted, brain tissue containing VIP-reactive neurons have been detected in numerous regions of the brain, including hypothalamic nuclei such as the suprachiasmatic nucleus (SCN) and Paraventricular Nucleus (PVN). First, this neuromodulator is pivotal in its regulation of food intake, with a 2022 rat study demonstrating that the appetite suppressing and metabolic effects of VIP are mediated through the PVN (Zaia et al., 2022). Second, VIP has been recognized for stimulating metabolic processes such as glycogenolysis and gluconeogenesis (Kalsbeek et al., 2010). Third, as is widely understood, microbes play a major part in the gut-brain axis, and VIP has been extensively studied in animal models to clarify its role in gut microbial regulation. By means of microbial neuroimmune mechanisms, it helps regulate intestinal motility (Bai et al., 2024). Besides that the balance between gut flora and epithelial cells depends on gut fucosylation. Through epithelium fucosylation, VIP generated by enteric neurons maintains gut microbiota, according to a study by Lei et al. (2022). Studies dating back to 1978 have also described this neuropeptide's role in the gut-brain axis. Results from a 1978 study by Ebeid et al., where VIP was examined in CSF in monkeys and dogs, demonstrate that peripheral elements, particularly liver function, heavily influence brain VIP processing. This indicates the gut-brain axis likely modulates both behavioral and intestinal activities (Ebeid et al., 1978).

Considering clinical relevance, VIP's therapeutic potential is made clear when we look at its role in GI disorders. In IBS, altered VIP levels (Iwasaki et al., 2019) have been associated with disordered gut motility and visceral hypersensitivity. IBD patients demonstrate dysregulated VIP signaling (Iwasaki et al., 2019), which likely exacerbates chronic inflammation and barrier dysfunction. Cancer research has shown that VIP can affect tumor growth and angiogenesis in some GI cancers (Moody et al., 2016), whereas in diabetes, VIP is involved in pancreatic beta-cell function and glucose homeostasis (Hou et al., 2022; Iwasaki et al., 2019).

In conclusion, VIP turns out to be a multifaceted neuropeptide with implications far beyond its origins as a "mere" vasodilator or hormone. Its widespread distribution in both nervous and digestive systems, along with its varied receptor-mediated activity, make VIP a key mediator of the gut-brain axis.

Ghrelin and gut-brain axis

Ghrelin is recognized as a peptide with a physiological glucoregulatory role, produced by the gastrointestinal tract, particularly A cells in the stomach and pancreas, as well as the brain (Dezaki et al., 2008). Ghrelin acts as an appetite-stimulating signal from the gastrointestinal tract to the brain, establishing communication through the afferent vagal pathway (Date et al., 2002). Ghrelin released from the stomach functions as an endocrine hormone, traveling via the bloodstream to its receptors in the area postrema, a part of the dorsal vagal complex (DVC) in the brain. These receptors play a role in activating vagal-vagal circuits, leading to the stimulation of efferent cholinergic vagal pathways, which enhance pancreatic enzyme secretion (Li et al., 2006).

Ghrelin levels rise before meals and decline after food intake, indicating its role in regulating the hunger-satiety cycle (Cummings et al., 2001). It also plays a significant role in body weight regulation and metabolism by stimulating hunger and modulating feeding behavior (Castañeda et al., 2010). Moreover, the secretion of gut peptides, including ghrelin and leptin, can be influenced by the gut microbiota. These hormones play a role in the gut-brain communication axis, contributing to energy homeostasis, appetite regulation, and even cognitive function (Abildinova et al., 2024)..In particular, ghrelin impacts the hippocampus and reduces neuroinflammation, leading to improved cognitive performance (Abildinova et al., 2024; Lai et al., 2024).

In a randomized, single-blind, crossover controlled trial conducted in 2012 by Greisa Vila et al., the effect of B-type natriuretic peptide (BNP) injection on appetite-regulating hormones (such as ghrelin), hunger, and satiety was investigated in 10 healthy participants. The participants received either a placebo or BNP-32 at a dose of 3.0 pmol/kg/min via continuous intravenous infusion for 4 hours in two separate sessions. Appetite-regulating peptide concentrations, including ghrelin, were measured hourly, and hunger and satiety were assessed using visual analog scales (VAS). The results showed that the fasting-induced increase in total and acylated ghrelin concentrations was inhibited by BNP. Additionally, compared to placebo, BNP reduced hunger and increased satiety. As an appetite-regulating hormone that rises during fasting and is associated with hunger stimulation, ghrelin's modulation by BNP demonstrated anorectic (appetite-suppressing) effects, suggesting the existence of a heart-gut-brain axis. This modulation of ghrelin by BNP could have therapeutic potential in treating patients with heart failure and obesity (Vila et al., 2012).

In 2005, H.P. Simonian and colleagues evaluated the effects of sham feeding and actual food intake on plasma ghrelin and pancreatic polypeptide (PP) concentrations in 15 healthy individuals. Serial blood samples were collected every 5 minutes for 30 minutes during sham feeding and 30 minutes after real food intake, with plasma concentrations of the two hormones measured using radioimmunoassay. The results indicated that sham feeding increased plasma ghrelin from 1730 ± 237 to 1917 ± 269 pg/mL and PP from 417 ± 50 to 841 ± 97 pg/mL, whereas real food intake raised PP from 782 ± 88 to 1710 ± 119 pg/mL but had no effect on ghrelin levels. These findings suggest that sham feeding, via efferent vagal pathways, increases ghrelin and PP secretion, while real food intake only elevates PP, confirming the role of efferent vagal pathways in regulating ghrelin and the gut-brain axis. The efferent vagal pathway during sham feeding enhances ghrelin and PP secretion, whereas during real food intake, only PP levels increase, with no change in plasma ghrelin. This study underscores the role of efferent vagal pathways in regulating ghrelin and the gut-brain axis (Simonian et al., 2005).

In a 2014 study by Maria Papastamataki and colleagues, preprandial levels of gut-brain axis hormones, including glucagon-like peptide-1 (GLP-1), glucose-dependent insulinotropic peptide (GIP), active amylin, acylated ghrelin (AG), leptin, pancreatic polypeptide (PP), and peptide tyrosine-tyrosine (PYY), were measured in 47 children with celiac disease (CD)—comprising 12 untreated (UCD), 22 treated with a gluten-free diet (TCD), and 13 with celiac disease and type 1 diabetes (DCD)—along with 18 healthy controls (HC) using hormone-map-array technology before and after treatment. The results revealed a significant reduction in acylated ghrelin (AG) concentrations in children with celiac disease and type 1 diabetes (DCD). However, no significant differences in ghrelin levels were observed in the UCD and TCD groups compared to the control group. As the active form of ghrelin, acylated ghrelin regulates appetite and metabolism, and its notable reduction in the DCD group may result from the interaction between celiac disease and type 1 diabetes, indicating more pronounced alterations in the gut-brain axis in this group (Papastamataki et al., 2014).

In 2006, Ying Li and colleagues investigated the effect of intravenous ghrelin injection (at doses of 5, 10, and 25 nmol) on pancreatic protein secretion in 4–5 anaesthetized and conscious rats per experimental group. Atropine, hexamethonium, acute vagotomy, and perivagal capsaicin were used as pretreatments to assess the role of cholinergic vagal pathways. To examine the role of the area postrema in the response to ghrelin, this region was selectively lesioned, and c-Fos expression was evaluated using immunohistochemistry in the area postrema, nucleus tractus solitarius (NTS), and dorsal motor nucleus of the vagus (DMV). The results showed that ghrelin injection increased pancreatic protein secretion in anaesthetized rats from 125 ± 6 mg/hour to 186 ± 8, 295 ± 12, and 356 ± 11 mg/hour, while a 10-nanomole dose in conscious rats resulted in a 2.2-fold increase over baseline. Furthermore, the pancreatic secretion response to ghrelin was completely abolished by pretreatment with atropine, hexamethonium, or acute vagotomy, but perivagal capsaicin had no effect. The pancreatic secretion response to ghrelin was also eliminated following area postrema lesioning, while this lesion had no impact on secretion induced by bile-pancreatic juice diversion. Ghrelin significantly increased the number of c-Fos-expressing neurons in the area postrema, NTS, and DMV, an effect that was abolished after area postrema lesioning.

These findings indicate that ghrelin, as an endocrine hormone, stimulates pancreatic secretion via cholinergic vagal pathways and the area postrema, sending digestive signals to the brain and playing a key role in the gut-brain axis (Li et al., 2006).

In a 2017 cross-sectional study by Pendharkar SA et al., aimed at exploring the association of gut-brain axis peptides, particularly ghrelin, with abnormal glucose metabolism (AGM), 83 patients were examined, of whom 36% (30 individuals) developed AGM following acute pancreatitis (AP). Fasting levels of ghrelin and other glucoregulatory peptides were measured using ELISA kits. The results indicated that, unlike oxyntomodulin, glicentin, and VIP, ghrelin showed no significant association with AGM and appears to play no direct role in glucose regulation post-AP. Ghrelin levels were above the detectable minimum in over 90% of patients, suggesting its continuous presence in the bloodstream, even in the fasting state. The study proposes that ghrelin production may be influenced by non-dietary stimuli, such as brain signals or gut microbiota. However, no significant correlation was found between ghrelin and oxyntomodulin, glicentin, or VIP. Consequently, ghrelin is not considered a therapeutic candidate for diabetes related to exocrine pancreatic diseases, whereas the study highlights oxyntomodulin as a potential therapeutic target (Pendharkar et al., 2017).

In a 2025 study by Shaoyang Liu et al., the impact of electroacupuncture (EA) on the gut-brain axis was evaluated in 24 rats with type 2 diabetes mellitus (T2DM) induced by a high-fat diet and intraperitoneal streptozotocin (STZ) injection. The rats were divided into four groups: T2DM, EA, EA with PI3K inhibitor (EA + LY294002), and control. After 5 weeks, the PI3K/Akt pathway, gut-brain axis indices, gut microbiota, food intake, body weight, fasting blood glucose (FBG), lipid profile (TC, TG, HDL, LDL), serum insulin levels, and insulin resistance and sensitivity indices were assessed. The results demonstrated that electroacupuncture (EA) reduced ghrelin protein and mRNA expression in the colon and hypothalamus in the T2DM group compared to the control group via the PI3K/Akt pathway, an effect reversed by the PI3K inhibitor (LY294002). This led to improvements in food intake, body weight, fasting blood glucose, lipid profile, and insulin resistance and sensitivity indices, as well as enhanced gut microbiota structure, which appears to be linked to ghrelin regulation. As a gut-derived hormone regulating appetite and energy metabolism, the reduced ghrelin levels in the EA group likely contributed to improved metabolism by decreasing appetite. Electroacupuncture, through ghrelin modulation in the gut-brain axis and the PI3K/Akt pathway, improved feeding behavior, body weight, glucolipid metabolism, and insulin resistance in T2DM rats, positioning this method as a potential therapeutic approach (Liu et al., 2025). A schematic overview of the mechanism is provide in Fig. 3.4A.

Somatostatin and gut-brain axis

Somatostatin (SST) is a peptide hormone with diverse functions in the central nervous system (CNS), peripheral nervous system (PNS), and enteric nervous system (ENS) (Timmermans et al., 1990).

Somatostatin is secreted by endocrine delta (D) cells in the stomach, intestine, and pancreas and, through various autocrine, paracrine, endocrine, and exoendocrine pathways, plays a role in processes including the secretion of hormones such as inhibiting the secretion of insulin, glucagon, and gastrin, reducing gastric acidity and inhibiting pancreatic digestive enzymes, intestinal motility, and inflammatory responses (Chen et al., 2002).

SST with inhibitory effect and 2–3 minutes half-life has five SST-receptors (SSTRs) in the GPCR superfamily. SST negatively affect the release of various hormones and cell proliferation via activation of its cognate receptors. The SSTR1-SSTR5 are binded with inhibitory G protein Gi/o and have role in motility, mucous and hormone secretion, blood vessels contractility, inflammation responses, and microbiotal flora (Stengel et al., 2011).

Studies have shown that capsaicin-sensitive sensory nerves can produce anti-anxiety and anti-inflammatory

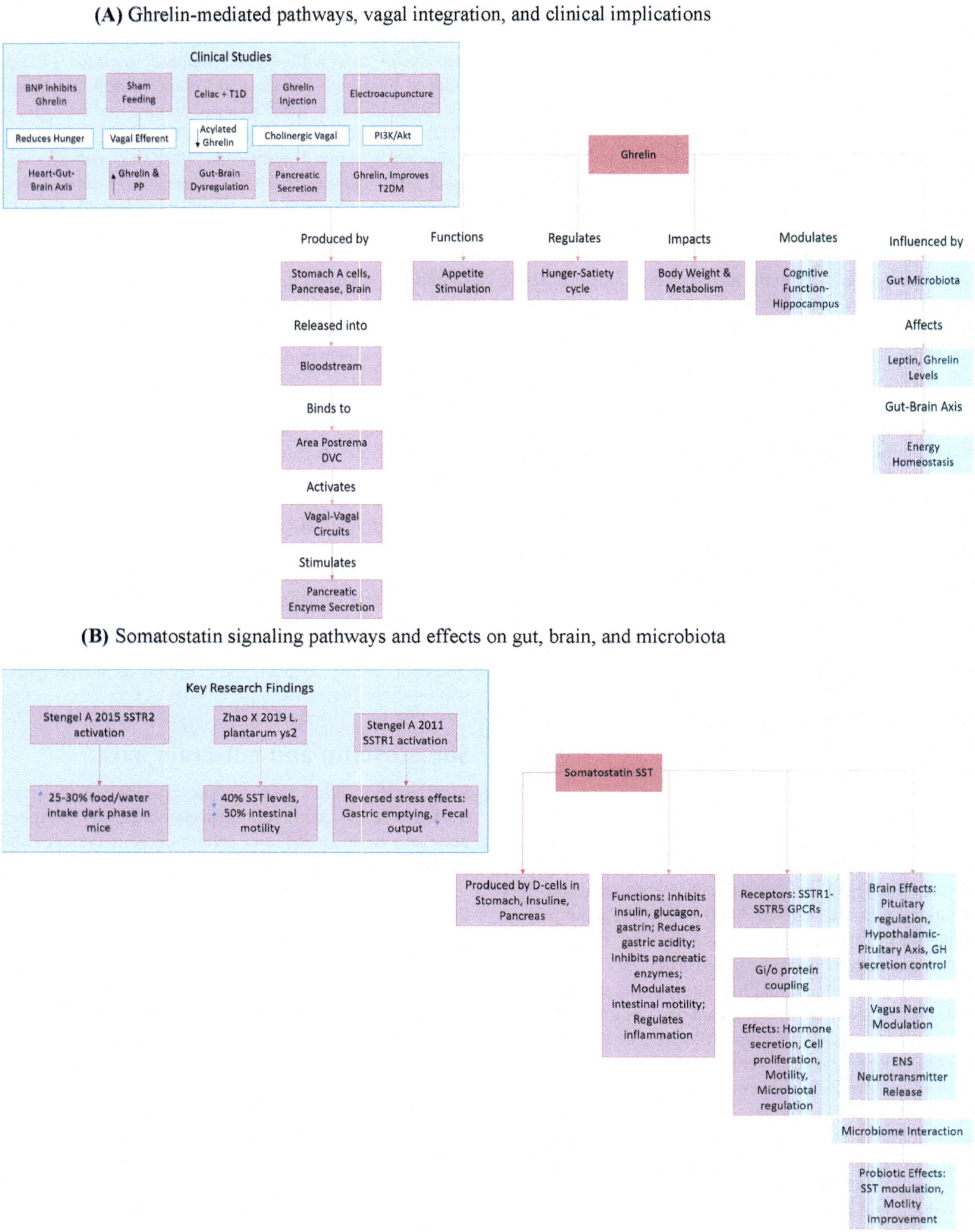

FIGURE 3.4 Hormonal modulation of the gut–brain axis: Somatostatin and Ghrelin. (A) Ghrelin-mediated pathways, vagal integration, and clinical implications. (B) Somatostatin signaling pathways and effects on gut, brain, and microbiota.

effects by secreting somatostatin via SSTR4 (Gribble & Reimann, 2016).

Somatostatin, as a neopeptide, can also bind to its receptors on human peripheral T cells, leading to stimulation or enhancement of T cell functions that are essential for maintaining health and fighting disease (Kántás et al., 2019).

The effect of somatostatin on the brain-gut axis is that somatotropic cells in the pituitary gland regulate digestive function through the bloodstream and nervous system through the secretion of somatostatin and the effect on the hypothalamic-pituitary axis (Timmermans et al., 1990).

For example, in regulating the body's metabolism and homeostasis to regulate growth hormone (GH) secretion, Pome neurons in the hypothalamus stimulate somatostatin neurons (Zhao et al., 2019).

Additionally, there is a feedback loop between the brain and the gut, in which the vagus nerve and local neurotransmitters of the ENS regulate the release of somatostatin in the digestive system (Timmermans et al., 1990).

In 2015, Stengel A et al. studied brain-gut interactions and the role of central activation of somatostatin in increasing the stimulation of food and water intake in the dark phase in rodents via the SSTR2 receptor.

Their study was conducted on approximately 30 mice (10 mice × 3 groups) in control, agonist and SST2 antagonist treatment groups, which underwent intracerebral injection of specific somatostatin receptor 2 agonists (such as octreotide) and antagonists, and feeding behaviors and water consumption were measured with cages equipped with an automatic recording system in the light and dark phases of the day, and then hormonal analysis (such as corticosterone) was performed to assess stress. The results showed that without significant changes in stress levels, sst2 activation led to a 25%–30% increase in food and water consumption in the dark phase, indicating an independent role of sst2 in regulating feeding behavior through interaction with the brain-gut axis (Engevik et al., 2020).

In their 2019 study, X. Zhao et al. investigated the role of somatostatin in regulating intestinal motility along with the effect of probiotic Lactobacillus plantarum ys2 in the treatment of activated carbon-induced constipation in male Kunming rats. For this purpose, 20 rats in the control and treatment groups underwent activated carbon-induced constipation, and the treatment group received L. plantarum ys2 orally at a daily dose of 1×10^9 CFU per rat (Zhao et al., 2019).

During the study, intestinal motility was recorded by manometry and serum SST levels were measured by ELISA, somatostatin receptor (SSTR) gene expression was measured by qPCR, and SSTR protein expression was measured by immunohistochemistry in intestinal tissue. The results of the study showed that treatment with L. plantarum ys2 reduced the time to the first black stool and improved intestinal peristalsis by about 50%. Serum SST levels were also reduced by about 40% in the treatment group, and SSTR gene and protein expression in the intestine were also reduced.

The results indicate that intestinal motility improved by reducing SST, indicating the inhibitory role of somatostatin in neuro-digestive interactions. It seems that L. plantarum ys2, by regulating the intestinal microbiome, reduced the production of SST by enteroendocrine cells and modulated its inhibitory effect on intestinal motility. Therefore, probiotics can play a prominent role in the treatment of constipation through hormonal modulation such as SST and digestive disorders (Cani et al., 2019).

In a study by Andreas Stengel et al., in 2011, the role of the somatostatin receptor 1 (sst1) in regulating gastric and colonic function following acute stress was investigated. This study was conducted on 32 male C57BL/6 mice that were randomly divided into 4 groups of 8 mice each: control, treatment with a specific sst1 agonist (S-406-062), treatment with somatostatin-28, and a stress group (tail suspension or CRF injection). Injections were administered intracerebrospinal (icv). Gastric emptying volume using phenol red marker, fecal output, and colonic transit time were measured throughout the study, and plasma corticosterone levels were also assessed to assess the stress response. The results of the study showed that acute stress resulted in an approximately 50% reduction in gastric emptying and a 2- to 3-fold increase in fecal output (Stengel et al., 2011).

Injection of the agonist sst1 or somatostatin-28 significantly reversed these changes and improved digestive function without significant changes in corticosterone levels. Thus, the effects of the sst1 receptor on digestive function under stress are independent of the hypothalamic-pituitary-adrenal (HPA) hormonal axis. Activation of sst1 appears to modulate gastrointestinal responses to stress via the brain-gut axis by inhibiting the release or action of CRF in the brain. These findings suggest the importance of the sst1 receptor in regulating gastrointestinal motility under stress conditions, as well as the therapeutic role of its agonists in stress-related gastrointestinal disorders, such as irritable bowel syndrome (Shamsi et al., 2021). Fig. 3.4B illustrates a summarized representation of mechanism.

Mini-dictionary of terms

- **Gut–Brain Axis:**

A bidirectional communication system linking the gastrointestinal tract and the brain via neural, hormonal, and immune pathways, regulating appetite, mood, metabolism, and homeostasis.

- **GLP-1 (Glucagon-Like Peptide-1):**

An incretin hormone derived from proglucagon, secreted by intestinal L-cells, which enhances insulin secretion, inhibits glucagon release, and signals to the brain to regulate appetite.

- **Vagus Nerve:**

A major component of the parasympathetic nervous system that transmits afferent and efferent signals between the gut and the brain, mediating visceral sensory input and autonomic control.

- **Somatostatin (SST):**

A peptide hormone that inhibits the secretion of several other hormones, including insulin, glucagon, and growth hormone; it also modulates gut motility and neural signaling.

- **Insulin Resistance:**

A pathological condition in which cells fail to respond effectively to insulin, leading to impaired glucose uptake and often contributing to metabolic disorders such as type 2 diabetes.

- **Enteroendocrine Cells (EECs):**

Specialized hormone-secreting cells within the intestinal epithelium that detect luminal nutrients and release gut peptides like GLP-1, GIP, and CCK to influence digestion and neural pathways.

- **Dorsal Vagal Complex (DVC):**

A brainstem structure comprising the nucleus tractus solitarius (NTS), area postrema, and dorsal motor nucleus of the vagus, essential for integrating gut-derived sensory input.

- **Ghrelin:**

A hunger-stimulating hormone produced primarily in the stomach and pancreas, which acts on the hypothalamus and vagal circuits to increase appetite and regulate energy balance.

- **Microbiota-Derived Metabolites:**

Bioactive compounds such as short-chain fatty acids (SCFAs) produced by gut microbiota, which influence intestinal integrity, immune responses, and neuroendocrine signaling.

- **GLUT2 (Glucose Transporter Type 2):**

A transmembrane glucose transporter involved in hepatic and intestinal glucose sensing; plays a critical role in activating neural circuits in the hepatoportal system.

Summary points

- The pancreas regulates metabolic homeostasis through endocrine hormones including insulin, glucagon, somatostatin, ghrelin, and VIP.
- Pancreatic hormones interact with the gut-brain axis via vagal, spinal, and hypothalamic pathways to influence glucose metabolism and appetite regulation.
- Insulin acts centrally in the brain to modulate hepatic glucose output and contributes to neuroplasticity and cognitive function.
- Glucagon and its derivatives, such as GLP-1 and oxyntomodulin, modulate both pancreatic function and central appetite centers via GLP-1 receptors.
- Somatostatin inhibits insulin, glucagon, and digestive hormone release, and regulates feeding behavior via specific brain receptors like SSTR2 and SSTR4.
- Ghrelin stimulates appetite and pancreatic enzyme secretion via afferent vagal signaling and plays a role in hippocampal neuroinflammation.
- VIP modulates glucagon and insulin secretion and influences gut motility, mucosal immunity, and energy balance via VPAC receptors.
- Dysregulation of the gut-brain-pancreas axis is implicated in type 2 diabetes, obesity, cognitive decline, and gastrointestinal motility disorders.

Disclosure of AI use: During the preparation of this chapter DeepSeek-R1 was used to improve readability and language.

References

Abildinova, G. Z., Benberin, V. V., Vochshenkova, T. A., Afshar, A., Mussin, N. M., Kaliyev, A. A., et al. (2024). The gut-brain-metabolic axis: Exploring the role of microbiota in insulin resistance and cognitive function. *Frontiers in Microbiology, 15*, 1463958.

Agibalova, T., Hempel, A., Maurer, H. C., Ragab, M., Ermolova, A., Wieland, J., et al. (2024). Vasoactive intestinal peptide promotes secretory differentiation and mitigates radiation-induced intestinal injury. *Stem Cell Research & Therapy, 15*(1), 348.

Alarcón, C., Lincoln, B., & Rhodes, C. J. (1993). The biosynthesis of the subtilisin-related proprotein convertase PC3, but no that of the PC2 convertase, is regulated by glucose in parallel to proinsulin biosynthesis in rat pancreatic islets. *Journal of Biological Chemistry, 268*(6), 4276–4280.

Alavi, S. E., & Shahmabadi, H. E. (2021). GLP-1 peptide analogs for targeting pancreatic beta cells. *Drug Discovery Today, 26*(8), 1936–1943.

Avruch, J. (1998). Insulin signal transduction through protein kinase cascades. *Molecular and Cellular Biochemistry, 182*(1–2), 31–48.

Bai, X., De Palma, G., Boschetti, E., Nishiharo, Y., Lu, J., Shimbori, C., et al. (2024). Vasoactive intestinal polypeptide plays a key role in the microbial-neuroimmune control of intestinal motility. *Cellular and Molecular Gastroenterology and Hepatology, 17*(3), 383–398.

Bessac, A., Cani, P. D., Meunier, E., Dietrich, G., & Knauf, C. (2018). Inflammation and gut-brain axis during type 2 diabetes: Focus on the crosstalk between intestinal immune cells and enteric nervous system. *Frontiers in Neuroscience, 12*, 725.

Burcelin, R., Crivelli, V., Perrin, C., Da Costa, A., Mu, J., Kahn, B. B., et al. (2003). GLUT4, AMP kinase, but not the insulin receptor, are required for hepatoportal glucose sensor-stimulated muscle glucose utilization. *Journal of Clinical Investigation, 111*(10), 1555–1562.

Burcelin, R., Da Costa, A., Drucker, D., & Thorens, B. (2001). Glucose competence of the hepatoportal vein sensor requires the presence of an activated glucagon-like peptide-1 receptor. *Diabetes, 50*(8), 1720–1728.

Burcelin, R., Dolci, W., & Thorens, B. (2000). Glucose sensing by the hepatoportal sensor is GLUT2-dependent: In vivo analysis in GLUT2-null mice. *Diabetes, 49*(10), 1643–1648.

Cani, P. D., Van Hul, M., Lefort, C., Depommier, C., Rastelli, M., & Everard, A. (2019). Microbial regulation of organismal energy homeostasis. *Nature Metabolism, 1*(1), 34–46.

Castañeda, T. R., Tong, J., Datta, R., Culler, M., & Tschöp, M. H. (2010). Ghrelin in the regulation of body weight and metabolism. *Frontiers in Neuroendocrinology, 31*(1), 44–60.

Chen, A., Ganor, Y., Rahimipour, S., Ben-Aroya, N., Koch, Y., & Levite, M. (2002). The neuropeptides GnRH-II and GnRH-I are produced by human T cells and trigger laminin receptor gene expression, adhesion, chemotaxis and homing to specific organs. *Nature Medicine, 8*(12), 1421–1426.

Csajbók, É. A., & Tamás, G. (2016). Cerebral cortex: A target and source of insulin? *Diabetologia, 59*(8), 1609–1615.

Cummings, D. E., Purnell, J. Q., Frayo, R. S., Schmidova, K., Wisse, B. E., & Weigle, D. S. (2001). A preprandial rise in plasma ghrelin levels suggests a role in meal initiation in humans. *Diabetes, 50*(8), 1714–1719.

Dang, X., Xu, M., Liu, D., Zhou, D., & Yang, W. (2020). Assessing the efficacy and safety of fecal microbiota transplantation and probiotic VSL#3 for active ulcerative colitis: A systematic review and meta-analysis. *PLoS One, 15*(3), e0228846.

Date, Y., Murakami, N., Toshinai, K., Matsukura, S., Niijima, A., Matsuo, H., et al. (2002). The role of the gastric afferent vagal nerve in ghrelin-induced feeding and growth hormone secretion in rats. *Gastroenterology, 123*(4), 1120–1128.

de Lartigue, G., & Diepenbroek, C. (2016). Novel developments in vagal afferent nutrient sensing and its role in energy homeostasis. *Current Opinion in Pharmacology, 31*, 38–43.

Dezaki, K., Sone, H., & Yada, T. (2008). Ghrelin is a physiological regulator of insulin release in pancreatic islets and glucose homeostasis. *Pharmacology & Therapeutics, 118*(2), 239–249.

Dias, J. A. R., Alves, L. L., Barros, F. A. L., Cordeiro, C. A. M., Meneses, J. O., Santos, T. B. R., et al. (2022). Comparative effects of using a single strain probiotic and multi-strain probiotic on the productive performance and disease resistance in *Oreochromis niloticus*. *Aquaculture, 550*, 737855.

Ebeid, A. M., Smith, A., Escourrou, J., Murray, P., & Fischer, J. E. (1978). Increased immunoreactive vasoactive intestinal peptide in the cerebro-spinal fluid (CSF) of dogs and monkeys in hepatic failure. *Journal of Surgical Research, 25*(6), 538–541.

Engevik, A. C., Kaji, I., & Goldenring, J. R. (2020). The physiology of the gastric parietal cell. *Physiological Reviews, 100*(2), 573–602.

Escobar, Y. H., O'Piela, D., Wold, L. E., & Mackos, A. R. (2022). Influence of the microbiota-gut-brain axis on cognition in Alzheimer's disease. *Journal of Alzheimer's Disease, 87*(1), 17–31.

Farilla, L., Bulotta, A., Hirshberg, B., Li Calzi, S., Khoury, N., Noushmehr, H., et al. (2003). Glucagon-like peptide 1 inhibits cell apoptosis and improves glucose responsiveness of freshly isolated human islets. *Endocrinology, 144*(12), 5149–5158.

Ferreira, L. S. S., Fernandes, C. S., Vieira, M. N. N., & De Felice, F. G. (2018). Insulin resistance in Alzheimer's disease. *Frontiers in Neuroscience, 12*, 830.

Green, C., Zaman, V., Blumenstock, K., Banik, N. L., & Haque, A. (2025). Dysregulation of metabolic peptides in the gut–brain axis promotes hyperinsulinemia, obesity, and neurodegeneration. *Biomedicines, 13*(1), 132.

Gribble, F. M., & Reimann, F. (2016). Enteroendocrine cells: Chemosensors in the intestinal epithelium. *Annual Review of Physiology, 78*, 277–299.

Grill, H. J. (2010). Leptin and the systems neuroscience of meal size control. *Frontiers in Neuroendocrinology, 31*(1), 61–78.

Hales, C. M., Carroll, M. D., Fryar, C. D., & Ogden, C. L. (2020). Prevalence of obesity and severe obesity among adults: United States, 2017-2018. *NCHS Data Brief, 360*, 1–8.

Holst, J. J. (1997). Enteroglucagon. *Annual Review of Physiology, 59*, 257–271.

Holst, J. J., & Deacon, C. F. (2005). Glucagon-like peptide-1 mediates the therapeutic actions of DPP-IV inhibitors. *Diabetologia, 48*(4), 612–615.

Hoosein, N. M., & Gurd, R. S. (1984). Identification of glucagon receptors in rat brain. *Proceedings of the National Academy of Sciences of the United States of America, 81*(14), 4368–4372.

Hou, X., Yang, D., Yang, G., Li, M., Zhang, J., Zhang, J., et al. (2022). Therapeutic potential of vasoactive intestinal peptide and its receptor VPAC2 in type 2 diabetes. *Frontiers in Endocrinology, 13*, 984198.

Hupe-Sodmann, K., McGregor, G. P., Bridenbaugh, R., Göke, R., Göke, B., Thole, H., et al. (1995). Characterisation of the processing by human neutral endopeptidase 24.11 of GLP-1(7-36) amide and comparison of the substrate specificity of the enzyme for other glucagon-like peptides. *Regulatory Peptides, 58*(3), 149–156.

Iwasaki, M., Akiba, Y., & Kaunitz, J. D. (2019). Recent advances in vasoactive intestinal peptide physiology and pathophysiology: Focus on the gastrointestinal system. *F1000Research, 8*. F1000 Faculty Rev-629.

Kántás, B., Börzsei, R., Szőke, É., Bánhegyi, P., Horváth, Á., Hunyady, Á., et al. (2019). Novel drug-like somatostatin receptor 4 agonists are potential analgesics for neuropathic pain. *International Journal of Molecular Sciences, 20*(24), 6245.

Kalsbeek, A., Bruinstroop, E., Yi, C., Klieverik, L., La Fleur, S., & Fliers, E. (2010). Hypothalamic control of energy metabolism via the autonomic nervous system. *Annals of the New York Academy of Sciences, 1212*(1), 114–129.

Kaufman, B. A., Li, C., & Soleimanpour, S. A. (2015). Mitochondrial regulation of β-cell function: Maintaining the momentum for insulin release. *Molecular Aspects of Medicine, 42*, 91–104.

Kleinridders, A., Ferris, H. A., Cai, W., & Kahn, C. R. (2014). Insulin action in brain regulates systemic metabolism and brain function. *Diabetes, 63*(7), 2232–2243.

Koote, R. S., Levin, E., Salojärvi, J., Smits, L. P., Hartstra, A. V., Udayappan, S. D., et al. (2017). Improvement of insulin sensitivity after lean donor feces in metabolic syndrome is driven by baseline intestinal microbiota composition. *Cell Metabolism, 26*(4), 611-9.e6.

Kreymann, B., Ghatei, M. A., Burnet, P., Williams, G., Kanse, S., Diani, A. R., et al. (1989). Characterization of glucagon-like peptide-1-(7-36)amide in the hypothalamus. *Brain Research, 502*(2), 325–331.

Kwoji, I. D., Aiyegoro, O. A., Okpeku, M., & Adeleke, M. A. (2021). Multi-strain probiotics: Synergy among isolates enhances biological activities. *Biology, 10*(4).

Lai, T. T., Tsai, Y. H., Liou, C. W., Fan, C. H., Hou, Y. T., Yao, T. H., et al. (2024). The gut microbiota modulate locomotion via vagus-dependent glucagon-like peptide-1 signaling. *NPJ Biofilms Microbiomes, 10*(1), 2.

Larsen, P. J., Tang-Christensen, M., Holst, J. J., & Orskov, C. (1997). Distribution of glucagon-like peptide-1 and other preproglucagon-derived peptides in the rat hypothalamus and brainstem. *Neuroscience, 77*(1), 257–270.

Lei, C., Sun, R., Xu, G., Tan, Y., Feng, W., McClain, C. J., et al. (2022). Enteric VIP-producing neurons maintain gut microbiota homeostasis through regulating epithelium fucosylation. *Cell Host & Microbe, 30* (10), 1417–1434 e8.

Lewis, G. F., & Brubaker, P. L. (2021). The discovery of insulin revisited: Lessons for the modern era. *Journal of Clinical Investigation, 131*(1).

Li, Y., Wu, X., Zhao, Y., Chen, S., & Owyang, C. (2006). Ghrelin acts on the dorsal vagal complex to stimulate pancreatic protein secretion. *American Journal of Physiology - Gastrointestinal and Liver Physiology, 290*(6), G1350–G1358.

Liu, S., Zhuang, S., Li, R., & Duan, H. (2025). Electroacupuncture modulates the gut-brain axis via the PI3K/Akt pathway to improve feeding behavior, body weight, glucolipid metabolism, and reduce insulin resistance in T2DM rats. *Diabetology & Metabolic Syndrome, 17*(1), 122.

Llewellyn-Smith, I. J., Reimann, F., Gribble, F. M., & Trapp, S. (2011). Preproglucagon neurons project widely to autonomic control areas in the mouse brain. *Neuroscience, 180*, 111–121.

Müller, T. D., Finan, B., Clemmensen, C., DiMarchi, R. D., & Tschöp, M. H. (2017). The new biology and pharmacology of glucagon. *Physiological Reviews, 97*(2), 721–766.

Manzaneque-López, M. C., Sánchez-López, C. M., Pérez-Bermúdez, P., Soler, C., & Marcilla, A. (2023). Dietary-derived exosome-like nanoparticles as bacterial modulators: Beyond MicroRNAs. *Nutrients, 15*(5).

Matheeussen, V., Baerts, L., De Meyer, G., De Keulenaer, G., Van der Veken, P., Augustyns, K., et al. (2011). Expression and spatial heterogeneity of dipeptidyl peptidases in endothelial cells of conduct vessels and capillaries. *Biological Chemistry, 392*(3), 189–198.

Moody, T. W., Nuche-Berenguer, B., & Jensen, R. T. (2016). Vasoactive intestinal peptide/pituitary adenylate cyclase activating polypeptide, and their receptors and cancer. *Current Opinion in Endocrinology Diabetes and Obesity, 23*(1), 38–47.

Nakagawa, A., Satake, H., Nakabayashi, H., Nishizawa, M., Furuya, K., Nakano, S., et al. (2004). Receptor gene expression of glucagon-like peptide-1, but not glucose-dependent insulinotropic polypeptide, in rat nodose ganglion cells. *Autonomic Neuroscience, 110*(1), 36–43.

Nauck, M. A., Heimesaat, M. M., Orskov, C., Holst, J. J., Ebert, R., & Creutzfeldt, W. (1993). Preserved incretin activity of glucagon-like peptide 1 [7-36 amide] but not of synthetic human gastric inhibitory polypeptide in patients with type-2 diabetes mellitus. *Journal of Clinical Investigation, 91*(1), 301–307.

Obici, S., Feng, Z., Karkanias, G., Baskin, D. G., & Rossetti, L. (2002). Decreasing hypothalamic insulin receptors causes hyperphagia and insulin resistance in rats. *Nature Neuroscience, 5*(6), 566–572.

Papastamataki, M., Papassotiriou, I., Bartzeliotou, A., Vazeou, A., Roma, E., Chrousos, G. P., et al. (2014). Incretins, amylin and other gut-brain axis hormones in children with coeliac disease. *European Journal of Clinical Investigation, 44*(1), 74–82.

Pendharkar, S. A., Asrani, V. M., Murphy, R., Cutfield, R., Windsor, J. A., & Petrov, M. S. (2017). The role of gut-brain axis in regulating glucose metabolism after acute pancreatitis. *Clinical and Translational Gastroenterology, 8*(1), e210.

Plamboeck, A., Holst, J. J., Carr, R. D., & Deacon, C. F. (2005). Neutral endopeptidase 24.11 and dipeptidyl peptidase IV are both mediators of the degradation of glucagon-like peptide 1 in the anaesthetised pig. *Diabetologia, 48*(9), 1882–1890.

Rodriguez-Rodriguez, P., Sandebring-Matton, A., Merino-Serrais, P., Parrado-Fernandez, C., Rabano, A., Winblad, B., et al. (2017). Tau hyperphosphorylation induces oligomeric insulin accumulation and insulin resistance in neurons. *Brain, 140*(12), 3269–3285.

Roh, E., Song, D. K., & Kim, M.-S. (2016). Emerging role of the brain in the homeostatic regulation of energy and glucose metabolism. *Experimental and Molecular Medicine, 48*(3), e216–e.

Sasaki, H., Ebitani, I., Tominaga, M., Yamatani, K., Yawata, Y., & Hara, M. (1980). Glucagon-like substance in the canine brain. *Endocrinologia Japonica, 27*(Suppl. 1), 135–140.

Schwartz, M. W., Woods, S. C., Porte, D., Jr., Seeley, R. J., & Baskin, D. G. (2000). Central nervous system control of food intake. *Nature, 404*(6778), 661–671.

Semaniakou, A., Chappe, F., Anini, Y., & Chappe, V. (2021). VIP reduction in the pancreas of F508del homozygous CF mice and early signs of cystic fibrosis related diabetes (CFRD). *Journal of Cystic Fibrosis, 20*(5), 881–890.

Shamsi, B. H., Chatoo, M., Xu, X. K., Xu, X., & Chen, X. Q. (2021). Versatile functions of somatostatin and somatostatin receptors in the gastrointestinal system. *Frontiers in Endocrinology, 12*, 652363.

Simonian, H. P., Kresge, K. M., Boden, G. H., & Parkman, H. P. (2005). Differential effects of sham feeding and meal ingestion on ghrelin and pancreatic polypeptide levels: Evidence for vagal efferent stimulation mediating ghrelin release. *Neuro-Gastroenterology and Motility, 17*(3), 348–354.

Stengel, A., Goebel-Stengel, M., Wang, L., Larauche, M., Rivier, J., & Taché, Y. (2011). Central somatostatin receptor 1 activation reverses acute stress-related alterations of gastric and colonic motor function in mice. *Neuro-Gastroenterology and Motility, 23*(6), e223–e236.

Timmermans, J.-P., Scheuermann, D. W., Stach, W., Adriaensen, D., & De Groodt-Lasseel, M. H. (1990). Distinct distribution of CGRP-, enkephalin-, galanin-, neuromedin U-, neuropeptide Y-, somatostatin-, substance P-, VIP-and serotonin-containing neurons in the two submucosal ganglionic neural networks of the porcine small intestine. *Cell and Tissue Research, 260*(2), 367–379.

Valente, R., Coppola, A., Scandavini, C. M., Halimi, A., Magnusson, A., Lauro, A., et al. (2024). Interactions between the exocrine and the endocrine pancreas. *Journal of Clinical Medicine, 13*(4), 1179.

Vasiljević, J., Torkko, J. M., Knoch, K. P., & Solimena, M. (2020). The making of insulin in health and disease. *Diabetologia, 63*(10), 1981–1989.

Vila, G., Grimm, G., Resl, M., Heinisch, B., Einwallner, E., Esterbauer, H., et al. (2012). B-type natriuretic peptide modulates ghrelin, hunger, and satiety in healthy men. *Diabetes, 61*(10), 2592–2596.

Volčanšek, Š., Koceva, A., Jensterle, M., Janež, A., & Muzurović, E. (2025). Amylin: From mode of action to future clinical potential in diabetes and obesity. *Diabetes Therapy*, 1–21.

Vu, J. P., Larauche, M., Flores, M., Luong, L., Norris, J., Oh, S., et al. (2015). Regulation of appetite, body composition, and metabolic hormones by vasoactive intestinal polypeptide (VIP). *Journal of Molecular Neuroscience, 56*(2), 377–387.

Wachsmuth, H. R., Weninger, S. N., & Duca, F. A. (2022). Role of the gut–brain axis in energy and glucose metabolism. *Experimental and Molecular Medicine, 54*(4), 377–392.

Washington, M. C., Raboin, S. J., Thompson, W., Larsen, C. J., & Sayegh, A. I. (2010). Exenatide reduces food intake and activates the enteric nervous system of the gastrointestinal tract and the dorsal vagal complex of the hindbrain in the rat by a GLP-1 receptor. *Brain Research, 1344*, 124–133.

Yamamoto, H., Kishi, T., Lee, C. E., Choi, B. J., Fang, H., Hollenberg, A. N., et al. (2003). Glucagon-like peptide-1-responsive catecholamine neurons in the area postrema link peripheral glucagon-like peptide-1 with central autonomic control sites. *Journal of Neuroscience, 23*(7), 2939–2946.

Yang, B. Y., Zhai, G., Gong, Y. L., Su, J. Z., Peng, X. Y., Shang, G. H., et al. (2018). Different physiological roles of insulin receptors in mediating nutrient metabolism in zebrafish. *American Journal of Physiology. Endocrinology and Metabolism, 315*(1), E38–e51.

Zaia, C. T. B. V., Uchôa, E. T., Dos Santos, A. A. T., de Andrade Ribeiro, R. C., Batista, A. C. S., Crespigio, J., et al. (2022). Vaso-active intestinal peptide promotes hypophagia and metabolic changes: Role of paraventricular hypothalamic nucleus and nitric oxide. *Brain Research Bulletin, 189*, 102–110.

Zhao, X., Qian, Y., Li, G., Yi, R., Park, K.-Y., & Song, J.-L. (2019). Lactobacillus plantarum YS2 (yak yogurt Lactobacillus) exhibited an activity to attenuate activated carbon-induced constipation in male Kunming mice. *Journal of Dairy Science, 102*(1), 26–36.

Chapter 4

Somatostatin: A crossroad between pancreatic and neural pathways

Mohammad Rahmanian[1, 2], Sana Esamaeilzadeh Gharajeh[3], Amiryousef Zahedi[4], Hossein Gharedaghi[5], Hana Abbasian[6] and Mahsa Kamali[7]

[1]*Gastroenterology and Liver Diseases Research Center, Research Institute for Gastroenterology and Liver Diseases, Shahid Beheshti University of Medical Sciences, Tehran, Iran;* [2]*Student Research Committee, School of Medicine, Shahid Beheshti University of Medical Sciences, Tehran, Iran;* [3]*Faculty of Veterinary Medicine, Islamic Azad University, Tabriz Branch, Tabriz, Iran;* [4]*Department of Internal Medicine, Caucasus Medical Center (Vian Health Holding), Tbilisi, Georgia;* [5]*School of Medicine, Zanjan University of Medical Science, Zanjan, Iran;* [6]*Harvard Medical School, Boston, MA, United States;* [7]*Universal Scientific Education and Research Network (USERN), Tehran, Iran*

Abbreviations

ACTH adrenocorticotropic hormone
AD Alzheimer's disease
APP amyloid precursor protein
Aβ amyloid-β
BBB blood–brain barrier
BLA basolateral amygdala
BPD bipolar disorder
Ca^{2+} calcium ion
cAMP cyclic adenosine monophosphate
CNS central nervous system
COX-2 cyclooxygenase-2
CRH corticotropin-releasing hormone
CRHR1 corticotropin-releasing hormone receptor 1
CSF cerebrospinal fluid
DM diabetes mellitus
ERK extracellular signal–regulated kinase
Fc fragment crystallizable (region of antibody)
FFAR4 free fatty acid receptor 4 (GPR120)
GABA gamma-aminobutyric acid
GHRH growth hormone-releasing hormone
GLP-1 glucagon-like peptide-1
GPCR G protein-coupled receptor
HD Huntington's disease
HPA hypothalamic–pituitary–adrenal
IL-1β interleukin-1 beta
LAR long-acting release
LPS lipopolysaccharide
LTP long-term potentiation
MAPK mitogen-activated protein kinase
MDD major depressive disorder
mRNA messenger ribonucleic acid
NEN neuroendocrine neoplasm
NET neuroendocrine tumor
NF-κB nuclear factor kappa B
NMDA N-methyl-D-aspartate
PET/CT positron emission tomography/computed tomography
PI3K/AKT phosphatidylinositol 3-kinase/protein kinase B pathway
PKC protein kinase C
pNET pancreatic neuroendocrine tumor
PRRT peptide receptor radionuclide therapy
PV parvalbumin
SGLT2 sodium-glucose cotransporter 2
SRIH somatotropin release-inhibiting hormone (somatostatin)
SST somatostatin
SSTR somatostatin receptor
TNF-α tumor necrosis factor alpha
VIP vasoactive intestinal peptide

Introduction to somatostatin

Somatostatin (SST) is a short inhibitory peptide that exists primarily as two bioactive forms—the 14-residue SST-14 and the N-terminally extended SST-28—. These forms are produced from a common preprohormone and conserved across vertebrates, suggesting an ancient role as a pleiotropic inhibitory neuroendocrine signal. The SST isoforms share a central pharmacophore that mediates receptor binding while their differing N-terminal lengths and tissue distributions confer distinct stability and local concentrations in neural and peripheral compartments (Burgus et al., 1973; Esch et al., 1980). These structural variations underpin both systemic hormonal effects and finely graded paracrine actions within micro-organs. Recent reviews have further emphasized the conserved structural basis of

Hormones in Neuroscience. https://doi.org/10.1016/B978-0-443-30263-3.00022-1

SST biology and highlighted newly emerging roles of its receptor system in endocrine regulation (Zhang et al., 2025).

SST is synthesized and secreted from anatomically and functionally distinct pools: hypothalamic neurons that regulate secretion from the pituitary, enteroendocrine cells diffusely dispersed within the gastrointestinal mucosa and pancreatic δ-cells within the islets of Langerhans of the pancreas. Pancreatic δ-cells, though numerically minor, display complex morphologies with long processes enabling extensive contacts with α- and β-cells; together these sources establish both circulating and tightly localized somatostatinergic tone that governs endocrine and exocrine physiology (Rorsman & Huising, 2018).

Cellular effects of somatostatin are transduced by a family of five G-protein–coupled receptors (SSTR1–5). These receptors principally couple to inhibitory Gi/o proteins to reduce adenylate cyclase activity, lower cAMP, activate G protein–gated inward rectifier K+ channels and suppress voltage-gated Ca2+ entry and exocytosis. The tissue-specific expression profiles of SSTR subtypes have important functional consequences: in human islets, SSTR2 has emerged as the major functional receptor mediating inhibition of insulin and glucagon secretion. In addition, recent high-resolution structural studies have uncovered subtype-selective mechanisms of ligand recognition and distinct binding modes of agonists and antagonists, thus paving the way for therapeutic targeting (Chen et al., 2022; Kailey et al., 2012; Zhao et al., 2022).

Following these studies, cryo-EM studies performed in 2024 resulted in high-resolution structures of SSTR1 and SSTR3 and delivered unprecedented insights into selective ligand recognition and activation processes (Wang et al., 2024).

Inside the pancreatic islets, somatostatin acts as a major paracrine inhibitor of hormone secretion and practically dampens pulses of insulin and glucagon, therefore regulating their pulsatility and preventing indiscriminate co-secretion. SST also reduces selected exocrine secretion and exerts regulatory actions on islet blood flow and local innervation. Autonomic and paracrine negative feedback systems add depth to this regulatory scheme: a notable case in point is urocortin-3 produced from beta cells to stimulate the release of somatostatin from δ-cells and create a negative feedback circuit responsible for terminating insulin discharge following a glucose challenge; impairments of this circuit are observed in diabetic islets and are likely to exaggerate dysglycemia (van der Meulen et al., 2015; Vergari et al., 2020).

Somatostatin secretion is tightly controlled by nutrient and neuronal signals. Glucose and a few selected amino acids and fatty acids alter δ-cell electrical activity and stimulus–secretion coupling; incretins such as GLP-1 adjust SST secretion and consequently influence downstream islet hormone balance. Autonomic and paracrine local mediators also control δ-cell output, so SST secretion is an intermediary for metabolic, luminal and neural information streams (Braun, 2014; Rorsman & Huising, 2018; Vergari et al., 2020).

From a clinical point of view, perturbations in somatostatin signaling have various implications. Excessive or misplaced somatostatinergic tone contributes to defective counter-regulatory glucagon release in diabetes, somatostatin receptors are exploited as diagnostic markers and therapeutic targets in neuroendocrine tumors, and alterations in cortical somatostatinergic interneuron function have been linked to mood disorders and neurodegenerative disease pathophysiology. The intersecting molecular pharmacology, islet physiology and translational potential of SST make it a compelling subject at the interface of pancreatic endocrinology and neuroscience (Braun, 2014; Rorsman & Huising, 2018; Song, Yoon, & Lee, 2021; van der Meulen et al., 2015). Collectively, this chapter integrates molecular, cellular and clinical perspectives to inform both ongoing research and patient care. This introduction sets the stage for the subsequent detailed discussions and clinical ramifications (Fig. 4.1).

Somatostatin in the pancreas

Somatostatin (SST) is produced by pancreatic δ-cells, with SST-14 as the predominant local peptide. Within islets it acts as a fast, paracrine "brake" that tempers both glucagon and insulin release and stabilizes moment-to-moment glucose control (Hauge-Evans et al., 2009; van der Meulen et al., 2015). Human islet studies indicate receptor selectivity across endocrine cell types: α-cells are dominated by SSTR2 signaling, whereas β-cells are also predominantly SSTR2-mediated in humans; expression levels vary by species and metabolic state but the α-cell SSTR2 predominance is a consistent finding (Kailey et al., 2012; Rorsman & Huising, 2018).

What drives δ-cells? Nutrients play a role, but β-cell-derived cues are pivotal. Urocortin-3 released from β-cells engages δ-cell receptors to trigger SST secretion, forming a local negative-feedback loop in which rising β-cell activity evokes δ-cell SST that restrains further insulin (and α-cell glucagon) output (van der Meulen et al., 2015). Ghrelin can selectively activate δ-cells, and autonomic inputs further tune δ-cell excitability, underscoring their integrator role at the endocrine–neural interface (DiGruccio et al., 2016; Gylfe & Gilon, 2014). Recent work also highlights specialized microdomains and close paracrine interactions between β- and δ-cells that shape the timing and strength of paracrine inhibition (van der Meulen et al., 2015).

Functional consequences are state-dependent. During hyperglycaemia, δ-cell SST helps suppress glucagon to

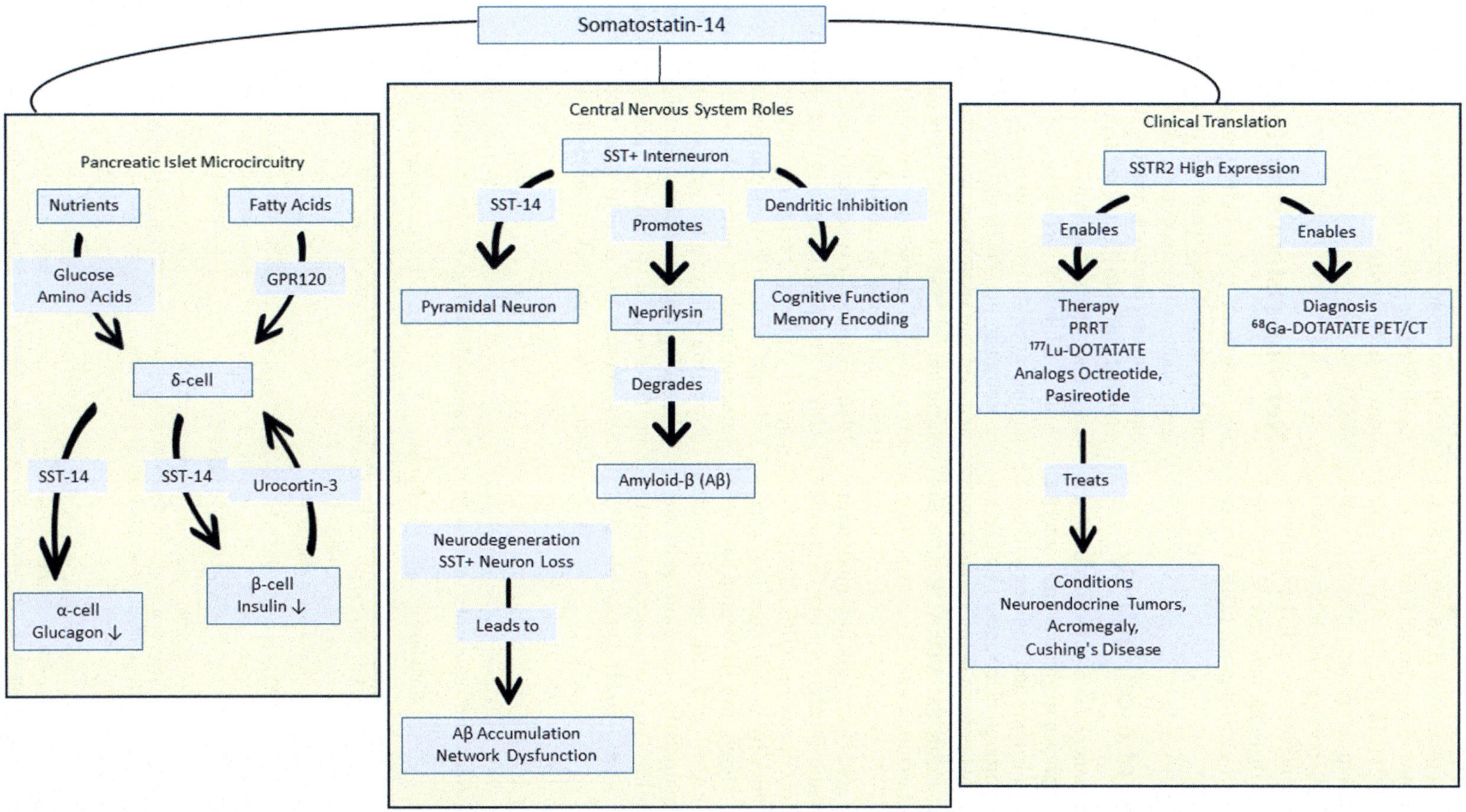

FIGURE 4.1 Schematic representation of the multifaceted roles of somatostatin-14 (SST-14). The diagram illustrates its involvement in (A) pancreatic islet microcircuitry, where SST-14 regulates glucagon and insulin secretion; (B) central nervous system functions, including modulation of pyramidal neurons, neprilysin activity, amyloid-β degradation, and cognitive processes, with implications in neurodegeneration; and (C) clinical translation, where high SSTR2 expression enables both diagnostic imaging with ˆ68Ga-DOTATATE PET/CT and therapeutic applications such as PRRT with ˆ177Lu-DOTATATE and somatostatin analogs for conditions including neuroendocrine tumors, acromegaly, and Cushing's disease.

prevent counterproductive hepatic glucose output; during falling glucose, excessive SSTR2 signaling on α-cells can blunt counter-regulation. Early clinical studies now test selective SSTR2 antagonism to restore glucagon responses in type 1 diabetes without provoking broad islet hypersecretion (Hoffman et al., 2023). In β-cells, SST signaling restrains cAMP and calcium entry and thereby curbs exocytosis; this restraint collaborates with other paracrine signals, including mechanisms such as SGLT2-insulin effects and GLP-1 pathways that can raise δ-cell SST and indirectly dampen glucagon (Vergari et al., 2019; Ørgaard & Holst, 2017) (Table 4.1).

Nutrient regulation of δ-cell activity

Somatostatin secretion by pancreatic δ-cells is modulated by circulating nutrients in addition to glucose and paracrine signals. Amino acids such as leucine and arginine stimulate δ-cell somatostatin output through electrogenic uptake via cationic and neutral amino acid transporters, together with oxidative metabolism that raises the ATP:ADP ratio and promotes membrane depolarization and exocytosis (Rorsman & Huising, 2018).

Long-chain free fatty acids exert distinct modulatory effects on δ-cells. Pancreatic δ-cells express G-protein-coupled fatty-acid receptors including GPR120/FFAR4; activation of these receptors by unsaturated long-chain fatty acids (for example oleate and linoleate) has been shown to lower δ-cell cAMP and calcium signaling, thereby suppressing somatostatin secretion and indirectly potentiating insulin and arginine-stimulated glucagon release (Croze et al., 2021; Du et al., 2022; Zhao, 2022).

These nutrient-sensing pathways enable δ-cells to integrate amino-acid and lipid cues with glucose-dependent inputs, refining paracrine control of α- and β-cell hormone release during mixed-nutrient states and identifying possible targets for pharmacological modulation of islet function (Rorsman & Huising, 2018; Zhao, 2022).

Nervous system somatostatin

Somatostatin (SST/SRIF)-secreting interneurons are broadly represented within the mammalian brain, including areas such as the neocortex, hippocampal formation (i.e., hippocampus, dentate gyrus, entorhinal cortex, subiculum), amygdala, hypothalamus, median eminence, preoptic area, brainstem, and somatosensory areas (Martel et al., 2012). Within inhibitory networks, SST is co-expressed with gamma-aminobutyric acid (GABA), the primary inhibitory neurotransmitter buffering microcircuits against hyperexcitability and stabilizing them (Esclapez & Houser, 1995). Secretion of SST is Ca^{2+}-dependent and involves the exocytosis of dense-core vesicles that need high-FN activity and robust intracellular Ca^{2+} transients (Epelbaum, 1986). Empirical experiments have shown that NMDA and AMPA receptor activity evoke secretion of peptides, thereby linking excitatory input with SRIF secretion (Fontana et al., 1996). Upon secretion, SST operates by paracrine mode by interacting through local receptors and decreasing presynaptic neurotransmitter output, thereby influencing transmission of signals and dendritic processing (Moore et al., 1988). Current experiments suggest that SST neurons are diverse cells that contain diverse transcriptomic and electrophysiological subtypes whose

TABLE 4.1 Physiological roles and regulatory mechanisms of pancreatic somatostatin (δ-cells).

Target/process	Cellular/signaling mechanism	Physiological effect
β-cells (insulin secretion)	Paracrine SST binds β-cell SSTRs (predominantly SSTR2) → reduces cAMP and Ca^{2+} signaling; urocortin-3 evokes δ-cell feedback to limit insulin secretion	Attenuation of insulin release and modulation of pulsatile insulin secretion
α-cells (glucagon secretion)	SST acts on α-cell SSTRs (SSTR2/5), suppressing exocytosis; GLP-1 and insulin can increase δ-cell SST to inhibit glucagon	Reduction in glucagon output, especially postprandially; contributes to glucose homeostasis
Amino-acid sensing (e.g., leucine, arginine)	Electrogenic uptake via amino-acid transporters and oxidative metabolism increases ATP:ADP ratio → membrane depolarization and exocytosis	Stimulates δ-cell SST release that shapes timing and magnitude of paracrine inhibition
Long-chain free fatty acids (palmitate, oleate, linoleate, stearate)	Activate fatty-acid receptors (e.g., GPR120/FFAR4) on δ-cells → modulate cAMP/Ca^{2+} signaling and suppress SST release in some contexts	Alters paracrine balance by reducing SST-mediated inhibition, thereby modulating insulin and glucagon responses to mixed nutrients

functionality is subject to influences from processes such as sleep/wake cycling, attention, learning, and predictive coding (Park et al., 2024).

Cognitive, behavioral, and neurodegenerative processes

SST-positive interneurons are the foundation of memory encoding by compartment-particular dendritic inhibition that organizes hippocampal ensembles (Honoré et al., 2021). Exogenous SST treatment potentiates mossy fiber–CA3 LTP, while deletion of SST diminishes Schaffer collateral–CA1 LTP (Matsuoka et al., 1991). In Alzheimer's disease (AD), there is a consistent feature of the degeneration of SST peptide and SST + neurons that contributes to loops of worsening pathology (Grouselle et al., 1998; Rossor et al., 1980). Loss of SST is correlated with increased levels of Aβ, while treatment with SST decreases tau phosphorylation at Ser262, indicative of neuroprotective effects (Saito et al., 2005). A Pharmacological Reviews article published in 2024 concludes that restorative SST signaling may become a treatment strategy against AD-related protein aggregation and synaptic dysfunction (Aspet, 2024).

Interactions with other neurotransmitters

SST also interacts with both excitatory and inhibitory networks. Stimulation of SSTR2 reduces glutamate release, providing seizure protection; overall, SST reduces excitatory transmission in hypothalamus, hippocampus, basal forebrain, and neocortex (Patel, 1999). Moreover, SST may inhibit GABA release presynaptically and reduce inhibitory currents of basal forebrain cholinergic neurons (Patel, 1999). Dopaminergic modulation has also been documented: infusion of SST increases striatal dopamine release by mechanisms that are presynaptic, changing locomotion comparably with amphetamine (Hathway et al., 2002). These results demonstrate the integrative nature of SST's action by balancing multiple neurotransmitter systems.

Somatostatin is an interface between the nervous and endocrine systems. SST was first discovered in 1973 as a growth hormone–inhibiting factor and named SRIH (Brazeau et al., 1973). There are two active isoforms: SST-14, which is most prominent in the brain, and SST-28, most represented in pancreas and intestine (Brazeau et al., 1973; Song et al., 2021). This bimodal expression permits SST to harmonize central and peripheral physiology by circulating after release in the blood. At the functional level, numerous endocrine and exocrine secretions are inhibited by SST (Brazeau et al., 1973). Stress response is suppressed by SST at the CNS by reducing levels of ACTH, catecholamines, and glucocorticoids (Brown et al., 1984; Stengel & Taché, 2017). Subtype-selective receptor actions also connect SST with the control of metabolism: SSTR5 with insulin secretion, SSTR2 with glucagon and immune response, and SSTR3 with growth control (Song et al., 2021). In doing so, SST bridges nervous activity with systemic metabolic homeostasis.

Somatostatin receptors (SSTR1–5) and signaling pathways

All SSTRs contain seven transmembrane GPCRs interacting with SST-14/28 by nanomolar affinity (Song et al., 2021). SSTR1 is rich on SST interneurons as an autoreceptor, while SSTR2 is the most common cortex subtype of deep excitatory neurons (Møller et al., 2017; Song et al., 2021). Traditionally, SSTRs inhibit adenylyl cyclase by coupling with Gi/Go proteins, lowering cAMP, activating GIRK K^+ channels, and reducing voltage-gated Ca^{2+} channels, leading to robust neuron inhibition (Song et al., 2021). In addition, SST signaling stimulates PLC, MAPK cascades, nitric oxide synthase, and Na^+/H^+ exchangers, extending its cellular effects (Epelbaum, 1986; Song et al., 2021). These concurrent mechanisms demonstrate SST's fine-tuning of excitability and secretion within neural and peripheral tissues.

Clinical relevance: imaging, PRRT, and treatment

Imaging and PRRT: The guidance from EANM/IAEA/SNMMI standardized protocols for PRRT using ^{177}Lu-DOTATATE, reflecting its importance in neuroendocrine tumor (NET) management (Bodei et al., 2013). Comparative reviews in 2025 analyzed DOTATATE, DOTATOC, and DOTANOC, describing differences in receptor affinity and clinical performance (Wijaya et al., 2025).

Somatostatin analogs: Clinical reports feature advancements with pasireotide LAR for acromegaly with evidence of safety and efficacy from real life (Favero et al., 2024). A systematic review noted pasireotide's metabolic effects of hyperglycemia and strategies of combining it with pegvisomant for maximum control with minimum adverse effects (Tóth et al., 2025).

Pancreatic disease

Somatostatin (SST) is a cyclic polypeptide that regulates pancreatic endocrine and exocrine functions by modulating insulin and glucagon secretion through five G protein-coupled receptors. Its effects are predominantly inhibitory, mediated by reductions in intracellular calcium and cAMP levels and increased potassium efflux. Dysregulation of SST signaling, through altered secretion or receptor expression, is associated with pancreatic disorders (O'Toole & Sharma, 2023; Periferakis et al., 2024). The contribution of increased levels of circulating glucagon and decreased levels of SST in type 1 and type 2 diabetes has been recognized as a contributor to hyperglycemia (Gylfe & Gilon, 2014; Makam et al., 2022). However, the exact extent of its involvement in the pathogenesis of diabetes mellitus in humans is not yet fully understood. Moreover, experimental studies suggested that dysregulated SST signaling particularly through overactivity of somatostatin receptor 2 (SSTR2) may cause loss glucose counter regulation in diabetes. Importantly, SSTR2 antagonists are highlighted as promising investigational agents that restore glucagon counter regulation in DM type 1 (Hoffman et al., 2024).

Pancreatic neuroendocrine tumors (pNETs) are rare and part of the diverse family of neuroendocrine neoplasms (NENs). SSTRs, particularly SSTR2 and SSTR5, are highly expressed in NENs and can be targeted by endogenous somatostatin and synthetic analogs such as octreotide and lanreotide, making them valuable for both diagnostic and therapeutic applications (Hu et al., 2021). Studies have demonstrated that low or absent SSTR2 expression assessed using immunohistochemical, associated with significantly poorer survival outcomes, indicating that SSTR2 expression may serve as a clinically useful biomarker for guiding therapeutic decisions and estimating future survival in pNET patients (Hu et al., 2021; Wang et al., 2017) (Table 4.2).

Neurological disorders

SST is expressed by a specific subset of GABAergic interneurons and serves as a molecular marker for these inhibitory interneurons. It plays a critical role in modulating cortical circuitry and regulating cognitive processes. Therefore, reduced SST expression is a defining feature of several neurological disorders (Song, Yoon, & Lee, 2021).

Multiple studies have consistently demonstrated that SST expression particularly SST2, SST4, and SST5 is markedly decreased in both the brain tissue and cerebrospinal fluid (CSF) of patients with Alzheimer's disease (AD) (Kumar, 2005). Moreover, total SST levels and the number of SST-expressing neurons in the frontal cortex and hippocampus of AD patients were reduced to less than 30% of those observed in healthy controls (Davies et al., 1980). Notably, in the hippocampus, the decline in SST-positive interneurons and SST mRNA expression precedes the alteration in other neurotransmitter systems, including GABAergic, glutamatergic, and cholinergic markers, as well as the loss of principal neurons. In addition, there is compelling evidence suggesting a linear correlation between SST deficiency and amyloid beta (Aβ), indicating that a reduction in SST may facilitate or exacerbate Aβ accumulation and the pathogenesis of AD (Ramos et al., 2006). A novel AD treatment mechanism, combining SST with a blood-brain barrier (BBB) transporter and an Fc fragment, was implemented in a murine model overexpressing amyloid precursor protein (APP). It was observed that the treatment significantly increased neprilysin levels, the primary enzyme responsible for degrading Aβ, which correlated with a decline in various Aβ forms (Metzendorf et al., 2025).

In schizophrenia (SCZ), hyperactivity of the ventral hippocampus has been reported, and it is proposed that this was an outcome of the loss of inhibitory interneurons, including parvalbumin-positive (PV+) and somatostatin-positive (SST+) subtypes. Reduced SST expression has been detected not only in the CSF of SCZ patients but also within key brain regions such as the hippocampus, thalamic reticular nucleus, and various cortical areas (Hoftman et al., 2015; Konradi et al., 2011). In addition, reduced SST levels in the lateral amygdala have been associated with neurochemical alterations identified in postmortem studies. Consistent with this, an SCZ mouse model carrying a duplication of the genomic region corresponding to human chromosome 16p11.2 exhibits decreased SST expression, accompanied by disrupted functional connectivity between the hippocampus and both

TABLE 4.2 Role of somatostatin in pancreatic diseases.

Disorder category	Key findings on SST	Mechanistic insights	Therapeutic/clinical implications
Pancreatic diseases	↓ SST levels and ↑ glucagon contributes to hyperglycemia in T1DM and T2DM	Dysregulated SST signaling via overactive SSTR2 impairs glucose counter-regulation	SSTR2 antagonists show promise in restoring glucagon response in T1DM
Pancreatic neuroendocrine tumors (pNETs)	High SSTR2/SSTR5 expression enables targeting by octreotide and lanreotide	Loss/low SSTR2 expression correlates with poor survival	SSTR2 expression is a biomarker for prognosis and treatment stratification

the orbitofrontal cortex and the amygdala (Bristow et al., 2020). These findings demonstrate the role of SST dysfunction in the pathophysiology of SCZ, specifically in relation to impaired inhibitory control and circuit-level connectivity.

The neuropeptide SST2 and SST4 are also abundantly expressed in brain regions implicated in seizure generation, primarily the hippocampus. This region's SST expression and release are dynamically regulated by seizure activity. Moreover, SST-expressing interneurons located in the hilus of the dentate gyrus are especially vulnerable to seizure-induced neuronal loss. Both in vivo and in vitro studies indicate that the loss of SST signaling in the dentate may contribute to the development of epilepsy and increase susceptibility to seizures. SST also regulates activity of CA1 and CA3 subregions of the hippocampus, representing its role as a homeostatic modulator within hippocampal circuits (Tallent & Qiu, 2008). Experimental models have demonstrated that SST possesses strong antiepileptic properties, with its primary site of action localized to the hippocampus. In contrast, some studies found SST-positive interneurons typically inhibitory, paradoxically promote seizure activity in the dysplastic brain, likely due to a switch in neurotransmitter phenotype (from GABA to glutamate release) (Zheng et al., 2023).

SST has been shown to exert a neuroprotective role in a Parkinson's disease (PD) rat model by reducing dopaminergic neuronal loss provoked by lipopolysaccharide (LPS) exposure. Furthermore, SST considerably suppressed microglial activation, oxidative stress, and inflammatory markers such as TNF-α, IL-1β, and COX-2. These findings alleviates neuroinflammation by inhibiting the NF-κB signaling pathway (Bai et al., 2015). Interestingly, a recent study on common and divergent signatures in PD, multiple system atrophy, and progressive supranuclear palsy showed that the most significant neuronal loss in PD occurred in inhibitory SST-expressing neurons (Dick et al., 2025).

The effect of SST reduction, especially of SST2 and SST3 in the pathophysiology of Major Depressive Disorder (MDD) is currently ambiguous. However, SST levels are decreased in the CSF of individuals with MDD, and these levels return to normal in patients who have been treated (Song, Yoon, & Lee, 2021). It is notable that, SST is concurrently released with GABA and plays important physiological and neuronal roles. Infusion of SST into rodent corticolimbic brain regions has demonstrated anxiolytic and antidepressant effects (Engin & Treit, 2009; Yeung et al., 2011). Furthermore, a recent study on mice revealed that high-frequency activation of basolateral amygdala (BLA) GABA neurons expressing SST and PV attenuated depressive and anxiety behaviors, while low-frequency activation had no significant effect (Asim et al., 2024).

Surprisingly, polymorphism in the SST5 receptor gene is one of the crucial candidates in the pathophysiology of BPD (Nyegaard et al., 2002). Studies in BPD patients revealed a decline in SST levels in the hippocampus and caudal entorhinal cortex (67, 68). Moreover, postmortem brain analyses of gene expression in BPD patients revealed alterations in nine genes, including SST (69). Decreased SST expression was also observed in the amygdala of BPD, indicating a disturbance of fear and anxiety regulation in these patients (70).

Experimental studies in Huntington disease (HD) mice also observed a reduction of PV+, SST+, and cholinergic interneurons. Notably, SSTR1 and SSTR5 were identified in double knockout mice (71–73). Additionally, postmortem human analysis and genetic manipulation in animal models revealed that the hypothalamus is affected in HD via altered neuropeptides. SST is one of the main factors that detects reduced hypothalamic nuclei, which regulate behavior and emotion (74). However, the underlying mechanism of SST effects on HD remains unclear, and further studies should be performed (Table 4.3).

Other hormones, endocrine tissues, and pathways

Somatostatin places δ-cells in a position as regional regulators of islet output, inhibiting α/β secretion to maintain the insulin–glucagon balance under normal and abnormal metabolic conditions (75). β- and α-cells of human and mouse islets almost entirely express SSTR2, allowing δ-cell signals to modulate both hormone outputs through a shared inhibitory receptor pathway (76). Intra-islet interactions also focus the scope of this network: pancreatic polypeptide has an inhibitory effect on δ-cell release, while somatostatin has an inhibitory effect on pancreatic polypeptide release, thus entering a reciprocal regulatory scheme over endocrine flux (77). Fasting-induced ghrelin stimulates δ-cells directly and hastens somatostatin release, thus islets modulate insulin levels while influencing glucagon during low nutrient supply (78). In general, δ-cell microcircuits act as modulatory brakes to prevent over-rapid signaling across the insulin–glucagon–PP–ghrelin axis (79).

Pancreatic mechanisms are integrated with brain and gut function through neuropeptide networks. In the cortex and hippocampus, somatostatin and neuropeptide-Y are commonly co-localized within GABAergic interneurons, enabling phase-locked inhibition and potentially modulating subsequent autonomic outflow (80). VIP interneurons act predominantly on somatostatin-expressing cells, creating a disinhibitory pattern that adaptively modulates cortical throughput and state-dependent processing (81). Under stress conditions, the corticotrope

TABLE 4.3 Neurological disorders associated with somatostatin dysfunction.

Neurological disorder	Somatostatin change	Pathophysiological role/effect	Main SSTR(s) involved
Alzheimer's disease (AD)	↓ SST in cortex & hippocampus (<30% of controls)	Promotes amyloid-β accumulation	SSTR2, SSTR4, SST5
Schizophrenia (SCZ)	↓ SST in CSF, hippocampus, thalamus, cortex	Loss of inhibitory interneurons disrupts connectivity	SSTR2
Epilepsy	Loss of SST + interneurons in hippocampus	Increases seizure susceptibility; excitatory switch in dysplastic brain	SSTR2, SSTR4
Parkinson's disease (PD)	↓ SST neurons, ↓ neuroprotection	Increased dopaminergic loss & neuroinflammation	SSTR1, SST2, SST3, SST4, SSTR5
Major depressive disorder (MDD)	↓ SST in CSF (restores with treatment)	Linked to anxiety and depression pathophysiology	SSTR2, SSTR3
Bipolar disorder (BPD)	↓ SST in hippocampus, entorhinal cortex, amygdala	Disturbed fear/anxiety regulation & circadian rhythm	SSTR5
Huntington (HD)	↓ PV+, SST+, cholinergic interneurons	Hypothalamic dysfunction; altered interneuron networks	SSTR1, SSTR5

SSTR5 reduces CRH responsiveness through CRHR1 downregulation and consequently limits HPA-axis sensitivity during adversity (82). Central interactions extend to peripheral systems via established brain-gut links, through which brain and pituitary somatostatin pathways modulate gastrointestinal barrier features, secretion, and movement (83). Hypothalamic regulators with opposing actions also determine the timing of growth hormone release: GHRH exerts release-promoting actions while somatostatin exerts inhibition, with collaborative modulation of somatotroph output on an ultradian and circadian level (84).

Beyond the pancreas, the endocrine actions of somatostatin are ubiquitous. Initially discovered in the hypothalamus, the gastrointestinal tract forms a significant source and prime target, providing neuromodulatory and enteroendocrine control of movement, secretion, acid formation, and barrier function (85). At the pituitary, SSTR2/SSTR5 activation inhibits release of growth hormone and has potential therapeutic value in reducing secretion from somatotroph adenomas, effectively translating receptor biology to clinical value (86). In experimental preparations from thyroid C-cells, the native peptide reduces calcitonin mRNA transcription and release, thus implicating calcium-hormone level modulation directly (87). Adrenocortical preparations demonstrate reduced steroidogenic output in the presence of multi-receptor analogs such as SOM230, highlighting the sensitivity of the adrenal cortex to somatostatin inhibition (88). Immune compartments accordingly demonstrate analogous responsiveness: human T-cell effector function and metabolism are reduced via an SSTR3–GSK3 checkpoint, accordingly implicating peptide signals in immune-metabolome regulation (76).

Mechanistically, conserved receptor pharmacology provides these distributed effects with a foundation. Somatostatin receptors are class-A GPCRs coupled with Gi/o, decreasing adenylyl cyclase activity, decreasing cAMP/PKA drive, and restricting Ca^{2+} influx with membrane hyperpolarization (85). Receptor activation crosses MAPK modules: Gi-dependent Ras activation and Gβγ sequestration reshape ERK1/2 signaling, time-locking rapid secretory control with longer-term growth/trophic programs (89). Simultaneous PI3K/AKT tuning with tyrosine phosphatase recruitment introduces additional levers over growth and survival branches, while C-terminal (de)phosphorylation closely gates signaling efficiency and desensitization (90). Through these hubs, endocrine and neuronal contexts are able to transduce a single ligand with tissue-specific outcomes without compromising network coherence (91). Physiological state then skews the network. Fasting raises circulating ghrelin, boosts δ-cell drive, and enhances somatostatin outflow that modulates islet hormones during nutrient deprivation (78, 92). Stress inputs are damped centrally as SSTR5 reduces corticotrope responsiveness to CRH, blunting ACTH/corticosterone surges (82). Circadian signals define peptide tone as photoperiod and light exposure regulate somatostatin expression in clock circuits, shifting timing of endocrine and behavioral rhythms (93). Together, these interactions describe somatostatin as an adaptive anchor that synchronizes pancreatic, neuropeptidergic, pituitary, gastrointestinal, thyroid, adrenal, and immune actions through GPCR-focused crosstalk, reconciling secretion and growth signals with feeding status, burden of stress, inflammation, and time-of-day architecture (83).

Clinical applications

Therapeutic applications of somatostatin analogs cover a wide range of conditions. In acromegaly, long-interval lanreotide (120 mg every >4 weeks) maintains biochemical control in most stable patients (94), while guidelines recommend switching to pasireotide LAR if control is inadequate (95). Oral octreotide capsules are also effective for maintenance in those previously responsive to injectables (96). For neuroendocrine tumors, lanreotide prolongs progression-free survival in metastatic entero-pancreatic disease (97), and octreotide LAR delays progression in midgut NETs (98). In gastrointestinal bleeding, best-practice advice is to start vasoactive agents such as octreotide promptly after suspicion or diagnosis of variceal hemorrhage (99), and long-acting octreotide reduces transfusion needs in chronic angiodysplasia (100). In Cushing's disease, subcutaneous pasireotide rapidly lowers urinary free cortisol in phase 3 trials (101), and after pancreatic surgery, prophylactic use of somatostatin analogs decreases pancreatic fistula rates and hospital stays (102).

Diagnostic applications

Diagnostic applications of somatostatin receptor imaging have advanced considerably. Appropriate-use criteria recommend that SSTR-PET should replace ^{111}In-pentetreotide (OctreoScan) wherever available for neuroendocrine tumor evaluation (103). Evidence from systematic reviews confirms that ^{68}Ga-DOTATATE PET/CT offers superior diagnostic accuracy and staging compared with ^{111}In-octreotide and conventional imaging methods (104). Furthermore, prospective studies demonstrate that ^{68}Ga-DOTATATE not only increases diagnostic accuracy but also reduces radiation exposure, influencing treatment decisions in a substantial proportion of patients (105).

Therapeutic innovations

Therapeutic breakthroughs based on the targeting of somatostatin receptors have transformed the management of neuroendocrine tumors. ^{177}Lu-DOTATATE substantially extends progression-free survival compared to high-dose octreotide in midgut NETs (106). Revised procedure guidelines also emphasize that PRRT maximizes tumor control as well as symptom relief among carefully selected patients (107). Consensus statements also endorse the use of SSTR-PET to optimize appropriate candidate selection for PRRT, notably among patients who have G3 NETs (108). Moreover, European perspectives also emphasize the need to optimize and individualize PRRT techniques to achieve the best outcomes (109).

Research applications

Applications of somatostatin research provide significant insights into the integration of neuroendocrine, GPCR regulation, and the brain–gut axis. Somatostatin is a potent inhibitory neuromodulator within endocrine, exocrine, and neural networks (110). Its biological activities are transduced through five GPCR subtypes (SSTR1-5), forming the classical paradigm of inhibitory GPCR signal transduction (111). SST/SSTR pathways intersect central and gastrointestinal processes, forming a controllable system that enables investigation of the brain–gut axis (112). At the cellular level, subtypes differ in β-arrestin recruitment as well as endosomal mobilization, supporting the concept of subtype-selective signaling (113). Moreover, SST acting upon Gi-coupled receptors inhibits proliferation, both cytostatically and cytotoxically, revealing dimensions of cancer biology (114). In cortical networks, somatostatin-expressing interneurons form a major inhibitory cohort, allowing systems-level analysis of neural networks (115).

Future directions (metabolic disease, neurodegeneration, immune modulatory)

Somatostatin stands at the crossroads of metabolic control, neural networks, and immunity, thus, making it an interesting target for potential future translational study. When it comes to metabolic disease, it is urgent to develop our knowledge regarding delta-cell physiology and somatostatin receptor (SSTR) signaling. Selective stimulation of receptor subtypes, especially SSTR2 and SSTR5, could help achieve a normal level of glucagon counter-regulation and maintain insulin release in a more proper way without triggering uncontrolled islet hypersecretion. Single-cell multi-omics in the context of human-relevant models, including islet organoids and perfused pancreas systems, will aid in the understanding of the heterogeneity of δ-cells and their response patterns to various metabolic conditions. These methods should be supplemented by creating non-invasive biomarkers and imaging modalities to track the function of delta cells and the expression of receptors in the body. Finally, combining somatostatin-based interventions with current treatment plans, e.g., GLP-1 receptor agonists or SGLT2 inhibitors, may result in new dimensions of more specific metabolic treatment.

Future studies in the context of neurodegeneration should be based on findings that somatostatin, in combination with neprilysin, promotes its activity and thus decreases the amyloid burden. Blood–brain barrier–penetrant somatostatin mimetics or biologics that enhance the activity of neprilysin are promising agents in the treatment of

Alzheimer's disease and related disorders. Concurrently, genetic, pharmacological, or trophic methods of protecting or restoring somatostatin-positive interneurons can potentially preserve network oscillations and cognitive performance. The development of biomarkers, especially cerebrospinal fluid tests and imaging devices that indicate somatostatin–cerebral receptor occupancy, will play an essential role in monitoring disease progression and target occupancy. Notably, it is imperative that systemic endocrine effects of any central somatostatin therapy should be well monitored to guarantee safety in regard to glucose homeostasis and pituitary functions.

The immunomodulatory effect of somatostatin is also an area that should be explored more. Receptor subtype mapping on immune cells, such as microglia, macrophages, and lymphocytes, will help to understand the effect of somatostatin on major inflammatory pathways. This understanding can be used to make rational decisions regarding the use of receptor-selective agonists or antagonists in diseases with neuroinflammation or autoimmunity. Preclinical research can investigate the ability of somatostatin analogs to suppress microglial activity and the release of cytokines, and combination therapy with current immunotherapies may result in synergistic effects without side effects. In addition, somatostatin has a potentially unrecognized role in the immune microenvironment of pancreatic islets, which opens the potential for new interventions in autoimmune diabetes. In each of the three areas, advancements will be based on the identification of receptor subtype-selective compounds with desirable pharmacokinetic and safety properties, and the implementation of human-relevant cellular and organoid models that will close the translational gap between animal and human studies. Mechanistic endpoints, including biomarker assays and functional receptor imaging, need to be added to future clinical trials to allow precision medicine treatments. With its capacity to modify metabolic, neural, and immune pathways at the same time, somatostatin still is an attractive model for innovative treatments to deal with the interplay of endocrinological, neurological, and immunological disorders.

Mini-dictionary of terms

Amyloid-β (Aβ): A peptide derived from amyloid precursor protein whose accumulation in the brain is strongly associated with Alzheimer's disease pathology.

Cytokines: Small signaling proteins released by immune cells that regulate inflammation, cell communication, and immune responses in health and disease.

G protein–coupled receptors (GPCRs): A large family of cell surface receptors that transmit extracellular signals into intracellular responses via heterotrimeric G proteins.

Interneurons: Specialized inhibitory nerve cells that regulate the excitability and synchronization of neural circuits, often critical for learning and memory.

Microglia: Resident immune cells of the central nervous system responsible for pathogen defense, clearance of debris, and regulation of neuroinflammation.

Neprilysin: A membrane-bound endopeptidase that degrades peptides such as amyloid-β, thereby playing a protective role in neurodegenerative conditions.

Organoids: Three-dimensional, stem cell–derived culture systems that mimic the structure and function of human organs, useful for disease modeling.

Paracrine signaling: A form of cell-to-cell communication in which a cell secretes factors that act locally on nearby cells.

Receptor subtype selectivity: The ability of a drug or ligand to preferentially bind to specific receptor subtypes within a receptor family, enhancing therapeutic precision and reducing side effects.

Tau protein: A microtubule-associated protein that stabilizes neuronal structure; abnormal aggregation of tau leads to neurofibrillary tangles in several neurodegenerative disorders.

Summary points

- Somatostatin regulates pancreatic islet function by modulating insulin and glucagon secretion through δ-cell activity and somatostatin receptor (SSTR) signaling.
- Selective SSTR subtype targeting provides therapeutic potential to restore glucagon counter-regulation and regulate insulin release in diabetes.
- In the central nervous system, somatostatin-positive interneurons contribute to cognitive function and neuroprotection by regulating network activity and enhancing amyloid-degrading pathways.
- Loss of somatostatin signaling is associated with neurodegenerative disorders like Alzheimer's disease, Parkinson's disease, and Huntington's disease.
- Somatostatin has immunomodulatory actions that affect the production of cytokines and the activation of immune cells, with potential effects on autoimmune and neuroinflammatory conditions.
- Recent developments in receptor-selective ligands and brain-penetrant analogs allow new therapeutic approaches.
- Recent developments in receptor-selective ligands, brain-penetrant analogs allow new routes to therapy.
- Organoid models of humans, e.g., islet organoids and neuronal cultures, are essential tools for understanding δ-cell and interneuron heterogeneity.
- Noninvasive imaging modalities and biomarker-based approaches are key to tracking somatostatin receptor

activity and guiding translational efforts toward clinical application.

- Further academic research must attempt to combine somatostatin-based approaches with established metabolic, neurological, and immunological therapeutic interventions to optimize clinical outcomes.

Declaration of AI assistance

ChatGPT and QuillBot were utilized to aid in the final language polishing of this manuscript. The authors thoroughly reviewed and edited all sections afterward and bear complete responsibility for the final version of the work.

Conflicts of interest

The authors declare that they have no competing interests.

References

Asim, M., Wang, H., Waris, A., & He, J. (2024). Basolateral amygdala parvalbumin and cholecystokinin-expressing GABAergic neurons modulate depressive and anxiety-like behaviors. *Translational Psychiatry, 14*(1), 418.

Aspet. (2024). Somatostatin: Linking cognition and Alzheimer's disease to therapeutic development. *Pharmacological Reviews, 76*(1), 25–47.

Bai, L., Zhang, X., Li, X., Liu, N., Lou, F., Ma, H., Luo, X., & Ren, Y. (2015). Somatostatin prevents lipopolysaccharide-induced neurodegeneration in the rat substantia nigra by inhibiting the activation of microglia. *Molecular Medicine Reports, 12*(1), 1002–1008.

Bodei, L., Mueller-Brand, J., Baum, R. P., Pavel, M. E., Hörsch, D., O'Dorisio, M. S., O'Dorisio, T. M., Howe, J. R., Cremonesi, M., Kwekkeboom, D. J., & Zaknun, J. J. (2013). The joint IAEA, EANM, and SNMMI practical guidance on peptide receptor radionuclide therapy (PRRNT) in neuroendocrine tumours. *European Journal of Nuclear Medicine and Molecular Imaging, 40*(5), 800–816. https://doi.org/10.1007/s00259-012-2330-6. PMID: 23389427; PMCID: PMC3622744.

Braun, M. (2014). The somatostatin receptor in human pancreatic β-cells. *Vitamins & Hormones, 95*, 165–193. https://doi.org/10.1016/b978-0-12-800174-5.00007-7

Brazeau, P., et al. (1973). Hypothalamic polypeptide that inhibits the secretion of immunoreactive pituitary growth hormone. *Science, 179* (4068), 77–79.

Bristow, G. C., Thomson, D. M., Openshaw, R. L., Mitchell, E. J., Pratt, J. A., Dawson, N., & Morris, B. J. (2020). 16p11 duplication disrupts hippocampal-orbitofrontal-amygdala connectivity, revealing a neural circuit endophenotype for schizophrenia. *Cell Reports, 31*(3).

Brown, M. R., et al. (1984). Central nervous system actions of somatostatin to inhibit stress-induced ACTH secretion. *Endocrinology, 114* (5), 1545–1549.

Burgus, R., Ling, N., Butcher, M., & Guillemin, R. (1973). Primary structure of somatostatin, a hypothalamic peptide that inhibits the secretion of pituitary growth hormone. *Proceedings of the National Academy of Sciences of the United States of America, 70*(3), 684–688. https://doi.org/10.1073/pnas.70.3.684

Chen, L. N., Wang, W. W., Dong, Y. J., Shen, D. D., Guo, J., Yu, X., Qin, J., Ji, S. Y., Zhang, H., Shen, Q., He, Q., Yang, B., Zhang, Y., Li, Q., & Mao, C. (2022). Structures of the endogenous peptide- and selective non-peptide agonist-bound SSTR2 signaling complexes. *Cell Research, 32*(8), 785–788. https://doi.org/10.1038/s41422-022-00669-z

Croze, M. L., Flisher, M. F., Guillaume, A., Tremblay, C., Noguchi, G. M., Granziera, S., Vivot, K., Castillo, V. C., Campbell, S. A., Ghislain, J., Huising, M. O., & Poitout, V. (2021). Free fatty acid receptor 4 inhibitory signaling in delta cells regulates islet hormone secretion in mice. *Molecular Metabolism, 45*, 101166. https://doi.org/10.1016/j.molmet.2021.101166

Davies, P., Katzman, R., & Terry, R. D. (1980). Reduced somatostatin-like immunoreactivity in cerebral cortex from cases of Alzheimer disease and Alzheimer senile dementa. *Nature, 288*(5788), 279–280.

Dick, F., Johanson, G. A. S., Tysnes, O.-B., Alves, G., Dölle, C., & Tzoulis, C. (2025). Brain proteome profiling reveals common and divergent signatures in Parkinson's disease, multiple system atrophy, and progressive supranuclear palsy. *Molecular Neurobiology, 62*(3), 2801–2816.

DiGruccio, M. R., Mawla, A. M., Donaldson, C. J., Noguchi, G. M., Vaughan, J., Cowing-Zitron, C., van der Meulen, T., & Huising, M. O. (2016). Comprehensive alpha, beta and delta cell transcriptomes reveal that ghrelin selectively activates delta cells and promotes somatostatin release from pancreatic islets. *Molecular Metabolism, 5*(7), 449–458. https://doi.org/10.1016/j.molmet.2016.04.007

Du, Y. Q., Sha, X. Y., Cheng, J., Wang, J., Lin, J. Y., An, W. T., Pan, W., Zhang, L. J., Tao, X. N., Xu, Y. F., Jia, Y. L., Yang, Z., Xiao, P., Liu, M., Sun, J. P., & Yu, X. (2022). Endogenous lipid-GPR120 signaling modulates pancreatic islet homeostasis to different extents. *Diabetes, 71*(7), 1454–1471. https://doi.org/10.2337/db21-0794

Engin, E., & Treit, D. (2009). Anxiolytic and antidepressant actions of somatostatin: The role of sst2 and sst3 receptors. *Psychopharmacology, 206*(2), 281–289.

Epelbaum, J. (1986). Somatostatin in the central nervous system: Physiology and pathophysiology. *Progress in Neurobiology, 27*(1), 63–100.

Esch, F., Böhlen, P., Ling, N., Benoit, R., Brazeau, P., & Guillemin, R. (1980). Primary structure of ovine hypothalamic somatostatin-28 and somatostatin-25. *Proceedings of the National Academy of Sciences of the United States of America, 77*(11), 6827–6831. https://doi.org/10.1073/pnas.77.11.6827

Esclapez, M., & Houser, C. R. (1995). Somatostatin neurons colocalize with GABA in the rat hippocampus. *Journal of Comparative Neurology, 356*(4), 601–617.

Favero, V., et al. (2024). Long-term real-world outcomes with pasireotide LAR in acromegaly. *Frontiers in Endocrinology, 15*, 1423362.

Fontana, A., et al. (1996). NMDA and AMPA receptor activation stimulates somatostatin release in the hippocampus. *European Journal of Neuroscience, 8*(12), 2519–2526.

Grouselle, D., et al. (1998). Loss of somatostatin in Alzheimer's disease is associated with amyloid pathology. *Neurobiology of Disease, 5*(5), 438–445.

Gylfe, E., & Gilon, P. (2014). Glucose regulation of glucagon secretion. *Diabetes Research and Clinical Practice, 103*(1), 1–10. https://doi.org/10.1016/j.diabres.2013.11.019

Hathway, G. J., et al. (2002). Exogenous somatostatin increases striatal dopamine release via presynaptic mechanisms. *Neuroscience, 115*(3), 951–960.

Hauge-Evans, A. C., King, A. J., Carmignac, D., Richardson, C. C., Robinson, I. C., Low, M. J., Christie, M. R., Persaud, S. J., & Jones, P. M. (2009). Somatostatin secreted by islet delta-cells fulfills multiple roles as a paracrine regulator of islet function. *Diabetes, 58* (2), 403–411. https://doi.org/10.2337/db08-0792

Hoffman, E. G., D'Souza, N. C., Liggins, R. T., & Riddell, M. C. (2023). Pharmacologic inhibition of somatostatin receptor 2 to restore glucagon counterregulation in diabetes. *Frontiers in Pharmacology, 14*, 1295639. https://doi.org/10.3389/fphar.2023.1295639

Hoffman, E. G., D'Souza, N. C., Liggins, R. T., & Riddell, M. C. (2024). Pharmacologic inhibition of somatostatin receptor 2 to restore glucagon counterregulation in diabetes. *Frontiers in Pharmacology, 14*, 1295639.

Hoftman, G. D., Volk, D. W., Bazmi, H. H., Li, S., Sampson, A. R., & Lewis, D. A. (2015). Altered cortical expression of GABA-related genes in schizophrenia: Illness progression vs developmental disturbance. *Schizophrenia Bulletin, 41*(1), 180–191.

Honoré, E., et al. (2021). Somatostatin interneurons regulate hippocampal network dynamics and memory. *Nature Neuroscience, 24*(5), 684–695.

Hu, Y., Ye, Z., Wang, F., Qin, Y., Xu, X., Yu, X., & Ji, S. (2021). Role of somatostatin receptor in pancreatic neuroendocrine tumor development, diagnosis, and therapy. *Frontiers in Endocrinology, 12*, 679000.

Kailey, B., van de Bunt, M., Cheley, S., Johnson, P. R., MacDonald, P. E., Gloyn, A. L., Rorsman, P., & Braun, M. (2012). SSTR2 is the functionally dominant somatostatin receptor in human pancreatic β- and α-cells. *American Journal of Physiology. Endocrinology and Metabolism, 303*(9), E1107–E1116. https://doi.org/10.1152/ajpendo.00207.2012

Konradi, C., Yang, C. K., Zimmerman, E. I., Lohmann, K. M., Gresch, P., Pantazopoulos, H., Berretta, S., & Heckers, S. (2011). Hippocampal interneurons are abnormal in schizophrenia. *Schizophrenia Research, 131*(1–3), 165–173.

Kumar, U. (2005). Expression of somatostatin receptor subtypes (SSTR1–5) in Alzheimer's disease brain: An immunohistochemical analysis. *Neuroscience, 134*(2), 525–538.

Møller, R. S., et al. (2017). Somatostatin receptor distribution and function in cortical interneurons. *Cerebral Cortex, 27*(1), 93–105.

Makam, A. A., Biswas, A., Kothegala, L., & Gandasi, N. R. (2022). Setting the stage for insulin granule dysfunction during type-1-diabetes: Is ER stress the culprit? *Biomedicines, 10*(11), 2695.

Martel, G., et al. (2012). Somatostatin interneurons and their role in cortical circuits. *Frontiers in Neural Circuits, 6*, 12.

Matsuoka, H., et al. (1991). Somatostatin enhances long-term potentiation in the hippocampus. *Brain Research, 558*(2), 137–140.

Metzendorf, N. G., Godec, A., Petrovic, A., Chourlia, A., Napoleone, A., Syvänen, S., Rofo, F., & Hultqvist, G. (2025). Somatostatin therapy, neprilysin activation, and amyloid beta reduction: A novel approach for Alzheimer's treatment. *Biomedicine & Pharmacotherapy, 189*, 118325.

Moore, S. D., et al. (1988). Somatostatin modulates synaptic transmission via presynaptic and postsynaptic actions. *Journal of Neuroscience, 8* (12), 4672–4686.

Nyegaard, M., Børglum, A., Bruun, T., Collier, D., Russ, C., Mors, O., Ewald, H., & Kruse, T. (2002). Novel polymorphisms in the somatostatin receptor 5 (SSTR5) gene associated with bipolar affective disorder. *Molecular Psychiatry, 7*(7), 745–754.

Ørgaard, A., & Holst, J. J. (2017). The role of somatostatin in GLP-1-induced inhibition of glucagon secretion in mice. *Diabetologia, 60* (9), 1731–1739. https://doi.org/10.1007/s00125-017-4315-2

O'Toole, T. J., & Sharma, S. (2023). Physiology, somatostatin. In *StatPearls [Internet]*. StatPearls Publishing.

Park, E., Mosso, M. B., & Barth, A. L. (2024). Neocortical somatostatin neurons: Diversity, state-dependence, and roles in learning. *Trends in Neurosciences, 47*(2), 140–155.

Patel, Y. C. (1999). Somatostatin and its receptor family. *Frontiers in Neuroendocrinology, 20*(3), 157–198.

Periferakis, A., Tsigas, G., Periferakis, A.-T., Tone, C. M., Hemes, D. A., Periferakis, K., Troumpata, L., Badarau, I. A., Scheau, C., & Caruntu, A. (2024). Agonists, antagonists and receptors of somatostatin: Pathophysiological and therapeutical implications in neoplasias. *Current Issues in Molecular Biology, 46*(9), 9721–9759.

Ramos, B., Baglietto-Vargas, D., del Rio, J. C., Moreno-Gonzalez, I., Santa-Maria, C., Jimenez, S., Caballero, C., Lopez-Tellez, J. F., Khan, Z. U., & Ruano, D. (2006). Early neuropathology of somatostatin/NPY GABAergic cells in the hippocampus of a PS1× APP transgenic model of Alzheimer's disease. *Neurobiology of Aging, 27* (11), 1658–1672.

Rorsman, P., & Huising, M. O. (2018). The somatostatin-secreting pancreatic δ-cell in health and disease. *Nature Reviews Endocrinology, 14*(7), 404–414. https://doi.org/10.1038/s41574-018-0020-6

Rossor, M. N., et al. (1980). Somatostatin in Alzheimer's disease. *The Lancet, 1*(8181), 92–93.

Saito, T., et al. (2005). Somatostatin regulates brain amyloid-β peptide Aβ42 through modulation of proteolytic degradation. *Nature Medicine, 11*(4), 434–439.

Song, J., et al. (2021). The role of somatostatin in the brain and periphery: Mechanisms and clinical relevance. *Experimental & Molecular Medicine, 53*, 328–338.

Song, Y. H., Yoon, J., & Lee, S. H. (2021). The role of neuropeptide somatostatin in the brain and its application in treating neurological disorders. *Experimental and Molecular Medicine, 53*(3), 328–338. https://doi.org/10.1038/s12276-021-00580-4

Stengel, A., & Taché, Y. (2017). Role of brain somatostatin in stress response regulation. *Frontiers in Neuroscience, 11*, 231.

Tallent, M. K., & Qiu, C. (2008). Somatostatin: An endogenous antiepileptic. *Molecular and Cellular Endocrinology, 286*(1–2), 96–103.

Tóth, M., et al. (2025). Effects of pasireotide on lipid and glucose metabolism in acromegaly: A systematic review. *Journal of Endocrinological Investigation, 48*(2), 215–229.

van der Meulen, T., Donaldson, C. J., Cáceres, E., Hunter, A. E., Cowing-Zitron, C., Pound, L. D., Adams, M. W., Zembrzycki, A., Grove, K. L., & Huising, M. O. (2015). Urocortin3 mediates somatostatin-dependent negative feedback control of insulin secretion. *Nature Medicine, 21*(7), 769–776. https://doi.org/10.1038/nm.3872

Vergari, E., Denwood, G., Salehi, A., Zhang, Q., Adam, J., Alrifaiy, A., Wernstedt Asterholm, I., Benrick, A., Chibalina, M. V., Eliasson, L., Guida, C., Hill, T. G., Hamilton, A., Ramracheya, R., Reimann, F., Rorsman, N. J. G., Spilliotis, I., Tarasov, A. I.,

Walker, J. N., ... Briant, L. J. B. (2020). Somatostatin secretion by Na(+)-dependent Ca(2+)-induced Ca(2+) release in pancreatic delta-cells. *Nature Metabolism, 2*(1), 32–40. https://doi.org/10.1038/s42255-019-0158-0

Vergari, E., Knudsen, J. G., Ramracheya, R., Salehi, A., Zhang, Q., Adam, J., Asterholm, I. W., Benrick, A., Briant, L. J. B., Chibalina, M. V., Gribble, F. M., Hamilton, A., Hastoy, B., Reimann, F., Rorsman, N. J. G., Spiliotis, I. I., Tarasov, A., Wu, Y., Ashcroft, F. M., & Rorsman, P. (2019). Insulin inhibits glucagon release by SGLT2-induced stimulation of somatostatin secretion. *Nature Communications, 10*(1), 139. https://doi.org/10.1038/s41467-018-08193-8

Wang, Y., Wang, W., Jin, K., Fang, C., Lin, Y., Xue, L., Feng, S., Zhou, Z., Shao, C., & Chen, M. (2017). Somatostatin receptor expression indicates improved prognosis in gastroenteropancreatic neuroendocrine neoplasm, and octreotide long-acting release is effective and safe in Chinese patients with advanced gastroenteropancreatic neuroendocrine tumors. *Oncology Letters, 13*(3), 1165–1174.

Wang, Y., Xu, Y., Wang, Y., Zhang, J., Chen, L., He, X., Fan, W., Wu, K., Hu, W., Cheng, X., Yang, G., Xu, H. E., Zhuang, Y., & Sun, S. (2024). Selective ligand recognition and activation of somatostatin receptors SSTR1 and SSTR3. *Proceedings of the National Academy of Sciences of the United States of America, 121*(41). https://doi.org/10.1073/pnas.2400298121. e2400298121.

Wijaya, R. R., et al. (2025). Comparison of radiolabeled somatostatin analogues for imaging gastroenteropancreatic neuroendocrine neoplasms. *Annals of Nuclear Medicine, 39*(1), 1–14.

Yeung, M., Engin, E., & Treit, D. (2011). Anxiolytic-like effects of somatostatin isoforms SST 14 and SST 28 in two animal models (Rattus norvegicus) after intra-amygdalar and intra-septal microinfusions. *Psychopharmacology, 216*(4), 557–567.

Zhang, B., Xue, L., & Wu, Z. B. (2025). Structure and function of somatostatin and its receptors in endocrinology. *Endocrine Reviews, 46*(1), 26–42. https://doi.org/10.1210/endrev/bnae022

Zhao, Y. F. (2022). Free fatty acid receptors in the endocrine regulation of glucose metabolism: Insight from gastrointestinal-pancreatic-adipose interactions. *Frontiers in Endocrinology, 13*, 956277. https://doi.org/10.3389/fendo.2022.956277

Zhao, W., Han, S., Qiu, N., Feng, W., Lu, M., Zhang, W., Wang, M., Zhou, Q., Chen, S., Xu, W., Du, J., Chu, X., Yi, C., Dai, A., Hu, L., Shen, M. Y., Sun, Y., Zhang, Q., Ma, Y., ... Zhao, Q. (2022). Structural insights into ligand recognition and selectivity of somatostatin receptors. *Cell Research, 32*(8), 761–772. https://doi.org/10.1038/s41422-022-00679-x

Zheng, Y., Xu, C., Sun, J., Ming, W., Dai, S., Shao, Y., Qiu, X., Li, M., Shen, C., & Xu, J. (2023). Excitatory somatostatin interneurons in the dentate gyrus drive a widespread seizure network in cortical dysplasia. *Signal Transduction and Targeted Therapy, 8*(1), 186.

Chapter 5

Feeding and fasting and impact on pancreatic hormones: insulin and beyond

Salma Younas[1], Sweta Sahu[2], Smruti Karale[3], Muhammad Nafees[4, 5] and Soha Farooq[6]

[1]Department of Pharmacy, University of the Punjab, Lahore, Pakistan; [2]Department of Medicine, JJM Medical College, Davanagere, Karnataka, India; [3]Department of Medicine, GMC Kolhapur, Kolhapur, Maharashtra, India; [4]Department of Pharmacy, University of Faisalabad, Faisalabad, Pakistan; [5]Jinnah Institute of Medical Sciences (JIMS), Faisalabad, Pakistan; [6]Department of Pharmacy, Shifa Tameer e Millat University, Islamabad, Pakistan

Introduction

Pancreatic hormones play a central role in systemic metabolism by tightly regulating blood glucose and energy storage. Insulin, secreted from pancreatic β-cells, promotes glucose uptake and storage in liver, muscle, and fat, thereby lowering blood sugar and driving anabolic processes (El Sayed & Mukherjee, 2023). Its counter-hormone glucagon, from α-cells, raises blood glucose by stimulating hepatic glycogen breakdown and gluconeogenesis, thus mobilizing energy during fasting. Other islet hormones such as somatostatin, pancreatic polypeptide (PP), and amylin provide additional layers of regulation, modulating the secretion and actions of insulin and glucagon and influencing appetite and digestion. Together, these hormones maintain glucose homeostasis through opposing actions: in the fed (postprandial) state, insulin dominates and glucagon is suppressed, whereas in the fasted state insulin falls and glucagon rises (Yoo et al., 2021).

The cyclic transition between feeding and fasting states has deep evolutionary and physiological significance. Early humans did not have continuous food availability; instead, periods of food intake alternated with short-term fasting or longer starvation. As a result, metabolic pathways evolved to be metabolically flexible, switching between glucose utilization in the fed state and fatty acid/ketone utilization in fasting (Palmer & Clegg, 2022). The feeding–fasting cycle is synchronized with circadian rhythms and energy demands: feeding triggers an anabolic cascade for storage and growth, whereas fasting triggers catabolic pathways to release stored fuels. This rhythmic hormonal balance had survival value by optimizing energy use during feast-famine cycles. Insulin and glucagon are key hormonal signals in this adaptation, and their precise regulation is critical for energy homeostasis across different nutritional states. The fact that insulin's actions are opposed by glucagon (and other counterregulatory hormones) highlights the bihormonal control of metabolism; indeed, both insulin deficiency and glucagon excess contribute to metabolic disorders like diabetes (Yoo et al., 2021).

This chapter focuses on insulin and beyond, that is, on insulin and other pancreatic hormones—to explore how feeding versus fasting states impact their secretion and actions. While insulin is the best-known pancreatic hormone, a full understanding requires examining its interplay with glucagon, somatostatin, amylin, and pancreatic polypeptide, especially as the body transitions between nutrient abundance and scarcity. Our aim is to provide a comprehensive review of: (1) pancreatic hormone physiology and interactions; (2) the dynamic changes in these hormones during feeding (fed state) versus fasting (fasted state); and (3) the implications of these hormonal patterns for metabolism and disease.

Physiology of the pancreas and pancreatic hormones

The pancreas has a dual role, with an exocrine portion (acinar tissue secreting digestive enzymes) and an endocrine portion (islets of Langerhans) scattered throughout. The islets are clusters of endocrine cells that secrete hormones directly into the bloodstream. Each islet contains several cell types, each defined by the hormone it produces. The main islet cell types include: β-cells (beta cells), which

Hormones in Neuroscience. https://doi.org/10.1016/B978-0-443-30263-3.00038-5

comprise the majority (~60%–70%) and secrete insulin (as well as the co-hormone amylin); α-cells (~20%) producing glucagon; δ-cells (delta cells, ~5%–10%) producing somatostatin; PP cells (F cells) producing pancreatic polypeptide; and a small number of ε-cells producing ghrelin (El Sayed & Mukherjee, 2023).

Insulin is a peptide hormone synthesized as proinsulin in β-cells and stored in secretory granules. It is released primarily in response to elevated blood glucose. Mechanistically, glucose uptake into β-cells (via GLUT2 transporters) and its metabolism raises the ATP/ADP ratio, causing closure of ATP-sensitive K^+ channels, cell membrane depolarization, Ca^{2+} influx, and exocytosis of insulin granules. This glucose-stimulated insulin secretion is biphasic: an acute first phase (release of pre-stored insulin, within minutes) and a slower second phase (continued release and new synthesis) to manage postprandial glycemia. Besides glucose, other nutrients can stimulate insulin release as well—amino acids (such as arginine and leucine) and free fatty acids in the blood modestly induce insulin secretion (El Sayed & Mukherjee, 2023).

Glucagon is a 29-amino-acid peptide hormone produced by α-cells. Its secretion is essentially the mirror opposite of insulin's: glucagon is released when blood glucose is low (hypoglycemia) and is suppressed when glucose is high (hyperglycemia) (Yoo et al., 2021). The primary physiological trigger for glucagon release is a drop in circulating glucose (as occurs during fasting), which threatens brain fuel supply. In addition, a rise in circulating amino acids after a protein-rich meal strongly stimulates glucagon secretion. This protein-induced glucagon release is thought to prevent hypoglycemia when insulin is secreted in response to dietary protein—glucagon ensures that blood sugar is maintained by promoting hepatic glucose output from the absorbed amino acids. Glucagon secretion is also enhanced by sympathetic nervous activity and circulating catecholamines (β-adrenergic stimulation of α-cells), and by exercise and stress (as part of the counterregulatory hormone ensemble). Inhibitors of glucagon release include high glucose and insulin levels (insulin has a paracrine suppressive effect on neighboring α-cells as well as somatostatin from δ-cells. When secreted, glucagon primarily targets the liver, binding to G_s-coupled receptors to increase cAMP and activate protein kinase A, thereby stimulating hepatic glycogenolysis and gluconeogenesis (El Sayed & Mukherjee, 2023). The result is an increase in glucose release into the bloodstream. Glucagon also promotes lipolysis and fatty acid oxidation in the liver, leading to ketone body production during prolonged fasting (El Sayed & Mukherjee, 2023). Notably, glucagon's actions are largely confined to the liver in humans—muscle and adipose tissues are not direct targets (they lack glucagon receptors), so glucagon's systemic effects are mostly indirect via raising glucose and fatty acid availability. In essence, glucagon acts to oppose insulin: during fasting, glucagon becomes dominant, raising blood glucose and fueling the body by tapping into energy reserves (Yoo et al., 2021).

Somatostatin is a universal inhibitory hormone produced by δ-cells of the pancreatic islets (as well as by neurons and gastrointestinal D-cells). In the islet, somatostatin acts locally to dampen the secretion of both insulin and glucagon, helping prevent excessive hormone output. Paradoxically, glucagon itself stimulates somatostatin release from δ-cells (a negative feedback loop) (El Sayed & Mukherjee, 2023). Nutrient stimuli that affect somatostatin secretion are less well characterized; somatostatin output tends to increase postprandially (as suggested by rises in somatostatin after mixed meals) (Gutniak et al., 1986), possibly due to high glucose and amino acids or gut signals. The role of somatostatin in islet physiology is primarily modulatory: by inhibiting insulin and glucagon (and itself being stimulated by those same hormones), it contributes to the fine-tuning of islet responses and prevents wild swings in hormone levels. Somatostatin also has systemic actions to inhibit gastrointestinal hormones and slow gut motility, but within the pancreas its paracrine effect is key—acting via somatostatin receptors on α- and β-cells to restrain their secretory activity (El Sayed & Mukherjee, 2023).

Pancreatic polypeptide (PP)

Pancreatic polypeptide is secreted by specialized PP (F) cells in the islets, mainly located in the pancreatic head. It is released primarily in response to meal ingestion, especially protein-rich meals, and via vagal (parasympathetic) stimulation during feeding. PP secretion has a characteristic biphasic pattern during a meal: an early phase mediated by cephalic vagal input and a later phase as nutrients (particularly amino acids) reach the intestine. Interestingly, PP release is not strongly correlated with the caloric content or macronutrient composition of the meal—for example, one study noted that PP output is "largely unrelated to meal calories and macronutrient balance," underscoring the dominance of neural regulation. Once in circulation, PP targets the brain and gastrointestinal tract. It binds to Y4 receptors in the hypothalamus and brainstem to promote satiety and reduce food intake, and it slows gastric emptying and pancreatic exocrine secretion (Rose, 2019). Through these actions, PP is thought to signal a state of fullness and aid in the termination of eating. Its secretion continues for hours after a meal, and baseline PP levels are higher in the postprandial state. During prolonged fasting, PP levels tend to rise as well (after 24–48 h without food), possibly as an adaptive signal to conserve resources and inhibit appetite (Villanueva et al., 1978).

Amylin is a peptide hormone co-packaged and co-secreted with insulin from β-cells, roughly in a fixed molar ratio to insulin. Like insulin, it is released in response to nutrient stimuli (primarily glucose) during feeding (Rose, 2019). Amylin's actions complement those of insulin: it suppresses glucagon secretion from α-cells (via intra-islet paracrine action) El Sayed & Mukherjee (2023), slows gastric emptying, and promotes satiety by acting on the area postrema in the brain. By delaying gastric emptying and reducing meal size, amylin helps prevent rapid spikes in blood glucose after eating El Sayed & Mukherjee (2023). Thus, as a partner to insulin, amylin contributes to postprandial glucose control and signals meal satisfaction (it is considered an "anorexigenic" hormone). In people with diabetes who lack sufficient insulin, amylin is also deficient, which partly explains excessive glucagon and rapid gastric emptying seen in diabetes. Synthetic amylin analogs (pramlintide) are used therapeutically to replace this hormone's function. In normal physiology, every meal triggers a rise in both insulin and amylin; this tandem release curtails both blood glucose elevation and ongoing hunger, illustrating how β-cells have a dual output to manage nutrient influx (Rose, 2019).

Intra-islet hormonal interactions

The endocrine pancreas is an intricate paracrine network—the hormones described above do not act in isolation, but rather influence each other's secretion within the islet to stabilize metabolic control. Notably, insulin (and co-secreted amylin) directly inhibit glucagon release from α-cells; when β-cells secrete insulin in response to high glucose, the neighboring α-cells sense insulin and are restrained, helping ensure glucagon is low during feeding. Glucagon, in turn, stimulates insulin secretion—this may seem counterintuitive given their opposite effects, but it serves to prevent hyperglycemia from unchecked glucagon action. Glucagon is elevated (as in a high-protein meal or early fasting), it raises blood glucose which then triggers some insulin; additionally, glucagon can directly signal β-cells to secrete insulin (glucagon is an insulin secretagogue) (El Sayed & Mukherjee, 2023).

Table 5.1 shows rimary pancreatic hormones, their cellular origin within the islets of Langerhans, and major metabolic actions in target tissues.

Feeding and pancreatic hormone dynamics

The shift into the fed state (postprandial period) triggers a well-orchestrated change in pancreatic hormone levels. As nutrients are absorbed into the bloodstream, insulin secretion surges while glucagon and other counter-regulatory hormones are suppressed. These changes ensure that incoming dietary glucose is efficiently stored and utilized, and prevent blood sugar from spiking excessively. Several mechanisms contribute to the dynamics of hormone release after a meal, including direct nutrient effects, the incretin system, and neural signals.

Upon food intake—especially a carbohydrate-rich meal—the pancreatic β-cells respond within minutes by releasing insulin. In healthy individuals, the plasma insulin level begins rising in the first few minutes after glucose enters the circulation, reflecting the first phase insulin release of pre-stored insulin granules. This first phase peaks about 5–10 min into the meal absorption. It is followed by a sustained second phase of insulin secretion, in which β-cells release newly synthesized insulin over the next 1–2 h, helping to dispose of prolonged glucose influx. The net result is that blood insulin concentration may

TABLE 5.1 Pancreatic hormones and primary functions.

Hormone	Cell of origin	Primary action	Major targets
Insulin (Da Silva Xavier, 2018)	β-cell (~60% of human islet cells)	↑Glucose uptake; ↑glycogen/lipid/protein synthesis	Muscle, adipose, liver
Glucagon (Da Silva Xavier, 2018)	α-cell (~30%)	↑Hepatic glucose output (glycogenolysis, gluconeogenesis)	Liver
Amylin (IAPP) (Mietlicki-Baase, 2016)	β-cell (co-secreted with insulin)	↓Gastric emptying; ↓Postprandial glucagon; ↑Satiety	Brainstem/hypothalamus, GI, α-cell
Somatostatin (Da Silva Xavier, 2018)	δ-cell (~10%)	↓Insulin and ↓glucagon (paracrine brake)	α-/β-cells, GI tract
Pancreatic polypeptide (PP) (Zhu et al., 2023)	PP (F) cell	↑Satiety via Y4 receptor; biphasic post-meal rise; vagal marker	Hypothalamus/brainstem; GI pancreas
Ghrelin (Shahid & Singh, 2022)	ε-cell (minor) and stomach	↑Hunger; ↓Insulin secretion; ↑GH	Hypothalamus; β-cell

increase several-fold above fasting levels, typically peaking around 30–60 min after a meal and gradually declining back toward baseline by about 2–3 h post-meal (in a person with normal glucose tolerance). The magnitude of this insulin spike depends on the meal's composition and size: meals high in carbohydrates elicit the most robust insulin responses, whereas protein or fat-rich meals provoke smaller insulin elevations. For example, a study comparing mixed meals found that a high-carbohydrate meal (62% carbs) caused a significantly greater insulin response than an isocaloric low-carb, high-fat meal (Gutniak et al., 1986; Karimi et al., 2024).

The gastrointestinal tract plays an integral endocrine role in the fed-state hormone response through the incretin effect. Incretins are hormones secreted by intestinal enteroendocrine cells in response to nutrients in the gut, and they augment islet hormone secretion. The two primary incretins are GLP-1 (glucagon-like peptide-1) from L-cells in the ileum/colon and GIP (glucose-dependent insulinotropic polypeptide) from K-cells in the duodenum/jejunum. During a meal, GLP-1 and GIP are released within minutes (often beginning while nutrients are still being digested, partly via neural reflexes). GLP-1 and GIP act in a glucose-dependent manner, meaning they amplify insulin release only when blood glucose is elevated (preventing inappropriate insulin in hypoglycemia). Besides insulin, incretins also influence glucagon: GLP-1 strongly inhibits glucagon release from α-cells when glucose is high (Fisman & Tenenbaum, 2021). Studies have shown that GLP-1 and GIP have "opposite" effects on glucagon: GLP-1 increases insulin and suppresses glucagon, whereas GIP increases insulin and can increase glucagon simultaneously (Yoo et al., 2021). Meanwhile, GLP-1 acts as both an insulin secretagogue and a brake on α-cells to avoid postprandial hyperglycemia. Incretins also have extrapancreatic actions: GLP-1 slows gastric emptying and enhances satiety (so less food intake), and GIP has receptors in adipose tissue (promoting fat storage). However, their central role in the feeding response is to coordinate with pancreatic islets: boosting insulin output and modulating glucagon so that blood glucose is controlled after meals. In healthy metabolism, the incretin system is highly effective—for example, infusion of GLP-1 in humans amplifies insulin and nearly shuts off the glucagon if glucose is elevated (Yoo et al., 2021). In type 2 diabetes, the incretin effect is often blunted (GIP responsiveness is reduced and GLP-1 secretion may be lower) (Fisman & Tenenbaum, 2021), contributing to inadequate insulin release and glucagon overshoot after meals.

A hallmark of the well-fed state is that glucagon levels decline or remain low while insulin is high. In healthy individuals, as blood glucose rises and insulin pours out after a meal, α-cells are inhibited (via direct glucose effect, insulin's paracrine action, and the influence of GLP-1). Consequently, glucagon concentrations in plasma fall in the early postprandial period relative to fasting levels. This appropriate suppression of glucagon prevents an unnecessary output of glucose from the liver when plenty of glucose is coming from the gut. For example, during an oral glucose tolerance test or a mixed meal, glucagon is transiently suppressed in non-diabetic individuals. In fact, insulin actively suppresses components of the counter-regulatory response—for instance, high insulin levels signal the hypothalamus that glucose is plentiful, reducing sympathetic output and adrenal epinephrine release. Likewise, growth hormone (GH), which surges during fasting or overnight, is blunted in the daytime after carbohydrate meals (hyperglycemia and somatostatin feedback both suppress GH release). Amylin, co-secreted with insulin, also contributes to glucagon suppression: amylin from β-cells has been shown to inhibit α-cell glucagon release, particularly postprandially (El Sayed & Mukherjee, 2023). This provides a secondary brake on glucagon in the presence of rising glucose. Somatostatin levels in plasma may increase somewhat after a meal (as suggested by elevated somatostatin-like immunoreactivity after feeding), which could further restrain both insulin and glucagon; however, any somatostatin rise is balanced such that insulin still predominates sufficiently to handle the meal (Gutniak et al., 1986).

Not all meals are equal in hormonal response—the mix of macronutrients (carbohydrate, fat, protein) profoundly influences how much insulin and glucagon are released. The body tailors the islet response to the nutrient profile to maintain balance. Carbohydrates (especially sugars and starches) cause the most pronounced insulin secretion. Glucose from carbs is a direct stimulus to β-cells; additionally, carbs trigger robust incretin release (both GIP and GLP-1) from the gut, which further boosts insulin. Thus, a high-carb meal leads to a high insulin peak and strongly suppressed glucagon (since glucose and insulin levels are high) (Gutniak et al., 1986). Ingestion of isolated protein or amino acids can cause a paradoxical rise in glucagon despite some insulin release. In one classic study, a low-carb, high-fat/protein meal in healthy subjects induced significantly greater glucagon release than a high-carb meal. The high-carb meal, conversely, produced a much larger insulin response and kept glucagon low (in healthy people) (Gutniak et al., 1986; Petroff & Greco, 2020).

The characteristic insulin and glucagon responses to carbohydrate-, protein-, and fat-rich meals are depicted in Fig. 5.1, and key comparative findings are summarized in Table 5.2.

Table 5.3 shows directional changes of key pancreatic hormones during feeding and fasting and their main metabolic consequences.

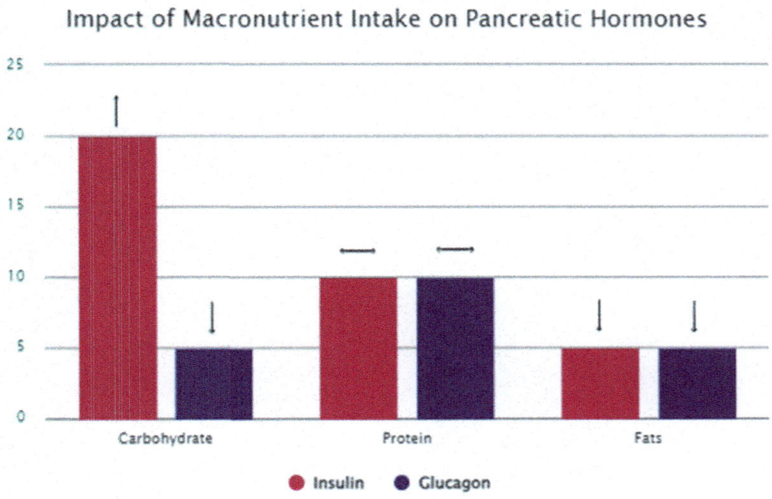

FIGURE 5.1 Impact of macronutrient intake on pancreatic hormones.

TABLE 5.2 Macronutrient-specific effects on pancreatic hormones.

Nutrient/meal	Insulin	Glucagon	Other hormonal notes	Evidence
High-carbohydrate mixed meal (Aulinger & D'Alessio, 2023)	↑↑ insulin AUC; strong incretin drive	↓ (suppressed)	Typical post-meal profile	Review/data on incretin effect (≈58%–79% insulin contribution)
Protein added to carbohydrate (Thomas et al., 2024)	↑ insulin AUC; ↓ glucose AUC	↔/↑ (amino acid–driven)	Blunts glycemic excursion	Human meal tests
High-protein or high-fat versus high-carb (Parvaresh Rizi et al., 2018)	Smaller glucose rise; stronger GLP-1/PYY; greater ghrelin suppression	Variable	Satiety hormones more favorable than HC	Crossover meal study (lean and obese)
High-fat or high-protein (isocaloric, lean adults) (Schuppelius et al., 2024)	–	↑ Glucagon and ↑GIR versus low-fat diet	–	Twin analysis (post-hoc)
Single macronutrient versus mixed (Göbl et al., 2022)	–	Temporal associations with glucose/ insulin differ	–	Controlled macronutrient tests

These reciprocal shifts in insulin, glucagon, and related hormones between the fed and fasted states are summarized schematically in Fig. 5.2.

Fasting and pancreatic hormone dynamics

Within just a few hours of the last caloric intake, as blood glucose begins to decline toward baseline, insulin secretion tapers off. The β-cells sense the lower glucose and reduce insulin release to only a basal trickle. This decrease in insulin is significant: plasma insulin levels during an overnight fast may drop to a small fraction of post-meal peaks. As insulin falls, its suppression of hepatic glucose output is lifted—allowing the liver to start releasing glucose from its stores. Simultaneously, glucagon secretion begins to increase. Even in an overnight fast (~8–12 h without food), α-cell activity rises enough to keep blood glucose from dipping too low. Glucagon levels in the early morning tend to be higher than in the fed evening prior, reflecting reduced glucose and insulin inhibition overnight (Karimi et al., 2024; Yoo et al., 2021). Growth hormone (GH) secretion is pulsatile and normally surges at night; fasting amplifies GH secretion after about 12–24 h, as low

TABLE 5.3 Direction of key hormones: feeding versus fasting.

Hormone	Feeding (post-meal)	Fasting (post-absorptive)	Key outcome
Insulin (Silver et al., 2018)	↑↑ (peaks ~30–60 min; near baseline by ~2–3 h)	↓	Glucose disposal/storage versus suppression in fasting
Glucagon (Geary, 2017)	↓ (suppressed by glucose/insulin/GLP-1)	↑ (maintains HGP; ketogenesis support)	Switch liver to production mode
Incretins (GLP-1/GIP) (Aulinger & D'Alessio, 2023)	↑ (major share of insulin response)	↓	Potentiate insulin; GLP-1 suppresses glucagon when glucose high
Amylin (Mietlicki-Baase, 2016)	↑ (co-secreted)	↓	Slows gastric emptying; reins in post-meal glucagon
PP (Lonovics et al., 1981)	↑ (biphasic; vagal + intestinal phases)	↑ with prolonged fasting (autonomic marker)	Satiety; pancreatic/GI regulation

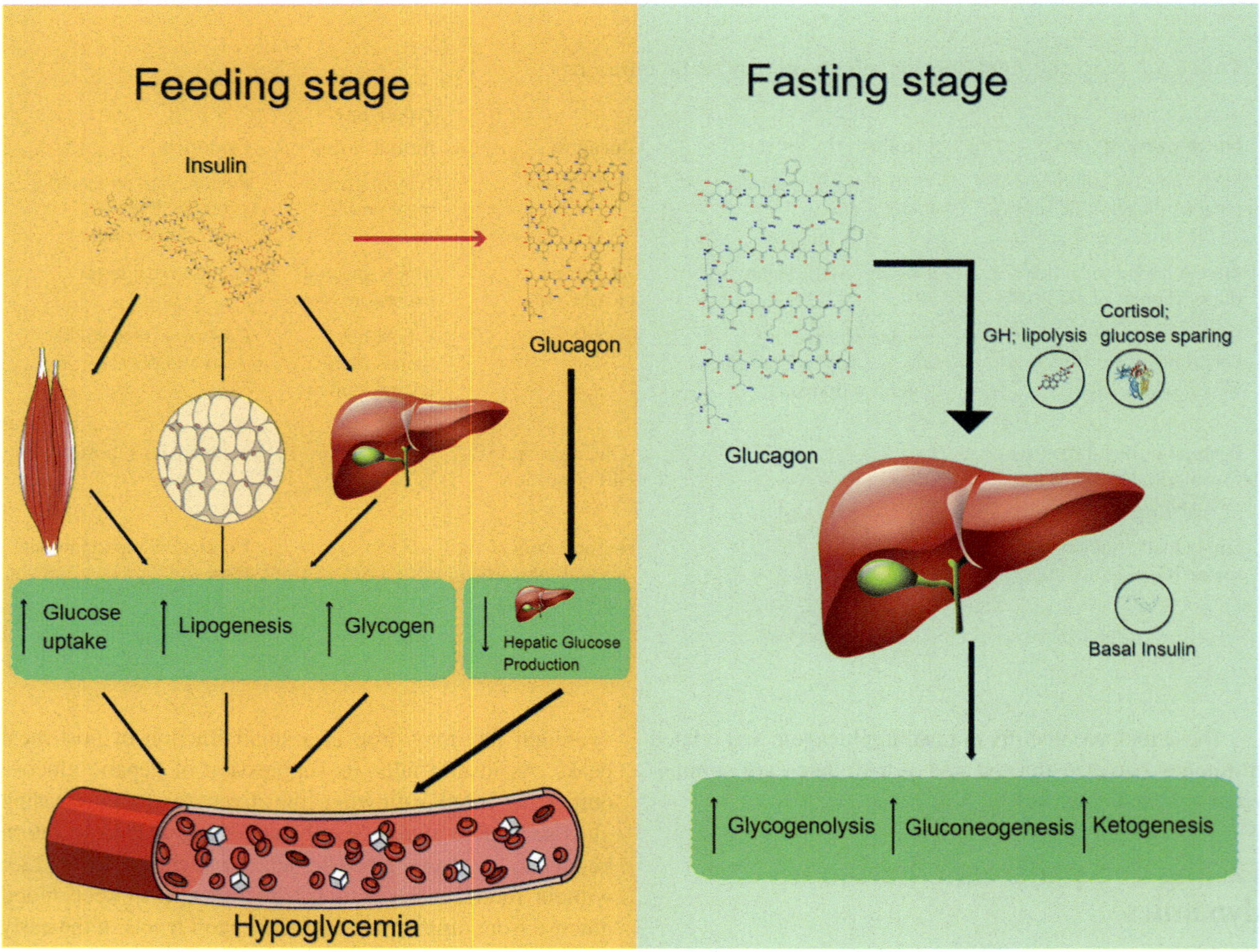

FIGURE 5.2 Feeding versus fasting hormonal axis.

insulin and low IGF-1 feedback remove inhibition on GH release (Espelund et al., 2005). GH during fasting helps mobilize fat and reduce muscle glucose uptake, thus sparing glucose for critical organs. In short-term fasting (up to 1 day), however, the dominant acute changes are low insulin and rising glucagon, since hepatic glycogen stores are still available for use. Studies show that during a 24-hour fast in humans, insulin levels fall markedly and glucagon levels rise, causing a several-fold increase in the glucagon:insulin ratio (Karimi et al., 2024). This hormonal shift enhances glycogenolysis and gluconeogenesis, while reducing peripheral glucose uptake and glycogen synthesis (Karimi et al., 2024). The result is maintenance of blood glucose within a safe range (usually lower than post-meal, but still euglycemic ~70–90 mg/dL) entirely by internal production. Also during short-term fasting, adipose tissue lipolysis is activated as insulin's antilipolytic effect wanes—leading to rising free fatty acid (FFA) levels in blood. Elevated FFAs become an alternate fuel for muscle and other tissues, sparing glucose. This is facilitated by the drop in insulin (which normally suppresses hormone-sensitive lipase in fat cells) and the presence of catecholamines and GH, which actively stimulate lipolysis. Therefore, even by 12–18 h of fasting, one can measure a significant increase in plasma FFA and ketone bodies begin to appear at low levels, indicating that fat is being utilized for energy. In summary, short-term fasting induces a "glucagon-dominant" hormonal profile: insulin is low, glucagon is relatively high, and this drives hepatic glucose output and fat breakdown to sustain metabolism in the absence of dietary fuel (Yoo et al., 2021).

The first 12–24 h of fasting, liver glycogenolysis is the major source of blood glucose, and glucagon is critical for this process. Humans typically store about 80–100 g of glycogen in the liver after a meal; overnight, roughly half of that may be depleted under glucagon's influence. By ~24 h, gluconeogenesis (making new glucose from lactate, glycerol, and amino acids) becomes increasingly important as glycogen stores wane. Glucagon strongly augments gluconeogenesis by increasing substrate availability (stimulating lipolysis yields glycerol, and proteolysis yields amino acids like alanine) and by upregulating key gluconeogenic enzymes. An interesting manifestation of glucagon's dominance is seen in experiments: infusing low-dose glucagon in a fasting individual rapidly raises blood glucose, whereas blocking glucagon can cause hypoglycemia during fasting. Even in the presence of low insulin, glucagon ensures that hepatic glucose production (HGP) matches glucose utilization by tissues. In healthy people, fasting glucagon levels are elevated enough to roughly double the hepatic glucose output compared to what it would be without glucagon (Huang et al., 2023).

When fasting continues beyond ~24–48 h, the metabolic state enters a starvation phase. By this time, liver glycogen is largely depleted and gluconeogenesis from non-carbohydrate sources (amino acids from muscle protein, glycerol from fat, lactate from RBCs) becomes the primary source of glucose. Ketone bodies generated from fatty acids sharply increase in concentration and start to be utilized by the brain, which reduces its demand for glucose (glucose requirements drop from ~120 g/day to perhaps 30–50 g/day as ketosis progresses). The hormonal milieu in prolonged fasting shows some further adaptations. Glucagon remains elevated, continuing to drive gluconeogenesis and maintain the liver in a catabolic mode. In fact, the insulin: glucagon ratio stays very low, which favors ketone production and ureagenesis (to excrete nitrogen from protein breakdown) (Karimi et al., 2024). Cortisol levels tend to be higher in prolonged fasting, which helps induce enzymes for gluconeogenesis and facilitates muscle protein breakdown to supply amino acids. Growth hormone secretion becomes markedly elevated after 1–2 days without food studies have shown a 5- to 10-fold increase in GH after 24–48h of fasting (Horne et al., 2024), which might seem counterintuitive since GH is an anabolic hormone; however, GH in fasting has a metabolic role: it promotes lipolysis and ketogenesis, and it also induces a degree of peripheral insulin resistance to spare glucose for the brain (Tong et al., 2016). Now, considering pancreatic hormones beyond insulin and glucagon: Pancreatic polypeptide (PP) appears to play a role in extended fasting. Normally a postprandial hormone, PP interestingly shows an increase after the first day of fasting. Research in humans indicates that plasma PP levels begin to rise approximately 24 h into a fast and remain elevated thereafter (Villanueva et al., 1978). The exact reason is not fully elucidated, but it may be related to increased vagal output or other adaptive responses. Elevated PP during starvation could serve to reduce appetite and gastric motility (helping the individual cope with prolonged absence of food) (Rose, 2019). PP is also considered a marker of parasympathetic activity; its rise might reflect changes in autonomic balance during fasting (somewhat paradoxically, since feeding stimulates vagus—but possibly the body cycles through phases of vagal activity even during fasting, or there may be a sensitization of PP release mechanisms). Additionally, high PP might help restrain insulin secretion further (PP can inhibit pancreatic secretion exocrine and perhaps endocrine to some extent) and promote efficient energy usage. Somatostatin from δ-cells may also have a role during extended fasting, though data are limited. It's plausible that somatostatin secretion from the islets could increase as a response to prolonged α-cell activity, thereby providing a check on glucagon. On the other hand, evidence suggests that during chronic fasting, somatostatin tone in some tissues decreases to allow heightened GH release (Córdoba-Chacón et al., 2011; Karimi et al., 2024).

Fasting does not only involve pancreatic and counter-regulatory hormones; it also engages gut-derived signals and neural circuits that communicate the body's energy status. One prominent hormone in this context is ghrelin, often dubbed the "hunger hormone." Ghrelin is produced mainly by the stomach, but also in small amounts by pancreatic ε-cells, and its levels rise during fasting. Prior to habitual mealtimes (when the stomach is empty), ghrelin surges and triggers the sensation of hunger via hypothalamic feeding centers (Rose, 2019). It also has endocrine effects: ghrelin inhibits insulin secretion from β-cells (both directly via islet paracrine action and indirectly by antagonizing insulin's effects) (El Sayed & Mukherjee, 2023). The gut–brain axis also encompasses behavioral adaptation: as fasting continues, the brain increases orexigenic (appetite-stimulating) signals like Neuropeptide Y (NPY) and AgRP in the hypothalamus, driven by low insulin and leptin. These central changes can indirectly influence pancreatic function via autonomic output. Pancreatic polypeptide (PP) is actually an interesting marker of vagal tone—in fasting beyond a day, elevated PP levels suggest a change in autonomic regulation (Villanueva et al., 1978). Some studies propose using PP responses to sham feeding as a test of vagal function (Witteman et al., 1994). PP itself also feeds back to the brain's appetite centers (like Y4 receptors in brainstem) to reduce hunger, so the fasting-induced PP increase might help temper appetite after the initial extreme hunger of the first 24 h (Rose, 2019).

Insulin and beyond: Broader hormonal interplay

Amylin is a peptide co-secreted with insulin from pancreatic β-cells during meals. It complements insulin's actions by promoting meal termination and preventing postprandial glucose spikes. Amylin acts on the central nervous system (area postrema and hypothalamus) to enhance satiety and reduce food intake, for example by inhibiting orexigenic (appetite-stimulating) neurons and slowing gastric emptying. It also modestly suppresses glucagon release after meals, curtailing excessive hepatic glucose output (Chan et al., 2022). Notably, amylin has adiposity-signaling properties alongside leptin and insulin—it sensitizes the brain to leptin's weight-reducing effects. A deficiency or resistance to amylin's effects (as observed in experimental models) can lead to larger meal sizes and weight gain, underscoring its role in energy homeostasis (Boyle et al., 2022).

Somatostatin, produced by pancreatic δ-cells, serves as a local inhibitory regulator within the islets. It is released in response to nutrient stimuli and acts on neighboring β- and α-cells to inhibit both insulin and glucagon secretion. This paracrine "brake" prevents over-secretion of insulin or glucagon, helping to fine-tune the islet hormonal output. Somatostatin thereby maintains a balance between anabolic and catabolic drives during feeding–fasting transitions. Beyond the pancreas, somatostatin has broad inhibitory effects on gastrointestinal function (slowing gut motility and enzyme secretion) and on other hormones (it suppresses growth hormone, thyrotropin, and gut peptides), reflecting its role in coordinating digestive and metabolic processes (Chan et al., 2022; Kellard et al., 2020).

Pancreatic polypeptide (PP) is secreted by the PP (F) cells of the pancreas, primarily in the postprandial state. Its levels surge after food intake—rising 8- to 10-fold above fasting baseline following a meal—especially in response to protein-rich meals and vagal stimulation. PP plays a role in signaling satiety and regulating appetite: it reduces food intake in humans when administered exogenously, and physiologically it correlates with meal termination. In healthy individuals, PP exhibits a clear circadian pattern, with higher circulating levels during daytime eating hours and low levels overnight. Intriguingly, PP dynamics reflect energy status—conditions of overnutrition (obesity or hyperphagia) are associated with blunted PP responses, whereas anorexia nervosa and prolonged fasting show elevated baseline PP levels (Batterham et al., 2003). This coordinated interplay ensures blood glucose remains in a narrow range. Importantly, these systemic hormones also feed back on the pancreas and other organs—e.g., chronically elevated cortisol (as in Cushing's or chronic stress) induces insulin resistance and beta-cell dysfunction, whereas GH excess can antagonize insulin's action on muscle (Kim & Park, 2017; Tavares et al., 2022).

Pathophysiological perspectives

In obesity and insulin-resistant states, normal hormonal rhythms linked to meals are often blunted or dysregulated. Many individuals with obesity engage in frequent eating or prolonged daily caloric intake, effectively shortening fasting periods and keeping insulin levels chronically elevated. This persistent hyperinsulinemia can downregulate insulin receptors and promote insulin resistance, creating a vicious cycle of "anabolic drive" and weight gain. Concurrently, obesity is characterized by hyperglucagonemia—in fact, fasting glucagon levels are inappropriately high in many obese and type 2 diabetic patients (Stern et al., 2019). The normal reciprocal relationship between insulin and glucagon is impaired: instead of high insulin suppressing glucagon, individuals with obesity/T2D often maintain elevated glucagon even in the fed state. This aberrant combination (too much glucagon alongside insufficient insulin action) is a core defect in the "bihormonal hypothesis" of diabetes. According to this concept, hyperglycemia in diabetes results from both inadequate insulin and excessive glucagon; the liver is thus simultaneously under-insulinized and over-glucagonized, driving overproduction of glucose (Rix et al., 2019). A classic manifestation is that during a meal tolerance test, instead of the normal suppression, plasma glucagon may paradoxically rise or remain elevated in type 2 diabetes (Stern et al., 2019).

TABLE 5.4 Dysregulation of feeding–fasting cycles in metabolic disease.

Condition	Hormonal profile	Clinical consequence
Type 2 diabetes (Geary, 2017)	Post-meal glucagon not adequately suppressed; blunted incretin effect	Exaggerated postprandial hyperglycemia; higher HGP despite feeding
Obesity/insulin resistance (Stern et al., 2019)	Fasting-induced rise in glucagon blunted; "pseudo-fed" signals persist	Impaired switching to fat use; metabolic inflexibility. (Lean vs. obese humans/mice data)
Metabolic syndrome (Bang et al., 2023)	Lower fasting glucagon-to-insulin ratio (GIR 14.0 ± 9.7 vs. 17.3 ± 10.3 in non-MetS)	Early marker of reduced metabolic flexibility; worse with more MetS components

Insulin resistance—whether due to obesity, lipotoxicity, or genetics—means that insulin's target tissues respond suboptimally, so the pancreas compensatorily secretes more insulin (hyperinsulinemia). Despite high circulating insulin, key tissues (liver, muscle, adipose) don't adequately take up glucose, and critically, the liver continues producing glucose. In healthy physiology, insulin inhibits α-cells (directly and via paracrine somatostatin), but in insulin-resistant states this inhibition falters. The result is inadequate suppression of glucagon when blood sugar is high. Postprandially, instead of dropping, glucagon may remain normal or even elevate in those with type 2 diabetes (Stern et al., 2019). The liver, therefore, receives mixed signals—insulin says "store glucose," but glucagon still says "release glucose." This discordance contributes to the elevated fasting glucose and exaggerated postprandial glucose peaks in diabetes. Additionally, insulin resistance is often accompanied by hepatic steatosis (fatty liver), which has been linked to aberrant glucagon responses (termed "glucagon resistance" at the liver) (Li et al., 2025). The liver of an insulin-resistant individual may become less sensitive to insulin's suppressive effect on glucose production, yet remain responsive to glucagon's stimulation—a combination that strongly favors hyperglycemia. Thus, impaired glucagon suppression is now recognized as an early feature of dysglycemia and metabolic syndrome (Laurenti et al., 2021).

At the far ends of the feeding-fasting spectrum, the endocrine system undergoes profound adaptations. During prolonged fasting (beyond 24–48 h), insulin levels drop to barely discernible basal concentrations, reflecting near-complete shutdown of β-cell secretion. This is accompanied by a marked rise in counter-regulatory hormones: glucagon increases (especially in the early days of fasting) to maintain blood glucose via hepatic glycogenolysis and gluconeogenesis (Rix et al., 2019). Growth hormone also climbs significantly with fasting, reaching levels several-fold higher than fed-state by the second day of fasting. GH in fasting helps mobilize fat (lipolysis) and spare muscle protein, partly by inducing peripheral insulin resistance to ensure glucose is reserved for vital organs (Horne et al., 2024). Cortisol tends to elevate as fasting becomes extended, peaking in the morning to support gluconeogenesis and blood pressure; indeed, fasting and psychological stress both raise cortisol, which in turn raises blood glucose (Batterham et al., 2003; Chan et al., 2022), aligning with the idea that PP is a marker of satiation that is diminished when food intake is excessive. Ketogenic diets (KDs), which are very low in carbohydrates and high in fats (with moderate protein), induce a physiological fasting-like state while food is still being consumed. By severely restricting carbohydrate, KDs keep insulin levels low and glucagon relatively elevated, thereby promoting ketone body formation similar to fasting. The insulin: glucagon ratio during a well-formulated ketogenic diet resembles that of fasting—low ratio—which drives lipolysis and ketogenesis (Paoli et al., 2023; Rix et al., 2019). Many individuals on ketogenic diets experience a mild rise in fasting glucagon and significantly lower fasting insulin, contributing to improved insulin sensitivity and weight loss. However, there can be adaptations: for instance, long-term ketogenic feeding in rodents has been noted to reduce β-cell mass and alter α-cell function (perhaps due to chronic low insulin signaling) (Ellenbroek et al., 2014; Kaja, 2025). Clinically, ketogenic diets often lead to lower HbA1c and reduced glycemic variability in type 2 diabetes, but some studies report impaired glucose tolerance on a carb challenge (likely because of physiologic insulin suppression and low de novo insulin response after long-term carb abstinence) (Kinzig et al., 2010; Merovci et al., 2024).

Table 5.4 shows altered pancreatic hormone profiles and clinical manifestations in metabolic and nutritional disorders.

Experimental and clinical evidence

Animal studies—mechanisms of pancreatic hormone adaptation: Controlled experiments in animals have been invaluable for unraveling how pancreatic hormones adapt to feeding and fasting. In rodents, researchers can mimic human feeding schedules or induce obesity/diabetes and then directly observe islet function. For example, studies in lean versus obese rodents highlight differential hormonal responses to fasting: lean mice fasting for 24 h show the expected rise in glucagon and fall in insulin, whereas obese (diet-induced obese) mice fail to upregulate glucagon during fasting. Instead, obese mice maintain suprabasal insulin and actually drop their glucagon output after prolonged fasting. This animal model mirrors the blunting of glucagon dynamics seen in obese humans. Pancreas perfusion experiments (isolating the pancreas or islets and perfusing with various substrates) confirm intrinsic islet defects—in obese mice, α-cells do not ramp up glucagon secretion upon low glucose, implicating intrinsic resistance to usual stimuli (Stern et al., 2019). Another line of animal research involves genetic models: by knocking out or altering specific hormones/receptors, one can see how the system compensates. Mice with somatostatin signaling knocked down (or with delta-cell ablation) exhibit hypersecretion of glucagon, especially on high-fat diets (Kellard et al., 2020). In one study, RAMP1 knockout mice had increased adiposity and gained excessive weight on a high-fat diet compared to wild-type mice. This aligns with amylin's normal function in curbing appetite and limiting weight gain. Conversely, transgenic rodents overexpressing amylin or treated with amylin analogs eat less and resist diet-induced obesity (Boyle et al., 2022). There are also α-cell knockout or glucagon receptor knockout models—glucagon receptor–null mice, for instance, have lower blood glucose and extreme α-cell hyperplasia (since lack of glucagon feedback leads to continuous α-cell stimulation). Such models confirm glucagon's role in preventing hypoglycemia and show that chronic absence of glucagon signaling pushes the pancreas to produce many more α-cells (demonstrating an intrinsic glucagon-insulin axis feedback). These "meal-fed" rats have insulin surges that become entrained to feeding time, and their tissues respond better to insulin—likely due to daily fasting recovery periods that restore insulin receptor sensitivity (Wiley & Leveille, 1970).

Rigorous clinical studies in humans have tested how structured feeding or fasting interventions impact pancreatic hormones and metabolic health. One notable example is trials of intermittent fasting (IF). In a randomized controlled trial, adults with metabolic syndrome were assigned to either an intermittent fasting regimen (24-hour water fast once or twice per week) or a standard diet. After 6 months, the IF group showed significantly improved insulin sensitivity: HOMA-IR (a measure of insulin resistance) dropped substantially, and fasting insulin levels were ~30% lower than baseline, despite minimal weight loss. This suggests that periodic fasting can reset insulin dynamics, improving how the body responds to insulin independent of just losing weight. Other trials have examined time-restricted feeding (TRF), such as eating all meals in a 6–8 h window earlier in the day (e.g., 8:00 am–2:00 pm feeding, and fasting for ~18 h until next morning). Early TRF (aligned with circadian rhythms) has been shown to enhance the diurnal insulin profile—one study found that after 5 weeks of early TRF, participants had markedly lower 24-hour insulin levels and improved β-cell responsiveness during an oral glucose test (Horne et al., 2024). They also exhibited lower evening blood glucose, reflecting improved glucose tolerance later in the day, which is typically a time of reduced insulin sensitivity. Such findings highlight that when we eat can influence metabolic hormones as much as what we eat. Controlled overfeeding or underfeeding studies offer additional insights: for instance, short-term overfeeding (to deliberately induce insulin resistance) causes fasting insulin and glucagon to both rise, as the body "anticipates" and tries to store excess nutrients (Colling et al., 2023).

A randomized trial in 2022 compared a 5:2 intermittent fasting diet (5 days normal diet, 2 non-consecutive days of ~500 kcal) against standard drug therapy in type 2 diabetics. Strikingly, after 16 weeks, the fasting regimen achieved better glycemic control (lower HbA1c) than the group on metformin and an SGLT2 inhibitor, despite the IF group taking no glucose-lowering medications (Guo et al., 2024). Furthermore, around 15% of the IF group reached normal blood glucose levels (partial diabetes remission) off medication, whereas virtually none did in the drug-treated group. Another study from China reported that an intermittent calorie restriction diet (with three consecutive days of very low-calorie intake per week) led to diabetes remission in nearly 50% of participants after 3 months, and many maintained normal blood sugar a year later without medications (Yang et al., 2022).

Table 5.5 showing representative animal and human studies examining how structured feeding or fasting regimens alter pancreatic hormone secretion and metabolic outcomes.

Future directions

Future research in feeding and fasting physiology is rapidly advancing toward precision metabolic profiling. The fasting glucagon-to-insulin ratio (GIR) has emerged as a promising marker of metabolic flexibility, with lower values linked to insulin resistance and metabolic syndrome. Next-generation assays using multiplex and ultrasensitive platforms will enable simultaneous measurement of multiple pancreatic and gut hormones, uncovering subtle hormonal imbalances long before glucose abnormalities appear. Continuous monitoring technologies may soon extend beyond glucose to real-time tracking of insulin, glucagon, or cortisol, transforming metabolic care.

TABLE 5.5 Representative clinical/experimental studies on feeding–fasting timing.

Study design and population	Intervention	Key hormonal/metabolic findings
Early time-restricted feeding (eTRF), overweight men, randomized crossover (Sutton et al., 2018).	6-h eating window (early) versus control	↓ 24-h insulin and improved β-cell responsiveness; improved insulin sensitivity despite prolonged pre-test fast
eTRF trials overview (review) (Jamshed et al., 2019).	4-day and 5-week eTRF studies	↓ Insulin levels; ↑ fat oxidation; ↓ ghrelin; improved glycemic control in prediabetes
10-h TRE, T2D outpatients, 12 weeks (Che et al., 2021).	Time-restricted eating	↓ FPG; ↑ insulin sensitivity; ↓ diabetes meds dose
Mixed-meal tests, diabetes versus control (Geary, 2017).	Standard meals/OGTT	Impaired postprandial glucagon suppression in T2D; contributes to hyperglycemia
Mechanistic review/assessment of incretin effect (Aulinger & D'Alessio, 2023).	Clamp/OGTT comparisons	Oral glucose elicits substantially greater insulin than IV at matched glycemia (incretin effect)

Another major frontier is the gut microbiome, which modulates insulin and glucagon responses through metabolites such as short-chain fatty acids and bile acids. Aligning feeding patterns with microbial rhythms could enhance hormonal balance and treatment outcomes.

Personalized nutrition and therapeutic fasting will integrate genetics, microbiome data, and circadian profiles to tailor interventions for each individual. Finally, preventive endocrinology will expand to include dynamic hormone testing and chrononutrition guidance. Together, these innovations will move metabolic medicine from reactive treatment to proactive precision care, detecting, predicting, and preventing disease by tracking and restoring the body's natural hormonal rhythms.

Conclusion

Feeding and fasting are coordinated hormonal states that maintain metabolic balance. In feeding, insulin promotes glucose uptake and nutrient storage, while amylin and gut hormones enhance satiety and suppress glucagon, cortisol, and growth hormone. During fasting, insulin declines and glucagon rises, stimulating glucose production, fat breakdown, and ketone formation to sustain energy. This daily oscillation between anabolic and catabolic phases preserves flexibility in fuel use. When disrupted by constant feeding or impaired fasting, it leads to insulin resistance, fat accumulation, and hyperglycemia. Restoring hormonal rhythm through time-restricted eating, intermittent fasting, or drugs that replicate fasting physiology improves metabolic control. Integrating insights from endocrinology, nutrition, and circadian biology enables personalized strategies that align meal timing with hormonal cycles. Maintaining this natural feeding-fasting rhythm is essential for preventing obesity, diabetes, and related metabolic disorders.

Summary points

- Feeding and fasting induce distinct metabolic states that are coordinated by pancreatic hormones, primarily insulin and glucagon, alongside somatostatin, amylin, and pancreatic polypeptide (PP).
- In the fed state, nutrient absorption stimulates biphasic insulin secretion, promotes amylin release, and suppresses glucagon through direct glucose effects, intra-islet signaling, and incretin hormones such as GLP-1 and GIP, facilitating glucose uptake, glycogen synthesis, and nutrient storage while enhancing satiety and slowing gastric emptying.
- PP and somatostatin further modulate appetite regulation, gut motility, and islet hormone output. During fasting, declining glucose availability reduces insulin secretion and increases glucagon release, shifting metabolism toward hepatic glycogenolysis, gluconeogenesis, lipolysis, and eventually ketogenesis.
- Prolonged fasting also engages growth hormone, cortisol, ghrelin, and rising PP levels, which support fuel mobilization, appetite signaling, and metabolic adaptation. Disruptions in these coordinated hormonal rhythms such as persistent hyperinsulinemia, impaired glucagon suppression, or altered amylin and PP responses, contribute to obesity, insulin resistance, and type 2 diabetes.
- Understanding the integrated actions of pancreatic hormones across feeding–fasting cycles provides insight into metabolic flexibility and disease mechanisms. It also supports therapeutic strategies such as time-restricted eating, intermittent fasting, and dietary interventions designed to restore physiological insulin–glucagon balance and improve metabolic health.

References

Aulinger, B. A., & D'Alessio, D. A. (2023). Assessment of the incretin effect in healthy subjects: Concordance between clamp and OGTT methods. *AJP Endocrinology and Metabolism, 325*, E412–E420. https://doi.org/10.1152/ajpendo.00104.2022

Bang, J., Lee, S. A., Koh, G., & Yoo, S. (2023). Association of glucagon to insulin ratio and metabolic syndrome in patients with type 2 diabetes. *Journal of Clinical Medicine, 12*. https://doi.org/10.3390/jcm12185806, 5806–6.

Batterham, R. L., Le Roux, C. W., Cohen, M. A., Park, A. J., Ellis, S. M., Patterson, M., et al. (2003). Pancreatic polypeptide reduces appetite and food intake in humans. *Journal of Clinical Endocrinology and Metabolism, 88*, 3989–3992. https://doi.org/10.1210/jc.2003-030630

Boyle, C. N., Zheng, Y., & Lutz, T. A. (2022). Mediators of amylin action in metabolic control. *Journal of Clinical Medicine, 11*, 2207. https://doi.org/10.3390/jcm11082207

Chan, K., Wong, F. S., & Pearson, J. A. (2022). Circadian rhythms and pancreas physiology: A review. *Frontiers in Endocrinology, 13*. https://doi.org/10.3389/fendo.2022.920261

Che, T., Yan, C., Tian, D., Zhang, X., Liu, X., & Wu, Z. (2021). Time-restricted feeding improves blood glucose and insulin sensitivity in overweight patients with type 2 diabetes: A randomised controlled trial. *Nutrition and Metabolism, 18*. https://doi.org/10.1186/s12986-021-00613-9

Colling, C., Bredella, M. A., Fazeli, P. K., Pachón-Peña, G., Singh, R. J., Rosen, C. J., et al. (2023). Changes in serum cortisol levels after 10 days of overfeeding and fasting. *AJP Endocrinology and Metabolism, 324*, E506–E513. https://doi.org/10.1152/ajpendo.00181.2022

Córdoba-Chacón, J, Gahete, M. D., Castaño, J. P., Kineman, R. D., & Luque, R. M. (2011). Somatostatin and its receptors contribute in a tissue-specific manner to the sex-dependent metabolic (fed/fasting) control of growth hormone axis in mice. *American Journal of Physiology-Endocrinology and Metabolism, 300*, E46–54. https://doi.org/10.1152/ajpendo.00514.2010

Da Silva Xavier, G. (2018). The cells of the islets of langerhans. *Journal of Clinical Medicine, 7*, 54. https://doi.org/10.3390/jcm7030054

El Sayed, S. A., & Mukherjee, S. (2023). *Physiology, Pancreas*. https://www.ncbi.nlm.nih.gov/books/NBK459261/.

Ellenbroek, J. H., van Dijck, L., Töns, H. A., Rabelink, T. J., Carlotti, F., Ballieux, B. E. P. B., et al. (2014). Long-term ketogenic diet causes glucose intolerance and reduced β- and α-cell mass but no weight loss in mice. *American Journal of Physiology. Endocrinology and Metabolism, 306*, E552–E558. https://doi.org/10.1152/ajpendo.00453.2013

Espelund, U., Hansen, T. K., Højlund, K., Beck-Nielsen, H., Clausen, J. T., Hansen, B. S., et al. (2005). Fasting unmasks a strong inverse association between ghrelin and cortisol in serum: Studies in obese and normal-weight subjects. *The Journal of Clinical Endocrinology and Metabolism, 90*. https://doi.org/10.1210/jc.2004-0604, 741–6.

Fisman, E. Z., & Tenenbaum, A. (2021). The dual glucose-dependent insulinotropic polypeptide (GIP) and glucagon-like peptide-1 (GLP-1) receptor agonist tirzepatide: a novel cardiometabolic therapeutic prospect. *Cardiovascular Diabetology, 20*. https://doi.org/10.1186/s12933-021-01412-5

Geary, N. (2017). Postprandial suppression of glucagon secretion: A puzzlement. *Diabetes, 66*, 1123–1125. https://doi.org/10.2337/dbi16-0075

Göbl, C., Morettini, M., Salvatori, B., Alsalim, W., Kahleova, H., Ahrén, B., et al. (2022). Temporal patterns of glucagon and its Relationships with glucose and insulin following ingestion of different classes of macronutrients. *Nutrients, 14*, 376. https://doi.org/10.3390/nu14020376

Guo, L., Xi, Y., Jin, W., Yuan, H., Qin, G., Chen, S., et al. (2024). A 5:2 intermittent fasting meal replacement diet and glycemic control for adults with diabetes. *JAMA Network Open, 7*. https://doi.org/10.1001/jamanetworkopen.2024.16786. e2416786–6.

Gutniak, M., Grill, V., & Efendlc, S. (1986). Effect of composition of mixed meals–low-versus high-carbohydrate content–oninsulin, glucagon, and somatostatin release in healthy humans and in patients with NIDDM. *Diabetes Care, 9*. https://doi.org/10.2337/diacare.9.3.244, 244–9.

Horne, B. D., Anderson, J. L., May, H. T., Bair, T. L., Le, V. T., Iverson, L., et al. (2024). Insulin resistance reduction, intermittent fasting, and human growth hormone: Secondary analysis of a randomized trial. *Npj Metabolic Health and Disease, 2*. https://doi.org/10.1038/s44324-024-00025-2

Huang, W., Xie, C., Wewer Albrechtsen, N. J., Jones, K. L., Horowitz, M., Rayner, C. K., et al. (2023). The "early" postprandial glucagon response is related to the rate of gastric emptying in type 2 diabetes. *Peptides, 161*, 170941. https://doi.org/10.1016/j.peptides.2023.170941

Jamshed, H., Beyl, R. A., Della Manna, D. L., Yang, E. S., Ravussin, E., & Peterson, C. M. (2019). Early time-restricted feeding improves 24-Hour glucose levels and affects markers of the circadian clock, aging, and autophagy in humans. *Nutrients, 11*, 1234. https://doi.org/10.3390/nu11061234

Kaja, F. (2025). Ketones and insulin: A paradoxical interplay with implications for glucose metabolism. *Journal of the Endocrine Society, 9*. https://doi.org/10.1210/jendso/bvaf101

Karimi, R., Yanovich, A., Fawzy, E., & Cleven, A. (2024). Adaptive effects of endocrine hormones on metabolism of macronutrients during fasting and starvation: A scoping review. *Metabolites, 14*. https://doi.org/10.3390/metabo14060336, 336–6.

Kellard, J. A., Nils, J., Rorsman, G., Hill, T. G., Armour, S. L., van, P. R., et al. (2020). Reduced somatostatin signalling leads to hypersecretion of glucagon in mice fed a high-fat diet. *Molecular Metabolism, 40*. https://doi.org/10.1016/j.molmet.2020.101021, 101021–1.

Kim, S. H., & Park, M. J. (2017). Effects of growth hormone on glucose metabolism and insulin resistance in human. *Annals of Pediatric Endocrinology & Metabolism, 22*. https://doi.org/10.6065/apem.2017.22.3.145, 145–52.

Kinzig, K. P., Honors, M. A., & Hargrave, S. L. (2010). Insulin sensitivity and glucose tolerance are altered by maintenance on a ketogenic diet. *Endocrinology, 151*, 3105–3114. https://doi.org/10.1210/en.2010-0175

Li, S., Liu, J., Yang, C., Jin, Y., Zhou, Y., Zhao, H., … Chen, Y. (2025). Impaired glucagon suppression accompanied by liver fat accumulation mediated worse blood glucose control in T2D patients. *The Journal of Clinical Endocrinology and Metabolism, 303*. https://doi.org/10.1210/clinem/dgaf303

Laurenti, M.C., Egan, A.M., Wismayer, D.S., Cobelli, C., Man, C.D., Vella, A. (2021). 259-OR: The effect of impaired glucagon suppression on endogenous glucose production is independent of insulin action. Diabetes, 70. 10.2337/db21-259-or.

Lonovics, J., Devitt, P., Watson, L.C., Rayford, P.L., Thompson, J.C. (1981). Pancreatic polypeptide. A review. *Archives of Surgery (Chicago, Ill.: 1960)*, *116*(10), 1256–1264. 10.1001/archsurg.1981.01380220010002

Merovci, A., Finley, B., Hansis-Diarte, A., Neppala, S., Abdul-Ghani, M. A., Cersosimo, E., et al. (2024). Effect of weight-maintaining ketogenic diet on glycemic control and insulin sensitivity in obese T2D subjects. *BMJ Open Diabetes Research & Care, 12*. https://doi.org/10.1136/bmjdrc-2024-004199. e004199–9.

Mietlicki-Baase, E. G. (2016). Amylin-mediated control of glycemia, energy balance, and cognition. *Physiology and Behavior, 162*, 130–140. https://doi.org/10.1016/j.physbeh.2016.02.034

Palmer, B. F., & Clegg, D. J. (2022). Metabolic flexibility and its impact on health outcomes. *Mayo Clinic Proceedings, 97*(4), 761–776. https://doi.org/10.1016/j.mayocp.2022.01.012

Paoli, A., Bianco, A., Moro, T., João, F. M., & de, C. (2023). The effects of ketogenic diet on insulin sensitivity and weight loss, which came first: The chicken or the egg? *Nutrients, 15*. https://doi.org/10.3390/nu15143120, 3120–0.

Parvaresh Rizi, E., Loh, T. P., Baig, S., Chhay, V., Huang, S., Caleb Quek, J., … Khoo, C. M. (2018). A high carbohydrate, but not fat or protein meal attenuates postprandial ghrelin, PYY and GLP-1 responses in Chinese men. *Plos One, 13*(1), Article e0191609. https://doi.org/10.1371/journal.pone.0191609

Petroff, B. K., & Greco, D. S. (2020). *Endocrine glands and their function* (pp. 394–427). Elsevier EBooks. https://doi.org/10.1016/b978-0-323-55227-1.00034-x.

Rix, I., Nexøe-Larsen, C., Bergmann, N. C., Lund, A., & Knop, F. K. (2019). *Glucagon physiology*. National Library of Medicine. https://www.ncbi.nlm.nih.gov/books/NBK279127/.

Rose, A. J. (2019). Role of peptide hormones in the adaptation to altered dietary protein intake. *Nutrients, 11*(9), 1990. https://doi.org/10.3390/nu11091990

Schuppelius, B., Schüler, R., Pivovarova-Ramich, O., Hornemann, S., Busjahn, A., Machann, J., … Pfeiffer, A. F. H. (2024). Alterations in glucagon levels and the glucagon-to-insulin ratio in response to high dietary fat or protein intake in healthy lean adult twins: A post hoc analysis. *Nutrients, 16*(22), 3905. https://doi.org/10.3390/nu16223905

Shahid, Z., & Singh, G. (2022, October 3). *Physiology, islets of langerhans*. Retrieved from Nih.gov website: https://www.ncbi.nlm.nih.gov/books/NBK542302/

Silver, B., Ramaiya, K., Andrew, S. B., Fredrick, O., Bajaj, S., Kalra, S., … Makhoba, A. (2018). EADSG guidelines: Insulin therapy in diabetes. *Diabetes Therapy, 9*(2), 449–492. https://doi.org/10.1007/s13300-018-0384-6

Stern, J. H., Smith, G. I., Chen, S., Unger, R. H., Klein, S., & Scherer, P. E. (2019). Obesity dysregulates fasting-induced changes in glucagon secretion. *Journal of Endocrinology, 243*, 149–160. https://doi.org/10.1530/JOE-19-0201

Sutton, E. F., Beyl, R., Early, K. S., Cefalu, W. T., Ravussin, E., & Peterson, C. M. (2018). Early time-restricted feeding improves insulin sensitivity, blood pressure, and oxidative stress Even without weight loss in men with prediabetes. *Cell Metabolism, 27*, 1212–1221.e3. https://doi.org/10.1016/j.cmet.2018.04.010

Tavares, M.R., Frazao, R., & Donato, J. (2022). Understanding the role of growth hormone in situations of metabolic stress. *The Journal of Endocrinology*, *256*(1), JOE–22-0159. 10.1530/JOE-22-0159

Thomas, M. S., Wolever, A. Z., Koecher, K., & Fei, A. Y. (2024). The effect of adding protein to a carbohydrate meal on postprandial glucose and insulin responses: A systematic review and meta-analysis of acute controlled feeding trials. *Journal of Nutrition, 154*(9). https://doi.org/10.1016/j.tjnut.2024.07.011

Tong, J., Davis, H. W., Gastaldelli, A., & D'Alessio, D. (2016). Ghrelin impairs prandial glucose tolerance and insulin secretion in healthy humans despite increasing GLP-1. *The Journal of Clinical Endocrinology and Metabolism, 101*(6), 2405–2414. https://doi.org/10.1210/jc.2015-4154

Villanueva, M. L., Hedo, J. A., & Marco, J. (1978). Fluctuations of human pancreatic polypeptide in plasma: Effect of normal food ingestion and fasting. *Experimental Biology and Medicine, 159*(2), 245–248. https://doi.org/10.3181/00379727-159-40324

Wiley, J. H., & Leveille, G. A. (1970). Significance of insulin in the metabolic adaptation of rats to meal ingestion. *The Journal of Nutrition, 100*, 1073–1080. https://doi.org/10.1093/jn/100.9.1073

Witteman, B. J., Edwards-Teunissen, K., Hopman, W. P., Jebbink, M. C., Masclee, A. A., Lamers, C. B., & Jansen, J. B. (1994). Nutrient-specific effects of modified sham feeding on pancreatic polypeptide release. *European Journal of Clinical Nutrition, 48*(8), 556–560.

Yang, X., Zhou, J., Shao, H., Huang, B., Kang, X., Wu, R., et al. (2022). Effect of an intermittent calorie-restricted diet on type 2 diabetes remission: A randomized controlled trial. *Journal of Clinical Endocrinology and Metabolism, 108*. https://doi.org/10.1210/clinem/dgac661

Yoo, S., Kim, D., & Koh, G. (2021). The change in glucagon following meal ingestion is associated with glycemic control, but not with incretin, in people with diabetes. *Journal of Clinical Medicine, 10* (11), 2487. https://doi.org/10.3390/jcm10112487

Zhu, W., Tanday, N., Flatt, P. R., & Irwin, N. (2023). Pancreatic polypeptide revisited: Potential therapeutic effects in obesity-diabetes. *Peptides, 160*, 170923. https://doi.org/10.1016/j.peptides.2022.170923

Further reading

Alaeddin, N., Jongejan, R. M. S., Stingl, J. C., Yolanda, P. R. P., Breteler, M. M. B., et al. (2023). Over- and undertreatment with levothyroxine—findings of the population-based Rhineland study. *Deutsches Ärzteblatt International*. https://doi.org/10.3238/arztebl.m2023.0192

AlAwaji, M. I., & Alhamwy, R. H. (2023). The impact of hypothyroidism on the quality of life of adults in Riyadh, Saudi Arabia. *Cureus, 15*. https://doi.org/10.7759/cureus.37636

Apostu, D., Lucaciu, O., Oltean-Dan, D., Mureşan, A.-D., Moisescu-Pop, C., Maxim, A., et al. (2020). The influence of thyroid pathology on osteoporosis and fracture risk: A review. *Diagnostics, 10*, 149. https://doi.org/10.3390/diagnostics10030149

Bernal, J., Guadaño-Ferraz, A., & Morte, B. (2003). Perspectives in the study of thyroid hormone action on brain development and function. *Thyroid, 13*, 1005–1012. https://doi.org/10.1089/105072503770867174

Bruinstroop, E., van der Spek, A., & Boelen, A. (2023). Role of hepatic deiodinases in thyroid hormone homeostasis and liver metabolism, inflammation and fibrosis. *European Thyroid Journal, 12*. https://doi.org/10.1530/etj-22-0211

Chambers, T., Anney, R., Taylor, P. N., Teumer, A., Peeters, R. P., Medici, M., et al. (2020). Effects of thyroid status on regional brain volumes: A diagnostic and genetic imaging study in UK biobank. *Journal of Clinical Endocrinology and Metabolism, 106*, 688–696. https://doi.org/10.1210/clinem/dgaa903

Contributors, W. (2025). *Levothyroxine*. Wikipedia.

Cooke, G. E., Mullally, S., Correia, N., O'Mara, S. M., & Gibney, J. (2014). Hippocampal volume is decreased in adults with hypothyroidism. *Thyroid, 24*, 433–440. https://doi.org/10.1089/thy.2013.0058

Cooper, H. E., Kaden, E., Halliday, L. F., Bamiou, D.-E., Mankad, K., Peters, C., et al. (2019). White matter microstructural abnormalities in children with severe congenital hypothyroidism. *NeuroImage. Clinical, 24*, 101980. https://doi.org/10.1016/j.nicl.2019.101980

de Escobar, G. M., Obregon, M. J., & del Rey, F. E. (1987). Fetal and maternal thyroid hormones. *Hormone Research, 26*, 12–27. https://doi.org/10.1159/000180681

Drugbank. (2023). *Levothyroxine. Godrugbankcom*. https://go.drugbank.com/drugs/DB00451.

Effraimidis, G., Watt, T., & Feldt-Rasmussen, U. (2021). Levothyroxine therapy in elderly patients with hypothyroidism. *Frontiers in Endocrinology, 12*. https://doi.org/10.3389/fendo.2021.641560

Eghtedari, B., & Correa, R. (2023). *Levothyroxine*. National Library of Medicine. https://www.ncbi.nlm.nih.gov/books/NBK539808/.

Ettleson, M. D., Karavolos, K., Burnett-Bowie, S.-A. M., Powell, L. H., & Janssen, I. (2024). The association between hypothyroidism and cognitive function change in women across the menopause transition: The study of Women's health across the nation (SWAN). *Thyroid*. https://doi.org/10.1089/thy.2024.0358

Freitas, B. C. G., Gereben, B., Castillo, M., Kalló, I., Zeöld, A., Egri, P., et al. (2010). Paracrine signaling by glial cell–derived triiodothyronine activates neuronal gene expression in the rodent brain and human cells. *Journal of Clinical Investigation, 120*, 2206–2217. https://doi.org/10.1172/jci41977

Głombik, K., Detka, J., Bobula, B., Bąk, J., Kusek, M., Tokarski, K., et al. (2021). Contribution of hypothyroidism to cognitive impairment and hippocampal synaptic plasticity regulation in an animal model of depression. *International Journal of Molecular Sciences, 22*, 1599. https://doi.org/10.3390/ijms22041599

Gaitonde, D. Y., Rowley, K. D., & Sweeney, L. B. (2012). Hypothyroidism: An update. *American Family Physician, 86*, 244–251.

Garber, J., Cobin, R., Gharib, H., Hennessey, J., Klein, I., Mechanick, J., et al. (2012). Clinical practice guidelines for hypothyroidism in adults: Cosponsored by the American association of clinical endocrinologists and the American thyroid association. *Endocrine Practice, 18*, 988–1028. https://doi.org/10.4158/ep12280.gl

Geest, van, Gunhanlar, N., Groeneweg, S., & Visser, W. E. (2021). Monocarboxylate transporter 8 deficiency: From pathophysiological understanding to therapy development. *Frontiers in Endocrinology, 12*. https://doi.org/10.3389/fendo.2021.723750

Guadaño-Ferraz, A., María, J. E., Rausell, E., & Bernal, J. (1999). Expression of type 2 iodothyronine deiodinase in hypothyroid rat brain indicates an important role of thyroid hormone in the development of specific primary sensory systems. *Journal of Neuroscience, 19*, 3430–3439. https://doi.org/10.1523/jneurosci.19-09-03430.1999

Ishii, S., Amano, I., & Koibuchi, N. (2021). The role of thyroid hormone in the regulation of cerebellar development. *Endocrinology and Metabolism, 36*, 703–716. https://doi.org/10.3803/enm.2021.1150

Kumar, M., Singh, S., Modi, S., Rana, P., D'souza, M., Sekhri, T., et al. (2024). Cortical gray matter thickness and volume changes and their association with memory functions in hyperthyroid patients. *Neuroendocrinology, 114*, 348–355. https://doi.org/10.1159/000536027

Kumar, M., Singh, S., Rana, P., D'souza, M., Kumaran, S. S., Sekhri, T., et al. (2025). Neurometabolite and cognitive changes in hypothyroid patients in response to treatment: In-vivo ^{1}H MRS study. *Neurochemistry International, 182*, 105915. https://doi.org/10.1016/j.neuint.2024.105915

Lekurwale, V., Acharya, S., Shukla, S., & Kumar, S. (2023). Neuropsychiatric manifestations of thyroid diseases. *Cureus, 15*. https://doi.org/10.7759/cureus.33987

Lischinsky, J. E., Skocic, J., Clairman, H., & Rovet, J. (2016). Preliminary findings show maternal hypothyroidism may contribute to abnormal cortical morphology in offspring. *Frontiers in Endocrinology, 7*. https://doi.org/10.3389/fendo.2016.00016

Liu, H., Li, W., Zhang, W., Sun, S., & Chen, C. (2022). Levothyroxine: Conventional and novel drug delivery formulations. *Endocrine Reviews, 44*, bnac030. https://doi.org/10.1210/endrev/bnac030

Martin, J. V., & Sarkar, P. K. (2023). Nongenomic roles of thyroid hormones and their derivatives in adult brain: Are these compounds putative neurotransmitters? *Frontiers in Endocrinology, 14*, 1210540. https://doi.org/10.3389/fendo.2023.1210540

Nillni, E. A. (2010). Regulation of the hypothalamic thyrotropin releasing hormone (TRH) neuron by neuronal and peripheral inputs. *Frontiers in Neuroendocrinology, 31*, 134–156. https://doi.org/10.1016/j.yfrne.2010.01.001

Nuguru, S., Rachakonda, S., Sripathi, S., Khan, M., Patel, N., & Meda, R. (2022). Hypothyroidism and depression: A narrative review. *Cureus, 14*, e28201. https://doi.org/10.7759/cureus.28201

Oppenheimer, J. H., & Schwartz, H. L. (1997). Molecular basis of thyroid hormone-dependent brain development. *Endocrine Reviews, 18*, 462–475. https://doi.org/10.1210/edrv.18.4.0309

Pak, K., Kim, M., Kim, K., Kim, B. H., Kim, S.-J., & Kim, I. J. (2020). Cerebral glucose metabolism and cerebral blood flow in thyroid dysfunction: An activation likelihood estimation Meta-analysis. *Scientific Reports, 10*. https://doi.org/10.1038/s41598-020-58255-5

Rastogi, M. V., & LaFranchi, S. H. (2010). Congenital hypothyroidism. *Orphanet Journal of Rare Diseases, 5*. https://doi.org/10.1186/1750-1172-5-17

Remaud, S., GothiÃ©, J.-D., Morvan-Dubois, G., & Demeneix, B. A. (2014). Thyroid hormone signaling and adult neurogenesis in mammals. *Frontiers in Endocrinology, 5*. https://doi.org/10.3389/fendo.2014.00062

Ritchie, M., & Yeap, B. B. (2015). Thyroid hormone: Influences on mood and cognition in adults. *Maturitas, 81*, 266–275. https://doi.org/10.1016/j.maturitas.2015.03.016

Rousset, B., Dupuy, C., Miot, F., & Dumont, J. (2000). *Chapter 2 thyroid hormone synthesis and secretion*. PubMed https://pubmed.ncbi.nlm.nih.gov/25905405/.

Şorodoc, V., Constantin, M., Asaftei, A., Lionte, C., Ceasovschih, A., Sîrbu, O., et al. (2023). The use of intravenous immunoglobulin in the treatment of Hashimoto's encephalopathy: Case based review. *Frontiers in Neurology, 14*. https://doi.org/10.3389/fneur.2023.1243787

Salas-Lucia, F. (2024). Mapping thyroid hormone action in the human brain. *Thyroid, 34*, 815–826. https://doi.org/10.1089/thy.2024.0120

Samuels, M. H. (2014). Psychiatric and cognitive manifestations of hypothyroidism. *Current Opinion in Endocrinology Diabetes and Obesity, 21*, 377–383. https://doi.org/10.1097/med.0000000000000089

Satkarjit, K. J., Shenoy, S., Randhawa, A. S., & Singh, A. P. (2025). Impaired sustained attention in patients with subclinical hypothyroidism. *PubMed, 23*, 45–52. https://doi.org/10.3121/cmr.2025.1925

Singh, S., Trivedi, R., Singh, K., Kumar, P., Shankar, L. R., & Khushu, S. (2014). Diffusion tensor tractography in hypothyroidism and its correlation with memory function. *Journal of Neuroendocrinology, 26*, 825–833. https://doi.org/10.1111/jne.12193

Taylor, P. N., Medici, M. M., Hubalewska-Dydejczyk, A., & Boelaert, K. (2024). Hypothyroidism. *The Lancet, 404*, 1347–1364. https://doi.org/10.1016/s0140-6736(24)01614-3

Toll, A., Blanco-Hinojo, L., Berge, D., Manzano, A., Abidi, K. E., Perez-Solà, V., et al. (2023). Relationship between thyroid-stimulating hormone, BDNF levels, and hippocampal volume in antipsychotic-naïve first-episode psychosis patients. *Frontiers in Psychiatry, 14*. https://doi.org/10.3389/fpsyt.2023.1301714

Uchida, K., & Suzuki, M. (2021). Congenital hypothyroidism and brain development: Association with other psychiatric disorders. *Frontiers in Neuroscience, 15*. https://doi.org/10.3389/fnins.2021.772382

Uchida, K., Yonezawa, M., Nakamura, S., Kobayashi, T., & Machida, T. (2005). Impaired neurogenesis in the growth-retarded mouse is reversed by T3 treatment. *NeuroReport, 16*, 103–106. https://doi.org/10.1097/00001756-200502080-00005

Waliszewska-Prosół, M., Bladowska, J., Budrewicz, S., Sąsiadek, M., Dziadkowiak, E., & Ejma, M. (2021). The evaluation of Hashimoto's thyroiditis with event-related potentials and magnetic resonance spectroscopy and its relation to cognitive function. *Scientific Reports, 11*. https://doi.org/10.1038/s41598-021-82281-6

Wang, Q., Jiang, Y., Lv, H., Lu, Q., Tao, S., Qin, R., et al. (2022). Association of maternal mild hypothyroidism with offspring neurodevelopment in TPOAb-Negative women: A prospective cohort study. *Frontiers in Endocrinology, 13*. https://doi.org/10.3389/fendo.2022.884851

Zhang, Y., Yang, Y., Tao, B., Lv, Q., Lui, S., & He, L. (2021). Gray matter and regional brain activity abnormalities in subclinical hypothyroidism. *Frontiers in Endocrinology, 12*. https://doi.org/10.3389/fendo.2021.582519

Zhao, C., Wang, Y., Xiao, L., & Li, L. (2022). Effect of levothyroxine on older patients with subclinical hypothyroidism: A systematic review and meta-analysis. *Frontiers in Endocrinology, 13*. https://doi.org/10.3389/fendo.2022.913749

Zinck, J. C., Isaksen Jonas, L., Ahlberg, G., Salling, O. M., Nygaard, B., Ellervik, C., et al. (2023). Interaction between DIO2 polymorphisms and LT4 treatment on psychological well-being outcomes and cardiovascular risk factors: A population-based cross-sectional study in UK biobank. *Endocrine Abstracts*. https://doi.org/10.1530/endoabs.92.op-04-01

Section 2

Focused areas of neurological pathology and physiology

Chapter 6

Type 2 diabetes and the increased risk of dementia

Ehsan Noroozi[1], Zeinab Kolahdouzan[2], Sayedeh-Fatemeh Sadat-Madani[3], Fatemeh Naseri Rad[4], Amir Abdi[5] and Alaleh Alizadeh[6]

[1]*Department of Internal Medicine, Shahid Labbafi Nezhad Hospital, Shahid Beheshti University of Medical Science, Tehran, Iran;* [2]*School of Medicine, Kashan University of Medical Sciences, Isfahan, Iran;* [3]*School of Medicine, Isfahan University of Medical Sciences, Isfahan, Iran;* [4]*Student Research Committee, Faculty of Medicine, Iran University of Medical Sciences, Tehran, Iran;* [5]*Student Research Committee, School of Medicine, Tehran Medical Sciences, Islamic Azad University, Tehran, Iran;* [6]*Faculty of Medicine, Mashhad Branch, Islamic Azad University, Mashhad, Iran*

List of abbreviations

AD Alzheimer's disease
AD[+DM] Alzheimer's disease with diabetes mellitus
AGEs advanced glycation end products
AMPK AMP-activated protein kinase
BBB blood-brain barrier
CNS central nervous system
CVD cerebrovascular disease
DPP-4 dipeptidyl peptidase-4
DrD diabetes-related dementia
GLP-1 RAs glucagon-like peptide-1 receptor agonists
MRI magnetic resonance imaging
NF-κB nuclear factor-kappa B
NMDAR N-methyl-D-aspartate receptor
PGC-1α peroxisome proliferator-activated receptor gamma coactivator-1 alpha
PPAR-γ peroxisome proliferator-activated receptor gamma
PSD-95 postsynaptic density protein-95
RAGE receptor for advanced glycation end products
rCBF regional cerebral blood flow
ROS reactive oxygen species
SGLT2is sodium-glucose cotransporter-2 inhibitors
SPECT single photon emission computed tomography
T2DM type 2 diabetes mellitus
TNF-α tumor necrosis factor-alpha
VaD vascular dementia
WHO World Health Organization

Introduction

The most common cause of dementia is Alzheimer's disease (AD). The main risk factors that have been identified for dementia by World Health Organization (WHO): (I) Age (Fukasawa et al.), high blood pressure, (III) high blood sugar (diabetes), (IV) Obesity or overweight, (V) smoking, (VI) Drinking too much alcohol, (VII) physical inactivity, (VIII) social isolation, and (IX) depression. In 2021, 57 million individuals and their caregivers suffered from dementia, which is more than 60 percent of these cases in low-income and middle-income areas. Dementia has not only an impact on patients with changes in their motivation, mood, behavior, and control but also it has a considerable physical, social, and psychological effect on caregivers, families, and society (World Health Organization, 2023).

This chapter delves into what we know so far about type 2 diabetes and dementia connection. It reviews population-level evidence and then leads to etiologies. The overall aim is to give clinicians and policy planners a concise, use-ready map of an evolving field.

Pathophysiology of diabetes-related dementia

As illustrated in Fig. 6.1, diabetes-related dementia arises from a complex interplay of metabolic, vascular, and neurodegenerative pathways, with insulin resistance and Advanced Glycation End Products (AGE)-Receptor for Advanced Glycation End Products (RAGE) signaling serving as pivotal contributors to neuronal damage. These mechanisms are summarized in Table 6.1, which provides an overview of their descriptions, impacts on brain function, and supporting evidence. The pathophysiologic milieu consists of a complex array of interconnected

Hormones in Neuroscience. https://doi.org/10.1016/B978-0-443-30263-3.00003-8

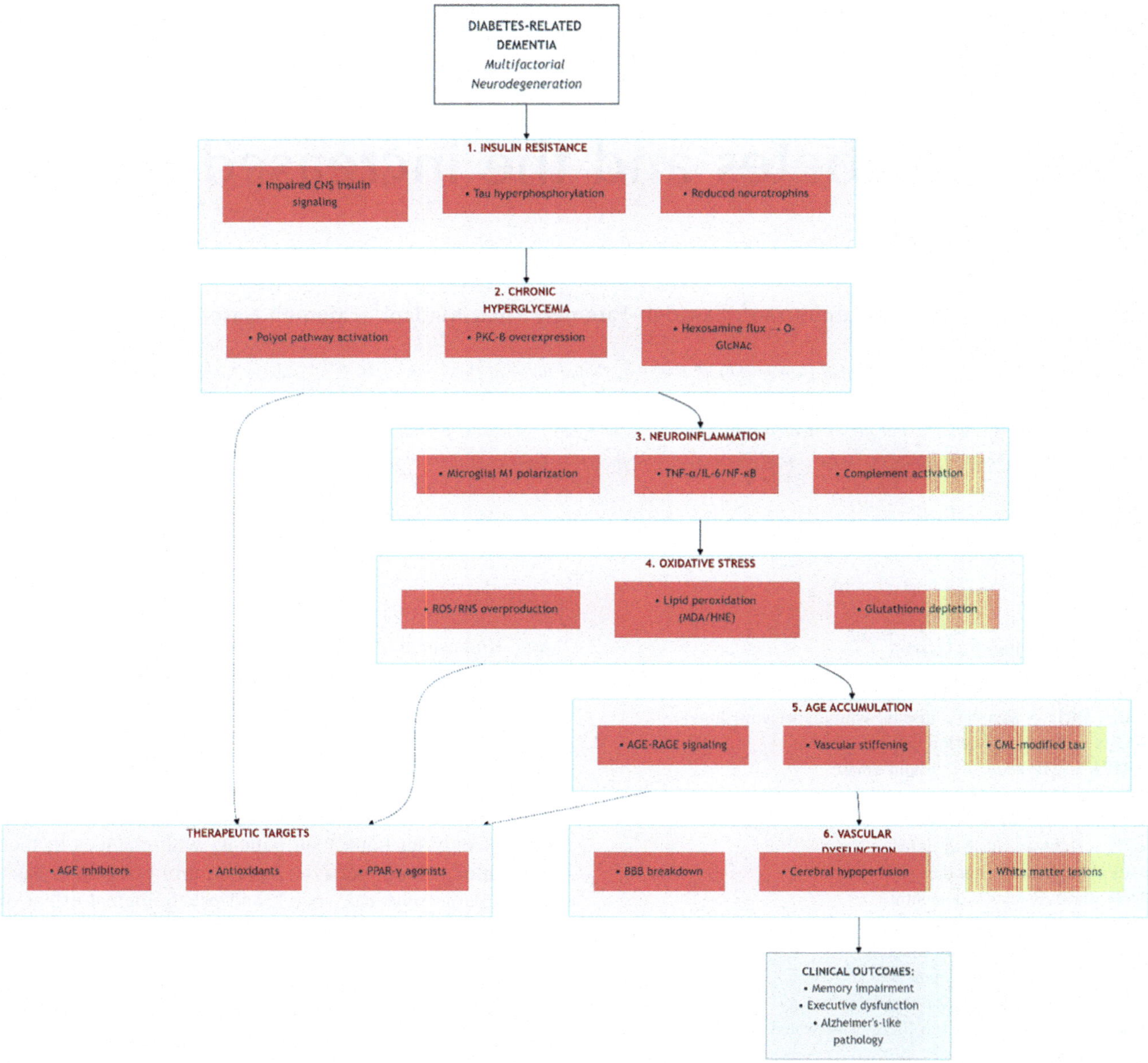

FIGURE 6.1 Multifactorial pathophysiology of diabetes-related dementia. The diagram summarizes key interconnected mechanisms, including insulin resistance, chronic hyperglycemia, neuroinflammation, oxidative stress, AGE accumulation, and vascular dysfunction, culminating in synaptic collapse and cognitive decline.

mechanisms whose cumulative impact adversely affects neuronal well-being and cognitive performance (Cholerton et al., 2016; Platzbecker et al., 2025).

Figure 6.2 provides a visual representation of how these mechanisms progress from type 2 diabetes to dementia, illustrating the interconnected pathways leading to cognitive decline. These mechanisms include chronic hyperglycemia-induced cellular damage, central nervous system (CNS) insulin resistance, microvascular disease, chronic inflammation, oxidative stress, and the accumulation of advanced glycation end products (Khalid et al., 2022; Mosquera-Sulbarán & Hernández-Fonseca, 2022; Singh et al., 2022). Recognition of these pathways is central to the process of establishing directed therapeutic strategies and achieving improved clinical results in the growing number of individuals with diabetes and cognitive impairment (Singh et al., 2022).

Insulin resistance and CNS dysfunction

Central insulin resistance is one of the keystone mechanisms in the pathogenesis of diabetes-associated dementia. The brain, traditionally an insulin-insensitive organ, actually demonstrates dispersed expression of insulin receptors

TABLE 6.1 Summary of pathophysiological mechanisms linking type 2 diabetes to dementia.

Mechanism	Description	Key effects on brain	References
Insulin resistance	Impaired insulin signaling in the brain, particularly in the hippocampus and cortex	Disrupted glucose uptake, synaptic plasticity, and increased tau phosphorylation	Caroline et al. (2020), Csajbók and Tamás (2016), Milstein and Ferris (2021); Spinelli et al. (2019)
Chronic hyperglycemia	Prolonged high glucose levels leading to neurotoxicity	Polyol pathway activation, osmotic stress, and protein kinase C-mediated inflammation	Hwang et al. (2017), Kadir et al. (2022), Morozova et al. (2024), Shao and Bayraktutan (2014), Tomlinson and Gardiner (2008)
Neuroinflammation	Chronic low-grade inflammation with elevated cytokines (e.g., IL-6, TNF-α)	Microglial activation, complement dysregulation, and synaptic loss	Abbatecola et al. (2024), Ciudin and Hernández (2024), De Felice and Ferreira (2014), Phosat et al. (2017), Ramaglia et al. (2021), Xu et al. (2024), Zhilian Li et al. (2023)
Oxidative stress	Increased reactive oxygen species and reduced antioxidant defenses	Lipid peroxidation, protein carbonylation, and mitochondrial dysfunction	Ahmad et al. (2017), Chen et al. (2022), Dawi et al. (2025), Hecker and Wagner (2018), Kim et al. (2024), Kruman et al. (1997), Sekhar et al. (2010), Weinberg Sibony et al. (2024), Zarkovic et al. (2013)
Advanced glycation end products (AGEs)	Non-enzymatic protein modifications by glucose	Neuroinflammation, NF-κB activation, and cerebrovascular dysfunction	Bhattacharya et al. (2023), Chen and Yu (2023), Duca et al. (2016), Fu et al. (1996), Hanyu (2019), Khalid et al. (2022), Kuzan et al. (2022), Mosquera-Sulbarán and Hernández-Fonseca (2022), Shaw et al. (2002), Singh et al. (2024), Stammers et al. (2020)
Vascular dysfunction	Endothelial dysfunction and microvascular damage	Impaired cerebral blood flow, blood-brain barrier disruption, and hypoxic neuronal damage	Almutairi et al. (2016), Biessels and Despa (2018), Dutta et al. (2022), Feng and Gao (2024), Padovani et al. (2025), van Sloten et al. (2020), Xiong et al. (2017)
Mitochondrial dysfunction	Impaired mitochondrial structure and bioenergetic failure	Reduced ATP production, increased ROS, and neuronal apoptosis	Alkhalifa et al. (2025), Belosludtsev et al. (2020), Chen et al. (2023), Xiao Liang (2024), Zhang et al. (2023)
Synaptic dysfunction	Reduced synaptic protein expression and impaired neurotransmission	Compromised long-term potentiation, memory deficits, and neural network disruption	Biessels and Despa (2018), Collingridge (2025), Iarkov et al. (2021), Kim and Song (2022), Singh et al. (2022), Wang et al. (2014)

This table outlines the major mechanisms contributing to diabetes-related cognitive impairment, detailing their descriptions; effects on brain function, and supporting references; Abbreviations, IL-6 (Interleukin-6); TNF-α (Tumor Necrosis Factor-alpha), NF-κB (Nuclear Factor-kappa B); ROS (Reactive Oxygen Species).

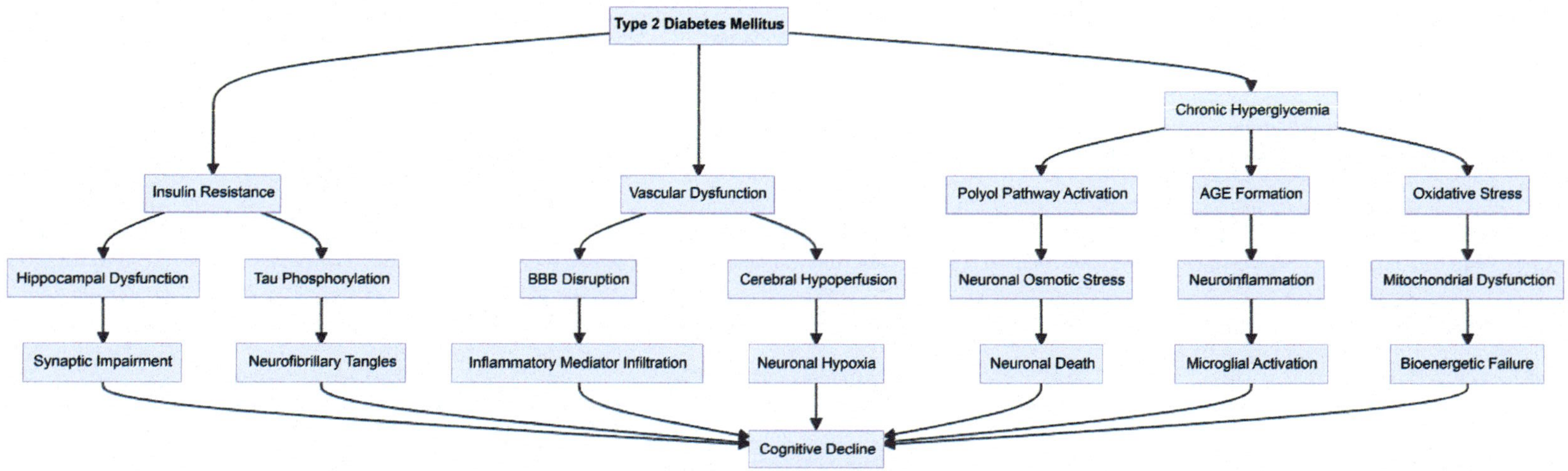

FIGURE 6.2 Progression from type 2 diabetes to dementia. This flowchart illustrates the sequential pathophysiological pathways linking type 2 diabetes mellitus (T2DM) to cognitive decline, highlighting key mechanisms such as chronic hyperglycemia, insulin resistance, vascular dysfunction, and their downstream effects on neuronal integrity and cognitive function. Abbreviations: AGE (advanced glycation end products), BBB (blood-brain barrier).

in cortical as well as subcortical regions, with the highest density in the hippocampus, hypothalamus, and cerebral cortex. Central insulin signaling is what governs neuronal glucose uptake, synaptic plasticity, neurotrophin expression, and memory-consolidating processes.

Central insulin resistance also impinges on the regulation of tau protein phosphorylation, resulting in neurofibrillary tangle formation observed in AD pathology (Caroline et al., 2020).

Chronic hyperglycemia and neurotoxicity

Extended hyperglycemia has immediate neurotoxicity through several cellular mechanisms.

Hyperglycemia also amplifies protein kinase C activation, triggering inflammatory cascades and vascular dysfunction. Amplified protein kinase C signaling increases inflammatory mediator production (Kadir et al., 2022) and increases endothelial dysfunction (Shao & Bayraktutan, 2014).

Oxidative stress markers and cellular damage

Oxidative stress is a common pathway in the pathogenesis of diabetes-related dementia. Many markers of oxidative damage show consistent elevation in affected patients, reflecting overloaded antioxidant protective mechanisms and increased generation of Reactive Oxygen Species (ROS) (Kim et al., 2024).

Advanced glycation end products (AGEs) and neurodegeneration

AGEs are a major pathophysiologic process in diabetes-related dementia. They are irreversible protein alterations caused by non-enzymatic glycosylation of reducing sugars with amino groups in proteins, lipids, and nucleic acids. During diabetes, elevated levels of blood glucose over a prolonged period increase the production of AGEs in the body, including the brain (Khalid et al., 2022; Mosquera-Sulbarán & Hernández-Fonseca, 2022).

AGEs exert pathological effects through a multiplicity of mechanisms. RAGE is widely expressed in brain tissue, particularly in neurons, microglia, and endothelial cells.

AGE cross-linking within the extracellular matrix alters tissue mechanical properties and deteriorates vascular function. AGE-modified collagen has increased cross-linking and reduced elasticity and contributes to cerebrovascular dysfunction. This AGE deposition in the vasculature compromises cerebral blood flow regulation and allows for Blood-Brain Barrier (BBB) dysfunction (Duca et al., 2016).

Vascular dysfunction and cerebrovascular pathology

Activation of matrix metalloproteinases leads to breakdown of tight junction proteins, contributing to loss of barrier function.

Autoregulation of the brain is also impaired in diabetes, lowering the ability of the brain to maintain constant flow of blood during fluctuations in blood pressure. This loss affects white matter regions in particular, resulting in subcortical ischemic change and executive dysfunction characteristic of vascular cognitive impairment (Dutta et al., 2022).

Mitochondrial dysfunction and bioenergetic failure

Mitochondrial DNA damage accumulates in diabetes-associated dementia, both a manifestation of higher oxidative stress exposure and compromised DNA repair mechanisms. This damage impacts genes encoding electron transport chain proteins, inducing a vicious cycle of worsening mitochondrial function and bioenergetic collapse.

Mitochondrial biogenesis is impaired by the reduced expression and activity of peroxisome proliferator-activated receptor gamma coactivator-1α (PGC-1α). This transcriptional coactivator plays a role in mitochondrial gene expression and biogenesis, and its dysfunction explains decreased mitochondrial density and function in brain tissue.

The mitochondrial permeability transition pore is more open in diabetic dementia, promoting cytochrome *c* release and apoptotic neuronal loss. Increased sensitivity is the result of oxidative post-translational modification of pore subunits and altered calcium homeostasis (Belosludtsev et al., 2020; Zhang et al., 2023).

Synaptic dysfunction and neural network disruption

Diabetes-associated dementia is typified by progressive synaptic dysfunction preceding overt neuronal loss. Synaptic proteins such as synaptophysin, postsynaptic density protein-95 (PSD-95), and N-methyl-D-aspartate receptor (NMDAR) subunits exhibit reduced expression in the affected brain regions. This reduction of synaptic proteins correlates with the degree of cognitive deterioration and disruption of functional connectivity (Biessels & Despa, 2018; Wang et al., 2014).

Neurotransmitter systems are disturbed in dementia of diabetes. There is cholinergic impairment by diminished choline acetyltransferase activity and synthesis of acetylcholine. The cholinergic deficiency strongly impacts attention and memory performance. Glutamatergic transmission is also disturbed by receptor desensitization and reduced synaptic plasticity.

Diabetic dementia pathophysiology encompasses a number of crucial interlinked mechanisms that collectively impinge on brain structure and function. Elucidation of these pathways is central to the development of mechanism-targeted therapeutic modalities and the improvement of clinical outcomes. Opportunities for future research will involve the search for early biomarkers, the development of mechanism-based therapies, and the deployment of prevention strategies to stem the rising tide of diabetic cognitive impairment.

The pathophysiologic mechanisms of diabetic dementia are multifactorial and will need multi-targeted interventions to treat inflammation, oxidative stress, atherosclerosis, and metabolic disturbances simultaneously. Early intensive treatment on the basis of glucose control, cardiovascular risk reduction, and neuroprotection offers the best hope for prevention or delay of diabetic cognitive impairment.

Subgroups of dementia associated with diabetes mellitus based on brain imaging

Type 2 Diabetes Mellitus (T2DM) is recognized as a significant risk factor for cognitive decline and different types of dementia, including AD and Vascular Dementia (VaD). Previous studies have investigated the vascular and neurodegenerative consequences of diabetes. Advanced neuroimaging technologies can be used for diagnosis of diabetes-associated cognitive syndromes. Diabetes-related dementia (DrD), emerge from DM-specific pathophysiological processes, is different from classic AD or Cerebrovascular Disease (CVD) mechanisms.

Neuroimaging modalities such as magnetic resonance imaging (MRI) and single photon emission computed tomography (SPECT) have been utilized in defining the structural and functional cerebral changes that distinguish different dementia subtypes in diabetic populations. In all types of dementia, SPECT provides high specificity for AD. MRI can detect cerebrovascular lesions, including white matter hyperintensities and infarctions, and SPECT is sensitive to perfusion changes in key cognitive regions such as the parietotemporal and posterior cingulate cortices, which are hallmarks of AD-related neurodegeneration (Bonte et al., 2006; Dougall et al., 2004).

Longitudinal neuroimaging of DrD

Longitudinal studies emphasized this distinction. In a follow-up SPECT investigation involving 18 patients with DrD and 29 with probable AD and diabetes mellitus (AD [+DM]), the two groups showed different Regional Cerebral Blood Flow (rCBF) trajectories in a 3-year period. At baseline, AD[+DM] patients exhibited hypoperfusion in the parietotemporal and posterior cingulate cortices that are patterns related to AD. In contrast, DrD patients showed hypoperfusion in the lateral and medial frontal lobes, indicative of non-AD cortical involvement. AD [+DM] patients showed worsening parietotemporal deficits over time, while DrD patients still showed isolated frontal lobe involvement, with no encroachment into AD-signature regions (Fukasawa et al., 2015).

Clinical and diagnostic implications

Based on these imaging and molecular findings, DrD is a distinct and predominantly metabolic form of dementia. Its pathogenesis is related to insulin resistance, oxidative stress, and chronic inflammation, with secondary contributions from tau-related neurodegeneration. This supports the need for diagnostic guidelines that incorporate clinical symptoms and neuropsychological profiles and neuroimaging features and molecular biomarkers to distinguish DrD from other dementias.

Endogenous and environmental factors

There are a number of parallels between T2DM and dementia, especially AD. They're connected by several shared risk factors, including oxidative stress, chronic inflammation, compromised mitochondrial health, unhealthy diets, lack of physical activity, and exposure to environmental toxins (Fig. 6.3). The hippocampus, the part of the brain that helps us form and store memories, is certainly vulnerable to this kind of damage. Diabetes-related blood vessel damage can reduce blood flow to the brain, which in turn worsens memory problems and slows down thinking (Basta et al., 2004; Ding et al., 2008; Geert Jan Biessels et al., 2014; Hassing et al., 2004; Quaegebeur et al., 2011).

Environmental factors (modifiable and non-modifiable)

Age, sex, genetic predisposition, and brain injuries are among the risk factors for dementia that cannot be changed. However, approximately 40% of dementia cases are linked to modifiable factors such as midlife depression, inactivity, obesity, smoking, diabetes, and hypertension (Livingston et al., 2020).

Clinical evidence and therapeutic potential for anti-diabetic drugs and dementia

Research into possible shared therapeutic approaches has been spurred by the rising prevalence of dementia and

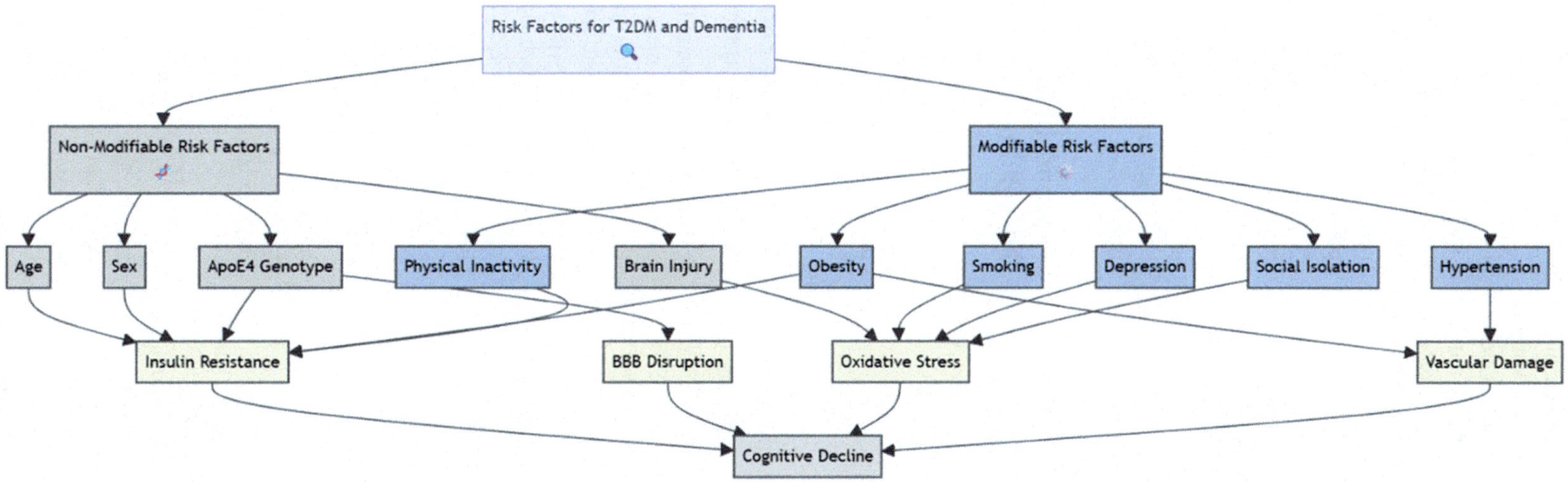

FIGURE 6.3 Environmental and genetic risk factors linking type 2 diabetes and dementia. This diagram categorizes non-modifiable and modifiable risk factors for type 2 diabetes mellitus (T2DM) and dementia, illustrating their contributions to cognitive decline through mechanisms such as insulin resistance, vascular damage, and oxidative stress. Abbreviations: ApoE4 (apolipoprotein E4), BBB (blood-brain barrier).

diabetes mellitus worldwide. Table 6.2 provides a comprehensive summary of the neuroprotective effects of anti-diabetic medications, detailing their mechanisms, clinical evidence, and supporting references. According to new research, some anti-diabetic drugs may affect neurodegenerative processes and cognitive function in a number of ways. With an emphasis on their pharmacological mechanisms and corroborating clinical evidence, this section investigates the neuroprotective potential of several anti-diabetic medication classes (Kuate Defo et al., 2024; Li et al., 2024).

Figure 6.4 illustrates the major classes of anti-diabetic medications and their proposed mechanisms of action in dementia pathophysiology. Insulin sensitizers (metformin and thiazolidinediones) demonstrate AMP-Activated Protein Kinase (AMPK) activation and Peroxisome Proliferator-Activated Receptor Gamma (PPAR-γ) mediated effects. Incretin-based therapies (GLP-1 receptor agonists and Dipeptidyl Peptidase-4 (DPP-4) inhibitors) show BBB penetration and amyloid clearance properties. SGLT2 inhibitors exhibit metabolic effects through ketone utilization, while intranasal insulin provides direct CNS delivery.

Insulin sensitizers: Metformin and thiazolidinediones (TZDs)

Multimodal neuroprotection with metformin

The most commonly prescribed oral anti-diabetic medication, metformin, has shown benefits for brain health in addition to glycemic control. Numerous neuroprotective pathways are started when the medication activates AMPK, improving insulin sensitivity and mitochondrial function. This drug Decreases tau phosphorylation and Aβ aggregation via AMPK-mediated pathways, and reduces oxidative damage in neuronal cells by lowering the production of ROS in the mitochondria, and inhibiting Nuclear Factor-Kappa B (NF-κB). Furthermore, by improving endothelial nitric oxide synthase activity, it improves cerebrovascular function and may raise cerebral perfusion (Kuate Defo et al., 2024; Tang et al., 2025).

TZDs (pioglitazone, rosiglitazone)

In dementia research, they have produced mixed results, despite the potential advantages of their PPAR-γ agonist activity, microglial activity modulation, and anti-inflammatory qualities (Hierro-Bujalance et al., 2020; Zhao et al., 2023). PPAR-γ agonists enhance insulin signaling, reduce Aβ production, and improve cerebral blood flow (Li et al., 2024). It is notable that they have been restricted due to safety concerns, and possible cardiovascular risks e.g., fluid retention, and heart failure (Zhao et al., 2023).

Incretin-based therapies

Glucagon-like peptide-1 receptor agonists (GLP-1 RAs)

Liraglutide, exenatide, and semaglutide are incretin-based drugs that show a number of characteristics related to neurodegeneration. They cross the BBB, binding to hippocampal and cortical GLP-1 receptors, leading to promote neurogenesis, synaptic plasticity, and reduce Aβ and tau pathology. By inhibiting pro-inflammatory cytokine release and microglial activation, these substances lessen neuroinflammatory reactions. By upregulating the expression of brain-derived neurotrophic factor, they also improve synaptic plasticity. In Alzheimer's models, preclinical research shows decreased tau and amyloid-beta pathology, indicating the possibility of disease modification (Fessel, 2024).

Dipeptidyl peptidase-4 inhibitors (sitagliptin, linagliptin)

These drugs increase endogenous GLP-1 levels, enhancing neuroprotection. There are mixed results on dementia; some studies report cognitive benefits, while others show no significant effect (Angelopoulou & Piperi, 2018; Jiang et al., 2024).

Sodium-glucose cotransporter-2 inhibitors (SGLT2i); new approaches to brain health

Empagliflozin, dapagliflozin, and canagliflozin are renal glucose-lowering medications that may improve neurological function in a number of ways. These medications enhance cerebral perfusion by improving systemic vascular function. They may lessen oxidative stress in neural tissues by improving mitochondrial efficiency. Moreover, SGLT2 inhibitors may provide indirect protection against VaD by lowering the risk of stroke. They improve endothelial function, reduced oxidative stress, increased ketone body utilization, providing an alternative energy source for neurons.

Although there are still few studies on dementia in humans, preclinical research yields encouraging findings. Preclinical studies suggest neuroprotection, but clinical data remain limited (Youn et al., 2024).

Complex neurological effects of insulin therapy

Insulin treatment seems to have a dose-and route-dependent effect on the nervous system. Because it avoids systemic effects and delivers insulin directly to the CNS, intranasal administration exhibits particular promise. Insulin can facilitate Aβ clearance and improves synaptic function (Long et al., 2022).

TABLE 6.2 Neuroprotective effects of anti-diabetic medications in dementia.

Drug class	Example drugs	Mechanism of action	Clinical evidence	References
Insulin sensitizers	Metformin, pioglitazone, rosiglitazone	AMPK activation, PPAR-γ agonism, reduced Aβ/tau pathology, improved cerebral perfusion	Observational studies show reduced dementia risk; mixed RCT results	Dai et al. (2022), Hierro-Bujalance et al. (2020), Huang et al. (2023), Kruczkowska et al. (2025), Kuate Defo et al. (2024), Li et al. (2024), Tang et al. (2025), Zhao et al. (2023)
GLP-1 receptor agonists	Liraglutide, exenatide, semaglutide	BBB penetration, neurogenesis, reduced Aβ/tau pathology, anti-inflammatory effects	Cognitive stabilization in early AD; reduced hippocampal atrophy in trials	Fessel (2024), Teixeira et al. (2025)
DPP-4 inhibitors	Sitagliptin, linagliptin	Increased endogenous GLP-1, reduced Aβ/tau pathology	Mixed results; some studies show cognitive benefits, others no effect	Angelopoulou and Piperi (2018), Jiang et al. (2024)
SGLT2 inhibitors	Empagliflozin, dapagliflozin, canagliflozin	Improved vascular function, reduced oxidative stress, ketone utilization	Preclinical neuroprotection; limited clinical data on cognitive outcomes	Youn et al. (2024)
Insulin therapy	Intranasal insulin	Direct CNS delivery, enhanced Aβ clearance, improved synaptic function	Favorable but mixed results in MCI trials; hypoglycemia risks with systemic insulin	Long et al. (2022)

This table summarizes the major classes of anti-diabetic drugs, their mechanisms of action in mitigating dementia risk; clinical evidence, and supporting references; Abbreviations, AMPK (AMP-activated Protein Kinase); PPAR-γ (Peroxisome Proliferator-Activated Receptor-gamma), Aβ (Amyloid-beta); BBB (Blood-Brain Barrier), GLP-1 (Glucagon-like Peptide-1); DPP-4 (Dipeptidyl Peptidase-4), SGLT2 (Sodium-Glucose Cotransporter-2); CNS (Central Nervous System), AD (Alzheimer's Disease); RCT (Randomized Controlled Trial), MCI (Mild Cognitive Impairment).

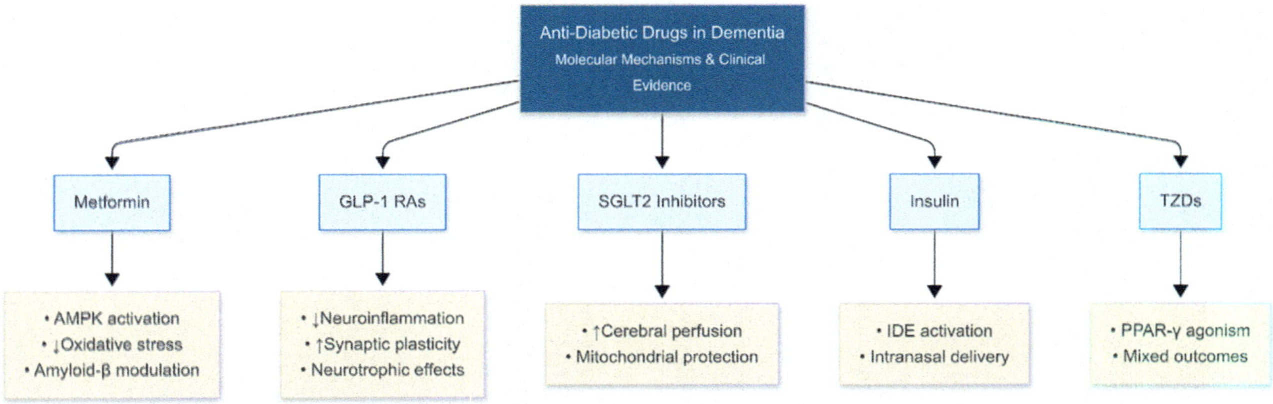

FIGURE 6.4 Neuroprotective mechanisms of anti-diabetic drugs in dementia.

Key points

Early intervention can be designed by treating prediabetes with metformin or GLP-1 RAs which may delay cognitive decline. Combination therapies with metformin + GLP-1 RAs may synergistically target insulin resistance and neurodegeneration. The challenges of this field include heterogeneity in studies (drug regimens, follow-up duration, cognitive assessments, safety concerns like fluid retention and cardiovascular complications after TZDs, hypoglycemia following insulin, and dehydration in elderly after SGLT2i (Li et al., 2024; Long et al., 2022; Youn et al., 2024; Zhao et al., 2023)).

Prospects for the future

Personalized medicine-based therapies can apply genetic (e.g., APOE ε4 status) and metabolic profiling that will select more effective drugs for every person There is a lack of appliable biomarkers. Despite clinicians need validated markers (e.g., neuroimaging, CSF Aβ/tau) to track cognitive outcomes.

Research gaps include long-term cognitive outcomes in non-diabetic populations, and mechanistic studies on SGLT2i and DPP-4 inhibitors. Establishing therapeutic efficacy will require the creation of specialized clinical trials with longer follow-up times and standardized cognitive outcomes. Future studies should also look into the best ways to dose patients and possible combination treatments that target several pathways in the cognitive decline linked to diabetes.

Conclusion

The complex relationship between T2DM and dementia is a vital research frontier in neuroscience with significant implications for clinical practice and health policy. The multifactorial pathophysiology of diabetes-associated dementia ranging from insulin resistance, chronic hyperglycemia, neuroinflammation, oxidative stress, AGEs, and cerebrovascular dysfunction highlights the intricacy of this relationship. These processes culminate in synaptopathy, damaging neural circuits and expediting cognitive loss, especially in susceptible brain areas like the hippocampus. Promisingly, anti-diabetic medications like metformin, GLP-1 receptor agonists, and SGLT2 inhibitors exhibit neuroprotective effects by modulating insulin signaling, reducing neurotoxic protein buildup, and enhancing cerebral vascular health. Precision medicine via genetic and metabolic profiling promises an attractive means to maximize treatment success. Synthesizing population-level data with mechanistic findings, this chapter presents a holistic framework for the clinician and policymaker to navigate the changing landscape of diabetes-related dementia, making the case for integrated solutions to reduce its increasing global burden.

Mini-dictionary of terms

- **Advanced Glycation End Products (AGEs):** Irreversible protein alterations resulting from a non-enzymatic reaction between reducing sugars and the amino groups of proteins, lipids, and nucleic acids.
- **AMP-activated Protein Kinase (AMPK):** A central metabolic regulator that, when activated, improves sensitivity and mitochondrial function in neuronal cells.
- **Blood-Brain Barrier (BBB):** A semipermeable barrier of endothelial cells that separates the brain from the rest of the body's circulation. In diabetes-related dementia, its function can be compromised due to vascular damage.
- **Neurogenesis:** The process by which new neurons are formed in the brain.
- **Neuroinflammation:** Chronic, low-grade inflammation within the brain characterized by elevated cytokines

and microglial activation, which contributes to synaptic loss and neuronal damage.

- **Oxidative Stress:** An imbalance between the production of reactive oxygen species (ROS) and the body's ability to counteract or detoxify their harmful effects. This process is a common pathway in diabetes-related dementia.
- **Synaptic Plasticity:** The ability of synapses to strengthen or weaken over time in response to increases or decreases in their activity. This process is disturbed in diabetic dementia.
- **Tau Protein:** A protein that helps stabilize microtubules in neurons. In Alzheimer's disease pathology, hyperphosphorylation of this protein leads to the formation of neurofibrillary tangles.

Summary points

- There is a strong connection between Type 2 diabetes mellitus (T2DM) and an increased risk of dementia, which is mediated by several common pathophysiological processes.
- Key mechanisms linking T2DM to dementia include brain insulin resistance, chronic hyperglycemia, neuroinflammation, oxidative stress, and vascular dysfunction.
- Anti-diabetic drugs like metformin, GLP-1 receptor agonists, and SGLT2 inhibitors may have neuroprotective effects by improving insulin sensitivity, reducing toxic protein buildup, and enhancing cerebral vascular health.
- Imaging technologies like MRI and SPECT can help differentiate between various dementia subtypes in diabetic patients, with diabetes-related dementia (DrD) showing distinct patterns from classic Alzheimer's disease.
- The progression of diabetes-related dementia involves a complex interplay of metabolic, vascular, and neurodegenerative pathways that lead to compromised neuronal function and cognitive loss.
- Approximately 40% of dementia cases are linked to modifiable factors such as a lack of physical activity, obesity, smoking, and diabetes.
- Personalized medicine, which considers an individual's genetic and metabolic profile, offers a promising approach for selecting more effective anti-diabetic drugs for dementia prevention.

Declaration of AI use

During the preparation of this work, the authors used ChatGPT, from OpenAI, to improve readability and correct grammar errors. After using this tool/service, the authors reviewed and edited the content as needed and take full responsibility for the content of the publication.

References

Abbatecola, A. M., Arosio, B., Cerasuolo, M., Auriemma, M. C., Di Meo, I., Langiano, E., & Rizzo, M. R. (2024). Common neurodegenerative pathways in brain aging, cognitive decline, type 2 diabetes and metabolic syndrome. *Journal of Gerontology and Geriatrics, 72*(1), 43–49. https://doi.org/10.36150/2499-6564-n691

Ahmad, W., Ijaz, B., Shabbiri, K., Ahmed, F., & Rehman, S. (2017). Oxidative toxicity in diabetes and Alzheimer's disease: Mechanisms behind ROS/RNS generation. *Journal of Biomedical Science, 24*(1), 76. https://doi.org/10.1186/s12929-017-0379-z

Alkhalifa, A. E., Alkhalifa, O., Durdanovic, I., Ibrahim, D. R., & Maragkou, S. (2025). Oxidative stress and mitochondrial dysfunction in Alzheimer's disease: Insights into pathophysiology and treatment. *Biochimica et Biophysica Acta, 2*(2), 17.

Almutairi, M. M. A., Gong, C., Xu, Y. G., Chang, Y., & Shi, H. (2016). Factors controlling permeability of the blood–brain barrier. *Cellular and Molecular Life Sciences, 73*(1), 57–77. https://doi.org/10.1007/s00018-015-2050-8

Angelopoulou, E., & Piperi, C. (2018). DPP-4 inhibitors: A promising therapeutic approach against Alzheimer's disease. *Annals of Translational Medicine, 6*(12), 255. https://doi.org/10.21037/atm.2018.04.41

Basta, G., Schmidt, A. M., & De Caterina, R. (2004). Advanced glycation end products and vascular inflammation: Implications for accelerated atherosclerosis in diabetes. *Cardiovascular Research, 63*(4), 582–592.

Belosludtsev, K. N., Belosludtseva, N. V., & Dubinin, M. V. (2020). Diabetes mellitus, mitochondrial dysfunction and Ca2+-dependent permeability transition pore. *International Journal of Molecular Sciences, 21*(18), 6559.

Bhattacharya, R., Alam, M. R., Kamal, M. A., Seo, K. J., & Singh, L. R. (2023). *AGE-RAGE axis culminates into multiple pathogenic processes: A central road to neurodegeneration*. https://doi.org/10.3389/fnmol.2023.1155175. Vols. 16, 2023.

Biessels, G. J., & Despa, F. (2018). Cognitive decline and dementia in diabetes mellitus: Mechanisms and clinical implications. *Nature Reviews Endocrinology, 14*(10), 591–604. https://doi.org/10.1038/s41574-018-0048-7

Biessels, G. J., Strachan, M. W., Visseren, F. L., Kappelle, L. J., & Whitmer, R. A. (2014). Dementia and cognitive decline in type 2 diabetes and prediabetic stages: Towards targeted interventions. *Lancet Diabetes & Endocrinology, 2*(3), 246–255.

Bonte, F. J., Harris, T. S., Hynan, L. S., Bigio, E. H., & White, C. L., 3rd (2006). Tc-99m HMPAO SPECT in the differential diagnosis of the dementias with histopathologic confirmation. *Clinical Nuclear Medicine, 31*(7), 376–378. https://doi.org/10.1097/01.rlu.0000222736.81365.63

Caroline, W., Kathrin, B., Robert, W., Stephanie, K., Hubert, P., Günter, S., … Martin, H. (2020). Insulin sensitivity predicts cognitive decline in individuals with prediabetes. *BMJ Open Diabetes Research & Care, 8*(2), e001741. https://doi.org/10.1136/bmjdrc-2020-001741

Chen, J. J., Thiyagarajah, M., Song, J., Chen, C., Herrmann, N., Gallagher, D., … Lanctôt, K. L. (2022). Altered central and blood glutathione in Alzheimer's disease and mild cognitive impairment: A meta-analysis. *Alzheimer's Research and Therapy, 14*(1), 23. https://doi.org/10.1186/s13195-022-00961-5

Chen, M., Yan, R., Luo, J., Ning, J., Zhou, R., & Ding, L. (2023). The role of PGC-1α-Mediated mitochondrial biogenesis in neurons.

Neurochemical Research, 48(9), 2595–2606. https://doi.org/10.1007/s11064-023-03934-8

Chen, Y., & Yu, Y. (2023). Tau and neuroinflammation in Alzheimer's disease: Interplay mechanisms and clinical translation. *Journal of Neuroinflammation, 20*(1), 165. https://doi.org/10.1186/s12974-023-02853-3

Cholerton, B., Baker, L. D., Montine, T. J., & Craft, S. (2016). Type 2 diabetes, cognition, and dementia in older adults: Toward a precision health approach. *Diabetes Spectrum, 29*(4), 210–219. https://doi.org/10.2337/ds16-0041

Ciudin, A., & Hernández, C. (2024). Chapter 14—Diabetes-Related cognitive impairment and dementia. In D. Mauricio, & N. Alonso (Eds.), *Chronic complications of diabetes mellitus* (pp. 215–230). Academic Press.

Collingridge, G. L. (2025). Long-term potentiation in the hippocampus: From magnesium to memory. *Neuroscience, 578*, 126–131. https://doi.org/10.1016/j.neuroscience.2024.11.069

Csajbók, É. A., & Tamás, G. (2016). Cerebral cortex: A target and source of insulin? *Diabetologia, 59*(8), 1609–1615. https://doi.org/10.1007/s00125-016-3996-2

Dai, J., Ports, K. D., Corrada, M. M., Odegaard, A. O., O'Connell, J., & Jiang, L. (2022). Metformin and dementia risk: A systematic review with respect to time related biases. *Journal of Alzheimer's Disease Reports, 6*(1), 443–459. https://doi.org/10.3233/adr-220002

Dawi, J., Misakyan, Y., Affa, S., Kades, S., Narasimhan, A., Hajjar, F., … Venketaraman, V. (2025). Oxidative stress, glutathione insufficiency, and inflammatory pathways in type 2 diabetes mellitus: Implications for therapeutic. *Intervention, 13*(1), 18.

De Felice, F. G., & Ferreira, S. T. (2014). Inflammation, defective insulin signaling, and mitochondrial dysfunction as common molecular denominators connecting type 2 diabetes to Alzheimer disease. *Diabetes, 63*(7), 2262–2272. https://doi.org/10.2337/db13-1954

Ding, J., Patton, N., Deary, I. J., Strachan, M. W., Fowkes, F., Mitchell, R. J., & Price, J. F. (2008). Retinal microvascular abnormalities and cognitive dysfunction: A systematic review. *British Journal of Ophthalmology, 92*(8), 1017–1025.

Dougall, N. J., Bruggink, S., & Ebmeier, K. P. (2004). Systematic review of the diagnostic accuracy of 99mTc-HMPAO-SPECT in dementia. *American Journal of Geriatric Psychiatry, 12*(6), 554–570. https://doi.org/10.1176/appi.ajgp.12.6.554

Duca, L., Blaise, S., Romier, B., Laffargue, M., Gayral, S., El Btaouri, H., … Maurice, P. (2016). Matrix ageing and vascular impacts: Focus on elastin fragmentation. *Cardiovascular Research, 110*(3), 298–308. https://doi.org/10.1093/cvr/cvw061

Dutta, B. J., Singh, S., Seksaria, S., Das Gupta, G., & Singh, A. (2022). Inside the diabetic brain: Insulin resistance and molecular mechanism associated with cognitive impairment and its possible therapeutic strategies. *Pharmacological Research, 182*, 106358. https://doi.org/10.1016/j.phrs.2022.106358

Feng, L., & Gao, L. (2024). *The role of neurovascular coupling dysfunction in cognitive decline of diabetes patients*. https://doi.org/10.3389/fnins.2024.1375908. Vols. 18, 2024.

Fessel, J. (2024). All GLP-1 agonists should, theoretically, cure Alzheimer's dementia but dulaglutide might be more effective than the others. *Journal of Clinical Medicine, 13*(13). https://doi.org/10.3390/jcm13133729

Fu, M.-X., Requena, J. R., Jenkins, A. J., Lyons, T. J., Baynes, J. W., & Thorpe, S. R. (1996). The advanced glycation end product, <em>N</em>ϵ-(carboxymethyl)lysine, is a product of both lipid peroxidation and glycoxidation reactions (*). *Journal of Biological Chemistry, 271*(17), 9982–9986. https://doi.org/10.1074/jbc.271.17.9982

Fukasawa, R., Hanyu, H., Shimizu, S., Kanetaka, H., Sakurai, H., & Ishii, K. (2015). Identification of diabetes-related dementia: Longitudinal perfusion SPECT and amyloid PET studies. *Journal of the Neurological Sciences, 349*(1–2), 45–51. https://doi.org/10.1016/j.jns.2014.12.023

Hanyu, H. (2019). Diabetes-related dementia. In Y. Nakabeppu, & T. Ninomiya (Eds.), *Diabetes mellitus: A risk factor for Alzheimer's disease* (pp. 147–160). Singapore: Springer Singapore.

Hassing, L. B., Hofer, S. M., Nilsson, S. E., Berg, S., Pedersen, N. L., McClearn, G., & Johansson, B. (2004). Comorbid type 2 diabetes mellitus and hypertension exacerbates cognitive decline: Evidence from a longitudinal study. *Age and Ageing, 33*(4), 355–361.

Hecker, M., & Wagner, A. H. (2018). Role of protein carbonylation in diabetes. *Journal of Inherited Metabolic Disease, 41*(1), 29–38. https://doi.org/10.1007/s10545-017-0104-9

Hierro-Bujalance, C., Infante-Garcia, C., Del Marco, A., Herrera, M., Carranza-Naval, M. J., Suarez, J., … Garcia-Alloza, M. (2020). Empagliflozin reduces vascular damage and cognitive impairment in a mixed murine model of Alzheimer's disease and type 2 diabetes. *Alzheimer's Research & Therapy, 12*(1), 40. https://doi.org/10.1186/s13195-020-00607-4

Huang, K. H., Tsai, Y. F., Lee, C. B., Gau, S. Y., Tsai, T. H., Chung, N. J., & Lee, C. Y. (2023). The correlation between metformin use and incident dementia in patients with new-onset diabetes mellitus: A population-based study. *Journal of Personalized Medicine, 13*(5). https://doi.org/10.3390/jpm13050738

Hwang, J. J., Jiang, L., Hamza, M., Dai, F., Belfort-DeAguiar, R., Cline, G., … Sherwin, R. S. (2017). The human brain produces fructose from glucose. *JCI Insight, 2*(4), e90508. https://doi.org/10.1172/jci.insight.90508

Iarkov, A., Mendoza, C., & Echeverria, V. (2021). *Cholinergic receptor modulation as a target for preventing dementia in Parkinson's disease*. https://doi.org/10.3389/fnins.2021.665820. Vols. 15, 2021.

Jiang, X., Li, J., Yao, X., Ding, H., Gu, A., & Zhou, Z. (2024). Neuroprotective effects of dipeptidyl peptidase 4 inhibitor on Alzheimer's disease: A narrative review. *Frontiers in Pharmacology, 15*. https://doi.org/10.3389/fphar.2024.1361651, 2024.

Kadir, R. R. A., Alwjwaj, M., & Bayraktutan, U. (2022). Protein kinase C-β distinctly regulates blood-brain barrier-forming capacity of brain microvascular endothelial cells and outgrowth endothelial cells. *Metabolic Brain Disease, 37*(6), 1815–1827. https://doi.org/10.1007/s11011-022-01041-1

Khalid, M., Petroianu, G., & Adem, A. (2022). Advanced glycation end products and diabetes mellitus. *Mechanisms and Perspectives, 12*(4), 542.

Kim, S., Jung, U. J., & Kim, S. R. (2024). Role of oxidative stress in blood–brain barrier disruption and. *Neurodegenerative Diseases, 13*(12), 1462.

Kim, H. K., & Song, J. (2022). Hypothyroidism and diabetes-related dementia: Focused on neuronal dysfunction. *Insulin Resistance, and Dyslipidemia, 23*(6), 2982.

Kruczkowska, W., Gałęziewska, J., Buczek, P., Płucienník, E., Kciuk, M., & Śliwińska, A. (2025). Overview of metformin and neurodegeneration: A comprehensive review. *Pharmaceuticals, 18* (4). https://doi.org/10.3390/ph18040486

Kruman, I., Bruce-Keller, A. J., Bredesen, D., Waeg, G., & Mattson, M. P. (1997). Evidence that 4-hydroxynonenal mediates oxidative stress-induced neuronal. *Apoptosis, 17*(13), 5089–5100. https://doi.org/10.1523/JNEUROSCI.17-13-05089.1997

Kuate Defo, A., Bakula, V., Pisaturo, A., Labos, C., Wing, S. S., & Daskalopoulou, S. S. (2024). Diabetes, antidiabetic medications and risk of dementia: A systematic umbrella review and meta-analysis. *Diabetes, Obesity and Metabolism, 26*(2), 441–462. https://doi.org/10.1111/dom.15331

Kuzan, A., Królewicz, E., Kustrzeba-Wójcicka, I., Lindner-Pawłowicz, K., & Sobieszczańska, M. (2022). How diabetes and other comorbidities of elderly patients and their treatment influence levels of glycation products. *International Journal of Environmental Research and Public Health, 19*(12), 7524.

Li, Z., Lin, C., Cai, X., Lv, F., Yang, W., & Ji, L. (2024). Anti-diabetic agents and the risks of dementia in patients with type 2 diabetes: A systematic review and network meta-analysis of observational studies and randomized controlled trials. *Alzheimer's Research and Therapy, 16*(1), 272. https://doi.org/10.1186/s13195-024-01645-y

Li, Z., Wu, H., Luo, Y., & Tan, X. (2023). Correlation of serum complement factor 5a level with inflammatory response and cognitive function in patients with Alzheimer's disease of different severity. *BMC Neurology, 23*(1), 319. https://doi.org/10.1186/s12883-023-03256-w

Livingston, G., Huntley, J., Sommerlad, A., Ames, D., Ballard, C., Banerjee, S., … Cooper, C. (2020). Dementia prevention, intervention, and care: 2020 report of the lancet commission. *The Lancet, 396* (10248), 413–446.

Long, C., Han, X., Yang, Y., Li, T., Zhou, Q., & Chen, Q. (2022). Efficacy of intranasal insulin in improving cognition in mild cognitive impairment or dementia: A systematic review and meta-analysis. *Frontiers in Aging Neuroscience, 14*. https://doi.org/10.3389/fnagi.2022.963933, 2022.

Milstein, J. L., & Ferris, H. A. (2021). The brain as an insulin-sensitive metabolic organ. *Molecular Metabolism, 52*, 101234. https://doi.org/10.1016/j.molmet.2021.101234

Morozova, M. P., Savinkova, I. G., & Gorbacheva, L. R. (2024). Peculiarities of brain cell functioning in hyperglycemia and diabetes mellitus. *Journal of Evolutionary Biochemistry and Physiology, 60* (4), 1531–1545. https://doi.org/10.1134/S0022093024040215

Mosquera-Sulbarán, J. A., & Hernández-Fonseca, J. P. (2022). Advanced glycation end products in diabetes. In V. B. Patel, & V. R. Preedy (Eds.), *Biomarkers in diabetes* (pp. 1–25). Cham: Springer International Publishing.

Padovani, A., Galli, A., Bazzoli, E., Tolassi, C., Caratozzolo, S., Gumina, B., … Pilotto, A. (2025). The role of insulin resistance and APOE genotype on blood–brain barrier integrity in Alzheimer's disease. *Alzheimer's and Dementia, 21*(2), e14556. https://doi.org/10.1002/alz.14556

Phosat, C., Panprathip, P., Chumpathat, N., Prangthip, P., Chantratita, N., Soonthornworasiri, N., … Kwanbunjan, K. (2017). Elevated C-reactive protein, interleukin 6, tumor necrosis factor alpha and glycemic load associated with type 2 diabetes mellitus in rural thais: A cross-sectional study. *BMC Endocrine Disorders, 17*(1), 44. https://doi.org/10.1186/s12902-017-0189-z

Platzbecker, A. L., Gronewold, J., Schramm, S., Moebus, S., Stang, A., Schmidt, B., … Group, t. H. N. R. S. I. (2025). Sex- and age-specific effect of known type 2 diabetes mellitus on incident mild cognitive impairment five years later: Results from the population-based Heinz Nixdorf. *Recall Study, 17*(2), e70130. https://doi.org/10.1002/dad2.70130

Quaegebeur, A., Lange, C., & Carmeliet, P. (2011). The neurovascular link in health and disease: Molecular mechanisms and therapeutic implications. *Neuron, 71*(3), 406–424.

Ramaglia, V., Dubey, M., Malpede, M. A., Petersen, N., de Vries, S. I., Ahmed, S. M., … Huitinga, I. J. A. N. (2021). Complement-associated loss of CA2 inhibitory synapses in the demyelinated hippocampus impairs memory. *Acta Neuropathologica, 142*(4), 643–667.

Sekhar, R. V., McKay, S. V., Patel, S. G., Guthikonda, A. P., Reddy, V. T., Balasubramanyam, A., & Jahoor, F. (2010). Glutathione synthesis is diminished in patients with uncontrolled diabetes and restored by dietary supplementation with cysteine and glycine. *Diabetes Care, 34*(1), 162–167. https://doi.org/10.2337/dc10-1006

Shao, B., & Bayraktutan, U. (2014). Hyperglycaemia promotes human brain microvascular endothelial cell apoptosis via induction of protein kinase C-βI and prooxidant enzyme NADPH oxidase. *Redox Biology, 2*, 694–701. https://doi.org/10.1016/j.redox.2014.05.005

Shaw, J. N., Baynes, J. W., & Thorpe, S. R. (2002). Nε-(carboxymethyl) lysine (CML) as a biomarker of oxidative stress in long-lived tissue proteins. In D. Armstrong (Ed.), *Oxidative stress biomarkers and antioxidant protocols* (pp. 129–137). Totowa, NJ: Humana Press.

Singh, A., Ansari, V. A., Mahmood, T., Hasan, S. M., Wasim, R., Maheshwari, S., … Research, M. (2024). Targeting abnormal tau phosphorylation for Alzheimer's therapeutics. *Harmone and Metabolic Research, 56*(07), 482–488.

Singh, D. D., Shati, A. A., Alfaifi, M. Y., Elbehairi, S. E. I., Han, I., Choi, E. H., & Yadav, D. K. (2022). Development of dementia in type 2 diabetes patients: Mechanisms of insulin resistance and antidiabetic drug development. *Cells, 11*(23). https://doi.org/10.3390/cells11233767

Spinelli, M., Fusco, S., & Grassi, C. (2019). *Brain insulin resistance and hippocampal plasticity: Mechanisms and biomarkers of cognitive decline*. https://doi.org/10.3389/fnins.2019.00788. Vols. 13, 2019.

Stammers, M., Ivanova, I. M., Niewczas, I. S., Segonds-Pichon, A., Streeter, M., Spiegel, D. A., & Clark, J. (2020). Age-related changes in the physical properties, cross-linking, and glycation of collagen from mouse tail tendon. *Journal of Biological Chemistry, 295*(31), 10562–10571. https://doi.org/10.1074/jbc.RA119.011031

Tang, C., Hao, J., Tao, F., Feng, Q., Song, Y., & Zeng, B. (2025). Association of metformin use with risk of dementia in patients with type 2 diabetes: A systematic review and meta-analysis. *Diabetes, Obesity and Metabolism, 27*(4), 1992–2001. https://doi.org/10.1111/dom.16192

Teixeira, L. C. R., Luizon, M. R., & Gomes, K. B. (2025). Exploring the role of GLP-1 receptor agonists in Alzheimer's disease: A review of preclinical and clinical evidence. *Receptor, 4*(1), 2.

Tomlinson, D. R., & Gardiner, N. J. (2008). Glucose neurotoxicity. *Nature Reviews Neuroscience, 9*(1), 36–45. https://doi.org/10.1038/nrn2294

van Sloten, T. T., Sedaghat, S., Carnethon, M. R., Launer, L. J., & Stehouwer, C. D. (2020). Cerebral microvascular complications of type 2 diabetes: Stroke, cognitive dysfunction, and depression. *The Lancet. Diabetes and Endocrinology, 8*(4), 325–336.

Wang, J., Gong, B., Zhao, W., Tang, C., Varghese, M., Nguyen, T., … Pasinetti, G. M. (2014). Epigenetic mechanisms linking diabetes and synaptic impairments. *Diabetes, 63*(2), 645–654. https://doi.org/10.2337/db13-1063

Weinberg Sibony, R., Segev, O., Dor, S., & Raz, I. (2024). Overview of oxidative stress and inflammation in diabetes. *Journal of Diabetes, 16* (10), e70014. https://doi.org/10.1111/1753-0407.70014

World Health Organization. (2023). *Dementia.*

Xiao Liang, K. (2024). Interplay of mitochondria and diabetes: Unveiling novel therapeutic strategies. *Mitochondrion, 75*, 101850. https://doi.org/10.1016/j.mito.2024.101850

Xiong, L., Liu, X., Shang, T., Smielewski, P., Donnelly, J., Guo, Z.-n., … Wong, K. S. (2017). Impaired cerebral autoregulation: Measurement and application to stroke. *Journal of Neurology, Neurosurgery and Psychiatry, 88*(6), 520–531. https://doi.org/10.1136/jnnp-2016-314385

Xu, W., Wang, X., Hou, X., Yang, Y., Ma, R., Lv, R., & Yin, Q. (2024). *The role of microglia in the pathogenesis of diabetic-associated cognitive dysfunction.* https://doi.org/10.3389/fendo.2023.1246979. Vols. 14, 2023.

Youn, Y. J., Kim, S., Jeong, H. J., Ah, Y. M., & Yu, Y. M. (2024). Sodium-glucose cotransporter-2 inhibitors and their potential role in dementia onset and cognitive function in patients with diabetes mellitus: A systematic review and meta-analysis. *Frontiers in Neuroendocrinology, 73*, 101131. https://doi.org/10.1016/j.yfrne.2024.101131

Zarkovic, N., Cipak, A., Jaganjac, M., Borovic, S., & Zarkovic, K. (2013). Pathophysiological relevance of aldehydic protein modifications. *Journal of Proteomics, 92*, 239–247. https://doi.org/10.1016/j.jprot.2013.02.004

Zhang, Z., Huang, Q., Zhao, D., Lian, F., Li, X., & Qi, W. (2023). The impact of oxidative stress-induced mitochondrial dysfunction on diabetic microvascular complications. *Frontiers in Endocrinology, 14*, 1112363. https://doi.org/10.3389/fendo.2023.1112363

Zhao, H., Zhuo, L., Sun, Y., Shen, P., Lin, H., & Zhan, S. (2023). Thiazolidinedione use is associated with reduced risk of dementia in patients with type 2 diabetes mellitus: A retrospective cohort study. *Journal of Diabetes, 15*(2), 97–109. https://doi.org/10.1111/1753-0407.13352

Chapter 7

The interstitial fluid dynamics and diabetes mellitus

Rukeye Tuerxun, Kaito Takabayashi, Takafumi Kitagawa, Rui Zou, Koyo Mizuta and Koji Kamagata
Department of Radiology, Juntendo University Graduate School of Medicine, Tokyo, Japan

List of abbreviations

ALPS analysis along the perivascular space
AQP4 aquaporin-4
EPVS enlarged perivascular spaces
FW free water
ISF interstitial fluid
NGM normal glucose metabolism
PVS perivascular spaces
T2DM type 2 diabetes mellitus

Introduction

Perivascular spaces (PVS), also referred to as Virchow–Robin spaces, are an essential anatomical feature of the brain that participate in the clearance of metabolic waste products and facilitate cerebrospinal fluid (CSF)–interstitial fluid (ISF) exchange (Mestre et al., 2020; Wardlaw et al., 2020). The concept of the glymphatic system has provided a unifying framework for understanding how PVS, aquaporin-4 (AQP4) channels on astrocytic end-feet, and vascular pulsatility collaborate to maintain brain homeostasis (Iliff et al., 2012; Jessen et al., 2015). Glymphatic dysfunction has been increasingly implicated in neurodegenerative disorders such as Alzheimer's disease (AD), Parkinson's disease, and cerebral small vessel disease (SVD) (Nedergaard & Goldman, 2020; Rasmussen et al., 2018).

Type 2 diabetes mellitus (T2DM) is a global health crisis, characterized by chronic hyperglycemia, insulin resistance, and vascular complications (ElSayed et al., 2023). Epidemiological data have consistently linked T2DM to an increased risk of cognitive impairment, vascular dementia, and AD (Biessels & Despa, 2018; Cheng et al., 2012). However, the underlying mechanisms remain incompletely understood. A growing body of evidence suggests that disruption of glymphatic clearance may be a critical mediator (Kamagata et al., 2022; Zhang et al., 2021). Prediabetes (Pre-DM), defined by impaired glucose tolerance or mildly elevated glycated hemoglobin (HbA1c), represents an intermediate stage of dysglycemia (Bansal, 2015; Davidson & Kahn, 2016; Echouffo-Tcheugui & Selvin, 2021). Emerging findings indicate that even in Pre-DM, subtle changes in PVS and glymphatic function can already be observed, implying that metabolic brain vulnerability begins before clinical diabetes develops (Sun et al., 2018; Tuerxun et al., 2024).

Figure 7.1 Glymphatic system overview.

Anatomy and imaging of the perivascular unit

PVS are fluid-filled spaces that accompany penetrating arterioles and venules as they descend from the subarachnoid space into brain parenchyma (Wardlaw et al., 2020). Within these compartments, CSF flows inward along arterial PVS, mixes with ISF through AQP4 channels on astrocytic end-feet and drains outward along venous PVS toward meningeal and cervical lymphatics (Iliff et al., 2012; Jessen et al., 2015; Rasmussen et al., 2018). This circulation is thought to underlie the brain's "self-cleaning" system, supporting metabolic clearance, nutrient distribution, and immune surveillance (Nedergaard & Goldman, 2020).

The structural integrity of this "perivascular unit," composed of astrocytes, pericytes, and endothelial cells, is highly sensitive to metabolic stress. In metabolic diseases such as diabetes, oxidative stress, hyperglycemia, and accumulation of advanced glycation end-products can disrupt AQP4 polarization, induce endothelial inflammation, and promote arteriosclerosis (Biessels & Despa, 2018). As a result, the patency of PVS and CSF–ISF exchange are impaired (Wardlaw et al., 2020). In diabetes, these pathological changes may lead to enlargement of

Hormones in Neuroscience. https://doi.org/10.1016/B978-0-443-30263-3.00033-6

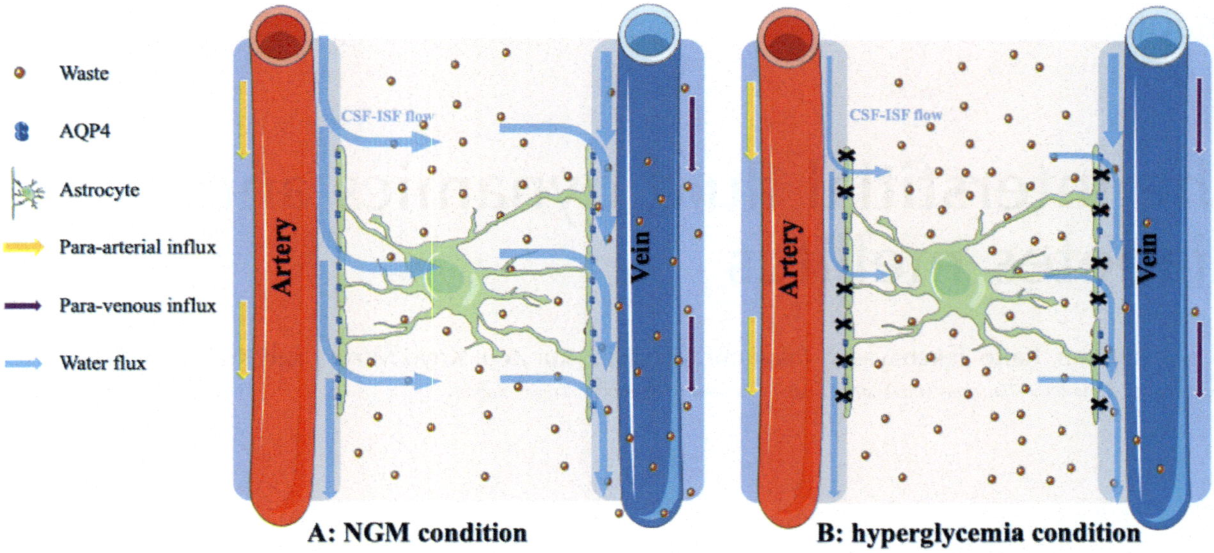

FIGURE 7.1 Comparison of glymphatic system activity between normal glucose metabolism and hyperglycemic states. (A) Under normal glucose metabolism (NGM), cerebrospinal fluid (CSF) passes through aquaporin-4 (AQP4) water channels located on astrocytic end-feet, mixes with interstitial fluid (ISF), and facilitates the clearance of metabolic waste from the brain. (B) In contrast, under hyperglycemic conditions, astrocytic damage leads to a reduction in AQP4 channel expression. As a result, CSF–ISF exchange within the brain parenchyma is impaired, causing the accumulation of metabolic waste products. *Reproduced from Tuerxun et al. Front. Aging Neurosci. (2024) 16:1362457, CC BY 4.0.*

PVS, stagnation of fluid, and accumulation of toxic metabolites (Kamagata et al., 2022; Zhang et al., 2016, 2021).

To visualize these structural and functional alterations, neuroimaging plays a crucial role. Structural PVS assessment is typically achieved with T2-weighted Magnetic Resonance Imaging (MRI). Visual rating scales (e.g., in basal ganglia, centrum semiovale, hippocampus) or automated segmentation quantify EPVS burden (Wardlaw et al., 2020; Zhang et al., 2016, 2021). Advances in deep-learning algorithms now enable objective volumetric measurement with high reproducibility. Furthermore, functional assessment has been pioneered by DTI-ALPS, which evaluates water diffusivity along medullary veins near the lateral ventricles. A higher ALPS index reflects efficient diffusivity along PVS, whereas reductions indicate impaired glymphatic transport (Kamagata et al., 2022). Modified ALPS (mALPS) indices and free water (FW) mapping provide complementary measures of perivascular fluid dynamics (Taoka et al., 2022; Taoka, Fukusumi, et al., 2017; Taoka, Masutani et al., 2017; Taoka & Naganawa, 2020; Taoka & Naganawa, 2021).

Diabetes and perivascular alterations: Evidence from MRI and DTI-ALPS

Several neuroimaging studies have reported greater PVS burden in patients with T2DM. For example, hippocampal EPVS have been associated with poorer memory performance in T2DM, linking structural alterations to cognitive outcomes (Zhang et al., 2016, 2021). In other cohorts, T2DM patients exhibited more severe basal ganglia and centrum semiovale EPVS compared with non-diabetic controls, often coexisting with white matter hyperintensities (WMH) and lacunes (Wardlaw et al., 2020). These findings suggest that diabetes-related vascular and metabolic stress may contribute to perivascular enlargement and glymphatic dysfunction (Table 7.1).

The impact of diabetes on PVS is further magnified by coexistent vascular risk factors such as hypertension, hyperlipidemia, and obesity. Studies of patients with both T2DM and hypertension demonstrate particularly high PVS burden, reflecting compounded microvascular stress (Biessels & Despa, 2018; Cheng et al., 2012). Pathological studies also support these imaging findings, showing thickened arteriolar walls, basement membrane disruption, and perivascular fibrosis in diabetic brains (Wardlaw et al., 2020). Although most evidence is cross-sectional, longitudinal data indicate that greater EPVS burden predicts future WMH progression and cognitive decline (Rasmussen et al., 2018). In diabetic cohorts, progressive enlargement of PVS has been observed alongside worsening executive and memory function, supporting its potential role as an early biomarker of neurovascular compromise (Someya et al., 2019, 2021; Tuerxun et al., 2024) (Table 7.2).

Beyond these structural alterations, DTI-ALPS offers dynamic evidence of glymphatic dysfunction in diabetes. The ALPS index quantitatively assesses water diffusivity along perivascular pathways, providing a functional marker of glymphatic activity. In our investigation of 70

TABLE 7.1 Structural MRI findings: EPVS burden in diabetes.

Study/Condition	Population	PVS region	EPVS severity	Cognitive/Structural outcome	Related vascular/ Metabolic factor
Zhang et al. (2021)	T2DM (n = 72)	Hippocampus	↑ EPVS	Poorer memory	–
Kamagata et al. (2022)	MCI ± T2DM	Basal ganglia	↑ EPVS in T2DM	Lower cognitive score	–
Taoka, Masutani, et al. (2017)	SVD + diabetes	Centrum semiovale	Significant	Executive dysfunction	–
T2DM + hypertension	–	–	Severe	↑ WMH, lacunes	Hypertension
T2DM + obesity	–	–	Moderate–Severe	↑ Brain atrophy	Obesity
T2DM + dyslipidemia	–	–	Mild	Co-localized WMH	Dyslipidemia

EPVS, Enlarged Perivascular Spaces; *MCI*, Mild Cognitive Impairment; *PVS*, Perivascular Spaces; *SVD*, Small Vessel Disease; *T2DM*, Type 2 Diabetes Mellitus; *WMH*, White Matter Hyperintensity.

TABLE 7.2 Comparison of ALPS indices and insulin resistance across glycemic groups.

Group	ALPS index (mean ± SD)	HOMA-IR (mean ± SD)	Statistical significance (*P*-value)
NGM	1.58 ± 0.12	1.1 ± 0.5	–
Pre-DM	1.45 ± 0.09	2.3 ± 0.7	$P < .01$ versus NGM
T2DM	1.42 ± 0.11	2.6 ± 0.8	$P < .01$ versus NGM; NS versus Pre-DM

ALPS Index, Analysis along the Perivascular Space Index; HOMA-IR, Homeostasis Model Assessment of Insulin Resistance; NGM, Normal Glucose Metabolism; Pre-DM, Prediabetes; T2DM, Type 2 Diabetes Mellitus; NS, Not Significant.

elderly participants from the Bunkyo Health Study, ALPS indices were significantly lower in both T2DM, and Pre-DM groups compared with individuals with normal glucose metabolism (NGM). Notably, no significant difference was observed between T2DM and Pre-DM, indicating that glymphatic impairment may already emerge in the prediabetic stage. Furthermore, the ALPS index correlated inversely with Homeostasis Model Assessment of Insulin Resistance (HOMA-IR), offering direct evidence that metabolic dysfunction contributes to altered ISF dynamics (Tuerxun et al., 2024).

Consistent with our findings, reductions in ALPS index have also been reported in AD, mild cognitive impairment, cerebral amyloid angiopathy, and SVD (Jessen et al., 2015; Li et al., 2021; Tang et al., 2022; Taoka, Fukusumi, et al., 2017; Taoka, Masutani et al., 2017). These observations underscore that impaired glymphatic function represents a shared mechanism linking metabolic and neurodegenerative pathology. Importantly, lower ALPS indices have been associated with amyloid-β accumulation and faster cognitive decline, further supporting the biological plausibility of glymphatic dysfunction as a mediator between diabetes and dementia (Sims-Robinson et al., 2010; Talbot, 2014; Zhao et al., 2008).

Methodologically, reproducibility studies have demonstrated moderate-to-good test–retest reliability of ALPS, although variability may arise from acquisition parameters, ROI placement, and susceptibility artifacts (Kamagata et al., 2022; Saito, Kamagata et al., 2023; Taoka et al., 2022; Tatekawa et al., 2023). Recent consensus emphasizes the need for harmonized acquisition protocols and complementary metrics—such as EPVS volume and free-water mapping—to enhance the interpretability of glymphatic imaging. Together, these multimodal MRI markers provide convergent evidence that diabetes impairs both the structural and functional integrity of the perivascular system.

Mechanistic pathways linking diabetes to glymphatic dysfunction

Astrocytic and endothelial dysfunction

Insulin resistance damages astrocytic end-feet, leading to mislocalization or loss of AQP4 channels. This diminishes CSF influx and ISF efflux, compromising glymphatic circulation. Animal models of diabetes consistently show AQP4 downregulation and impaired amyloid clearance

(Benveniste et al., 2019; Iliff et al., 2013; Iliff & Nedergaard, 2013; McMurray et al., 2016; Rehman & Akash, 2017).

Diabetes accelerates endothelial injury through oxidative stress, inflammation, and reduced nitric oxide bioavailability. Stiffened arteries lose their pulsatility, a critical driver of glymphatic flow (Jessen et al., 2015; McMurray et al., 2016; Rehman & Akash, 2017). Pathological studies confirm arteriolosclerosis and basement membrane thickening in diabetic patients, consistent with impaired perivascular pumping.

Inflammation, pericyte injury, and amyloid accumulation

Figure 7.2 ALPS index calculation.

Chronic hyperglycemia activates microglia and perivascular macrophages, promoting neuroinflammation (Giwa et al., 2012; Kim et al., 2018; Prasad et al., 2014; Roriz-Filho et al., 2009). Pericyte degeneration compromises BBB integrity and expands PVS volume. These changes synergize with astrocytic and endothelial dysfunction to impair clearance.

Insulin resistance interferes with insulin-degrading enzyme activity, reducing amyloid-β clearance. Coupled with glymphatic impairment, this accelerates amyloid deposition, tau aggregation, and neurodegeneration, explaining the higher dementia risk in diabetic populations (Ho et al., 2004; Sims-Robinson et al., 2010; Talbot, 2014; Zhang et al., 2016, 2021).

Sleep disruption, autonomic imbalance, and blood pressure variability—common in diabetes—may further reduce glymphatic efficiency. Sleep, in particular, enhances glymphatic clearance, and diabetic sleep disturbances could exacerbate impairment (Saito, Hayakawa et al., 2023) (Table 7.3).

Clinical implications and translational perspectives

Clinical utility of PVS-based imaging

Reduced ALPS indices in Pre-DM highlight the potential for PVS-based imaging to identify at-risk individuals before clinical diabetes onset. Such early markers may complement conventional vascular risk scores and guide preventive strategies (Jessen et al., 2015; Jiang et al., 2017; Kamagata et al., 2022).

Furthermore, ALPS and EPVS measures may serve as surrogate endpoints for lifestyle or pharmacological interventions. Improved insulin sensitivity through exercise, dietary modification, or medications such as metformin and GLP-1 receptor agonists could theoretically restore

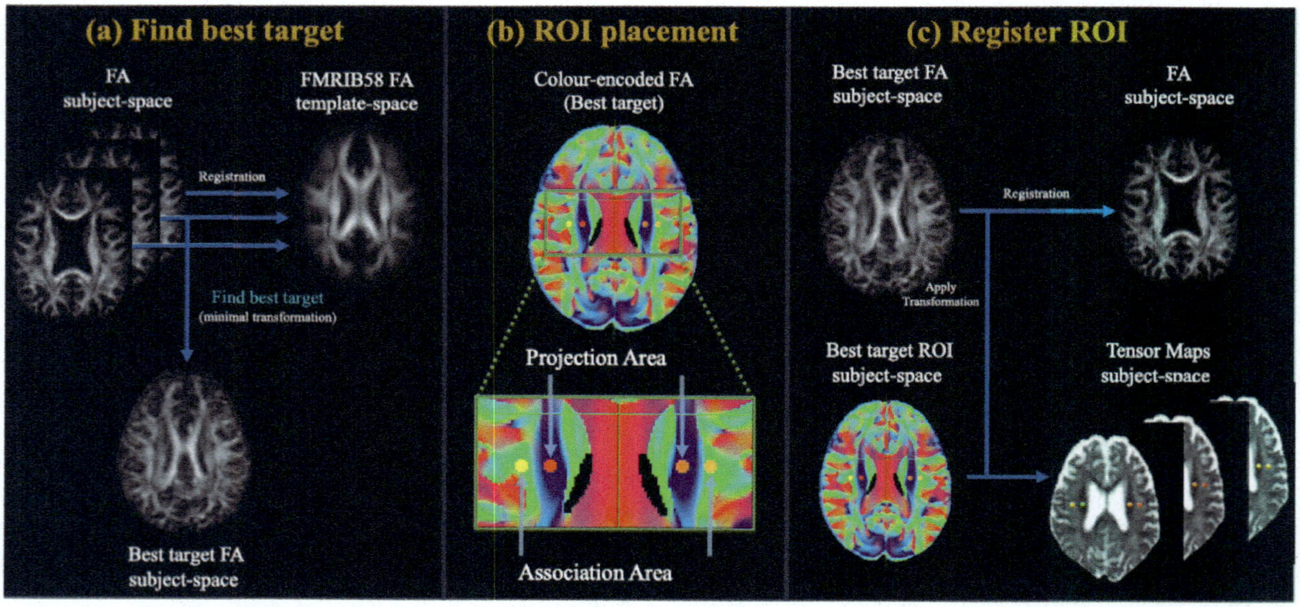

FIGURE 7.2 Flowchart of the study process. (A) Initially, the FA maps of all participants underwent a linear followed by non-linearly registration to the high-definition FMRIB58_FA standard-space reference image. Subsequently, based on the criterion of minimal deformation, specifically, the lowest sum of squared discrepancies, we have selected the appropriate participant to determine the region-of-interest (ROI). (B) Employing the native color-coded FA map from the chosen participant, we set ROIs measuring 5 mm in diameter next to the lateral ventricle bodies within both projection and association regions across both hemispheres. In the projection region, the domain fibers are oriented along the z-axis, standing perpendicular to the x- and y-axes. Conversely, in the association region, principal fibers are oriented along the y-axis, perpendicular to the x- and z-axes. (C) After this, the delineated ROIs underwent registration with an individual FA map. *Reproduced from Tuerxun et al. Front. Aging Neurosci. (2024) 16:1362457, CC BY 4.0.*

TABLE 7.3 Mechanistic links between diabetes and glymphatic dysfunction.

Mechanistic pathway	Key cellular component	Functional impact	Potential intervention	Target biomarker	Supporting references
AQP4 mislocalization	Astrocytes	↓ CSF influx, ↓ clearance	Metformin	AQP4, EPVS	Iliff et al. (2012), Tuerxun et al. (2024)
Endothelial dysfunction	Vascular endothelium	↓ Pulsation, vascular stiffness	GLP-1 agonist	ALPS	McMurray et al. (2016), Kamagata et al. (2022)
Inflammation/ Pericyte loss	Pericytes, Pericytes, microglia	BBB leakage, ↑ PVS	Exercise	ALPS, EPVS	Prasad et al. (2014), Someya et al. (2021)

ALPS, Analysis along the Perivascular Space; *AQP4*, Aquaporin-4; *BBB*, Blood–Brain Barrier; *CSF*, Cerebrospinal Fluid; *EPVS*, Enlarged Perivascular Spaces; *GLP-1*, Glucagon-Like Peptide-1; *PVS*, Perivascular Space.

TABLE 7.4 Methodological challenges and future research priorities.

Challenge	Proposed solution	Related future direction
ROI placement variability	Use of automated pipelines	Standardization of ALPS
Single-center data	Multi-center collaboration	Enhanced generalizability
Interpretation ambiguity	Combine ALPS + EPVS + FW	Multimodal biomarker development

ALPS, Analysis along the Perivascular Space; *EPVS*, Enlarged Perivascular Spaces; *FW*, Free Water.; *ROI*, Region of Interest.

glymphatic function. Future clinical trials should incorporate PVS metrics as exploratory biomarkers (Lee et al., 2016; Someya et al., 2021; Talbot, 2014).

In addition, Integration with multimodal biomarkers including amyloid/tau PET, plasma neurofilament light (NfL), glial fibrillary acidic protein (GFAP), and cognitive testing could yield comprehensive risk profiles. Such multimodal approaches may improve prediction of dementia in diabetic patients (Kamagata et al., 2022; Ke et al., 2022; Li et al., 2021).

Relevance beyond diabetes

The insights gained from diabetes extend to broader contexts of vascular-metabolic brain injury. PVS measures could inform risk stratification in obesity, metabolic syndrome, and aging, highlighting their translational reach (Biessels et al., 2014; Buysschaert et al., 2015; Jessen et al., 2015; Nguyen et al., 2007).

Methodological considerations and standardization needs

Acquisition: DTI-ALPS indices are sensitive to diffusion gradient schemes and b-values. Standardized acquisition protocols are essential for reproducibility (Andersson & Sotiropoulos, 2016; Tournier et al., 2019) (Table 7.4).

Analysis: Manual ROI placement can introduce bias. Automated and semi-automated pipelines are under development to enhance consistency (Kamagata et al., 2022; Taoka, Fukusumi, et al., 2017).

Interpretation: ALPS reflects perivascular diffusivity, not direct CSF flow (Iliff et al., 2012, 2013; Iliff & Nedergaard, 2013; Taoka & Naganawa, 2021). Complementary markers such as EPVS volume and FW fraction should be integrated (Kamagata et al., 2022; Tang et al., 2022).

Generalizability: Many studies are single-center with modest sample sizes. Multi-center collaborations and harmonized pipelines are needed to confirm findings (Saito, Kamagata et al., 2023; Tuerxun et al., 2024).

Future directions

Large-scale longitudinal studies tracking PVS and ALPS changes in diabetes across disease stages and interventions (Taoka et al., 2024; Taoka, Masutani et al., 2017).

Mechanistic studies integrating animal models, high-field MRI, and molecular assays to clarify causal pathways (Lv et al., 2023).

Interventional trials testing whether insulin-sensitizing therapies (e.g., GLP-1 agonists, SGLT2 inhibitors) improve glymphatic metrics and cognitive outcomes (Mei et al., 2024; Sasaki et al., 2024; Youn et al., 2024).

Technological advances in automated PVS quantification, deep-learning segmentation, and standardized ALPS pipelines (Huang et al., 2024; Pham et al., 2022; Taoka, Masutani et al., 2017).

Multimodal biomarker integration combining MRI, PET, and blood-based biomarkers for precision risk profiling (Drzezga & Barthel, 2025; Ma et al., 2025; Okazawa et al., 2024).

Mini-dictionary of terms

AQP4: Astrocytic water channel critical for CSF–ISF exchange in glymphatic circulation.

Blood–brain barrier (BBB): A selective permeability barrier that regulates entry of molecules into the brain parenchyma.

Diffusion tensor imaging (DTI): MRI technique quantifying water diffusion to assess microstructural and fluid dynamics.

FW: An imaging marker representing extracellular water fraction, often increased in edema or inflammation.

Glymphatic system: Brain clearance pathway involving CSF influx, ISF efflux, and PVS transport.

Homeostasis model assessment of insulin resistance (HOMA-IR): An index derived from glucose and insulin levels to quantify insulin resistance.

PVS: Fluid-filled channels surrounding cerebral vessels, visible on T2-weighted MRI.

Summary points

Diabetes is consistently associated with greater PVS burden and reduced glymphatic function.

Our study showed that both T2DM and Pre-DM patients have significantly lower ALPS indices compared with NGM, with an inverse correlation between ALPS and insulin resistance.

Hippocampal and basal ganglia EPVS are enlarged in T2DM and correlate with cognitive impairment.

PVS and ALPS measures integrate structural and dynamic aspects of glymphatic dysfunction, offering early biomarkers of brain vulnerability.

Insulin resistance drives astrocytic AQP4 loss, vascular stiffness, and impaired amyloid clearance, converging on glymphatic failure.

Non-invasive imaging biomarkers may guide early prevention, risk stratification, and monitoring of therapeutic interventions in diabetes.

Methodological standardization and automated analysis are critical for clinical translation.

Future research should emphasize longitudinal, interventional, and multimodal biomarker studies.

References

Andersson, J. L. R., & Sotiropoulos, S. N. (2016). An integrated approach to correction for off-resonance effects and subject movement in diffusion MR imaging. *NeuroImage, 125*, 1063–1078.

Bansal, N. (2015). Prediabetes diagnosis and treatment: A review. *World Journal of Diabetes, 6*(2), 296–303.

Benveniste, H., Liu, X., Koundal, S., Sanggaard, S., Lee, H., & Wardlaw, J. (2019). The glymphatic System and waste clearance with brain aging: A review. *Gerontology, 65*(2), 106–119.

Biessels, G. J., & Despa, F. (2018). Cognitive decline and dementia in diabetes mellitus: Mechanisms and clinical implications. *Nature Reviews Endocrinology, 14*(10), 591–604.

Biessels, G. J., Strachan, M. W., Visseren, F. L., Kappelle, L. J., & Whitmer, R. A. (2014). Dementia and cognitive decline in type 2 diabetes and prediabetic stages: Towards targeted interventions. *Lancet Diabetes & Endocrinology, 2*(3), 246–255.

Buysschaert, M., Medina, J. L., Bergman, M., Shah, A., & Lonier, J. (2015). Prediabetes and associated disorders. *Endocrine, 48*(2), 371–393.

Cheng, G., Huang, C., Deng, H., & Wang, H. (2012). Diabetes as a risk factor for dementia and mild cognitive impairment: A meta-analysis of longitudinal studies. *Internal Medicine Journal, 42*(5), 484–491.

Davidson, M. B., & Kahn, R. A. (2016). A reappraisal of prediabetes. *Journal of Clinical Endocrinology and Metabolism, 101*(7), 2628–2635.

Drzezga, A., & Barthel, H. (2025). Imaging and fluid biomarkers of alzheimer disease: Complementation rather than competition. *Journal of Nuclear Medicine, 66*(Suppl. 2), S32–S44.

Echouffo-Tcheugui, J. B., & Selvin, E. (2021). Prediabetes and what it means: The epidemiological evidence. *Annual Review of Public Health, 42*, 59–77.

ElSayed, N. A., Aleppo, G., Aroda, V. R., Bannuru, R. R., Brown, F. M., Bruemmer, D., Collins, B. S., Hilliard, M. E., Isaacs, D., Johnson, E. L., Kahan, S., Khunti, K., Leon, J., Lyons, S. K., Perry, M. L., Prahalad, P., Pratley, R. E., Seley, J. J., Stanton, R. C., Gabbay, R. A., & A. on behalf of the American Diabetes. (2023). 2. Classification and diagnosis of diabetes: Standards of care in diabetes-2023. *Diabetes Care, 46*(Suppl. 1), S19–S40.

Giwa, M. O., Williams, J., Elderfield, K., Jiwa, N. S., Bridges, L. R., Kalaria, R. N., Markus, H. S., Esiri, M. M., & Hainsworth, A. H. (2012). Neuropathologic evidence of endothelial changes in cerebral small vessel disease. *Neurology, 78*(3), 167–174.

Ho, L., Qin, W., Pompl, P. N., Xiang, Z., Wang, J., Zhao, Z., Peng, Y., Cambareri, G., Rocher, A., Mobbs, C. V., Hof, P. R., & Pasinetti, G. M. (2004). Diet-induced insulin resistance promotes amyloidosis in a transgenic mouse model of Alzheimer's disease. *The FASEB Journal, 18*(7), 902–904.

Huang, P., Liu, L., Zhang, Y., Zhong, S., Liu, P., Hong, H., Wang, S., Xie, L., Lin, M., Jiaerken, Y., Luo, X., Li, K., Zeng, Q., Cui, L., Li, J., Chen, Y., Zhang, R., & Alzheimer's Disease Neuroimaging, I. (2024). Development and validation of a perivascular space segmentation method in multi-center datasets. *NeuroImage, 298*, 120803.

Iliff, J. J., Lee, H., Yu, M., Feng, T., Logan, J., Nedergaard, M., & Benveniste, H. (2013). Brain-wide pathway for waste clearance captured by contrast-enhanced MRI. *Journal of Clinical Investigation, 123*(3), 1299–1309.

Iliff, J. J., & Nedergaard, M. (2013). Is there a cerebral lymphatic system? *Stroke, 44*(6 Suppl. 1), S93–S95.

Iliff, J. J., Wang, M., Liao, Y., Plogg, B. A., Peng, W., Gundersen, G. A., Benveniste, H., Vates, G. E., Deane, R., Goldman, S. A., Nagelhus, E. A., & Nedergaard, M. (2012). A paravascular pathway facilitates CSF flow through the brain parenchyma and the clearance of interstitial solutes, including amyloid beta. *Science Translational Medicine, 4*(147), 147ra111.

Jessen, N. A., Munk, A. S., Lundgaard, I., & Nedergaard, M. (2015). The glymphatic system: A beginner's guide. *Neurochemical Research, 40* (12), 2583–2599.

Jiang, Q., Zhang, L., Ding, G., Davoodi-Bojd, E., Li, Q., Li, L., Sadry, N., Nedergaard, M., Chopp, M., & Zhang, Z. (2017). Impairment of the glymphatic system after diabetes. *Journal of Cerebral Blood Flow and Metabolism, 37*(4), 1326–1337.

Kamagata, K., Andica, C., Takabayashi, K., Saito, Y., Taoka, T., Nozaki, H., Kikuta, J., Fujita, S., Hagiwara, A., Kamiya, K., Wada, A., Akashi, T., Sano, K., Nishizawa, M., Hori, M., Naganawa, S., Aoki, S., & I. Alzheimer's Disease Neuroimaging. (2022). Association of MRI indices of glymphatic system with amyloid deposition and cognition in mild cognitive impairment and alzheimer disease. *Neurology, 99*(24), e2648–e2660.

Ke, Z., Mo, Y., Li, J., Yang, D., Huang, L., Yang, Z., Qin, R., Mao, C., Lv, W., Huang, Y., Hu, Z., Zhang, B., & Xu, Y. (2022). Glymphatic dysfunction mediates the influence of white matter hyperintensities on episodic memory in cerebral small vessel disease. *Brain Sciences, 12*(12).

Kim, Y. K., Nam, K. I., & Song, J. (2018). The glymphatic system in diabetes-induced dementia. *Frontiers in Neurology, 9*, 867.

Lee, J. E., Shin, D. W., Yun, J. M., Kim, S. H., Nam, Y. S., Cho, B., Lim, J. S., Jeong, H. Y., Kwon, H. M., & Park, J. H. (2016). Insulin resistance is a risk factor for silent lacunar infarction. *Stroke, 47*(12), 2938–2944.

Li, M., Kitamura, A., Beverley, J., Koudelka, J., Duncombe, J., Lennen, R., Jansen, M. A., Marshall, I., Platt, B., Wiegand, U. K., Carare, R. O., Kalaria, R. N., Iliff, J. J., & Horsburgh, K. (2021). Impaired glymphatic function and pulsation alterations in a mouse model of vascular cognitive impairment. *Frontiers in Aging Neuroscience, 13*, 788519.

Lv, C., Han, S., Sha, Z., Liu, M., Dong, S., Zhang, C., Li, Z., Zhang, K., Lu, S., Xu, Z., Bie, L., & Jiang, R. (2023). Cerebral glucagon-like peptide-1 receptor activation alleviates traumatic brain injury by glymphatic system regulation in mice. *CNS Neuroscience and Therapeutics, 29*(12), 3876–3888.

Ma, J., Chen, M., Liu, G. H., Gao, M., Chen, N. H., Toh, C. H., Hsu, J. L., Wu, K. Y., Huang, C. M., Lin, C. M., Fang, J. T., Lee, S. H., & Lee, T. M. C. (2025). Effects of sleep on the glymphatic functioning and multimodal human brain network affecting memory in older adults. *Molecular Psychiatry, 30*(5), 1717–1729.

McMurray, F., Patten, D. A., & Harper, M. E. (2016). Reactive oxygen species and oxidative stress in obesity-recent findings and empirical approaches. *Obesity, 24*(11), 2301–2310.

Mei, J., Li, Y., Niu, L., Liang, R., Tang, M., Cai, Q., Xu, J., Zhang, D., Yin, X., Liu, X., Shen, Y., Liu, J., Xu, M., Xia, P., Ling, J., Wu, Y., Liang, J., Zhang, J., & Yu, P. (2024). SGLT2 inhibitors: A novel therapy for cognitive impairment via multifaceted effects on the nervous system. *Translational Neurodegeneration, 13*(1), 41.

Mestre, H., Mori, Y., & Nedergaard, M. (2020). The brain's glymphatic system: Current controversies. *Trends in Neurosciences, 43*(7), 458–466.

Nedergaard, M., & Goldman, S. A. (2020). Glymphatic failure as a final common pathway to dementia. *Science, 370*(6512), 50–56.

Nguyen, T. T., Wang, J. J., & Wong, T. Y. (2007). Retinal vascular changes in pre-diabetes and prehypertension: New findings and their research and clinical implications. *Diabetes Care, 30*(10), 2708–2715.

Okazawa, H., Nogami, M., Ishida, S., Makino, A., Mori, T., Kiyono, Y., & Ikawa, M. (2024). PET/MRI multimodality imaging to evaluate changes in glymphatic system function and biomarkers of Alzheimer's disease. *Scientific Reports, 14*(1), 12310.

Pham, W., Lynch, M., Spitz, G., O'Brien, T., Vivash, L., Sinclair, B., & Law, M. (2022). A critical guide to the automated quantification of perivascular spaces in magnetic resonance imaging. *Frontiers in Neuroscience, 16*, 1021311.

Prasad, S., Sajja, R. K., Naik, P., & Cucullo, L. (2014). Diabetes mellitus and blood-brain barrier dysfunction: An overview. *J Pharmacovigil, 2*(2), 125.

Rasmussen, M. K., Mestre, H., & Nedergaard, M. (2018). The glymphatic pathway in neurological disorders. *The Lancet Neurology, 17*(11), 1016–1024.

Rehman, K., & Akash, M. S. H. (2017). Mechanism of generation of oxidative stress and pathophysiology of type 2 diabetes mellitus: How are they interlinked? *Journal of Cellular Biochemistry, 118*(11), 3577–3585.

Roriz-Filho, J. S., Sa-Roriz, T. M., Rosset, I., Camozzato, A. L., Santos, A. C., Chaves, M. L., Moriguti, J. C., & Roriz-Cruz, M. (2009). (Pre)diabetes, brain aging, and cognition. *Biochimica et Biophysica Acta, 1792*(5), 432–443.

Saito, Y., Hayakawa, Y., Kamagata, K., Kikuta, J., Mita, T., Andica, C., Taoka, T., Uchida, W., Takabayashi, K., Tuerxun, R., Mahemuti, Z., Yoshida, S., Kitagawa, T., Arai, T., Suzuki, A., Sato, K., Nishizawa, M., Akashi, T., Shimoji, K., Wada, A., & Aoki, S. (2023). Glymphatic system impairment in sleep disruption: Diffusion tensor image analysis along the perivascular space (DTI-ALPS). *Japanese Journal of Radiology, 41*(12), 1335–1343.

Saito, Y., Kamagata, K., Andica, C., Taoka, T., Tuerxun, R., Uchida, W., Takabayashi, K., Owaki, M., Yoshida, S., Yamazaki, K., Naganawa, S., & Aoki, S. (2023). Multisite harmonization of diffusion tensor image analysis along the perivascular space using the COMBined association test. *Japanese Journal of Radiology, 41*(10), 1072–1083.

Saito, Y., Kamagata, K., Andica, C., Uchida, W., Takabayashi, K., Yoshida, S., Nakaya, M., Tanaka, Y., Kamio, S., Sato, K., Nishizawa, M., Akashi, T., Shimoji, K., Wada, A., & Aoki, S. (2023). Glymphatic system impairment in corticobasal syndrome: Diffusion tensor image analysis along the perivascular space (DTI-ALPS). *Japanese Journal of Radiology, 41*(11), 1226–1235.

Saito, Y., Kamagata, K., Andica, C., Uchida, W., Takabayashi, K., Yoshida, S., Nakaya, M., Tanaka, Y., Kamiyo, S., Sato, K., Nishizawa, M., Akashi, T., Shimoji, K., Wada, A., & Aoki, S. (2023). Reproducibility of automated calculation technique for diffusion tensor image analysis along the perivascular space. *Japanese Journal of Radiology, 41*(9), 947–954.

Sasaki, K., Fujita, H., Sato, T., Kato, S., Takahashi, Y., Takeshita, Y., Kanda, T., Saito, T., Saido, T. C., Hattori, S., Hozumi, Y., Yamada, Y., & Waki, H. (2024). GLP-1 receptor signaling restores aquaporin 4 subcellular polarization in reactive astrocytes and promotes amyloid beta clearance in a mouse model of Alzheimer's disease. *Biochemical and Biophysical Research Communications, 741*, 151016.

Sims-Robinson, C., Kim, B., Rosko, A., & Feldman, E. L. (2010). How does diabetes accelerate Alzheimer disease pathology? *Nature Reviews Neurology, 6*(10), 551–559.

Someya, Y., Tamura, Y., Kaga, H., Nojiri, S., Shimada, K., Daida, H., Ishijima, M., Kaneko, K., Aoki, S., Miida, T., Hirayama, S., Konishi, S., Hattori, N., Motoi, Y., Naito, H., Kawamori, R., & Watada, H. (2019). Skeletal muscle function and need for long-term care of urban elderly people in Japan (the Bunkyo Health Study): A prospective cohort study. *BMJ Open, 9*(9), e031584.

Someya, Y., Tamura, Y., Kaga, H., Sugimoto, D., Kadowaki, S., Suzuki, R., Aoki, S., Hattori, N., Motoi, Y., Shimada, K., Daida, H., Ishijima, M., Kaneko, K., Nojiri, S., Kawamori, R., & Watada, H. (2021). Insulin resistance and muscle weakness are synergistic risk factors for silent lacunar infarcts: The Bunkyo Health Study. *Scientific Reports, 11*(1), 21093.

Sun, B. L., Wang, L. H., Yang, T., Sun, J. Y., Mao, L. L., Yang, M. F., Yuan, H., Colvin, R. A., & Yang, X. Y. (2018). Lymphatic drainage system of the brain: A novel target for intervention of neurological diseases. *Progress in Neurobiology, 163–164*, 118–143.

Talbot, K. (2014). Brain insulin resistance in Alzheimer's disease and its potential treatment with GLP-1 analogs. *Neurodegenerative Disease Management, 4*(1), 31–40.

Tang, J., Zhang, M., Liu, N., Xue, Y., Ren, X., Huang, Q., Shi, L., & Fu, J. (2022). The association between glymphatic system dysfunction and cognitive impairment in cerebral small vessel disease. *Frontiers in Aging Neuroscience, 14*, 916633.

Taoka, T., Fukusumi, A., Miyasaka, T., Kawai, H., Nakane, T., Kichikawa, K., & Naganawa, S. (2017). Structure of the medullary veins of the cerebral hemisphere and related disorders. *RadioGraphics, 37*(1), 281–297.

Taoka, T., Ito, R., Nakamichi, R., Kamagata, K., Sakai, M., Kawai, H., Nakane, T., Abe, T., Ichikawa, K., Kikuta, J., Aoki, S., & Naganawa, S. (2022). Reproducibility of diffusion tensor image analysis along the perivascular space (DTI-ALPS) for evaluating interstitial fluid diffusivity and glymphatic function: CHanges in Alps index on multiple conditiON acquIsition eXperiment (CHAMONIX) study. *Japanese Journal of Radiology, 40*(2), 147–158.

Taoka, T., Ito, R., Nakamichi, R., Nakane, T., Kawai, H., & Naganawa, S. (2024). Diffusion tensor image analysis ALong the perivascular space (DTI-ALPS): Revisiting the meaning and significance of the method. *Magnetic Resonance in Medical Sciences, 23*(3), 268–290.

Taoka, T., Masutani, Y., Kawai, H., Nakane, T., Matsuoka, K., Yasuno, F., Kishimoto, T., & Naganawa, S. (2017). Evaluation of glymphatic system activity with the diffusion MR technique: Diffusion tensor image analysis along the perivascular space (DTI-ALPS) in Alzheimer's disease cases. *Japanese Journal of Radiology, 35*(4), 172–178.

Taoka, T., & Naganawa, S. (2020). Glymphatic imaging using MRI. *Journal of Magnetic Resonance Imaging, 51*(1), 11–24.

Taoka, T., & Naganawa, S. (2021). Imaging for central nervous system (CNS) interstitial fluidopathy: Disorders with impaired interstitial fluid dynamics. *Japanese Journal of Radiology, 39*(1), 1–14.

Tatekawa, H., Matsushita, S., Ueda, D., Takita, H., Horiuchi, D., Atsukawa, N., Morishita, Y., Tsukamoto, T., Shimono, T., & Miki, Y. (2023). Improved reproducibility of diffusion tensor image analysis along the perivascular space (DTI-ALPS) index: An analysis of reorientation technique of the OASIS-3 dataset. *Japanese Journal of Radiology, 41*(4), 393–400.

Tournier, J. D., Smith, R., Raffelt, D., Tabbara, R., Dhollander, T., Pietsch, M., Christiaens, D., Jeurissen, B., Yeh, C. H., & Connelly, A. (2019). MRtrix3: A fast, flexible and open software framework for medical image processing and visualisation. *NeuroImage, 202*, 116137.

Tuerxun, R., Kamagata, K., Saito, Y., Andica, C., Takabayashi, K., Uchida, W., Yoshida, S., Kikuta, J., Tabata, H., Naito, H., Someya, Y., Kaga, H., Miyata, M., Akashi, T., Wada, A., Taoka, T., Naganawa, S., Tamura, Y., Watada, H., Kawamori, R., & Aoki, S. (2024). Assessing interstitial fluid dynamics in type 2 diabetes mellitus and prediabetes cases through diffusion tensor imaging analysis along the perivascular space. *Frontiers in Aging Neuroscience, 16*, 1362457.

Wardlaw, J. M., Benveniste, H., Nedergaard, M., Zlokovic, B. V., Mestre, H., Lee, H., Doubal, F. N., Brown, R., Ramirez, J., MacIntosh, B. J., Tannenbaum, A., Ballerini, L., Rungta, R. L., Boido, D., Sweeney, M., Montagne, A., Charpak, S., Joutel, A., Smith, K. J., Black, S. E., & D. colleagues from the Fondation Leducq Transatlantic Network of Excellence on the Role of the Perivascular Space in Cerebral Small Vessel. (2020). Perivascular spaces in the brain: Anatomy, physiology and pathology. *Nature Reviews Neurology, 16*(3), 137–153.

Youn, Y. J., Kim, S., Jeong, H. J., Ah, Y. M., & Yu, Y. M. (2024). Sodium-glucose cotransporter-2 inhibitors and their potential role in dementia onset and cognitive function in patients with diabetes mellitus: A systematic review and meta-analysis. *Frontiers in Neuroendocrinology, 73*, 101131.

Zhang, L., Chopp, M., Zhang, Y., Xiong, Y., Li, C., Sadry, N., Rhaleb, I., Lu, M., & Zhang, Z. G. (2016). Diabetes mellitus impairs cognitive function in middle-aged rats and neurological recovery in middle-aged rats after stroke. *Stroke, 47*(8), 2112–2118.

Zhang, W., Zhou, Y., Wang, J., Gong, X., Chen, Z., Zhang, X., Cai, J., Chen, S., Fang, L., Sun, J., & Lou, M. (2021). Glymphatic clearance function in patients with cerebral small vessel disease. *NeuroImage, 238*, 118257.

Zhao, W. Q., De Felice, F. G., Fernandez, S., Chen, H., Lambert, M. P., Quon, M. J., Krafft, G. A., & Klein, W. L. (2008). Amyloid beta oligomers induce impairment of neuronal insulin receptors. *The FASEB Journal, 22*(1), 246–260.

Chapter 8

Brain insulin receptor gene networks, behavior and disease risk over the life course

Angela Marcela Jaramillo-Ospina and Patricia Pelufo Silveira
Department of Psychiatry, Faculty of Medicine and Health Sciences, McGill University, Ludmer Centre for Neuroinformatics and Mental Health, Douglas Hospital Research Centre, Montreal, QC, Canada

List of abbreviations

ADHD attention-deficit/hyperactivity disorder
BIBO basal influences on the baby's development
CEBQ child eating behavior questionnaire
CRP C-reactive protein
ELA early-life adversity
ePRS expression-based polygenic score
ePRS-IR expression-based polygenic score for the insulin receptor gene network
ERK extracellular signal-regulated kinase
FOXO forkhead box O
GSK3 glycogen synthase kinase 3
GWAS genome-wide association studies
HBCS helsinki birth cohort study
HDL-C high-density lipoprotein cholesterol
HPC hippocampus
IGFBP-1 insulin-like growth factor binding protein 1
IR-A insulin receptor isoform A
IR-B insulin receptor isoform B
IST information sampling task
LDL-C low-density lipoprotein cholesterol
MAPK mitogen-activated protein kinase
MAVAN maternal adversity, vulnerability and neurodevelopment
mPFC medial prefrontal cortex
MRI magnetic resonance imaging
mTOR mammalian target of rapamycin
NAc nucleus accumbens
NCDs chronic non-communicable diseases
NKT number knowledge task
PFC prefrontal cortex
PI3K/Akt phosphoinositide 3-kinase/Akt pathway
PRS polygenic risk score
SAGE study of addiction, genetics and environment repository
SAT subcutaneous adipose tissue
STR striatum
T2D type 2 diabetes
VAT visceral adipose tissue
VTA ventral tegmental area

Introduction

Prevalent diseases such as T2D and cardiovascular disease, as well as their major risk factors (obesity and metabolic syndrome) represent major global health challenges. Their complex origins arise from the cumulative effects of multiple genetic variants interacting with environmental exposures across the lifespan. Traditional genomic approaches have advanced our understanding of genetic liability, but remain limited in covering the mechanisms that contribute to disease risk. Expression-based polygenic scores (ePRS) address this gap by integrating gene co-expression networks grounded in biologically relevant pathways, thereby offering a functional approach to genetic profiling. A key example is the ePRS for the insulin receptor gene network (ePRS-IR), developed for the mesocorticolimbic and hippocampal regions. This score reflects variations in the brain insulin signaling, a pathway implicated in both metabolic regulation and neurobehavioral processes, and highlights a potential mechanism linking early-life adversity, a relevant environmental risk factor for the development of adult chronic diseases, to detrimental health outcomes across the lifespan.

Hormones in Neuroscience. https://doi.org/10.1016/B978-0-443-30263-3.00021-X

Expression-based polygenic scores: A biologically informed genomic profiling approach

NCDs, such as T2D (Auton et al., 2015), cardiovascular diseases (Kathiresan & Srivastava, 2012), metabolic syndrome (Lind, 2019), and obesity (Loos & Yeo, 2022), are considered complex polygenic conditions. This means that their appearance is driven by the small cumulative effects of multiple genetic variants, as well as their interaction with the environment over the lifespan. Traditional genomic approaches, including genome-wide association studies (GWAS), have identified many significant genetic variants associated with these conditions. However, the underlying biological mechanisms of these conditions remain only partially understood, and effective treatments that account for individual differences are still lacking.

Polygenic risk scores (PRS) are typically computed as the sum of risk alleles weighted by their effect sizes derived from GWAS. These scores summarize an individual's genetic risk for a specific polygenic condition or trait. PRSs are the gold-standard method for evaluating genetic risk (Miguel & Silveira, 2025). PRSs are based on the statistical association described in a GWAS, and therefore do not account for the biological relationships among the variants or other functional contexts (Wray & Goddard, 2010). This limits their utility in capturing the molecular pathways and processes that contribute to disease risk.

Expression-based polygenic scores (ePRS) represent a novel approach to genomic profiling. Here, the emphasis is the relation or co-expression between multiple genes or proteins that are likely involved in a biological function (Miguel & Silveira, 2025). ePRS are therefore informed by gene networks, which are sets of co-expressed genes within a specific tissue or brain region, anchored to a gene of interest and based on their contribution to biologically meaningful gene networks, allowing ePRS to better reflect underlying biological processes. This approach is based on the biologically plausible assumption that gene networks, rather than isolated genes, represent the relevant unit of influence during development (Hari Dass et al., 2019). ePRS shifts the focus from disease states to biological variability, capturing subtle individual variations in endophenotypes among individuals in the general population. This makes ePRS particularly suited for use in community and pediatric samples, where the goal is to identify early biological markers of risk before the onset of clinical symptoms. As such, ePRS enables the investigation of individual variations in biological mechanisms that may underlie vulnerability to disease later in life, providing a powerful tool for advancing developmental and translational science (Hari Dass et al., 2019; Miguel & Silveira, 2025).

Central insulin signaling mechanisms: effects on behavioral and health outcomes

Insulin is a polypeptide hormone with critical metabolic functions required for normal growth and development. Beyond its peripheral effects, insulin also plays important regulatory roles within the central nervous system.

Although pancreatic β-cells are the primary source of central insulin, local central synthesis also occurs in neurons (Kuwabara et al., 2011) and astrocytes (Pitt et al., 2017). Once secreted, insulin enters the circulation, where it acts on peripheral tissues and crosses the blood-brain barrier via a specific, saturable transport system located on capillary endothelial cells (Yazdani et al., 2019). Within the brain, insulin binds to its receptors, which exist in two high-affinity isoforms: isoform A (IR-A) and isoform B (IR-B) (Fernandez & Torres-Alemán, 2012). Both isoforms are expressed in glial cells, whereas neurons exclusively express IR-A (Pomytkin et al., 2018). Upon ligand binding, insulin receptors function as tyrosine kinases, initiating two major intracellular signaling cascades: the phosphoinositide 3-kinase (PI3K)/AKT pathway and the mitogen-activated protein kinase (MAPK) pathway (LeRoith et al., 2021; White, 2003).

The PI3K/AKT pathway phosphorylates AKT, which in turn regulates downstream targets such as glycogen synthase kinase-3 (GSK3), the mammalian target of rapamycin (mTOR), and the transcription factor forkhead box O (FOXO). These signaling proteins are critical for neurodevelopment, synaptic plasticity, and neurodegenerative processes (Kim et al., 2006; Salcedo-Tello et al., 2011; Stoica et al., 2011). The MAPK pathway, in contrast, results in phosphorylation of extracellular signal-regulated kinase (ERK), which translocates to the nucleus to activate transcription factors and additional protein kinases. Dysregulation of these pathways impairs glucose homeostasis, disrupts energy metabolism, alters redox regulation, compromises neurogenesis and synaptogenesis, and increases oxidative stress and neurodegeneration (Duarte, 2015).

Insulin receptor isoforms are widely distributed across brain regions, including the ventral tegmental area (VTA), striatum (STR), hippocampus (HPC), hypothalamus, amygdala, and prefrontal cortex (PFC). Through receptor activation, insulin modulates intracellular signaling cascades, gene expression, and neuronal activity in dopaminergic, serotonergic, GABAergic and glutamatergic systems (Alberry & Silveira, 2023; Batra et al., 2021). These neuromodulatory effects contribute to the regulation of key cognitive and behavioral processes, such as incentive salience (Figlewicz et al., 2008), impulsivity (Schoffelmeer et al., 2011), mood regulation (Kleinridders et al.,

2015), decision-making (Mucellini et al., 2017), energy balance, and eating behavior (Morton et al., 2006; Mucellini et al., 2017).

Specifically, glucose and insulin regulate the expression of dopamine-related genes, thereby modulating dopaminergic system function in brain regions such as the midbrain, including the VTA and substantia nigra. This regulation is considered a critical feature in the pathophysiology of several psychiatric disorders, including schizophrenia, bipolar disorders, and major depression (Figlewicz et al., 2003; Lozovsky et al., 1981; Mansur et al., 2018; Saller & Chiodo, 1980). Beyond gene expression, insulin also influences dopaminergic signaling at the circuit level: in the nucleus accumbens (NAc) and medial prefrontal cortex (mPFC), insulin modulates molecular responses to dopamine, ultimately shaping dopamine-dependent behaviors such as value attribution and incentive salience (Figlewicz et al., 1994, 2006). Supporting this evidence, a mouse model of brain-specific knockout of the insulin receptor exhibits anxiety and depressive-like behaviors (Kleinridders et al., 2015), while intranasal insulin administration has been shown to reduce palatable food consumption (Tiedemann et al., 2017).

Although the insulin-dopamine interaction has received the most attention, other neurotransmitter systems are also influenced by insulin. For example, peripheral insulin administration enhances serotonin synthesis (Martin et al., 2021) and increases expression of multiple GABA-A receptor subunits in both the hypothalamus (Cai et al., 2021) and mPFC (Trujeque-Ramos et al., 2018). In addition, insulin resistance in the HPC alters glutamate receptor phosphorylation, impairing hippocampal-dependent learning (Grillo et al., 2015). In the NAc, insulin further regulates glutamatergic excitatory transmission through synaptic mechanisms (Fetterly et al., 2021).

Together, these findings underscore insulin's broad neuromodulatory role, extending beyond dopamine to influence multiple neurotransmitter systems critical for cognition, emotion, and learning/motivation.

Gene networks and ePRS

Most studies in neurobiology focus on the effect of single genes or proteins on phenotypical traits. However, in each biological process, genes or proteins do not work in isolation. Take, for example, a synapse. There are multiple proteins involved in the process of neurotransmission: calcium channels, proteins involved in vesicles formation and neurotransmitters release, receptors, postsynaptic signaling cascades, and specialized presynaptic transporters. Studies focusing on variations of a single component of the neurotransmission process have only a partial view. There is a paucity of methods that consider how these multiple components work together within this biological process, and how the combined variation in these components can affect outcomes influenced by this neurobiological mechanisms.

This was the main driver for the creation of the ePRS (Barth et al., 2024; Silveira & Meaney, 2023). The technique was built from the idea that genes act not in isolation but within networks, making the gene network the key biological unit. Based on this principle, we developed an innovative genomic profiling approach that characterizes networks of co-expressed genes in specific tissues or brain regions (Barth et al., 2024; Silveira & Meaney, 2023). These co-expression networks are identified using large RNA sequencing datasets, typically anchored around a target gene of interest. Because networks vary across tissues, this method captures both gene–gene interactions and region-specific relevance.

From this framework, we created the expression-based polygenic score (ePRS), which integrates: i) gene co-expression within a tissue, ii) tissue-specific expression levels, iii) individual genetic variation, and iv) tissue-specific allele–expression associations (from GTEx). SNPs in network genes are annotated, filtered for redundancy (linkage disequilibrium clumpling), and weighted by their effect sizes before being summed into a single score. The result is a functional measure of how genetic variation shapes the activity of a specific gene network. Unlike other approaches, such as pharmagenic enrichment scores (Reay et al., 2020), which focus on predefined druggable pathways, ePRS derives directly from genome-wide co-expression networks and their regulatory relationships. This allows for a more biologically grounded and tissue-specific assessment. Combining ePRS with clinical and experimental studies provides a powerful way to understand individual differences in behavior, disease vulnerability, and responses to treatment.

The brain insulin receptor gene networks

We have previously calculated ePRS for the brain insulin receptor gene networks (ePRS-IR) in the meso-corticolimbic and the hippocampal regions using a step-wise approach (Hari Dass et al., 2019; Jaramillo-Ospina, Dalle Molle et al., 2025; Selenius et al., 2021; Selenius, Silveira, Bonsdorff, et al., 2024; Selenius, Silveira, Haapanen, et al., 2024). First, we identified genes co-expressed with the insulin receptor in the PFC and STR, or the HPC, in mice using RNA sequencing data from GeneNetwork (Mulligan et al., 2017), and only those genes with an absolute value of the co-expression correlation $r>|0.5|$ were selected. Ortholog human genes were identified using the BiomaRt package. Second, due to our interest in genes that are relevant during early developmental phases, Brainspan

(Miller et al., 2014) was used to filter genes with at least a 1.5-fold change difference in the expression during fetal development and childhood compared to adulthood, in the same brain regions. The final list of genes included 263 genes for the mesocorticolimbic ePRS-IR and 488 genes for the hippocampal ePRS-IR. Third, the chromosomal start and end positions for all genes in the final gene list were extracted using the BiomaRt package (Durinck et al., 2005, 2009). Next, all SNPs corresponding to each gene in the network were included obtaining the final SNP list. This SNP list was generated prior to applying the human cohort filter step. Fourth, the identified SNP list was merged with the genotype data from the human cohort of interest, retaining only the SNPs present in both. The resulting SNPs were then pruned using linkage disequilibrium clumping ($r2>0.2$ within 500 kb region) to generate the final cohort-specific SNP list. Each SNP was weighted by the GTEx-derived tissue-specific effect size (PFC or HPC) and summarized into a single score. Before summarizing the weighted SNPs in the mesocorticolimbic ePRS-IR or the hippocampal ePRS-IR, the weighted SNPs were multiplied by ± 1 depending on the sign of the correlation coefficient between the insulin receptor gene expression and the expression of each of the other genes in the network in the original RNA sequencing data in mice (Jaramillo-Ospina, Elgbeili, et al., 2025) (Fig. 8.1).

Brain specificity of the mesocorticolimbic ePRS-IR has been supported by the stronger co-expression of its genes in the PFC and STR, compared to genes linked to the top SNP from the fasting insulin GWAS (Hari Dass et al., 2019). In addition, the mesocorticolimbic and hippocampal ePRS-IR brain region specificity has been indicated by performing gene ontology enrichment analyses. Results showed that while statistically significant enrichment for pathways involved in cell cycle regulation and process networks for neurogenesis/axonal guidance, transcription and translation for the mesocorticolimbic ePRS-IR, pathways involved in development and transcription, as well as process networks involved in cell division, were revealed for the hippocampal ePRS-IR (Hari Dass et al., 2019).

Compelling evidence for the brain region specificity of the mesocorticolimbic and hippocampal ePRS-IR is also reflected in their differential associations with behavioral and metabolic outcomes across analyses. For example, considering the modulation of dopaminergic neurotransmission by insulin acting on the mesocorticolimbic system, and how dopamine influences decision-making processes (Chakroun et al., 2023), it was hypothesized that variations in the mesocorticolimbic ePRS-IR would be associated with behaviors linked to decision-making and impulse control. To test that, we initially studied children from the Maternal Adversity, Vulnerability and Neurodevelopment (MAVAN) community-based cohort placed in Montreal (Quebec) and Hamilton (Ontario), Canada. We investigated if variations in the mesocorticolimbic ePRS-IR were linked to MAVAN children performance on the Information Sampling Task (IST) applied at 6 years of age. This task is designed to measure how much information a child requires before making a decision under uncertainty. In a typical version, children see a grid of hidden stimuli (e.g., colored cards), and they may uncover, one by one, some of those hidden pieces before selecting their guess of which color is displayed in majority. The key behavioral outcome is how many samples a child draws before they commit to a choice, revealing their tolerance for uncertainty or impulsivity in decision making. In neuroscience and psychology studies, this task helps reveal individual differences in exploration versus exploitation, risk attitudes, and how neural circuits compute value of information. It was observed that a lower mesocorticolimbic ePRS-IR, but not hippocampal ePRS-IR, was significantly associated with reduced certainty when making decisions, predicting greater impulsivity specifically in 6-year-old boys. This association was not observed in girls (Hari Dass et al., 2019).

Moreover, given the clinical overlap between impulsivity and risk for addiction (Kozak et al., 2019; Mitchell & Potenza, 2014), and based on the results observed in children, it was tested whether the mesocorticolimbic ePRS-IR would predict risk for addiction in men. Indeed, a lower mesocorticolimbic ePRS-IR was significantly associated with a higher number of addiction comorbidities in adult men from the Study of Addiction, Genetics and Environment (SAGE) repository. Addiction to alcohol, cocaine, marijuana, nicotine, opiates, as well as other drugs, was included in the count of addictions comorbidities. A lower mesocorticolimbic ePRS-IR was also associated with greater probability of alcohol dependence in men. The same associations were not observed in women (Hari Dass et al., 2019). These findings provided additional evidence for the previously described links between impulsivity, conduct disorders, and drug addiction (Levy et al., 2014; Rosenström et al., 2018). Moreover, insulin function has been associated with the risk of drug addiction, where diminished insulin sensitivity correlates with fewer endogenous dopamine receptor in the ventral striatum (O'Dell & Nazarian, 2016). This reinforces the notion that metabolic processes are involved in dysfunctions of the mesocorticolimbic system implicated in drug dependence (Hari Dass et al., 2019).

Supporting our hypothesis that the hippocampal ePRS-IR connects metabolic and cognitive processes, evidence shows that patients with Alzheimer's disease exhibit reduced insulin receptor expression and altered components of the insulin signaling in the HPC (Talbot et al., 2012). In addition, down-regulation of insulin signaling has been shown to impair memory through inhibition of serine phosphorylation in the insulin receptor (Bomfim

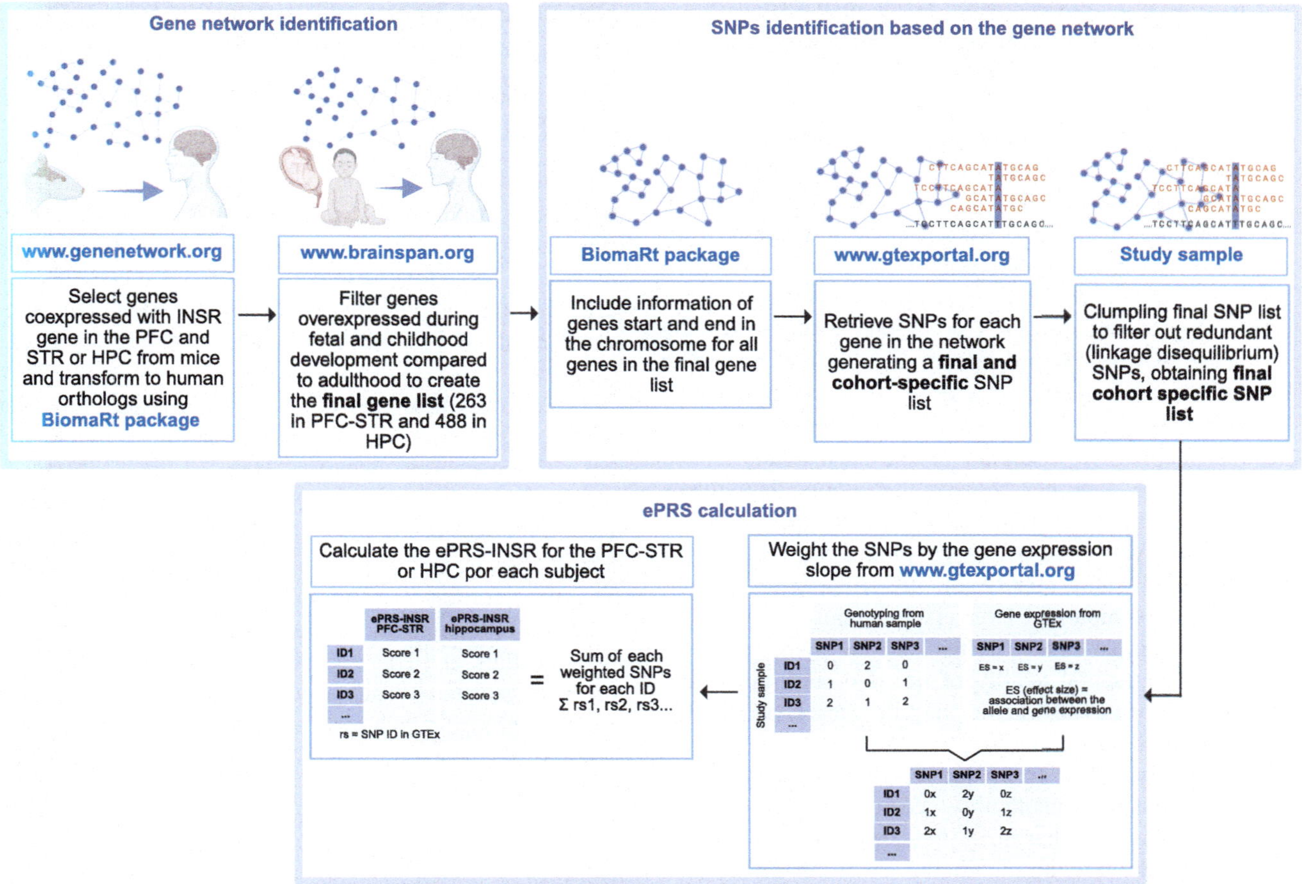

FIGURE 8.1 Calculation of the expression-based polygenic score (ePRS) from the genes co-expressed with the insulin receptor gene in the prefrontal cortex and striatum or hippocampus. Gene network identification: GeneNetwork was used to identify a list of genes co-expressed with the insulin receptor gene in prefrontal cortex (PFC) and striatum (STR) or hippocampus (HPC) in mice. These genes were mapped to their human orthologs using the BiomarRt package. BrainSpan was used to identify genes overexpressed in the PFC and STR or HPC during fetal development and up to 5 years of age, compared with adult samples. Both gene lists were merged, and only genes present in both were retained to generate the final gene list. **SNPs identification based on the gene network:** Star and end chromosomal position for all genes in the final gene list were extracted from the BiomaRt package. The resulting SNP list was clumpled to retain only independent SNPs, yielding the final list of SNPs for the study sample of interest. **ePRS-IR calculation:** For each SNP in the final list, the number of effect alleles at a given SNP carrier by each participant (rs1, rs2 …) was multiplied by the estimated effect size of the genotype-gene expression association from GTEx. The weighted SNPs values were further multiplied by ±1 depending on the sign of the correlation coefficient between the insulin receptor gene expression and the expression of each of the other genes in the network in the original RNA sequencing data in mice. The sum of these values across all SNPs provides the mesocorticolimbic (PFC-STR) or the hippocampal (HPC) ePRS-IR.

et al., 2012). Aligned with these findings, in children from the MAVAN cohort, variations in the hippocampal ePRS-IR (but not the mesocorticolimbic ePRS-IR) were linked to cognitive performance evaluated using the Number Knowledge Task (Lawn et al., 2023) at 4 years of age. Here, a higher hippocampal ePRS-IR was associated with poorer performance, reflecting reduced cognitive ability, in both boys and girls. Moreover, in adult participants from the multi-side GEN-ADA study, a higher hippocampal ePRS-IR was linked to a greater prevalence of Alzheimer's disease (Hari Dass et al., 2019). Taken together, these findings highlight the brain region specificity of the ePRS-IR across the lifespan: the mesocorticolimbic score links insulin-related mechanisms to impulsive decision-making and addiction risk emerging in childhood through adulthood, whereas the hippocampal score connects insulin pathways to cognitive vulnerability in childhood and Alzheimer's disease in later life.

After the initial ePRS-IR study, interest in using this brain region specific estimation of the insulin receptor function emerged from multiple groups. For example, the Helsinki Birth Cohort Study (HBCS) is a population-based study of adults with a mean age of 61 years, of whom approximately 56% are women. Within this cohort, the association between mesocorticolimbic and hippocampal ePRS-IR, cardiometabolic risk factors, and T2D was examined. Results indicated that in men, a higher mesocorticolimbic ePRS-IR was associated with increased levels of HDL-C and decreased levels of IGFBP-1 (Selenius, Silveira, Bonsdorff, et al., 2024). In contrast, a higher hippocampal ePRS-IR was linked to adverse metabolic outcomes particularly in women. Women with elevated hippocampal ePRS-IR showed greater body fat, higher triglycerides, total cholesterol, LDL-C, glucose and insulin levels, together with reduced HDL-C levels. In men, the hippocampal ePRS-IR was associated with increased 120-min post–oral glucose tolerance test insulin levels. When both sexes were considered together, a higher hippocampal ePRS-IR predicted greater risk of prediabetes and T2D (Selenius, Silveira, Bonsdorff, et al., 2024). In the same cohort, it was also explored whether mesocorticolimbic or hippocampal ePRS-IR scores modified the relationship between depressive symptom profiles and health-related quality of life. Findings showed that among individuals with low hippocampal ePRS-IR, the presence of melancholic-type depressive symptoms was associated with better mental quality of life compared to non-melancholic symptoms. Conversely, among those with high hippocampal ePRS-IR, melancholic symptoms predicted poorer mental quality of life. No such modifying effects were observed when testing the mesocorticolimbic ePRS-IR (Selenius et al., 2021).

Longitudinal data from the HBCS provided insights into aging-related vulnerability. For example, the frailty index is a measure that quantifies an individual's health vulnerability by tallying the number of health deficits that they have, relative to a total set of possible deficits. These deficits can include symptoms, diseases, signs, disabilities, and laboratory abnormalities. The more deficits a person accumulates, the higher their frailty index and the greater their vulnerability to adverse outcomes like mortality, disability, and decline in health.

In adult women with a mean age of 61 years, a higher hippocampal ePRS-IR was significantly associated with an elevated frailty index, both at baseline and after 12 years of follow-up. These associations were not observed with the mesocorticolimbic ePRS-IR, underscoring the stronger link between hippocampal insulin-related pathways and late-life frailty (Selenius, Silveira, Haapanen, et al., 2024). In summary, findings from the HBCS show that the hippocampal ePRS-IR is consistently associated with adverse cardiometabolic outcomes, poorer mental quality of life, and greater frailty, highlighting brain region- and sex-specific pathways across health domains.

Findings across diverse cohorts demonstrate the brain region specificity of the ePRS-IR, with the mesocorticolimbic score linked to impulsive decision-making and addiction risk in males, and the hippocampal score associated with adverse metabolic, cognitive, and aging outcomes, particularly in women, underscoring sex and age-specific effects across the lifespan (Table 8.1).

The role of brain insulin receptor gene networks in moderating the effect of early adversity on eating behavior and metabolism

Early-life adversity (ELA), occurring during prenatal and childhood periods, both critical windows for development, can exert long-lasting effects on physiology and increase the risk of behavioral disorders as well as metabolic disturbances throughout life (Bellis et al., 2019; Boynton-Jarrett et al., 2012; Entringer et al., 2008; McKay et al., 2022).

ELA exerts profound effects on the brain and other biological systems, increasing vulnerability to psychiatric and cardiometabolic disorders later in life (Alberry & Silveira, 2023; Batra et al., 2021) and their comorbidity (Gómez-Ilescas & Silveira, 2025). Insulin may play a critical role in this link, as its central actions influence synaptic plasticity, neurotransmission, and executive functions in brain areas such as the PFC, NAc, amygdala, hypothalamus, and HPC, areas particularly sensitive to the impact of adversity. This provides a potential mechanistic pathway through which early adversity contributes to adult physical and mental disease risk (Alberry & Silveira, 2023; Batra et al., 2021).

In this context, child behavior (e.g., cognitive performance, impulsive decision-making, eating behavior) is theorized as important mediators of the effect of ELA on health. For example, we hypothesize that small changes in eating behavior induced by ELA contribute to an increased risk of higher adiposity and metabolic disease in the long term (Scudine et al., 2024). A question that emerged was whether variations in brain insulin function would moderate the effect of ELA on eating behavior, suggesting that brain insulin receptor gene networks were relevant mechanistic processes linking ELA to adult disease. Using our ePRS reflecting functional variation of the insulin receptor gene network, we explored whether individual variation in the expression of the mesocorticolimbic ePRS-IR moderates the association between ELA and childhood eating behaviors in participants of the MAVAN cohort. Adversity exposure during early life was used as a cumulative adversity score calculated adding one point for the presence of each of the environmental adverse components: smoking during pregnancy, low household income, lack of money, high marital strain, presence of physical or sexual domestic violence, pregnancy anxiety, maternal disorganized attachment, poor family function, postnatal maternal mental health symptoms, birth size $<$10th or $>$90th percentile, and gestational age $\leq$37 weeks (Silveira et al., 2017). A higher cumulative adversity score reflected a higher adverse environment. The outcome, eating behavior, was measured using the Child Eating Behavior

TABLE 8.1 Mesocorticolimbic and hippocampal expression-based polygenic risk score for insulin receptor gene networks: associations with behavioral, metabolic, and inflammatory outcomes across cohorts.

Cohort	Life stage	ePRS-IR	Association	References
Mesocorticolimbic ePRS-IR				
MAVAN (Canada)	Childhood	Low[a]	Greater impulsivity: 6-year-old boys	Hari Dass et al. (2019)
		Low	Higher ELA → higher satiety responsiveness, lower enjoyment of food, lower food approach/food avoidance: 6-year-old girls	Jaramillo-Ospina, Dalle Molle et al. (2025)
		High	Higher ELA → higher desire to drink, higher food approach score, higher food approach/food avoidance ratio, lower enjoyment of food: 4-year-old children	
		High	Higher ELA → higher food responsiveness, higher food approach score, higher food approach/food avoidance ratio: 6-year-old children	
		High	Higher ELA → was associated with higher desire to drink: 4-year-old boys	
		High	Higher ELA → higher food approach and food approach/food avoidance ratio: 4-year-old girls	
		High	Higher ELA → higher desire to drink: 6-year-old girls	
BIBO (The Netherlands)	Childhood	High	Higher ELA → higher desire to drink: 6-year-old children	Jaramillo-Ospina, Dalle Molle et al. (2025)
SAGE (USA)	Adulthood	Low[a]	Higher number of addiction comorbidities: Adult men Greater probability of alcohol dependence: Adult men	Hari Dass et al. (2019)
HBCS (Finland)	Late adulthood	High	Increased HDL-C and decreased IGFBP-1: Adult men	Selenius, Silveira, Bonsdorff, et al. (2024)
UK Biobank (United Kingdom)	Adulthood	High	Higher ELA → Greater risk of metabolic syndrome, increased abdominal SAT and VAT volumes: Adult women and men	Jaramillo-Ospina, Elgbeili, et al. (2025)
		High	Higher ELA → Greater risk of metabolic syndrome, increased VAT volumes, greater percentage of energy derived from fat: adult women	
Hippocampal ePRS-IR				
MAVAN (Canada)	Childhood	High[a]	Reduced cognitive ability without sex differences	Hari Dass et al. (2019)
GEN-ADA (USA)	Adulthood	High[a]	Greater prevalence of Alzheimer's disease	Hari Dass et al. (2019)
HBCS (Finland)	Late adulthood	High	Increased risk of prediabetes and diabetes mellitus	Selenius, Silveira, Bonsdorff, et al. (2024)

Continued

TABLE 8.1 Mesocorticolimbic and hippocampal expression-based polygenic risk score for insulin receptor gene networks: associations with behavioral, metabolic, and inflammatory outcomes across cohorts.—cont'd

Cohort	Life stage	ePRS-IR	Association	References
		High	Increased body fat, triglycerides, total cholesterol, LDL-C, glucose and insulin levels, and reduced HDL-C levels in women	
		High	Increased 120-min post-oral glucose tolerance test insulin levels in men	
		Low	Melancholic-type depressive symptoms predicted better mental quality of life	Selenius et al. (2021)
		High	Melancholic symptoms predicted poorer mental quality of life	
		High	Elevated frailty index at baseline and after 12 years of follow-up	Selenius, Silveira, Haapanen, et al. (2024)

BIBO: Basal Influences on the Baby's Development; ELA: early-life adversity; HBCS: Helsinki Birth Cohort Study; HDL-C: High-density lipoprotein cholesterol; IGFBP-1: Insulin-like growth factor binding protein 1; LDL-C: Low-density lipoprotein cholesterol; MAVAN: Maternal Adversity, Vulnerability and Neurodevelopment; SAGE: Study of Addiction, Genetics and Environment; SAT: Subcutaneous adipose tissue; VAT: Visceral adipose tissue.

[a]*Comparisons between the mesocorticolimbic and hippocampal ePRS-IR derived in the study by Hari Dass et al. (2019) and those subsequently reported by Selenius et al. and Jaramillo-Ospina et al. should be interpreted with caution, as the current ePRS-IR is weighted using GTEx data, whereas the earlier version was weighted using genome-wide association study data on attention-deficit/hyperactivity disorder (ADHD), thereby limiting the comparability of high and low scores between studies.*

Questionnaire (CEBQ), which is a parent-reported questionnaire applied when the participants were four and 6 years of age. The CEBQ measures eight subscales, four evaluate a tendency to overeat (food approach), and four assess a tendency toward avoiding eating (food avoidance). Food approach, food avoidance scores and the ratio between them were calculated (Jaramillo-Ospina, Dalle Molle et al., 2025).

This study showed a significant interaction effect between ELA and the mesocorticolimbic ePRS-IR on eating behavior in childhood. Specifically, only in the high mesocorticolimbic ePRS-IR group, a greater early adversity score was significantly associated with a higher desire to drink, higher food approach score and food approach/food avoidance ratio. Additionally, only in the high mesocorticolimbic ePRS-IR group, a higher early adversity score was significantly associated with a lower enjoyment of food. These interactions were observed at age 4. At age 6, only in the high mesocorticolimbic ePRS-IR group, a greater early adversity score was significantly associated with a higher food responsiveness, a higher food approach score, and a higher food approach/food avoidance ratio (Jaramillo-Ospina, Dalle Molle et al., 2025).

Moreover, sex-specific interactions were observed; in 4-year-old boys, a higher early adversity score was significantly associated with a higher desire to drink, only in the high mesocorticolimbic ePRS-IR group. Also, in 4-year-old girls, a higher early adversity score was significantly associated with a higher food approach score and with a higher food approach/food avoidance ratio, only in those in the high mesocorticolimbic ePRS-IR group. In addition, in 6-year-old girls, only in the high mesocorticolimbic ePRS-IR group, a higher early adversity score was associated with a higher desire to drink, while in those in the low ePRS-IR group, a higher early adversity score was associated with a higher satiety responsiveness, as well as lower enjoyment of food and food approach/food avoidance ratio (Jaramillo-Ospina, Dalle Molle et al., 2025). Results were replicated using data from the Basal Influences on the Baby's Development (BIBO), an ongoing prospective longitudinal project that recruited participants from The Netherlands. In BIBO, a significant moderation of ePRS-IR in the association between the early adversity score and the desire to drink score in the whole cohort at 6 years of age was observed (Jaramillo-Ospina, Dalle Molle et al., 2025). Overall, the study confirmed that the mesocorticolimbic ePRS-IR moderated the effect of ELA on eating behavior in boys and girls, in which children with a high ePRS-IR were more sensitive to the effects of ELA on food approach/overeating. This suggested that individuals from the high ePRS-IR group exposed to ELA are at risk for altered eating behavior early during childhood, with potential metabolic consequences over the life course.

Indeed, we were able to confirm this metabolic vulnerability. We examined whether individual variation in the expression of the mesocorticolimbic ePRS-IR moderated the association between ELA and abdominal subcutaneous (SAT) and visceral (VAT) adipose tissue volumes, risk of metabolic syndrome, and food intake in adult women and men from the large-scale biomedical dataset, the UK Biobank. Early life environmental exposure was characterized using a composite score that included both adverse and positive experiences during the prenatal period, birth, and early childhood. To calculate this score, positive points (+1) were assigned for each supportive/enriching experience and negative points (−1) for each adversity experience, with the final score obtained by summing these values. More negative values indicate a predominantly adverse environment, whereas more positive values represent a supportive/enriching environment. The outcomes assessed were abdominal SAT and VAT volumes, estimated using Magnetic Resonance Imaging (MRI), risk of metabolic syndrome, which included C-reactive protein (CRP) levels; and food intake, derived from participants' 24-hour dietary recalls (Jaramillo-Ospina, Elgbeili, et al., 2025).

We observed a significant interaction effect between early life environmental exposure and the mesocorticolimbic ePRS-IR across all outcomes. Specifically, in the high ePRS-IR group, greater early adversity (reflected by lower environmental exposure score) was significantly associated with higher SAT and VAT volumes and a higher risk of metabolic syndrome. Moreover, the association of greater early adversity with higher VAT volume, increased risk of metabolic syndrome, and higher percentage of energy derived from fat was especially strong and statistically significant in women compared with men (Jaramillo-Ospina, Elgbeili, et al., 2025).

Overall, these studies demonstrate that greater exposure to adversity is associated with increased abdominal SAT and VAT accumulation, higher risk of metabolic syndrome, and more obesogenic food intake and eating behaviors in both adults and children with a high mesocorticolimbic ePRS-IR (Table 8.1). In other words, the detrimental effects of adversity on risk factors for non-communicable diseases during childhood and adulthood are particularly evident among individuals with a high ePRS-IR. Thus, the ePRS-IR may help identify those who are especially vulnerable to the effects of early life adversity, which can manifest as risk factors for non-communicable diseases and, in turn, contribute to an increased risk of cardiovascular outcomes later in life (Doom et al., 2024; Hübel et al., 2021; Munkholm et al., 2016). These findings highlight a gene–environment interaction in the development of non-communicable disease risk factors, mediated by insulin receptor co-expression pathways within mesocorticolimbic circuits.

Insulin as a key mechanism linking ELA to altered behavior and disease risk

Insulin has been proposed as a central mechanism linking ELA to altered behavior and increased risk of disease in adulthood. Importantly, the co-occurrence of metabolic and psychiatric disorders within the same individual is highly prevalent, with ELA representing a common developmental risk factor (Gómez-Ilescas & Silveira, 2025; Silveira et al., 2025) and variations in insulin as a potential mediator of these associations. Given their critical roles in reward processing, self-regulation, value attribution (Alberry & Silveira, 2023; Batra et al., 2021), learning, and memory (Riedel & Micheau, 2001) and their particular sensitivity to ELA (Dalle Molle et al., 2022; Fabricius et al., 2008), the PFC, STR and HPC have been identified as key brain regions linking early life stress to long-term behavioral and metabolic outcomes. Importantly, their involvement in brain-insulin signaling pathways points to a neurobiological mechanism through which early adversity may disrupt brain-body communication, thereby increasing disease vulnerability (e.g. Barth et al., 2024). Altered insulin signaling not only disrupts glucose homeostasis in the periphery but also reshapes dopamine- and reward-related circuits, with sex-specific effects. These neurobiological changes contribute to increased vulnerability to impulsive decision-making, altered eating behaviors increasing food approach and intake of palatable foods, increased abdominal fat deposition and comorbid metabolic and psychiatric disorders in adulthood, underscoring insulin's central role in developmental programming of brain–body communication.

Perspectives in ePRS-IR

A central question for future research is whether altered insulin function in individuals exposed to ELA affects the transcriptional landscape of the PFC and STR, and in turn contributes to maladaptive behavioral phenotypes. To address this, studies examining whether peripheral insulin modulates PFC and STR gene expression profiles across developmental windows in animal models of ELA are warranted (e.g., Miguel et al., 2025).

Future directions for the use of the ePRS include testing its applicability across diverse genetic backgrounds and evaluating the hypothesis that the ePRS can identify individuals at risk for disease in early life. In addition, subsequent studies will examine its predictive value for outcomes such as excessive weight gain, low-grade inflammation, and altered eating behaviors in younger cohorts, as well as chronic non-communicable disease risk factors in older cohorts.

Beyond the investigation of metabolic and inflammatory biomarkers, another area of interest is the examination of brain structure and function. For example, variations in brain volume and resting-state functional connectivity can be investigated in individuals varying in ELA exposure and ePRS-IR backgrounds, to identify whether the gene-environment interaction alters structural and functional aspects in regions related to the behavioral outcomes (e.g., eating, decision-making, cognitive aspects). This approach will allow us to assess not only the neural regions most sensitive to early life stress but also to identify individuals who are potentially more vulnerable or more strongly affected.

As we discussed previously, while traditional PRS aggregate the effects of multiple SNPs identified by GWAS to provide an overall estimate of genetic susceptibility (Loos & Yeo, 2022), ePRS extend this framework by integrating functional, tissue-specific gene co-expression networks that capture additional sources of variability (Hari Dass et al., 2019). Building on this complementary perspective, we hypothesize that combining GWAS-derived PRS with ePRS-IR will improve predictive accuracy by reflecting both broad genetic liability and functionally relevant molecular networks. Future studies could test this integration across multiple human cohorts, spanning diverse developmental stages and backgrounds, to evaluate their joint contribution to metabolic and mental health outcomes.

Although ePRS overcomes some limitations of conventional PRS by incorporating tissue specificity, gene functionality, and individual-level calculation, it still faces significant barriers to clinical translation. Interpretation and communication of probabilistic risk estimates remain challenging for clinicians and patients, often generating anxiety, fatalism, or, misplaced reassurance. Confidence has further been undermined by the premature commercialization of unvalidated direct-to-consumer genetic tests. Moreover, the lack of large-scale prospective studies demonstrating clear clinical benefits, together with ongoing concerns about genetic discrimination, calls for caution and stronger safeguards before implementation in healthcare (Torkamani et al., 2018). Nonetheless, current evidence suggests that ePRS offers particular value for studying gene-environment interactions in community-based cohorts and at the individual level. By integrating tissue-specific functional networks with genetic variation, ePRS can disentangle complex developmental trajectories and provide complementary insights beyond traditional GWAS-based scores, positioning ePRS as a promising tool for advancing personalized medicine.

Conclusion

Early-life adversity heightens vulnerability to metabolic and mental health disorders across the lifespan, partly through its interaction with genetic predisposition. Insulin plays a central role in this link by shaping brain–body communication in key regions such as the prefrontal cortex, striatum, and hippocampus, which are especially sensitive to early adversity. Expression-based polygenic scores offer a novel approach to capture tissue-specific biological processes and identify early markers of risk before chronic conditions take place, providing a powerful tool to advance developmental and translational science toward prevention and early intervention.

Other hormones

Whilst this chapter reviews the work related to the expression-based polygenic scores from the brain insulin receptor gene networks, other central and peripheral gene networks are shown to be relevant to identify individuals more vulnerable to the long-term effects of ELA. For example, we have described that variations in the expression of the PFC-leptin receptor (LepR) gene network similarly moderates the effect of ELA on eating behavior outcomes (de Lima et al., 2022). Moreover, a liver-LepR gene network moderates the effect of ELA on mental health outcomes in different cohorts (de Lima et al., 2025). Future studies focusing on multiple gene networks and their cumulative effects will bring more understanding of their individual as well as combined contribution.

Applications of the material

This chapter reviews the application of novel polygenic metrics estimating individual variations in the expression of the brain insulin receptor co-expression networks. It is suggested that these novel methods for genomic profiling, informed by biological function, can capture part of the variance associated with gene-early environmental adversity interactions effects on the risk for adult disease. The authors argue that the consideration of multiple levels of information (relation between genes, tissue specific gene expression, genotype, and early environment risk) should be considered when evaluating risk profiles and this could inform the development of precision medicine approaches.

Mini-dictionary of terms

- **Polygenic Risk Score.** A PRS summarizes the genetic individual's genetic risk of developing a specific disease base on SNPs that show statistical association with the condition in population level studies.
- **Expression-based polygenic score.** An ePRS is a novel genomic approach that summarizes individual variations in the expression of gene co-expression networks within a specific brain region or tissue, providing greater biological relevance.
- **Co-expression gene network.** Group of genes that are expressed together within a specific tissue or brain region, typically identified based on a gene of interest and their shared contribution to biologically meaningful processes.
- **Early-life adversity.** It refers to adverse events experienced from the prenatal period through childhood that affect physical and mental development and increase disease risk across the lifespan.

Summary points

- Early-life adversity increases the risk of metabolic and mental health problems across the lifespan.
- The negative effects of early-life adversity are pronounced in individuals with increased genetic susceptibility.
- The expression-based polygenic score for the insulin receptor gene network (ePRS-IR) identifies individuals at greater risk of developing metabolic and mental diseases across the lifespan.
- ePRS are derived from co-expressed genes within specific tissues or brain regions, providing stronger biological relevance than traditional risk scores.
- ePRS-IR in mesocorticolimbic brain regions is associated with risk-related behaviors across the lifespan and moderates the link between early-life adversity and behavioral, metabolic and mental health outcomes.
- ePRS-IR in the hippocampus is associated with risk-related behaviors across the lifespan as well as vulnerability to Alzheimer's disease, metabolic dysfunction, and mental health problems.
- These results strengthen the evidence that insulin is a potential mechanism linking early-life adversity with metabolic and mental health issues across the lifespan.

References

Alberry, B., & Silveira, P. P. (2023). Brain insulin signaling as a potential mediator of early life adversity effects on physical and mental health. *Neuroscience and Biobehavioral Reviews, 153*, 105350. https://doi.org/10.1016/j.neubiorev.2023.105350

Auton, A., Brooks, L. D., Durbin, R. M., Garrison, E. P., Kang, H. M., Korbel, J. O., … Abecasis, G. R. (2015). A global reference for human genetic variation. *Nature, 526*(7571), 68–74. https://doi.org/10.1038/nature15393

Barth, B., Arcego, D. M., de Mendonça Filho, E. J., de Lima, R. M. S., Parent, C., Dalmaz, C., Portella, A. K., Pokhvisneva, I., Meaney, M. J., & Silveira, P. P. (2024). Striatal dopamine gene network moderates the effect of early adversity on the risk for adult psychiatric and cardiometabolic comorbidity. *Scientific Reports, 14*(1), 27349. https://doi.org/10.1038/s41598-024-78465-5

Barth, B., de Mendonça Filho, E. J., Arcego, D. M., Pokhvisneva, I., Meaney, M. J., & Silveira, P. P. (2024). Expression based polygenic scores—A gene network perspective to capture individual differences in biological processes. *bioRxiv*. https://doi.org/10.1101/2024.03.01.583008, 2024.2003.2001.583008.

Batra, A., Latsko, M., Portella, A. K., & Silveira, P. P. (2021). Early adversity and insulin: Neuroendocrine programming beyond glucocorticoids. *Trends in Endocrinology and Metabolism, 32*(12), 1031–1043. https://doi.org/10.1016/j.tem.2021.09.003

Bellis, M. A., Hughes, K., Ford, K., Ramos Rodriguez, G., Sethi, D., & Passmore, J. (2019). Life course health consequences and associated annual costs of adverse childhood experiences across Europe and North America: A systematic review and meta-analysis. *The Lancet Public Health, 4*(10), e517–e528. https://doi.org/10.1016/S2468-2667(19)30145-8

Bomfim, T. R., Forny-Germano, L., Sathler, L. B., Brito-Moreira, J., Houzel, J. C., Decker, H., … De Felice, F. G. (2012). An anti-diabetes agent protects the mouse brain from defective insulin signaling caused by Alzheimer's disease-associated Aβ oligomers. *Journal of Clinical Investigation, 122*(4), 1339–1353. https://doi.org/10.1172/jci57256

Boynton-Jarrett, R., Rosenberg, L., Palmer, J. R., Boggs, D. A., & Wise, L. A. (2012). Child and adolescent abuse in relation to obesity in adulthood: The Black Women's Health Study. *Pediatrics, 130*(2), 245–253. https://doi.org/10.1542/peds.2011-1554

Cai, W., Zhang, X., Batista, T. M., García-Martín, R., Softic, S., Wang, G., … Kahn, C. R. (2021). Peripheral insulin regulates a broad network of gene expression in Hypothalamus, hippocampus, and nucleus accumbens. *Diabetes, 70*(8), 1857–1873. https://doi.org/10.2337/db20-1119

Chakroun, K., Wiehler, A., Wagner, B., Mathar, D., Ganzer, F., van Eimeren, T., … Peters, J. (2023). Dopamine regulates decision thresholds in human reinforcement learning in males. *Nature Communications, 14*(1), 5369. https://doi.org/10.1038/s41467-023-41130-y

Dalle Molle, R., de Mendonca Filho, E. J., Minuzzi, L., Machado, T. D., Reis, R. S., Rodrigues, D. M., … Silveira, P. P. (2022). Thrifty-eating behavior phenotype at the food court—Programming goes beyond food preferences. *Frontiers in Endocrinology, 13*, 882532. https://doi.org/10.3389/fendo.2022.882532

de Lima, R. M. S., Barth, B., Arcego, D. M., de Mendonça Filho, E. J., Patel, S., Wang, Z., … Silveira, P. P. (2025). Peripheral leptin receptor gene network modulates the impact of childhood adversity on mental health disorders. *Brain, Behavior, and Immunity, 129*, 585–594. https://doi.org/10.1016/j.bbi.2025.06.035

de Lima, R. M. S., Barth, B., Mar Arcego, D., de Mendonça Filho, E. J., Patel, S., Wang, Z., … Silveira, P. P. (2022). Leptin receptor co-expression gene network moderates the effect of early life adversity on eating behavior in children. *Communications Biology, 5*(1), 1092. https://doi.org/10.1038/s42003-022-03992-8

Doom, J. R., Deer, L. K., Mickel, T., Infante, A., & Rivera, K. M. (2024). Eating behaviors as pathways from early childhood adversity to adolescent cardiometabolic risk. *Health Psychology, 43*(6), 448–461. https://doi.org/10.1037/hea0001340

Duarte, J. M. N. (2015). Metabolic alterations associated to brain dysfunction in diabetes. *Aging and disease, 6*(5), 304–321. https://doi.org/10.14336/ad.2014.1104

Durinck, S., Moreau, Y., Kasprzyk, A., Davis, S., De Moor, B., Brazma, A., & Huber, W. (2005). BioMart and bioconductor: A powerful link between biological databases and microarray data analysis. *Bioinformatics, 21*(16), 3439–3440. https://doi.org/10.1093/bioinformatics/bti525

Durinck, S., Spellman, P. T., Birney, E., & Huber, W. (2009). Mapping identifiers for the integration of genomic datasets with the R/Bioconductor package biomaRt. *Nature Protocols, 4*(8), 1184–1191. https://doi.org/10.1038/nprot.2009.97

Entringer, S., Wüst, S., Kumsta, R., Layes, I. M., Nelson, E. L., Hellhammer, D. H., & Wadhwa, P. D. (2008). Prenatal psychosocial stress exposure is associated with insulin resistance in young adults. *American Journal of Obstetrics and Gynecology, 199*(5). https://doi.org/10.1016/j.ajog.2008.03.006, 498.e491-497.

Fabricius, K., Wörtwein, G., & Pakkenberg, B. (2008). The impact of maternal separation on adult mouse behaviour and on the total neuron number in the mouse hippocampus. *Brain Structure and Function, 212*(5), 403–416. https://doi.org/10.1007/s00429-007-0169-6

Fernandez, A. M., & Torres-Alemán, I. (2012). The many faces of insulin-like peptide signalling in the brain. *Nature Reviews Neuroscience, 13*(4), 225–239. https://doi.org/10.1038/nrn3209

Fetterly, T. L., Oginsky, M. F., Nieto, A. M., Alonso-Caraballo, Y., Santana-Rodriguez, Z., & Ferrario, C. R. (2021). Insulin bidirectionally alters NAc glutamatergic transmission: Interactions between insulin receptor activation, endogenous opioids, and glutamate release. *Journal of Neuroscience, 41*(11), 2360–2372. https://doi.org/10.1523/jneurosci.3216-18.2021

Figlewicz, D. P., Bennett, J. L., Aliakbari, S., Zavosh, A., & Sipols, A. J. (2008). Insulin acts at different CNS sites to decrease acute sucrose intake and sucrose self-administration in rats. *American Journal of Physiology—Regulatory, Integrative and Comparative Physiology, 295*(2), R388–R394. https://doi.org/10.1152/ajpregu.90334.2008

Figlewicz, D. P., Bennett, J. L., Naleid, A. M., Davis, C., & Grimm, J. W. (2006). Intraventricular insulin and leptin decrease sucrose self-administration in rats. *Physiology and Behavior, 89*(4), 611–616. https://doi.org/10.1016/j.physbeh.2006.07.023

Figlewicz, D. P., Evans, S. B., Murphy, J., Hoen, M., & Baskin, D. G. (2003). Expression of receptors for insulin and leptin in the ventral tegmental area/substantia nigra (VTA/SN) of the rat. *Brain Research, 964*(1), 107–115. https://doi.org/10.1016/s0006-8993(02)04087-8

Figlewicz, D. P., Szot, P., Chavez, M., Woods, S. C., & Veith, R. C. (1994). Intraventricular insulin increases dopamine transporter mRNA in rat VTA/substantia nigra. *Brain Research, 644*(2), 331–334. https://doi.org/10.1016/0006-8993(94)91698-5

Gómez-Ilescas, A., & Silveira, P. P. (2025). Early adversity and the co-morbidity between metabolic disease and psychopathology. *Journal of Physiology (Cambridge)*. https://doi.org/10.1113/jp285927

Grillo, C. A., Piroli, G. G., Lawrence, R. C., Wrighten, S. A., Green, A. J., Wilson, S. P., … Reagan, L. P. (2015). Hippocampal insulin resistance impairs spatial learning and synaptic plasticity. *Diabetes, 64*(11), 3927–3936. https://doi.org/10.2337/db15-0596

Hübel, C., Herle, M., Santos Ferreira, D. L., Abdulkadir, M., Bryant-Waugh, R., Loos, R. J. F., … Micali, N. (2021). Childhood overeating is associated with adverse cardiometabolic and inflammatory profiles in adolescence. *Scientific Reports, 11*(1), 12478. https://doi.org/10.1038/s41598-021-90644-2

Hari Dass, S. A., McCracken, K., Pokhvisneva, I., Chen, L. M., Garg, E., Nguyen, T. T. T., … Silveira, P. P. (2019). A biologically-informed polygenic score identifies endophenotypes and clinical conditions associated with the insulin receptor function on specific brain regions. *EBioMedicine, 42*, 188–202. https://doi.org/10.1016/j.ebiom.2019.03.051

Jaramillo-Ospina, A. M., Dalle Molle, R., Patel, S., Kelly, S., Pokhvisneva, I., de Weerth, C., & Silveira, P. P. (2025). A mesocorticolimbic insulin receptor gene co-expression network moderates the association between early life adversity and food approach eating behaviour style in childhood. *Appetite, 204*, 107762. https://doi.org/10.1016/j.appet.2024.107762

Jaramillo-Ospina, A. M., Elgbeili, G., Patel, S., Pokhvisneva, I., & Silveira, P. P. (2025). Brain insulin receptor gene network shapes risk for metabolic disease after early-life stress in women. *Communications Biology, 8*(1), 1372. https://doi.org/10.1038/s42003-025-08750-0

Kathiresan, S., & Srivastava, D. (2012). Genetics of human cardiovascular disease. *Cell, 148*(6), 1242–1257. https://doi.org/10.1016/j.cell.2012.03.001

Kim, M.-S., Pak, Y. K., Jang, P.-G., Namkoong, C., Choi, Y.-S., Won, J.-C., … Lee, K.-U. (2006). Role of hypothalamic Foxo1 in the regulation of food intake and energy homeostasis. *Nature Neuroscience, 9* (7), 901–906. https://doi.org/10.1038/nn1731

Kleinridders, A., Cai, W., Cappellucci, L., Ghazarian, A., Collins, W. R., Vienberg, S. G., … Kahn, C. R. (2015). Insulin resistance in brain alters dopamine turnover and causes behavioral disorders. *Proceedings of the National Academy of Sciences, 112*(11), 3463–3468. https://doi.org/10.1073/pnas.1500877112

Kozak, K., Lucatch, A. M., Lowe, D. J. E., Balodis, I. M., MacKillop, J., & George, T. P. (2019). The neurobiology of impulsivity and substance use disorders: Implications for treatment. *Annals of the New York Academy of Sciences, 1451*(1), 71–91. https://doi.org/10.1111/nyas.13977

Kuwabara, T., Kagalwala, M. N., Onuma, Y., Ito, Y., Warashina, M., Terashima, K., … Asashima, M. (2011). Insulin biosynthesis in neuronal progenitors derived from adult hippocampus and the olfactory bulb. *EMBO Molecular Medicine, 3*(12), 742–754. https://doi.org/10.1002/emmm.201100177

Lawn, J. E., Ohuma, E. O., Bradley, E., Idueta, L. S., Hazel, E., Okwaraji, Y. B., … Prendergast, A. (2023). Small babies, big risks: Global estimates of prevalence and mortality for vulnerable newborns to accelerate change and improve counting. *The Lancet, 401*(10389), 1707–1719. https://doi.org/10.1016/S0140-6736(23)00522-6

LeRoith, D., Holly, J. M. P., & Forbes, B. E. (2021). Insulin-like growth factors: Ligands, binding proteins, and receptors. *Molecular Metabolism, 52*, 101245. https://doi.org/10.1016/j.molmet.2021.101245

Levy, S., Katusic, S. K., Colligan, R. C., Weaver, A. L., Killian, J. M., Voigt, R. G., & Barbaresi, W. J. (2014). Childhood ADHD and risk for substance dependence in adulthood: A longitudinal, population-based study. *PLoS One, 9*(8), e105640. https://doi.org/10.1371/journal.pone.0105640

Lind, L. (2019). Genome-wide association study of the metabolic syndrome in UK biobank. *Metabolic Syndrome and Related Disorders, 17*(10), 505–511. https://doi.org/10.1089/met.2019.0070

Loos, R. J. F., & Yeo, G. S. H. (2022). The genetics of obesity: From discovery to biology. *Nature Reviews Genetics, 23*(2), 120–133. https://doi.org/10.1038/s41576-021-00414-z

Lozovsky, D., Saller, C. F., & Kopin, I. J. (1981). Dopamine receptor binding is increased in diabetic rats. *Science, 214*(4524), 1031–1033. https://doi.org/10.1126/science.6458088

Mansur, R. B., Fries, G. R., Subramaniapillai, M., Frangou, S., De Felice, F. G., Rasgon, N., … McIntyre, R. S. (2018). Expression of dopamine signaling genes in the post-mortem brain of individuals with mental illnesses is moderated by body mass index and mediated by insulin signaling genes. *Journal of Psychiatric Research, 107*, 128–135. https://doi.org/10.1016/j.jpsychires.2018.10.020

Martin, H., Bullich, S., Guiard, B. P., & Fioramonti, X. (2021). The impact of insulin on the serotonergic system and consequences on diabetes-associated mood disorders. *Journal of Neuroendocrinology, 33*(4), e12928. https://doi.org/10.1111/jne.12928

McKay, M. T., Kilmartin, L., Meagher, A., Cannon, M., Healy, C., & Clarke, M. C. (2022). A revised and extended systematic review and meta-analysis of the relationship between childhood adversity and adult psychiatric disorder. *Journal of Psychiatric Research, 156*, 268–283. https://doi.org/10.1016/j.jpsychires.2022.10.015

Miguel, P. M., Barth, B., Batra, A., Alves, M. B., Arcego, D. M., Alberry, B., … Silveira, P. P. (2025). Insulin modulates mPFC gene expression and emotional behavior in a sex-specific manner following fetal growth restriction. *bioRxiv*. https://doi.org/10.1101/2025.08.14.669951, 2025.2008.2014.669951.

Miguel, P. M., & Silveira, P. P. (2025). Using functional genomics in the understanding of gene-environment interaction effects on mental health. In C. R. Martin, V. R. Preedy, V. B. Patel, & R. Rajendram (Eds.), *Handbook of the biology and pathology of mental disorders* (pp. 1–27). Cham: Springer Nature Switzerland.

Miller, J. A., Ding, S.-L., Sunkin, S. M., Smith, K. A., Ng, L., Szafer, A., … Lein, E. S. (2014). Transcriptional landscape of the prenatal human brain. *Nature, 508*(7495), 199–206. https://doi.org/10.1038/nature13185

Mitchell, M. R., & Potenza, M. N. (2014). Addictions and personality traits: Impulsivity and related constructs. *Current Behavioral Neuroscience Reports, 1*(1), 1–12. https://doi.org/10.1007/s40473-013-0001-y

Morton, G., Cummings, D., Baskin, D., Barsh, G., & Schwartz, M. (2006). Central nervous system control of food intake and body weight. *Nature, 443*(7109), 289–295.

Mucellini, A. B., da Fonseca, N. K.d. O., Manfro, G. G., & Silveira, P. P. (2017). Hippocampal insulin resistance and altered food decision-making as players on obesity risk. *Neuroscience and Biobehavioral Reviews, 77*, 165–176.

Mulligan, M. K., Mozhui, K., Prins, P., & Williams, R. W. (2017). GeneNetwork: A toolbox for systems genetics. *Methods in Molecular Biology, 1488*, 75–120. https://doi.org/10.1007/978-1-4939-6427-7_4

Munkholm, A., Olsen, E. M., Rask, C. U., Clemmensen, L., Rimvall, M. K., Jeppesen, P., … Skovgaard, A. M. (2016). Early predictors of eating problems in preadolescence—A prospective birth cohort study. *Journal of Adolescent Health, 58*(5), 533–542. https://doi.org/10.1016/j.jadohealth.2016.01.006

O'Dell, L. E., & Nazarian, A. (2016). Enhanced vulnerability to tobacco use in persons with diabetes: A behavioral and neurobiological framework. *Progress In Neuro-Psychopharmacology & Biological Psychiatry, 65*, 288–296. https://doi.org/10.1016/j.pnpbp.2015.06.005

Pitt, J., Wilcox, K. C., Tortelli, V., Diniz, L. P., Oliveira, M. S., Dobbins, C., … Klein, W. L. (2017). Neuroprotective astrocyte-derived insulin/insulin-like growth factor 1 stimulates endocytic processing and extracellular release of neuron-bound Aβ oligomers. *Molecular Biology of the Cell, 28*(20), 2623–2636. https://doi.org/10.1091/mbc.e17-06-0416

Pomytkin, I., Costa-Nunes, J. P., Kasatkin, V., Veniaminova, E., Demchenko, A., Lyundup, A., ... Strekalova, T. (2018). Insulin receptor in the brain: Mechanisms of activation and the role in the CNS pathology and treatment. *CNS Neuroscience and Therapeutics, 24*(9), 763–774. https://doi.org/10.1111/cns.12866

Reay, W. R., Atkins, J. R., Carr, V. J., Green, M. J., & Cairns, M. J. (2020). Pharmacological enrichment of polygenic risk for precision medicine in complex disorders. *Scientific Reports, 10*, 879. https://doi.org/10.1038/s41598-020-57795-0

Riedel, G., & Micheau, J. (2001). Function of the hippocampus in memory formation: Desperately seeking resolution. *Progress In Neuro-Psychopharmacology & Biological Psychiatry, 25*(4), 835–853. https://doi.org/10.1016/s0278-5846(01)00153-1

Rosenström, T., Torvik, F. A., Ystrom, E., Czajkowski, N. O., Gillespie, N. A., Aggen, S. H., ... Reichborn-Kjennerud, T. (2018). Prediction of alcohol use disorder using personality disorder traits: A twin study. *Addiction, 113*(1), 15–24. https://doi.org/10.1111/add.13951

Salcedo-Tello, P., Ortiz-Matamoros, A., & Arias, C. (2011). GSK3 function in the brain during development, neuronal plasticity, and neurodegeneration. *International Journal of Alzheimer's Disease, 2011*(1), 189728. https://doi.org/10.4061/2011/189728

Saller, C. F., & Chiodo, L. A. (1980). Glucose suppresses basal firing and haloperidol-induced increases in the firing rate of central dopaminergic neurons. *Science, 210*(4475), 1269–1271. https://doi.org/10.1126/science.6254155

Schoffelmeer, A. N. M., Drukarch, B., De Vries, T. J., Hogenboom, F., Schetters, D., & Pattij, T. (2011). Insulin modulates cocaine-sensitive monoamine transporter function and impulsive behavior. *Journal of Neuroscience, 31*(4), 1284–1291. https://doi.org/10.1523/jneurosci.3779-10.2011

Scudine, K. G. O., Castelo, P. M., Hoppe, J. P. M., Portella, A. K., & Silveira, P. P. (2024). Early influences on development of sensory perception and eating habits. *Advances in Nutrition, 15*(12), 100325. https://doi.org/10.1016/j.advnut.2024.100325

Selenius, J. S., Silveira, P. P., Bonsdorff, M. V., Lahti, J., Koistinen, H., Koistinen, R., ... Wasenius, N. S. (2024). Biologically informed polygenic scores for brain insulin receptor network are associated with cardiometabolic risk markers and diabetes in women. *Diabetes & Metabolism J.* https://doi.org/10.4093/dmj.2023.0039

Selenius, J. S., Silveira, P. P., Haapanen, M. J., von Bonsdorff, M., Lahti, J., Eriksson, J. G., & Wasenius, N. S. (2024). The brain insulin receptor gene network and associations with frailty index. *Age and Ageing, 53*(5). https://doi.org/10.1093/ageing/afae091

Selenius, J. S., Silveira, P. P., Salonen, M., Kautiainen, H., von Bonsdorff, M., Kajantie, E., ... Wasenius, N. S. (2021). The relationship between health-related quality of life and melancholic depressive symptoms is modified by brain insulin receptor gene network. *Scientific Reports, 11*(1), 21588. https://doi.org/10.1038/s41598-021-00631-w

Silveira, P. P., & Meaney, M. J. (2023). Examining the biological mechanisms of human mental disorders resulting from gene-environment interdependence using novel functional genomic approaches. *Neurobiology of Disease, 178*, 106008. https://doi.org/10.1016/j.nbd.2023.106008

Silveira, P. P., Pokhvisneva, I., Parent, C., Cai, S., Rema, A. S. S., Broekman, B. F., ... Meaney, M. J. (2017). Cumulative prenatal exposure to adversity reveals associations with a broad range of neurodevelopmental outcomes that are moderated by a novel, biologically informed polygenetic score based on the serotonin transporter solute carrier family C6, member 4 (SLC6A4) gene expression. *Development and Psychopathology, 29*(5), 1601–1617.

Silveira, P. P., Portella, A. K., Wang, Z., Hoppe, J. P. M., Pokhvisneva, I., & Meaney, M. (2025). Early life adversity and the comorbidity of psychopathology and cardiometabolic disorders: Insights from the UK Biobank. *BMJ Public Health, 3*(2), e003051. https://doi.org/10.1136/bmjph-2025-003051

Stoica, L., Zhu, P. J., Huang, W., Zhou, H., Kozma, S. C., & Costa-Mattioli, M. (2011). Selective pharmacogenetic inhibition of mammalian target of Rapamycin complex I (mTORC1) blocks long-term synaptic plasticity and memory storage. *Proceedings of the National Academy of Sciences of the United States of America, 108*(9), 3791–3796. https://doi.org/10.1073/pnas.1014715108

Talbot, K., Wang, H. Y., Kazi, H., Han, L. Y., Bakshi, K. P., Stucky, A., ... Arnold, S. E. (2012). Demonstrated brain insulin resistance in Alzheimer's disease patients is associated with IGF-1 resistance, IRS-1 dysregulation, and cognitive decline. *Journal of Clinical Investigation, 122*(4), 1316–1338. https://doi.org/10.1172/jci59903

Tiedemann, L. J., Schmid, S. M., Hettel, J., Giesen, K., Francke, P., Büchel, C., & Brassen, S. (2017). Central insulin modulates food valuation via mesolimbic pathways. *Nature Communications, 8*(1), 16052. https://doi.org/10.1038/ncomms16052

Torkamani, A., Wineinger, N. E., & Topol, E. J. (2018). The personal and clinical utility of polygenic risk scores. *Nature Reviews Genetics, 19*(9), 581–590. https://doi.org/10.1038/s41576-018-0018-x

Trujeque-Ramos, S., Castillo-Rolón, D., Galarraga, E., Tapia, D., Arenas-López, G., Mihailescu, S., & Hernández-López, S. (2018). Insulin regulates GABA(A) receptor-mediated tonic currents in the prefrontal cortex. *Frontiers in Neuroscience, 12*, 345. https://doi.org/10.3389/fnins.2018.00345

White, M. F. (2003). Insulin signaling in health and disease. *Science, 302*(5651), 1710–1711. https://doi.org/10.1126/science.1092952

Wray, N. R., & Goddard, M. E. (2010). Multi-locus models of genetic risk of disease. *Genome Medicine, 2*(2), 10. https://doi.org/10.1186/gm131

Yazdani, S., Jaldin-Fincati, J. R., Pereira, R. V. S., & Klip, A. (2019). Endothelial cell barriers: Transport of molecules between blood and tissues. *Traffic, 20*(6), 390–403. https://doi.org/10.1111/tra.12645

Chapter 9

Linking exosomes and neuropathy in diabetics conditions

Lamiaa A. Ahmed[1] and Khaled F. Al-Massri[2]

[1]*Pharmacology and Toxicology Department, Faculty of Pharmacy, Cairo University, Cairo, Egypt;* [2]*Pharmacy and Biotechnology Department, Faculty of Medicine and Health Sciences, University of Palestine, Gaza, Palestine*

List of abbreviations

ACTA1 actin alpha 1 (skeletal muscle)
AD-SC adipose-derived stem cell
BM-MSC bone marrow-mesenchymal stem cell
DM diabetes mellitus
DNMT1 DNA (Cytosine-5)-methyltransferase 1
DPN diabetic peripheral neuropathy
DRG dorsal root ganglion
ECH electroconductive hydrogel
ECH-Exos electroconductive hydrogel loaded with BM-MSCs-derived exosomes
ERK extracellular signal-regulated kinase
EVs extracellular vesicles
Exos exosomes
FT3 free triiodothyronine
GSH glutathione
HIF1α hypoxia-inducible factor 1-alpha
HO-1 heme oxygenase-1
MDA malondialdehyde
MEK mitogen-activated protein kinase kinase
miR-130a-3p microRNA-130a-3p
miRNA microRNA
MMP mitochondrial membrane potential
MNCV motor nerve conduction velocity
NRF2 nuclear factor erythroid 2-related factor 2
OH hydroxyvitamin
PNI peripheral nerve injury
SC Schwann cell
SC-Exos Sc-derived Exos
SIRT1 sirtuin 1
SOD superoxide dismutase
T2D type 2 diabetes
TOMM20 translocase of outer mitochondrial membrane 20
TRAF6 tumor necrosis factor receptor–associated factor 6

Introduction

Diabetes mellitus (DM), the ninth leading cause of death worldwide, has exhibited a rising incidence and prevalence in recent years (Liu et al., 2020; Ye et al., 2023). It is estimated that the burden of DM will continue to rise in the upcoming years, partly due to unfavorable changes in lifestyle, with over 7000 per 100,000 individuals affected by type 2 DM in 2030 (Khan et al., 2020; Zhu et al., 2022). Diabetes is considered to be one of the major causes of neuropathy (Wu et al., 2021). Diabetic peripheral neuropathy (DPN) is one of the most common complications of diabetes and the main cause of non-traumatic amputation, with no ideal treatment (Selvarajah et al., 2019). DPN is a disabling symptom and one of the most common complications of DM, affecting more than 50% of patients (Pop-Busui et al., 2022). Resulting from damage to distal sensory nerves, DPN manifests clinically with sensory and motor symptoms, including limb numbness, pain, tingling, hyperalgesia, impaired motor coordination, and ataxia, which is difficult to reverse. It also reduces patient quality of life, increases morbidity, and negatively affects the economic burden of the disease (Ahmed & Al-Massri, 2024; Callaghan et al., 2012; Wang et al., 2020).

Chronic inflammation, immune system upregulation, glial cell activation, antioxidant system disruption, blood flow impairment, and alteration in expression of ion channels are among the mechanisms that contribute to axonal degeneration and segmental demyelination, leading to the development of DPN (Javed et al., 2015; Wang et al., 2020). Hyperglycemia plays a key role in triggering mechanisms that lead to DPN. Chronic hyperglycemia

Hormones in Neuroscience. https://doi.org/10.1016/B978-0-443-30263-3.00009-9

activates the immune system, which induces neural cell damage through chronic inflammation, vascular dysfunction, and oxidative stress. In DM, antioxidant defense is compromised due to inflammatory status, lipid peroxidation, and the accumulation of advanced glycation end products, resulting in oxidative stress, nerve injury, and apoptosis. Additionally, blood flow impairment in DM contributes to vascular dysfunction, compromised blood-nerve barrier integrity, and subsequent ischemic nerve injury, further exacerbating neuropathic damage. Current therapeutic strategies for the management of DPN are not satisfying since available treatment options are associated with dose-limiting side effects and systemic adverse events, primarily alleviate symptoms without addressing the underlying pathology, and result in symptom reversal following discontinuation (Al-Massri et al., 2020; Coy-Dibley et al., 2025). Thus, exploring underlying pathobiological mechanisms and discovering novel treatment strategies remains crucial for DPN prevention, early diagnosis, and better management.

Exosomes (Exos) are bilayer nanovesicles originating from cellular endosomes with a 30–200 nm diameter (Wang et al., 2020). Exos are excreted by almost all cell types and carry lipid, protein, nucleic acid, and other metabolite cargoes to target cells, thus mediating cell-to-cell interactions and affecting the function of the recipient cell by regulating biological signaling pathways (Zhang et al., 2019). Although Exos were previously thought to only remove cell wastes and debris, in recent years, a growing body of evidence has revealed a key role for these structures in the pathogenesis of various disorders (Ahmed & Al-Massri, 2024). In addition, these vesicles have also emerged as promising therapeutic options due to their lower risk of immunogenicity, higher biocompatibility, lower toxicity, and ability to cross biological barriers such as microvascular networks and blood-brain barrier (Konala et al., 2016; Zhao et al., 2019).

Although emerging evidence suggests that Exos are involved in the development and treatment of diabetes and its complications, including retinopathy and nephropathy, their role in the pathogenesis, diagnosis, and treatment of diabetes-induced neuropathy remains insufficiently characterized (Chen et al., 2021; Niu et al., 2022; Zhu et al., 2022). Therefore, the present chapter aims to provide a comprehensive review of the multifaceted roles of Exos in DPN. In this chapter, we first summarized different approaches for isolation, characterization, purification, and preservation of Exos. We also demonstrated the utilization of various cells-derived Exos in DPN. Furthermore, we examined recent research focused on modified Exos and their practical implementation in peripheral neuropathy and focusing on enhancement of the therapeutic efficiency of Exos. Herein, we also identified problems and limitations that should be addressed to advance the clinical utilization of Exos in DPN.

Insulin and diabetic peripheral neuropathy

Since neurons are not sensitive to insulin, insulin does not directly regulate the activity of the central and peripheral nervous systems. Insulin is indirectly involved in the pathogenesis of DPN through regulation of hyperglycemia. Additionally, there is mounting evidence that insulin plays a direct role in development of DPN, being as it act as a neurotrophic factor that supports peripheral nerves (Agrawal et al., 2021; Grote & Wright, 2016). A study conducted by Shetter et al. (2011) demonstrated that insulin receptors are expressed by dorsal root ganglion (DRG) on the basal lamina and plasma membrane of SCs where SC dysfunction is a key component in DPN pathophysiology and insulin can impact DPN through modifying SC function.

Insulin increases nerve conduction velocity, decreases demyelination, and enhances axonal development, all of which can improve DPN (Huang et al., 2005). Ozaki et al. (2013) revealed that treatment of Wistar Bonn Kobori rats that developed diabetes on their own, with insulin can enhance the nerve conduction velocity compared to control group. Additionally, the sciatic and tibial nerves showed improvements in axonal distortion and myelin expansion in the same study (Ozaki et al., 2013). The sensory nerve conduction velocity of WBN/Kob rats was also improved after a 6-week treatment regimen with honey and insulin (Sirisha et al., 2021). Researchers observed that insulin enhanced nerve function in the sciatic nerve and improved demyelination caused by low lipoprotein lipase production in hyperglycemic Schwann cells (SCs) (Rachana et al., 2018). Previous investigations demonstrated that myelin development can be influenced by insulin, insulin receptors, and insulin resistance. Research conducted by Brussee et al. (2004) revealed that insulin can have a direct impact on DPN, where phosphatidylinositol 3-kinase-protein kinase B signaling pathway is one of the downstream signaling pathways that insulin receptor activation can use to enhance axon growth (Boucher et al., 2014; Soltoff et al., 1992). Myelin sheath development and SC differentiation are both enhanced by activating the protein kinase B signaling pathway (Ogata et al., 2004). Importantly, insulin reveals better outcomes in type 1 diabetes with DPN (Aghanoori et al., 2017). Insulin has limited outcomes on DPN in type 2 diabetes (T2D) due to the development of insulin resistance with changed phosphorylation status of insulin receptor substrate proteins, and impaired activation of axonal growth-related pathways (Grote et al., 2011; Guo et al., 2011). Thus, more research is still needed to further investigate the role of other related molecules in insulin signaling to discover new therapeutic targets or approaches to manage DPN.

Complex mechanisms contribute to the development of diabetic neuropathy including polyol, glycolysis, hexosamine, and advanced glycation end-product pathways that are activated in a chain reaction due to hyperglycemia, dyslipidemia, and insulin resistance (Cheng et al., 2024). Endoplasmic reticulum stress, mitochondrial malfunction, DNA damage, and increased levels of inflammatory factors are all outcomes of activation of these pathways, leading to amplified oxidative stress and inflammatory signaling (Baum et al., 2021; Stino et al., 2020). Understanding the detailed pathogenic mechanisms of diabetic neuropathy and developing efficient treatments are, hence, of the utmost importance.

Other hormones, endocrine tissues, and pathways

An increased risk of DPN development has been linked to a decline in peripheral thyroid hormone sensitivity in elderly patients with T2D and adequate thyroid function. One reliable indicator of DPN is free triiodothyronine (FT3)/free tetraiodothyronine ration. Furthermore, pathology involving large and tiny nerve fibers may be impacted by a decline in peripheral thyroid hormone sensitivity (Fei et al., 2025b). On the other hand, Sun et al. (2024) found that individuals with reduced levels of central and peripheral thyroid hormone sensitivity had a lower likelihood of developing DPN. Xu et al. (2025) also found that euthyroid individuals with T2D who had low FT3 are more likely to develop DPN. Reduced FT3 levels may impact peripheral neuropathy through specifically influencing the ulnar nerve's motor and sensory fiber conduction velocity and the tibial nerve's motor fiber conduction velocity (Chen et al., 2024). According to another study, thyroid-stimulating hormone can worsen oxidative stress, iron overload, and mitochondrial dysfunction via inhibiting the nuclear factor erythroid 2-related factor 2 (NRF2)/glutathione peroxidase four signaling pathway, which in turn causes ferroptosis in Scs. These results shed new light on the mechanistic relationship between thyroid dysfunction and the development of DPN (Fei et al., 2025a).

Additionally, DPN is associated with low 25-hydroxyvitamin D (25[OH]D) levels in patients with T2D. The chance of developing DPN increases when serum levels of 25(OH) D fall below 19.94 ng/mL, since 25(OH) D acts as a protective factor against DPN. According to Chen et al. (2025), regular assessments of 25(OH) D levels could help with DPN early diagnosis and prompt intervention. According to Al-Joda et al. (2024), neuropathy risk factors can be better evaluated by understanding the correlations between age, HbA1C, parathyroid hormone, C-peptide, and vitamin D2 levels (Al-Joda et al., 2024). Other factors can contribute to the development of neuropathy in diabetes, and thus the previously mentioned biomarkers can have limited diagnostic usefulness in this regard. Further research is required to elucidate the detailed intricate mechanisms underlying neuropathy and to discover more potential diagnostic markers for early intervention and management.

Applications of various cells-derived exosomes in DPN

Exos derived from diverse cell types, including stem cells, SCs, plasma, and keratinocytes as shown in Fig. 9.1, hold significant therapeutic potential for DPN by promoting nerve regeneration, providing neuroprotection, and offering metabolic support (Tajabadi et al., 2025).

Stem cells-derived exosomes

Bone marrow-mesenchymal stem cell (BM-MSC) is a type of stem cell with multidifferentiation ability. The cell is considered a desirable seed cell in tissue engineering due to its low immunogenicity, easy obtainability, and strong self-renewal ability (Ahmed & Al-Massri, 2024). Meanwhile, it has been found that Exos secreted by the stem cells can regulate the function of tissue cells and effectively repair the damaged tissues including peripheral nerve injury and spinal cord (Wang, Lu, et al., 2024; Wang, Zhao, et al., 2024; Yu et al., 2021).

BM-MSC-Exo can achieve marked therapeutic effects in treating DM complications, which may be related to its ability to lower blood glucose, improve renal function, and inhibit the Janus kinase 2/signal transducer and activator of transcription 3 pathway activity (Ahmed & Al-Massri, 2024; Wang et al., 2021). Another study demonstrated that the delivery of MSC-Exo reduced the inflammatory response and promoted neurovascular remodeling and functional recovery of DPN in diabetic mice (Fan et al., 2020).

On the other hand, adipose-derived stem cell (AD–SC)–Exos were found to improve high glucose-induced oxidative stress injury and accelerate diabetic foot ulcer wound healing via regulating Kelch-like ECH-associated protein 1 (Keap1)/Nrf2 axis (Tian et al., 2025). AD–SC–Exos promoted the regeneration of the myelin sheath by inhibiting SCs autophagy in rat model of sciatic nerve injury. This study revealed that AD–SC–Exos promoted the regeneration of myelin sheath by modulating the autophagy of injured SCs via miRNA-26b downregulation of karyopherin subunit alpha 2 (Kpna2) (Yin et al., 2021). Cao et al. (2014) assessed the mechanisms of human AD–SC–Exos in DPN,

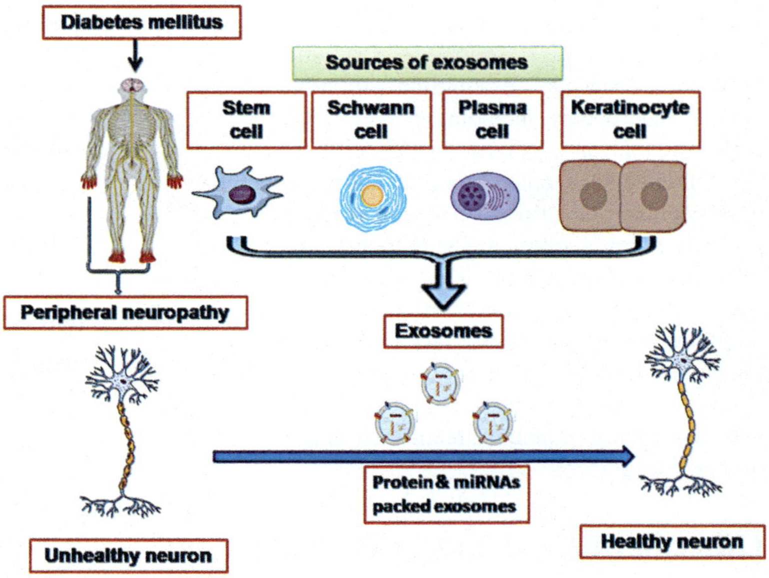

FIGURE 9.1 Therapeutic potential of exosomes in diabetic neuropathy. Exosomes derived from stem cells, Schwann cells, plasma, and keratinocytes exhibit therapeutic effects in diabetic peripheral neuropathy. These exosomes contribute to nerve repair, reduce inflammation, and promote functional recovery, highlighting their potential as a treatment strategy for neuropathic complications in diabetes.

demonstrating their role in modulating inflammatory and apoptotic pathways, significantly improving nerve function, and alleviating neuropathic pain.

Schwann cells-derived exosomes

Myelin-forming SCs, the most abundant cells in the peripheral nervous system, interact with axons and blood vessels to regulate peripheral nerve function (Corfas et al., 2004; Feldman et al., 2017; Lehmann & Höke, 2010). Dysfunction of these interactions contributes to the development of DPN (Gonçalves et al., 2017). Patients with T2D have shown to exhibit increased susceptibility to neuropathy when they have polymorphisms in miR-146a, miR-27a, and miR-124a genes (Wang et al., 2015). Treatment of diabetic mice with miR-146a mimics also reduced DPN by targeting proinflammatory genes, interleukin-1 receptor–associated kinase, and tumor necrosis factor receptor–associated factor 6 (TRAF6) (Liu et al., 2017). Another study in type 1 diabetic mice with DPN showed significantly reduced levels of let-7i and miR-27a, among other miRs (Kobayashi & Zochodne, 2018). On the other hand, intranasal administration of exogenous let-7i miRNA substantially has shown to decrease DPN (Cheng et al., 2015).

Importantly, Exos derived from high-glucose–challenged SCs promoted the development of DPN in T2D mice (Jia et al., 2018), while SC-Exos substantially enhanced the regeneration of injured peripheral axons in a rat model of sciatic nerve injury (Lopez-Verrilli et al., 2013). Treatment of T2D mice with SC-Exos robustly ameliorated DPN by reducing peripheral nerve damage (Wang et al., 2020). Zhu et al. also examined the neuroprotective role of SC-Exos in mitigating apoptosis of DRG neurons through regulating the inositol-requiring enzyme 1 alpha pathway, demonstrating their utility in preventing neuronal damage (Zhu et al., 2021). Notably, administration of SC-Exos restored sciatic nerve function by replenishing key miRNAs, promoting myelination, reducing inflammatory markers, and improving neuropathic symptoms (Wang, Lu, et al., 2024; Wang, Zhao, et al., 2024).

Plasma-derived exosomes

Obviously, miRNA expression in plasma-derived Exos of healthy rats was significantly different from that of age-matched DPN rats. By injection of plasma-derived Exos into DPN rats, the thermal sensitivity and motor ability of rats were increased, and the nerve conduction speed was accelerated in addition to myelin regeneration of the sciatic nerve, increased distal local blood perfusion, and enhanced neuromuscular junction and muscle spindle innervation (Li et al., 2023). Specifically, miR-20b-3p was enriched in Exos of healthy plasma and was found to be re-upregulated in the sciatic nerve of DPN rats after plasma-derived Exos treatment (Li et al., 2023). Moreover, miR-20b-3p agomir improved DPN symptoms to a level similar to plasma-derived Exos with alleviation of autophagy impairment induced by high glucose in SCs miR-20b-3p has demonstrated to reduce the amount of phosphorylated signal transducer and activator of transcription 3 levels, which negatively regulated autophagy processes (Li et al., 2023).

Keratinocyte-derived exosomes

Keratinocytes, the most abundant epidermal cell type, are closely positioned to cutaneous nerve terminals, suggesting the possibility of bidirectional communication (Coy-Dibley et al., 2025). A recent study revealed a potential novel mode of communication between keratinocytes and DRG neurons in the skin, revealing a possible role of keratinocytes in contributing to axonal degeneration that underlies neuropathic pain in DPN (Coy-Dibley et al., 2025). This study suggested the role of keratinocyte-derived extracellular vesicles (EVs) in the communication between keratinocytes and DRG neurons (Coy-Dibley et al., 2025), recommending a novel strategy for treating DPN by focusing on the skin rather than the entire body and overcoming challenges with current DPN treatments, including their systemic off-target effects (Attal & Bouhassira, 2021; Finnerup et al., 2015; Shinu et al., 2022). Hence, methods could be utilized to develop and deliver improved topical treatments for DPN and other nervous system diseases (Coy-Dibley et al., 2025).

Collectively, Exos derived from various cell types hold significant therapeutic promise for DPN by promoting nerve regeneration, neuroprotection, and metabolic support (Tajabadi et al., 2025). Among them, SC-Exos have been the most extensively studied, demonstrating robust efficacy in nerve repair and neuropathic symptom relief (Lopez-Verrilli et al., 2013; Wang et al., 2020; Zhu et al., 2021). Similarly, MSC-Exos exhibit potent antiinflammatory and neuroprotective properties, with engineered variants further augmenting their therapeutic potential (Mathew et al., 2023). AD–SC–Exos also contribute to neuroprotection and nerve regeneration by modulating apoptotic and inflammatory pathways (Zhou et al., 2022). While direct comparative studies remain limited, SC-Exos and MSC-Exos currently exhibit the most promising therapeutic profiles for DPN treatment.

Modified Exos and their practical implementation in DPN

Exos represent a promising emerging therapeutic strategy for diabetic neuropathy, offering minimal immunogenicity, the ability to cross biological barriers, and the potential for precise delivery of therapeutic agents. However, studies to date have been mostly preclinical, primarily involving rodent models where their long-term effects on humans are not yet well understood (Yorek, 2022). Regarding Exo engineering approaches, the functional molecule can be either displayed on the surface of Exos or inserted into its lumen using two approaches either parental cell-based or direct Exo engineering as demonstrated in Fig. 9.2. Importantly, genetic engineering is used for therapeutic molecules loading into the lumen or expressing them on the surface of Exos in parental cell-based engineering approach. On the other side, several chemical and mechanical methods including bioconjugation, extrusion, sonication, and electroporation in the postisolation Exos engineering approach (Jafari et al., 2020).

Engineered Exos have garnered attention for their ability to enhance targeting specificity, therapeutic longevity, and safety profile. Compared to native Exos, engineered variants can be modified to improve therapeutic loading efficiency and optimize tissue-specific targeting, potentially offering greater precision in treatment (Sun et al., 2023). However, native Exos derived from the patient's cells show a lower risk of immune response. The choice between engineered and native Exos depends on treatment objectives and balancing factors such as targeting accuracy, safety, and long-term therapeutic effects (Tajabadi et al., 2025). Collectively, these findings highlight the transformative potential of Exos-based therapies in diabetic neuropathy, underscoring the necessity for further research and clinical implementation.

Engineered Exos and nanoparticles offer innovative therapeutic approaches for diabetic neuropathy management as shown in Table 9.1. These studies have utilised Exos from several sources, including Schwann-like cells derived from BM-MSCs, AD-SCs, and astrocytes. Recent advancements have pioneered engineered Exos and nanoparticle-based strategies to enhance therapeutic outcomes (Abbasi et al., 2025). Singh et al. (2021) demonstrated that Exos from BM-MSCs, when stimulated electrically, effectively targeted the sciatic nerve and significantly ameliorated DPN. Additionally, combining polymeric nerve guidance conduits with Exos has demonstrated improved nerve regeneration in diabetic neuropathy models (Singh et al., 2021). Furthermore, Luo et al. (2019) investigated the efficacy of nanoparticle-microRNA-146a-5p polyplexes, which were revealed to reduce sciatic nerve inflammation and apoptosis, underscoring the potential of nanoparticle-Exos integration in precision therapeutics. BM-MSC-Exos in combination with polypyrrole nanoparticles encapsulating liposomes, also showed marked neuroprotective and antioxidant potential, providing a potential effective approach for managing DPN (Singh et al., 2021).

Additionally, Kasimu et al. (2022) evaluated Exos enriched with miR-125a-5p in astrocytes, which targeted the TRAF6 pathway and significantly attenuated peripheral neuropathy in mice with T2D, further advancing the benefits of miRNA-mediated mechanisms. Shan et al. (2022) also reported that Exos containing sirtuin 1 (SIRT1) from BM-MSCs preserved mitochondrial function and attenuated symptoms of peripheral neuropathy by sustaining cellular energy homeostasis. Furthermore, Chen et al. (2023) investigated adipose-derived stem cell Exos delivering miR-130a-3p, which demonstrated neuroprotective effects by regulating the DNA methyltransferase 1/NRF2/hypoxia inducible factor 1α/skeletal muscle actin alpha 1 axis, presenting a potential therapeutic strategy.

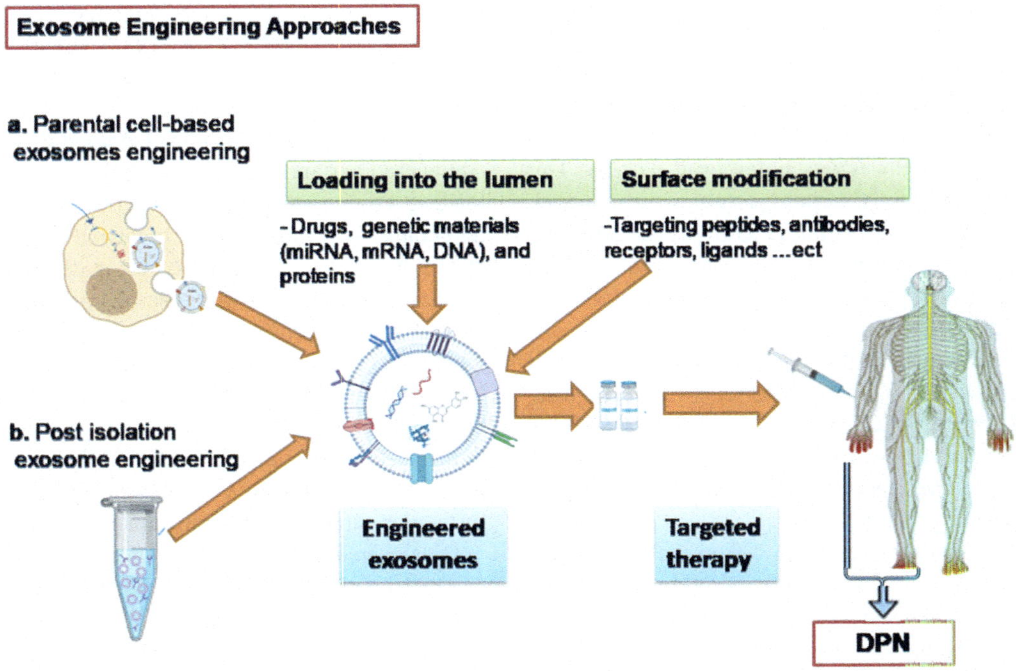

FIGURE 9.2 Exosome engineering strategies. Therapeutic molecules can be loaded into exosomes either by modifying the parental cells genetically before exosome production or through direct post-isolation methods, including chemical and mechanical approaches. These strategies enhance the therapeutic potential of exosomes for targeted delivery.

Compared to native Exos, treatment of DPN with engineered mesenchymal stromal cell-derived Exos enriched with microRNA-146a provided amplified therapeutic efficacy with significant increases in nerve conduction velocity, reduced sensitivity to thermal and mechanical stimuli, enhanced neurovascular function, and reduced inflammatory monocytes and activated macrophages (Fan et al., 2021). Yang et al. (2023) also applied electroconductive hydrogels loaded with Exos derived from BM-SCs to treat diabetic peripheral nerve damage with promotion of SCs adhesion and migration, regulation of M2 macrophage polarization, and enhancement of functional recoveryand myelinated axon regeneration through controlling the mitogen-activated protein kinase/extracellular signal-regulated kinase (MEK/ERK) pathway (Yang et al., 2023). Moreover, implantation of Exos derived from human umbilical cord MSC-encapsulated Gelfoam improved spinal cord injury-induced motor dysfunction and neuropathic pain, possibly through its capabilities to improve nerve regeneration and remyelination in addition to its antiinflammatory and antiapoptotic activities (Poongodi et al., 2024). Overall, Exos could serve as a promising therapeutic alternative for conventional therapies in DPN management.

Challenges and future perspectives

Critical gaps exist despite substantial advancements in utilizing Exo in DN and understanding some of its mechanisms or functions. Exos are pivotal in neuroinflammation, oxidative stress, and apoptosis, all contributing to DPN pathogenesis (Mosquera-Heredia et al., 2021). However, the exact mechanisms underlying these processes are not yet fully elucidated (Liu et al., 2019). Future studies should examine more oligodendrocyte dysfunction and explore the therapeutic potential of SC-derived Exos, particularly their small RNA components where miRNA alterations during aging have impact on inflammation, neurovascular remodeling, and functional recovery in DPN which necessitate further investigations (Tajabadi et al., 2025).

Importantly, a major obstacle to clinical implementation of Exo-based diagnostics is the limitations of current detection and analysis technologies. The process is often hindered by time-consuming methodologies, challenges ensuring consistent quality control, and substantial variability across laboratories. Establishing a comprehensive library of exosomal patterns associated with certain diseases is crucial for reducing diagnostic inaccuracies (Lai et al., 2022). Moreover, the scalability and reproducibility of Exos extraction techniques are insufficient. Emerging technologies, such as microfluidics-based isolation, offer promising solutions with high sensitivity, improved recovery rates, and low sample requirements, potentially enabling personalized diagnostics and therapeutic interventions (Oushyani Roudsari et al., 2024). Standardized protocols and scalable production techniques are also essential for integrating Exos into clinical research and practice (Chen & Li, 2025). Other ethical challenges

TABLE 9.1 Examples of the therapeutic efficacy of engineered MSC-EXOs in diabetic peripheral neuropathy.

No.	*Title of study*	*Method*	*Outcomes*	*References*
1.	Treatment of diabetic peripheral neuropathy with engineered mesenchymal stromal cell-derived exosomes enriched with microRNA-146a provide amplified therapeutic efficacy	Engineered MSC-Exos were loaded with miR-146a (exo-146a) in diabetic (db/db) mice with DPN.	Significant increase in nerve conduction velocity, reduced sensitivity to thermal and mechanical stimuli, enhanced neurovascular function, and reduced inflammatory monocytes and activated macrophages with promotion of M2 polarization and decreased adhesion molecule expression in endothelial cells.	Fan et al. (2021)
2.	Transplantation of engineered exosomes derived from bone marrow mesenchymal stromal cells ameliorate diabetic peripheral neuropathy under electrical stimulation	Exos derived from bone marrow mesenchymal stromal cells were fused with polypyrrole nanoparticles containing liposomes to deliver both cues in a single delivery vehicle.	Significant improvement of MNCV and compound muscle action potential with enhanced nerve regeneration, supported muscle recovery, and modulated gene expression related to nerve repair.	Singh et al. (2021)
3.	SIRT1-enriched exosomes derived from bone marrow mesenchymal stromal cells alleviate peripheral neuropathy via conserving mitochondrial function	Exos were prepared from bone marrow mesenchymal stromal cells that had been transduced with SIRT1 lentivirus (exo-SIRT1).	Significant improvement of nerve conduction velocity, morphological, and behavioral changes with elevated TOMM20 and Nrf2/HO-1 expression, reduced MDA levels, increased GSH and SOD activity, and increased MMP.	Shan et al. (2022)
4.	Exosomes-loaded electroconductive nerve dressing for nerve regeneration and pain relief against diabetic peripheral nerve injury	An electroconductive hydrogels (ECH) was loaded with BM-SCs-Exos (ECH-Exos) as electroconductive nerve dressing to treat diabetic PNI and achieve functional recovery and pain relief while avoiding the cumbersome implantation process.	Enhancement of myelinated axonal regrowth and relief of inflammatory pain through regulation of the MEK/ERK pathway, thereby improvement of muscle denervation atrophy and facilitation of functional restoration.	Yang et al. (2023)
5.	Delivery of miR-130a-3p through AD–SC–secreted EVs protects against diabetic peripheral neuropathy via DNMT1/NRF2/HIF1α/ACTA1 axis	miR-130a-3p was delivered to AD-SCs and thus their derived EVs, inhibiting SCs apoptosis and promoting their proliferation under high-glucose environment.	Promotion of angiogenesis through activated NRF2/HIF1α/ACTA11 axis, providing a potential treatment against DPN.	Chen et al. (2023)

ACTA1, Actin alpha 1 (Skeletal Muscle); AD-SCs, Adipose-derived stem cells; DPN, Diabetic peripheral neuropathy; DNMT1, DNA (Cytosine-5)-methyltransferase 1; ECH, Electroconductive hydrogel; ECH-Exos, Electroconductive hydrogel loaded with BMSCs-derived exosomes; Exos, Exosomes; ERK, Extracellular signal-regulated kinase; EVs, Extracellular vesicles; GSH, Glutathione; HO-1, Heme oxygenase-1; HIF1α, Hypoxia-inducible factor 1-alpha; MDA, Malondialdehyde; miR-130a-3p, microRNA-130a-3p; MMP, Mitochondrial membrane potential; MEK, Mitogen-activated protein kinase kinase; MNCV, Motor nerve conduction velocity; NRF2, Nuclear factor erythroid 2–related factor 2; PNI, Peripheral nerve injury; SC, Schwann cell; SIRT1, Sirtuin 1; SOD, Superoxide dismutase; TOMM20, Translocase of outer mitochondrial membrane 20.

remain, including concerns about stem cell sourcing, patient safety, equitable access to treatment, and long-term monitoring. Addressing these issues requires transparent collaboration among researchers, clinicians, ethicists, and regulatory authorities to safeguard patient welfare while advancing research and adhering to rigorous ethical standards (Bosholm et al., 2023).

The previously mentioned findings underscore the potential of Exos in treating neurological diseases including DPN. Advancing Exo-based therapeutics requires further investigation into their molecular mechanisms, refinement of isolation techniques, and targeted engineering to enhance specificity and efficacy (Sadeghi et al., 2023). Importantly, gene therapy and genomic profiling innovations could further improve Exos utilization both in diagnostics and treatments. However, rigorous clinical trials are first essential to ensure safety, large-scale applicability, and regulatory approval. Interestingly, Exo-based therapies are well-suited for precision medicine, leveraging genetic and environmental factors to provide tailored interventions (Minocha & Sardana, 2023).

Conclusion

Exos have emerged as critical players in DPN pathogenesis, diagnosis, and treatment, holding substantial promise for the clinical management of this debilitating condition. By mediating intercellular communication and modulating key molecular pathways, Exos contribute to the progression of DPN and serve as valuable biomarkers for early detection and monitoring. Importantly, their therapeutic applications are equally promising, with Exo-based strategies demonstrating significant neuroprotection and regenerative potential through axonal repair and modulation of neuroinflammation. Despite these advancements, several barriers, including the standardization of isolation techniques, scalability of production, and ethical considerations, still persist as obstacles for successful clinical translation. Future research should prioritize optimizing Exos engineering, enhancing delivery specificity, and conducting robust clinical trials to maximize the therapeutic potential of Exos in precision medicine for DPN.

Mini-dictionary of terms

- Exosome. Tiny membrane-bound vesicles (30–150 nm) released by cells that carry proteins, lipids, and RNAs to communicate with other cells and influence their behavior.
- Keratinocytes. The predominant cell type in the epidermis responsible for forming the skin barrier and participate in wound healing.
- Peripheral Neuropathy. A disease or degenerative state of the peripheral nerves in which motor, sensory, or vasomotor nerve fibers may be affected and is marked by muscle weakness and atrophy, pain, and numbness
- MicroRNA. A short segment of RNA that suppresses gene expression by binding to complementary segments of messenger RNA and interfering with the formation of proteins during translation.
- Insulin Resistance. Reduced sensitivity to insulin which results in decreased activity of the body's insulin-dependent processes such as glucose uptake, lipolysis, and inhibition of glucose production by the liver.
- Schwann Cells. The principal glial cells of the peripheral nervous system that insulate axons by forming a protective myelin sheath around larger ones, increasing nerve signal conduction speed.

Summary points

- Insulin supports nerve growth and myelination and its signaling impairment contributes to DPN.
- Low thyroid hormones and vitamin D levels may increase DPN risk with other factors.
- Exosomes offer a novel approach for diagnosis and management of DPN
- Exosomes alleviate neurovascular dysfunction and axonal demyelination
- Exosomes mediate antiapoptotic, antiremodeling, and antiinflammatory actions
- Utilization of modified or engineered exosomes enhance their actions compared to native ones

Highlights

- Exosomes offer a novel approach for diagnosis and management of DPN
- Exosomes alleviate neurovascular dysfunction and axonal demyelination
- Exosomes mediate antiapoptotic, antiremodeling, and antiinflammatory actions
- Utilization of modified or engineered exosomes enhance their actions compared to native ones

Acknowledgments

Everyone who contributed to the manuscript was placed as an author in the chapter.

Funding

This research received no specific grant from public, commercial, or not-for-profit funding agencies.

Conflict of interest

The authors declare no conflict of interest.

Ethical statement

No animal experiments were carried out for this chapter.

References

Abbasi, R., Alamdari-Mahd, G., Maleki-Kakelar, H., Momen-Mesgin, R., Ahmadi, M., Sharafkhani, M., & Rezaie, J. (2025). Recent advances in the application of engineered exosomes from mesenchymal stem cells for regenerative medicine. *European Journal of Pharmacology, 989*, 177236.

Aghanoori, M. R., Smith, D. R., Roy Chowdhury, S., Sabbir, M. G., Calcutt, N. A., & Fernyhough, P. (2017). Insulin prevents aberrant mitochondrial phenotype in sensory neurons of type 1 diabetic rats. *Experimental Neurology, 297*, 148–157.

Agrawal, R., Reno, C. M., Sharma, S., Christensen, C., Huang, Y., & Fisher, S. J. (2021). Insulin action in the brain regulates both central and peripheral functions. *American Journal of Physiology. Endocrinology and Metabolism, 321*, E156–E163.

Ahmed, L. A., & Al-Massri, K. F. (2024). Exploring the role of mesenchymal stem cell-derived exosomes in diabetic and chemotherapy-induced peripheral neuropathy. *Molecular Neurobiology, 61*, 5916–5927.

Al-Joda, B. M. S., Al-Hindy, H. A. A. M., & Mousa, M. J. (2024). Exploring the role of vitamin D2, parathyroid hormone, and C-peptides as biomarkers in diabetic neuropathy development. *Medical Journal of Babylon, 21*, 438–443.

Al-Massri, K. F., Ahmed, L. A., & El-Abhar, H. S. (2020). Mesenchymal stem cells in chemotherapy-induced peripheral neuropathy: A new challenging approach that requires further investigations. *Journal of Tissue Engineering and Regenerative Medicine, 14*, 108–122.

Attal, N., & Bouhassira, D. (2021). Advances in the treatment of neuropathic pain. *Current Opinion in Neurology, 34*, 631–637.

Baum, P., Toyka, K. V., Blüher, M., Kosacka, J., & Nowicki, M. (2021). Inflammatory mechanisms in the pathophysiology of diabetic peripheral neuropathy (dn)-new aspects. *International Journal of Molecular Sciences, 22*, 10835.

Bosholm, C. C., Zhu, H., Yu, P., Cheng, K., Murphy, S. V., McNutt, P. M., & Zhang, Y. (2023). Therapeutic benefits of stem cells and exosomes for sulfur-mustard-induced tissue damage. *International Journal of Molecular Sciences, 24*, 9947.

Boucher, J., Kleinridders, A., & Kahn, C. R. (2014). Insulin receptor signaling in normal and insulin-resistant states. *Cold Spring Harbor Perspectives in Biology, 6*, a009191.

Brussee, V., Cunningham, F. A., & Zochodne, D. W. (2004). Direct insulin signaling of neurons reverses diabetic neuropathy. *Diabetes, 53*, 1824–1830.

Callaghan, B. C., Cheng, H. T., Stables, C. L., Smith, A. L., & Feldman, E. L. (2012). Diabetic neuropathy: Clinical manifestations and current treatments. *The Lancet Neurology, 11*, 521–534.

Cao, T., Hao, T., Xiao, D., Zhang, W. F., Ji, P., Jia, Y. H., … Tao, K. (2024). Effect and mechanism of human adipose-derived stem cell exosomes on diabetic peripheral neuropathy. *Zhonghua Shao Shang yu Chuang Mian xiu fu za zhi, 40*, 240–248.

Chen, S., Chen, J., Zheng, S., Yan, Y., Huang, M., & Chen, N. (2025). Correlation between serum 25-hydroxyvitamin D levels and peripheral neuropathy in type 2 diabetes mellitus. *Frontiers in Endocrinology, 16*, 1541388.

Chen, H., & Li, Q. (2025). Recent advances in scalable exosome production: Challenges and innovations. *Chinese Journal of Plastic and Reconstructive Surgery, 7*, 149–163.

Chen, J., Li, G., Liu, X., Chen, K., Wang, Y., Qin, J., & Yang, F. (2023). Delivery of mir-130a-3p through adipose-derived stem cell-secreted EVS protects against diabetic peripheral neuropathy via dnmt1/nrf2/hif1α/acta1 axis. *Molecular Neurobiology, 60*, 3678–3694.

Chen, Y., Sun, L., Chen, M., Zhang, H., Song, B., Wang, H., Jiang, A., Zhang, L., Li, S., Wang, J., Wang, W., & Zhang, H. (2024). Lower free triiodothyronine is a risk factor of diabetic peripheral neuropathy in patients with type 2 diabetes mellitus. *Diabetes, Metabolic Syndrome and Obesity, 17*, 4407–4415.

Chen, J., Zhang, Q., Liu, D., & Liu, Z. (2021). Exosomes: Advances, development and potential therapeutic strategies in diabetic nephropathy. *Metabolism: Clinical and Experimental, 122*, 154834.

Cheng, Y., Chen, Y., Li, K., Liu, S., Pang, C., Gao, L., Xie, J., Wenjing, L. V., Yu, H., & Deng, B. (2024). How inflammation dictates diabetic peripheral neuropathy: An enlightening review. *CNS Neuroscience and Therapeutics, 30*, e14477.

Cheng, C., Kobayashi, M., Martinez, J. A., Ng, H., Moser, J. J., Wang, X., Singh, V., Fritzler, M. J., & Zochodne, D. W. (2015). Evidence for epigenetic regulation of gene expression and function in chronic experimental diabetic neuropathy. *Journal of Neuropathology and Experimental Neurology, 74*, 804–817.

Corfas, G., Velardez, M. O., Ko, C. P., Ratner, N., & Peles, E. (2004). Mechanisms and roles of axon-Schwann cell interactions. *Journal of Neuroscience, 24*, 9250–9260.

Coy-Dibley, J., Jayaraj, N. D., Ren, D., Pacifico, P., Belmadani, A., Wang, Y.-Z., Gebis, K. K., Savas, J. N., Paller, A. S., & Miller, R. J. (2025). Keratinocyte-derived extracellular vesicles in painful diabetic neuropathy. *Neurobiology of Pain, 17*, 100176.

Fan, B., Chopp, M., Zhang, Z. G., & Liu, X. S. (2021). Treatment of diabetic peripheral neuropathy with engineered mesenchymal stromal cell-derived exosomes enriched with microRNA-146a provide amplified therapeutic efficacy. *Experimental Neurology, 341*, 113694.

Fan, B., Li, C., Szalad, A., Wang, L., Pan, W., Zhang, R., Chopp, M., Zhang, Z. G., & Liu, X. S. (2020). Mesenchymal stromal cell-derived exosomes ameliorate peripheral neuropathy in a mouse model of diabetes. *Diabetologia, 63*, 431–443.

Fei, S., Guo, L., & Pan, Q. (2025a). 248-OR: Thyroid-stimulating hormone aggravates ferroptosis in Schwann cells via suppression of the nrf2/gpx4 pathway—Implications for diabetic peripheral neuropathy. *Diabetes, 74*, 248–OR.

Fei, S. J., Luo, J. Y., Wang, W. H., Guo, L. X., & Pan, Q. (2025b). Correlation between diabetic peripheral neuropathy and thyroid hormone sensitivity in elderly patients with type 2 diabetes mellitus. *World Journal of Diabetes, 16*, 98897.

Feldman, E. L., Nave, K. A., Jensen, T. S., & Bennett, D. L. H. (2017). New horizons in diabetic neuropathy: Mechanisms, bioenergetics, and pain. *Neuron, 93*, 1296–1313.

Finnerup, N. B., Attal, N., Haroutounian, S., McNicol, E., Baron, R., Dworkin, R. H., Gilron, I., Haanpää, M., Hansson, P., Jensen, T. S., Kamerman, P. R., Lund, K., Moore, A., Raja, S. N., Rice, A. S., Rowbotham, M., Sena, E., Siddall, P., Smith, B. H., & Wallace, M. (2015). Pharmacotherapy for neuropathic pain in adults: A systematic review and meta-analysis. *The Lancet Neurology, 14*, 162–173.

Gonçalves, N. P., Vægter, C. B., Andersen, H., Østergaard, L., Calcutt, N. A., & Jensen, T. S. (2017). Schwann cell interactions with axons and microvessels in diabetic neuropathy. *Nature Reviews Neurology, 13*, 135–147.

Grote, C. W., Morris, J. K., Ryals, J. M., Geiger, P. C., & Wright, D. E. (2011). Insulin receptor substrate 2 expression and involvement in neuronal insulin resistance in diabetic neuropathy. *Experimental Diabetes Research, 2011*, 212571.

Grote, C. W., & Wright, D. E. (2016). A role for insulin in diabetic neuropathy. *Frontiers in Neuroscience, 10*, 581.

Guo, G., Kan, M., Martinez, J. A., & Zochodne, D. W. (2011). Local insulin and the rapid regrowth of diabetic epidermal axons. *Neurobiology of Disease, 43*, 414–421.

Huang, T. J., Verkhratsky, A., & Fernyhough, P. (2005). Insulin enhances mitochondrial inner membrane potential and increases ATP levels through phosphoinositide 3-kinase in adult sensory neurons. *Molecular and Cellular Neuroscience, 28*, 42–54.

Jafari, D., Shajari, S., Jafari, R., Mardi, N., Gomari, H., Ganji, F., Forouzandeh Moghadam, M., & Samadikuchaksaraei, A. (2020). Designer exosomes: A new platform for biotechnology therapeutics. *BioDrugs, 34*, 567–586.

Javed, S., Petropoulos, I. N., Alam, U., & Malik, R. A. (2015). Treatment of painful diabetic neuropathy. *Therapeutic Advances in Chronic Disease, 6*, 15–28.

Jia, L., Chopp, M., Wang, L., Lu, X., Szalad, A., & Zhang, Z. G. (2018). Exosomes derived from high-glucose–stimulated Schwann cells promote development of diabetic peripheral neuropathy. *The FASEB Journal, 32*, 6911.

Kasimu, A., Apizi, X., Talifujiang, D., Ma, X., Fang, L., & Zhou, X. (2022). miR-125a-5p in astrocytes attenuates peripheral neuropathy in type 2 diabetic mice through targeting TRAF6. *Endocrinologia, Diabetes y Nutricion, 69*, 43–51.

Khan, M. A. B., Hashim, M. J., King, J. K., Govender, R. D., Mustafa, H., & Al Kaabi, J. (2020). Epidemiology of type 2 diabetes—Global burden of disease and forecasted trends. *Journal of Epidemiology and Global Health, 10*, 107–111.

Kobayashi, M., & Zochodne, D. W. (2018). Diabetic neuropathy and the sensory neuron: New aspects of pathogenesis and their treatment implications. *Journal of Diabetes Investigation, 9*, 1239–1254.

Konala, V. B., Mamidi, M. K., Bhonde, R., Das, A. K., Pochampally, R., & Pal, R. (2016). The current landscape of the mesenchymal stromal cell secretome: A new paradigm for cell-free regeneration. *Cytotherapy, 18*, 13–24.

Lai, J. J., Chau, Z. L., Chen, S. Y., Hill, J. J., Korpany, K. V., Liang, N. W., Lin, L. H., Lin, Y. H., Liu, J. K., Liu, Y. C., Lunde, R., & Shen, W. T. (2022). Exosome processing and characterization approaches for research and technology development. *Advanced Science, 9*, e2103222.

Lehmann, H. C., & Höke, A. (2010). Schwann cells as a therapeutic target for peripheral neuropathies. *CNS & Neurological Disorders: Drug Targets, 9*, 801–806.

Li, J., Wu, G., Li, W., Zhou, X., Li, W., Xu, X., Xu, K., Cao, R., & Cui, S. (2023). Plasma exosomes improve peripheral neuropathy via miR-20b-3p/Stat3 in type I diabetic rats. *Journal of Nanobiotechnology, 21*, 447.

Liu, W., Bai, X., Zhang, A., Huang, J., Xu, S., & Zhang, J. (2019). Role of exosomes in central nervous system diseases. *Frontiers in Molecular Neuroscience, 12*, 240.

Liu, X. S., Fan, B., Szalad, A., Jia, L., Wang, L., Wang, X., Pan, W., Zhang, L., Zhang, R., Hu, J., Zhang, X. M., Chopp, M., & Zhang, Z. G. (2017). MicroRNA-146a mimics reduce the peripheral neuropathy in type 2 diabetic mice. *Diabetes, 66*, 3111–3121.

Liu, J., Ren, Z. H., Qiang, H., Wu, J., Shen, M., Zhang, L., & Lyu, J. (2020). Trends in the incidence of diabetes mellitus: Results from the global burden of disease study 2017 and implications for diabetes mellitus prevention. *BMC Public Health, 20*, 1415.

Lopez-Verrilli, M. A., Picou, F., & Court, F. A. (2013). Schwann cell-derived exosomes enhance axonal regeneration in the peripheral nervous system. *Glia, 61*, 1795–1806.

Luo, Q., Feng, Y., Xie, Y., Shao, Y., Wu, M., Deng, X., Yuan, W. E., Chen, Y., & Shi, X. (2019). Nanoparticle-microRNA-146a-5p polyplexes ameliorate diabetic peripheral neuropathy by modulating inflammation and apoptosis. *Nanomedicine: Nanotechnology, Biology and Medicine, 17*, 188–197.

Mathew, B., Acha, L. G., Torres, L. A., Huang, C. C., Liu, A., Kalinin, S., Leung, K., Dai, Y., Feinstein, D. L., Ravindran, S., & Roth, S. (2023). MicroRNA-based engineering of mesenchymal stem cell extracellular vesicles for treatment of retinal ischemic disorders: Engineered extracellular vesicles and retinal ischemia. *Acta Biomaterialia, 158*, 782–797.

Minocha, N., & Sardana, S. (2023). Exosome and other extracellular vesicles in gene therapy and personalized care. *Current Pharmacogenomics and Personalized Medicine, 20*(3), 135–145.

Mosquera-Heredia, M. I., Morales, L. C., Vidal, O. M., Barceló, E., Silvera-Redondo, C., Vélez, J. I., & Garavito-Galofre, P. (2021). Exosomes: Potential disease biomarkers and new therapeutic targets. *Biomedicines, 9*, 1061.

Niu, S. R., Hu, J. M., Lin, S., & Hong, Y. (2022). Research progress on exosomes/microRNAs in the treatment of diabetic retinopathy. *Frontiers in Endocrinology, 13*, 935244.

Ogata, T., Iijima, S., Hoshikawa, S., Miura, T., Yamamoto, S., Oda, H., Nakamura, K., & Tanaka, S. (2004). Opposing extracellular signal-regulated kinase and Akt pathways control Schwann cell myelination. *Journal of Neuroscience, 24*, 6724–6732.

Oushyani Roudsari, Z., Esmaeili, Z., Nasirzadeh, N., Heidari Keshel, S., Sefat, F., Bakhtyari, H., & Nadri, S. (2024). Microfluidics as a promising technology for personalized medicine. *BioImpacts, 15*, 29944.

Ozaki, K., Yamano, S., Matsuura, T., & Narama, I. (2013). Insulin-ameliorated peripheral motor neuropathy in spontaneously diabetic WBN/Kob rats. *Journal of Veterinary Medical Science, 750*, 1323–1328.

Poongodi, R., Yang, T. H., Huang, Y. H., Yang, K. D., Chen, H. Z., Chu, T. Y., Wang, T. Y., Lin, H. C., & Cheng, J. K. (2024). Stem cell exosome-loaded Gelfoam improves locomotor dysfunction and neuropathic pain in a rat model of spinal cord injury. *Stem Cell Research and Therapy, 15*, 143.

Pop-Busui, R., Ang, L., Boulton, A. J. M., Feldman, E. L., Marcus, R. L., Mizokami-Stout, K., Singleton, J. R., & Ziegler, D. (2022). Diagnosis and treatment of painful diabetic peripheral neuropathy. *ADA Clinical Compendia, 2022*, 1–32.

Rachana, K. S., Manu, M. S., & Advirao, G. M. (2018). Insulin-induced upregulation of lipoprotein lipase in Schwann cells during diabetic peripheral neuropathy. *Diabetes and Metabolic Syndrome, 12*, 525–530.

Sadeghi, S., Tehrani, F. R., Tahmasebi, S., Shafiee, A., & Hashemi, S. M. (2023). Exosome engineering in cell therapy and drug delivery. *Inflammopharmacology, 31*, 145–169.

Selvarajah, D., Kar, D., Khunti, K., Davies, M. J., Scott, A. R., Walker, J., & Tesfaye, S. (2019). Diabetic peripheral neuropathy: Advances in diagnosis and strategies for screening and early intervention. *Lancet Diabetes & Endocrinology, 7*, 938–948.

Shan, L., Zhan, F., Lin, D., Feng, F., Chen, X., Li, X., & Yang, D. (2022). SIRT1-enriched exosomes derived from bone marrow mesenchymal stromal cells alleviate peripheral neuropathy via conserving mitochondrial function. *Journal of Molecular Neuroscience, 72*, 2507–2516.

Shetter, A. R., Muttagi, G., & Sagar, C. B. (2011). Expression and localization of insulin receptors in dissociated primary cultures of rat Schwann cells. *Cell Biology International, 35*, 299–304.

Shinu, P., Morsy, M. A., Nair, A. B., Mouslem, A. K. A., Venugopala, K. N., Goyal, M., Bansal, M., Jacob, S., & Deb, P. K. (2022). Novel therapies for the treatment of neuropathic pain: Potential and pitfalls. *Journal of Clinical Medicine, 11*, 3002.

Singh, A., Raghav, A., Shiekh, P. A., & Kumar, A. (2021). Transplantation of engineered exosomes derived from bone marrow mesenchymal stromal cells ameliorate diabetic peripheral neuropathy under electrical stimulation. *Bioactive Materials, 6*, 2231–2249.

Sirisha, A., Gaur, G. S., Pal, P., Bobby, Z., Balakumar, B., & Pal, G. K. (2021). Effect of honey and insulin treatment on oxidative stress and nerve conduction in an experimental model of diabetic neuropathy Wistar rats. *PLoS One, 16*, e0245395.

Soltoff, S. P., Rabin, S. L., Cantley, L. C., & Kaplan, D. R. (1992). Nerve growth factor promotes the activation of phosphatidylinositol 3-kinase and its association with the Trk tyrosine kinase. *Journal of Biological Chemistry, 267*, 17472–17477.

Stino, A. M., Rumora, A. E., Kim, B., & Feldman, E. L. (2020). Evolving concepts on the role of dyslipidemia, bioenergetics, and inflammation in the pathogenesis and treatment of diabetic peripheral neuropathy. *Journal of the Peripheral Nervous System, 25*, 76–84.

Sun, Y., Sun, F., Xu, W., & Qian, H. (2023). Engineered extracellular vesicles as a targeted delivery platform for precision therapy. *Tissue Engineering and Regenerative Medicine, 20*, 157–175.

Sun, M., Yu, L., Zhao, X., Ren, L., Liu, X., Gang, X., & Wang, G. (2024). Correlation between thyroid hormone sensitivity and diabetic peripheral neuropathy in euthyroid patients with type 2 diabetes mellitus. *Scientific Reports, 14*, 19603.

Tajabadi, Z., Dadkhah, P. A., Gholami Chahkand, M. S., Esmaeilpour Moallem, F., Karimi, M. A., Amini-Salehi, E., & Karimi, M. (2025). Exploring the role of exosomes in diabetic neuropathy: From molecular mechanisms to therapeutic potential. *Biomedicine and Pharmacotherapy, 185*, 117959.

Tian, C., Xu, H., Li, C., Gao, J., Zhang, H., Wang, P., Wang, Y., Liu, D., Jing, L., Wang, B., Liu, W., Liao, D., Li, J., Shen, K., & Hu, D. (2025). ADSC exosomes improve high glucose induced fibroblast oxidative stress injury and accelerate DFU wound healing via regulating Keap1/Nrf2 axis. *Cellular Signalling, 134*, 111936.

Wang, S., Bao, L., Fu, W., Deng, L., & Ran, J. (2021). Protective effect of exosomes derived from bone marrow mesenchymal stem cells on rats with diabetic nephropathy and its possible mechanism. *American Journal of Translational Research, 13*, 6423–6430.

Wang, T. T., Chen, Y. J., Sun, L. L., Zhang, S. J., Zhou, Z. Y., & Qiao, H. (2015). Affection of single-nucleotide polymorphisms in miR-27a, miR-124a, and miR-146a on susceptibility to type 2 diabetes mellitus in Chinese Han people. *Chinese Medical Journal, 128*, 533–539.

Wang, L., Chopp, M., Szalad, A., Lu, X., Zhang, Y., Wang, X., Cepparulo, P., Lu, M., Li, C., & Zhang, Z. G. (2020). Exosomes derived from Schwann cells ameliorate peripheral neuropathy in type 2 diabetic mice. *Diabetes, 69*, 749–759.

Wang, L., Lu, X., Szalad, A., Liu, X. S., Zhang, Y., Wang, X., Golembieski, W. A., Powell, B., Mccann, M., Lu, M., Chopp, M., & Zhang, Z. G. (2024a). Schwann cell-derived exosomes ameliorate peripheral neuropathy induced by ablation of dicer in Schwann cells. *Frontiers in Cellular Neuroscience, 18*, 1462228.

Wang, H., Zhao, C., Rong, Q., Cao, J., Chen, H., Li, R., Zhang, B., & Xu, P. (2024b). The role of exosomes from mesenchymal stem cells in spinal cord injury: A systematic review. *International Journal of Stem Cells, 17*, 236–252.

Wu, C. J., Huang, K. J., Tsai, Y. C., Yeh, T. P., Hsieh, C. F., & Wang, Y. J. (2021). Peripheral neuropathy: Comparison of symptoms and severity between colorectal cancer survivors and patients with diabetes. *Clinical Journal of Oncology Nursing, 25*, 395–403.

Xu, B. E., Yang, X., Ma, Y., Jiang, Y., Jiang, Y., Li, X., … Bian, H. (2025). Association between free triiodothyronine and diabetic peripheral neuropathy in euthyroid patients with type 2 diabetes. *Journal of Clinical Endocrinology and Metabolism, 111*, 119–128.

Yang, Q., Su, S., Liu, S., Yang, S., Xu, J., Zhong, Y., Yang, Y., Tian, L., Tan, Z., Wang, J., Yu, Z., Shi, Z., & Liang, F. (2023). Exosomes-loaded electroconductive nerve dressing for nerve regeneration and pain relief against diabetic peripheral nerve injury. *Bioactive Materials, 26*, 194–215.

Ye, J., Wu, Y., Yang, S., Zhu, D., Chen, F., Chen, J., Ji, X., & Hou, K. (2023). The global, regional and national burden of type 2 diabetes mellitus in the past, present and future: A systematic analysis of the global burden of disease study 2019. *Frontiers in Endocrinology, 14*, 1192629.

Yin, G., Yu, B., Liu, C., Lin, Y., Xie, Z., Hu, Y., & Lin, H. (2021). Exosomes produced by adipose-derived stem cells inhibit Schwann cells autophagy and promote the regeneration of the myelin sheath. *The International Journal of Biochemistry and Cell Biology, 132*, 105921.

Yorek, M. (2022). Treatment for diabetic peripheral neuropathy: What have we learned from Nimal models? *Current Diabetes Reviews, 18*. e040521193121.

Yu, T., Xu, Y., Ahmad, M. A., Javed, R., Hagiwara, H., & Tian, X. (2021). Exosomes as a promising therapeutic strategy for peripheral nerve injury. *Current Neuropharmacology, 19*, 2141–2151.

Zhang, Z. G., Buller, B., & Chopp, M. (2019). Exosomes—Beyond stem cells for restorative therapy in stroke and neurological injury. *Nature Reviews Neurology, 15*(4), 193–203.

Zhao, T., Sun, F., Liu, J., Ding, T., She, J., Mao, F., Xu, W., Qian, H., & Yan, Y. (2019). Emerging role of mesenchymal stem cell-derived exosomes in regenerative medicine. *Current Stem Cell Research and Therapy, 14*, 482–494.

Zhou, N., Xu, Z., Li, X., Ren, S., Chen, J., Xiong, H., Wang, C., Guo, J., Kang, Y., Chen, Z., Li, W., Yang, X., Zhang, X., & Xu, X. (2022).

Schwann cell-derived exosomes induce the differentiation of human adipose-derived stem cells into Schwann cells. *Frontiers in Molecular Biosciences, 8*, 835135.

Zhu, Y., Han, S., Li, X., Gao, Y., Zhu, J., Yang, X., & Xu, L. (2021). Paeoniflorin effect of Schwann cell-derived exosomes ameliorates dorsal root ganglion neurons apoptosis through ire1α pathway. *Evidence-Based Complementary and Alternative Medicine, 2021*, 6079305.

Zhu, M., Wu, J., & Gao, J. (2022). Exosomes for diabetes syndrome: Ongoing applications and perspective. *Biomaterials Science, 10*, 2154–2171.

Chapter 10

The divalent metal transporter 1 (DMT1), neuroprotection and insulin

Rosaria Ingrassia
Section of Biotechnologies, Department of Molecular and Translational Medicine, University of Brescia, Brescia, Italy

Introduction

Divalent metal transporter 1 (DMT1) (Nramp2/SLC11A2, solute carrier family 11 member 2) (Burdo et al., 2001; Fleming et al., 1997; Gunshin et al., 1997) is the major importer of ferrous iron and responsible for the uptake of nontransferrin-bound iron transport (NTBI) and divalent metals. DMT1 complement its function to the parallel receptor-mediated endocytosis of transferrin-bound iron (TBI) involved in dietary iron uptake leading to its transport to the peripheral tissues, after uptake through the apical membrane of the enterocytes in the acidic duodenal milieu (Garrick et al., 2003). The duodenal NTBI absorption of iron and heavy metals is influenced by systemic iron homeostasis, also regulated by inflammation (Ganz, 2018), that may lead to iron dyshomeostasis, accumulation in peripheral tissues, and to the possible development of neurodegenerative diseases. DMT1, in fact, was found to be widely expressed throughout the rat brain, in the olfactory epithelium, olfactory bulb, hippocampus, cerebellum, and ventral area (Gunshin et al., 1997), and highest DMT1 expression in the rat brain was also found in neurons in the striatum, cerebellum, thalamus, ependymal cells lining the third ventricle, and vascular cells throughout the brain (Burdo et al., 2001), also up-regulated in several neurodegenerative diseases with alteration of the intracellular transport of heavy metals (Ingrassia et al., 2019). Interestingly, the expression of DMT1 in pancreatic β cells (Hansen et al., 2012), regulated by interleuchin-1β, with the presence of Insulin receptors in the central nervous system (Patel et al., 2023), makes this an interesting paradigm to evaluate the interference between iron and glucose metabolism at the crossroads of iron overload-dependent neurodegeneration with altered glucose metabolism.

DMT1 and neurodegeneration

DMT1 is finely regulated at the transcriptional level by the 5′ and 3′ alternative splicing in several tissues (Hubert & Hentze, 2002). The 5′ prime splicing drives the expression of the alternative promoters 1A and 1B, the former responsive to hypoxia while the latter responsive to NF-kB (Lis et al., 2005; Paradkar & Roth, 2006). Conversely, the 3′ alternative transcripts, with or without an iron-responsive element (IRE), are subjected to the post-transcriptional regulation due to the IRE/IRP (Iron Responsive Element/Iron Responsive Protein) secondary to intracellular iron levels (Hubert & Hentze, 2002), leading to the expression of four alternative DMT1 mRNAs, named 1A(+)IRE, 1A(−)IRE, 1B(+)IRE, and 1B(−)IRE, according to the 5′ and 3′ alternative transcripts. This wide range of expressed transcripts was indeed well evaluated also by Hubert and Hentze, in terms of isoform-specific tissue distribution, showing an interesting tissue-specific pattern of DMT1 expression, that participate in iron regulation (Hubert & Hentze, 2002), where the expression of the isoforms by IRE-dependent post-transcriptional regulation or by alternative promoters show peculiar downstream functional mechanisms in strict relationship to conditions of both iron overload or iron deficiency, with a differential pattern of tissue-specific expression, with 1A/DMT1 predominantly expressed in kidney and duodenum and 1B/DMT1 ubiquitously expressed in the peripheral organs and brain. A further complexity of DMT1 depends on two differently glycosylated components: the immature, partially glycosylated, endo H-sensitive, and the mature, fully glycosylated, PNGaseF-sensitive component, with molecular masses of 60 and 90 kDa, respectively (Tabuchi et al., 2002). Interestingly, the differential promoters of DMT1 showed a peculiar responsiveness to different stimuli. In fact, hypoxia induces the transcription factors HIF-1α and NF-κB, found to

Hormones in Neuroscience. https://doi.org/10.1016/B978-0-443-30263-3.00005-1

differentially activate the 5′ regulatory regions of 1A–1B/DMT1 (Li et al., 2008; Lis et al., 2005; Paradkar & Roth, 2006; Qian et al., 2011; Shah et al., 2009; Wang et al., 2010). A broad up-regulation of DMT1 expression was found in the substantia nigra of PD cases as well as in animal and cellular models of PD (Burdo et al., 2001; Salazar et al., 2008; Zhang et al., 2009). To this end, previous studies have clearly described the brain distribution of the principal iron transporters, highlighting the significant role of both the transferrin receptor (TfR) in the central nervous system development (Levy et al., 1999), and divalent metal transporter-1 (DMT1) (Gunshin et al., 1997). However, while TfR was shown to lack a complete match in the substantia nigra for the binding of transferrin in the basal ganglia, with no difference between PD patients and control subjects (Faucheux et al., 1993), the up-regulation of DMT1 and iron were clearly described in the substantia nigra of post-mortem human brain slices, in the PD model of 1-methyl-4-phenyl-1,2,3,6-tetrahydropyridine (MPTP) intoxication in mice (Salazar et al., 2008) and in the mice model of neurodegeneration with parkinsonism, the NFkB/c-Rel knockout mice (Baiguera et al., 2012) with reduced expression of the dopamine transporter (DAT) in the striatum of aged mice and increased expression of DMT1 in both the substantia nigra pars compacta and striatum, and iron staining in the ventral tegmental area. Moreover, NF-κB transcriptional activation was found to specifically regulate both (−)IRE and (+)IRE 1B/DMT1 isoforms in neuronal cells exposed to sodium-nitro-prusside (Paradkar & Roth, 2006), and the (−)IRE DMT1 isoform in MES23.5 cells exposed to MPP+ (Xu et al., 2010), while the role of DMT1 during ischemic neurodegeneration was initially highlighted only by indirect evidence of NF-kB-mediated DMT1 regulation in an ischemic model of MCAO (Yang et al., 2011), where tanshinone IIA, a natural compound reported to inhibit NF-κB activity (Jang et al., 2006), downregulated DMT1 expression, iron and brain infarct volume. Later findings, indeed, showed the epigenetic regulation during post-ischemic neurodegeneration due to the activation of the dimer NF-κB p50/RelA (Sarnico et al., 2009) and requires specific RelA acetylation on the Lys310 residue (Lanzillotta et al., 2010). Inhibition of both 1B/(−)IRE DMT1 expression and promoter activation during post-ischemic neurodegeneration leads to protection from cell death with downregulation of intracellular iron levels, exerted through RelA-Lys310 hypo-acetylation in both in vitro and in vivo models (Ingrassia et al., 2012), and further showed to be exacerbated or significantly reduced by 1B/(−)IRE DMT1 overexpression or (−)IRE DMT1 RNA silencing, respectively, during in vitro ischemia in differentiated human neuroblastoma. Thus, the described mechanism may represent a therapeutic mechanistic target during post-ischemic neurodegeneration in stroke patients.

DMT1, neurodegeneration and insulin signaling

It is noteworthy that neuroinflammation is the prodromal event for several neurodegenerative diseases, like Parkinson's, Alzheimer and ischemia, with the involvement of the NFkB signaling, downstream to cytokine exposure, not only implicated in the metabolic syndrome leading to insulin resistance (Arkan et al., 2005), but also to the NFkB-dependent activation of DMT1 (Ingrassia et al., 2019), via both central and peripheral pathways (Fig. 10.1). Interestingly, DMT1 regulation by insulin was described in Langerhans cells (Hansen et al., 2012), where IL-1β induced NFkB transcriptional activity and DMT1 expression in rat and human islets, and the DMT1 conditional knockout in beta-cells improved glucose-stimulated insulin secretion, known the role of iron in the pancreatic beta-cells, necessary for the correct insulin synthesis and processing (Marku et al., 2021). Previous interesting evidence, both in patients and in the human hepatoma cell line HepG2, showed that insulin resistance, evaluated by homeostatic model assessment (HOMA)-insulin resistance, significantly decreased after diet, and iron depletion by chelation with desferrioxamine increased by twofold the 125I-insulin-specific receptor binding, whereas iron addition decreased insulin binding (Fargion et al., 2005). Noteworthy, in insulin-producing cells, the NF-kB/p65 DNA binding was reduced by small-molecule inhibitors of HDAC lysine deacetylase, leading to the abrogation of the inflammation-dependent NF-kB transcriptional activation, with the protection of β-cells from inflammation-induced apoptosis (Li et al., 2020). In this regard, further evidence showed the activation of the NFkB inducing kinase (NIK) in β cell islets with their consequent failure, as established in obesity, associated with diabetes, also induced in a

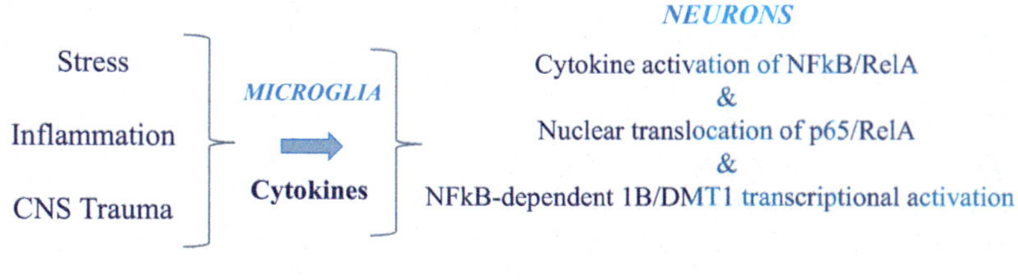

FIGURE 10.1 Neuroinflammation and neurodegeneration through activation of 1B/DMT1 expression.

mice model at a young age (Lundh et al., 2010). Importantly, not only the impaired function of glucose transporters is described in neurodegenerative diseases such as Alzheimer's, Parkinson's, Huntington's disease, GLUT1-deficiency syndrome, stroke, and traumatic brain injury (Szablewski, 2025), but also insulin mRNA and insulin were detected in the CNS, in several regions, including the hypothalamus, hippocampus, cerebral and cerebellar cortex, and olfactory bulb with specific roles of neuroprotection, control of growth hormone secretion, and regulation of appetite and neuronal glucose uptake, with impairment of insulin signaling also associated with memory decline in humans (Dakic et al., 2023; Milstein et al., 2021). Insulin not only has a described role in the hypothalamus, but also at the level of mesolimbic dopamine neurons in the midbrain, and in the striatum, with significant expression of insulin receptors (InsRs) (Patel et al., 2019), and acute exposure to insulin induced the dopamine-related transmission in basal ganglia with consequent activation of striatal insulin receptors that amplifies axonal dopamine release in rodent brain slices. Indeed, insulin and insulin-like growth factor 1 (IGF-1) have been shown to play a key role in the CNS by regulating neuronal growth and plasticity with multiple effects in the brain, including neuroprotection in cerebral ischemia and stroke patients by intranasal treatment as an effective route of administration (Lioutas et al., 2015), bypassing the blood–brain barrier and maximizing the distribution to the CNS without systemic side effects and metabolic influence (Zorina et al., 2023). Interestingly, increased insulin in the striatum stimulated dopamine release (Patel et al., 2023). However, particular attention must be undertaken to the discussion of the protective role of insulin, in an overall complex picture where studies about systemic insulin signaling also described short-term calories restriction as able to lower blood insulin with attenuated Aβ plaques in early adulthood (Patel et al., 2005). Lower insulin signaling was also associated with increased antioxidant mechanisms and improved autophagy linked to anti-inflammatory effects (Green et al., 2022), in both clinical trials and experimental studies. Noteworthy, in this respect, further evaluation highlighted a differential systemic response in terms of insulin secretion, in trials of short-term versus long-term fasting in young males, where prolonged fasting of 36-hours was associated with increased serum insulin levels during the refeeding test, in comparison of 12-hour fasting (Jørgensen et al., 2021), thus highlighting the overall complexity of this picture. We also need to recall the significant findings of improved insulin secretion, after short-term fasting, shown in the β cells conditional knockout of DMT1, further sustained by the finding in human and rat β cells of DMT1 upregulation by interleukin 1-β stimulation, through activation of the proinflammatory signaling via NF-kB (Hansen et al., 2012), underlining the complex relationship between iron and glucose metabolism.

DMT1, neurodegeneration and mechanisms of insulin neuroprotection

Moreover, insulin was shown to inhibit the noxious NFkB signaling through downregulation of IkB expression during in vitro ischemia in cardiomyocytes, where insulin targets the NFkB signaling with protection from ischemic cell death (Díaz et al., 2015). In this relationship, we need to consider that the iron-overload cardiomyopathy is a diffused risk of mortality in patients with secondary iron overload or with early onset forms of genetic hemochromatosis (Wood, 2009), and iron overload decreased insulin sensitivity in primary neonatal rat cardiomyocytes with reduced insulin-stimulated glucose uptake, associated with the decrease in Akt phosphorylation (Sung et al., 2019). Moreover, it becomes of interest the convergence of the insulin and leptin signaling pathways on the IRS/PI3K/Akt axis previously described in midbrain dopaminergic neurons (Khanh et al., 2014). In fact, neuroprotection after ischemic neuronal injury, both in vitro and in vivo, was exerted by exogenous leptin administration, known to activate INS2 receptor as a common mechanism with Insulin, in a c-Rel-dependent manner, in primary cortical neurons of wild type mice, compared to the knockout model of Parkinsonism NFkB/cRel, the subunit of NFkB with protective role against inflammation, thus highlighting the link between insulin/leptin, NF-kB, and ischemia (Valerio et al., 2009). Thus, leptin mediates neuroprotection during cerebral ischemia through the protective subunit cRel/NFkB. Given the involvement of leptin signaling, acting on NFkB regulation as shown in cortical neurons of the cRel knockout model of Parkinsonism when exposed to in vitro ischemia (Valerio et al., 2009), in association to the epigenetic regulation of NFkB (Lanzillotta et al., 2010; Sarnico et al., 2009) also found to target the 1B/DMT1 promoter and its expression during brain ischemia (Ingrassia et al., 2012), we need to mention the results about DMT1 regulation in beta-cells of rat and human islets (Hansen et al., 2012), showing the inflammatory cascade of interleuchin-1β and NFkB targeting DMT1 and insulin secretion. In primary neuronal cultures during in vitro ischemia, insulin was also shown to exert neuroprotection by direct targeting of the NFkB-dependent 1B/DMT1 promoter (Fenaroli et al., 2024), bearing a NFkB cis-acting element (Paradkar & Roth, 2006), possibly through the already shown NFkB/cRel-specific protective mechanism (Valerio et al., 2009). In fact, in both mouse primary cortical neurons and in differentiated human neuroblastoma cells, it was shown the direct targeting of 1B/DMT1 promoter downregulated by insulin treatment with neuroprotection from cell death during oxygen glucose deprivation (OGD) and by acute ferrous iron (Fenaroli et al., 2024). In this relationship, remarkable results were found in the human trial with intranasal insulin in stroke patients (Lioutas et al., 2015) and in cerebral ischemia (Zorina et al., 2023), thus supporting for the targeting by insulin of neurodegenerative pathologies with iron overload (Ingrassia et al., 2019), as highlighted in Fig. 10.2. Thus, iron and glucose, physiologically essential elements when at the equilibrium, may induce damage to the central nervous system and to the dopaminergic system in Parkinson disease as a downstream effect triggered by neuroinflammatory events (Ingrassia et al., 2019). To this end, it is worth mentioning that the intranasal treatment with a low dose of insulin in a rat model of Parkinson disease, after a 6-OHDA-induced lesion lasting for 14 days, determined a significant decrease in the 6-OHDA-induced motor dysfunction and dopaminergic cell death (Iravanpour et al., 2021). Interestingly, the results later described by Novak et al. (2019) in a trial on Parkinson and Multiple System Atrophy patients, showed the rescue of both motor and cognitive symptoms following intranasal Insulin administration, which represent an important synthesis of the underpinning mechanism related to the glucose-iron paradigm, in relationship to the iron overload condition, leading to ferroptosis, present in neurodegenerative pathologies associated to the progression of type 2 diabetes mellitus (T2DM) (Miao et al., 2023).

Several clinical evidence highlighted a link between PD and Type 2 Diabetes defining the nigrostriatal insulin resistance with the influence of high fat diet on dopamine function and increased iron deposition in the substantia nigra of the high fat group of rats, found by Magnetic resonance imaging (Morris et al., 2011). Moreover, the increase of iron and glucose are known to be cross regulated because of ferroptosis involvement in the impairment of glucose-stimulated insulin secretion (GSIS) in the pathogenesis of diabetes (Sha et al., 2021). In this relationship, in the hemochromatosis patients the risk of type 2 diabetes is increased and associated with abnormal iron homeostasis and iron overload, decreased insulin secretion, insulin resistance, and metabolic syndrome (Barton & Acton, 2017). Interestingly, in this regard, it was recently found by MRI the presence of brain iron deposition in newly diagnosed diabetic patients with hyperglycemia (Pirozzi et al., 2026). Importantly, the regulation of systemic iron metabolism is exerted through the complex mechanisms developed by living organisms to acquire, distribute and sequester iron, through to the liver-peptide hormone hepcidin, secreted by hepatocytes to control iron homeostasis (Ganz, 2005), by dietary iron absorption and recycling by macrophages. Indeed, hyperglycemia induces secretion of hepcidin (Moreno-Navarrete et al., 2025), which blocks the intracellular iron export by downregulation of ferroportin, the only iron exporter known so far (Nemeth et al., 2004), with consequent cellular iron overload and cell death (Aigner et al., 2013), mechanism also involved in iron overload disorders like hemochromatosis (Papanikolaou et al., 2005).

Noteworthy, hepcidin knockout mice do not develop hyperglycemia (Ramey et al., 2007). These mice, in fact, showed increased pancreatic iron deposits but only in the acini and the β-cell function was unaffected, with a normal secretion of insulin, thus controlling hyperglycemia.

This increasing body of evidence is in relationship with the described iron overload and DMT1 upregulation both in the substantia nigra of post-mortem slices of Parkinson patients (Salazar et al., 2008) and in the caudate-putamen in the aged mice of the NFkB/cRel model of Parkinsonism (Baiguera et al., 2012) with reduced expression of the dopamine transporter in the striatum (Fig. 10.3).

Of note, Parkinson's patients, like hemochromatosis patients, both with an iron overload condition, may develop diabetes (König & Outeiro, 2024) and growing literature based on animal and in vitro models suggest that diabetes mellitus may induce neurodegeneration by making the nigrostriatal pathway more vulnerable and less resilient (De Pablo-Fernandez et al., 2018; Pagano et al., 2018). For instance, dysregulation of glucose metabolism can appear years before motor symptoms, suggesting the crucial role of hyperglycemia in facilitating the onset of Parkinson (Dunn et al., 2014). Indeed, insulin receptors are present in the basal ganglia and substantia nigra (Athauda & Foltynie, 2016), and insulin has a key role in the

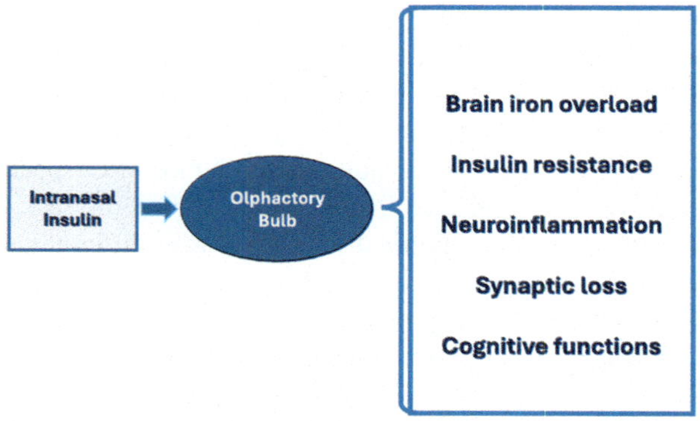

FIGURE 10.2 Intranasal insulin administration with protection against neurodegenerative events.

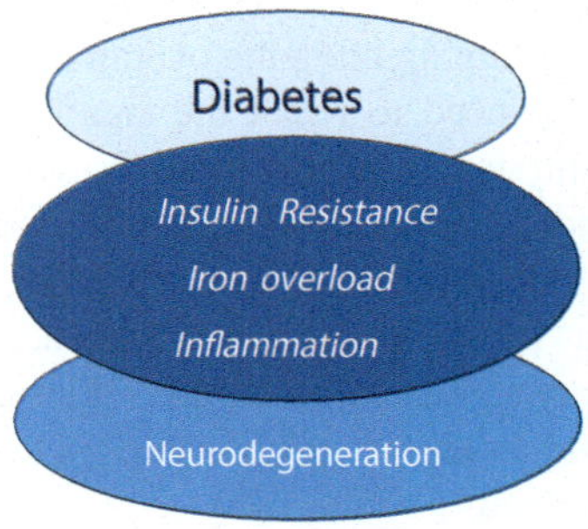

FIGURE 10.3 Risk factors and potential therapeutic targets.

regulation of the dopaminergic transmission (Ferrario et al., 2016; Liu & Borgland, 2019). Hence, new therapeutic targets acting both on metabolic changes and neurodegeneration may arise, being that insulin has demonstrated protective effects on dopaminergic neurons in experimental models (Komici et al., 2021).

Moreover, ongoing clinical trials have shown that patients with Parkinson and also Alzheimer disease respond to the treatment with GLP-1 receptor agonist, being glucagon-like peptide 1 (GLP-1) the incretin secreted in response to nutrients with a peripheral and a central action, including stimulation of glucose-dependent insulin secretion and insulin biosynthesis (Monney et al., 2023).

Furthermore, in support of these findings, exercise is later emerged as good practice in Parkinson disease (Agley et al., 2024), and we may recall that physical exercise induces the translocation of Glut4 to the cellular membrane of myocytes increasing glucose uptake (Fig. 10.4) (Richter & Hargreaves, 2013), thus counteracting the hyperglycemic endothelial damage.

We may conclusively suggest that controlling glycemia and serum insulin could be a good practice, as a possible preventive strategy of support for diabetes and in patients with the insurgence of movement disorders, particularly because of the shown effectiveness of intranasal insulin treatment in the recovery of motor and cognitive symptoms in Parkinson's patients, with no side effects (Novak et al., 2019), and in the reduction of infarction area and injury to the CNS in both animal models and stroke patients (Lioutas et al., 2015; Zorina et al., 2023). Thus, an emerging regulatory mechanism between altered glucose response in the CNS and dopaminergic damage is highlighted, also considering the significant finding that intraventricular injection of insulin in the rat ventral tegmental area/substantia nigra (VTA/SN) can increase the expression of DAT mRNA, as demonstrated by in situ hybridization (Figlewicz et al., 1994).

Finally, we may envisage powerful treatment with intranasal insulin as a strategy to reach a double target with efficacy. Insulin treatment could, indeed, be helpful in several neurodegenerative diseases by two parallel mechanisms that need to be better defined, like reducing glucose levels and lowering excessive iron not only through systemic hepcidin regulation, but also by epigenetic control of the NFkB-dependent 1B/DMT1 promoter (Fenaroli et al., 2024), with possible influence also on the expression of DAT (Figlewicz et al., 1994), thus highlighting a protective mechanism for the dopaminergic system, in relationship to the recovery of neurological symptoms demonstrated in the trial with intranasal insulin in Parkinson's patients (Novak et al., 2019).

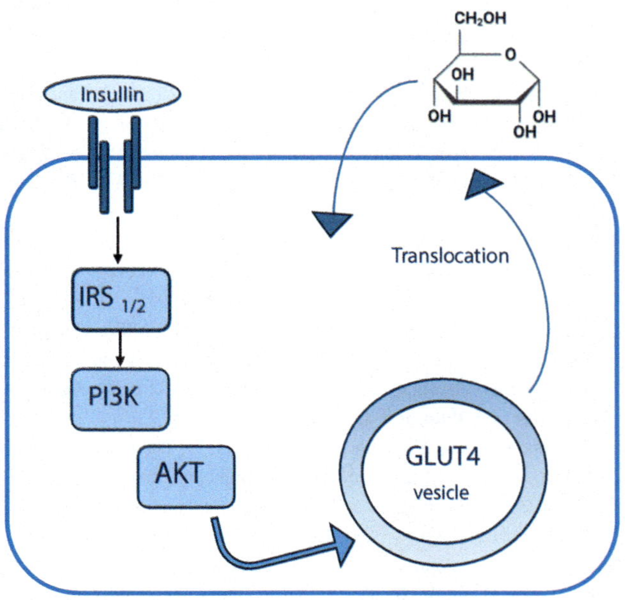

FIGURE 10.4 Exercise-dependent Glut4 translocation to the myocyte's cellular membrane.

References

Agley, L., Hartley, P., & Lafortune, L. (2024). Exploring the experiences, priorities and preferences of people living with Parkinson's on exercise and physical activity promotion in the UK. *PLoS One, 19*(6), e0304223.

Aigner, E., Felder, T. K., Oberkofler, H., Hahne, P., Auer, S., Soyal, S., Stadlmayr, A., Schwenoha, K., Pirich, C., Hengster, P., Datz, C., & Patsch, W. (2013). Glucose acts as a regulator of serum iron by increasing serum hepcidin concentrations. *Journal of Nutritional Biochemistry, 24*(1), 112–117.

Arkan, M. C., Hevener, A. L., Greten, F. R., Maeda, S., Li, Z. W., Long, J. M., Wynshaw-Boris, A., Poli, G., Olefsky, J., & Karin, M. (2005). IKK-beta links inflammation to obesity-induced insulin resistance. *Nature Medicine, 11*, 191–198.

Athauda, D., & Foltynie, T. (2016). Insulin resistance and Parkinson's disease: A new target for disease modification? *Progress in Neurobiology, 145*, 98–120.

Baiguera, C., Alghisi, M., Pinna, A., Bellucci, A., De Luca, M. A., Frau, L., Morelli, M., Ingrassia, R., Benarese, M., Porrini, V., et al. (2012). Late-onset parkinsonism in NF-kB/c-Rel-deficient mice. *Brain, 135*(Pt 9), 2750–2765.

Barton, J. C., & Acton, R. T. (2017). Diabetes in HFE hemochromatosis. *Journal of Diabetes Research, 2017*, 9826930.

Burdo, J. R., Menzies, S. L., Simpson, I. A., Garrick, L. M., Garrick, M. D., Dolan, K. G., Haile, D. J., Beard, J. L., & Connor, J. R. (2001). Distribution of divalent metal transporter 1 and metal transport protein 1 in the normal and Belgrade rat. *Journal of Neuroscience Research, 66*(6), 1198–1207.

Díaz, A., Humeres, C., González, V., Gómez, M. T., Montt, N., Sanchez, G., Chiong, M., & García, L. (2015). Insulin/NFkB protects against ischemia-induced necrotic cardiomyocyte death. *Biochemical and Biophysical Research Communications, 467*, 451–457.

Dakic, T., Jevdjovic, T., Lakic, I., Ruzicic, A., Jasnic, N., Djurasevic, S., Djordjevic, J., & Vujovic, P. (2023). The expression of insulin in the central nervous system: What have we learned so far? *International Journal of Molecular Sciences, 24*, 6586.

De Pablo-Fernandez, E., Goldacre, R., Pakpoor, J., Noyce, A. J., & Warner, T. T. (2018). Association between diabetes and subsequent Parkinson disease: A record-linkage cohort study. *Neurology, 91*(2), e139–e142.

Dunn, L., Allen, G. F. G., Mamais, A., et al. (2014). Dysregulation of glucose metabolism is an early event in sporadic Parkinson's disease. *Neurobiology of Aging, 35*(5), 1111–1115.

Fargion, S., Dongiovanni, P., Guzzo, A., Colombo, S., Valenti, L., & Fracanzani, A. L. (2005). Iron and insulin resistance. *Alimentary Pharmacology & Therapeutics, 22*(Suppl. 2), 61–63.

Faucheux, B. A., Hirsch, E. C., Villares, J., Selimi, F., Mouatt-Prigent, A., Javoy-Agid, F., Hauw, J. J., & Agid, Y. (1993). Distribution of 125I-ferrotransferrin binding sites in the mesencephalon of control subjects and patients with Parkinson's disease. *Journal of Neurochemistry, 60*, 2338–2341.

Fenaroli, F., Valerio, A., & Ingrassia, R. (2024). Ischemic neuroprotection by insulin with down-regulation of divalent metal transporter 1 (DMT1) expression and ferrous iron-dependent cell death. *Biomolecules, 14*(7), 856.

Ferrario, C. R., Labouèbe, G., Liu, S., et al. (2016). Homeostasis meets motivation in the battle to control food intake. *Journal of Neuroscience, 36*(45), 11469–11481.

Figlewicz, D. P., Szot, P., Chavez, M., Woods, S. C., & Veith, R. C. (1994). Intraventricular insulin increases dopamine transporter mRNA in rat VTA/substantia nigra. *Brain Research, 644*(2), 331–334.

Fleming, M. D., Trenor, C. C., Su, M. A., Foernzler, D., Beier, D. R., Dietrich, W. F., & Andrews, N. C. (1997). Microcytic anaemia mice have a mutation in Nramp2, a candidate iron transporter gene. *Nature Genetics, 16*, 383–386.

Ganz, T. (2005). Cellular iron: Ferroportin is the only way out. *Cell Metabolism, 1*(3), 155–157.

Ganz, T. (2018). Iron and infection. *International Journal of Hematology, 107*, 7–15.

Garrick, M. D., Dolan, K. G., Horbinski, C., Ghio, A. J., Higgins, D., Porubcin, M., Moore, E. G., Hainsworth, L. N., Umbreit, J. N., Conrad, M. E., et al. (2003). DMT1: A mammalian transporter for multiple metals. *Biometals, 16*, 41–54.

Green, C. L., Lamming, D. W., & Fontana, L. (2022). Molecular mechanisms of dietary restriction promoting health and longevity. *Nature Reviews Molecular Cell Biology, 23*(1), 56–73.

Gunshin, H., Mackenzie, B., Berger, U. V., Gunshin, Y., Romero, M. F., Boron, W. F., Nussberger, S., Gollan, J. L., & Hediger, M. A. (1997). Cloning and characterization of a mammalian proton-coupled metal-ion transporter. *Nature, 388*, 482–488.

Hansen, J. B., Tonnesen, M. F., Madsen, A. N., Hagedorn, P. H., Friberg, J., Grunnet, L. G., Heller, R. S., Nielsen, A.Ø., Størling, J., Baeyens, L., Anker-Kitai, L., Qvortrup, K., Bouwens, L., Efrat, S., Aalund, M., Andrews, N. C., Billestrup, N., Karlsen, A. E., Holst, B., Pociot, F., & Mandrup-Poulsen, T. (2012). Divalent metal transporter 1 regulates iron-mediated ROS and pancreatic β cell fate in response to cytokines. *Cell Metabolism, 16*(4), 449–461.

Hubert, N., & Hentze, M. W. (2002). Previously uncharacterized isoforms of divalent metal transporter DMT1: Implications for regulation and cellular function. *Proceedings of the National Academy of Sciences of the United States of America, 99*, 12345–12350.

Ingrassia, R., Garavaglia, B., & Memo, M. (2019). DMT1 expression and iron levels at the crossroads between aging and neurodegeneration. *Frontiers in Neuroscience, 13*, 575.

Ingrassia, R., Lanzillotta, A., Sarnico, I., Benarese, M., Blasi, F., Borgese, L., Bilo, F., Depero, L., Chiarugi, A., Spano, P. F., et al. (2012). 1B/(−)IRE DMT1 expression during brain ischemia contributes to cell death mediated by NF-kB/RelA acetylation at Lys310. *PLoS ONE, 7*, e38019.

Iravanpour, F., Dargahi, F., Rezaei, M., Haghani, M., Heidari, R., Valian, N., & Ahmadiani, A. (2021). Intranasal insulin improves mitochondrial function and attenuates motor deficits in a rat 6-OHDA model of Parkinson's disease. *CNS Neuroscience & Therapeutics, 27*, 308–319.

Jørgensen, S. W., Hjort, L., Gillberg, L., Justesen, L., Madsbad, S., Brøns, C., & Vaag, A. A. (2021). Impact of prolonged fasting on insulin secretion, insulin action, and hepatic versus whole body insulin secretion disposition indices in healthy young males. *American Journal of Physiology - Endocrinology and Metabolism, 320*(2), E281–E290.

Jang, S. I., Kim, H. J., Kim, Y. J., Jeong, S. I., & You, Y. O. (2006). Tanshinone IIA inhibits LPS-induced NF-kappaB activation in RAW 264.7 cells: Possible involvement of the NIK-IKK, ERK1/2, p38 and JNK pathways. *European Journal of Pharmacology, 542*(1–3), 1–7.

König, A., & Outeiro, T. F. (2024). Diabetes and Parkinson's disease: Understanding shared molecular mechanisms. *Journal of Parkinson's Disease, 14*(5), 917–924.

Khanh, D. V., Choi, Y. H., Moh, S. H., Kinyua, A. W., & Kim, K. W. (2014). Leptin and insulin signaling in dopaminergic neurons: Relationship between energy balance and reward system. *Frontiers in Psychology, 5*, 846.

Komici, K., Femminella, G. D., Bencivenga, L., Rengo, G., & Pagano, G. (2021). Diabetes mellitus and Parkinson's disease: A systematic review and meta-analyses. *Journal of Parkinson's Disease, 11*(4), 1585–1596. https://doi.org/10.3233/JPD-212725

Lanzillotta, A., Sarnico, I., Ingrassia, R., Boroni, F., Branca, C., Benarese, M., Faraco, G., Blasi, F., Chiarugi, A., Spano, P., et al. (2010). The acetylation of RelA in Lys310 dictates the NF-kB-dependent response in post-ischemic injury. *Cell Death & Disease, 1*, e96.

Levy, J. E., Jin, O., Fujiwara, Y., Kuo, F., & Andrews, N. C. (1999). Transferrin receptor is necessary for development of erythrocytes and the nervous system. *Nature Genetics, 21*, 396–399.

Li, Z., Lai, Z., Ya, K., Fang, D., Ho, Y. W., et al. (2008). Correlation between the expression of divalent metal transporter 1 and the content of hypoxia-inducible factor-1 in hypoxic HepG2 cells. *Journal of Cellular and Molecular Medicine, 12*(2), 569–579.

Li, X., Wu, Y., Song, Y., Ding, N., Lu, M., Jia, L., Zhao, Y., Liu, M., & Chen, Z. (2020). Activation of NF-kB-Inducing kinase in islet β cells causes β cell failure and diabetes. *Molecular Therapy, 28*, 2430–2441.

Lioutas, V. A., Alfaro-Martinez, F., Bedoya, F., Chung, C. C., Pimentel, D. A., & Novak, V. (2015). Intranasal insulin and insulin-like growth factor 1 as neuroprotectants in acute ischemic stroke. *Translational Stroke Research, 6*, 264–275.

Lis, A., Paradkar, P. N., Singleton, S., Kuo, H. C., Garrick, M. D., et al. (2005). Hypoxia induces changes in expression of isoforms of the divalent metal transporter (DMT1) in rat pheochromocytoma (PC12) cells. *Biochemical Pharmacology, 69*(11), 1647–1655.

Liu, S., & Borgland, S. L. (2019). Insulin actions in the mesolimbic dopamine system. *Experimental Neurology, 320*, 113006.

Lundh, M., Christensen, D. P., Rasmussen, D. N., Mascagni, P., Dinarello, C. A., Billestrup, N., Grunnet, L. G., & Mandrup-Poulsen, T. (2010). Lysine deacetylases are produced in pancreatic beta cells and are differentially regulated by proinflammatory cytokines. *Diabetologia, 53*, 2569–2578.

Marku, A., Galli, A., Marciani, P., Dule, N., Perego, C., & Castagna, M. (2021). Iron metabolism in pancreatic beta-cell function and dysfunction. *Cells, 10*(11), 2841.

Miao, R., Fang, X., Zhang, Y., Wei, J., Zhang, Y., & Tian, J. (2023). Iron metabolism and ferroptosis in type 2 diabetes mellitus and complications: Mechanisms and therapeutic opportunities. *Cell Death & Disease, 14*(3), 186.

Milstein, J. L., Heather, A., & Ferris, H. A. (2021). The brain as an insulin-sensitive metabolic organ. *Molecular Metabolism, 52*, 101234.

Monney, M., Jornayvaz, F. R., & Gariani, K. (2023). GLP-1 receptor agonists effect on cognitive function in patients with and without type 2 diabetes. *Diabetes Metab, 49*(5), 101470.

Moreno-Navarrete, J. M., Rodrigues, I. G., & Fernández-Real, J. M. (2025). The impact of iron homeostasis in insulin-sensitive tissues and gut microbiome on obesity-driven metabolic disorders. *Advances in Experimental Medicine and Biology, 1480*, 253–269.

Morris, J. K., Bomhoff, G. L., Gorres, B. K., Davis, V. A., Kim, J., Lee, P. P., Brooks, W. M., Gerhardt, G. A., Geiger, P. C., & Stanford, J. A. (2011). Insulin resistance impairs nigrostriatal dopamine function. *Experimental Neurology, 231*(1), 171–180.

Nemeth, E., Tuttle, M. S., Powelson, J., Vaughn, M. B., Donovan, A., Ward, D. M., Ganz, T., & Kaplan, J. (2004). Hepcidin regulates cellular iron efflux by binding to ferroportin and inducing its internalization. *Science, 306*(5704), 2090–2093.

Novak, P., Maldonado, D. A. A. P., & Novak, V. (2019). Safety and preliminary efficacy of intranasal insulin for cognitive impairment in Parkinson's disease and multiple system atrophy: A double-blinded placebo-controlled pilot study. *PLoS ONE, 14*, e0214364.

Pagano, G., Polychronis, S., Wilson, H., et al. (2018). Diabetes mellitus and Parkinson disease. *Neurology, 90*(19), e1654–e1662.

Papanikolaou, G., Tzilianos, M., Christakis, J. I., Bogdanos, D., Tsimirika, K., MacFarlane, J., Goldberg, Y. P., Sakellaropoulos, N., Ganz, T., & Nemeth, E. (2005). Hepcidin in iron overload disorders. *Blood, 105*(10), 4103–4105.

Paradkar, P. N., & Roth, J. A. (2006). Post-translational and transcriptional regulation of DMT1 during P19 embryonic carcinoma cell differentiation by Retinoic acid. *Biochemical Journal, 394*, 173–183.

Patel, J. C., Carr, K. D., & Rice, M. E. (2023). Actions and consequences of insulin in the striatum. *Biomolecules, 13*(3), 518.

Patel, N. V., Gordon, M. N., Connor, K. E., Good, R. A., Engelman, R. W., Mason, J., Morgan, D. G., Morgan, T. E., & Finch, C. E. (2005). Caloric restriction attenuates Abeta-deposition in Alzheimer transgenic models. *Neurobiology of Aging, 26*(7), 995–1000.

Patel, J. C., Stouffer, M. A., Mancini, M., Nicholson, C., Carr, K. D., & Rice, M. E. (2019). Interactions between insulin and diet on striatal dopamine uptake kinetics in rodent brain slices. *European Journal of Neuroscience, 49*, 794–804.

Pirozzi, M. A., Esposito, F., Trojsi, F., Passaniti, C., Di Meo, I., Auriemma, M. C., Chianese, A., Cerasuolo, M., Paolisso, G., Cirillo, M., & Rizzo, M. R. (2026). In newly diagnosed diabetes, hyperglycemia is associated with increased brain iron deposition, measured by MRI-based Quantitative Susceptibility Mapping (QSM). *Journal of Advanced Research, 81*, 521–533.

Qian, Z. M., Wu, X. M., Fan, M., Yang, L., Du, F., et al. (2011). Divalent metal transporter 1 is a hypoxia-inducible gene. *Journal of Cellular Physiology, 226*(6), 1596–1603.

Ramey, G., Faye, A., Durel, B., Viollet, B., & Vaulont, S. (2007). Iron overload in Hepc1(-/-) mice is not impairing glucose homeostasis. *FEBS Letters, 581*(5), 1053–1057.

Richter, E. A., & Hargreaves, M. (2013). Exercise, GLUT4, and skeletal muscle glucose uptake. *Physiological Reviews, 93*(3), 993–1017.

Salazar, J., Mena, N., Hunot, S., Prigent, A., Alvarez-Fischer, D., et al. (2008). Divalent metal transporter 1 (DMT1) contributes to neurodegeneration in animal models of Parkinson's disease. *Proceedings of the National Academy of Sciences of the United States of America, 105*(47), 18578–18583.

Sarnico, I., Lanzillotta, A., Boroni, F., Benarese, M., Alghisi, M., et al. (2009). NF-kappaB p50/RelA and c-Rel-containing dimers: Opposite regulators of neuron vulnerability to ischemia. *Journal of Neurochemistry, 108*(2), 475–485.

Sha, W., Hu, F., Xi, Y., Chu, Y., & Bu, S. J. (2021). Mechanism of ferroptosis and its role in type 2 diabetes mellitus. *Diabetes Research, 2021*, 9999612.

Shah, Y. M., Matsubara, T., Ito, S., Yim, S. H., & Gonzalez, F. J. (2009). Intestinal hypoxia-inducible transcription factors are essential for iron absorption following iron deficiency. *Cell Metabolism, 9*, 152–164.

Sung, H. K., Song, E., Jahng, J. W. S., Pantopoulos, K., & Sweeney, G. (2019). Iron induces insulin resistance in cardiomyocytes via regulation of oxidative stress. *Scientific Reports, 9*(1), 4668.

Szablewski, L. (2025). Human glucose transporters in health and selected neurodegenerative diseases. *International Journal of Molecular Sciences, 26*(15), 7392.

Tabuchi, M., Tanaka, N., Nishida-Kitayama, J., Ohno, H., & Kishi, F. (2002). Alternative splicing regulates the subcellular localization of divalent metal transporter 1 isoforms. *Molecular Biology of the Cell, 13*(12), 4371–4387.

Valerio, A., Dossena, M., Bertolotti, P., Boroni, F., Sarnico, I., Faraco, G., Chiarugi, A., Frontini, A., Giordano, A., Liou, H. C., et al. (2009). Leptin is induced in the ischemic cerebral cortex and exerts neuroprotection through NF-kappaB/c-Rel-dependent transcription. *Stroke, 40*, 610–617.

Wang, D., Wang, L. H., Zhao, Y., Lu, Y. P., & Zhu, L. (2010). Hypoxia regulates the ferrous iron uptake and reactive oxygen species level via divalent metal transporter 1 (DMT1) Exon1B by hypoxia-inducible factor-1. *IUBMB Life, 62*(8), 629–636.

Wood, J. C. (2009). History and current impact of cardiac magnetic resonance imaging on the management of iron overload. *Circulation, 120*, 1937–1939.

Xu, H., Jiang, H., Wang, J., & Xie, J. (2010). Rg1 protects the MPP+-treated MES23.5 cells via attenuating DMT1 up-regulation and cellular iron uptake. *Neuropharmacology, 58*(2), 488–494.

Yang, L., Zhang, B., Yin, L., Cai, B., Shan, H., et al. (2011). Tanshinone IIA prevented brain iron dyshomeostasis in cerebral ischemic rats. *Cellular Physiology and Biochemistry, 27*(1), 23–30.

Zhang, S., Wang, J., Song, N., Xie, J., & Jiang, H. (2009). Up-regulation of divalent metal transporter 1 is involved in 1-methyl-4-phenylpyridinium (MPP(+))-induced apoptosis in MES23.5 cells. *Neurobiology of Aging, 30*(9), 1466–1476.

Zorina, I. I., Avrova, N. F., Zakharova, I. O., & Shpakov, A. O. (2023). Prospects for the use of intranasally administered insulin and insulin-like growth factor-1 in cerebral ischemia. *Biochemistry (Moscow), 88*, 374–391.

Chapter 11

Peripheral nerves in diabetic neuropathy: Features and methods

Rajeswari Aghoram[1], Saranya B. Gomathy[1] and Stalin Viswanathan[2]
[1]Department of Neurology, Jawaharlal Institute of Postgraduate Medical Education and Research, Puducherry, India; [2]Department of General Medicine, Jawaharlal Institute of Postgraduate Medical Education and Research, Puducherry, India

List of abbreviations

ACE Angiotensin Converting Enzyme
ADA American Diabetes Association
BMI Body mass index
BP Blood pressure
CAN Cardiovascular autonomic neuropathy
CARTs Cardiovascular reflex tests
CCM Corneal confocal microscopy
CKD Chronic kidney disease
CMAP Compound muscle action potential
COMPASS Composite Autonomic Symptom Score
CRP C reactive protein
DAN Diabetic autonomic neuropathy
DBP Diastolic blood pressure
DM Diabetes mellitus
DN Diabetic neuropathy
DPN Diabetic peripheral neuropathy
DSPN Diabetic sensorimotor polyneuropathy
EMG Electromyography
ESR Erythrocyte Sedimentation Rate
GABA Gamma amino butyric acid
GAN Gastrointestinal autonomic neuropathy
HRV Heart rate variability
IENFD Intraepidermal nerve fiber density
MIBG Meta-iodobenzylguanidine
MRN Magnetic Resonance Neurography
NCS Nerve conduction studies
PREP Pain-related Evoked Potential
QST Quantitative Sensory Testing
SBP Systolic blood pressure
SFN Small Fiber Neuropathy
SNAP Sensory nerve action potential
SRAR Sural-to-radial amplitude ratio
SSR Sympathetic skin response
T1DM Type 1 diabetes mellitus
T2DM Type 2 diabetes mellitus
TCSS Toronto Clinical Scoring System
TIN Treatment-induced Neuropathy
UAN Urogenital autonomic neuropathy
VLCAD Acyl CoA dehydrogenase, very long chain
VPT Vibration perception threshold
WPN Whole Plantar Nerve

Introduction

Diabetic peripheral neuropathy (DPN) is a complication of type 2 diabetes mellitus (T2DM), type 1 diabetes mellitus (T1DM) and can also affect individuals with prediabetes. DPN is defined as "the presence of symptoms and/or signs of peripheral nerve dysfunction in people with diabetes after the exclusion of other causes" (Ang et al., 2022; Jensen et al., 2021). Painful DPN is "pain as a direct consequence of abnormalities in the peripheral somatosensory system in people with diabetes" (Rosenberger et al., 2020). Typical DPN is "a symmetrical length-dependent sensorimotor polyneuropathy attributable to chronic hyperglycaemia, associated metabolic derangements, cardiovascular risk covariates, and micro vessel alteration" (Tesfaye et al., 2010). Apart from producing pain, DPN can impair gait and balance leading to falls and fractures. The pain itself contributes to depression, sleep disruption and reduced quality of life. It is also a risk factor for diabetic foot ulcers; that are chronic, non-healing, and often result in amputations. In many countries, diabetes and diabetic neuropathy is a leading cause of non-traumatic amputations resulting in considerable morbidity and economic burden. Diabetic neuropathy is associated with increased white matter lesions, cognitive impairment and is also an independent predictor of all-cause mortality. Thus, it poses both direct and indirect costs to both the individual and the system. In this chapter we will discuss clinical, electrophysiological and some emerging technologies for the evaluation of DPN.

Hormones in Neuroscience. https://doi.org/10.1016/B978-0-443-30263-3.00032-4

Epidemiology

Diabetes is the most common cause of peripheral neuropathy in the world (Freeman, 2009; Yang et al., 2025). The reported prevalence of DPN ranges from 6% to 60%, depending on the type of study, the definitions adopted, setting (higher prevalence in developing countries) and the methods used. For instance, in a large cohort of diabetics from Denmark, using a clinical screening instrument 18% had possible DPN while 23.8% had abnormal nerve conduction studies and 32.6% had intraepidermal nerve fiber abnormalities and all these increased over a 5 year follow up (Gylfadottir et al., 2020).

Large fiber DPN is seen in 40%–60% of patients. It results in sensory and motor deficits (Yang et al., 2025). DPN usually begins 5–10 years after diagnosis of diabetes (Oh, 2020) but may be present in 20%–30% at the time of diagnosis (Pop-Busui et al., 2022). Small fiber neuropathy (SFN) is seen in 25% of patients with DPN (Oh, 2020). Of all the various types distal sensorimotor polyneuropathy (DSPN) is the most common and best studied and is often referred to as typical diabetic neuropathy (Jensen et al., 2021; Pop-Busui et al., 2017). Diabetic autonomic neuropathy (DAN) can coexist with DSPN in 50%. Cardiac autonomic neuropathy (CAN) is associated with increased morbidity and mortality in diabetics. Its prevalence varies from 44% among T1DM to 60% among those with T2DM (Ang et al., 2022; Pop-Busui, 2023). The prevalence of gastrointestinal autonomic neuropathy (GAN) ranges from 30% to 50% (Yang et al., 2025). Urogenital autonomic neuropathy (UAN) is seen in up to 40% of diabetics, and erectile dysfunction in 75% of males (Yang et al., 2025). In women, reduced libido, vaginal dryness, dyspareunia, and decreased sexual arousal occur in 30%–50% (Yang et al., 2025). Sudomotor dysfunction is observed in 30% of patients, while pupillary abnormalities may be seen in 25%.

Risk factors

DSPN is associated with age, smoking, blood pressure, high body mass index (BMI), lipids, duration of diabetes, sedentary activity, hypertension, alcohol use and depression (Pop-Busui et al., 2017; Ziegler, 2023). Chronic kidney disease, elevated triglycerides, and smoking are additional associations for CAN (Ang et al., 2022). Among women recent UTI, and microvascular complications are risk factors for UAN (Dillon et al., 2024). Painful DPN is associated with female gender, poor glycemic control, high BMI, and renal dysfunction (Yang et al., 2025). Genetic factors, such as gain-of-function mutations in voltage-gated sodium channels may also contribute to painful DPN (Yang et al., 2025).

Classification

Diabetic neuropathy is an umbrella term for many types of neuropathies that are related to diabetes mellitus (Oh, 2020). The classification of DPN (Fig. 11.1) may be based on clinical features, type of fiber affected, based on the system affected, based on the diagnostic criteria etc (Lee & Won, 2025). Though, all of these systems have their utility and proponents, they have significant overlap.

- **Distal sensorimotor polyneuropathy (DSPN):** The Toronto clinical criteria, accepted worldwide, classifies DSPN as—possible in the presence of signs or symptoms; probable when both are present and confirmed when associated with nerve conduction study abnormalities (Kallinikou et al., 2019; Oh, 2020). The distal legs and feet are involved symmetrically, which progress proximally in a "stocking and glove" pattern.
- **Small fiber neuropathy (SFN):** This generally presents as painful neuropathy due to affection of the small fibers. There is considerable overlap between SFN, DSPN and Autonomic neuropathies.
- **Autonomic neuropathies:** These are classified based on the organ system involved such as Cardiac autonomic neuropathy, genito-urinary autonomic neuropathy, sudomotor autonomic neuropathy etc. It is often associated with small fiber neuropathy (Yang et al., 2025).
- **Mononeuropathies:** Cranial nerve palsies are uncommon, with an incidence of 1% (Samakidou et al., 2021). The oculomotor and facial nerves are commonly involved, followed by the abducens and trochlear nerves (Samakidou et al., 2021). Most cases are unilateral and present with a subacute onset. Facial nerve palsy is associated with sparing of taste sensation (Samakidou et al., 2021). About 1/3 of patients may have an entrapment syndrome, which is often asymptomatic. Median neuropathy is the commonest, seen in about 5.8% of patients. Other nerves commonly involved are the ulnar nerve, radial nerve and common peroneal nerve. These generally present with mixed motor and sensory symptoms. Lateral femoral cutaneous neuropathy is a pure sensory syndrome, causing pain and sensory loss over the lateral thigh (Clinical Overview: Diabetic Peripheral Neuropathy, 2021).
- **Radiculopathies/Plexopathies:** Thoracic radiculoplexus neuropathy presents with girdle-like distribution of pain affecting the lower or upper abdominal wall associated with paresthesia, anhidrosis and muscle weakness (Samakidou et al., 2021). Motor, sensory and autonomic nerve fibers are involved. It primarily affects T2DM (Samakidou et al., 2021). Cervical radiculoplexus neuropathy presents with pain and weakness of the arm or shoulder due to involvement of the

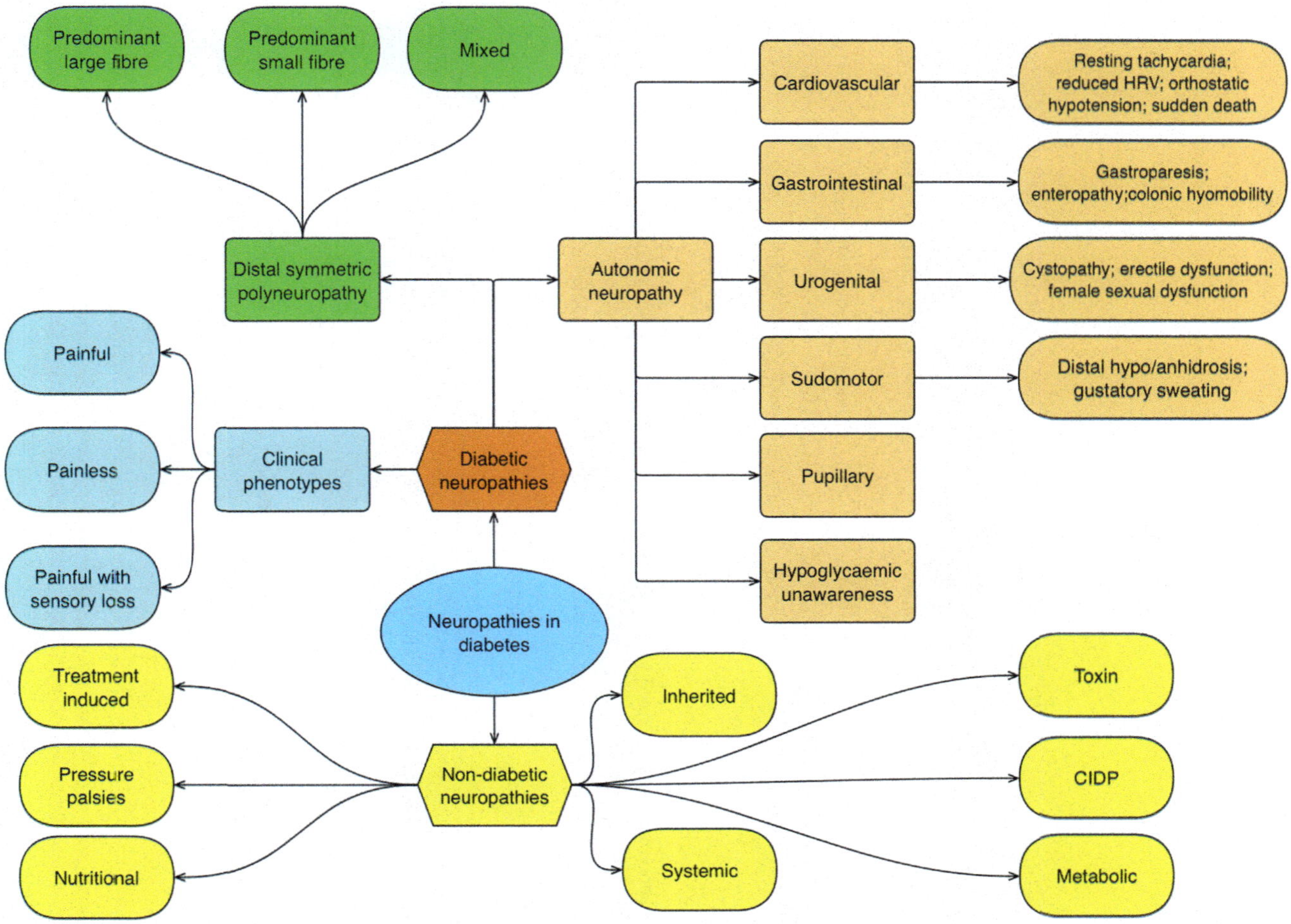

FIGURE 11.1 Classification and differential diagnosis of diabetic neuropathy.

brachial plexus and C7 trunk (Samakidou et al., 2021). Lumbosacral radiculoplexus neuropathy (diabetic amyotrophy/Bruns-Garland syndrome) affects 1% of middle-aged T2DM, usually with an abrupt onset, asymmetrical deep pain in the thigh that progresses to motor weakness and atrophy (Samakidou et al., 2021).

- **Treatment-induced neuropathy:** In treatment-induced neuropathy (previously insulin neuritis; TIN), symptoms emerge rapidly within 2 months following a sudden improvement in glycemic control through diet or therapy. The prevalence of TIN in individuals with T1DM is 11% (Oh, 2020). Symptoms typically involve continuous burning pain in the soles of the feet.

Screening

DPN unlike diabetic retinopathy and nephropathy, does not have an effective screening and treatment program to reduce progression (Burgess et al., 2021). Currently available screening instruments rely heavily on the examiners' interpretations, while bedside tests rely on patients' subjective responses. The American Diabetes Association (ADA) recommends that patients with T1DM (duration >5 years) and T2DM should be assessed annually for DSPN using a comprehensive history and a simple clinical examination (American Diabetes Association Professional Practice Committee, 2025; Burgess et al., 2021). Pin prick, vibration using a 128Hz tuning fork and a 10g monofilament test is recommended for screening of DPN (Burgess et al., 2021). An abnormal monofilament test carries a 15% 3-year risk of ulcers or limb amputation. Assessment methods, such as vibration perception testing, pinprick, thermal threshold, and light touch, demonstrate specificity ranging from 77% to 96% and positive predictive values (PPV) between 95% and 98%.

Popular instruments for screening and assessing severity of diabetic neuropathy are given in Table 11.1. These help detect and follow patients (Oh, 2020). Michigan neuropathy screening instrument (MSNI) and Neuropathy Disability Score (NDS) are most often used. The

TABLE 11.1 Commonly used scoring systems for DPN[a].

Screening tool	Components	Cutoffs	Sensitivity/ Specificity (%)	Comments
Michigan diabetic neuropathy score	Clinical examination + NCS	Higher scores more abnormal	80–88/85–90	Comprehensive; time-consuming
Michigan neuropathy screening instrument	Symptom questionnaire + foot inspection + clinical examination	≥3	75–80/80–95	Validated; most widely used.
Diabetic neuropathy symptom score	Only 4 symptoms	>1	79/28	Screening tool for DPN
Diabetic neuropathy examination score	Examination score	>3	80/95	No special instruments, hierarchical scoring system
Neuropathy symptom score	Symptoms scored based on severity	>1 abnormal; >5 moderate symptoms	75–90/78–90	Quick screening
Neuropathy disability score	Based on sensory and motor examination	>3 abnormal (can be score 0–10 or 0–28)	65/91	Validated; routine use
TNS	8 components, scored 0–4	Total = 32; ≥5 suggestive	84–86/84–90	Comprehensive
Toronto clinical scoring system	Symptoms, reflexes and sensory tests;	>5	80–88/80–90	Both large fiber and small fibers are evaluated; longer time required
Douleur neuropathique 4 questions (DN4)	7 symptoms; 3 findings	≥4 makes neuropathy likely	80/92	Suitable for clinical practice
Neuropathy pain questionnaire	Weighted discriminant score of 12 questions	>0 abnormal; >0.52 stricter definition	66/74	Useful for screening

[a]*Rosenberger et al. (2020); Oh (2020); Kallinikou et al. (2019); Sun et al. (2020); Perkins et al. (2001); Dias et al. (2015); Sommer et al. (2011); Meijer et al. (2002); Carmichael et al. (2021); Zelman et al. (2005); Cornblath et al. (1999); Feldman et al. (1994); Sletten et al. (2012); Zilliox et al. (2011).*

Toronto Clinical Neuropathy Score (TCNS) provides a semi-quantitative clinical grading (0–19 points), incorporating symptoms, reflexes, and sensory deficits. TCNS correlates strongly with nerve conduction studies (NCS) results and is widely used for disease staging and monitoring in clinical trials. Specific questionnaires also exist for autonomic symptoms like COMPASS-31.

Evaluation and diagnosis

There is no single gold standard test for confirming DPN (Jensen et al., 2021). Hierarchical systems are utilized in research, while confirmatory tests are often unnecessary in clinical practice as clinical neuropathy can be diagnosed without abnormal NCS (Jensen et al., 2021). A thorough history-taking with a physical exam is necessary (Dillon et al., 2024). Important factors in history include age, type and duration of diabetes, renal and thyroid conditions, benign prostatic hyperplasia, urinary infections, social determinants, and medication use (Ziegler, 2023). Painful DPN can also be graded (Jensen et al., 2021).

Differential diagnosis

The differential diagnosis for DN includes alcohol use, toxin abuse, hypothyroidism, chronic kidney disease, paraproteinemia, and B12 deficiency (Pop-Busui et al., 2022). Blood tests HbA1c, serum electrolytes, creatinine and blood urea nitrogen, serum B12, folate, TSH, fasting lipids, liver function tests, thiamine, erythrocyte sedimentation rate, C-reactive protein, Angiotensin converting enzyme levels and protein immunoelectrophoresis are often needed (Pop-Busui et al., 2022). Alcohol neuropathy is sensory dominant and slowly progressive. Chronic kidney disease neuropathy is sensorimotor (Freeman, 2009). Vitamin B12 deficiency is large fiber more than small fiber neuropathy and may coexist with myelopathy (Freeman, 2009). Chronic inflammatory demyelinating neuropathy (CIDP) has proximal more than distal weakness with sensory loss and areflexia (Freeman, 2009). Approximately half of patients with DPN may have an additional underlying cause of peripheral neuropathy (Freeman, 2009). Anemia, thyroid and adrenal disease may contribute to CAN symptoms (Dillon et al., 2024). UAN symptoms may mimic benign prostatic hyperplasia (Dillon et al., 2024).

Clinical evaluation

Sensory: Small fibers induce symptoms early, and patients typically present with burning or aching pain, or dysesthesia, mostly at night, often accompanied by hyperalgesia and/or allodynia (Pop-Busui et al., 2022). Reduced sensation due to small and large fiber involvement may lead to slippage of footwear or Charcot's joints (Ziegler, 2023). The feet may be numb or insensate. Other symptoms may also include unsteadiness, falls, ulcers, and painless burns (Pop-Busui et al., 2022). With progression the upper limbs may become involved. Truncal nerves when involved result in sensory deficits in the anterior chest or abdominal wall.

Autonomic: Orthostatic symptoms and falls due to CAN, are more frequent during fever or after large meals. Diuretics, beta blockers or ACE inhibitors may exacerbate this (Elsevier Point of Care, 2025). Gastroparesis is the most common manifestation of GAN (Pop-Busui, 2023). Nausea, vomiting, early satiety, bloating, dyspepsia, constipation, diarrhea, esophageal motility, and fecal incontinence are the other typical symptoms of GAN (Pop-Busui, 2023). Erectile dysfunction can exist in between 1/4th and 2/3rd of patients (Pop-Busui, 2023). In Genito-urinary autonomic neuropathy lower urinary tract symptoms including storage and voiding symptoms (urgency, urge incontinence, frequency, hesitancy, slow stream, dribbling and nocturia) along with sexual dysfunction may occur in both sexes (Elsevier Point of Care, 2025; Pop-Busui, 2023). Reduced libido, erectile dysfunction and abnormal ejaculation are the key elements in male sexual dysfunction while in women, reduced libido, inadequate lubrication and dyspareunia are suggestive of sexual dysfunction (Elsevier Point of Care, 2025). Lack of sweating due to hypoglycaemia unawareness is a slowly progressive illness. Patients do not experience hypoglycaemic symptoms such as palpitations, diaphoresis or light-headedness (Oh, 2020). Silent ischemia refers to episodes of myocardial ischemia that occur without anginal pain and pupillary involvement is generally asymptomatic (Elsevier Point of Care, 2025).

Motor: Varying degrees of distal predominant wasting and weakness are seen in typical DSPN, however proximal involvement can be seen in diabetic plexopathies like diabetic amyotrophy or in the distribution of particular nerves in diabetic mononeuropathy.

Examination: Inspection of the lower limbs includes color, fissuring, hyperkeratosis, calluses, subcutaneous bleeding, hypohidrosis, blistering, dry skin, foot ulcers, signs of infection, healed foot lesions, and foot deformities, as well as footwear (Burgess et al., 2021; Ziegler, 2023). Palpation of the feet involves assessing temperature, pedal pulses, sweating, foot deformities such as Charcot's arthropathy, muscle power, and gait (Burgess et al., 2021). Sensory examination includes vibration testing using a 128Hz tuning fork on the great toe, proprioception, testing with a 10g monofilament, and testing of the ankle jerk (Oh, 2020; Willits et al., 2015). Small fiber function should be assessed by either pinprick or temperature sensation (Pop-Busui et al., 2022). High-risk foot requires the metatarsal heads and distal hallux of the first, third and fifth toes of each foot be assessed using a monofilament (Pop-Busui et al., 2022). Other methods of sensory or autonomic testing that can be performed bedside and their sensitivity

TABLE 11.2 Extended bedside tests for DPN.

Method	Sensitivity (95% confidence interval)	Specificity (95% confidence interval)
Monofilament[a]	0.53(0.32–0.74)	0.88(0.78–0.94)
Neuropad[b]	0.86(0.79–0.91)	0.65(0.51–0.76)
Ipswich touch test[b]	0.77(0.69–0.85)	0.96(0.93–0.98)
Tuning fork[a,c]	0.01–0.19	0.87–0.99
Biesthesiometer[a,c]	0.6–0.81	0.64–0.76
Vibratip[d]	0.79	0.82
Nerve check[c]	0.75–0.89	0.66–0.81

[a]*Compared to Nerve conduction studies.*
[b]*Compared to variable combination of clinical evaluation and nerve conduction studies.*
[c]*Range of estimates from various studies.*
[d]*In comparison to neuresthesiometer.*

and specificity are in Table 11.2 (Mogilevskaya et al., 2024).

Resting tachycardia may be the first parasympathetic sign (Oh, 2020). Orthostatic hypotension and postural orthostatic tachycardia may occur (Ziegler, 2023). Abdominal examination may show a succussion splash, hardened stools in the rectum or poor anal sphincter tone (Ziegler, 2023). A distended bladder may be palpable (Elsevier Point of Care, 2025). For pupillary responses, pupillometry and pharmacological response to pilocarpine can be used (Yang et al., 2025).

Electrophysiological evaluation

Diabetic neuropathy involves damage to all three types of nerve fibers: sensory, motor, and autonomic. Small fiber neuropathy (SFN) affects unmyelinated C and thinly myelinated Aδ fibers, impairing pain and temperature sensation. Large fiber neuropathy involves myelinated Aβ fibers, causing sensory and motor deficits. Around half of all DPN patients exhibit mixed fiber involvement (Yang et al., 2025).

Nerve conduction study (NCS)

The NCS protocol for diabetic neuropathy is designed to evaluate large-fiber peripheral nerve function. Both motor and sensory nerves in the upper and lower limbs are tested to assess the pattern and severity of neuropathy. The motor nerve studies typically include the peroneal nerve, the tibial nerve, the median nerve, and the ulnar nerve. The key parameters measured are distal motor latency, compound muscle action potential (CMAP) amplitude, conduction velocity, and F-wave latency.

The sensory nerve studies include the sural nerve, superficial peroneal nerve, median sensory nerve and ulnar sensory nerve. Parameters assessed are sensory nerve action potential (SNAP) amplitude and sensory conduction velocity. Among these, the sural SNAP amplitude is the most sensitive early indicator of diabetic neuropathy and has been considered the gold standard.

Early findings include reduced or absent sensory nerve action potentials (sural and superficial peroneal), with relatively preserved conduction velocities. As neuropathy progresses, there is a gradual reduction in CMAP amplitudes of motor nerves (especially peroneal and tibial), mild-to-moderate slowing of motor and sensory conduction velocities, prolonged distal motor latencies, and delayed or absent F-waves and H-reflexes. Motor conduction velocities, particularly in the peroneal nerve, have prognostic significance. Slower velocities are associated with higher risks of foot ulceration and mortality. Sensory abnormalities generally precede motor involvement. Findings of focal nerve compressions may produce electrophysiologic patterns resembling mononeuropathy. Nevertheless, several studies suggest that such findings are integral to the DSPN spectrum and may reflect predisposition to polyneuropathy.

Unilateral NCS are generally sufficient, as DSPN is usually symmetric. If clinical asymmetry is observed, alternative diagnoses should be considered, and more extensive testing may be warranted. Finally, the interpretation should describe the distribution (symmetric vs. asymmetric), type (axonal, demyelinating, or mixed), and severity (mild, moderate, severe) of neuropathic involvement (Perkins et al., 2002).

The sural nerve, being more distal is affected earlier than radial nerve in DSPN. By comparing the sural-to-radial amplitude ratio helps normalize for technical and physiological factors such as age, height, or skin temperature, thereby improving diagnostic consistency. A reduced ratio (<0.21) may indicate early axonal loss even when absolute sural amplitudes are normal (Guo et al., 2015).

Since DSPN is a length-dependent process, there is early involvement of the medial and lateral plantar nerves which are distal branches of the tibial nerve. Thus, these nerves may provide evidence of early neuropathic change (Galiero et al., 2021; Kartheka et al., 2024). In DSPN, prolongation of sensory latency and marked reduction or absence of SNAP amplitude are common even in patients with normal sural nerve responses. Plantar nerve conduction studies show a sensitivity of 85%–95% and specificity of 75%–90% for detecting early DSPN (Galiero et al., 2021). However, technical challenges and the lack of normative values can make accurate interpretation difficult (Uluc et al., 2008).

Sympathetic skin response (SSR): While NCS primarily evaluates large myelinated fibers, SSR provides complementary information on autonomic and small-fiber function, which often precedes large-fiber impairment in DM. It assesses the electrical potential changes in the skin triggered by sympathetic sudomotor activity. SSR abnormalities were reported in 83.3% of diabetic patients in one study (Liu et al., 2024). Overall, SSR represents a sensitive, non-invasive, and practical electrophysiologic measure for early detection of diabetic peripheral and autonomic neuropathy. Integrating SSR into routine neurophysiologic evaluation may therefore improve early diagnosis and timely intervention of DPN (Braune & Horter, 1996).

Electromyography (EMG)

Electromyographic (EMG) examination in patients with DSPN is generally considered supplementary and exploratory, rather than part of routine NCS. EMG is most valuable when alternative diagnoses—such as radiculopathy, motor neuron disease, or inflammatory myopathy——are strongly suspected and need to be excluded. When performed in DSPN, abnormal findings like increased insertional activity, spontaneous activity like fibrillations and reduced recruitment are typically observed in distal muscles, particularly in patients with moderate to severe neuropathy (Lamontagne & Buchthal, 1970). Chronic axonal loss and subsequent reinnervation are reflected by remodeled motor unit potentials, characterized by increased duration, amplitude, and polyphasic morphology of the moto unit action potentials.

Advantages

NCS provide objective, quantitative, and reproducible assessment of peripheral nerve function in DSPN. They show lower variability than clinical exams or quantitative sensory testing and correlate well with both structural and clinical severity of neuropathy. It can detect changes related to short-term metabolic fluctuations as well as long-term hyperglycaemia, and motor nerve conduction velocities can predict serious complications such as foot ulcers and increased mortality. Their standardization makes them essential for accurate diagnosis, particularly when the clinical picture is unclear or alternative neuropathies are suspected (Tankisi et al., 2019).

Limitations

- **Limited sampling of nerves:** As discussed above in only 4-5 subcutaneous mixed nerves are studied in routine NCS. Thus involvement of other nerves or involvement of more proximal portions is not evaluated (Perkins & Bril, 2014).
- **Focus on DSPN:** NCS and its abnormalities have been best studied in the context of DSPN and forms part of the diagnostic criteria. However, its role in other types of DPN is less well established.
- **Focus on large fibers:** NCS primarily evaluate large myelinated fibers, which mediate vibration, proprioception, and motor function, and do not directly assess small unmyelinated fibers responsible for pain, temperature, and autonomic function. Nevertheless, NCS remain clinically relevant because most patients with DSPN have diffuse nerve fiber involvement, and progressive loss typically affects both large and small fibers.
- **Accuracy and reference ranges:** Interpretation of NCS can be limited by overlap between normal and abnormal values, and results are influenced by anthropometric factors such as height, gender, and BMI. Establishing accurate reference ranges is challenging.
- **Assessment of small fibers and autonomic nerves:** Conventional NCS cannot evaluate small fiber or autonomic dysfunction. Tests like sympathetic skin response (SSR), but their reproducibility is limited, and they often do not correlate with symptoms.

Role of EP studies in diabetes

Diagnosis: Screening tools may detect neuropathy in clinics, but definitive diagnosis of DSPN currently requires NCS. These tests confirm the presence and severity of DSPN, delineate its extent, and help identify features suggesting alternative diagnoses—such as:

- marked conduction slowing with preserved amplitudes (demyelinating process),
- normal sensory but abnormal motor findings (motor neuropathy),
- normal NCS with absent reflexes and high vibration thresholds (polyradiculopathy),
- or normal NCS with intact reflexes but sensory loss (central lesion).

Staging: In mild cases, overlap with normal reference values can limit sensitivity, especially in small fiber predominant neuropathies. As severity increases, distal sensory and motor responses are lost in the lower limbs and eventually in upper limbs. Disease extent is gauged by the number and distribution of affected nerves. Isolated abnormalities suggest mononeuropathy, while diffuse abnormalities confirm DSPN.

In research, NCS are central to standardized definitions of DSPN. The Toronto Consensus Panel established diagnostic criteria incorporating symptoms, signs, and NCS findings:

- *Possible DSPN*: symptoms or signs only
- *Probable DSPN*: both symptoms and signs
- *Confirmed DSPN*: abnormalities on NCS in addition to symptoms and/or signs

Monitoring: NCS parameters change gradually in DSPN, making them valuable for longitudinal monitoring. Minimal changes over 12 months are typical; rapid decline suggests poor glycemic control or superimposed conditions like CIDP. NCS based outcomes are often preferred in trials assessing DPN progression or therapeutic response as it reliable, objective and reproducible (Yu, 2021).

Emerging methods for evaluation of DPN

The conventional methods of screening and diagnosis for diabetic neuropathy have poor sensitivity and miss about a third of patients with DPN. Often abnormalities in the nerve structure and function predate clinical findings. Thus, newer techniques may help early identification and provide a window to intervene. Broadly they can be classified into structural methods, functional methods, chemical biomarkers and imaging methods.

Structural methods

As a group, these techniques assess the unmyelinated C fibers using morphometric methods. They are often abnormal even in diabetic individuals without clinical neuropathy. However, these abnormalities are not specific to DM.

- **Intraepidermal nerve fiber density (IENFD):** The skin is richly innervated by axons belonging to unmyelinated c-fibers which can be lost in DPN (Samolis et al., 2025). Spot skin biopsies are stained for protein gene product 9.5 and immunoflourescence is used to estimate the density of these nerve fibers. Though considered the gold standard for SFN diagnosis, there is a high variability due to site of biopsy, age, gender etc. In comparison with NCS this method has a sensitivity of 51.1% (43.7–58.5) and a specificity of 90% (79.5–96.2) for DSPN (Gylfadottir et al., 2020).
- **Corneal confocal microscopy (CCM):** The cornea is innervated by small fibers whose distribution can be directly observed as the cornea is transparent using CCM. It is a rapid, non-invasive in vivo technique. The corneal nerves appear as hyper-refractile beaded structures. Corneal nerve fiber density, nerve fiber length, branch density and tortuosity are lowest in individuals with diabetic neuropathy; followed by diabetics without neuropathy compared to normal controls (Gad et al., 2022). Its sensitivity is comparable to quantitative sensory testing for diagnosis of DSPN (Ferdousi et al., 2020). This method has been used to diagnose, rate severity and to monitor effects of treatment in DPN (Zhang et al., 2024). However this is a contact technique and uses expensive equipment which limits its applicability and accessibility.

Functional methods

- **Quantitative sensory testing (QST):** This includes a panel of tests that quantifies thresholds for perception and pain to a variety of standardised stimuli. It quantifies both loss of sensation and increased sensitivity. Standardised protocols like the German research networks for neuropathic pain are used (van Driel et al., 2024). This technique requires active patient engagement and it evaluates the entire somatosensory system from the peripheral nerve to the central nervous system (Kahn, 1992). When compared to nerve conduction studies the sensitivity of cold threshold is reported to be 27%–98%, heat threshold 22%–98% and vibration 58%–84% with a specificity of 51%–86% (Yu, 2021). However as for the structural methods QST can be abnormal even in diabetic individuals without clinical or electrophysiological evidence of diabetic neuropathy (Kahn, 1992; Yu, 2021).
- **Pain related evoked potentials (PREP):** In this the peripheral nociceptive fibers (Aδ and C) are stimulated and the response of the brain is recorded using electroencephalograph signals over the vertex. Intracutaneous low intensity current, contact thermode or laser can be used to stimulate the nociceptive fibers alone. Compared to healthy adults the resultant PREP latencies and/or amplitudes are affected in individuals with neuropathy and the degree of abnormality may correlate with the severity (Bubenzer et al., 2023). In DPN laser evoked potentials have been reported to have a sensitivity of 78% and specificity of 81% compared to IENFD (Marshall et al., 2021). Such evoked potentials are non-invasive, easy to perform and objective, but may cause some discomfort, are not widely available and are expensive. Also these potentials tend to habituate and are dependent on patients attention (Marshall et al., 2021).
- **Microneurography:** In this technique an intraneural needle electrode is inserted into a peripheral nerve like the superficial peroneal nerve and action potentials recorded (Jin et al., 2021). A "sawtooth" pattern is

elicited when there is spontaneous activity. Sensitivity of the nerve can also be assessed. This technique is invasive and time consuming (Marshall et al., 2021).

- **Autonomic function testing:** The autonomic nerves system efferents are small unmyelinated C fibers. Various testing methods are used depending upon the target system. Sudomotor tests have been widely used for evaluation of DPN (Akbar et al., 2023). The quantitative sudomotor axon reflex testing quantifies the amount of sweat produced to fixed quantities of acetylcholine delivered using iontophoresis. This is non-invasive but may be uncomfortable and time consuming. It has a sensitivity of 52%–80% and specificity of more than 80% for diagnosis of SFN. Direct indirect test, electrochemical skin conductance test, and thermoregulatory sweat test are other tests of sudomotor activity apart from SSR discussed previously. Specific autonomic testing like heart rate variability for cardiac autonomic neuropathy, urodynamic studies for genitorurinary, breath tests for gastrointestinal neuropathy also exist and further discussion of these tests is beyond the scope of this chapter.
- **Rate dependent depression:** This examines the spinal inhibitory systems by utilizing the H wave, an electrophysiological equivalent of the ankle reflex. Loss of GABAergic neurons in the spinal cord results in increased pain sensitivity and a loss of rate dependent depression of the H wave (Jin et al., 2021; Marshall et al., 2021). This does not evaluate peripheral nerves but the central mechanisms behind painful diabetic neuropathy.
- **Nerve excitability testing:** This is similar to standard nerve conduction studies but utilizes both submaximal and supramaximal stimulation using standardized excitability protocols. This provides an insight into the axon membrane and ion channel properties at the stimulation site (Marshall et al., 2021). These abnormalities often precede abnormalities in standard nerve conduction studies (Bae et al., 2011). Changes in glycemic control have shown to result in changes in the nerve excitability. These tests can be conducted using standard NCS equipment but relevant normative ranges are not available and it does not test small fibers (Marshall et al., 2021).

Imaging methods

- **Ultrasound:** In DPN the cross sectional area of the nerve increases and there is blurring of the characteristic "honeycomb" appearance. Echogenicity reduces due to increased water in the nerves and vascularity measured using doppler is increased but may reduce with worsening arteriosclerosis (Huang & Wu, 2021). It appears that ultrasound may show abnormalities before NCS but severity is better rated with NCS (Tandon et al., 2021). Ultrasound is a particularly useful tool to differentiate diabetic neuropathy from CIDP (Heiling et al., 2024) Focussed ultrasound may also be used as a therapeutic tool to improve nerve microcirculation (Tan et al., 2020). Shear wave sonoelastography which gives an indication of fibrosis is a newer technological addition to traditional ultrasound which may have a higher sensitivity (90%) and specificity (85%) compared to cross sectional area alone (sensitivity 60%; specificity 75%) for diagnosis of DPN (Dikici et al., 2017). The ability to simultaneously image muscles and quantify atrophy or neurogenic changes is another advantage for ultrasound.
- **Magnetic Resonance Neurography (MRN):** The advent of 3T and higher field strengths of MRI has resulted in fascicular level resolution of the peripheral nerve. MRN can also image more proximal nerves such as that involved in diabetic amyotrophy unlike Ultrasound. Larger nerve cross sectional area, T2 weighted hyperintense and hypointense lesions, evidence of nerve edema, reduced fractional anisotropy and increased apparent diffusion coefficient were reported in DPN in comparison to controls (Evans et al., 2021). Simultaneous imaging of the muscles can be done. Cost, specific imaging parameters, nerve choice and duration of the testing are limitations (Zhang & Zhang, 2023).
- **Others:** I-123 sympathetic MIBG scintigraphy and 11C-HED-PET are used to detect cardiac sympathetic activity in CAN (Pop-Busui, 2023).

Chemical biomarkers

Diabetic neuropathy can be characterized as resulting from a combination of the effects of the oxidative stress, advance glycation end products, depletion of nitric oxide, abnormal protein kinase C activity and reduced neurotrophic factors. Thus, chemical signals can be potential biomarkers of DPN (Adki & Kulkarni, 2023). Though a detailed description of these biomarkers is beyond the scope of this chapter, they can be broadly classified as:

- **Inflammatory biomarkers** like monocyte chemoattractant protein-1, Vascular endothelial growth factor, transient receptor potential vanilloid 1 and nuclear factor kappa-light chain enhancer of activated B cells.
- **Oxidative biomarkers** like nuclear factor erythroid derived 2 and adiponectin,
- **Enzyme biomarkers** like nicotnamide adnine dinucleotide phosphate oxidase, ceruloplasmin, heme oxygenase 1 dipeptidyl peptidase 4 and polyADP ribose polymerase
- **Miscellaneous** like sirtuin 1, caviolin1, microRNAs, metastatsis associated lung adenomcarcinoma transcript 1

TABLE 11.3 Point of care devices for DPN.

Name of instrument	Principle	Sensitivity, specificity
Neuroquick	Subjective thermal perception using airflow	Good correlation with quantitative sensory testing thresholds
NC stat, DPN check	Hand held evaluation of sural SNAP	88%–94%, 82%–94%
Sudoscan	Electrochemical skin conductance measured to estimate sudomotor function	77%–82%, 61%–86%

Digital and wearable technologies

- **Point of care and wearable devices:** Multiple point of care devices have become available for evaluation of diabetic neuropathy. In general most of these devices have been evaluated in non-primary care settings. These work on various principles and are summarized in Table 11.3 (Ananthakumar et al., 2016). While multiple methods are available their reliability, ability to monitor progress or relationship to clinically important outcomes is not known.
- **Artificial and machine learning (AI/ML) technologies:** There is a move toward automating and utilizing AI/ML technologies to obtain, classify and interpret diagnostic studies such as CCM and imaging studies. In a study from China utilizing data of 1544 diabetic patients using automated feature selection yielded algorithms with area under the curve of 0.811, 95% CI: 0.776–0.843 in the testing cohort for detection of DPN (Sun et al., 2025). These technologies can be used to power a multiomic approach to DPN as well. Direct therapeutic applications also exist as in the development of DPN footwear with realtime sensors which facilitate pressure offloading and thus prevent foot ulcers (Okaduwa et al., 2024).

Conclusion

In conclusion many existing and emerging techniques can help profile DPN. Unfortunately, their potential is limited by the lack of gold standard for definition of DPN. There is considerable overlap in presentation. For instance, around 50% of those with DSPN will have associated SFN and autonomic neuropathy. Structural and functional abnormalities in the nerves often predates the of overt clinical or NCS manifestations. Many of the newer techniques detect this and hence comparison with standard clinical or NCS becomes meaningless. On the other hand, such early identification may provide a window to intervene and prevent further damage.

In essence, we feel an approach with identification and quantification of symptoms, signs and the various types of nerve fibers involved, associated risk factors and various biomarkers and their integration by leveraging AI/ML technologies to provide a complex endophenotype may offer the best approach to diagnosis, screening and monitoring of diabetic neuropathy. Such methods can provide real time support to the clinician in providing a personalized approach to managing and prognosticating patients with DPN. However, such algorithms will need to be developed, tested and validated in large real-world cohorts.

Mini-dictionary of terms

1. *Nerve conduction studies*: These are lab evaluation where the peripheral nerves are stimulated using current and the responses recorded from the muscles or the nerve itself as compound muscle action potential or sensory nerve action potential.
2. *Electromyography*: These are lab evaluation methods where by needle electrodes are used to interrogate the muscle functions directly.
3. *Artificial intelligence/machine learning*: These refer to a group of techniques that leverage advances in computing technology to crunch and identify patterns in big databases that can be used for predictions, prognostication or as clinical decision support systems.
4. *Nerve fiber types*: Broadly the peripheral nerves can be grouped according to their diameters as large fibers Aα, Aβ which are myelinated and responsible for the motor activity as well as proprioception, touch and pressure sensations; Aγ fibers are smaller and are motor to muscle spindles while Aδ which are thinly myelinated carry fast pain, B fibers are preganglionic for the autonomic

nervous system and the unmyelinated C fibers have the smallest diameter and they carry pain, temperature and post-ganglionic autonomic functions.

5. *Autonomic functions*: These group together both the sympathetic and parasympathetic systems.

Summary points

1. Diabetic neuropathy is an important complication of diabetes and can cause significant morbidity and economic burden.
2. It is a very heterogenous condition and can have a variable presentations.
3. Current classifications and diagnostic criteria are inadequate to fully describe the entire spectrum leading to confusion and missed opportunities.
4. Clinical evaluation centers around symptoms and examination of the sensori-motor and autonomic systems.
5. Electrophysiological evaluations in the nerve conduction labs interrogate the distal nerves and predominatly signify large fiber involvement.
6. Many emerging methods exist that can detect abnormalities even in subclincal neuropathies. Their role in routine diagnostic, screening or monitoring pathways is yet to be elucidated.
7. Developing complex endophenotypes including not only clinical and electrophysiological but other newer and emerging biomarkers by leveraging AI/ML technologies may be the way forward for both diagnosis, screening, prognostication and therapy of diabetic neuropathy.

References

Adki, K. M., & Kulkarni, Y. A. (2023). Biomarkers in diabetic neuropathy. *Archives of Physiology and Biochemistry, 123*(2), 460–475.

Akbar, M., Wandy, A., Soraya, G. V., Goysal, Y., Lotisna, M., & Basri, M. I. (2023). Sudomotor dysfunction in diabetic peripheral neuropathy (DPN) and its testing modalities: A literature review. *Heliyon, 9*(7), e18184.

American Diabetes Association Professional Practice Committee. (2025). Pharmacologic approaches to glycemic treatment: Standards of care in diabetes-2025. *Diabetes Care*. https://doi.org/10.2337/dc25-S009

Ananthakumar, T., Heneghan, C., Price, C. P., Van den Bruel, A., & Plüddemann, A. (2016). *Point-of-care devices for detecting diabetic polyneuropathy: Horizon Scan Report 0046*. Oxford: NIHR Diagnostic Evidence Co-operative.

Ang, L., Mizokami-Stout, K., Eid, S. A., et al. (2022). The conundrum of diabetic neuropathies—Past, present, and future. *Journal of Diabetes and Its Complications, 36*(11), 108334.

Bae, J. S., Kim, O. K., & Kim, J. M. (2011). Altered nerve excitability in subclinical/ealry diabetic neuropathy: Evidence for early neurovascular process in diabetes mellitus. *Diabetes Research and Clinical Practice, 91*(2), 183–189.

Braune, H. J., & Horter, C. (1996). Sympathetic skin response in diabetic neuropathy: A prospective clinical and neurophysiological trial on 100 patients. *Journal of the Neurological Sciences, 138*(1–2), 120–124.

Bubenzer, L. J., Konsolke, L., Enax-Krumova, E., Eberhardt, F., Tegenthoff, M., Höffken, O., & Özgül, Ö. S. (2023). Pain-related evoked potentials with concentric surface electrodes in patients and healthy subjects: A systematic review. *Brain Structure and Function, 228*(7), 1581–1594.

Burgess, J., Frank, B., Marshall, A., et al. (2021). Early detection of diabetic peripheral neuropathy: A focus on small nerve fibres. *Diagnostics, 11*(2), 165.

Carmichael, J., Fadavi, H., Ishibashi, F., Shore, A. C., & Tavakoli, M. (2021). Advances in screening, early diagnosis and accurate staging of diabetic neuropathy. *Frontiers in Endocrinology, 12*, 671257.

Clinical overview: Diabetic peripheral neuropathy. In. *ClinicalKey*, (2021)). Elsevier. http://www.clinicalkey.com.

Cornblath, D. R., Chaudhry, V., Carter, K., Lee, D., Seysedadr, M., Miernicki, M., & Joh, T. (1999). Total neuropathy score: Validation and reliability study. *Neurology, 53*(8), 1660–1664.

Dias, L. S., Nienov, O. H., Parisi, M. C. R., & Schmid, H. (2015). Sensitivity and specificity of neuropathy diabetes score, neuropathy symptoms score, diabetic neuropathy score and esthesiometry compared with the gold standards Michigan neuropathy screening instrument (MNSI) and Beck depression inventory (BDI). *Diabetology & Metabolic Syndrome, 7*(S1), A199.

Dikici, A. S., Ustabasioglu, F. E., Delil, S., Nalbantoglu, M., Korkmaz, B., Bakan, S., Kula, O., Uzun, N., Mihmanli, I., & Kantarci, F. (2017). Evaluation of the tibial nerve with shear-wave elastography: A potential sonographic method for the diagnosis of diabetic peripheral neuropathy. *Radiology, 282*(2), 494–501.

Dillon, B. R., Ang, L., & Pop-Busui, R. (2024). Spectrum of diabetic neuropathy: New insights in diagnosis and treatment. *Annual Review of Medicine, 75*(1), 293–306.

Elsevier Point of Care. (2025). *Diabetic autonomic neuropathy*. Elsevier BV. https://elsevier.health/en-US/preview/diabetic-autonomic-neuropathy-co.

Evans, M. C., Wade, C., Hehenschurz-Schmidt, D., Lally, P., Ugwudike, A., Shah, K., Bangerter, N., Sharp, D. J., & Rice, A. S. C. (2021). Magnetic resonance imaging as a biomarker in diabetic and HIV associated peripheral neuropathy: A systematic review based narrative. *Frontiers in Neuroscience, 15*, 727311.

Feldman. (1994). A practical two-step quantitative clinical and electrophysiological assessment for the diagnosis and staging of diabetic neuropathy. *Diabetes Care*. https://doi.org/10.2337/diacare.17.11.1281

Ferdousi, M., Kalteniece, A., Azmi, S., Petropoulos, I. N., Worthington, A., D'Onofrio, L., et al. (2020). Corneal confocal microscopy compared with quantitative sensory testing and nerve conduction for diagnosing and stratifying the severity of diabetic peripheral neuropathy. *BMJ Open Diabetes Research & Care, 8*, e001801.

Freeman, R. (2009). Not all neuropathy in diabetes is of diabetic etiology: Differential diagnosis of diabetic neuropathy. *Current Diabetes Reports, 9*(6), 423–431.

Gad, H., Petropoulos, I. N., Khan, A., Ponirakis, G., MacDonald, R., Alam, U., & Malik, R. A. (2022). Corneal confocal microscopy for the diagnosis of diabetic peripheral neuropathy: A systematic review and meta-analysis. *Journal of Diabetes Investigation, 13*(1), 134–147.

Galiero, R., Ricciardi, D., Pafundi, P. C., Todisco, V., Tedeschi, G., Cirillo, G., & Sasso, F. C. (2021). Whole plantar nerve conduction study: A new tool for early diagnosis of peripheral diabetic neuropathy. *Diabetes Research and Clinical Practice, 176*, 108856.

Guo, Y., Palmer, J. L., Brown, X. S., & Fu, J. B. (2015). Sural and radial sensory responses in patients with sensory polyneuropathy. *Clinical Medical Reviews and Case Reports, 2*(3), 049.

Gylfadottir, S. S., Christensen, D. H., Nicolaisen, S. K., Andersen, H., Callaghan, B. C., Itani, M., Khan, K. S., Kristensen, A. G., Nielsen, J. S., Sindrup, S. H., Andersen, N. T., Jensen, T. S., Thomsen, R. W., & Finnerup, N. B. (2020). Diabetic polyneuropathy and pain, prevalence, and patient characteristics: A cross-sectional questionnaire study of 5,514 patients with recently diagnosed type 2 diabetes. *Pain, 161*(3), 574–583.

Heiling, B., Kneer, K., He, W., Lehmann, T., Müller, N., Kloos, C., Grimm, A., & Axer, H. (2024). Nerve ultrasound helps to distinguish CIDP patients with diabetes from patients with diabetic polyneuropathy. *Scientific Reports, 14*(1), 30504.

Huang, H., & Wu, S. (2021). Application of high-resolution ultrasound on diagnosing diabetic peripheral neuropathy. *Diabetes, Metabolic Syndrome and Obesity, 14*, 139–152.

Jensen, T. S., Karlsson, P., Gylfadottir, S. S., et al. (2021). Painful and non-painful diabetic neuropathy, diagnostic challenges and implications for future management. *Brain, 144*(6), 1632–1645.

Jin, H. Y., Moon, S. S., & Calcutt, N. A. (2021). Lost in translation? Measuring diabetic neuropathy in humans and animals. *Diabetes & Metabolism Journal, 45*(1), 27–42.

Kahn, R. (1992). Quantitative sensory testing. *Diabetes Care, 15*(s3), 1092–1095.

Kallinikou, D., Soldatou, A., Tsentidis, C., et al. (2019). Diabetic neuropathy in children and adolescents with type 1 diabetes mellitus: Diagnosis, pathogenesis, and associated genetic markers. *Diabetes Metabolism Research, 35*(7), e3178.

Kartheka, R., Aghoram, R., Faith, A. J., & Wadwekar, V. (2024). Cross-sectional study of the relationship between medial plantar nerve conduction studies and severity of diabetic neuropathy. *Annals of Indian Academy of Neurology, 27*(2), 183–187.

Lamontagne, A., & Buchthal, F. (1970). Electrophysiological studies in diabetic neuropathy. *Journal of Neurology Neurosurgery and Psychiatry, 33*(4), 442–452.

Lee, J. E., & Won, J. C. (2025). Clinical phenotypes of diabetic peripheral neuropathy: Implications for phenotypic-based therapeutics strategies. *Diabetes & Metabolism Journal, 49*(4), 542–564.

Liu, H., Tan, S., Ma, Z., Gao, Q., & Yang, W. (2024). Sympathetic skin response for early detection of type 2 diabetic peripheral neuropathy and nephropathy. *Journal of Diabetes Investigation, 15*(1), 106–112.

Marshall, A., Alam, U., Themistocleous, A., Calcutt, N., & Marshall, A. (2021). Novel and emerging electrophysiological biomarkers of diabetic neuropathy and painful diabetic neuropathy. *Clinical Therapeutics, 43*(9), 1441–1456.

Meijer, J. W., Smit, A. J., Sonderen, E. V., Groothoff, J. W., Eisma, W. H., & Links, T. P. (2002). Symptom scoring systems to diagnose distal polyneuropathy in diabetes: The diabetic neuropathy symptom score. *Diabetic Medicine, 19*(11), 962–965.

Mogilevskaya, M., Gaviria-Carrillo, M., Feliciano-Alfonso, J. E., Barragan, A. M., Calderon-Ospina, C. A., & Nava-Mesa, M. O. (2024). Diagnostic accuracy of screening tests for diabetic peripheral neuropathy: An umbrella review. *Journal of Diabetes Research, 2024*, 5902036.

Oh, J. (2020). Clinical spectrum and diagnosis of diabetic neuropathies. *Korean Journal of Internal Medicine, 35*(5), 1059–1069.

Okaduwa, S. I. R., Igiri, B. E., Tagang, J. I., Okaduwa, U. J., & Adeyi, A. O. (2024). Therapeutic footwear approach for management of neuropathic diabetic foot ulcer: Current challenges and focus on future perspectives. *Medicine in Novel Technology and Devices, 23*, 100311. https://doi.org/10.1016/j.medntd.2024.100311

Perkins, B., & Bril, V. (2014). Electrophysiologic testing in diabetic neuropathy. *Handbook of Clinical Neurology, 126*, 235–248.

Perkins, B. A., Ngo, M., & Bril, V. (2002). Symmetry of nerve conduction studies in different stages of diabetic polyneuropathy. *Muscle & Nerve, 25*(2), 212–217.

Perkins, B. A., Olaleye, D., Zinman, B., & Bril, V. (2001). Simple screening tests for peripheral neuropathy in the diabetes clinic. *Diabetes Care, 24*(2), 250–256.

Pop-Busui, R. (2023). Autonomic diabetic neuropathies: A brief overview. *Diabetes Research and Clinical Practice, 206*, 110762.

Pop-Busui, R., Ang, L., Boulton, A., et al. (2022). Diagnosis and treatment of painful diabetic peripheral neuropathy. *Compendia, 2022*(1), 1–32.

Pop-Busui, R., Boulton, A. J. M., Feldman, E. L., et al. (2017). Diabetic neuropathy: A position statement by the American Diabetes Association. *Diabetes Care, 40*(1), 136–154.

Rosenberger, D. C., Blechschmidt, V., Timmerman, H., Wolff, A., & Treede, R.-D. (2020). Challenges of neuropathic pain: Focus on diabetic neuropathy. *Journal of Neural Transmission, 127*(4), 589–624.

Samakidou, G., Eleftheriadou, I., Tentolouris, A., Papanas, N., & Tentolouris, N. (2021). Rare diabetic neuropathies: It is not only distal symmetrical polyneuropathy. *Diabetes Research and Clinical Practice, 177*, 108932.

Samolis, A., Troupis, T., Politis, C., Pantazis, N., Triantafyllou, G., Tsakotos, G., Tegos, T., Lazaridis, N., Natsis, K., & Piagkou, M. (2025). Intraepidermal nerve fiber density as an indicator of neuropathy predisposition: A systematic review with meta-analysis. *Diagnostics, 15*(11), 1311.

Sletten, D. M., Suarez, G. A., Low, P. A., Mandrekar, J., & Singer, W. (2012). Compass 31: A refined and abbreviated composite autonomic symptom score. *Mayo Clinic Proceedings, 87*(12), 1196–1201.

Sommer. (2011). A modified score to identify and discriminate neuropathic pain: A study on the German version of the Neuropathic Pain Symptom Inventory (NPSI). *BMC Neurol*. https://doi.org/10.1186/1471-2377-11-104

Sun, M., Sun, X., Wang, F., & Liu, L. (2025). Machine learning-based prediction of diabetic peripheral neuropathy: Model development and clinical validation. *Frontiers in Endocrinology, 16*, 1614657.

Sun, J., Wang, Y., Zhang, X., Zhu, S., & He, H. (2020). Prevalence of peripheral neuropathy in patients with diabetes: A systematic review and meta-analysis. *Primary Care Diabetes, 14*(5), 435–444.

Tan, J. S., Lin, C. C., & Chen, G. S. (2020). Vasomodulation of peripheral blood flow by focused ultrasound potentiates improvement of diabetic neuropathy. *BMJ Open Diabetes Research & Care, 8*(1), e001004.

Tandon, A., Khullar, T., Maheshwari, S., Bhatt, S., & Narang, S. (2021). High resolution ultrasound in subclinical diabetic neuropathy: A potential screening tool. *Ultrasound, 29*(3), 150–161.

Tankisi, H., Pugdahl, K., Beniczky, S., Andersen, H., & Fuglsang-Frederiksen, A. (2019). Evidence-based recommendations for examination and diagnostic strategies of polyneuropathy electrodiagnosis. *Clinical Neurophysiology Practice, 4*, 214–222.

Tesfaye, S., Boulton, A. J. M., Dyck, P. J., et al. (2010). Diabetic neuropathies: Update on definitions, diagnostic criteria, estimation of severity, and treatments. *Diabetes Care, 33*(10), 2285–2293.

Uluc, K., Isak, B., Borucu, D., Temucin, C. M., Cetinkaya, Y., Koytak, P. K., Tanridag, T., & Us, O. (2008). Medial plantar and dorsal sural nerve conduction studies increase the sensitivity in the detection of neuropathy in diabetic patients. *Clinical Neurophysiology, 119*(4), 880–885.

van Driel, M. E. C., Huygen, F. J. P. M., & Risjsdijk, M. (2024). Quantitiative sensory testing: A practical guide and clinical applications. *British Journal of Anaesthesia, 24*(9), 326–334.

Willits, I., Cole, H., Jones, R., Dimmock, P., Arber, M., Craig, J., & Sims, A. (2015). ViibraTip for testing vibration perception to detect diabetic peripheral neuropathy: A NICE medical technology guidance. *Applied Health Economics and Health Policy, 13*(4), 315–324. https://doi.org/10.1007/s40258-015-0181-6

Yang, Y., Zhao, B., Wang, Y., Lan, H., Liu, X., Hu, Y., & Cao, P. (2025). Diabetic neuropathy: Cutting-edge research and future directions. *Signal Transduction and Targeted Therapy, 10*(1), 132.

Yu, Y. (2021). Gold standard for diagnosis of DPN. *Frontiers in Endocrinology, 12*, 719356.

Zelman, D. C., Gore, M., Dukes, E., Tai, K. S., & Brandenburg, N. (2005). Validation of a modified version of the brief pain inventory for painful diabetic peripheral neuropathy. *Journal of Pain and Symptom Management, 29*(4), 401–410.

Zhang, C., Ding, M., Liang, X., & Zhuo, L. (2024). Application and progress of corneal confocal microscopy in the evaluation of diabetes-related peripheral neuropathy. *Endokrynologia Polska, 75*(4), 395–402.

Zhang, X., & Zhang, F. (2023). Peripheral neuropathy in diabetes: What can MRI do? *Diabetes, 72*(8), 1060–1069.

Ziegler, D. (2023). Pathogenetic treatments for diabetic peripheral neuropathy. *Diabetes Research and Clinical Practice, 206*, 110764.

Zilliox, L., Peltier, A. C., Wren, P. A., Anderson, A., Smith, A. G., Singleton, J. R., Feldman, E. L., Alexander, N. B., & Russell, J. W. (2011). Assessing autonomic dysfunction in early diabetic neuropathy: The survey of autonomic symptoms. *Neurology, 76*(12), 1099–1105.

Chapter 12

Linking brain insulin receptors, neural stem cells and synaptic plasticity: Potential implications for neurodegenerative disorders

Sneha Raut[1], Jaldeep Langhnoja[2] and Prakash Pillai[1]

[1]*Division of Neurobiology, Department of Zoology, Faculty of Science, The Maharaja Sayajirao University of Baroda, Vadodara, Gujarat, India;* [2]*Department of Radiation Oncology, University of California, Irvine, Medical Sciences I, Irvine, CA, United States*

List of abbreviations

AMPA α-amino-3-hydroxy-5-methyl-4-isoxazolepropionic acid (receptor)
BDNF Brain-derived neurotrophic factor
CAMKII Ca^{2+}/calmodulin-dependent protein kinase II
CREB cAMP response element-binding protein
FGF Fibroblast growth factor
FOXO Forkhead box O transcription factors
GLP-1 Glucagon-like peptide-1
GLUT 1/4 Glucose transporter type 1 and type 4
IGF Insulin-like growth factor
IL-6 Interleukin-6
IR Insulin receptor
JNK c-Jun N-terminal kinase
MAPK Mitogen-activated protein kinase
mTOR Mechanistic (mammalian) target of rapamycin
NMDA N-methyl-D-aspartate (receptor)
NR2B GluN2B subunit of the NMDA receptor
NSC Neural stem cell
PI3K Phosphoinositide 3-kinase
PKC Protein kinase C
PP1 Protein phosphatase one
PTP1B Protein tyrosine phosphatase 1B
S6K Ribosomal protein S6 kinase
SGZ Subgranular zone
SVZ Subventricular zone
TNF-α Tumor necrosis factor alpha
TrkA Tropomyosin receptor kinase A
TrkB Tropomyosin receptor kinase B
Wnt Wingless-related integration site signaling pathway

Introduction

Neurodegenerative disorders like Alzheimer's disease (AD), Parkinson's disease (PD) and Huntington's disease (HD) have been a growing global health challenge, affecting millions of individuals and are marked by progressive neuronal loss and impaired synaptic communication and a gradual decline in the cognitive and functional abilities. Increasing evidences have showed that these disorders are not solely driven by protein misfolding or cellular degeneration but also by disruptions in brain's metabolic and signaling activity (Hartl & Hayer-Hartl, 2009). However, these activities are largely irreversible and the current modes of treatment are insufficient as they mostly focus on relieving the symptoms of the patients and therefore, effective treatments are urgently required.

One of the early signs of these neurodegenerative disorders is the synaptic dysfunction. Disruptions of synaptic processes such as long-term potentiation (LTP) and long-term depression (LTD), compromise neural stability and is now recognised as a key factor of disease progression rather than it simply being a consequence of neuronal death (Andrade-Talavera & Rodríguez-Moreno, 2021).

In recent years, there has been a growing interest in research around the insulin signaling in brain. Traditionally, insulin was viewed as a peripheral hormone that controlled the glucose metabolism. However, it has now been observed that it also has significant functions in the central nervous system. Insulin receptors (IRs) are widely

Hormones in Neuroscience. https://doi.org/10.1016/B978-0-443-30263-3.00024-5

expressed in the brain, particularly in areas associated with memory, cognition and homeostasis. Recent evidences indicate that insulin resistance in the brain, characterized by impaired insulin signaling, leads to synaptic failure, disrupted metabolic activity and neurodegenerative pathology, particularly in AD (Kellar & Craft, 2020).

In addition to synaptic alterations, neural stem cell activity and adult neurogenesis have been identified as additional factors that are involved in brain plasticity and repair (Horgusluoglu et al., 2017). Adult NSCs survive in specific brain niches and produce neurons that integrate into the existing circuits, particularly within the hippocampus. In neurodegenerative disorders, NSC proliferation, differentiation and survival are impaired, limiting endogenous regeneration. Markedly, NSCs are sensitive to metabolic cues such as insulin and insulin-like growth factor (IGF) signaling, placing them at the junction of neural plasticity and metabolic regulation.

Despite significant advances in each of these areas, insulin receptor signaling, neural stem cell biology and synaptic plasticity are often studied separately. An increasing amount of research over the years, have observed that these systems are intricately linked. Therefore, it is important to understand how metabolic signaling interacts with synaptic and cellular plasticity.

This chapter summarizes the current understanding of the brain insulin receptors, neural stem cells and synaptic plasticity, with an emphasis on their interrelated roles in neurodegenerative disorders (Fig. 12.1).

Brain insulin receptors and neurodegenerative disorders

In the brain, insulin signaling has been observed to be a crucial regulator of synaptic plasticity, neuronal survival and cognitive function. Insulin functions as a neuromodulator that affects neurotransmission, learning, memory and neurogenesis in addition to its traditional metabolic functions. It is now known that disruption of brain insulin receptor signaling plays a major role in the pathophysiology of a number of neurodegenerative diseases where impaired insulin responsiveness is associated with synaptic dysfunction, metabolic stress and progressive neuronal loss. Understanding the structure, distribution and signaling mechanisms of brain receptors is therefore essential for clarifying their role in neurodegeneration and identifying potential therapeutic targets.

Insulin receptor structure and ligand

The insulin receptor (IR) is a transmembrane receptor tyrosine kinase exists as a heterodimer consisting of two α-chains and two β-chains, commonly represented as $(\alpha\beta)_2$ where the two extracellular α-chains and the two transmembrane β-chains are linked by disulfide bonds (Meyts et al., 2013). Alternative splicing produces two isoforms: IR-A and IR-B, where IR-A has higher affinity for IFGs and IR-B binds with insulin better and drives the glucose uptake in the cell (Escribano et al., 2017). When the receptor interacts with insulin, its ligand, it undergoes a conformational change that activates the intrinsic tyrosine kinase activity of the β subunits, leading to receptor autophosphorylation at multiple sites. This activation triggers a series of coordinated phosphorylation events and molecular interactions. These events drive Akt2-dependent recruitment of glucose transporter Glut4 to the cell surface, thereby improving the glucose uptake. Parallelly, MAPK signaling regulates transcriptional programs and promote lipid, protein and glycogen synthesis (Gonzalez & McGraw, 2006). The differential expression of insulin receptors in brain has functional implications. The IR-A isoform is predominantly expressed in neurons, where it supports the cellular growth, survival and synaptic maintenance and IR-B is more expressed in astrocytes and is mostly involved in metabolic regulation (Pomytkin et al., 2018). Alternative splicing or structural modifications of receptors may alter ligand binding kinetics and downstream signaling.

Insulin receptor distribution in the brain

Insulin receptors are present throughout the brain, although a bit unevenly. The hippocampus, hypothalamus, cerebral cortex, olfactory bulb, amygdala, cerebellum and choroid plexus are hotspots in cognition and metabolism hubs. Majority of IRs are found in neurons, particularly at the synapses and postsynaptic densities. IR-A, which is predominantly expressed in neurons has a lower molecular weight due to differential glycosylation (Belfiore et al., 2009). The hippocampal CA1/CA3 and dentate gyrus have high density of IRs, connecting to memory circuits. Astrocytes and microglia exhibit comparatively low levels of insulin receptor expression, predominantly the IR-B isoform (Chen et al., 2025). The colocalization of IRs with glucose transporters such as GLUT1 and GLUT4 suggests a role for insulin signaling in fine tuning local glucose utilization within the brain. Brain insulin originates from both local neuronal synthesis and peripherally secreted pancreatic β-cells via transport proteins across vascular endothelium (Milstein & Ferris, 2021). This dual source of insulin may support its critical roles in neuronal signal modulation, glucose uptake and synaptic plasticity in the CNS (Table 12.1).

Insulin receptor signaling pathways

Upon binding of insulin, the β subunits of insulin receptor undergo autophosphorylation, creating docking sites for adaptor proteins such as insulin receptor substrates 1 and 2

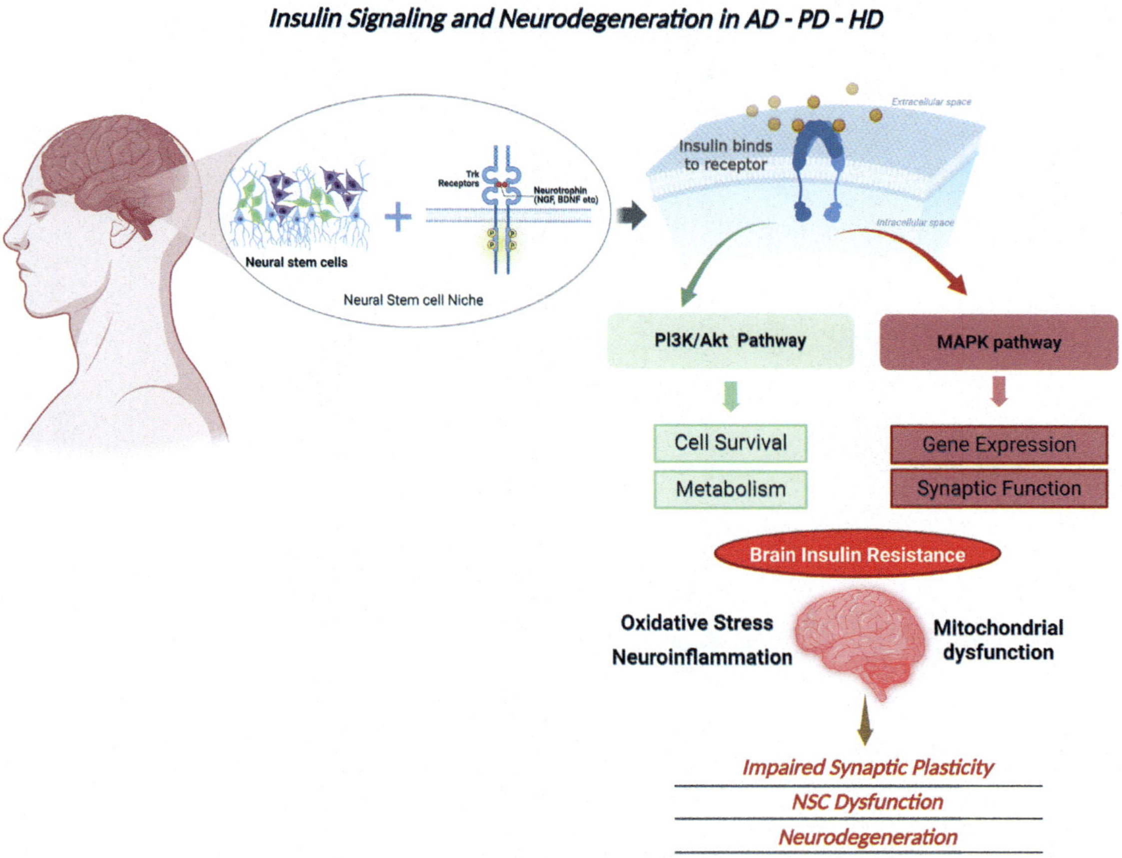

FIGURE 12.1 Insulin signaling and its role in neurodegeneration in Alzheimer's Disease (AD), Parkinson's Diseases (PD) and Huntington's Disease (HD). This schematic depicts the role of insulin signaling in the CNS. Within the NSC niche, insulin receptor binding activates key downstream signaling pathways which are crucial for survival, growth and functional regulation. Disruption of insulin signaling in brain leads to insulin resistance which ultimately results in neuronal dysfunction and mechanisms underlying neurodegenerative disorders. *Created in BioRender. at UCI, L. L. (2026) https://BioRender.com/xrpj8g2.*

(Virkamäki et al., 1999). This interaction triggers PI3K and MAPK pathways, the two main downstream signaling pathways (De Meyts, 2000). Neuronal metabolism, survival and synaptic maintenance are all heavily influenced by the PI3K-AKT axis. FOXO transcription factors (control quiescence and the stress responses), GSK3β (involved in tau phosphorylation) and the mechanistic target of rapamycin pathway, which controls the protein synthesis and cellular growth, are just a few of the substrates that activate AKT phosphorylates (Kumar & Bansal, 2022; Long et al., 2021). Additionally, Akt signaling facilitates GLUT4 translocation to the plasma membrane, which improves glucose absorption and supports energy requirement of neurons.

In parallel, insulin receptor activation engages MAPK/ERK pathways through adaptor proteins such as Grb2 and Sos, leading to Ras dependent ERK activation (Choi & Bai, 2023). This pathway mainly controls the expression of activity-dependent neuronal genes and activation of Elk1, commonly linked to neuronal function, which may further be amplified by crosstalk between IR or IGFR-1 signaling, especially during development and synaptic remodeling.

Hypothalamic ERK activity further regulates appetite and energy homeostasis while, hippocampal AKT signaling can support LTP and synaptic stability. Additionally, serine phosphorylation of IRS proteins, which results from long-term activation of kinases like S6K and JNK, reduces insulin signaling and increases central insulin resistance (Boura-Halfon & Zick, 2009). TNF-α and other neuroinflammatory mediators can aggravate AD pathophysiology by disrupting insulin signaling, leading to excessive mTOR activation, impaired autophagy and increased neuronal vulnerability due to reduce Akt activity. Therapeutic strategies targeting phosphatases such as PTP1B aim to restore insulin signaling and protect synaptic integrity in neurodegenerative disorders.

Effect of insulin resistance in neurodegeneration

Brain insulin resistance disrupts PI3K/Akt signaling, leading to pathological activation of downstream effectors that contribute to neurodegeneration. Unchecked GSK3-β

TABLE 12.1 Insulin receptor distribution across different cell types in brain.

Brain region	Cell types expressing insulin receptors	Functional notes	Species
Cerebral cortex	Pyramidal neurons, interneurons, astrocytes	Regulates synaptic plasticity, cognition, and glucose uptake in neurons and glia	Human, rodent
Hippocampus	CA1 and CA3 pyramidal neurons, dentate gyrus granule cells, astrocytes	Supports learning, memory, and synaptic modulation via insulin signaling	Human, rodent
Hypothalamus	Neurons in arcuate nucleus, paraventricular nucleus, and ventromedial hypothalamus	Central regulation of appetite, glucose metabolism, and energy balance	Human, rodent
Olfactory bulb	Mitral and granule cells	Modulates odor sensitivity and metabolic signaling	Rodent
Cerebellum	Purkinje neurons, Bergmann glia	Contributes to motor coordination and insulin-mediated neuroprotection	Rodent
Striatum	Medium spiny neurons, interneurons	Involved in dopamine regulation and reward processing	Human, rodent
Amygdala	Principal neurons, interneurons	Affects emotional regulation and stress response through insulin signaling	Rodent
Brainstem	Neurons in dorsal vagal complex, locus coeruleus	Regulates autonomic and metabolic responses	Rodent
Choroid plexus & endothelial cells	Epithelial and endothelial cells lining ventricles and capillaries	Mediates insulin transport across blood - brain barrier	Human, rodent

activation due to decreased AKT activity results in tau hyperphosphorylation and development of neurofibrillary tangles (Lee et al., 2024). Simultaneously, decreased insulin signaling reduces activity of insulin-degrading enzymes, which lowers the clearance of amyloid beta and promotes its accumulation (Farris et al., 2003). These molecular alterations result in synaptic failure, which is defined by decreased AMPA receptor trafficking, loss of dendritic spines and impaired LTP. These characteristics are frequently seen in models of PD and AD and are similar to phenotypes seen in mice with neuron-specific insulin receptor knockouts.

Insulin resistance also worsens oxidative stress and mitochondrial dysfunction. Persistent nuclear localization of FOXO TFs enhances stress related gene expression, while compromised mitochondrial bioenergetics further undermines neuronal viability (Maiese, 2017). In AD, these processes form a self-reinforcing cycle in which amyloid beta interferes with insulin receptor function, and worsening insulin resistance accelerates amyloid pathology. PD is associated with insulin signaling deficits in dopaminergic neurons of the substantia nigra, contributing to neuronal loss, while HD exhibits pronounced striatal degeneration linked to disrupted metabolic and synaptic signaling.

Synaptic plasticity is particularly vulnerable to impaired insulin signaling. Both LTP and LTD are compromised and adult hippocampal neurogenesis within the dentate gyrus is significantly reduced, further limiting network adaptability. Therapeutic interventions aimed at restoring insulin sensitivity, including metformin and GLP-1R agonists, have demonstrated partial recovery of Akt signaling and synaptic plasticity in preclinical models (Ríos et al., 2024). However, advanced age and chronic neuroinflammation significantly limits therapeutic efficacy, underscoring the complexity of reversing insulin resistance in the diseased brain. Collectively, these findings highlight how metabolic dysfunction destabilizes neural circuits and identify insulin signaling as a potential target for disease modifying interventions.

Neural stem cells in neurodegenerative disorders

Neural stem cells (NSCs) are an integral factor for brain development, plasticity and repair due to their capacity for self-renewal and differentiation into neurons and glial cells. NSCs are recognised as key factors in neurodegeneration and a comprehensive assessment of NSC regulation, dysfunction and potential in neurodegeneration can be achieved through understanding how NSCs respond to neurodegenerative cues and how it serves as a basis for investigating disease mechanisms and therapeutics strategies.

Neural stem cell niches

NSCs are the self-renewing, multipotent and largely undifferentiated cells, originating from the central nervous system (CNS). They have capacity to give rise to all major and important cells of the CNS such as neurons, astrocytes and oligodendrocytes. These are localized in specialized niche called subventricular zone (SVZ) and subgranular zone (SGZ) (Llorente et al., 2022), located in the two neurogenic regions called lateral ventricle and dentate gyrus of the hippocampus in an adult mammalian brain. Within these niches, NSCs receive signals that regulate their proliferation, differentiation and self-regulation, allowing the niche to maintain homeostasis modulate broader CNS regulation.

NSC development begins with proliferation, where the neuroepithelial cells or the radial cells first undergo a symmetric division and then asymmetric division to balance the self-renewal with production of neurons and intermediate progenitors (Kempermann et al., 2015). During this stage cells acquire regional identity, through transcription factors and spatial cues. In the neurogenic phase, these cells produce transit-amplifying progenitors that rapidly divide into neuroblasts, which then migrate and differentiate into neurons. As development proceeds, NSCs transition into gliogenic phase, producing astrocytes and then oligodendrocytes in a temporally ordered manner. A subset of embryonic NSC persists even in adulthood as quiescent or slowly cycling cells in SVZ or SGZ niches where they continue to undergo activation, proliferation, neuronal commitment, migration and maturation.

NSCs are also tightly regulated by extrinsic factors from their microenvironment. These growth factors (GFs) play a central role in controlling the proliferation, differentiation and lineage commitment. These signaling molecules are essential for both embryonic and adult neurogenesis, and understanding their mechanisms is critical for insights into brain development, plasticity, and potential regenerative therapies.

Role of NSCs in neurodegenerative disorders

In neurodegenerative disorders, NSC function becomes compromised or insufficient to compensate the speed of neuronal loss. Although NSCs often initiate compensatory proliferation in response to injury to degeneration, chronic disease-associated changes such as persistent inflammation, oxidative stress, vascular dysfunction or disrupted metabolic signaling can impair their capacity to generate functional neurons (Torrisi et al., 2025). This reduced availability of key GFs such as BDNF, IGF-1, FGF-2, etc., may further limit their function. Additionally, abnormal glial activation and ECM remodeling within the diseased niches can create a non-permissive environment for neuronal integration, thereby weakening the brain's intrinsic regenerative and repair mechanisms.

Link between insulin receptors and NSCs

NSC function within neurogenic niches is governed by extrinsic signals, including insulin receptor signaling and transcriptional regulation. IR activation additionally dictates lineage specification within the NSCs. IR-A isoform is primarily associated with cellular proliferation and survival, whereas IR-B is mainly involved in metabolic functions. It also initiates multiple downstream signaling pathways that regulate NSC maintenance and differentiation. The PI3K/Akt pathway facilitates the cell cycle progression and provides protection against apoptosis (Hossini et al., 2016). Simultaneously, the MAPK/ERK pathway influences differentiation and early lineage specification. Insulin signaling via PI3K pathway also influences FOXO transcription factor, balancing NSC quiescence, oxidative stress response and long-term cell preservation (Sakaguchi, 2024). Impairments in insulin signaling can result in reduced neuronal output and a shift toward glial fate adoption. Insulin also modulates the production of essential factors such as BDNF, VEGF and Wnt proteins, which are important for creating an environment conductive to neurogenesis.

Insulin receptor signaling in NSCs is a key factor in neurodegenerative disorders like AD, PD and HD. Impaired signaling reduces neurogenic potential and contributes to cognitive decline.

Synaptic plasticity and neurodegenerative disorders

Synaptic plasticity is the capacity of synapses to alter their strength in response to activity-underlies learning, memory, and adaptive behavior. By enabling dynamic neural reorganization, it integrates experience into lasting brain modifications. Yet this flexibility renders synapses highly susceptible to aging and neurodegenerative disruptions that precipitate cognitive and functional decline.

Molecular mechanisms underlying synaptic plasticity

Synaptic plasticity embodies the molecular conversion of transient neuronal activity into enduring modifications in synaptic efficacy. The concept that neurons coactivated through experience strengthen their connections-experimentally verified as long-term potentiation (LTP) and its counterpart, long-term depression (LTD)-forms the cellular basis of learning and memory. NMDA-type glutamate receptors serve as critical coincidence detectors, mediating calcium influx only when presynaptic glutamate release coincides with postsynaptic depolarization. The amplitude and duration of this calcium signal dictate plasticity polarity: rapid, high-amplitude transients activate CaMKII and PKC, enhancing AMPA receptor phosphorylation and membrane incorporation during LTP, whereas prolonged, moderate calcium activates phosphatases such as PP1 and calcineurin, driving receptor dephosphorylation and internalization in LTD (Yasuda et al., 2022) Concurrently, actin remodeling governed by Rho-family GTPases sculpts dendritic spine morphology. Sustained plasticity requires CREB-dependent transcription of Arc, c-fos, and BDNF, while BDNF-TrkB signaling via MAPK/ERK and PI3K-Akt-mTOR pathways consolidates structural and metabolic reinforcement of synaptic strength (Park & Poo, 2013).

Synaptic plasticity in memory and learning

Synaptic plasticity constitutes the fundamental cellular mechanism by which experiences reshape neural circuitry, enabling the encoding, storage, and retrieval of memories. Memory formation unfolds in temporally distinct stages: the encoding phase depends on early long-term potentiation (LTP), marked by rapid post-translational modification and activity-dependent AMPA receptor incorporation; the consolidation phase engages late LTP, requiring de novo gene transcription and synaptic remodeling driven by CREB and BDNF signaling (Kandel et al., 2014). Conversely, long-term depression (LTD) refines and reorganizes memory traces by weakening nonessential synapses. Experimental evidence strongly supports this link-NMDA receptor inhibition disrupts spatial learning, while behavioral paradigms such as fear conditioning evoke LTP-like potentiation (Ogden et al., 2014). Optogenetic manipulations further reveal that targeted induction or reversal of plasticity can generate or abolish specific memories. At the systems level, hippocampal replay during sleep reinforces neocortical connectivity, while neuromodulators-dopamine, acetylcholine, and norepinephrine-fine-tune synaptic gain, ensuring adaptive stabilization of experience-dependent neural representations through BDNF-mediated signaling.

Synaptic plasticity in aging and neurodegeneration

Advancing age and neurodegenerative pathology progressively compromise the molecular systems sustaining synaptic plasticity. Even during normative aging, hippocampal LTP and LTD exhibit attenuated magnitude and stability due to reduced NMDA receptor efficiency, notably diminished NR2B subunit expression, which limits calcium influx and downstream activation of CaMKII. Dysregulated AMPA receptor trafficking and impaired calcium buffering further constrain synaptic adaptability. Mitochondrial deficits which characterized by reduced biogenesis, decreased ATP output, and elevated reactive oxygen species which induce oxidative injury to synaptic proteins and cytoskeletal scaffolds essential for structural plasticity

(Mattson & Arumugam, 2018). Concurrently, chronic neuroinflammation driven by activated microglia and astrocytes releases cytokines (such as TNF-α, IL-1β, IL-6 etc.) that suppress LTP (Navakkode & Kennedy, 2024) and promote synapse loss. Transcriptional rigidity, marked by reduced CREB activity, BDNF expression, and epigenetic repression, limits adaptive gene expression. In Alzheimer's, amyloid-β and hyperphosphorylated tau disrupt receptor signaling and cytoskeletal stability; in Parkinson's and Huntington's disease, dopaminergic imbalance and impaired BDNF transport further degrade synaptic plasticity. Enhancing mitochondrial resilience, neurotrophic signaling, and anti-inflammatory pathways represents a key therapeutic frontier.

Regulators of synaptic plasticity: molecular, cellular, and transcriptional control

Synaptic plasticity is governed by an intricate hierarchy of molecular, cellular, and transcriptional regulators that collectively maintain neuronal adaptability and stability. At the molecular level, reversible post-translational modifications enable rapid and spatially confined modulation of synaptic strength. CaMKII serves as a central transducer of transient calcium influx, sustaining potentiation through AMPA receptor phosphorylation, while phosphatases such as PP1 and PP2A mediate counteracting dephosphorylation to restore synaptic equilibrium (Foley et al., 2021). The dynamic interplay between these opposing enzyme systems dictates whether neural activity elicits potentiation or depression. At the cellular level, calcium-dependent signaling orchestrates actin remodeling via Rho-family GTPases, linking biochemical cues to spine morphology. Long-term stabilization arises through transcriptional and epigenetic control: CREB activation induces immediate early genes including Arc and zif268, coordinating receptor trafficking and cytoskeletal organization. Histone acetylation, HDAC inhibition, and microRNA-mediated translational control further refine activity-dependent gene expression, ensuring experience-driven plasticity remains both persistent and homeostatically constrained (Kandel et al., 2014).

Alterations in synaptic plasticity in Alzheimer's, Parkinson's, and Huntington's diseases

Neurodegenerative disorders share a convergent hallmark, progressive impairment of synaptic plasticity that precedes neuronal loss and drives cognitive and motor decline. In Alzheimer's disease (AD), soluble amyloid-β oligomers perturb NMDA receptor mediated calcium signaling, favoring LTD and AMPA receptor internalization, while tau pathology destabilizes microtubules and disrupts axonal transport. Mitochondrial dysfunction and chronic neuroinflammation exacerbate oxidative injury, with synaptic loss showing the strongest correlation to cognitive deterioration. Therapeutic restoration of BDNF-TrkB signaling and modulation of glutamatergic transmission remain key experimental targets (De Strooper & Karran, 2016). In Parkinson's disease (PD), degeneration of nigrostriatal dopaminergic neurons abolishes Dand D2 receptor regulation of LTP and LTD, while α-synuclein aggregates impair vesicular release and mitochondrial energy dynamics (Rubenstein, 1971). In Huntington's disease (HD), mutant huntingtin alters transcriptional control and BDNF transport, weakening corticostriatal connectivity and promoting excitotoxic calcium overload. Across these disorders, disrupted calcium signaling, mitochondrial failure, kinase-phosphatase imbalance, and neuroinflammation form a shared pathological axis, underscoring synaptic restoration as a central therapeutic goal.

Connecting insulin receptors, synaptic plasticity, and neural stem cells

Insulin receptor signaling in the brain integrates metabolic cues with neurogenesis and synaptic plasticity. Through PI3K–AKT and MAPK–ERK pathways, insulin receptors regulate neural stem cell fate, neurotrophic signaling, and synaptic function, positioning disrupted insulin signaling as a central mechanism underlying cognitive decline and neurodegenerative vulnerability.

Insulin receptor regulation of neural stem cell proliferation and differentiation

Adult neural stem cells (NSCs) reside within specialized neurogenic niches which is the subgranular zone of the dentate gyrus and the subventricular zone lining the lateral ventricles, where they balance quiescence, proliferation, and differentiation to sustain lifelong neurogenesis (Llorente et al., 2022). Insulin receptor signaling functions as a key integrator of systemic metabolic cues with local developmental programs, thereby regulating this equilibrium. Ligand binding activates receptor autophosphorylation and initiates downstream cascades, primarily the AKT and ERK pathways. Within NSCs, AKT signaling promotes proliferation by stimulating mTOR-mediated anabolic metabolism and relieving cell-cycle inhibition through suppression of regulators such as p21 and p27. In parallel, ERK activation modulates lineage specification, with sustained signaling favoring neuronal differentiation via transcriptional regulators including CREB. Beyond these canonical mechanisms, insulin receptor engagement enhances neurotrophin signaling by upregulating BDNF, NGF, and their receptors TrkB and TrkA, reinforcing neuronal commitment and maturation (Rossetti et al., 2021). Conversely, impaired insulin receptor responsiveness, typical of brain insulin resistance, compromises both proliferative potential and differentiation accuracy, leading to reduced neurogenesis and aberrant lineage outcomes linked to cognitive decline and neurodegenerative pathology.

Insulin-mediated synaptic plasticity: role and function

Insulin receptors are strategically distributed across presynaptic terminals and postsynaptic densities, particularly at excitatory glutamatergic synapses, positioning insulin as a critical modulator of neurotransmission and circuit remodeling. Upon activation, insulin receptors enhance glutamatergic signaling through modulation of NMDA and AMPA receptor dynamics. Insulin potentiates NMDA receptor mediated calcium influx essential for long-term potentiation (LTP) while regulating AMPA receptor trafficking via subunit-specific insertion and endocytosis, thereby fine-tuning synaptic efficacy (de León-López et al., 2025). Downstream activation of AKT and ERK pathways promotes local protein synthesis and stabilization of postsynaptic scaffolding components, maintaining dendritic spine integrity and synaptic persistence. Experimental disruption of insulin signaling results in reduced synapse density and impaired experience-dependent plasticity. Physiological insulin action supports hippocampal learning, whereas insulin resistance commonly associated with aging, obesity, or type 2 diabetes, attenuates LTP and cognitive performance (Heni, 2024). Notably, insulin mediated mechanisms parallel those governing neural stem cell differentiation through shared neurotrophin dependent pathways, suggesting that receptor dysfunction may jointly compromise synaptic maintenance and adult neurogenesis, thereby exacerbating cognitive decline.

Role of insulin receptors in neural stem cell maintenance and regulation

Beyond their functions in proliferation and differentiation, insulin receptors are fundamental to the long-term preservation and functional stability of the neural stem cell (NSC) pool. Sustained NSC maintenance requires ongoing survival signaling, metabolic equilibrium, and responsiveness to niche, derived cues, all of which depend on active insulin receptor pathways. Through PI3K/AKT signaling, insulin receptors inhibit pro-apoptotic factors while supporting mitochondrial integrity and biosynthetic processes necessary for cellular longevity (Godoy-Parejo et al., 2019). Equally important, insulin receptor activity regulates neurotrophic signaling networks, modulating the expression of neurotrophins such as BDNF and NGF and their receptors TrkB and TrkA (Rossetti et al., 2021). This coordination establishes autocrine and paracrine loops that promote NSC survival, self-renewal, and adaptive plasticity. Disruption of insulin signaling reduces neurotrophin expression, impairs receptor availability, and destabilizes stem cell homeostasis, leading to premature quiescence or misdirected differentiation. Such alterations contribute to the age-related decline in neurogenesis and heightened vulnerability to neurodegenerative pathology. Unlike the transient effects on proliferation or differentiation, insulin mediated maintenance reflects a prolonged regulatory mechanism that sustains metabolic competence and survival signaling over extended periods, thereby safeguarding the NSC reservoir throughout the lifespan.

Crosstalk between insulin receptor signaling pathways and neurotrophic pathways

Insulin receptor signaling functions within an intricate regulatory network that interfaces with major neurotrophic systems, including BDNF-TrkB, NGF-TrkA, and IGF pathways. These interconnected cascades converge on shared intracellular effectors such as PI3K-AKT and MAPK-ERK, coordinating neurogenesis, neuronal survival, and synaptic plasticity (Chiu & Cline, 2010). During neural stem cell differentiation, insulin receptor activation modulates neurotrophin gene expression and receptor availability, thereby influencing lineage specification and maturation. In mature neurons, insulin signaling amplifies BDNF driven synaptic remodeling and dendritic spine formation by enhancing neurotrophin responsiveness (Toader et al., 2025). Conversely, insulin resistance attenuates neurotrophic signaling efficiency, contributing to synaptic weakening and accelerated neurodegeneration. Mechanistically, insulin receptors regulate neurotrophin transcription, processing of proneurotrophins, and trafficking of their receptors to synaptic membranes. They also modulate proteases and matrix-binding proteins that determine neurotrophin bioavailability. Importantly, this communication is bidirectional like neurotrophic signaling can reciprocally influence insulin receptor expression and sensitivity, aligning neuronal energy metabolism with plasticity demands. Collectively, these interactions position insulin receptors as central integrators linking metabolic and neurotrophic networks, essential for sustaining neural stem cell maintenance, synaptic adaptability, and cognitive resilience across the lifespan. Their dysfunction unites metabolic and neurodegenerative pathology, highlighting insulin signaling as a key therapeutic axis for brain health.

Existing and emerging therapeutics involving insulin receptors, neural stem cells, synaptic plasticity

Stem cell therapies aim to restore the tissue function by enhancing body's natural repair mechanisms through

TABLE 12.2 List of therapies targeting neurodegenerative disorders.

Disease	Therapy type	Target	Innovations	Key insights
AD	Amyloid-β clearing antibodies	Removes amyloid plaques, slows cognitive decline	Lecanemab, Donanemab	First disease-modifying biologics for AD
	Tau aggregation inhibitors	Prevents formation of neurofibrillary tangles	Microtubule associated inhibitor- TRx0237	Aims to reduce tau pathology, mixed results
	Multi-target small molecules	Combines antioxidant, anti-amyloid & cholinesterase inhibition	Novel hybrid compounds	All-in-one strategy to address multiple AD pathways
PD	Dopaminergic replacement	Restores dopamine neurotransmission	Levodopa, Pramipexole	Core symptomatic treatment for PD motor symptoms
	α-synuclein immunotherapy	Neutralizes toxic α-synuclein aggregates	Prasinezumab	Slows disease spread; promising but early
	Gene therapy	Delivers GDNF genes to improve dopamine function	AAV2-GDNF	Restores neural function via targeted delivery
HD	Gene silencing	Reduces mutant huntingtin (mHTT) protein expression	Tominersen	First targeted genetic approach for HD
ALS	Glutamate modulation	Lowers excitotoxic neuronal death	Riluzole, edaravone	Modest survival benefit, core standard of care
	RNA-targeted therapy	Small molecules stabilize RNA to prevent TDP-43 aggregation	Novel RNA modulators	Precision strategy for protein misfolding diseases
	Nanoparticle-based delivery	Enhances CNS drug delivery across BBB	Lipid nanoparticles	Overcomes key barrier in brain-targeted therapy
	CRISPR/Cas9 gene editing	Corrects mutations	CRISPR-Cas systems	Groundbreaking curative potential
	Anti-inflammatory & microglial modulation	Targets NLRP3 inflammasome and microglial overactivation	NLRP3 inhibitors, microglia reprogrammers	Aims to restore immune homeostasis in brain
	Neurotrophic factor therapy	Supports neuronal growth and survival	BDNF, GDNF infusion	Improves neuron resilience and repair
Pan-neurodegenerative	Stem cell–based regeneration	Replaces lost neurons and glia	Neural stem cell transplantation	Promising for tissue restoration, ongoing trials

replacement of damaged cells or modulation of tissue microenvironment. Therapeutic strategies for neurodegenerative disorders are now shifting from symptomatic management toward restoring cellular functions or target upstream disease-driving mechanisms. NSCs from induced pluripotent stem cells (iPSCs), embryonic stem cells (ESCs) or even fetal sources have shown therapeutic potential by migrating to the damaged areas, differentiating into neurons/glia and releasing beneficial factors like BDNF/VEGF resulting in improved motor skills in PD, reduced tau pathology in AD models and enhancing cognition. BDNF, FGF2, EGF or Wnt activators along with Notch inhibitors can be used to stimulate endogenous NSCs in regions like SVZ or SGZ. Emerging approaches include genetically engineered NSCs overexpressing GFs to improve survival, neuronal differentiation or modulation of inflammatory and metabolic pathways in degenerating brain (De Gioia et al., 2020). NSC-derived exosomes can also be used, which reduce inflammation, carry protective signals and enhance memory and motor functions without raising the risk of rejection (Isik et al., 2025).

Recently, IR signaling has gained attention due to insulin's prominent role in neuronal function. Intranasal insulin bypasses the BBB to enhance cognition, memory and hippocampal connectivity in individuals with mild cognitive impairment or early AD without adversely affecting the blood sugar levels (Hallschmid, 2021). Insulin sensitizers such as metformin and GLP-1 agonists, help correct insulin resistance in the brain and promote mitochondrial function.

A crucial third pillar is made up of treatments that directly target synaptic plasticity through the modulators of NMDA and AMPA receptors, neurotrophin mimetics like BDNF analogs and TrkB agonists and compounds that enhance mitochondrial function or activate CREB signaling (Chiu et al., 2022). Ketone therapies and mitochondrial enhancers like coenzyme Q10 and nicotinamide riboside improve neuronal bioenergetics and decrease oxidative damage while lifestyle interventions like exercise and intermittent fasting improve insulin signaling, raise BDNF levels, promote adult hippocampal neurogenesis and improve cognitive functions. These effects are reinforced by metabolic-based interventions that support both synapses and NSCs.

The development of multi-target or combination strategies that simultaneously enhance insulin signaling, increase NSC function and strengthen synaptic plasticity can be an important future direction. This integrative approach is exemplified by GLP-1 receptor agonists, which improve metabolic regulation, protect synapses and promote hippocampal neurogenesis. Graft survival, trophic factor production and integration into diseased circuits may be further enhanced by combining NSC-derived exosomes or stem cell transplantation with metabolic modulators like GLP-1 analogs, metformin or mitochondrial boosters (Song et al., 2025). Emerging approaches include the use of biomaterials and engineered scaffolds to improve NSC delivery, retention and differentiation in the diseased brain, CRISPR-based correction of monogenic neurodegenerative mutations, brain organoid platforms for high-throughput testing of combined metabolic, NSC-targeted and synaptic drugs as well as small molecules made to coordinate metabolic and synaptic pathways, are starting to offer a translational bridge between mechanistic understanding and clinical application (Table 12.2).

Conclusion

Neurodegenerative disorders arise from a complex interplay of metabolic dysfunction, synaptic failure and impaired regenerative capacity in the brain. Over the years studies have identified insulin resistance as a key contributor to the neurodegenerative disorders such as AD, PD and HD. Insulin resistance in the brain leads to disrupted neuronal function, intracellular signaling, stress responses, etc., and these metabolic alterations critically impair synaptic plasticity, affect neurotransmission and activity-dependent synaptic remodeling that are essential for cognitive function. Synaptic plasticity on the other hand, represents an early and central event in neurodegeneration and often precedes extensive neuronal loss. Impaired insulin signaling increases excitotoxity, neuroinflammatory responses and also cause accumulation of protein aggregates, collectively accelerating synapse loss. Parallelly, insulin resistance leads to suppression of NSC proliferation and differentiation through inflammatory mediators and dysregulated growth factor signaling. This reduced NSC activity causes limited synaptic remodeling and neuronal replacement, further causing a functional decline.

Thus, insulin resistance, synaptic plasticity impairment and NSC dysfunction, together constitute an interdependent network driving neurodegeneration. Therapeutic strategies which target insulin signaling pathway, NSC activation and synaptic stability may synergistically slow neurodegeneration.

Application of materials (250–750 words)

The concepts discussed in this chapter have important translational implications for understanding and managing neurodegenerative disorders. This framework provides a mechanistic link metabolic dysfunction and progressive neuronal and synaptic loss by integrating brain insulin receptor signaling with neural stem cells functions and synaptic plasticity. Such a perspective supports the development of therapeutic strategies aimed not only at symptom management but also at controlling the disease progression.

This work also discusses the important applications in therapeutic target identification. Impaired brain insulin signaling is emerging as a modifiable contributor to neurodegeneration, with the molecules of the insulin receptor pathway being recognized as the new promising drug targets. Interventions such as intranasal insulin, glucagon-like peptide-1 receptor agonists and modulators of downstream signaling pathways including PI3K-Akt and mTOR have shown promising potential in preserving synaptic function and neuronal survival. Understanding how these pathways regulate synaptic function may guide the direction for designing therapies to maintain cognitive function.

The contents of this chapter also discuss regenerative and stem cell-based approaches. NSCs respond to insulin-dependent metabolic signals and a disruption in this signaling impairs their neurogenic potential. Insights into insulin receptor mediated signaling can be leveraged to enhance endogenous neurogenesis or improve the efficacy of transplanted stem cells. Genetically engineered NSCs that enhance insulin or neurotrophic signaling may offer combined neuroprotective and regenerative benefits.

From a research point of view, adding metabolic parameters like insulin resistance to neurodegeneration experimental models could enhance therapeutic screening and disease relevance. Clinically, these ideas support the use of metabolic biomarkers for early intervention and emphasize the significance of metabolic health in brain function. Overall, this integrated framework creates new opportunities for translational studies that focus on metabolism-driven mechanisms of neurodegeneration.

Mini-dictionary of terms

1. Long-term depression

A persistent weakening of synaptic strength induced by low-frequency stimulation, involving AMPA receptor endocytosis and actin cytoskeleton remodeling, critical for learning and memory refinement.

2. Long-term potentiation

A sustained increase in synaptic efficacy triggered by high-frequency stimulation, dependent on NMDA receptor activation, calcium influx and AMPA receptor trafficking therefore, forming the basis of synaptic plasticity and memory.

3. Neurogenesis

The generation of new neurons from the neural stem cell or progenitor cells, persisting in the adult dentate gyrus and subventricular zone and essential for hippocampal-dependent learning and mood regulation.

4. Neurogenic niches (SVZ and SGZ)

Specialized microenvironments in the adult brain which house neural stem cells that monitor neurogenesis through intricate cell-cell signaling and vascular support. The subventricular zone (SVZ) line the lateral ventricles and generate neuroblasts that migrate to the olfactory bulbs whereas, the subgranular zone (SGZ) of the hippocampal dentate gyrus are responsible for producing granule neurons essential for learning and memory.

5. Neuroinflammation

Chronic activation of microglia and astrocytes releasing pro-inflammatory cytokines (TNF-α, IL-1β), worsening the neurodegeneration through synaptic loss and impaired plasticity in AD/PD/HD.

6. Synaptic plasticity

Activity dependent modification of synaptic strength and connectivity, encompassing LTP/LTD and structural changes like dendrite spine dynamics, underpinning learning, memory and circuit adaptability.

Summary points

- Synaptic dysfunction and impaired plasticity are early and central features of neurodegenerative disorders.
- Brain insulin receptors regulate key processes such as neuronal metabolism, synaptic functions and neural stem cell maintenance.
- Disruption of insulin signaling contributes to synaptic failure, reduced neurogenesis and progressive neuronal vulnerability.
- NSCs function is compromised in neurodegenerative disorders due to chronic inflammation, oxidative stress and altered growth factor signaling mechanisms.
- Converging evidence links brain insulin resistance to the pathogenesis of AD, PD and HD.
- Therapeutic strategies targeting insulin signaling, synaptic plasticity and neural stem cells may offer prospective disease modifying effects.
- Future interventions will most likely require multi-target approaches that integrate metabolic, inflammatory and regenerative mechanisms.

Declaration of usage of AI

During the preparation of this manuscript, the author used Perplexity AI to assist with the language refinement, clarity and structuring of specific sections such as *Application of Material*. After using the tool, the author reviewed and edited the content as needed and take full responsibility for the content of the publication.

References

Andrade-Talavera, Y., & Rodríguez-Moreno, A. (2021). Synaptic plasticity and oscillations in Alzheimer's disease: A complex picture of a multifaceted disease. *Frontiers in Molecular Neuroscience, 14.* https://doi.org/10.3389/fnmol.2021.696476

Belfiore, A., Frasca, F., Pandini, G., Sciacca, L., & Vigneri, R. (2009). Insulin receptor isoforms and insulin Receptor/insulin-like growth factor receptor hybrids in physiology and disease. *Endocrine Reviews, 30*(6), 586–623. https://doi.org/10.1210/er.2008-0047

Boura-Halfon, S., & Zick, Y. (2009). Phosphorylation of IRS proteins, insulin action, and insulin resistance. *American Journal of Physiology-Endocrinology and Metabolism, 296*(4), E581–E591. https://doi.org/10.1152/ajpendo.90437.2008

Chen, W., Liu, X., Muñoz, V. R., & Kahn, C. R. (2025). Loss of insulin signaling in microglia impairs cellular uptake of Aβ and neuroinflammatory response exacerbating AD-like neuropathology. *Proceedings of the National Academy of Sciences, 122*(21), Article e2501527122. https://doi.org/10.1073/pnas.2501527122

Chiu, S.-L., & Cline, H. T. (2010). Insulin receptor signaling in the development of neuronal structure and function. *Neural Development, 5,* 7. https://doi.org/10.1186/1749-8104-5-7

Chiu, Y.-J., Lin, T.-H., Chang, K.-H., Lin, W., Hsieh-Li, H. M., Su, M.-T., Chen, C.-M., Sun, Y.-C., & Lee-Chen, G.-J. (2022). Novel TRKB agonists activate TRKB and downstream ERK and AKT signaling to protect Aβ-GFP SH-SY5Y cells against Aβ toxicity. *Aging, 14*(18), 7568–7586. https://doi.org/10.18632/aging.204306

Choi, E., & Bai, X. (2023). Activation mechanism of the insulin receptor: A structural perspective. *Annual Review of Biochemistry, 92,* 247–272. https://doi.org/10.1146/annurev-biochem-052521-033250

De Gioia, R., Biella, F., Citterio, G., Rizzo, F., Abati, E., Nizzardo, M., Bresolin, N., Comi, G. P., & Corti, S. (2020). Neural stem cell transplantation for neurodegenerative diseases. *International Journal of Molecular Sciences, 21*(9), 3103. https://doi.org/10.3390/ijms21093103

de León-López, C. A. M., Carretero-Rey, M., & Khan, Z. U. (2025). AMPA receptors in synaptic plasticity, memory function, and brain diseases. *Cellular and Molecular Neurobiology, 45*(1), 14. https://doi.org/10.1007/s10571-024-01529-7

De Meyts, P. (2000). The insulin receptor and its signal transduction network. In K. R. Feingold, R. A. Adler, S. F. Ahmed, B. Anawalt, M. R. Blackman, G. Chrousos, E. Corpas, W. W. de Herder, K. Dhatariya, K. Dungan, E. Hamilton, J. Hofland, S. Jan de Beur, S. Kalra, G. Kaltsas, N. Kapoor, M. Kim, C. Koch, P. Kopp, & D. P. Wilson (Eds.), *Endotext.* MDText.com, Inc. http://www.ncbi.nlm.nih.gov/books/NBK378978/.

De Strooper, B., & Karran, E. (2016). The cellular phase of Alzheimer's disease. *Cell, 164*(4), 603–615. https://doi.org/10.1016/j.cell.2015.12.056

Escribano, O., Beneit, N., Rubio-Longás, C., López-Pastor, A. R., & Gómez-Hernández, A. (2017). The role of insulin receptor isoforms in diabetes and its metabolic and vascular complications. *Journal of Diabetes Research, 2017,* 1403206. https://doi.org/10.1155/2017/1403206

Farris, W., Mansourian, S., Chang, Y., Lindsley, L., Eckman, E. A., Frosch, M. P., Eckman, C. B., Tanzi, R. E., Selkoe, D. J., & Guénette, S. (2003). Insulin-degrading enzyme regulates the levels of insulin, amyloid β-protein, and the β-amyloid precursor protein intracellular domain in vivo. *Proceedings of the National Academy of Sciences of the United States of America, 100*(7), 4162–4167. https://doi.org/10.1073/pnas.0230450100

Foley, K., McKee, C., Nairn, A. C., & Xia, H. (2021). Regulation of synaptic transmission and plasticity by protein phosphatase 1. *Journal of Neuroscience, 41*(14), 3040–3050. https://doi.org/10.1523/JNEUROSCI.2026-20.2021

Godoy-Parejo, C., Deng, C., Liu, W., & Chen, G. (2019). Insulin stimulates PI3K/AKT and cell adhesion to promote the survival of individualized human embryonic stem cells. *Stem Cells, 37*(8), 1030–1041. https://doi.org/10.1002/stem.3026

Gonzalez, E., & McGraw, T. E. (2006). Insulin signaling diverges into Akt-dependent and -independent signals to regulate the recruitment/docking and the fusion of GLUT4 vesicles to the plasma membrane. *Molecular Biology of the Cell, 17*(10), 4484–4493. https://doi.org/10.1091/mbc.e06-07-0585

Hallschmid, M. (2021). Intranasal insulin for Alzheimer's disease. *CNS Drugs, 35*(1), 21–37. https://doi.org/10.1007/s40263-020-00781-x

Hartl, F. U., & Hayer-Hartl, M. (2009). Converging concepts of protein folding in vitro and in vivo. *Nature Structural & Molecular Biology, 16*(6), 574–581. https://doi.org/10.1038/nsmb.1591

Heni, M. (2024). The insulin resistant brain: Impact on whole-body metabolism and body fat distribution. *Diabetologia, 67*(7), 1181–1191. https://doi.org/10.1007/s00125-024-06104-9

Horgusluoglu, E., Nudelman, K., Nho, K., & Saykin, A. J. (2017). Adult neurogenesis and neurodegenerative diseases: A systems biology perspective. *American Journal of Medical Genetics. Part B, Neuropsychiatric Genetics, 174*(1), 93–112. https://doi.org/10.1002/ajmg.b.32429

Hossini, A. M., Quast, A. S., Plötz, M., Grauel, K., Exner, T., Küchler, J., Stachelscheid, H., Eberle, J., Rabien, A., Makrantonaki, E., & Zouboulis, C. C. (2016). PI3K/AKT signaling pathway is essential for survival of induced pluripotent stem cells. *PLoS ONE, 11*(5), e0154770. https://doi.org/10.1371/journal.pone.0154770

Isik, S., Osman, S., Yeman-Kiyak, B., Shamshir, S. R. M., & Sanchez, N. M. E. (2025). Advances in neurodegenerative disease therapy: Stem cell clinical trials and promise of engineered exosomes. *CNS Neuroscience & Therapeutics, 31*(9), e70577. https://doi.org/10.1111/cns.70577

Kandel, E. R., Dudai, Y., & Mayford, M. R. (2014). The molecular and systems biology of memory. *Cell, 157*(1), 163–186. https://doi.org/10.1016/j.cell.2014.03.001

Kellar, D., & Craft, S. (2020). Brain insulin resistance in Alzheimer's disease and related disorders: Mechanisms and therapeutic approaches. *The Lancet. Neurology, 19*(9), 758–766. https://doi.org/10.1016/S1474-4422(20)30231-3

Kempermann, G., Song, H., & Gage, F. H. (2015). Neurogenesis in the adult hippocampus. *Cold Spring Harbor Perspectives in Biology, 7*(9), a018812. https://doi.org/10.1101/cshperspect.a018812

Kumar, M., & Bansal, N. (2022). Implications of phosphoinositide 3-kinase-Akt (PI3K-Akt) pathway in the pathogenesis of Alzheimer's disease. *Molecular Neurobiology, 59*(1), 354–385. https://doi.org/10.1007/s12035-021-02611-7

Lee, E., Lee, Y., Yang, S., Gong, E. J., Kim, J., Ha, N.-C., Jo, D.-G., Mattson, M. P., & Lee, J. (2024). Akt-activated GSK3β inhibitory peptide effectively blocks tau hyperphosphorylation. *Archives of Pharmacal Research, 47*(10), 812–828. https://doi.org/10.1007/s12272-024-01513-1

Llorente, V., Velarde, P., Desco, M., & Gómez-Gaviro, M. V. (2022). Current understanding of the neural stem cell niches. *Cells, 11*(19), 3002. https://doi.org/10.3390/cells11193002

Long, H.-Z., Cheng, Y., Zhou, Z.-W., Luo, H.-Y., Wen, D.-D., & Gao, L.-C. (2021). PI3K/AKT signal pathway: A target of natural products in the prevention and treatment of Alzheimer's disease and Parkinson's disease. *Frontiers in Pharmacology, 12*. https://doi.org/10.3389/fphar.2021.648636

Maiese, K. (2017). Forkhead transcription factors: Formulating a FOXO target for cognitive loss. *Current Neurovascular Research, 14*(4), 415–420. https://doi.org/10.2174/1567202614666171116102911

Mattson, M. P., & Arumugam, T. V. (2018). Hallmarks of brain aging: Adaptive and pathological modification by metabolic states. *Cell Metabolism, 27*(6), 1176–1199. https://doi.org/10.1016/j.cmet.2018.05.011

Meyts, P. D., Sajid, W., Palsgaard, J., Theede, A.-M., Gauguin, L., Aladdin, H., & Whittaker, J. (2013). Insulin and IGF-I receptor structure and binding mechanism. In *Madame curie bioscience database*. Landes Bioscience. https://www.ncbi.nlm.nih.gov/books/NBK6192/.

Milstein, J. L., & Ferris, H. A. (2021). The brain as an insulin-sensitive metabolic organ. *Molecular Metabolism, 52*, 101234. https://doi.org/10.1016/j.molmet.2021.101234

Navakkode, S., & Kennedy, B. K. (2024). Neural ageing and synaptic plasticity: Prioritizing brain health in healthy longevity. *Frontiers in Aging Neuroscience, 16*. https://doi.org/10.3389/fnagi.2024.1428244

Ogden, K. K., Khatri, A., Traynelis, S. F., & Heldt, S. A. (2014). Potentiation of GluN2C/D NMDA receptor subtypes in the amygdala facilitates the retention of fear and extinction learning in mice. *Neuropsychopharmacology, 39*(3), 625–637. https://doi.org/10.1038/npp.2013.241

Park, H., & Poo, M. (2013). Neurotrophin regulation of neural circuit development and function. *Nature Reviews Neuroscience, 14*(1), 7–23. https://doi.org/10.1038/nrn3379

Pomytkin, I., Costa-Nunes, J. P., Kasatkin, V., Veniaminova, E., Demchenko, A., Lyundup, A., Lesch, K., Ponomarev, E. D., & Strekalova, T. (2018). Insulin receptor in the brain: Mechanisms of activation and the role in the CNS pathology and treatment. *CNS Neuroscience & Therapeutics, 24*(9), 763–774. https://doi.org/10.1111/cns.12866

Ríos, J. A., Bórquez, J. C., Godoy, J. A., Zolezzi, J. M., Furrianca, M. C., & Inestrosa, N. C. (2024). Emerging role of metformin in Alzheimer's disease: A translational view. *Ageing Research Reviews, 100*, 102439. https://doi.org/10.1016/j.arr.2024.102439

Rossetti, G., Ermer, J. A., Stentenbach, M., Siira, S. J., Richman, T. R., Milenkovic, D., Perks, K. L., Hughes, L. A., Jamieson, E., Xiafukaiti, G., Ward, N. C., Takahashi, S., Gray, N., Viola, H. M., Hool, L. C., Rackham, O., & Filipovska, A. (2021). A common genetic variant of a mitochondrial RNA processing enzyme predisposes to insulin resistance. *Science Advances, 7*(39), eabi7514. https://doi.org/10.1126/sciadv.abi7514

Rubenstein, D. (1971). Causes of viral gastroenteritis. *Clinical Science, 41*(2), 7P. https://doi.org/10.1042/cs041007p

Sakaguchi, M. (2024). The role of insulin signaling with FOXO and FOXK transcription factors. *Endocrine Journal, 71*(10), 939–944. https://doi.org/10.1507/endocrj.EJ24-0205

Song, D., Zou, X., Ma, D., Zhao, Y., Liu, T., Shen, B., & Cheng, O. (2025). The GLP1R agonist semaglutide inhibits reactive astrocytes and enhances the efficacy of neural stem cell transplantation therapy in Parkinson's disease mice. *Advanced Science, 12*(43), e17664. https://doi.org/10.1002/advs.202417664

Toader, C., Serban, M., Munteanu, O., Covache-Busuioc, R.-A., Enyedi, M., Ciurea, A. V., & Tataru, C. P. (2025). From synaptic plasticity to neurodegeneration: BDNF as a transformative target in medicine. *International Journal of Molecular Sciences, 26*(9), 4271. https://doi.org/10.3390/ijms26094271

Torrisi, F., Denaro, S., Ragonese, J., D'Aprile, S., Zappalà, A., & Parenti, R. (2025). Neural stem cells fate under neuroinflammatory conditions and oxidative stress response. *Frontiers in Cellular Neuroscience, 19*, 1736865. https://doi.org/10.3389/fncel.2025.1736865

Virkamäki, A., Ueki, K., & Kahn, C. R. (1999). Protein–protein interaction in insulin signaling and the molecular mechanisms of insulin resistance. *The Journal of Clinical Investigation, 103*(7), 931–943. https://doi.org/10.1172/JCI6609

Yasuda, R., Hayashi, Y., & Hell, J. W. (2022). CaMKII: A central molecular organizer of synaptic plasticity, learning and memory. *Nature Reviews Neuroscience, 23*(11), 666–682. https://doi.org/10.1038/s41583-022-00624-2

Section 3

Psychology and behavior as a consequence of changes in the pancreas or the pancreatic endocrine status or the effects of hormones

Chapter 13

Cognitive links with brain insulin signaling in metabolism: A new narrative

Habib Yaribeygi[1] and Behina Forouzanmehr[2]

[1]*Research Center of Physiology, Research Institute of Neurosciences, Semnan University of Medical Sciences, Semnan, Iran;* [2]*Functional Neurosurgery Research Center, Research Institute of Functional Neurosurgery, Shohada Tajrish Comprehensive Neurosurgical Center of Excellence, Shahid Beheshti University of Medical Sciences, Tehran, Iran*

List of abbreviations

AD Alzheimer's disease
Akt (Protein Kinase B)
AMPK AMP-activated protein kinase
Aβ Amyloid-β
BBB Blood-brain barrier
BDNF Brain-derived neurotrophic factor
BIR Brain insulin resistance
CNS Central nervous system
CSF Cerebrospinal fluid
FOXO Forkhead box O
GABA Gamma-Aminobutyric Acid
GLP-1 Glucagon-like peptide 1
GLUT4 Glucose transporter type 4
GSK-3β Glycogen synthase kinase-3β
HFD High-fat diet
ICV Intracerebroventricular
IGF-1 Insulin-like growth factor 1
IRs Insulin receptors
LTP Long-term potentiation
MAPK Mitogen-activated protein kinase
MCI Mild cognitive impairment
mTOR Mammalian target of rapamycin
PI3K/Akt Phosphoinositide 3-kinase/protein kinase B
PPAR-γ Peroxisome proliferator-activated receptor gamma
T2DM Type 2 diabetes mellitus

Introduction

Importance of brain insulin beyond glucose control

Insulin has long been characterized as a peripheral hormone secreted by pancreatic β-cells, primarily responsible for regulating blood glucose homeostasis (Jones et al., 2025). However, over the past 2 decades, a paradigm shift has occurred in our understanding of insulin's functions, particularly its role in the central nervous system (CNS) (Yaribeygi et al., 2023). The brain is now recognized not only as a target of insulin action but also as a key modulator of systemic energy balance (Yaribeygi et al., 2023). Unlike the traditional belief that insulin acts only in peripheral tissues such as muscle, liver, and adipose tissue, current evidence suggests that insulin in the brain regulates a wide range of functions unrelated to glucose uptake, especially in neuronal populations that are not insulin-dependent for glucose utilization (Scherer et al., 2021; Yaribeygi et al., 2023).

Insulin receptors are widely expressed in the CNS, especially in regions associated with cognition and emotion, including the hippocampus, cerebral cortex, amygdala, and hypothalamus (Foroozanmehr et al., 2025). Within these areas, insulin influences synaptic transmission, neuronal growth, plasticity, and the activity of neurotransmitter systems such as dopamine, glutamate, and GABA (Foroozanmehr et al., 2025; Yaribeygi et al., 2023). Additionally, insulin signaling in the brain is involved in appetite control, body weight regulation, and even reproductive behavior (Scherer et al., 2021). These discoveries have redefined insulin as a neuroactive peptide with broad regulatory effects, making it a key player in both metabolic and cognitive domains (Ahmad et al., 2022).

A brief overview of insulin's role in the brain and traditional metabolic concepts

Traditionally, metabolism and cognition have been studied as distinct physiological domains (Scherer et al., 2021). Metabolic regulation has been attributed to peripheral organs working in concert under the influence of hormones

Hormones in Neuroscience. https://doi.org/10.1016/B978-0-443-30263-3.00030-0

like insulin and glucagon, with minimal acknowledgment of central control. Conversely, cognitive neuroscience has historically focused on neuronal networks and neurotransmitter systems with limited reference to metabolic cues. However, this dichotomy is increasingly being challenged by research showing that the brain is both a regulator and a target of metabolic hormones (Scherer et al., 2021; Yang et al., 2024). Insulin crosses the blood-brain barrier (BBB) via receptor-mediated transport and activates intracellular signaling cascades such as the phosphoinositide 3-kinase/protein kinase B (PI3K/Akt) pathway and the mitogen-activated protein kinase (MAPK) pathway in neurons and glial cells (Yang et al., 2024). These signaling events are crucial not only for metabolic sensing but also for synaptic plasticity, long-term potentiation (LTP), and memory encoding (Yang et al., 2024).

The need to revisit the metabolic–cognitive interface

The intersection between metabolism and cognition has become increasingly relevant in light of epidemiological data linking metabolic disorders such as type 2 diabetes mellitus (T2DM), obesity, and metabolic syndrome with cognitive impairment, dementia, and Alzheimer's disease (Wang et al., 2024). These associations are not merely coincidental; instead, they are thought to stem from shared molecular mechanisms such as chronic low-grade inflammation, oxidative stress, mitochondrial dysfunction, and—crucially—brain insulin resistance. The term "type 3 diabetes" has been proposed to describe Alzheimer's disease, based on the observation that central insulin resistance precedes and potentially contributes to neurodegenerative changes (Michailidis et al., 2022). Insulin resistance in the brain impairs neuronal survival, synaptic function, and neurogenesis, while also promoting the accumulation of amyloid-beta and hyperphosphorylated tau, two pathological hallmarks of Alzheimer's disease (Kciuk et al., 2024; Michailidis et al., 2022).

Given these intricate relationships, it is no longer tenable to treat metabolic and cognitive dysfunction as isolated clinical entities. This chapter aims to contribute to that emerging model by synthesizing the current literature on the bidirectional links between brain insulin signaling and cognition, exploring underlying molecular pathways, and highlighting therapeutic interventions that address both metabolic and cognitive domains.

Insulin signaling in the brain: an overview

The CNS is now recognized as a metabolically active organ that not only responds to hormonal cues, but also plays a vital role in systemic energy regulation (Casimir et al., 2024). Contrary to early assumptions that the brain is insulin-insensitive, it is now established that insulin is transported across the BBB through a saturable, receptor-mediated mechanism, and exerts multiple effects on neuronal and glial cells (Casimir et al., 2024).

Distribution of insulin receptors in the brain

Insulin receptors (IRs) are widely distributed throughout the brain, though their density varies by region. High concentrations of IRs have been identified in the hypothalamus, hippocampus, olfactory bulb, amygdala, cerebellum, and cerebral cortex (Fig. 13.1). Notably, the hippocampus and prefrontal cortex—regions critical for memory and executive function—are among the most insulin-sensitive areas in the brain. The presence of IRs in both neurons and glia underscores the diverse cellular targets of insulin signaling in the CNS (Yaribeygi et al., 2023). These receptors are structurally and functionally similar to peripheral IRs (Zeng et al., 2016).

Major insulin signaling pathways in the brain

Two major intracellular pathways are activated by insulin in the brain:

1. **PI3K–Akt (Protein Kinase B) Pathway:** This pathway is critical for cell survival, growth, and metabolism. In neurons, PI3K-Akt signaling enhances synaptic plasticity, promotes LTP, and supports learning and memory. It also plays a role in glucose transporter translocation, particularly GLUT4, and in modulating mitochondrial activity and oxidative stress responses (Gabbouj et al., 2019).
2. **MAPK Pathway:** This pathway is involved in cell proliferation and differentiation, and in neurons, it contributes to synaptic remodeling, memory encoding, and neurite outgrowth (Zeng et al., 2016).

Of course, other downstream molecules such as glycogen synthase kinase-3β (GSK-3β), mammalian target of rapamycin (mTOR), and forkhead box O (FOXO) transcription factors also play important roles in regulating neuronal health, autophagy, and inflammation (McNay & Recknagel, 2011; Zeng et al., 2016).

Functional roles of central insulin signaling

Beyond molecular signaling, insulin in the brain exerts several physiological and cognitive effects:

- **Appetite and Energy Homeostasis:** In the hypothalamus, insulin inhibits orexigenic neurons (e.g., NPY/AgRP) and activates anorexigenic neurons (e.g., POMC/CART), thereby reducing food intake and

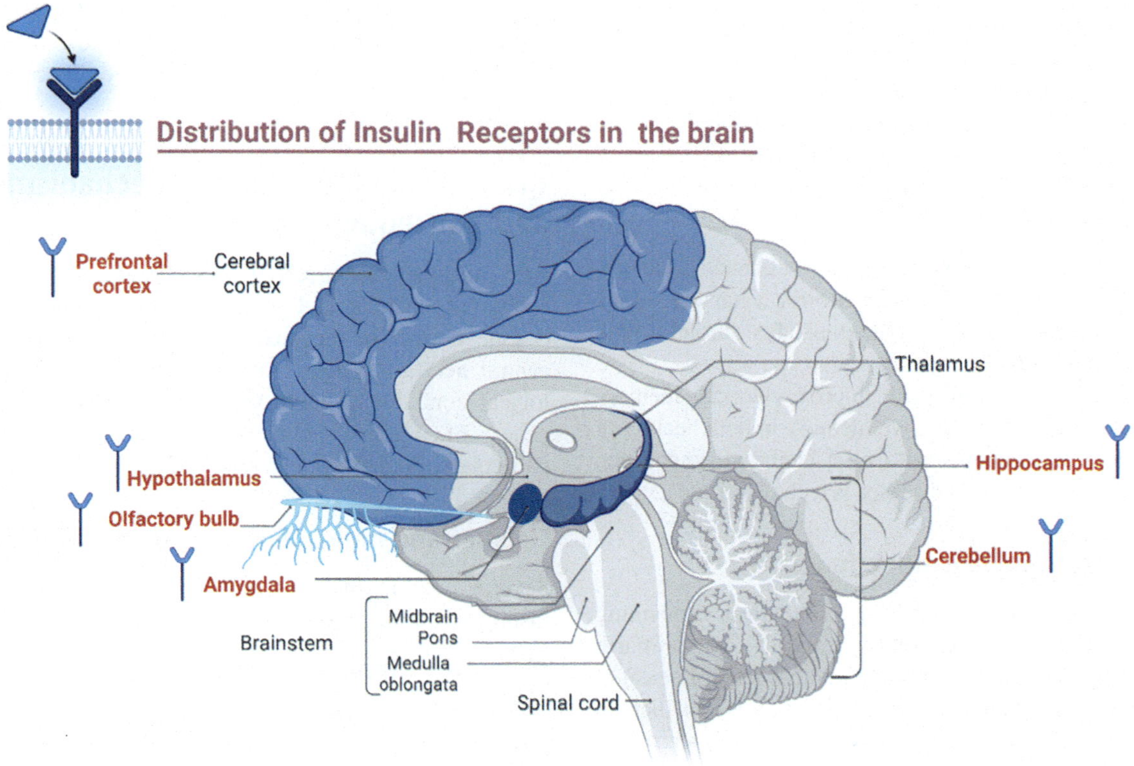

FIGURE 13.1 Distribution of insulin receptors in the brain.

influencing body weight regulation (Antasouras et al., 2024).

- **Neuroprotection and Plasticity:** Insulin promotes neuronal survival by inhibiting pro-apoptotic signaling and enhancing neurotrophic factors such as BDNF (brain-derived neurotrophic factor). It supports synaptogenesis and dendritic spine formation, processes essential for learning and memory (Nuñez et al., 2023).
- **Neurotransmission:** Insulin modulates several neurotransmitter systems, including increasing GABAergic activity, regulating dopamine turnover, and enhancing NMDA receptor-mediated glutamatergic transmission (Kleinridders et al., 2014).

Cognitive functions and insulin action

Several lines of experimental, clinical, and neuroimaging research suggest that insulin signaling within the CNS—especially in the hippocampus and prefrontal cortex—is essential for optimal learning, memory, and executive function (Yaribeygi et al., 2023).

Insulin and synaptic plasticity

Synaptic plasticity is a cellular mechanism underlying learning and memory (Kennedy, 2013). Insulin enhances synaptic plasticity by modulating neurotransmitter receptor trafficking, promoting dendritic spine formation, and regulating ion channel activity (Huang et al., 2010). Specifically, insulin has been shown to:

- **Increase NMDA receptor activity**, thereby enhancing long-term potentiation (LTP), which is considered a key neurophysiological correlate of memory consolidation (Barutçu et al., 2023; Kumar, 2011).
- **Modulate AMPA receptor trafficking**, which affects the strength and number of active synapses (Spinelli et al., 2019).
- **Promote dendritic branching and synaptogenesis**, increasing the structural complexity of neural circuits involved in memory storage (Palazzo et al., 2024).

Animal models with disrupted insulin signaling—such as brain-specific insulin receptor knockout (NIRKO) mice—exhibit reduced LTP, impaired spatial memory, and altered synaptic architecture, underscoring the critical role of insulin in plasticity (Palazzo et al., 2024; Spinelli et al., 2019).

Insulin's role in learning and memory

The hippocampus, a region densely populated with insulin receptors, plays a central role in spatial navigation, episodic memory, and contextual learning (Spinelli et al., 2019). Insulin administration, whether via intracerebroventricular (ICV) injection in rodents or intranasal delivery in humans, has been shown to improve memory performance in various cognitive domains:

- **In rodents**, hippocampal insulin enhances memory retention in tasks such as the Morris water maze, passive avoidance, and contextual fear conditioning (Tang et al., 2023).
- **In humans**, intranasal insulin improves working memory, verbal recall, and declarative memory in both healthy individuals and patients with mild cognitive impairment (MCI) or Alzheimer's disease (Wu et al., 2023).

Brain insulin and executive function

Executive functions—such as attention, planning, decision-making, and inhibitory control—are primarily governed by the prefrontal cortex, another insulin-sensitive brain region (Kullmann et al., 2020). Studies in diabetic and insulin-resistant individuals have reported impairments in attention and executive control, which correlate with reduced insulin sensitivity both peripherally and centrally (Kullmann et al., 2016, 2020).

Neuroimaging studies using fMRI and PET scans have demonstrated that intranasal insulin can increase activity in prefrontal cortical regions during cognitive tasks. These changes are paralleled by improvements in task performance, suggesting a causal relationship between insulin signaling and executive brain function (Kullmann et al., 2016, 2017).

Experimental and clinical evidence

Multiple lines of evidence support the cognitive role of insulin:

- **Animal models:** Central administration of insulin enhances performance in maze-based learning tasks. Conversely, induction of insulin resistance—whether genetically, pharmacologically, or through high-fat/high-sugar diets—leads to cognitive deficits and synaptic dysfunction (Lv et al., 2020).
- **Human studies:** Intranasal insulin therapy has been associated with improved cognition in patients with MCI and early Alzheimer's disease. Conversely, type 2 diabetes and obesity—conditions characterized by insulin resistance—are consistently linked to accelerated cognitive decline and increased risk of dementia (Fanelli et al., 2022; Wu et al., 2023).
- **Molecular insights:** Impaired insulin signaling disrupts multiple neuroprotective pathways, including BDNF production, mitochondrial integrity, and antioxidant defense, all of which are essential for cognitive resilience (He et al., 2025).

Brain insulin resistance: mechanisms and consequences

Brain insulin resistance (BIR) is defined as a diminished ability of neurons and glial cells to respond to insulin despite normal or elevated levels of the hormone in the central nervous system (Arnold et al., 2018). This pathological condition is increasingly implicated in the pathogenesis of cognitive decline and neurodegenerative diseases, particularly Alzheimer's disease (AD) (Yoon et al., 2023).

Cellular and molecular mechanisms of BIR

Several overlapping mechanisms contribute to impaired insulin signaling in the brain:

- **Impaired Insulin Receptor Signaling:**

Chronic hyperinsulinemia, often present in systemic insulin resistance, downregulates insulin receptor expression or leads to post-receptor defects such as impaired tyrosine phosphorylation and increased serine phosphorylation of insulin receptor substrates (IRS), particularly IRS-1 and IRS-2 (Copps & White, 2012).

- **Neuroinflammation:**

Elevated pro-inflammatory cytokines such as TNF-α, IL-1β, and IL-6 in the brain can interfere with insulin signaling by promoting serine phosphorylation of IRS proteins and activating stress kinases such as JNK and IKKβ (Lecca et al., 2022).

- **Oxidative Stress and Mitochondrial Dysfunction:**

Oxidative damage to neuronal mitochondria leads to reduced energy availability and impairs insulin signal transduction (Trigo et al., 2022).

- **Aβ and Tau Pathology:**

Amyloid-β (Aβ) oligomers and hyperphosphorylated tau—key pathological markers of AD—directly disrupt insulin receptor function and downstream signaling (Zheng & Wang, 2021).

Functional consequences of brain insulin resistance

The cognitive and structural consequences of BIR are extensive (Fig. 13.2):

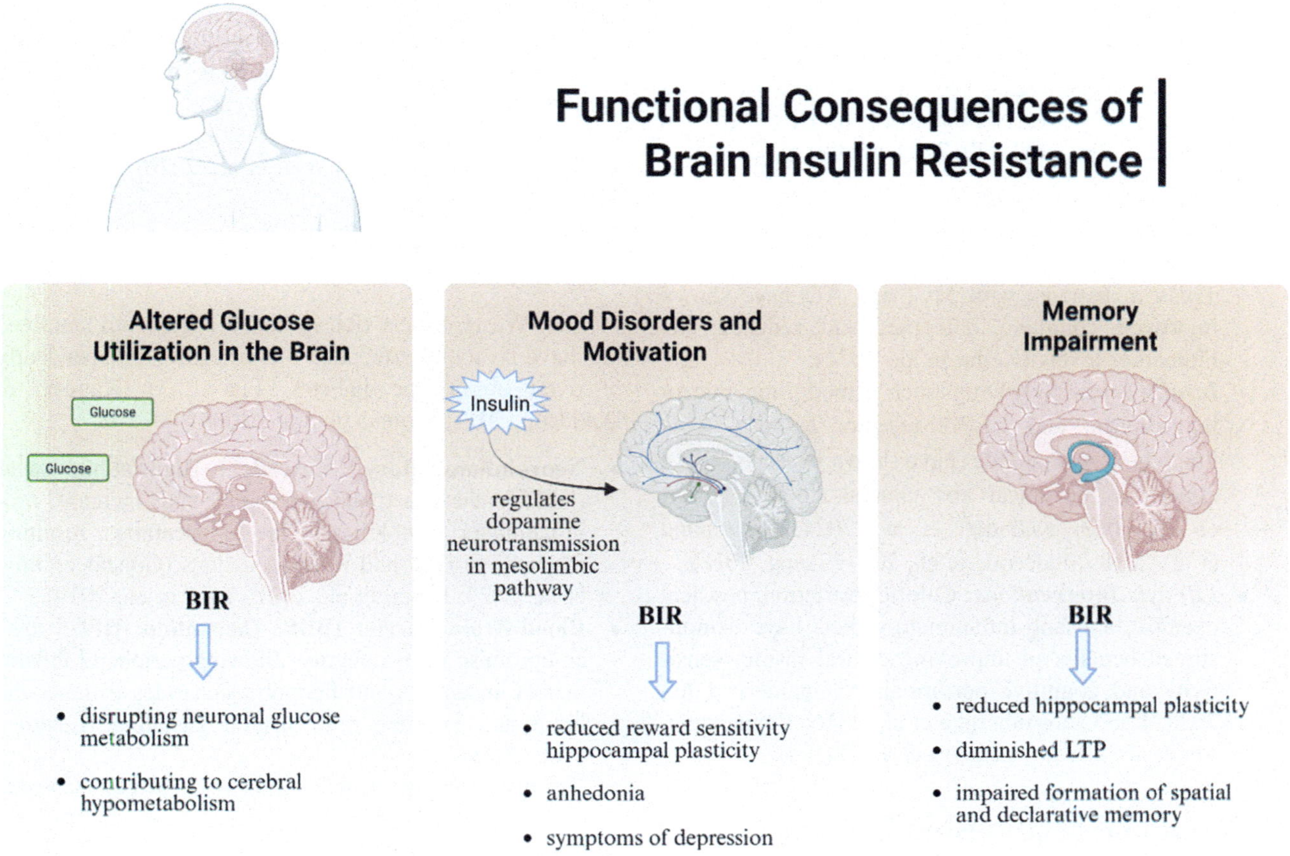

FIGURE 13.2 Functional consequences of brain insulin resistance.

- **Memory Impairment:**

BIR is associated with reduced hippocampal plasticity, diminished LTP, and impaired formation of spatial and declarative memory (Benedict & Grillo, 2018).

- **Mood Disorders and Motivation:**

Insulin also regulates dopamine neurotransmission in the mesolimbic pathway, and BIR has been linked to reduced reward sensitivity, anhedonia, and symptoms of depression—frequently observed in metabolic disorders and neurodegenerative conditions (Martin et al., 2021).

- **Altered Glucose Utilization in the Brain:**

BIR disrupts neuronal glucose metabolism, contributing to cerebral hypometabolism—a characteristic finding in FDG-PET scans of Alzheimer's patients (Blázquez et al., 2022).

Brain insulin resistance and Alzheimer's disease: "type 3 diabetes" hypothesis

An increasing body of evidence supports the concept of Alzheimer's disease as a form of diabetes localized to the brain—often referred to as "Type 3 Diabetes." This terminology reflects the overlap between insulin resistance and hallmark features of AD, including:

- Impaired insulin and IGF-1 signaling in the brain (Miao et al., 2025; Nguyen et al., 2020).
- Increased levels of Aβ and tau due to disrupted degradation and clearance (Nguyen et al., 2020).
- Diminished cerebral glucose metabolism and energy failure (Meng et al., 2025; Nguyen et al., 2020).
- Cognitive decline and neuronal loss parallel to features of diabetes-induced neurodegeneration (Michailidis et al., 2022).

Indeed, postmortem studies have demonstrated markedly reduced insulin receptor density and signaling efficacy in the brains of AD patients (Talbot, 2014). Furthermore, animal models of insulin resistance (e.g., HFD-induced obesity or streptozotocin-induced insulin deficiency) replicate many features of AD pathology (Hahm et al., 2020; Skovsø, 2014).

Clinical and therapeutic implications

Understanding brain insulin resistance opens new avenues for early detection and intervention in cognitive decline:

- **Biomarkers:**

Changes in CSF insulin levels, phosphorylated IRS proteins, and neuroimaging indicators of glucose metabolism may serve as early biomarkers of BIR (Mosconi et al., 2008; Mullins et al., 2017; Thomas et al., 2024).

- **Therapeutic Strategies:**
  - ***Intranasal Insulin Therapy*:** This method bypasses the BBB and directly delivers insulin to the brain. Trials in patients with MCI and AD have shown improved cognitive outcomes and reduced AD biomarker levels (Kellar et al., 2022).
  - ***Insulin Sensitizers*:** Drugs such as metformin, thiazolidinediones, and GLP-1 receptor agonists (e.g., liraglutide, semaglutide) have shown neuroprotective effects in preclinical and clinical studies (Ferrari et al., 2022; Kalinderi et al., 2024; Rotermund et al., 2018; Siddeeque et al., 2024; Tseng, 2018).
  - ***Lifestyle Interventions*:** Caloric restriction, physical exercise, and anti-inflammatory diets have demonstrated benefits in improving central insulin sensitivity and cognitive performance (Gualtieri et al., 2025; Van Cauwenberghe et al., 2016; Vandersmissen et al., 2025; Yaribeygi et al., 2019a).

Bidirectional relationship between metabolism and cognition

Impact of obesity, insulin resistance, and T2DM on cognitive performance

Obesity and insulin resistance—central components of metabolic syndrome—have consistently been associated with accelerated cognitive aging and an increased risk of neurodegenerative diseases (Abbasi et al., 2023; Komleva et al., 2020):

- **Obesity and Brain Structure:** Neuroimaging studies show that individuals with central obesity exhibit reductions in total brain volume, particularly in the prefrontal cortex and hippocampus—areas critical for executive function and memory. Adiposity-related inflammation and oxidative stress play pivotal roles in this neurodegeneration (Debette et al., 2010; Thirunavu et al., 2019; Weijie et al., 2024).
- **Insulin Resistance and Cognitive Impairment:** Insulin resistance impairs hippocampal synaptic plasticity, disrupts neurogenesis, and blunts insulin-mediated neurotransmitter regulation. Clinically, it is linked with poorer performance in verbal memory, attention, and information processing speed (Kleinridders et al., 2014; Spinelli et al., 2017, 2020).
- **T2DM as a Cognitive Risk Factor:** Type 2 diabetes is associated with a 50%–100% increased risk of developing dementia (Gudala et al., 2013). Hyperglycemia, microvascular injury, and insulin resistance synergistically contribute to neuronal dysfunction and structural brain damage. Glycemic variability and chronic hyperinsulinemia further exacerbate neuronal stress (Umegaki, 2014; Watt et al., 2020).

Influence of high-sugar/high-fat diets on cognitive decline

Modern Western diets rich in added sugars and saturated fats have profound effects on brain health—independently of overt obesity or diabetes (Fig. 13.3) (Kanoski & Davidson, 2011; Stranahan et al., 2008):

- **Neuroinflammatory Cascade:** High-fat/high-sugar (HFHS) diets activate microglia and increase pro-inflammatory cytokines in the hippocampus, resulting in impaired LTP and learning deficits (González Olmo et al., 2023; Patkar et al., 2021; Yang et al., 2019).
- **Blood-Brain Barrier (BBB) Disruption:** HFHS diets compromise BBB integrity, allowing peripheral inflammatory mediators and lipopolysaccharides to enter the brain and interfere with insulin signaling (Leidmaa et al., 2025).
- **Reward Circuit and Cognitive Control:** Chronic exposure to palatable energy-dense foods desensitizes dopamine pathways, diminishing cognitive flexibility and increasing impulsivity, thereby promoting a vicious cycle of poor dietary habits and executive dysfunction (Hartmann et al., 2019, 2020).
- **Animal Models:** Rodents fed HFHS diets show significant deficits in spatial memory (Morris water maze), increased anxiety-like behavior, and reduced brain insulin receptor expression, highlighting a direct causal link between diet and neurocognitive dysfunction (Abbott et al., 2019).

Effects of physical activity and dietary interventions on brain insulin sensitivity

Lifestyle interventions not only improve systemic metabolic health but also directly enhance cognitive performance and central insulin responsiveness (Fig. 13.4) (Kivipelto et al., 2018; Kullmann et al., 2022):

- **Exercise and Brain Insulin Signaling:** Aerobic and resistance exercise increase insulin receptor density and enhance PI3K/Akt signaling in the brain. Exercise-induced BDNF secretion supports synaptic remodeling and memory formation (Fang et al., 2013; Koo et al., 2013).
- **Neurocognitive Benefits of Exercise:** Regular physical activity has been associated with increased hippocampal volume, improved executive function, and

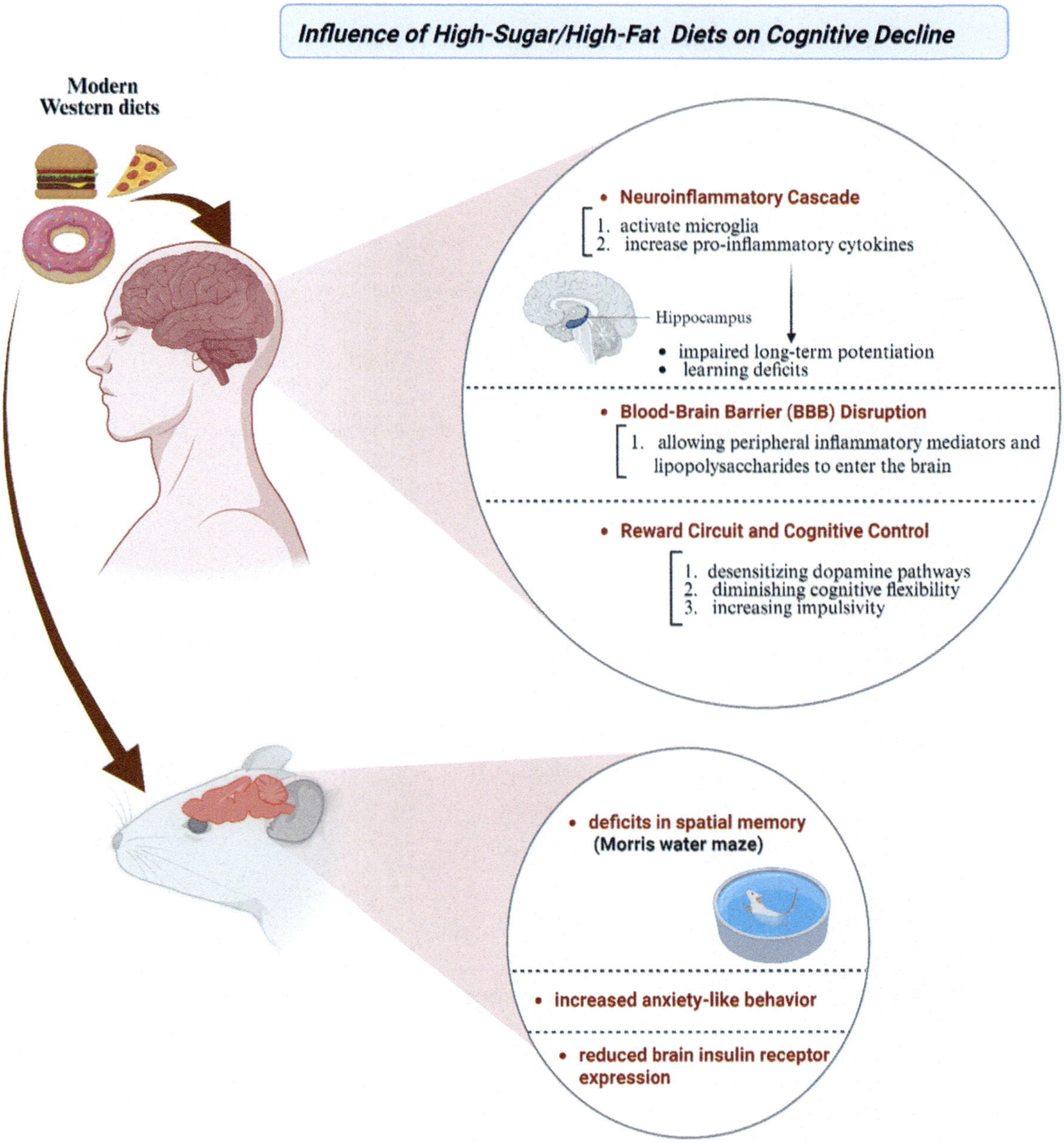

FIGURE 13.3 Effects of high-fat and high-sugar diets on cognitive performances.

delayed progression of cognitive decline, even in older adults with insulin resistance (Erickson et al., 2011).

- **Dietary Strategies:**
  - ***Mediterranean Diet*:** Rich in polyphenols, omega-3 fatty acids, and fiber, it reduces systemic and neural inflammation and is linked to better cognitive outcomes (Fekete et al., 2025).
  - ***Ketogenic Diet*:** By enhancing mitochondrial efficiency and reducing glycolytic stress, ketogenic diets may improve insulin signaling and protect against neurodegeneration (Gano et al., 2014).
  - ***Caloric Restriction and Intermittent Fasting*:** These approaches improve insulin sensitivity, activate sirtuins and autophagy, and support neuroplasticity and memory (Brocchi et al., 2022).
- **Combined Lifestyle Interventions:** Interventions that integrate physical activity with dietary modifications produce synergistic effects on cognitive function, brain

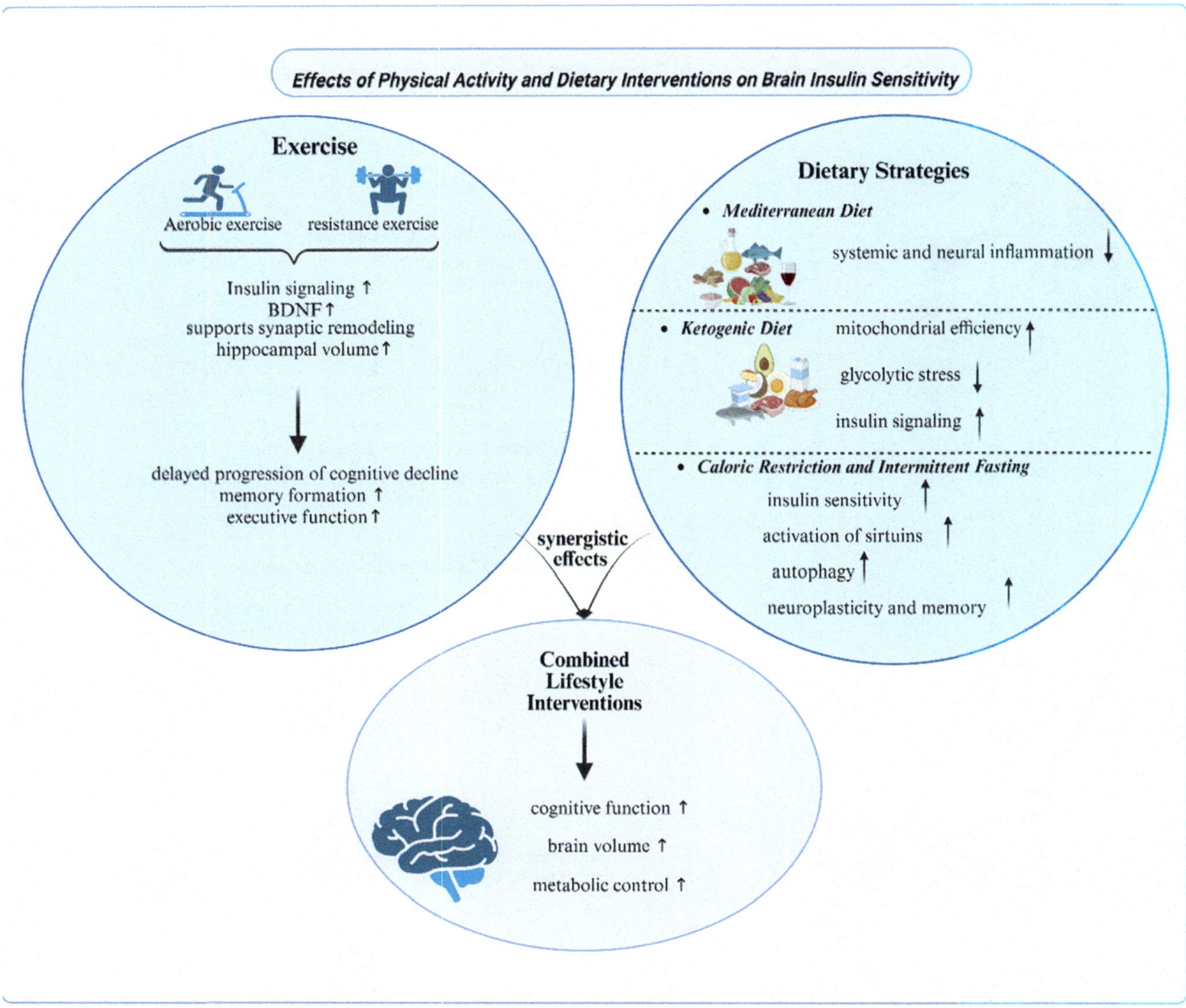

FIGURE 13.4 Effects of physical activities and dietary interventions on brain insulin sensitivity.

volume, and metabolic control—offering a powerful approach to prevent and reverse early cognitive deficits associated with metabolic syndrome (Popescu et al., 2025; Yaribeygi et al., 2019a; 2019b).

Therapeutic potentials: targeting brain insulin signaling to improve cognition

As brain insulin resistance emerges as a critical node in the intersection between metabolism and cognition, it presents a compelling therapeutic target Addressing impaired insulin signaling in the central nervous system may offer not only metabolic benefits but also substantial neurocognitive improvements.

A. ***Intranasal Insulin* Therapy:** allows insulin to bypass the BBB via the olfactory and trigeminal neural pathways, achieving central effects without systemic hypoglycemia [95, 96].
B. ***Antidiabetic Drugs with Neuroprotective Effects*:** Several antidiabetic agents (e.g., GLP-1 Receptor Agonists) have shown promise in modulating central insulin signaling and improving cognitive outcomes (Yaribeygi et al., 2021).
C. ***Lifestyle Interventions*:** Non-pharmacological strategies including physical exercise, dietary modifications and sleep & circadian rhythm regulation are essential adjuncts or even primary methods for enhancing brain insulin sensitivity (Malin et al., 2022; Sanguinetti et al., 2019; Stenvers et al., 2019).

Conclusion and future perspectives

Targeting brain insulin signaling represents a novel frontier in neurotherapeutics. As we unravel the molecular intricacies linking metabolic regulation to cognitive integrity, the potential for integrated treatment strategies—combining pharmacology, lifestyle, and neurotechnology—becomes increasingly apparent. Future research should aim to:

- Identify reliable biomarkers for early detection of brain insulin resistance.
- Clarify sex-, age-, and disease-specific responses to interventions.
- Design longitudinal clinical trials focusing on both prevention and treatment of cognitive decline via metabolic modulation.

Ultimately, reframing cognitive disorders through a metabolic lens may pave the way toward more effective, individualized interventions that bridge neurology and endocrinology.

Mini-dictionary of terms

- Ketogenic Diet: The ketogenic diet is a low-carb, high-fat diet that makes the body use fat for energy, producing ketones.
- Brain Insulin Resistance: Brain insulin resistance is a condition where the brain's cells respond poorly to insulin, impairing glucose metabolism and affecting cognitive functions.
- Mediterranean Diet: The Mediterranean diet is a plant-focused eating pattern rich in fruits, vegetables, whole grains, olive oil, and fish, promoting heart and brain health.
- Intranasal Insulin Therapy: Intranasal insulin therapy delivers insulin through the nose to directly target the brain, improving cognitive function and glucose regulation.
- Type 3 Diabetes: Type 3 diabetes refers to insulin resistance in the brain, often linked to Alzheimer's disease and cognitive decline.

Summary points

- Insulin in the brain regulates not only glucose metabolism but also synaptic plasticity, neuroprotection, appetite, and body weight.
- Brain insulin signaling involves key pathways such as PI3K–Akt and MAPK, influencing neuronal survival, learning, and memory.
- Insulin receptors are densely expressed in cognition-related regions like the hippocampus and prefrontal cortex.
- Brain insulin resistance (BIR) impairs synaptic function, neurogenesis, and cognitive performance, contributing to Alzheimer's disease.
- High-fat/high-sugar diets and obesity exacerbate BIR, neuroinflammation, and cognitive decline.
- Intranasal insulin therapy can bypass the BBB and improve memory, attention, and executive function in humans.
- Antidiabetic drugs like GLP-1 receptor agonists, metformin, and thiazolidinediones show neuroprotective effects via enhanced brain insulin signaling.
- Lifestyle interventions, including physical exercise, caloric restriction, and Mediterranean or ketogenic diets, improve brain insulin sensitivity and cognition.
- Brain-targeted nanocarriers, gene therapies, and combination approaches are emerging strategies to modulate central insulin signaling.
- Integrating metabolic and cognitive therapies may offer a novel, individualized approach to prevent and treat neurodegenerative diseases.

References

Abbasi, F., Robakis, T. K., Myoraku, A., Watson, K. T., Wroolie, T., & Rasgon, N. L. (2023). Insulin resistance and accelerated cognitive aging. *Psychoneuroendocrinology, 147*, 105944. https://doi.org/10.1016/j.psyneuen.2022.105944

Abbott, K. N., Arnott, C. K., Westbrook, R. F., & Tran, D. M. D. (2019). The effect of high fat, high sugar, and combined high fat-high sugar diets on spatial learning and memory in rodents: A meta-analysis. *Neuroscience & Biobehavioral Reviews, 107*, 399–421. https://doi.org/10.1016/j.neubiorev.2019.08.010

Ahmad, R. M. A. H., Ababneh, N. A., & Al-Domi, H. A. (2022). Brain insulin resistance as a mechanistic mediator links peripheral metabolic disorders with declining cognition. *Diabetes & Metabolic Syndrome: Clinical Research Reviews, 16*(4), 102468.

Antasouras, G., Dakanalis, A., Chrysafi, M., Papadopoulou, S. K., Trifonidi, I., Spanoudaki, M., Alexatou, O., Pritsa, A., Louka, A., & Giaginis, C. (2024). Could insulin be a better regulator of appetite/satiety balance and body weight maintenance in response to glucose exposure compared to sucrose substitutes? Unraveling current knowledge and searching for more appropriate choices. *Medical Science, 12*(2), 29.

Arnold, S. E., Arvanitakis, Z., Macauley-Rambach, S. L., Koenig, A. M., Wang, H. Y., Ahima, R. S., Craft, S., Gandy, S., Buettner, C., Stoeckel, L. E., Holtzman, D. M., & Nathan, D. M. (2018). Brain insulin resistance in type 2 diabetes and Alzheimer disease: Concepts and conundrums. *Nature Reviews Neurology, 14*(3), 168–181. https://doi.org/10.1038/nrneurol.2017.185

Barutçu, Ö., Süer, C., Dursun, N., Tufan, E., Gülpınar, E. A., & Tan, B. (2023). Insulin-induced long-term potentiation in the dentate gyrus of hippocampal formation. *Psychoneuroendocrinology, 157*, 106343. https://doi.org/10.1016/j.psyneuen.2023.106343

Benedict, C., & Grillo, C. A. (2018). Insulin resistance as a therapeutic target in the treatment of Alzheimer's disease: A state-of-the-art review. *Frontiers in Neuroscience, 12*, 215. https://doi.org/10.3389/fnins.2018.00215

Blázquez, E., Hurtado-Carneiro, V., LeBaut-Ayuso, Y., Velázquez, E., García-García, L., Gómez-Oliver, F., Ruiz-Albusac, J. M., Ávila, J., & Pozo, M.Á. (2022). Significance of brain glucose hypometabolism, altered insulin signal transduction, and insulin resistance in several neurological diseases. *Frontiers in Endocrinology, 13*, 873301.

Brocchi, A., Rebelos, E., Dardano, A., Mantuano, M., & Daniele, G. (2022). Effects of intermittent fasting on brain metabolism. *Nutrients, 14*(6). https://doi.org/10.3390/nu14061275

Casimir, P., Iwata, R., & Vanderhaeghen, P. (2024). Linking mitochondria metabolism, developmental timing, and human brain evolution. *Current Opinion in Genetics & Development, 86*, 102182.

Copps, K. D., & White, M. F. (2012). Regulation of insulin sensitivity by serine/threonine phosphorylation of insulin receptor substrate proteins IRS1 and IRS2. *Diabetologia, 55*(10), 2565–2582. https://doi.org/10.1007/s00125-012-2644-8

Debette, S., Beiser, A., Hoffmann, U., Decarli, C., O'Donnell, C. J., Massaro, J. M., Au, R., Himali, J. J., Wolf, P. A., Fox, C. S., & Seshadri, S. (2010). Visceral fat is associated with lower brain volume in healthy middle-aged adults. *Annals of Neurology, 68*(2), 136–144. https://doi.org/10.1002/ana.22062

Erickson, K. I., Voss, M. W., Prakash, R. S., Basak, C., Szabo, A., Chaddock, L., Kim, J. S., Heo, S., Alves, H., White, S. M., Wojcicki, T. R., Mailey, E., Vieira, V. J., Martin, S. A., Pence, B. D., Woods, J. A., McAuley, E., & Kramer, A. F. (2011). Exercise training increases size of hippocampus and improves memory. *Proceedings of the National Academy of Sciences of the United States of America, 108*(7), 3017–3022. https://doi.org/10.1073/pnas.1015950108

Fanelli, G., Mota, N. R., Salas-Salvado, J., Bullo, M., Fernandez-Aranda, F., Camacho-Barcia, L., Testa, G., Jimenez-Murcia, S., Bertaina-Anglade, V., & Franke, B. (2022). The link between cognition and somatic conditions related to insulin resistance in the UK biobank study cohort: A systematic review. *Neuroscience & Biobehavioral Reviews, 143*, 104927.

Fang, Z. H., Lee, C. H., Seo, M. K., Cho, H., Lee, J. G., Lee, B. J., Park, S. W., & Kim, Y. H. (2013). Effect of treadmill exercise on the BDNF-mediated pathway in the hippocampus of stressed rats. *Neuroscience Research, 76*(4), 187–194. https://doi.org/10.1016/j.neures.2013.04.005

Fekete, M., Varga, P., Ungvari, Z., Fekete, J. T., Buda, A., Szappanos, Á., Lehoczki, A., Mózes, N., Grosso, G., Godos, J., Menyhart, O., Munkácsy, G., Tarantini, S., Yabluchanskiy, A., Ungvari, A., & Győrffy, B. (2025). The role of the Mediterranean diet in reducing the risk of cognitive impairment, dementia, and Alzheimer's disease: A meta-analysis. *GeroScience, 47*(3), 3111–3130. https://doi.org/10.1007/s11357-024-01488-3

Ferrari, F., Moretti, A., & Villa, R. F. (2022). Incretin-based drugs as potential therapy for neurodegenerative diseases: Current status and perspectives. *Pharmacology & Therapeutics, 239*, 108277.

Foroozanmehr, B., Hemmati, M. A., Yaribeygi, H., Karav, S., Jamialahmadi, T., & Sahebkar, A. (2025). Parkinson's disease and brain insulin signaling: Mechanisms and potential role of GLP-1 mimetics. *Brain Research, 1862*, 149738.

Gabbouj, S., Ryhänen, S., Marttinen, M., Wittrahm, R., Takalo, M., Kemppainen, S., Martiskainen, H., Tanila, H., Haapasalo, A., & Hiltunen, M. (2019). Altered insulin signaling in Alzheimer's disease brain–special emphasis on PI3K-Akt pathway. *Frontiers in Neuroscience, 13*, 629.

Gano, L. B., Patel, M., & Rho, J. M. (2014). Ketogenic diets, mitochondria, and neurological diseases. *The Journal of Lipid Research, 55*(11), 2211–2228. https://doi.org/10.1194/jlr.R048975

González Olmo, B. M., Bettes, M. N., DeMarsh, J. W., Zhao, F., Askwith, C., & Barrientos, R. M. (2023). Short-term high-fat diet consumption impairs synaptic plasticity in the aged hippocampus via IL-1 signaling. *Npj Science of Food, 7*(1), 35. https://doi.org/10.1038/s41538-023-00211-4

Gualtieri, P., De Lorenzo, A., & Di Renzo, L. (2025). Mediterranean diet, inflammation, and neurovulnerability: Towards personalized nutritional strategies. *Nutrients, 17*(14), 2269. https://www.mdpi.com/2072-6643/17/14/2269.

Gudala, K., Bansal, D., Schifano, F., & Bhansali, A. (2013). Diabetes mellitus and risk of dementia: A meta-analysis of prospective observational studies. *Journal of Diabetes Investigation, 4*(6), 640–650. https://doi.org/10.1111/jdi.12087

Hahm, J. R., Jo, M. H., Ullah, R., Kim, M. W., & Kim, M. O. (2020). Metabolic stress alters antioxidant systems, suppresses the adiponectin receptor 1 and induces alzheimer's like pathology in mice brain. *Cells, 9*(1). https://doi.org/10.3390/cells9010249

Hartmann, H., Pauli, L. K., Janssen, L. K., Huhn, S., Ceglarek, U., & Horstmann, A. (2019). Differences in intake of high-fat high-sugar diet are related to variations in central dopamine in humans. *bioRxiv*, 839647. https://doi.org/10.1101/839647

Hartmann, H., Pauli, L. K., Janssen, L. K., Huhn, S., Ceglarek, U., & Horstmann, A. (2020). Preliminary evidence for an association between intake of high-fat high-sugar diet, variations in peripheral dopamine precursor availability and dopamine-dependent cognition in humans. *Journal of Neuroendocrinology, 32*(12), e12917.

He, Y., Sun, M., Qu, M., Lu, Y., Yang, H., Wang, R., Li, Y., Li, P., Mi, W., & Ma, Y. (2025). Brain insulin signaling pathway regulation of hippocampal neuroplasticity in neurocognitive disorders: Mechanisms and therapeutic implications. *Journal of Integrative Neuroscience, 24*(8), 39446.

Huang, C.-C., Lee, C.-C., & Hsu, K.-S. (2010). The role of insulin receptor signaling in synaptic plasticity and cognitive function. *Chang Gung Medical Journal, 33*(2), 115–125.

Jones, D. L., Kusinski, L. C., Gillies, C., & Meek, C. L. (2025). A critique of measurement of defective insulin secretion and insulin sensitivity as a precision approach to gestational diabetes. *Diabetologia, 68*(4), 752–765.

Kalinderi, K., Papaliagkas, V., & Fidani, L. (2024). GLP-1 receptor agonists: A new treatment in Parkinson's disease. *International Journal of Molecular Sciences, 25*(7), 3812.

Kanoski, S. E., & Davidson, T. L. (2011). Western diet consumption and cognitive impairment: Links to hippocampal dysfunction and obesity. *Physiology and Behavior, 103*(1), 59–68. https://doi.org/10.1016/j.physbeh.2010.12.003

Kciuk, M., Kruczkowska, W., Gałęziewska, J., Wanke, K., Kałuzińska-Kołat, Ż., Aleksandrowicz, M., & Kontek, R. (2024). Alzheimer's disease as type 3 diabetes: Understanding the link and implications. *International Journal of Molecular Sciences, 25*(22), 11955. https://www.mdpi.com/1422-0067/25/22/11955.

Kellar, D., Register, T., Lockhart, S. N., Aisen, P., Raman, R., Rissman, R. A., Brewer, J., & Craft, S. (2022). Intranasal insulin modulates cerebrospinal fluid markers of neuroinflammation in mild cognitive impairment and Alzheimer's disease: A randomized trial. *Scientific Reports, 12*(1), 1346. https://doi.org/10.1038/s41598-022-05165-3

Kennedy, M. B. (2013). Synaptic signaling in learning and memory. *Cold Spring Harbor Perspectives in Biology, 8*(2), a016824. https://doi.org/10.1101/cshperspect.a016824

Kivipelto, M., Mangialasche, F., & Ngandu, T. (2018). Lifestyle interventions to prevent cognitive impairment, dementia and Alzheimer disease. *Nature Reviews Neurology, 14*(11), 653–666. https://doi.org/10.1038/s41582-018-0070-3

Kleinridders, A., Ferris, H. A., Cai, W., & Kahn, C. R. (2014). Insulin action in brain regulates systemic metabolism and brain function. *Diabetes, 63*(7), 2232–2243. https://doi.org/10.2337/db14-0568

Komleva, Y., Chernykh, A., Lopatina, O., Gorina, Y., Lokteva, I., Salmina, A., & Gollasch, M. (2020). Inflamm-aging and brain insulin resistance: New insights and role of life-style strategies on cognitive and social determinants in aging and neurodegeneration. *Frontiers in Neuroscience, 14*, 618395. https://doi.org/10.3389/fnins.2020.618395

Koo, J. H., Kwon, I. S., Kang, E. B., Lee, C. K., Lee, N. H., Kwon, M. G., Cho, I. H., & Cho, J. Y. (2013). Neuroprotective effects of treadmill exercise on BDNF and PI3-K/Akt signaling pathway in the cortex of transgenic mice model of Alzheimer's disease. *Journal of Exercise Nutrition & Biochemistry, 17*(4), 151–160. https://doi.org/10.5717/jenb.2013.17.4.151

Kullmann, S., Goj, T., Veit, R., Fritsche, L., Wagner, L., Schneeweiss, P., Hoene, M., Hoffmann, C., Machann, J., Niess, A., Preissl, H., Birkenfeld, A. L., Peter, A., Häring, H. U., Fritsche, A., Moller, A., Weigert, C., & Heni, M. (2022). Exercise restores brain insulin sensitivity in sedentary adults who are overweight and obese. *JCI Insight, 7*(18). https://doi.org/10.1172/jci.insight.161498

Kullmann, S., Heni, M., Hallschmid, M., Fritsche, A., Preissl, H., & Häring, H. U. (2016). Brain insulin resistance at the crossroads of metabolic and cognitive disorders in humans. *Physiological Reviews, 96*(4), 1169–1209. https://doi.org/10.1152/physrev.00032.2015

Kullmann, S., Heni, M., Veit, R., Scheffler, K., Machann, J., Häring, H.-U., Fritsche, A., & Preissl, H. (2017). Intranasal insulin enhances brain functional connectivity mediating the relationship between adiposity and subjective feeling of hunger. *Scientific Reports, 7*(1), 1627. https://doi.org/10.1038/s41598-017-01907-w

Kullmann, S., Kleinridders, A., Small, D. M., Fritsche, A., Häring, H. U., Preissl, H., & Heni, M. (2020). Central nervous pathways of insulin action in the control of metabolism and food intake. *Lancet Diabetes & Endocrinology, 8*(6), 524–534. https://doi.org/10.1016/s2213-8587(20)30113-3

Kumar, A. (2011). Long-term potentiation at CA3-CA1 hippocampal synapses with special emphasis on aging, disease, and stress. *Frontiers in Aging Neuroscience, 3*, 7. https://doi.org/10.3389/fnagi.2011.00007

Lecca, D., Jung, Y. J., Scerba, M. T., Hwang, I., Kim, Y. K., Kim, S., Modrow, S., Tweedie, D., Hsueh, S. C., & Liu, D. (2022). Role of chronic neuroinflammation in neuroplasticity and cognitive function: A hypothesis. *Alzheimer's & Dementia, 18*(11), 2327–2340.

Leidmaa, E., Zimmer, A., Stein, V., & Gellner, A. K. (2025). Acute high-fat high-sugar diet rapidly increases blood-brain barrier permeability in mice. *The Journal of Nutrition, Health & Aging, 29*(7), 100574. https://doi.org/10.1016/j.jnha.2025.100574

Lv, H., Tang, L., Guo, C., Jiang, Y., Gao, C., Wang, Y., & Jian, C. (2020). Intranasal insulin administration may be highly effective in improving cognitive function in mice with cognitive dysfunction by reversing brain insulin resistance. *Cognitive Neurodynamics, 14*(3), 323–338.

Malin, S. K., Stewart, N. R., Ude, A. A., & Alderman, B. L. (2022). Brain insulin resistance and cognitive function: Influence of exercise. *Journal of Applied Physiology, 133*(6), 1368–1380.

Martin, H., Bullich, S., Guiard, B. P., & Fioramonti, X. (2021). The impact of insulin on the serotonergic system and consequences on diabetes-associated mood disorders. *Journal of Neuroendocrinology, 33*(4), e12928.

McNay, E. C., & Recknagel, A. K. (2011). Reprint of:'brain insulin signaling: A key component of cognitive processes and a potential basis for cognitive impairment in type 2 diabetes. *Neurobiology of Learning and Memory, 96*(4), 517–528.

Meng, X., Zhang, H., Zhao, Z., Li, S., Zhang, X., Guo, R., Liu, H., Yuan, Y., Li, W., & Song, Q. (2025). Type 3 diabetes and metabolic reprogramming of brain neurons: Causes and therapeutic strategies. *Molecular Medicine, 31*(1), 61.

Miao, J., Zhang, Y., Su, C., Zheng, Q., & Guo, J. (2025). Insulin-like growth factor signaling in Alzheimer's disease: Pathophysiology and therapeutic strategies. *Molecular Neurobiology, 62*(3), 3195–3225.

Michailidis, M., Moraitou, D., Tata, D. A., Kalinderi, K., Papamitsou, T., & Papaliagkas, V. (2022). Alzheimer's disease as type 3 diabetes: Common pathophysiological mechanisms between Alzheimer's disease and type 2 diabetes. *International Journal of Molecular Sciences, 23*(5), 2687.

Mosconi, L., Pupi, A., & De Leon, M. J. (2008). Brain glucose hypometabolism and oxidative stress in preclinical Alzheimer's disease. *Annals of the New York Academy of Sciences, 1147*, 180–195. https://doi.org/10.1196/annals.1427.007

Mullins, R. J., Mustapic, M., Goetzl, E. J., & Kapogiannis, D. (2017). Exosomal biomarkers of brain insulin resistance associated with regional atrophy in Alzheimer's disease. *Human Brain Mapping, 38*(4), 1933–1940. https://doi.org/10.1002/hbm.23494

Nguyen, T. T., Ta, Q. T. H., Nguyen, T. K. O., Nguyen, T. T. D., & Van Giau, V. (2020). Type 3 diabetes and its role implications in Alzheimer's disease. *International Journal of Molecular Sciences, 21*(9), 3165. https://www.mdpi.com/1422-0067/21/9/3165.

Nuñez, A., Zegarra-Valdivia, J., Fernandez de Sevilla, D., Pignatelli, J., & Torres Aleman, I. (2023). The neurobiology of insulin-like growth factor I: From neuroprotection to modulation of brain states. *Molecular Psychiatry, 28*(8), 3220–3230.

Palazzo, E., Marabese, I., Boccella, S., Belardo, C., Pierretti, G., & Maione, S. (2024). Affective and cognitive impairments in rodent models of diabetes. *Current Neuropharmacology, 22*(8), 1327–1343. https://doi.org/10.2174/1570159x22666240124164804

Patkar, O. L., Mohamed, A. Z., Narayanan, A., Mardon, K., Cowin, G., Bhalla, R., Stimson, D. H. R., Kassiou, M., Beecher, K., Belmer, A., Alvarez Cooper, I., Morgan, M., Hume, D. A., Irvine, K. M., Bartlett, S. E., Nasrallah, F., & Cumming, P. (2021). A binge high sucrose diet provokes systemic and cerebral inflammation in rats without inducing obesity. *Scientific Reports, 11*(1), 11252. https://doi.org/10.1038/s41598-021-90817-z

Popescu, I., Lenain, M., Rovini, E., & Ruthsatz, M. (2025). Lifestyle interventions and innovative approaches for the management of neurodegenerative disorders in older adults-state-of-the-art and future directions. *American Journal of Lifestyle Medicine, 19*(8), 1216. https://doi.org/10.1177/15598276251330170

Rotermund, C., Machetanz, G., & Fitzgerald, J. C. (2018). The therapeutic potential of metformin in neurodegenerative diseases. *Frontiers in Endocrinology, 9*, 400.

Sanguinetti, E., Guzzardi, M. A., Panetta, D., Tripodi, M., De Sena, V., Quaglierini, M., Burchielli, S., Salvadori, P. A., & Iozzo, P. (2019). Combined effect of fatty diet and cognitive decline on brain metabolism, food intake, body weight, and counteraction by intranasal insulin therapy in 3× tg mice. *Frontiers in Cellular Neuroscience, 13*, 188.

Scherer, T., Sakamoto, K., & Buettner, C. (2021). Brain insulin signalling in metabolic homeostasis and disease. *Nature Reviews Endocrinology, 17*(8), 468–483.

Siddeeque, N., Hussein, M. H., Abdelmaksoud, A., Bishop, J., Attia, A. S., Elshazli, R. M., Fawzy, M. S., & Toraih, E. A. (2024). Neuroprotective effects of GLP-1 receptor agonists in neurodegenerative disorders: A large-scale propensity-matched cohort study. *International Immunopharmacology, 143*, 113537. https://doi.org/10.1016/j.intimp.2024.113537

Skovsø, S. (2014). Modeling type 2 diabetes in rats using high fat diet and streptozotocin. *Journal of Diabetes Investigation, 5*(4), 349–358. https://doi.org/10.1111/jdi.12235

Spinelli, M., Fusco, S., & Grassi, C. (2019). Brain insulin resistance and hippocampal plasticity: Mechanisms and biomarkers of cognitive decline. *Frontiers in Neuroscience, 13*, 788. https://doi.org/10.3389/fnins.2019.00788

Spinelli, M., Fusco, S., & Grassi, C. (2020). Brain insulin resistance impairs hippocampal plasticity. *Vitamins & Hormones, 114*, 281–306. https://doi.org/10.1016/bs.vh.2020.04.005

Spinelli, M., Fusco, S., Mainardi, M., Scala, F., Natale, F., Lapenta, R., Mattera, A., Rinaudo, M., Li Puma, D. D., Ripoli, C., Grassi, A., D'Ascenzo, M., & Grassi, C. (2017). Brain insulin resistance impairs hippocampal synaptic plasticity and memory by increasing GluA1 palmitoylation through FoxO3a. *Nature Communications, 8*(1), 2009. https://doi.org/10.1038/s41467-017-02221-9

Stenvers, D. J., Scheer, F. A., Schrauwen, P., la Fleur, S. E., & Kalsbeek, A. (2019). Circadian clocks and insulin resistance. *Nature Reviews Endocrinology, 15*(2), 75–89.

Stranahan, A. M., Norman, E. D., Lee, K., Cutler, R. G., Telljohann, R. S., Egan, J. M., & Mattson, M. P. (2008). Diet-induced insulin resistance impairs hippocampal synaptic plasticity and cognition in middle-aged rats. *Hippocampus, 18*(11), 1085–1088. https://doi.org/10.1002/hipo.20470

Talbot, K. (2014). Brain insulin resistance in Alzheimer's disease and its potential treatment with GLP-1 analogs. *Neurodegenerative Disease Management, 4*(1), 31–40. https://doi.org/10.2217/nmt.13.73

Tang, L., Wang, Y., Gong, X., Xiang, J., Zhang, Y., Xiang, Q., & Li, J. (2023). Integrated transcriptome and metabolome analysis to investigate the mechanism of intranasal insulin treatment in a rat model of vascular dementia. *Frontiers in Pharmacology, 14*, 1182803.

Thirunavu, V., McCullough, A., Su, Y., Flores, S., Dincer, A., Morris, J. C., Cruchaga, C., Benzinger, T. L. S., & Gordon, B. A. (2019). Higher body mass index is associated with lower cortical amyloid-β burden in cognitively normal individuals in late-life. *Journal of Alzheimer's Disease, 69*(3), 817–827. https://doi.org/10.3233/jad-190154

Thomas, P., Leclerc, M., Evitts, K., Brown, C., Miller, W., Hanson, A. J., Banks, W. A., Gibbons, L., Domoto-Reilly, K., Jayadev, S., Li, G., Peskind, E., Young, J. E., Calon, F., & Rhea, E. M. (2024). Cerebrospinal fluid soluble insulin receptor levels in Alzheimer's disease. *Alzheimer's & Dementia, 16*(2), e12603. https://doi.org/10.1002/dad2.12603

Trigo, D., Avelar, C., Fernandes, M., Sa, J., & da Cruz e Silva, O. (2022). Mitochondria, energy, and metabolism in neuronal health and disease. *FEBS Letters, 596*(9), 1095–1110.

Tseng, C.-H. (2018). Pioglitazone reduces dementia risk in patients with type 2 diabetes mellitus: A retrospective cohort analysis. *Journal of Clinical Medicine, 7*(10), 306.

Umegaki, H. (2014). Type 2 diabetes as a risk factor for cognitive impairment: Current insights. *Clinical Interventions in Aging, 9*, 1011–1019. https://doi.org/10.2147/cia.S48926

Van Cauwenberghe, C., Vandendriessche, C., Libert, C., & Vandenbroucke, R. E. (2016). Caloric restriction: Beneficial effects on brain aging and Alzheimer's disease. *Mammalian Genome, 27*, 300–319.

Vandersmissen, J., Dewachter, I., Cuypers, K., & Hansen, D. (2025). The impact of exercise training on the brain and cognition in type 2 diabetes, and its physiological mediators: A systematic review. *Sports Medicine - Open, 11*(1), 42. https://doi.org/10.1186/s40798-025-00836-7

Wang, J., Rang, Y., & Liu, C. (2024). Effects of caloric restriction and intermittent fasting and their combined exercise on cognitive functioning: A review. *Current Nutrition Reports, 13*(4), 691–700.

Watt, C., Sanchez-Rangel, E., & Hwang, J. J. (2020). Glycemic variability and CNS inflammation: Reviewing the connection. *Nutrients, 12*(12), 3906. https://www.mdpi.com/2072-6643/12/12/3906.

Weijie, Z., Meng, Z., Chunxiao, W., Lingjie, M., Anguo, Z., Yan, Z., Xinran, C., Yanjiao, X., & Li, S. (2024). Obesity-induced chronic low-grade inflammation in adipose tissue: A pathway to Alzheimer's disease. *Ageing Research Reviews, 99*, 102402. https://doi.org/10.1016/j.arr.2024.102402

Wu, S., Stogios, N., Hahn, M., Navagnanavel, J., Emami, Z., Chintoh, A., Gerretsen, P., Graff-Guerrero, A., Rajji, T. K., & Remington, G. (2023). Outcomes and clinical implications of intranasal insulin on cognition in humans: A systematic review and meta-analysis. *PLoS One, 18*(6), e0286887.

Yang, X., Qiu, K., Jiang, Y., Huang, Y., Zhang, Y., & Liao, Y. (2024). Metabolic crosstalk between liver and brain: From diseases to mechanisms. *International Journal of Molecular Sciences, 25*(14), 7621.

Yang, Y., Zhong, Z., Wang, B., Xia, X., Yao, W., Huang, L., Wang, Y., & Ding, W. (2019). Early-life high-fat diet-induced obesity programs hippocampal development and cognitive functions via regulation of gut commensal Akkermansia muciniphila. *Neuropsychopharmacology, 44*(12), 2054–2064. https://doi.org/10.1038/s41386-019-0437-1

Yaribeygi, H., Atkin, S. L., Simental-Mendía, L. E., & Sahebkar, A. (2019a). Molecular mechanisms by which aerobic exercise induces insulin sensitivity. *Journal of Cellular Physiology, 234*(8), 12385–12392.

Yaribeygi, H., Butler, A. E., & Sahebkar, A. (2019b). Aerobic exercise can modulate the underlying mechanisms involved in the development of diabetic complications. *Journal of Cellular Physiology, 234*(8), 12508–12515.

Yaribeygi, H., Maleki, M., Butler, A. E., Jamialahmadi, T., & Sahebkar, A. (2023). Brain insulin signaling and cognition: Possible links. *EXCLI Journal, 22*, 237.

Yaribeygi, H., Rashidy-Pour, A., Atkin, S. L., Jamialahmadi, T., & Sahebkar, A. (2021). GLP-1 mimetics and cognition. *Life Sciences, 264*, 118645.

Yoon, J. H., Hwang, J., Son, S. U., Choi, J., You, S. W., Park, H., Cha, S. Y., & Maeng, S. (2023). How can insulin resistance cause

Alzheimer's disease? *International Journal of Molecular Sciences, 24*(4). https://doi.org/10.3390/ijms24043506

Zeng, Y., Zhang, L., & Hu, Z. (2016). Cerebral insulin, insulin signaling pathway, and brain angiogenesis. *Neurological Sciences, 37*(1), 9–16.

Zheng, M., & Wang, P. (2021). Role of insulin receptor substance-1 modulating PI3K/Akt insulin signaling pathway in Alzheimer's disease. *3 Biotech, 11*(4), 179.

Chapter 14

Insulin-linked synaptic modulation in Alzheimer's pathology: Emerging potential of nanofibers

Patrick R. Sellers, Binod Pokharel, Shanvanth Arnipalli and Ouliana Ziouzenkova
Department of Human Sciences, College of Human Education and Ecology, The Ohio State University, Columbus, OH, United States

List of abbreviations

AAC2– Amino acid–based nanofiber complex two
ACC1– Acetyl-CoA carboxylase 1
ACE– Angiotensin-converting enzyme
AD– Alzheimer's disease
AGE– Advanced glycation end product
APOE– Apolipoprotein E
APP– Amyloid precursor protein
AS160– Akt substrate of 160 kDa
ATP– Adenosine triphosphate
Aβ– Amyloid beta
BBB– Blood–brain barrier
BDNF– Brain-derived neurotrophic factor
CERAD– Consortium to Establish a Registry for Alzheimer's Disease
CP– Choroid plexus
CSF– Cerebrospinal fluid
DNA– Deoxyribonucleic acid
E/I– Excitatory/inhibitory balance
ECM– Extracellular matrix
ERK– Extracellular signal–regulated kinase
FDG– Fluorodeoxyglucose
fMRI– Functional magnetic resonance imaging
GABA– γ-Aminobutyric acid
GLT-1 (EAAT2)– Glutamate transporter 1 (Excitatory amino acid transporter 2)
GLUT1/3/4/8– Glucose transporter isoforms 1, 3, 4, 8
Grb2– Growth factor receptor-bound protein two
GSH– Glutathione
IDE– Insulin-degrading enzyme
IGF1/IGF2– Insulin-like growth factor 1/2
IGF1R– Insulin-like growth factor 1 receptor
IKKβ– IκB kinase beta
INS/Ins1/Ins2– Insulin gene (human INS, rodent Ins1/2)
IR-A/IR-B– Insulin receptor isoform A/isoform B
IRS1/2– Insulin receptor substrate 1/2
IR– Insulin receptor
JNK– c-Jun N-terminal kinase
Km– Michaelis constant (enzyme affinity measure)
L-LTP– Late-phase long-term potentiation
LTD– Long-term depression
NF-κB– Nuclear factor kappa-light-chain-enhancer of activated B cells
NFT– Neurofibrillary tangle
NOX– NADPH oxidase
pAkt– Phosphorylated Akt
PDX1 (IPF1)– Pancreatic and duodenal homeobox 1 (Insulin promoter factor 1)
PI3K– Phosphoinositide 3-kinase
PKC– Protein kinase C
PKD– Protein kinase D
PVN– Paraventricular nucleus (hypothalamic region)
RAC1– Ras-related C3 botulinum toxin substrate 1 (small GTPase)
RAGE– Receptor for advanced glycation end products
RNA– Ribonucleic acid
ROS– Reactive oxygen species
SHC– Src homology two domain–containing adaptor protein
SHP– SH2 domain-containing phosphatase
SLC2A4– Solute carrier family two member 4 (GLUT4)
SNIFF– Study of Nasal Insulin in the Fight against Forgetfulness (clinical trial)
SREBP2– Sterol regulatory element–binding protein two
Syn2/Syp– Synapsin II/Synaptophysin
TAK1– Transforming growth factor β–activated kinase 1
TNFα– Tumor necrosis factor alpha
TrkB– Tropomyosin receptor kinase B
ZO-1– Zonula occludens-1Acadvl

Hormones in Neuroscience. https://doi.org/10.1016/B978-0-443-30263-3.00031-2

Introduction

Alzheimer's disease (AD), distinguished from other types of dementia by the presence of amyloid plaques and neurofibrillary tangles (NFTs) of hyperphosphorylated tau in brain regions critical for memory (Table 14.1), is the leading cause of progressive cognitive decline globally. More than 57 million individuals are currently affected by AD worldwide, with projections estimating an increase to 152 million cases by 2050. This persistent and accelerating rise, coupled with a notable trend toward earlier disease onset has shifted research priorities toward early risk identification, encompassing environmental factors and complex genetic landscape (Eteleeb et al., 2024).

The focus of this chapter is on glucose and insulin as modifiable, diet-dependent risk factors for Alzheimer's disease (AD). Notably, 81% of patients with AD present with hyperglycemia or type 2 diabetes (Janson et al., 2004). AD is more than twice as common in T2D, with cerebral insulin resistance therefore termed "type 3 diabetes" (Blazquez et al., 2014). High risk, 79% risk is also attributed to heritable factors. Notably, among the 75 major genetic risk loci, several directly intersect with glucose metabolism including APP, ACE, and SLC2A4 (GLUT4), highlighting the convergence of genetic and metabolic risk in AD.

Pancreatic insulin is a peptide hormone that directs glucose via GLUT4 to immediate energy needs and anabolic processes, including proliferation, growth, repair, and the synthesis of proteins, lipids, nucleic acids, and lipid storage. Elevated glucose triggers pulsatile insulin release, synchronized with nutrient supply, to restore euglycemia, maintain osmotic balance, and limit the formation of advanced glycation end products (AGEs) that promote inflammation (Song et al., 2000). The convergent activation of inflammatory and insulin signaling cascades is a

TABLE 14.1 Selected characteristics of brain regions affected by AD. SynL-synaptic loss, IR-insulin receptor A or B, Epen+ – ependymal cells. Arrow represents the direction of arterial flow.

Brain region	Early pathology	Late pathology	Barrier	IR	*Ins*	GLUT4
FRONTAL CTX	SynL, judgment↓ Aβ+/NFT−	Syn/neuronL, exec↓ Aβ ++/NFT+	BBB	A > B	Low	Med-Hi
PARIETAL CTX	SynL, visuosp ↓ Aβ+/NFT−	Syn/neuronL, language ↓ Aβ ++/NFT+	BBB	A > B	Low	Med-Hi
INSULAR/CINGULATE	SyuL Aβ ± /NFT−	Syn/neuron/connectivity L Aβ + /NFT++	BBB	A > B	Mod	Med
TEMPORAL CTX	SynL Aβ+/NFT+	Syn/neuronL Aβ++/NFT++	BBB	A > B	Mod	Med-Hi
ENTORHINAL CTX	SynL Aβ−/NFT + in II	Syn/neuronL Aβ+/NFT + II	BBB	A > B	Low-mod	Med
HIPPOCAMPUS	SynL Atrophy Aβ−/NFT+	Syn/neuronL Aβ + /NFT++	BBB	A > B	Mod-Hi	High
PARAHIPPOCAMPAL	SynL Focal atrophy Aβ−/NFT+	Connectivity L Aβ + /NFT+	BBB	A > B	Low-mod	Med
AMYGDALA	Mild SynL Aβ−/NFT−	SynL/neuronL Aβ + /NFT+	BBB	A > B	Mod	Med
THALAMUS	Mild SynL Aβ−/NFT−	Syn/neuronL ±Aβ/NFT±	BBB	A ≥ B	Low	Low-mod
STRIATUM	Subtle synL Aβ−/NFT−	syn/neuronL ±Aβ/NFT±	BBB	A ~ B	Low	Low
PERIVENTRICULAR	Glial activation mild synL Aβ−/NFT−	Glial/neuronL ±Aβ/NFT±	CP	N/A	Min	Epen+
CHOROID plexus, (CP)	Inflammation barrier stress ±Aβ/±NFT±	Barrier fails ±Aβ/NFT± in CSF	CP	A > B	Very High	DevL
VENT. CSF COMP.	(N/A biomaxrker)	Barrier fails ±Aβ/NFT± in CSF	CP	A	From CP	Low

key driver of insulin resistance and compensatory hyperinsulinemia (Talbot et al., 2012).

Insulin signaling in the brain possesses autonomous features that interact cooperatively or competitively with peripheral tissues to maintain homeostasis. Insulin's functions are redirected from growth and storage toward neuronal survival, synaptic plasticity, and memory formation, with insulin receptors closely integrated with glucose transporters and metabolic pathways. These mechanisms are highly species-specific, limiting direct translation from animal AD models to humans.

Here, we examine cerebral insulin signaling, its progressive remodeling and age-related decline, with particular emphasis on synaptic modulation as an early dysfunction in AD. We conclude with perspectives on nanotherapeutic strategies for targeting AD.

The synaptic loss in IR-positive regions is a pathological feature across AD subtypes

The heterogeneity of AD pathogenesis was first recognized in 1906, with subtypes classified primarily by clinical features until the 1970s. Advances in immunohistochemistry and pathological diagnostics then established the CERAD and Braak staging systems, later expanded by imaging studies to delineate typical, limbic-predominant, hippocampal-sparing, and minimal atrophy subtypes based on Aβ plaques, NFT, and regional atrophy patterns.

Following the pioneering discoveries of genetic AD subtypes by the Gajdusek and Müller-Hill laboratories in 1987, distinct hereditary forms of AD were established through genetic characterization. Subsequent large-scale multiomics (Eteleeb et al., 2024) profiling of brains from both AD patients and cognitively intact individuals, revealed remarkable heterogeneity between heritable and non-heritable AD cases, as well as the paradoxical presence of Aβ plaques and NFTs in individuals without cognitive deficits.

To reflect AD heterogeneity in causality, manifestation, and progression, current 21st-century classification systems implement new AD subtypes grounded in the A/T/N (amyloid/, tau/, neurodegeneration) biomarker framework, as well as molecular signatures and functional network hubs. An example of this approach is a recent classification based on 1056 proteomic markers, which highlighted five major molecular AD subtypes (Tijms et al., 2024): 1: Proteins involved in neuronal hyperplasticity; 2: Innate immune activation; 3: RNA dysregulation; 4: Choroid plexus (CP) dysfunction; and 5: BBB impairment. It is now recognized that the idiosyncratic combination, intensity, and timing of these factors yield diverse neuropathological mechanisms, driving the clinical heterogeneity of AD subtypes. These multisystem assessments improve diagnosis, prognosis, and the potential for personalized therapy by reflecting the complexity of AD biology.

Despite this heterogeneity, synaptic plasticity and synaptic loss in IR-positive regions, specifically in areas shielded by the BBB or CP (Table 14.1) are consistent features of both early and late phases of AD, and their severity predicts earlier disease onset and increased risk of earlier mortality compared to AD cohorts without pronounced synaptic dysfunction (Eteleeb et al., 2024). Strikingly, robust markers of AD progression, such as CLU, SNCA/αSyn, IGF1, NRXN3, and YWHAZ were integrally linked to synaptic function, IR-dependent signaling, and barrier function. Combining measures of synaptic plasticity, synaptic loss, proteinopathy, and cerebral FDG-glucose uptake yields more accurate diagnostic and prognostic indicators of cognitive decline in AD than proteinopathy markers alone (Bloudek et al., 2011). Thus, preventing neuronal hypoglycemia, synaptic loss, and barrier dysfunction is central to AD prevention, with insulin metabolism providing a unique, modifiable means of regulation.

Autonomous regulation of cerebral insulin and glucose metabolism

Autonomic regulation of insulin and glucose metabolism in the brain supports trophic and anabolic pathways, as well as neuronal and synaptic responses. It is governed by two key mechanisms: (1) BBB- and CP-mediated entry of circulating pancreatic insulin and (2) intrinsic insulin expression within the brain.

Pancreatic insulin is the major source of cerebral insulin

Pancreatic β-cells express the highest levels of the human insulin gene (*INS*) under PDX1 (IPF1) control. Rodents express two insulin genes: *Ins2*, the ortholog of the human *INS*, and *Ins1* a rodent-specific retroposed gene sharing homologous regulatory regions with *Ins2*. Human INS is translated into preproinsulin, processed to proinsulin, and cleaved into C-peptide and mature insulin. Basal secretion is modulated by endocrine signals, while glucose-stimulated insulin secretion (GSIS) is triggered by glucokinase activity, ATP production, K_ATP channel closure, and calcium influx. Chronic hyperglycemia enhances PDX1-dependent *INS* expression and production. Together, these mechanisms sustain the major endocrine pool of circulating insulin.

Extrapancreatic sites, including bone marrow, the thymus, liver, and some lymphoid tissues, also express *INS*, but in the absence of PDX1 these tissues show impaired processing and lack glucose responsiveness, thereby contributing to tissue-specific functions.

Circulating pancreatic insulin is the main source of cerebral insulin, assessed in human cerebrospinal fluid (CSF) rather than brain tissue due to methodological challenges. The BBB maintains homeostatic separation between systemic and cerebral insulin, preventing acute peripheral fluctuations from directly affecting brain insulin concentrations (Rhea & Banks, 2021) (Fig. 14.1A). Despite the pancreatic origin of most insulin present in the brain, both human CSF and mouse brain contain 100-fold lower concentrations (Blazquez et al., 2014) than insulin concentrations (≥1 nM) typically used in in vitro neuronal culture studies. This discrepancy likely reflects insulin compartmentalization and accumulation in specific brain regions, unlike the broad systemic effects of circulating insulin (Bingham et al., 2002).

CP and BBB in autonomous cerebral insulin and glucose metabolism

The BBB maintains a gradient between cerebral and peripheral insulin and glucose, blunting acute hyperinsulinemic and hyperglycemic fluctuations (Fig. 14.2A) (Rhea & Banks, 2021). CP serves as another structure where insulin crosses the blood-CSF barrier to supply insulin to CSF and interstitial fluid (ISF). CSF/CP function contributes approximately 10% of cerebral insulin (blood and ISF), while BBB-mediated transport from the circulating pool is the main route (Gray et al., 2014) reducing concentrations from 10–100 fold. Even during peripheral insulin infusion at rates such as 0.3 $mU{\cdot}kg^{-1}{\cdot}min^{-1}$, only a modest (~15%) increase in insulin is observed in the high neuronal density regions, accompanied by **decreased** FDG glucose uptake in the motor cerebellar cortex (Bingham et al., 2002). The BBB expresses more IRB than IRA

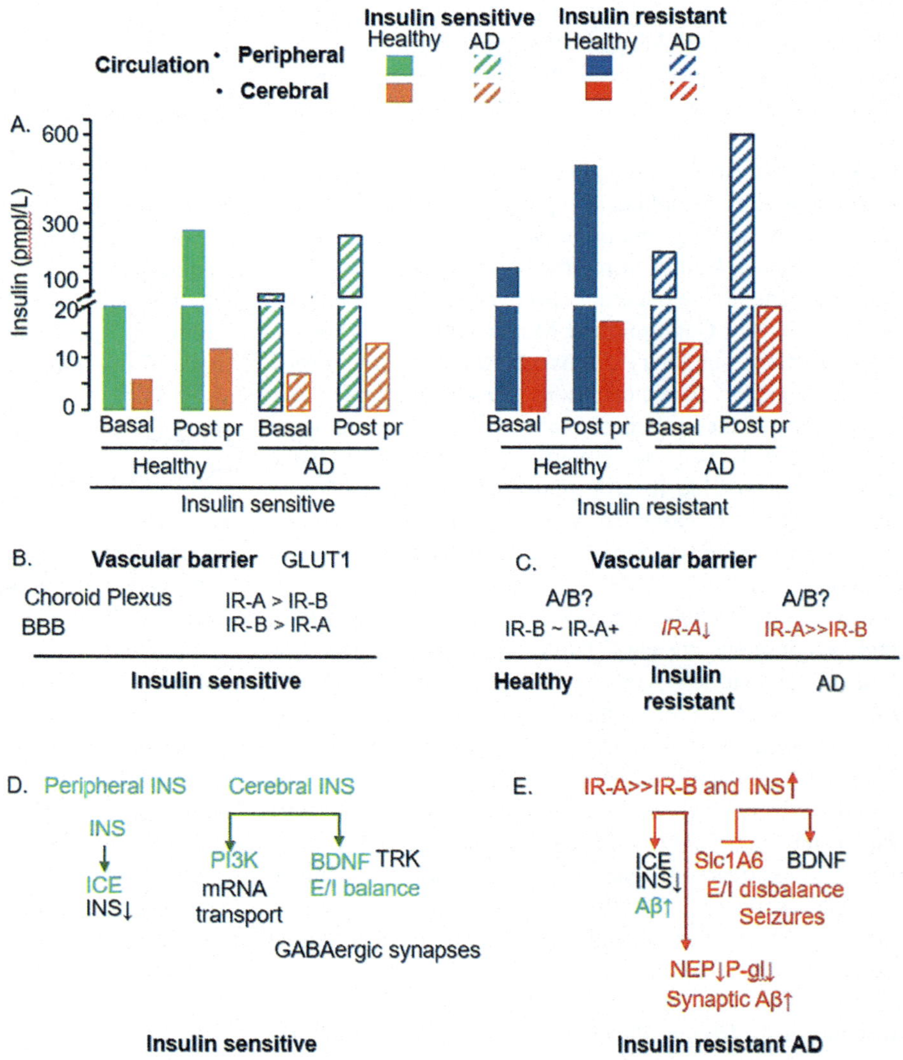

FIGURE 14.1 The differences in peripheral and cerebral metabolism of insulin. (A) Insulin concentrations in insulin-sensitive subjects with and without AD (left) and insulin resistant subjects with and without AD (right). (B) Glut 1 is the dominant BBB transporter. Intact BBB and CP express IR B > A (Watson et al., 2022), (C) insulin resistantce: BBB more IRA in the membrane (IRA+), less *IRA* mRNA expression is downregulated (*IRA*-); AD: IR-B is downregulated: Ratio IRA > IRB BBB become insulin resistant. (D) ICE is the major insulin clearance pathway in peripheral tissues. Cerebral insulinregulated pathways (Deng et al., 2024). (E) BBB under insulin resistance and AD (Leclerc et al., 2023) IR expression correlates positively with IDE, and negatively with Aβ. Neprilysin (NEP) cleaves Aβ, P-glycoprotein (P-gpl) responsible for Aβ clearance (Mohamed et al., 2016). C. AGE levels are shown as % to control based on published data.

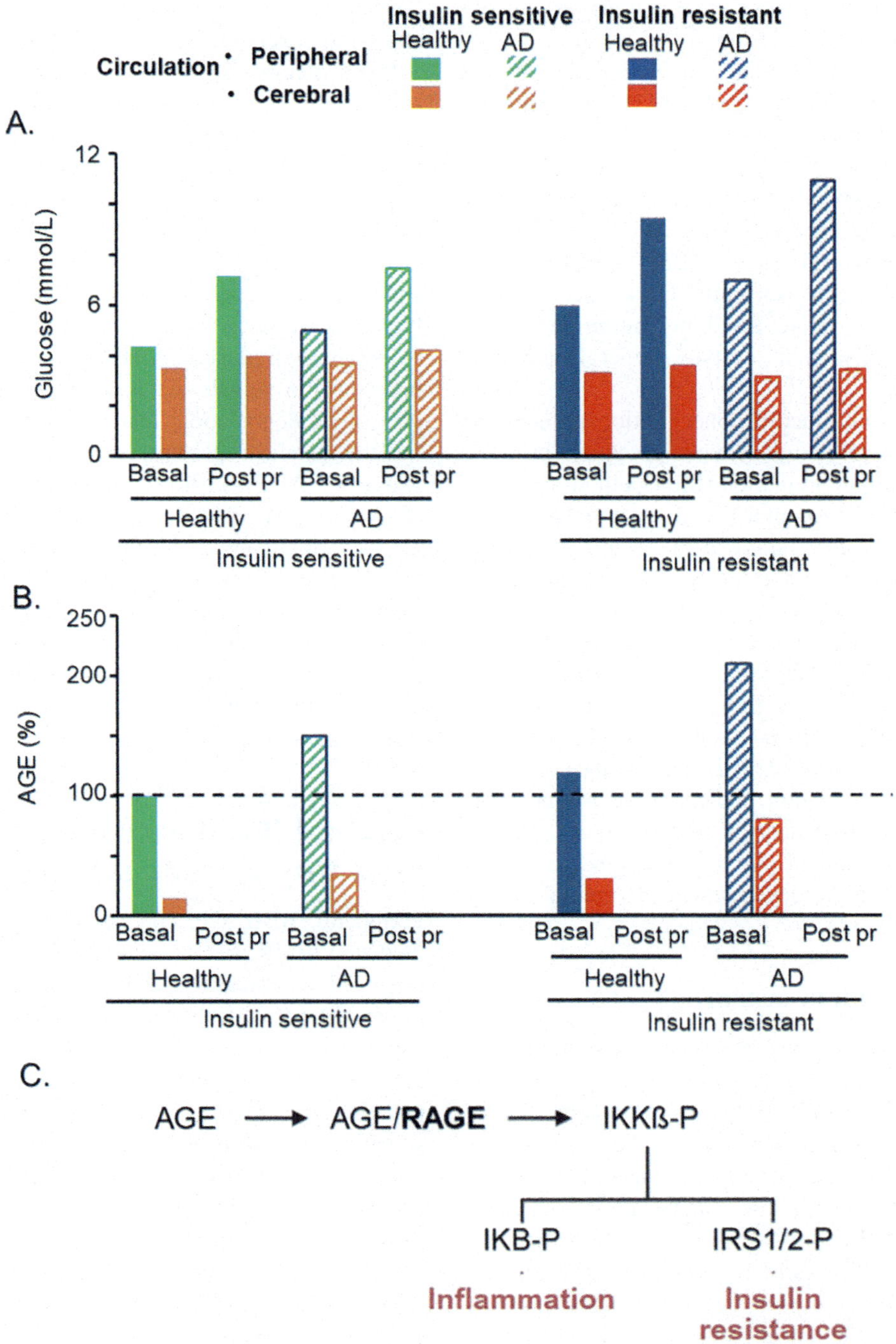

FIGURE 14.2 The differences in peripheral and cerebral metabolism of glucose and AGE in insulin-sensitive subjects with and without AD (left) and insulin resistant subjects with and without AD (right) A. Basal and postprandial glucose levels (Jarvis et al., 2023). B. AGE levels are shown as % to control (Dakterzada et al., 2023; Monacelli et al., 2014). C.AGE activate cognate RAGE receptor phosphorylating IKKβ (IKKβ-P). Activated IKKβ-P phosphorylate IκB for degradation to activate NFκB-dependent inflammatory pathways, whereas phosphorylation of IRS1/2 disable IR signaling.

isoforms, enabling saturable insulin transport that both balances and blunted dietary and pharmacological insulin spikes in peripheral blood, maintaining homeostatic control of insulin entry into the brain.

Insulin homeostasis is fundamentally important for cerebral glucose metabolism, limiting recruitment of GLUT4 regulated by insulin. BBB glucose uptake is predominantly mediated by GLUT1 (Fig. 14.1B) and its genetic deficiency causes severe reductions in glucose delivery and neuronal dysfunction.

Cerebral glucose is taken up predominantly through **insulin-independent** GLUT1 and GLUT3 relaying on facilitated diffusion effective at low cerebral glucose concentrations (Custodio et al., 2021). Therefore, brain glucose concentrations are significantly lower than in plasma of healthy humans and mice and lactate has the opposite trend (Nagalski et al., 2016). However, in hyperinsulinemic insulin resistant states, this glucose/lactate gradient may be reversed or diminished, often resulting in hypoglycemia despite hyperglycemia in circulation.

BBB control over glucose concentration reduces the formation of reactive non-enzymatic glycation products (AGEs) in the cerebral circulation (Monacelli et al., 2014) (Fig. 14.2B). The AGE receptor RAGE, expressed at BBB endothelium is a key contributor to EC insulin resistance, acting both through proinflammatory TNFα signaling and directly via the RAGE/IRS1 axis in EC or peripheral tissues (Fig. 14.2C). RAGE activation in neurons, microglia, and astrocytes amplify AD pathogenesis leasing to synaptic dysregulation, NFT accumulation, and Aβ influx (Johnson et al., 2023) demonstrated in RAGE deficiency mice protected from endothelial insulin resistance (Song et al., 2014), Aβ accumulation, and cognitive decline (Fang et al., 2018). Thus, BBB and CP integrity play a pivotal and causal role in brain insulin, glucose, and redox homeostasis, underlying the development of cerebral insulin resistance.

Cerebral expression of insulin

An additional autonomous characteristic of the brain is the regional expression of INS in humans and predominant expression of *Ins2* in mice in regions vulnerable to AD pathology (Table 14.1), with the highest expression detected in the CP in response to serotonin and intracellular Ca^{2+} levels, but not to glucose (Mazucanti et al., 2019) consistent with PDX1 expression only in the hypothalamus in adults. Instead, *Ins2* expression is under the control of WNT3α and GSK3, supressed by principal AD amyloid peptide Aβ1–42, while αSyn induces insulin secretion. Thus, AD proteopathy may provoke loss of intrinsic cerebral insulin synthesis.

Cerebral insulin biosynthesis is restricted to a small (3%–5%) subset of neurons and neuronal progenitors and resembles pancreatic expression of preproinsulin, which undergoes proteolytic cleavage to yield active insulin in response to glucose. Insulin expression has been most intensively investigated in subsets of GABAergic interneurons, where it modulates both GABAergic and glutamatergic synaptic transmission to maintain E/I balance in response to glucose through local adjustments in insulin concentration to prevent epileptiform discharges, seizures, and neurodegeneration.

Spaciotemporal, low-level autocrine and paracrine insulin production in the brain shapes both cerebral and systemic physiology. Its systemic impact is evident in hypothalamic PVN neurons, where insulin drives pituitary growth hormone release and body growth, deficits that are rescued by restoring neuronal insulin expression (Lee et al., 2020b).This finding underscores the unique paracrine and autocrine actions of cerebral insulin, governing brain functions while exerting systemic effects.

Whether produced by the BBB, CP, or in neural tissue, brain insulin signals via its cognate IR or is cleared by proteolytic degradation. Insulin is a high-affinity substrate for the insulin-degrading enzyme IDE (Km = 0.1 μM), whereas Aβ substrate has much lower affinity (Km = 2.7 μM). IDE degrades more than 50% of cerebral Aβ in normal conditions (Farris et al., 2003), but elevated insulin can competitively inhibit Aβ degradation, thus impairing its clearance (Fig. 14.1D**,** **E**). CNS insulin signaling is modulated by crosstalk with IGF1/IGF2 and their receptors, yet IR remains essential for synaptic structure, function, and plasticity, dependent on local insulin availability.

Cerebral IRs in excitatory and inhibitory synaptic systems

Peripheral, BBB, CP, and cerebral responses insulin are mediated by IR, a heterotetrameric tyrosine kinase receptor. IGF1 activates IR and insulin activates IGF1R, both at lower affinity than their cognate interactions. These contributions may be relevant, as IGF1 is linked to an AD subtype with higher astrocyte content and poorer cognitive outcomes (Eteleeb et al., 2024), though details of its signaling are discussed elsewhere (Pomytkin et al., 2018). Two major isoforms of IR-A and IR-B are generated by alternative splicing of exon 11 in the α-subunit and expressed in peripheral and CNS tissues (Table1).

IR-A, lacking exon 11, is the predominant isoform in neurons, although some neurons express IR-B alone or a combination of IR-A and IR-B, suggesting functional heterogeneity within neurons (Spencer et al., 2018). IR-B, including exon 11, is predominant in astrocytes (Gonzalez-Garcia et al., 2021) and is the predominant form in intact BBB endothelium (Fig. 14.1B). Other brain cell populations display heterogeneous expression patterns. The concentration of IR in the periphery and the brain is regulated independently, even under hyperinsulinemia or hypoinsulinemia (Rhea et al., 2022), which do not necessarily alter total IR expression or protein abundance in the brain changes in the IR-A/IR-B ratio and their subcellular distribution (Fig. 14.1C). Such isoform distributions and localization patterns are linked to normal synaptic function, whereas isoform shifts and aberrant localization are associated with insulin resistance–driven neurodegenerative conditions, including aging and AD.

Tripartite excitatory glutamatergic and inhibitory GABAergic synapses express in IR-A in pre-synaptic and postsynaptic neurons, while both type of neurones localize IR-A within the soma, shafts, and dendritic spine. These synapses are enwrapped by astrocytes, expressing IR-B.

Insulin activation of IR-A in cortical and hippocampal glutamatergic neurons stimulates BDNF production (Fig. 14.1D), which effects Trk receptors on same neurons (Deng et al., 2024). In glutamatergic neurons, presynaptic BDNF–TrkB signaling enhances synaptic strength by increasing glutamate release, while postsynaptic BDNF–TrkB activation promotes GLUT3 expression and trophic glucose utilization, resulting in improved synaptic plasticity. In GABAergic synapses, both presynaptic and postsynaptic BDNF–TrkB signaling enhance inhibitory synaptic strength and overall inhibitory input (via $GABA_A$ expression and GABA release, respectively), to balance E/I within neural circuits as well as synaptogenesis, and neuroplasticity.The critical role of induced BDNF secretion is underscored by the Val66Met polymorphism, which impairs BDNF trafficking and diminishes hippocampal integrity and episodic memory.

Insulin/IR-B signaling upregulates GLUT1 astrocytic glucose supply in astrocytes (Cai et al., 2018), energy demanding processes, including:

(1) astrocyte-neuron lactate shuttle nourishing glutamatergic and GABAergic neurons (Ardanaz et al., 2024);
(2) synthesis of the majority of brain cholesterol (Chen et al., 2023); trafficking of cholesterol-loaded ApoE (ApoE4) from astrocytes to neurons (Wang et al., 2021);
(3) release of gliotransmitters (e.g., ATP, glutamate, adenosine) that modulate synaptic activity at excitatory and inhibitory synapses (Ardanaz et al., 2024; Gonzalez-Garcia et al., 2021); and
(4) attenuation of excessive excitatory drive via increased surface expression and activity of the astrocytic glutamate transporter GLT-1 (EAAT2) (Cai et al., 2018).

Astrocytes' high glucose demand has an overarching impact on neuronal glucose metabolism. Astrocyte-specific GLUT1 deficiency leads to increased IR expression and enhanced glucose uptake in other brain cells and peripheral tissues and vice versa IR deficiency in astrocytes reduce *Glut1* expression (Tables 14.2–14.3). Genetic deficiency of the IR in astrocytes disrupts the canonical insulin–GLUT1 regulatory axis, which cannot be compensated by upregulation of GLUT3 or GLUT8, and leads to reduced

TABLE 14.2 Loss-of-function phenotypes in mouse models relevant to peripheral insulin signaling.

Peripheral			
Target	**Context**	**Phenotype**	**PMID**
IR	Adipose tissue	Lipodystrophy, insulin resistance, hyperglycemia	11602628
IR	Muscle	Muscle IR, elevated fat mass, normal glucose tolerance	9844629
IR	Whole body	Growth retardation, early death	12471165
IR inducible, Rescue IR-A, Rescue IR-B	Whole body	Glucose intolerance, insulin resistance Full rescue Partial rescue	27562101
IR	Hepatocyte spec.	Impaired glucose metabolism, hepatic cholesterol↑, 40%↓ in *Srebp2* and cholesterologenic genes 50%–90%↓ in the cholesterologenic genes	25963408, 10949030, 24516236
IR-B	β-cell spec.	Impaired glucose tolerance, hyperinsulinemia and hyperproinsulinemia ERK-SREBP1- dependent	39021804
IRS-1	Whole body	Growth retardation, mild insulin resistance, prolonged survival	12904469
IRS-2	Whole body	Small brain size, impaired synaptic plasticity, diabetes, hyperphosphorylated τ	12904469, 21955917
AKT2	Whole body	Developmental deficits, reduced organ size, GLUT1↓, neural apoptosis, type-2 diabetes-like phenotype	12843127 11387480
GLUT1	Whole body	Neonatal death, growth retardation, diabetes	14673082
GLUT3	Whole body	Reduced body weight and brain size, epileptiform activity	33482226

TABLE 14.3 Loss-of-function phenotypes in mouse models relevant to cerebral insulin signaling.

Cerebral			
Target	**Context**	**Phenotype**	**PMID**
IR	Hippocampus spec.	Normal bodyweight, body composition, glycemia, and insulinemia. Significantly impaired long-term memory, learning ability, and synaptic plasticity	26216852
IR	Neuron spec.	Normal brain size. Increased food intake, bodyweight, and adiposity. Hyperinsulinemia and dyslipidemia. Systemic insulin resistance. Impaired neuronal PI3K/Akt activity. Impaired synaptic plasticity	14981233, 11000114, 22661254
IR	Astrocyte spec.	Anxiety- and depressive-like behaviors hyperphosphorylated tau, impaired Aβ clearance, ↑ApoE expression. ↓Glucose influx, ↓*Glut1*	37186836 29664737 27518562
IRS-1	Neuron spec.	Growth retardation, cerebral and systemic mitochondrial dysfunction, impaired synaptic outgrowth	36812323, 33980693
IRS-2	Neuron spec.	Diabetes, systemic insulin resistance, incresaed food intake, enhanced memory, enhanced spine formation	21597043, 15467829
p110α	Neuron spec.	Postsynaptic structural losses, memory and learning deficits	36417513
p110β	Neuron spec.	Presynaptic dysfunction, loss of neurotransmission and synaptic plasticity, memory deficits	36417513
AKT1	Neuron spec.	Impaired synaptic plasticity, impaired protein synthesis, altered dendritic complexity, anxiety-like behavior, memory deficit	17077150, 33325370
AKT2	Neuron spec.	Reduced axonal length, impaired synaptic outgrowth	22509246
AKT3	Neuron spec.	Reduced brain size, abnormal brain development, impaired synaptic plasticity	15713641, 17077150
ERK2	Neuron spec.	Reduced cortical thickness and neurogenesis, impaired associative learning and memory	18596172
GLUT1	Astrocyte spec.	Normal behavior, enhanced glucose metabolism	40328784
GLUT3	Neuron spec.	Reduced dendritic spine density, impaired neurotransmission, altered behavior	30232223
GLUT4	Neuron spec.	Systemic glucose intolerance, hepatic insulin resistance, reduced brain glucose uptake	27797912

cholesterol synthesis in astrocytes, resulting in synaptic dysfunction and neurodegenerative disorders. Functional IR deficiency under insulin-resistant conditions reduces cholesterol synthesis and ApoE/ApoE4-mediated transport in HDL-like particles from astrocytes to neurons for incorporation into synaptic membranes, leading to lower synaptic cholesterol, disrupted lipid rafts, and increased membrane fluidity (Suzuki et al., 2010), disrupting synaptic function. This membrane milieu increases β- and γ-secretase activities, leading to elevated Aβ production (Wang et al., 2021). ApoE4-dependent remodeling of the synaptic membrane also promotes increased endocytosis, altering the proportion of IR-A in neurons. Thus, IR isoform signaling in neurons and astrocytes shapes synaptic architecture, composition, and dynamics, which chronic insulin resistance disrupts, rendering synapses vulnerable to metabolic and environmental stressors.

Synaptic signaling determinants of insulin sensitivity

The cerebral and peripheral IR signaling (Taniguchi et al., 2006), share signaling nodes; nevertheless, loss-of-function studies (Tables 14.2,14.3) revealed their unique functional outcomes in the brain cell population. The roles of IR-A and IR-B in specific phenotypes remain unclear. In liver-specific IR knockout mice, insulin resistance and glucose intolerance were rescued by IR-A overexpression, whereas IR-B had only moderate effects, highlighting differences in their downstream signaling. Independent on IR isoform insulin-activated IR undergoes conformational changes that promote autophosphorylation of specific tyrosine residues (IR^P), creating docking sites for subsequent phosphorylation of two adaptor protein families (Yunn et al., 2023). The central high-affinity adaptor protein is insulin receptor substrates (IRS), comprising of six members with IRS1 and two docking to IR^{P} (Fig. 14.3A). Although both IRS1 and IRS2 bind to IR^{P}, they mediate different responses (Tables 14.2,14.3). Both whole-body and neuronal IRS2 deficiency induce insulin resistance with tau hyperphosphorylation, reduced synaptic plasticity, and smaller brain size, whereas neuronal IRS1 deficiency primarily disrupts mitochondrial function. IRS binding and phosphorylation by IR^{P} assemble the PI3K/Akt/mTOR signaling node that mediates the metabolic branch of IR responses. Failure of IR^{P} IRS^{P} assembly defines brain insulin resistance (Arnold et al., 2018).

An alternative mitogenic signaling node involves the adaptor protein SHC binding to IR^{P} or IRS1/2. Phosphorylated SHC then recruits Grb2, linking IR^{P}–SHC^{P} to the Ras/Raf/MEK/ERK cascade.

Cerebral insulin signaling shares core IR features but varies by isoform, function, and localization, producing cell-specific responses through the IR^{P}IRS^{P}/PI3K^{P}/Akt^{P}/mTORC1/2^{P} cascade. Of the three PI3K classes, only class I ubiquitously expressed in neurons and astroglia, functioning as heterodimers of regulatory p85 (α, β, γ) and catalytical p110 (α, β, γ, δ) subunits. While p110α and p110β are broadly expressed, p110β is especially critical for presynaptic neurotransmitter synthesis and vesicle release (Sanchez-Castillo et al., 2022). p110α predominates in dendritic and postsynaptic compartments, where its loss causes structural deficits, while p110α drives the

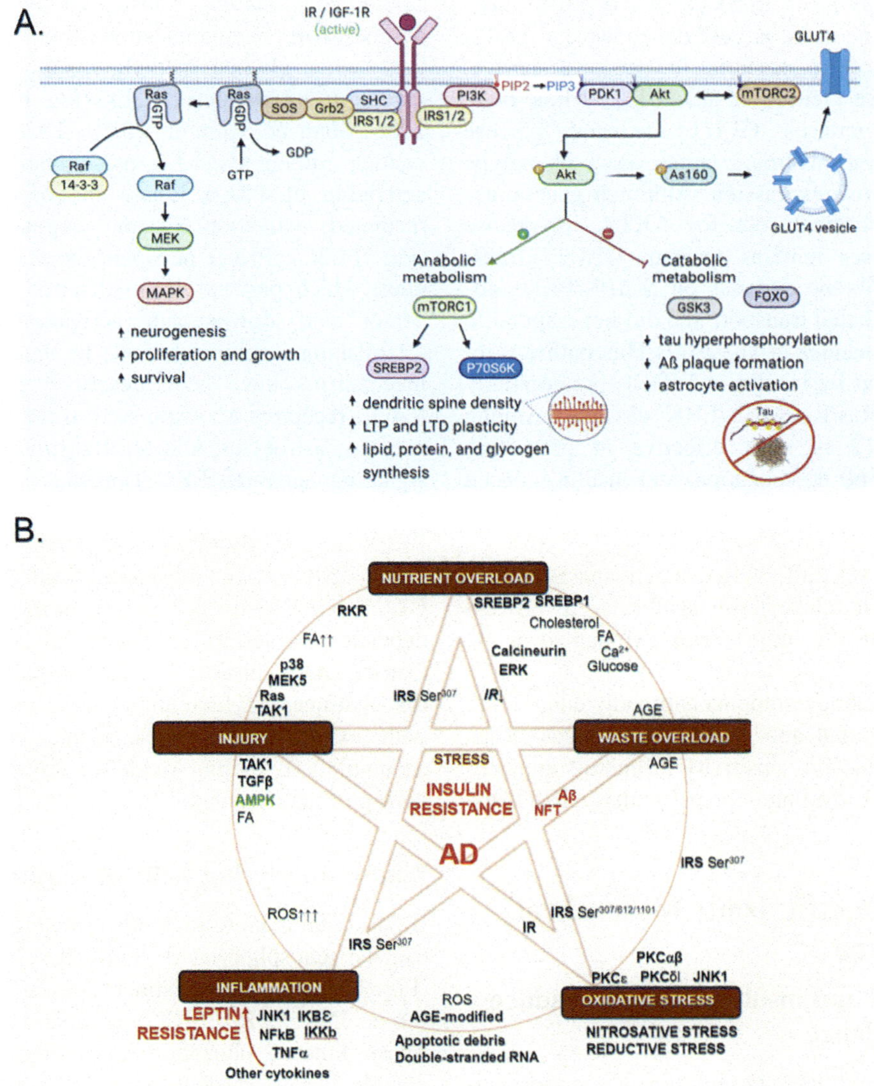

FIGURE 14.3 Insulin signaling pathway and its downstream targets (A), and integrated circuits of self-perpetuating pathways driving insulin resistance (B). Specific kinases inactivate IR or IRS through phosphorylation, amplifying insulin resistance. NF-κB activation induces cytokine expression, enhancing inflammation and promoting TNFα-mediated JNK phosphorylation, which contributes to LepR deficiency. Metabolites involved in the cycle are shown in regular font, and enzymes in bold.

Akt1–mTORC1 axis during LTP. Memory-related plasticity depends on mRNA transport from soma to synapses, regulated by PI3K p110β through phosphorylation of structural proteins and local translation via the p110β→Akt1/Akt3→mTORC1 axes. Disruption of these steps impairs LTP, leading to learning and memory deficits.

PI3K activates AKT isoforms, generating tissue-specific signaling responses (Table 14.2). Akt phosphorylation branches the cascade through kinases such as mTORC1, mTORC2, and GSKβ, or functional proteins such as AS160 to regulate GLUT4 and GLUT1 (Fig. 14.3) (Arnold et al., 2018). AKT1 and AKT3 are highly expressed in hippocampal and cortical neurons, whereas AKT2 is enriched in astrocytes. AKT1 is the most abundant brain isoform, dominating synaptic plasticity by supporting late-phase LTP and LTD, though Akt isoforms can partially compensate for one another (Tables 14.2–14.3).

Among their many roles, astrocytes exert critical neurotrophic effects on neurons, largely dependent on AKT2, their predominant isoform, essential for neuronal survival and insulin resistance (Jensen et al., 2010). Whole-body AKT2 deficiency reduces GLUT1-mediated glucose transport and increases neuronal apoptosis, a phenotype rescued by GLUT1 overexpression. Although genetic evidence confirms a critical role for AKT2, its precise regulation at synapses remains unclear. GLUT1 translocation to the membrane depends on MAPK/PKD activation of RAC1-mediated transport and did not respond to AKT inhibitors (Fernandez et al., 2017).This pathway requires joint activation by insulin and IGF1, preferentially inducing the SHP/Ras/Raf/MEK/ERK cascade. Among Akt isoforms, AKT2 is most effective in promoting glucose uptake, AS160 translocation, and insulin-induced membrane translocation in neuronal cultures and in vivo (Tables 14.2–14.3).

Activated Akt drives trafficking that activates SREBP2-dependent genes for cholesterol synthesis, a process inhibited by Aβ42 in the brain cortex (Mohamed et al., 2018).

Although IR signaling components broadly adapt brain cell functions to internal and external challenges (Rhea et al., 2022), brain insulin sensitivity is mainly impaired through IRS or AKT dysfunction, particularly IRS2 and AKT2 isoforms.

General and specific traits of cerebral insulin resistance

Glucose overload and insulin resistance induce metabolic discordance

Next, we focus on modifiable environmental stressors underlying insulin resistance, while genetic factors have been addressed (Semple et al., 2011). Disruption of insulin signaling is typically initiated by strong interfering signals from interconnected processes—dietary overload (glucose) → waste overload (AGE) ↔ oxidative stress (AGE-induced) ↔ inflammation ↔ injury—forming a self-perpetuating cycle (Fig. 14.3B). These processes are tightly interconnected, such that the activation of one stressor can propagate to others, collectively impairing key nodal points of insulin signaling:

1. The receptor–membrane interface
2. Adaptor proteins, particularly IRS
3. Kinases, particularly AKT

Dietary glucose overload is the most extensively studied and clinically relevant trigger of insulin resistance in Western dietary patterns, although excessive fat or protein intake also contributes to metabolic waste, redox imbalance, and chronic inflammation. Surplus glucose that cannot be efficiently transported by GLUT carriers undergoes non-enzymatic conversion into reactive byproducts such as glyoxal, methylglyoxal (MGO), and advanced glycation end products (AGEs) (Fig. 14.2B) (Cassese et al., 2008; Song & Schmidt, 2012). These products shift the cellular milieu toward a pro-oxidative state, leading to activation of PKCα kinase in conjunction with AGE-mediated activation of the cognate RAGE receptor (Fig. 14.2C). PKCα phosphorylates IRS1 on serine residues, which prevents its association with the insulin receptor and downstream activation of Akt, thereby establishing insulin resistance in skeletal muscle. Similar mechanisms have been described in the CNS, where RAGE receptors are expressed in the blood–brain barrier, neurons, astrocytes, and reactive microglia. AGE–RAGE signaling activates PKC isoforms, which phosphorylate IRS1 and impair insulin signaling. In the CNS specifically, PKCε can directly disable the insulin receptor (Nandipati et al., 2017). This episodic insulin resistance reduces GLUT4/GLUT3 trafficking in neurons and disrupts APOE-dependent cholesterol transfer from astrocytes to neurons. Consequently, insulin resistance establishes a **metabolic discordance**, in which intracellular hypoglycemia coexists with extracellular glucose excess and elevated AGE accumulation, fueling oxidative stress and inflammation with glucose overload.

Kinase interfering with IR sensitivity

IRS1 contains more than 30 serine/threonine residues that can be phosphorylated. Different stressors and states (Fig. 14.3B) activate distinct kinases, including PKC isoforms, JNK, IKKβ, ERK2, mTORC1/S6K1, and GSK3β. These kinases phosphorylate inhibitory residues, most notably Ser307 (rodent)/Ser312 (human) and Ser636/639 (Talbot et al., 2012). Although these phosphorylation events occur at different residues, their outcome is similar:

IRS-1 phosphorylation disrupts its interaction with the IR and impairs downstream Akt activation, thereby affecting both Akt-suppressed kinases such as GSK3β and AKT-regulated mTORC1/S6K1 (Fig. 14.3A). Insight into the importance of IRS1 inactivation for IR signaling was provided by the systematic analysis of AD brain tissue by Talbot et al. (2012), which revealed IRS1 phosphorylation at Ser636/639. This modification was closely associated with dietary overload (Arnold et al., 2018), mediated by glucose-induced ERK2 and amino acid–induced mTORC1/S6K1, as well as insulin resistance through the release of GSK3β suppression, and chronic inflammation via IKKβ and JNK1/2 activation. The convergence of chronic dietary overload with inflammatory signaling marks a critical inflection point, driving the shift from episodic, transient insulin resistance to the persistent state characteristic of brain insulin resistance in AD.

Inflammation and oxidative stress drivers of insulin resistance

Inflammation can serve as an independent entry point (Fig. 14.3B, Inflammation) for the development of autonomous cerebral or systemic insulin resistance in the brain, in both insulin-sensitive and insulin-resistant individuals. Cerebral injury (Gardner et al., 2014), infection (Bohn et al., 2023; Richmond-Rakerd et al., 2024), and acute or chronic systemic inflammation (Chen et al., 2017) are each associated with increased risk of incident AD or other dementias, with hazard ratios of roughly 1.7–2.93 in large cohorts—including ARIC (n~15,700) (Bohn et al., 2023), a nationwide New Zealand registry (n~1.74 million) (Richmond-Rakerd et al., 2024), and a chronic periodontitis cohort (n~18,672). Thus, irrespective of source, pathogen exposure, trauma, or metabolic disease, systemic inflammation confers increased AD risk.

Despite heterogeneous manifestations, inflammatory inputs converge on a shared the transcription factor NF-κB (Fig. 14.3B). NF-κB is activated when phosphorylated IKKβ ($IKK\beta^{P}$) phosphorylates IκBα, targeting it for degradation and releasing active NF-κB. IKKβ activation is driven by upstream kinases such as PKC during oxidative stress, TAK1 during injury or infection. cytokine signaling, and other TLR/pattern recognition/scavenger receptors pathways engaged by viral, metabolic, apoptotic debris, or toxic insults. Across contexts, activated IKKβ can phosphorylate IRS1/2 at inhibitory serine residues, inducing insulin resistance, curtailing GLUT4-dependent glucose uptake in adipose and muscle, and functionally preserving glucose availability for the BBB/CNS and immune compartments that rely predominantly on GLUT1.

NF-κB increases transcription of pro-inflammatory cytokines and ROS-generating enzymes (e.g., NOX, iNOS, COX-2), accelerating non-enzymatic glycoxidation of glucose into reactive dicarbonyls and AGEs, an effect markedly amplified under hyperglycemia in diabetes. ROS can also inactivate AKT2 via oxidation of Cys124, inducing an intramolecular disulfide and loss of kinase activity, thereby providing an alternative route to insulin resistance and weakening anti-apoptotic signaling.

Cerebral inflammation insulin resistance and amyloidogenic traits

At the BBB, AGE–RAGE–NF-κB signaling induces metalloproteinases that degrade tight-junction proteins, particularly claudin-5, occludin, and ZO-1, thereby weakening barrier integrity with aging and in AD prior to Aβ and NFT accumulation (Greene et al., 2019). In an AD mouse model, hippocampal delivery of claudin-5 rescued learning and memory, in part by normalizing inhibitory GABAergic neurotransmission. In humans, claudin-5 is the principal sealing tight-junction protein at the BBB; its degradation marks BBB disruption and is associated with E/I imbalance and cognitive decline across multiple neurodegenerative disorders.

In the brain, AGE–RAGE/NF-κB inflammatory axes have additional specific traits. Beyond sensing AGEs, RAGE mediates Aβ influx across the BBB (Deane, 2012), modulating parenchymal Aβ and providing an additional inflammatory stimulus. NF-κB activation remodels ECM by upregulating collagen I (Peng et al., 2016). In a hyperglycemic/hyperinsulinemic milieu (Fig. 14.1 and 14.2), elevated MGO, AGEs promote glycation of insulin (AGE–insulin) and collagen I, generating cross-linked fibrillar structures that seed Aβ aggregation and facilitate tau/NFT pathology. Chronic collagen I production combined with AGE modification impair vasculature in AD. Astrocytes, with high-capacity GLUT1-mediated glucose uptake, respond to D-ribose–derived AGEs via RAGE→NF-κB, producing a spectrum of inflammatory cytokines and driving exaggerated neuroinflammatory outputs with cognitive impairment in vivo. These astrocytic signals recruit and prime reactive microglia through parallel AGE- and cytokine-driven NF-κB cascades, further promoting Aβ aggregation and NFT burden in advanced AD.

Cerebral inflammation accompanied by insulin and leptin resistance

Inflammation activates also JNK1/2 kinase (Fig. 14.3B), which jointly inactivates IRS and leptin receptor (LepR) downstream of TNFα. LepR activated by adipokine leptin, integrates peripheral and CNS responses, coordinates leptin–insulin signaling in the hypothalamus to restrain appetite, promote energy expenditure, and regulate glucose homeostasis (Friedman, 2019). Intact insulin/IR stimulates leptin synthesis and secretion from adipose tissue, while insulin resistance and leptin resistance co-exist. In fact,

LepR signaling is required for insulin sensitivity; genetic deficiency in leptin (*ob/ob*) and or loss-of-function mutations LepR in mice (*db/db*) and in humans lead to central and peripheral insulin resistance, obesity culminating in T2D in *db/db* mice (Friedman, 2019) Leptin/LepR signaling supports catabolic glucose utilization via GLUT1 (Lee et al., 2020a) and complements insulin/IR, which favors glucose utilization via GLUT4, thereby maintaining metabolic homeostasis.

LepR is also expressed in the BBB endothelial cells, where leptin-activated LepR can modulate GLUT1-dependent glucose transport (Lee et al., 2020a). In the 5xFAD mouse model, LepR expression in the BBB and neurons declines with age and is accompanied by neuroinflammation, whereas astrocytes show a different trend consistent with their high glucose demand dependent on GLUT1.

LepR is expressed on GABAergic postsynaptic neurons in cortex and hippocampus. Leptin/LepR signaling via JAK2–PI3K modulates GABA release to POMC in a glucose-dependent manner (Lee et al., 2015) determining the strength of GABAergic synapses and glucose homeostasis. In addition, leptin strengthens GABAergic inhibitory postsynaptic currents in the hippocampus; β-PIX/CaMKK/ERK signaling mediates leptin-driven synaptogenesis. Consistent with insulin–leptin synergy in glucose regulation, *ob/ob* or *db/db* mice crossed with AD models exhibit worsened amyloid pathology and tau pathology, manifested as cognitive deficits without (*ob/ob*) or with vascular pathology (*db/db*).

AD general and specific traits of cerebral insulin signaling in AD

AD develops through a self-amplifying "vicious" cycle of insulin resistance (Fig. 14.3B), after which synaptic and vascular cells undergo structural and metabolic remodeling, inducing hypoglycemia and increased α/γ-secretase activity that generates Aβ. These vulnerable cells are further exposed to excess glucose and insulin through a permeable BBB (Fig. 14.1 and 14.2). Glucose surplus, combined with ROS in an oxidative–inflammatory milieu, elevates AGE and AGE-modified proteins and promotes Aβ and NFT aggregation, while excess insulin reduces Aβ cleavage by ICE. Notably, even insulin-sensitive patients show increased cerebral AGE despite only moderate elevations in glycemia (Fig. 14.2A,B). These waste products create an oxidative milieu and perpetuate inflammation, leading to insulin and leptin resistance. This disrupts not only IR signaling but also anti-apoptotic AKT2 activity, driving neurodegeneration.

These processes cannot be resolved because of metabolic discordance: intracellular hypoglycemia limits energy production and enzymatic defenses under chronic inflammation and oxidative stress. A meta-analysis of GSH levels in AD and MCI found that blood/intracellular GSH is lower in AD subjects compared to controls, implicating disrupted antioxidant homeostasis in AD patients (Chen et al., 2022), further amplifying oxidative stress. Chronic glucose deficiency impairs synaptic activity and metabolic processes like lipid synthesis and promotes methylation of genes involved in neurotrophic support, synaptic function, and myelination in AD (Fu & An, 2024).

The AD structural and metabolic environment alters responses to insulin arising from either hyperinsulinemia or pharmacological insulin treatment. In astrocytes, combined treatment with insulin and fructose induced insulin resistance resulted in decreased IR and Akt phosphorylation (pAkt S473) (Garwood et al., 2015). In AD, clearance of insulin by IDE becomes especially significant in the brain (Fig. 14.1E), where insulin outcompetes Aβ for degradation, thereby increasing Aβ accumulation. These metabolic effects raise concerns regarding the pharmacological effects of insulin therapy in the AD context.

In the APP-PS1 mouse model without diabetes, insulin treatment increased seizure-related mortality and was associated with brain mass loss, yet cognitive performance improved in surviving mice (Deng et al., 2024). These effects were attributed to insulin-stimulated *Bdnf*, which is linked to well-documented neurotrophic responses and cognitive benefits via Trk receptor activation. However, the inhibitory balance normally mediated by BDNF was blunted due to downregulation of *Slc1A6*, which clears excitatory glutamate to restore E/I balance. Slc1A6 deficiency is associated with epileptiform activity. Thus, insulin may pose specific risks in subjects with type 3 diabetes.

Intranasal insulin treatment has shown promise in partially delaying memory loss and improving plasticity, partially restoring insulin sensitivity in AD. To date, 12-month observations report episodes of hypoglycemia, headache, dizziness, and fatigue in a small proportion of patients, while epileptiform activity has not been evaluated. The SNIFF clinical trial (12 months) reported no clinically important adverse events and no worsening of Aβ/tau endpoints compared with placebo, although MRI analysis revealed a small hippocampal-volume reduction (Craft et al., 2020), suggesting neurodegeneration. These changes underscore that the metabolic and structural landscape of AD, shaped by chronic insulin and leptin resistance, cannot be rescued by insulin therapy alone and requires multi-target strategies.

Nanotechnology approach in AD

A peptide-based system termed AAC2 represents an emerging nano-therapeutic platform with significant potential for addressing the multifactorial pathology of AD (Lee et al., 2020a). AAC2 self-assembles into nanofibers capable of forming stable complexes with insulin (Lee et al., 2021). The resulting AAC2-insulin complex demonstrates improved pharmacokinetics, extending circulatory half-life while mitigating the risk of hypoglycemic events associated with insulin monotherapy in diabetic model (Lee et al., 2021). AAC2 enhances glucose uptake in endothelial cells of BBB, addressing the cerebral metabolic deficits characteristic of AD pathology (Deng et al., 2024). AAC2 administration sustain brain mass and reduced anxiety-like behaviors (Lee et al., 2020a). The therapeutic platform also demonstrates notable neuroprotective effects, preventing seizure-related mortality and promoting synaptic homeostasis through upregulation of inhibitory synaptic proteins Syn2 and Syp (Deng et al., 2024). This comprehensive approach targeting both metabolic dysfunction and excitotoxic mechanisms positions AAC2 as a versatile nanotherapeutic platform merits further investigation for complex neurological disorders.

Mini-dictionary of terms

- AGE (Advanced Glycation End products):
  Reactive molecules formed by non-enzymatic glycation of glucose with proteins such as insulin or collagen, or lipids; they promote oxidative stress, inflammation, and proteopathy in AD.
- BBB (Blood-Brain Barrier):
  Specialized endothelial interface that regulates entry of glucose, insulin, and other molecules into the brain; its dysfunction permits metabolic and inflammatory stressors to affect cerebral homeostasis.
- IDE (Insulin-Degrading Enzyme):
  Protease that normally clears insulin and Aβ; competition between these substrates contributes to impaired Aβ clearance in AD.
- IRS (Insulin Receptor Substrate):
  Adaptor proteins (IRS1/2) linking insulin receptor activation to downstream PI3K/Akt/mTOR signaling; aberrant phosphorylation of IRS promotes insulin resistance.
- RAGE (Receptor for Advanced Glycation End products):
  Cell-surface receptor binding AGEs and Aβ, activating NF-κB and inflammatory cascades that amplify AD pathology and vascular dysfunction.

Summary points

- The BBB safeguards autonomous cerebral insulin and glucose metabolism from postprandial fluctuations.
- Dietary glycemic overload damages insulin receptor–controlled BBB, leading to episodic cerebral hyperglycemia and hyperinsulinemia.
- Non-enzymatic glucose conversion into AGEs fuels oxidative stress and inflammation, driving insulin/leptin resistance and seeding proteopathy under chronic glucose overload.
- Glycemic overload, AGE accumulation, oxidative stress, inflammation, and injury form an interdependent circuit that drives aberrant phosphorylation, progressively disrupting insulin signaling and leading to chronic insulin resistance.
- Chronic insulin resistance alters synaptic structure and function, creating discordant glucose metabolism in which cerebral hypoglycemia coexists with systemic hyperglycemia.
- AD is driven by the same vicious cycle as insulin resistance, but in vulnerable synapses and metabolically discordant cells with altered membranes, disrupted intracellular trafficking, and remodeled insulin receptor landscapes, all under a progressively increasing load of oxidized waste products.
- AD risk centers on dysregulated insulin–glucose biology

Diet-driven hyperinsulinemia/hyperglycemia perpetuate neuronal insulin signaling toward pathology, causing synaptic remodeling, impaired plasticity/memory, by a self-perpetuating ‘vicious’ cycle (glycemic overload → AGEs → oxidative stress → inflammation → BBB injury).

- Nanotherapies represent a multi-target strategy capable of addressing multiple pathways simultaneously.

Acknowledge the use of AI

During the preparation of this work the authors used ChatGPT 5, from OpenAI, strictly to improve readability and language usage. Under no circumstances has it been used to draw insights from data, conclude or produce recommendations. After using this tool/service, the authors reviewed and edited the content as needed and take full responsibility for the content of the publication.

References

Ardanaz, C. G., de la Cruz, A., Minhas, P. S., Hernandez-Martin, N., Pozo, M. A., Valdecantos, M. P., et al. (2024). Astrocytic GLUT1 reduction paradoxically improves central and peripheral glucose homeostasis. *Science Advances, 10*, eadp1115.

Arnold, S. E., Arvanitakis, Z., Macauley-Rambach, S. L., Koenig, A. M., Wang, H. Y., Ahima, R. S., et al. (2018). Brain insulin resistance in type 2 diabetes and Alzheimer disease: concepts and conundrums. *Nature Reviews Neurology, 14*, 168–181.

Bingham, E. M., Hopkins, D., Smith, D., Pernet, A., Hallett, W., Reed, L., et al. (2002). The role of insulin in human brain glucose metabolism: An 18fluoro-deoxyglucose positron emission tomography study. *Diabetes, 51*, 3384–3390.

Blazquez, E., Velazquez, E., Hurtado-Carneiro, V., & Ruiz-Albusac, J. M. (2014). Insulin in the brain: Its pathophysiological implications for states related with central insulin resistance, type 2 diabetes and alzheimer's disease. *Frontiers in Endocrinology, 5*, 161.

Bloudek, L. M., Spackman, D. E., Blankenburg, M., & Sullivan, S. D. (2011). Review and meta-analysis of biomarkers and diagnostic imaging in Alzheimer's disease. *Journal of Alzheimer's Disease, 26*, 627–645.

Bohn, B., Lutsey, P. L., Misialek, J. R., Walker, K. A., Brown, C. H., Hughes, T. M., et al. (2023). Incidence of dementia following hospitalization with infection among adults in the Atherosclerosis risk in communities (ARIC) study cohort. *JAMA Network Open, 6*, e2250126.

Cai, W., Xue, C., Sakaguchi, M., Konishi, M., Shirazian, A., Ferris, H. A., et al. (2018). Insulin regulates astrocyte gliotransmission and modulates behavior. *Journal of Clinical Investigation, 128*, 2914–2926.

Cassese, A., Esposito, I., Fiory, F., Barbagallo, A. P., Paturzo, F., Mirra, P., et al. (2008). In skeletal muscle advanced glycation end products (AGEs) inhibit insulin action and induce the formation of multimolecular complexes including the receptor for AGEs. *Journal of Biological Chemistry, 283*, 36088–36099.

Chen, W., Huang, Q., Lazdon, E. K., Gomes, A., Wong, M., Stephens, E., et al. (2023). Loss of insulin signaling in astrocytes exacerbates alzheimer-like phenotypes in a 5xFAD mouse model. *Proceedings of the National Academy of Sciences of the U S A, 120*. e2220684120.

Chen, J. J., Thiyagarajah, M., Song, J., Chen, C., Herrmann, N., Gallagher, D., et al. (2022). Altered central and blood glutathione in Alzheimer's disease and mild cognitive impairment: A meta-analysis. *Alzheimer's Research & Therapy, 14*, 23.

Chen, C. K., Wu, Y. T., & Chang, Y. C. (2017). Association between chronic periodontitis and the risk of Alzheimer's disease: A retrospective, population-based, matched-cohort study. *Alzheimer's Research & Therapy, 9*, 56.

Craft, S., Raman, R., Chow, T. W., Rafii, M. S., Sun, C. K., Rissman, R. A., et al. (2020). Safety, efficacy, and feasibility of intranasal insulin for the treatment of mild cognitive impairment and alzheimer disease dementia: A randomized clinical trial. *JAMA Neurology, 77*, 1099–1109.

Custodio, T. F., Paulsen, P. A., Frain, K. M., & Pedersen, B. P. (2021). Structural comparison of GLUT1 to GLUT3 reveal transport regulation mechanism in sugar porter family. *Life Science Alliance, 4*.

Dakterzada, F., Jove, M., Cantero, J. L., Pamplona, R., & Pinoll-Ripoll, G. (2023). Plasma and cerebrospinal fluid nonenzymatic protein damage is sustained in Alzheimer's disease. *Redox Biology, 64*, 102772.

Deane, R. J. (2012). Is RAGE still a therapeutic target for Alzheimer's disease? *Future Medicinal Chemistry, 4*, 915–925.

Deng, Z., Lee, A., Lin, T., Taneja, S., Kowdley, D., Leung, J. H., et al. (2024). Amino acid compound 2 (AAC2) treatment counteracts insulin-induced synaptic gene expression and seizure-related mortality in a mouse model of Alzheimer's disease. *International Journal of Molecular Sciences, 25*.

Eteleeb, A. M., Novotny, B. C., Tarraga, C. S., Sohn, C., Dhungel, E., Brase, L., et al. (2024). Brain high-throughput multi-omics data reveal molecular heterogeneity in Alzheimer's disease. *PLoS Biology, 22*, e3002607.

Fang, F., Yu, Q., Arancio, O., Chen, D., Gore, S. S., Yan, S. S., et al. (2018). RAGE mediates Abeta accumulation in a mouse model of Alzheimer's disease via modulation of beta- and gamma-secretase activity. *Human Molecular Genetics, 27*, 1002–1014.

Farris, W., Mansourian, S., Chang, Y., Lindsley, L., Eckman, E. A., Frosch, M. P., et al. (2003). Insulin-degrading enzyme regulates the levels of insulin, amyloid beta-protein, and the beta-amyloid precursor protein intracellular domain in vivo. *Proceedings of the National Academy of Sciences of the U S A, 100*, 4162–4167.

Fernandez, A. M., Hernandez-Garzon, E., Perez-Domper, P., Perez-Alvarez, A., Mederos, S., Matsui, T., et al. (2017). Insulin regulates astrocytic glucose handling through cooperation with IGF-I. *Diabetes, 66*, 64–74.

Friedman, J. M. (2019). Leptin and the endocrine control of energy balance. *Nature Metabolism, 1*, 754–764.

Fu, J., & An, L. (2024). Histone methylation, energy metabolism, and Alzheimer's disease. *Aging and Disease, 16*, 2831–2858.

Gardner, R. C., Burke, J. F., Nettiksimmons, J., Kaup, A., Barnes, D. E., & Yaffe, K. (2014). Dementia risk after traumatic brain injury vs nonbrain trauma: The role of age and severity. *JAMA Neurology, 71*, 1490–1497.

Garwood, C. J., Ratcliffe, L. E., Morgan, S. V., Simpson, J. E., Owens, H., Vazquez-Villasenor, I., et al. (2015). Insulin and IGF1 signalling pathways in human astrocytes in vitro and in vivo; characterisation, subcellular localisation and modulation of the receptors. *Molecular Brain, 8*, 51.

Gonzalez-Garcia, I., Gruber, T., & Garcia-Caceres, C. (2021). Insulin action on astrocytes: From energy homeostasis to behaviour. *Journal of Neuroendocrinology, 33*, e12953.

Gray, S. M., Meijer, R. I., & Barrett, E. J. (2014). Insulin regulates brain function, but how does it get there? *Diabetes, 63*, 3992–3997.

Greene, C., Hanley, N., & Campbell, M. (2019). Claudin-5: Gatekeeper of neurological function. *Fluids and Barriers of the CNS, 16*, 3.

Janson, J., Laedtke, T., Parisi, J. E., O'Brien, P., Petersen, R. C., & Butler, P. C. (2004). Increased risk of type 2 diabetes in Alzheimer disease. *Diabetes, 53*, 474–481.

Jarvis, P. R. E., Cardin, J. L., Nisevich-Bede, P. M., & McCarter, J. P. (2023). Continuous glucose monitoring in a healthy population: Understanding the post-prandial glycemic response in individuals without diabetes mellitus. *Metabolism, 146*, 155640.

Jensen, P. J., Gunter, L. B., & Carayannopoulos, M. O. (2010). Akt2 modulates glucose availability and downstream apoptotic pathways during development. *Journal of Biological Chemistry, 285*, 17673–17680.

Johnson, R. J., Tolan, D. R., Bredesen, D., Nagel, M., Sanchez-Lozada, L. G., Fini, M., et al. (2023). Could Alzheimer's disease be a maladaptation of an evolutionary survival pathway mediated by intracerebral fructose and uric acid metabolism? *The American Journal of Clinical Nutrition, 117*, 455–466.

Leclerc, M., Bourassa, P., Tremblay, C., Caron, V., Sugere, C., Emond, V., et al. (2023). Cerebrovascular insulin receptors are defective in Alzheimer's disease. *Brain, 146*, 75–90.

Lee, D. K., Jeong, J. H., Chun, S. K., Chua, S., Jr., & Jo, Y. H. (2015). Interplay between glucose and leptin signalling determines the strength of GABAergic synapses at POMC neurons. *Nature Communications, 6*, 6618.

Lee, J., Kim, K., Cho, J. H., Bae, J. Y., O'Leary, T. P., Johnson, J. D., et al. (2020b). Insulin synthesized in the paraventricular nucleus of the hypothalamus regulates pituitary growth hormone production. *JCI Insight, 5*.

Lee, A., Mason, M. L., Lin, T., Kumar, S. B., Kowdley, D., Leung, J. H., et al. (2021). Amino acid nanofibers improve glycemia and confer cognitive therapeutic efficacy to bound insulin. *Pharmaceutics, 14*.

Lee, A., Sun, Y., Lin, T., Song, N. J., Mason, M. L., Leung, J. H., et al. (2020a). Amino acid-based compound activates atypical PKC and leptin receptor pathways to improve glycemia and anxiety like behavior in diabetic mice. *Biomaterials, 239*, 119839.

Mazucanti, C. H., Liu, Q. R., Lang, D., Huang, N., O'Connell, J. F., Camandola, S., et al. (2019). Release of insulin produced by the choroid plexis is regulated by serotonergic signaling. *JCI Insight, 4*.

Mohamed, L. A., Keller, J. N., & Kaddoumi, A. (2016). Role of P-glycoprotein in mediating rivastigmine effect on amyloid-beta brain load and related pathology in Alzheimer's disease mouse model. *Biochimica et Biophysica Acta, 1862*, 778–787.

Mohamed, A., Viveiros, A., Williams, K., & Posse de Chaves, E. (2018). Abeta inhibits SREBP-2 activation through Akt inhibition. *The Journal of Lipid Research, 59*, 1–13.

Monacelli, F., Borghi, R., Pacini, D., Serrati, C., Traverso, N., & Odetti, P. (2014). Pentosidine determination in CSF: A potential biomarker of Alzheimer's disease? *Clinical Chemistry and Laboratory Medicine, 52*, 117–120.

Nagalski, A., Kozinski, K., & Wisniewska, M. B. (2016). Metabolic pathways in the periphery and brain: Contribution to mental disorders? *The International Journal of Biochemistry & Cell Biology, 80*, 19–30.

Nandipati, K. C., Subramanian, S., & Agrawal, D. K. (2017). Protein kinases: Mechanisms and downstream targets in inflammation-mediated obesity and insulin resistance. *Molecular and Cellular Biochemistry, 426*, 27–45.

Peng, Y., Kim, J. M., Park, H. S., Yang, A., Islam, C., Lakatta, E. G., et al. (2016). AGE-RAGE signal generates a specific NF-kappaB RelA "barcode" that directs collagen I expression. *Scientific Reports, 6*, 18822.

Pomytkin, I., Costa-Nunes, J. P., Kasatkin, V., Veniaminova, E., Demchenko, A., Lyundup, A., et al. (2018). Insulin receptor in the brain: Mechanisms of activation and the role in the CNS pathology and treatment. *CNS Neuroscience & Therapeutics, 24*, 763–774.

Rhea, E. M., & Banks, W. A. (2021). A historical perspective on the interactions of insulin at the blood-brain barrier. *Journal of Neuroendocrinology, 33*, e12929.

Rhea, E. M., Banks, W. A., & Raber, J. (2022). Insulin resistance in peripheral tissues and the brain: A tale of two sites. *Biomedicines, 10*.

Richmond-Rakerd, L. S., Iyer, M. T., D'Souza, S., Khalifeh, L., Caspi, A., Moffitt, T. E., et al. (2024). Associations of hospital-treated infections with subsequent dementia: Nationwide 30-year analysis. *Nature Aging, 4*, 783–790.

Sanchez-Castillo, C., Cuartero, M. I., Fernandez-Rodrigo, A., Briz, V., Lopez-Garcia, S., Jimenez-Sanchez, R., et al. (2022). Functional specialization of different PI3K isoforms for the control of neuronal architecture, synaptic plasticity, and cognition. *Science Advances, 8*, eabq8109.

Semple, R. K., Savage, D. B., Cochran, E. K., Gorden, P., & O'Rahilly, S. (2011). Genetic syndromes of severe insulin resistance. *Endocrine Reviews, 32*, 498–514.

Song, F., Hurtado del Pozo, C., Rosario, R., Zou, Y. S., Ananthakrishnan, R., Xu, X., et al. (2014). RAGE regulates the metabolic and inflammatory response to high-fat feeding in mice. *Diabetes, 63*, 1948–1965.

Song, S. H., McIntyre, S. S., Shah, H., Veldhuis, J. D., Hayes, P. C., & Butler, P. C. (2000). Direct measurement of pulsatile insulin secretion from the portal vein in human subjects. *Journal of Clinical Endocrinology & Metabolism, 85*, 4491–4499.

Song, F., & Schmidt, A. M. (2012). Glycation and insulin resistance: Novel mechanisms and unique targets? *Arteriosclerosis, Thrombosis, and Vascular Biology, 32*, 1760–1765.

Spencer, B., Rank, L., Metcalf, J., & Desplats, P. (2018). Identification of insulin receptor splice variant B in neurons by in situ detection in human brain samples. *Scientific Reports, 8*, 4070.

Suzuki, R., Lee, K., Jing, E., Biddinger, S. B., McDonald, J. G., Montine, T. J., et al. (2010). Diabetes and insulin in regulation of brain cholesterol metabolism. *Cell Metabolism, 12*, 567–579.

Talbot, K., Wang, H. Y., Kazi, H., Han, L. Y., Bakshi, K. P., Stucky, A., et al. (2012). Demonstrated brain insulin resistance in Alzheimer's disease patients is associated with IGF-1 resistance, IRS-1 dysregulation, and cognitive decline. *Journal of Clinical Investigation, 122*, 1316–1338.

Taniguchi, C. M., Emanuelli, B., & Kahn, C. R. (2006). Critical nodes in signalling pathways: Insights into insulin action. *Nature Reviews Molecular Cell Biology, 7*, 85–96.

Tijms, B. M., Vromen, E. M., Mjaavatten, O., Holstege, H., Reus, L. M., van der Lee, S., et al. (2024). Cerebrospinal fluid proteomics in patients with Alzheimer's disease reveals five molecular subtypes with distinct genetic risk profiles. *Nature Aging, 4*, 33–47.

Wang, H., Kulas, J. A., Wang, C., Holtzman, D. M., Ferris, H. A., & Hansen, S. B. (2021). Regulation of beta-amyloid production in neurons by astrocyte-derived cholesterol. *Proceedings of the National Academy of Sciences of the U S A*, 118.

Watson, L. S., Wilken-Resman, B., Williams, A., DiLucia, S., Sanchez, G., McLeod, T. L., et al. (2022). Hyperinsulinemia alters insulin receptor presentation and internalization in brain microvascular endothelial cells. *Diabetes and Vascular Disease Research s, 19*, 14791641221118626.

Yunn, N. O., Kim, J., Ryu, S. H., & Cho, Y. (2023). A stepwise activation model for the insulin receptor. *Experimental & Molecular Medicine, 55*, 2147–2161.

Chapter 15

Sexual dysfunction and mood symptoms in patients with diabetes

Shubhangi Gupta[1], Anuradha Rathod[2] and Sreelakshmi Vaidyanathan[3]
[1]Department of Psychiatry, Seth VC Gandhi & MA Vora Municipal General Hospital, Mumbai, Maharashtra, India; [2]Department of Psychiatry, Government Medical College, Thane, Maharashtra, India; [3]Department of Psychiatry, All India Institute of Medical Sciences, Nagpur, Maharashtra, India

List of abbreviations

AOR Adjusted odds ratio
BBB Blood brain barrier
BDNF Brain derived neurotrophic factor
cAMP PKA cyclic adenosine monophosphate protein kinase A
CBT Cognitive behavior therapy
CPAP Continuous positive airway pressure
ED Erectile dysfunction
FSD Female sexual dysfunction
GLP 1 Glucagon like peptide 1
GnRH Gonadotropin releasing hormone
HPA axis Hypothalamo pituitary adrenal axis
HPG axis Hypothalamic pituitary gonadal axis
HPO axis Hypothalamic pituitary ovarian axis
ICD-11 International classification of diseases, 11th revision
IDO Indoleamine 2,3 dioxygenase
IL-6 Interleukin 6
JAK STAT Janus kinase signal transducer and activator of transcription
MAO Monoamine oxidase
MAPK Mitogen activated protein kinase
MDD Major depressive disorder
NDRI Norepinephrine dopamine reuptake inhibitor
NLPR3 Nucleotide binding domain, leucine rich containing family, pyrin domain containing-3
NO Nitric oxide
OSA Obstructive sleep apnea
PCOS Polycystic ovarian syndrome
PDE5i Phosphodiesterase type 5 inhibitors
PHQ9 Patient health questionnaire 9
PI3K Akt Phosphoinositide 3 kinase protein kinase B (Akt)
SGLT2 Sodium glucose cotransporter 2
SHBG Sex hormone binding globulin
SNRI Serotonin norepinephrine reuptake inhibitors
SSRI Selective serotonin reuptake inhibitors
TCA Tricyclic antidepressants
TNF-α Tumor necrosis factor alpha
VTA Ventral tegmental area

The pancreas plays a key role in the metabolic cycle of glucose, along with the modulation of various neural, behavioral and psychological processes. Several pancreatic hormones, such as insulin, glucagon and amylin, affect the central nervous system, leading to changes in cognition, emotional regulation, energy levels, and reproductive behavior. Diabetes mellitus, caused by dysregulation of pancreatic endocrine function, is now understood to present with symptoms beyond primary endocrine dysfunction, including mood disturbances and sexual dysfunctions (Shapiro, 2022). These manifestations result from a complex interplay among hormonal, neuronal, psychological, and metabolic pathways.

The etiology of these symptoms is multifactorial, stemming from the disease process as well as its psychological sequelae as a chronic medical illness. Physiologically, chronic hyperglycemia, fluctuating levels of glucose and insulin, and insulin resistance can, on one hand, lead to endothelial dysfunction and altered metabolism of sex steroids in both men and women, resulting in impaired sexual functioning. On the other hand, they also affect brain regions involved in mood regulation and reactivity to stress, predisposing one to depression and anxiety. The chronicity of illness, the need to take regular medications or injectables, and the effect on quality of life can further perpetuate the cycle. The exacerbation of metabolic dysfunction due to the impact of mood and anxiety symptoms on lifestyle and medication adherence completes this self-reinforcing vicious cycle.

This chapter aims to provide critical insight into the psychoneuroendocrinology of diabetes, with an in-depth

Hormones in Neuroscience. https://doi.org/10.1016/B978-0-443-30263-3.00019-1

review of the interconnections between pancreatic hormones and neural circuits responsible for mood and sexual behavior, a detailed exploration of the underlying biological and psychological mechanisms, and the development of an integrated approach to management.

Mood symptoms in diabetics

Introduction

A bidirectional relationship

Diabetes and mood disorders have a well-studied bidirectional, though temporally complex relationship, with depression increasing diabetes risk and vice versa (Sartorius, 2018). The main reason for such strong associations is their shared pathophysiology.

Rationale for integrated approach

Mood symptoms and diabetes have significant overlaps in metabolic and causative pathways. The shared neurobiological substrates (inflammation, insulin signaling, HPA axis) and the synergistic effect of improved mood on adherence, glycemic control, complications, and mortality underscore the need for combined metabolic and psychiatric interventions (Allen et al., 2021; Joseph & Golden, 2017; Leonard & Wegener, 2020; Slyepchenko et al., 2016).

A comprehensive overview of the emergence of mood symptoms in persons with diabetes, unique phenotypes presenting with varied symptoms, mechanisms underlying the brain-pancreas relationship, and hormonal pathways involved in mood symptoms in pancreatic dysfunction is provided below. Certain translational aspects and future research directions are also discussed.

Epidemiological trends and clinical phenotypes

Global prevalence patterns

The global burden of diabetes and depression is a significant public health issue (Sartorius, 2018). Depression prevalence in DM is around 35%. Significantly different depression rates in diabetics versus non-diabetics have been reported, with depression in T2DM at 17.6% versus 9.8% in controls. One-third of youth with T1DM have depressive symptoms (Buchberger et al., 2016), while in T2DM, the prevalence of depression is $\approx$ 10%–20% (Moulton et al., 2015; Semenkovich et al., 2015). In female patients with diabetes, the risk is much higher due to the higher prevalence of obesity, effects of sex hormones, and psychosocial causes (Kautzky-Willer et al., 2016). Diabetes patients in lower-middle-income countries have a greater prevalence of depression than patients in higher-income countries.

Anxiety in diabetes is also significantly more than in the normal population, with an AOR of 1.25. Diabetes distress is reported twice as much as clinical depression in this population (Tareen & Tareen, 2017).

Sizable gaps still exist in the literature, like limited high-quality data from lower-middle income countries, subjectivity between diabetes distress and MDD definitions, sparse T1DM studies, and few bipolar depression and anxiety prevalence studies (Cannon et al., 2024).

Type-specific differences

Both Type 1 and Type 2 diabetes mellitus are linked to a higher risk of MDD. In case of T2DM, depression rates go around 19%, ranging from 6.5% to 33%, while among the non-diabetic population, it's about 10.7%. In T1DM, depression is related primarily to psychological distress of lifelong insulin therapy and neurological impact of glucose fluctuations over time (Buchberger et al., 2016; Moulton et al., 2015). Whereas in T2DM, biological causes such as chronic inflammation, elevated cytokines, and a disrupted HPA axis are likely contributing factors (Moulton et al., 2015).

Thus, depression in diabetes is not one-size-fits-all. For T1DM patients, better glycemic stability and counseling may help, whereas in T2DM, there is a greater need to address biological factors such as inflammation and the hormonal stress response.

Clinical correlates

The presence of diabetes complications (neuropathy, nephropathy, retinopathy) is reflected in raised depression prevalence, and being on insulin therapy makes it 53% more likely to have depression (Yang et al., 2025), likely reflecting the increased severity of diabetes warranting insulin. A consistent Fasting glucose level $\geq$126 mg/dL is associated with an AOR of 2.51 for depression (Yang et al., 2025; Golden), which is concerning, as a large number of diabetes patients have glucose levels above this cutoff even with regular treatment. Persistent MDD is linked to a higher incidence of T2DM, one causative factor being long-term antidepressant use, which has been shown to raise T2DM risk independently. The overlap between somatic symptoms, like fatigue, sleep/appetite changes, is present both in diabetes and depression, which compounds diagnostic confusion. There is often overdiagnosis with the PHQ-9, for which Tataru suggests using a PHQ-9 cutoff of 12 instead of 10 to improve specificity in diabetes.

Neurobiological mechanisms and brain–pancreas axis

Central insulin resistance

Brain insulin receptors are densely located in the hippocampus, prefrontal cortex, hypothalamus, and striatum, which are key centers for emotion, memory, and reward (Grillo et al., 2015). Peripheral insulin enters the brain via BBB, but this transport is compromised in systemic insulin resistance (Grillo et al., 2015; Leonard & Wegener, 2020). Inside the brain, insulin primarily signals through the

PI3K-Akt-mTOR pathway, and in diabetes, this gets impaired, affecting neuroplasticity and neurotransmitter function (Fanelli et al., 2022; Grillo et al., 2015).

Synaptic plasticity, neurogenesis, and receptor sensitivity drop, mainly in the hippocampus (Grillo et al., 2015). Hippocampal insulin resistance is linked to memory loss and low mood (Grillo et al., 2015). Intranasal insulin bypasses BBB and improves mood, memory, and hypothalamic activity (Fanelli et al., 2022; Grillo et al., 2015).

Neurotransmitter imbalance

With insulin resistance, the dopamine reward system (VTA–NAc) suffers, with loss of insulin receptors leading to reduced D2/D3 activity (Grillo et al., 2015). Lowered receptor availability further disturbs dopamine turnover in reward circuits (Grillo et al., 2015; Leonard & Wegener, 2020), and this pathway disruption underlies anhedonia in diabetic depression (Grillo et al., 2015). Emergence of anxiety or depressive symptoms could also be attributed to a lack of insulin in astrocytes, contributing to further dopamine dysregulation (Grillo et al., 2015).

Insulin normally inhibits MAO-A/B and raises brain serotonin in models (Grillo et al., 2015). Low insulin levels in diabetes increase MAO expression, reducing serotonin and disrupting serotonergic tone (Fanelli et al., 2022). Cytokine-driven IDO activation diverts tryptophan from serotonin to kynurenine. Centrally, reduced serotonin explains weak SSRI response and mood instability. Sympathetic overactivity worsens insulin resistance and reduces insulin secretion (Joseph & Golden, 2017; Xu et al., 2024). Excess norepinephrine causes anxiety and hyperarousal (Joseph & Golden, 2017). Excess glutamate triggers oxidative damage (Allen et al., 2021). Restoring insulin signalling is reported to improve neurotransmitter balance (Fanelli et al., 2022; Grillo et al., 2015).

HPA axis dysregulation

Chronic stress and depression overactivate the HPA axis, raising cortisol and worsening insulin resistance (Leonard & Wegener, 2020; Xu et al., 2024). Cortisol and cytokine rise in depression further worsens insulin resistance (Leonard & Wegener, 2020; Xu et al., 2024). Cortisol also promotes fat deposition, worsening insulin function (Joseph & Golden, 2017; Leonard & Wegener, 2020). In T2DM and depression, the diurnal cortisol rhythm is flattened (Leonard & Wegener, 2020; Moulton et al., 2015), and Glucocorticoid Receptor defects reduce negative feedback, thereby prolonging the stress response (Allen et al., 2021). High cortisol levels correlate with poor metabolic control and predict relapse (Leonard & Wegener, 2020; Moulton et al., 2015).

Autonomic dysfunction

Autonomic dysfunction adds to metabolic stress. Sympathetic overactivity suppresses insulin secretion, increases glucose output, and fuels insulin resistance (Joseph & Golden, 2017; Xu et al., 2024). Poor Heart-Rate Variability, an indicator of vagal withdrawal, is common in both diabetes and depression (Joseph & Golden, 2017; Slyepchenko et al., 2016). Gastrointestinal symptoms from autonomic neuropathy correlate with anxiety and mood scores (Slyepchenko et al., 2016). Incretin signaling loss post-vagotomy worsens glycemic control (Lach et al., 2018; Slyepchenko et al., 2016).

Vagal tone improves with exercise and reduces inflammation (Joseph & Golden, 2017; Slyepchenko et al., 2016), and vagus nerve stimulation shows promise for mood and neuropathy (Slyepchenko et al., 2016). GLP-1RAs may have central mood benefits, though human data are sparse (Fanelli et al., 2022; Lach et al., 2018).

Neuroinflammation

Inflammation plays a central role. Elevated IL-6, TNF-α, and IL-1β cross the BBB, alter neural signaling, and trigger microglial activation (Fanelli et al., 2022; Leonard & Wegener, 2020). BBB is more permeable in diabetes and depression (Leonard & Wegener, 2020). Microglia release neurotoxic agents like quinolinic acid, damaging neurons (Leonard & Wegener, 2020). Cytokines impact the tryptophan-kynurenine pathway, as mentioned above.

Inflammation correlates with symptom severity and antidepressant non-response (Joseph & Golden, 2017; Leonard & Wegener, 2020). Drugs like metformin, pioglitazone and SGLT2 inhibitors may help in mood improvement by restoring insulin signaling and improving BDNF levels (Fang et al., 2020; Mansour et al., 2025). Metformin has anti-inflammatory CNS effects, and agents like omega-3 and NSAIDs help both mood and insulin sensitivity (Fang et al., 2020; Mansour et al., 2025).

Oxidative stress & mitochondrial dysfunction

Oxidative stress and mitochondrial dysfunction are shared features of depression and diabetes. Hyperglycemia impairs mitochondrial function and worsens brain insulin resistance (Allen et al., 2021; Xu et al., 2024). Hyperglycemia also increases ROS, damages mitochondria, and triggers ER stress (Allen et al., 2021; Xu et al., 2024). ROS causes lipid and DNA damage. Mitochondrial DNA activates the NLRP3 inflammasome, leading to inflammation (Allen et al., 2021; Tanaka et al., 2022). Depression features low ATP and poor mitochondrial efficiency (Allen et al., 2021; Tanaka et al., 2022). Chronic hyperglycemia leads to failed mitophagy and death of hippocampal neurons (Allen et al., 2021; Tanaka et al., 2022).

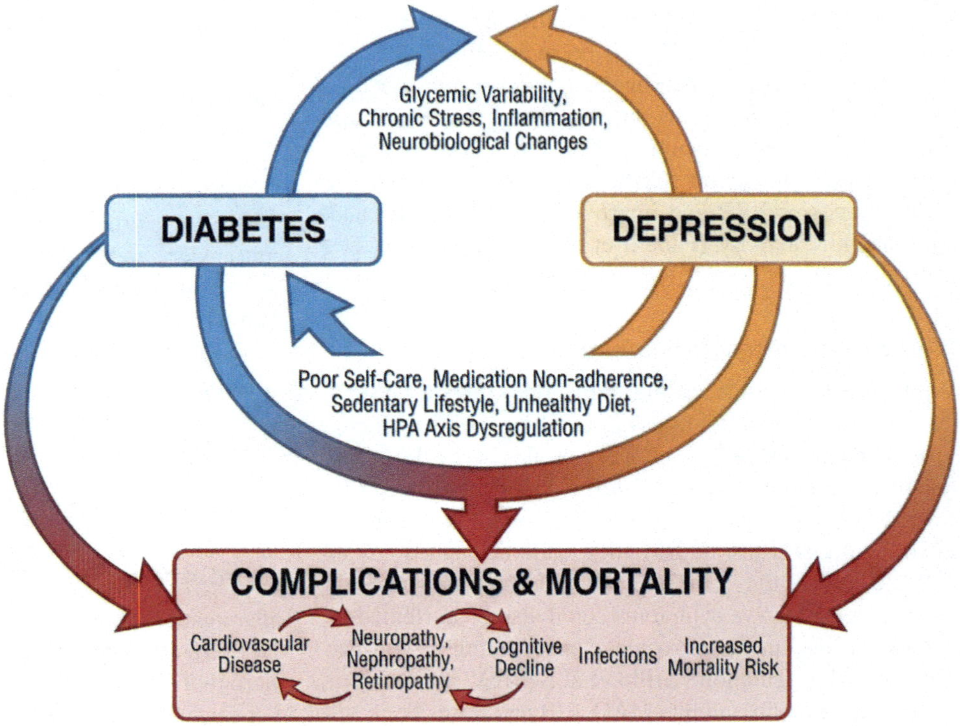

FIGURE 15.1 The bidirectional relationship between diabetes and mood disorders.

Depletion of neurotrophic factors

Reduced BDNF and IGF-1 are common to depression and T2DM (Leonard & Wegener, 2020; Moulton et al., 2015). BDNF supports hippocampal plasticity, and stress downregulates it via cortisol (Yang et al., 2020). Inflammation also reduces BDNF and TrkB signaling (Fanelli et al., 2022; Leonard & Wegener, 2020). Astrocytes and microglia release neurotrophins, but inflammation impairs this (Leonard & Wegener, 2020; Moulton et al., 2015).

Circadian disruptions

Sleep and circadian disruptions worsen mood and glycemic control. Clock gene misalignment impairs insulin secretion and mood regulation (Leonard & Wegener, 2020; Xu et al., 2024). Sleep loss increases inflammation, lowers insulin sensitivity, and raises depression risk (Leonard & Wegener, 2020). Light therapy and melatonin can improve outcomes (Leonard & Wegener, 2020; Xu et al., 2024).

Gut microbiome imbalance

The gut-brain axis links dysbiosis with insulin resistance and mood (Leonard & Wegener, 2020; Slyepchenko et al., 2016). A leaky gut allows LPS and cytokines into circulation, triggering neuroinflammation (Slyepchenko et al., 2016). GLP-1R agonists enhance neurogenesis and reduce depressive symptoms (Lach et al., 2018; Slyepchenko et al., 2016). FMT and probiotics improve insulin sensitivity and mood (Lach et al., 2018; Slyepchenko et al., 2016).

The interplay of these multiple factors in the human body is illustrated in Fig. 15.1.

Other pancreatic and neuroendocrine pathways

The role of other pancreatic hormones in mood regulation, along with glucose metabolism, is increasingly recognized in diabetes research as a potential target for improved diabetes management. These have been summarized in Table 15.1.

GLP-1 and GLP-1 receptor agonists

GLP-1 receptors are found in the pancreas as well as in brain regions such as the amygdala & hippocampus, which are involved in emotion, appetite, memory, and executive function. GLP-1 receptor agonists (liraglutide and semaglutide) may improve mood, motivation and cognition by improving neuroinflammation, neuroplasticity, and weight loss (Carmellini et al., 2025). Although there have been occasional reports of mood worsening or suicidal thoughts, overall, a neutral-beneficial effect has been found. Research is actively ongoing on GLP-1 agonists for neuropsychiatric indications beyond diabetes.

TABLE 15.1 Pancreatic and related hormones with receptor distribution, CNS effects, intracellular signaling pathways, and mood outcomes.

Hormone	Key receptor distribution	CNS effects	Major intracellular pathways	Mood-related outcomes
GLP-1	Pancreas; amygdala; hippocampus; prefrontal regions	Appetite regulation; emotion processing; cognition; neuroprotection	cAMP–PKA; PI3K–Akt; anti-inflammatory signaling	Generally neutral → beneficial mood effects; improved motivation/cognition; rare reports of mood worsening
Amylin	Pancreas; area postrema; hypothalamus	Satiety; reward modulation; eating behavior	cAMP pathways; interaction with calcitonin receptor	Possible improvements in appetite control and mood when combined with GLP-1; evidence still emerging
Leptin	Hypothalamus; VTA; hippocampus	Appetite, reward energy, motivation	JAK–STAT; PI3K; MAPK	Leptin resistance linked to low motivation, fatigue, depressive symptoms; improved with weight loss/activity
Thyroid hormones	Widespread—including cortex, limbic system	Metabolism, arousal, cognition, affect regulation	Nuclear receptor gene-transcription effects	Deficiency or fluctuations worsen mood/energy; replacement/supplement ation improves mood and metabolic control
Gonodal hormones	Limbic system; hypothalamus; cortex	Mood, stress-axis modulation, energy, libido	Estrogen receptor signaling; androgen receptor gene regulation	Deficiency or fluctuations worsen mood/energy; replacement/supplement ation improves mood and metabolic control

Amylin

Amylin works alongside insulin to regulate appetite and satiety, and synthetic analogs such as pramlintide are already used in diabetes management. Early work suggests amylin may also influence reward pathways and eating behavior, and when combined with GLP-1 agonists, may enhance weight- and mood-related outcomes. However, the psychiatric evidence base is still emerging (Alhomoud et al., 2025).

Leptin

Many people with Type 2 diabetes have elevated leptin levels but reduced central sensitivity to the hormone. This resistance affects appetite regulation and may also contribute to low energy, reduced motivation, and depressive symptoms through dopaminergic and stress-axis pathways (Friedman, 2019). Lifestyle interventions such as physical activity and weight loss may partially restore leptin sensitivity, and this may indirectly support mood regulation.

Thyroid and gonadal hormone overlap

Thyroid dysfunction is more common in people with diabetes, and both hypothyroidism and hyperthyroidism can worsen mood symptoms and glycemic control. Similarly, sex-hormone fluctuations across life stages, such as PCOS, perimenopause, or testosterone deficiency in men, may influence mood, energy, and insulin sensitivity (Taylor et al., 2023). Routine screening and timely treatment of thyroid or hormonal imbalance can support both metabolic and emotional outcomes.

A tabular representation of pancreatic and related hormones with receptor profiles & mood effects is given in Table 15.1.

Translational applications

Pharmacologic implications

Antidiabetic medications with CNS/mood impact

Insulin has indirect mood benefits by improving glycemic control. Intranasal insulin is being studied for cognitive enhancement & mood benefits, though it carries risks of hypoglycemia-related anxiety (Tabassum et al., 2024).

Metformin works by AMP Kinase activation, improves mitochondrial function and neuroplasticity, is anti-inflammatory (reduces IL-6 and TNF-α), and increases BDNF expression. GLP-1 Receptor Agonists (liraglutide, semaglutide, dulaglutide, exenatide) work similarly, with multiple studies showing a reduced risk of depression. Testing for direct impact in psychiatric indications is ongoing for both (Guo et al., 2023; Lach et al., 2018; Slyepchenko et al., 2016).

Antidepressants with glycemic impact

SSRIs are generally neutral or mildly beneficial for glycemic control. Sertraline has mixed evidence, with some studies showing worsening. Preferred are escitalopram or fluoxetine in diabetes with depression (Miedlich & Lamberti, 2025). SNRIs like Duloxetine, also FDA-approved for diabetic neuropathic pain, are generally neutral in glycemic effect and may help with comorbid pain and depression (Miedlich & Lamberti, 2025). Bupropion, an NDRI, is effective for anhedonia and fatigue commonly associated with diabetes, is weight-neutral or weight-loss promoting, and may improve glycemic control. Importantly, it has no sexual side effects; thus, it is the first choice for patients with diabetes-related sexual dysfunction (Miedlich & Lamberti, 2025).

TCAs are generally avoided in diabetes due to worsening of glycemic control, weight gain, and cardiac effects. The newer molecule Vortioxetine has limited data but is promising and may help cognition. Rapid-acting Ketamine/Esketamine, an emerging molecule for treatment-resistant depression, has limited data on glycemic effects (Miedlich & Lamberti, 2025).

Non-pharmacologic and integrative interventions

Psychotherapy

CBT adapted for Diabetes, which includes psychoeducation on the bidirectional mood-glucose relationship, encouraging behavioral activation, and challenging diabetes-related negative automatic thoughts, has given a sustained reduction in depressive symptoms. Telepsychiatry and the expansion of collaborative care platforms encourage more diabetes patients to seek CBT (Li et al., 2017).

Physical activity

Exercise improves emotional well-being and insulin sensitivity. The minimum recommendation is 150 min of aerobic activity per week, along with strength training (Colberg et al., 2016).

Nutrition

Diet affects metabolic state as well as mood. Focus on the Mediterranean/DASH-style diet is crucial for diabetes as it is associated with reduced inflammation & reduced depressive symptoms (Lassale et al., 2019). Micronutrients relevant to mood and inflammation are vitamin D, B vitamins and dietary fiber (Osimo et al., 2020).

Sleep and circadian regulation

Sleep disturbance directly worsens both mood and glycemic control (Duan et al., 2023). Practical sleep-focused strategies & basic sleep hygiene are essential. As OSA is common in T2DM, CPAP therapy improves mood, daytime functioning, and metabolic indices.

Fig. 15.2 represents an integrated management algorithm for co-morbid diabetes and depression.

Emerging research directions

Gut microbiome and mood regulation

Gut microbiome health significantly affects metabolism and mood, with dysbiosis commonly found in both diabetes and depression. A high-fiber diet improves microbiome health & leads to better mood regulation by various mechanisms, including reducing glucose spikes. Precision probiotics, such as specific strains of gut bacteria (Lactobacillus helvetica, Bifidobacterium longum for mood, Akkermansia muciniphila for metabolic health), could be the future course. Fecal microbiota transplantation (FMT) is being explored, though efficacy & safety trials are needed (Gao et al., 2023).

Sexual dysfiunction in diabetics

The intersection of metabolic and sexual health

Diabetes mellitus, defined as a chronic metabolic disorder, is known to affect multiple organ systems. However, sexual dysfunction, despite its prevalence and profound impact on quality of life, has historically occupied a peripheral position in diabetic care.

Defining sexual dysfunction in the clinical context

According to ICD-11, sexual dysfunction is a heterogeneous group of conditions characterized by difficulties in achieving personally satisfying and non-coercive sexual experiences. For a formal diagnosis of sexual dysfunction, the condition must meet three specific parameters:

1. **Frequency:** The dysfunction must occur frequently, although it may be absent on some occasions.
2. **Duration:** The symptoms must have been present for at least several months (unlike acute, situational sexual difficulties).
3. **Distress:** The dysfunction must be associated with clinically significant distress - highlights the psychological burden of the condition.

Global epidemiology

Data suggest that sexual impairment is a common experience for most patients living with diabetes. A comprehensive systematic review and meta-analysis by Gebeyehu et al. (2023), with data spanning from 2008 to 2022, provides an estimation of this burden.

Table 15.2 illustrates the comparative prevalence rates across key demographic and clinical subgroups.

The global prevalence of sexual dysfunction among diabetic patients was reported as 61.4%. This high prevalence underscores the systemic nature of diabetes with damage to the microvasculature and nervous system, compromising the highly vascularized and innervated genital organs.

The gender gap in prevalence

Males (65.9%) are more likely than females (58.8%) to report sexual dysfunction. The gender disparity is likely multifactorial. Biologically, male sexual function, specifically the mechanics of erection, is highly dependent on vascular integrity, and ED is often the first clinical sign of systemic endothelial dysfunction. Consequently, the threshold for symptomatic presentation in men may be lower than in women. In women, sexual arousal involves more complex, diffuse vascular engorgement and lubrication processes that may be less immediately obvious. Furthermore, diagnostic bias may play a role - FSD criteria are more ambiguous compared to ED, leading to a lesser frequency of screening.

Type 1 versus Type 2 diabetes mellitus

Persons with Type 2 DM (71%) are twice as likely to suffer from Sexual Dysfunction as compared to those with Type 1 DM (35.4%). T2DM is frequently part of a broader metabolic syndrome, characterized by central obesity, hypertension, and dyslipidemia. These comorbidities independently damage the vascular and nervous systems, compounding the effects of hyperglycemia. Additionally, the Type 2 demographic is generally older, introducing age-related gonadal decline into the clinical picture.

In contrast, T1DM typically has an earlier onset. While the prevalence is lower (35.4%), it remains alarmingly high for a generally younger population. This suggests that the mechanism in T1DM is driven more directly by the duration of hyperglycemia and insulin deficiency than by the cluster of metabolic comorbidities seen in Type 2.

The pathophysiological cascade: organic mechanisms-

The etiology of sexual dysfunction in diabetes is a multifactorial pathology where organic, psychological, and relational factors converge (Fig. 15.3). The organic component is driven by a triad of vascular, neurologic, and hormonal deteriorations, all of which are amplified by the toxic metabolic environment of chronic hyperglycemia, similar to how mood symptoms arise as was described above.

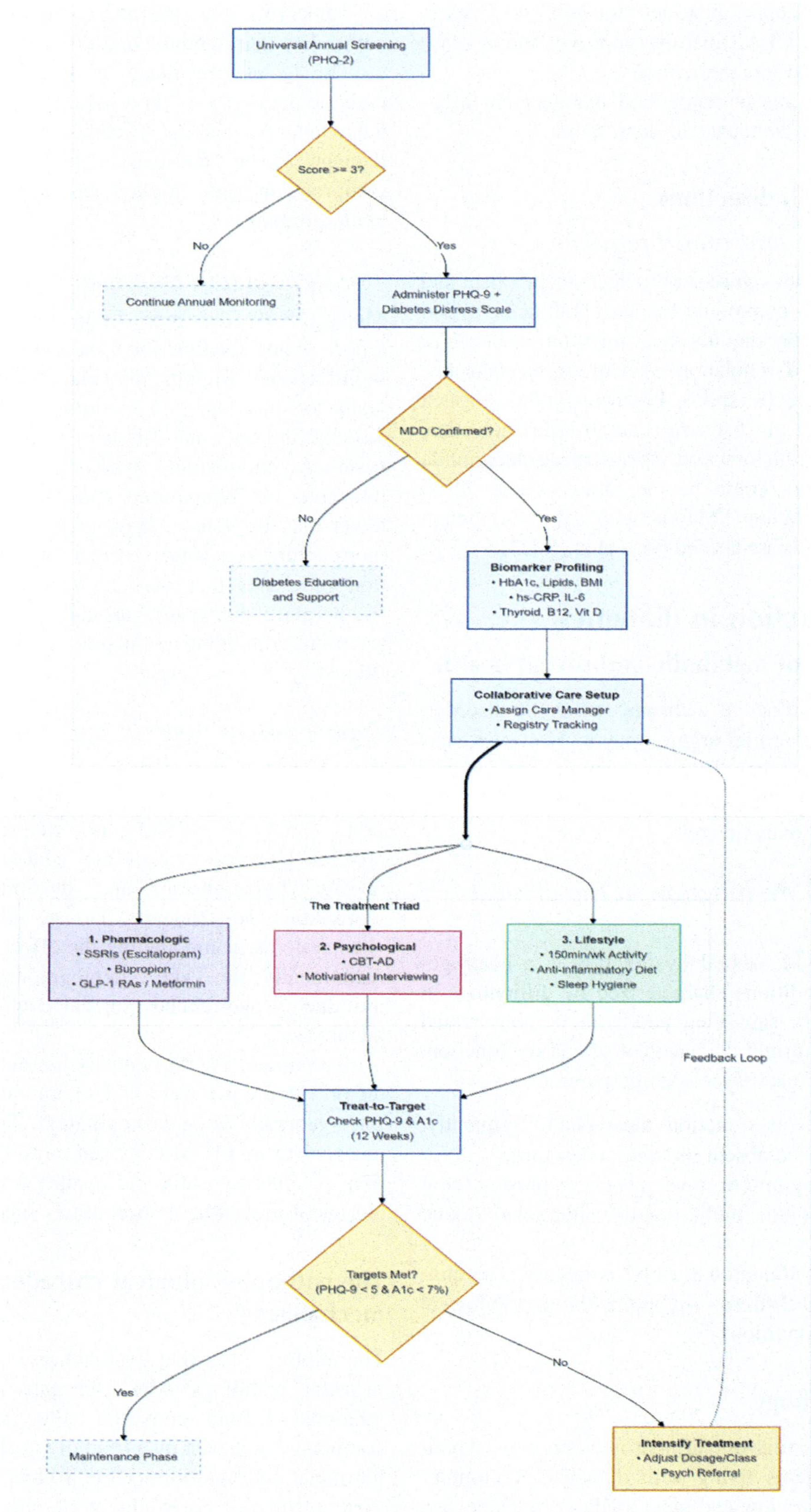

FIGURE 15.2 Integrated management algorithm for co-morbid diabetes and depression.

TABLE 15.2 Prevalence of sexual dysfunction in diabetic populations (Gebeyehu et al., 2023).

Demographic/Clinical group	Prevalence rate (%)	Clinical interpretation
Global population	61.4	Represents the aggregate burden, indicating SD is a majority complication.
Male patients	65.9	Reflects high susceptibility to vascular erectile dysfunction and easier objective diagnosis.
Female patients	58.8	Slightly lower reported rate, potentially due to underreporting or subjective diagnostic challenges.
Type 2 diabetes (T2DM)	71.0	Significantly higher prevalence driven by older age, obesity, and metabolic syndrome comorbidities.
Type 1 diabetes (T1DM)	35.4	Lower relative prevalence, likely associated with younger age of onset, though still clinically substantial.

Endothelial dysfunction and atherosclerosis

The vascular theory of diabetic sexual dysfunction states that the impairment of blood flow is the primary organic driver. Chronic hyperglycemia initiates a cascade of vascular damage that begins at the endothelial level of blood vessels.

Endothelial dysfunction and nitric oxide

In a healthy physiological state, sexual arousal triggers the release of NO from endothelial cells and nitrergic nerves. NO, a potent vasodilator, diffuses into the smooth muscle cells of the corpus cavernosum (in men) and the vaginal vasculature (in women), causing relaxation and rapid inflow of arterial blood, leading to erection and vaginal lubrication/engorgement. Chronic hyperglycemia in diabetes causes endothelial dysfunction, impairing the synthesis and bioavailability of NO, leading to inadequate relaxation of smooth muscles.

Atherosclerosis and macrovascular damage

Diabetes also accelerates atherosclerosis, leading to physical restriction of blood volume flowing into the pelvic region with reduced flow to the genitals, causing issues like ED and decreased lubrication/arousal. This vascular mechanism explains why ED is often considered a marker of coronary artery disease: the smaller penile arteries occlude earlier than the larger coronary arteries.

Neuropathic deterioration: the loss of sensation and control

Diabetic neuropathy affects both the autonomic and somatic nervous systems, which are critical for sexual function.

Autonomic neuropathy

The autonomic nervous system regulates the involuntary physiological responses related to sex, since it controls the shunting of blood, the relaxation of smooth muscle, and lubrication. Damage to these nerves uncouples the brain's sexual signals from the genital response.

Somatic neuropathy

Somatic nerves supply sensation to the genital skin. Chronic hyperglycemia damages these sensory nerves, leading to a reduction in tactile sensitivity, disrupting the sensory input required to maintain arousal and trigger orgasm. The damage impairs sensory input, arousal, and the psychological responses needed for erections, ejaculations, or vaginal lubrication. The loss of sensation can create a situation where the physical act fails to generate the expected pleasure, eventually leading to a secondary loss of libido.

Metabolic toxicity: the cellular environment

The diabetic metabolic environment acts as a continuous toxin to cellular health. Poor glycemic control, insulin resistance, and the accumulation of AGEs cause direct tissue and cellular damage.

With constantly elevated blood glucose levels, glucose molecules attach to proteins, forming AGEs. These compounds accumulate in the vessel walls and connective tissues everywhere, including the genitalia, leading to decreased elasticity of the tissues and reducing the expansions of penile and vaginal structures during arousal. They induce oxidative stress, which further scavenges NO, worsening the vascular defect (Coppola et al., 2025; Shindel & Lue, 2021).

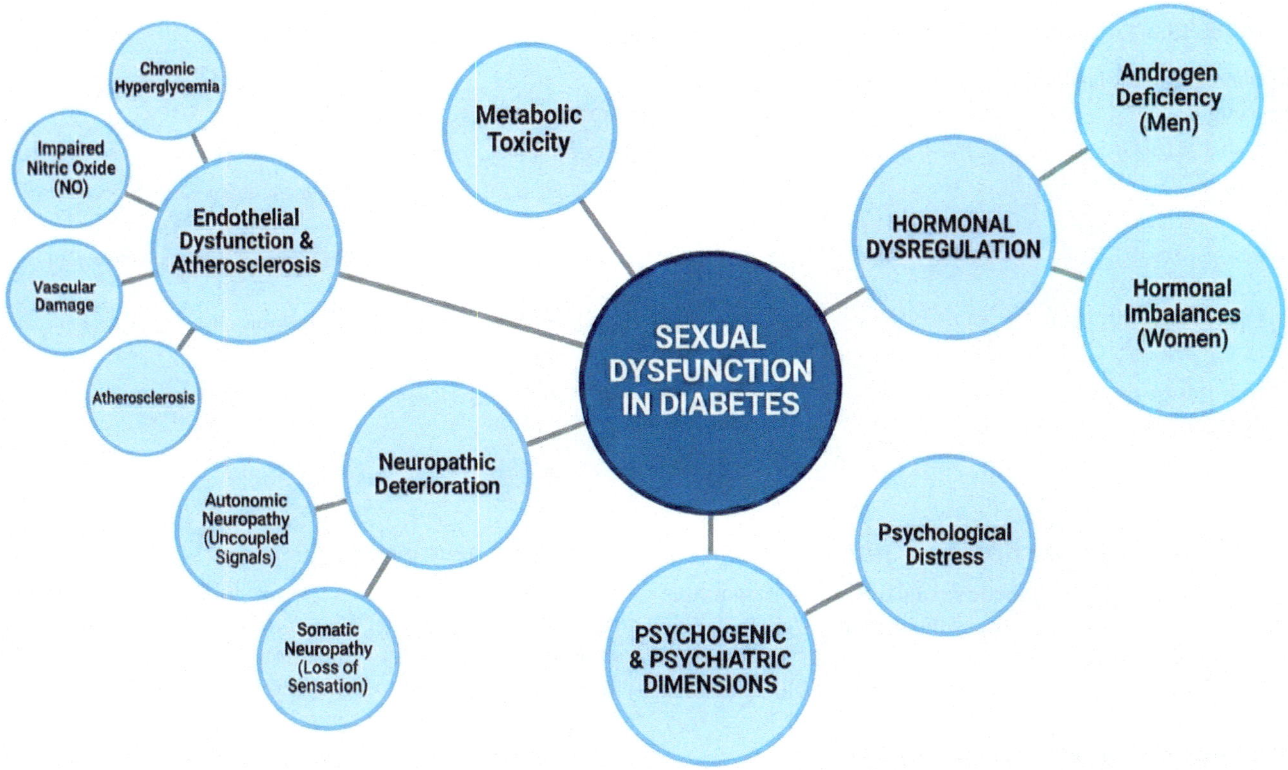

FIGURE 15.3 Multifactorial pathophysiology of sexual dysfunction in diabetes.

Hormonal dysregulation: the endocrine disruptions

Diabetes exerts a disruptive influence on the endocrine system, extending beyond insulin to affect the HPG axis. This hormonal dysregulation is a key, yet often overlooked, driver of sexual dysfunction.

Androgen deficiency in men

New evidence suggests that men with diabetes are at an increased risk of low serum testosterone levels, or hypogonadism (Ray & Pramanik, 2020). The etiology remains unclear but appears to be secondary to a decline in the levels of pituitary hormones that stimulate testicular production of testosterone. Additionally, insulin resistance is associated with reduced SHBG, thereby altering the fraction of bioavailable testosterone. Low levels of testosterone lead to a decline in libido and, directly or indirectly, contribute to ED. Testosterone is essential for the structural integrity of the penile tissue and the maintenance of NO synthase levels. Therefore, a deficiency compromises both the desire and the mechanical ability to engage in the act.

Hormonal imbalances in women

Diabetes disrupts the HPG axis, or, in the case of women, the HPO axis. In women with T1DM, this dysfunction leads to irregular menstruation and altered sex hormone levels, contributing to reduced libido and vaginal atrophy. In T2DM, insulin resistance and hyperinsulinemia affect the HPO axis, altering estrogen levels and, to a lesser extent, progesterone and testosterone levels. This disruption in the balance of estrogens and androgens leads to diminished sexual desire, vaginal dryness, and poor genital responsiveness.

Women with PCOS, a condition linked to insulin resistance, in which androgen excess is often seen, share similar cardiometabolic characteristics with men who have androgen insufficiency (Prosperi & Chiarelli, 2025).

The psychogenic and psychiatric dimensions

While the organic damage caused by diabetes is substantial, the psychological impact is equally profound. Both T1 and T2 DM patients must continuously exercise vigilance over their diets and glucose levels, which may lead to psychological distress. The higher prevalence of mood disorders in patients with diabetes can independently reduce sexual performance and libido. However, the relationship is distinctively circular, akin to a feedback loop, as illustrated in Fig. 15.4:

Thus, even if the medical/metabolic cause is treated (e.g., with medication), the anticipatory anxiety may maintain the dysfunction. High psychological distress

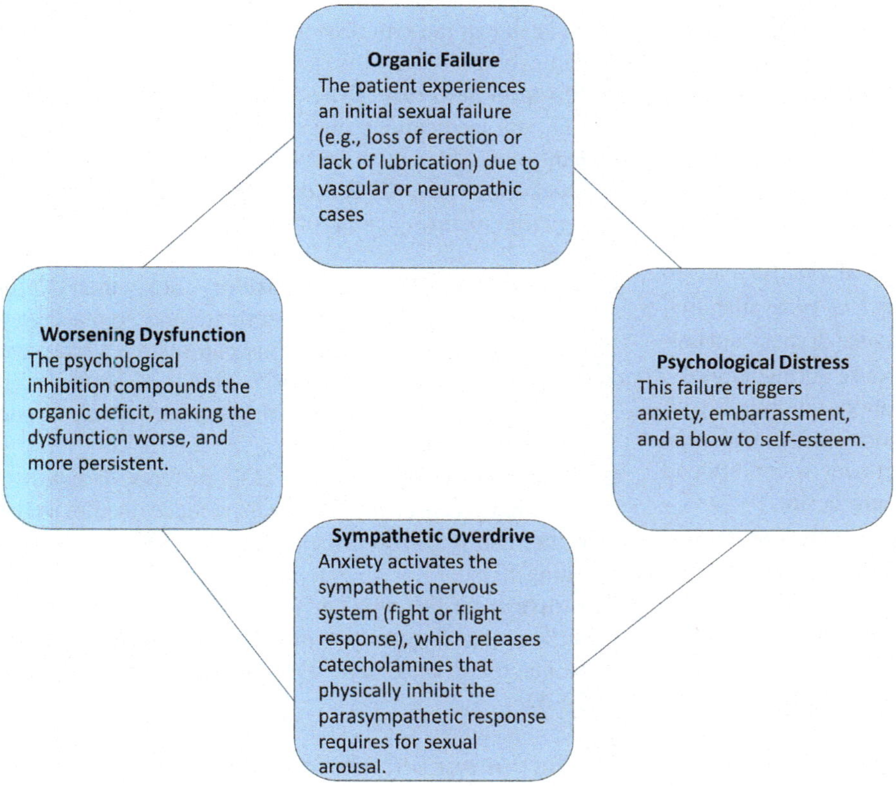

FIGURE 15.4 The psychogenic feedback loop of sexual dysfunction.

creates a self-sustaining, negatively spiraling cycle where sexual dysfunction worsens mood, and vice versa.

Role of other hormones

While insulin and testosterone are the primary focus, other critical hormones drive the "Diabetes-Depression-Sexual Dysfunction" cycle.

1. Prolactin-both high and low levels are deleterious to sexual function. High glucose/stress elevates prolactin, which in turn suppresses GnRH. Low prolactin has been linked to arteriogenic ED (Corona et al., 2014).
2. Oxytocin-it is crucial for sexual receptivity. Diabetic neuropathy and chronic stress can lower oxytocin levels, leading to a reduced desire for physical intimacy (Melis & Argiolas, 2021).
3. Cortisol- Chronic hyperglycemia keeps the body in a "sympathetic dominance" (fight-or-flight) state, resulting in chronically elevated cortisol levels. Cortisol is antagonistic to testosterone as it physically constricts blood flow to the genitals to shunt blood to muscles. Thus, reducing stress can be an effective means of restoring sexual function.
4. Adipose tissue as an endocrine organ-aromatase, present in high quantities in fat, actively converts testosterone to estradiol, which can exacerbate sexual dysfunction and lead to estrogenic symptoms like mood swings and loss of morning erections. Furthermore, high levels of inflammatory markers in fat, called adipokines, can directly damage the endothelium and prevent the release of NO.

Clinical manifestations

Gender influences the clinical presentation of sexual dysfunction, thus requiring tailored diagnostic inquiries.

In men, the primary manifestation is **Erectile Dysfunction,** the result of mechanistic synergy: reduced blood inflow leading to vascular occlusion, autonomic neuropathy preventing smooth muscle relaxation, and metabolic damage resulting in impaired penile tissue structure.

However, the spectrum of dysfunction is broader (He et al., 2021), including other ejaculatory disorders (like retrograde ejaculation - due to dysfunction of the internal sphincter of the bladder) (Mostafa & Abdel-Hamid, 2021) and symptoms of hypogonadism - testosterone deficiency (fatigue, decreased bone mineral density, and depressed mood).

FSD in diabetic women is less reported, but equally significant, affecting 20%–80% of women with T2DM.

- **Lubrication and Arousal:** High blood glucose leads to blood vessel damage and reduced NO bioavailability, affecting vaginal blood flow with vaginal dryness and failure of the genital engorgement response during arousal.
- **Dyspareunia:** The lack of lubrication, combined with potential vaginal atrophy from hormonal imbalances, leads to dyspareunia. This pain creates a conditioned aversion to sex, secondarily reducing desire.
- **Infection Risk:** High blood glucose levels alter vaginal pH and the environment, making it more prone to infections such as candidiasis. Recurrent infections cause discomfort and further discourage sexual activity.

Diagnostic frameworks and screening protocols

Effective management begins with identification. Healthcare providers should evaluate sexual health at the first admission and during follow-ups, integrating it into the standard diabetes assessment.

Active solicitation for sexual complaints is a must, as patients rarely volunteer such complaints. Validated questionnaires serve as non-threatening entry points for this conversation, e.g.,

- **Female Sexual Function Index (FSFI):** a self-report questionnaire consisting of 19 items developed to assess female sexual function across domains of desire, arousal, lubrication, orgasm, satisfaction, and pain (Rosen et al., 2000).
- **Brief Sexual Symptom Checklist (BSSC):** a concise tool suitable for rapid screening in busy clinics designed to identify key sexual symptoms, including libido, arousal, orgasm/ejaculatory issues, and satisfaction. The male version has formal psychometric validation (Romero Otero et al., 2018).

Comprehensive management strategies

Keeping in mind the multifactorial etiology of sexual dysfunction in diabetes, the management also needs to be multi-pronged.

Medical and pharmacological interventions

1. **Glycemic Control:** The foundation of all therapy is strict glycemic control. It is the first line of defense against the accumulation of AGEs and oxidative stress and can halt the progression of neuropathy and vascular damage.
2. **Phosphodiesterase Type 5 Inhibitors:** For men with ED, oral medications like sildenafil, tadalafil, or vardenafil are safe and effective. These drugs inhibit the breakdown of cGMP, thereby enhancing the NO signaling pathway and promoting vasodilation (Balhara et al., 2015).
3. **Testosterone Supplementation:** For men with confirmed hypogonadism, testosterone replacement therapy may be indicated. This can restore libido and improve the response to PDE5 inhibitors. However, it requires careful monitoring of hematocrit and prostate health (Kumari et al., 2023).
4. **Weight Management:** For T2DM patients, weight loss can improve testosterone levels, reduce insulin resistance, and improve vascular function, thereby ameliorating SD.
5. **Lubricants:** For women, vaginal lubricants and moisturizers are simple yet effective symptomatic treatments for dryness and dyspareunia (Shindel & Lue, 2021).

Psychotherapeutic and non-medical interventions

Recognizing the "psychological feedback loop," non-medical interventions are essential.

1. CBT is proven to improve sexual function and mental well-being, especially in women. It helps patients identify and resolve unhelpful thinking patterns, such as performance anxiety or negative body image related to diabetes (Mestre-Bach et al., 2022).
2. Sensate Focus is a specific sex therapy technique. It involves exercises in non-judgmental physical intimacy where the couple focuses on touch and sensation without the pressure of achieving an erection or orgasm. This rebuilding of connection helps reduce anxiety and breaks the behavior that fuels dysfunction (Huang et al., 2024).

Conclusion

Diabetes mellitus is a condition with a complex interaction of metabolic, endocrine, neurological and psychological dimensions, whose nuances are often missed due to its standard view as a purely endocrine condition. Factors such as impaired transmission of neurotransmitters—primarily dopamine and serotonin, disruption in insulin transport across the BBB and overactive HPA axis, form the basis of the pathophysiology of both mood symptoms and sexual dysfunction in diabetes. These also have a bidirectional interaction with symptoms such as depression, anxiety and diabetes—related distress, impairing sexual functioning and sexual dysfunction leading to poor self-esteem, performance anxiety and perpetuating mood symptoms. This picture is further compounded by the level of metabolic control, which mood symptoms can influence. It can amplify endothelial damage, autonomic neuropathy, hormonal imbalance and chronic inflammation, in turn contributing to the pathophysiology of both mood symptoms and sexual dysfunction, thereby completing the vicious cycle. Knowledge of these associated domains, early identification and holistic management are essential in every case of diabetes.

The emerging evidence further advocates such holistic management with findings suggestive of central neuroprotective and antidepressant-like effects of molecules like metformin, GLP-1 receptor agonists, SGLT2 inhibitors, etc., through modulation of insulin signaling, neurotropic factors—BDNF and inflammatory pathways. Psychotherapeutic interventions and lifestyle modifications tailored to diabetes, such as Cognitive Behavior Therapy, optimization of circadian cycles and sleep, physical activity, etc., have proven to improve glycemic outcomes and mood. This requires integrated care with a management framework involving endocrinologists, psychiatrists, and behavioral therapists, thus creating an archetype that addresses the entire biopsychosocial model of the illness.

Mood and sexual dysfunction must thus be considered as core clinical presentations of a dysfunctional pancreas–brain connection, rather than secondary complications of diabetes. The concept of the illness needs to be re-envisioned as a psycho-neuro-endocrine condition, with a paradigm shift in practice and research, to make management personalized, multidisciplinary, and to improve the individual's overall quality of life.

Mini-dictionary of terms

1. **Dyspareunia:** Persistent genital pain associated with sexual intercourse, often resulting from vaginal dryness and atrophy, which can lead to a secondary loss of sexual desire.
2. **Retrograde Ejaculation:** A male sexual disorder caused by nerve damage where semen enters the bladder instead of exiting through the urethra due to a malfunctioning internal sphincter.
3. **Sensate Focus:** A psychotherapeutic technique involving non-judgmental physical touch exercises designed to reduce performance anxiety and rebuild intimacy without the immediate goal of erection or orgasm.
4. **Major Depressive Disorder:** comprising persistent sadness of mood, with other significant symptoms like anhedonia and neurovegetative symptoms.
5. **Subsyndromal/subthreshold depression**, having clinically significant mood symptoms but not meeting full criteria for MDD
6. **Anxiety disorders,** like generalized anxiety, panic, diabetes-specific fears (hypoglycemia anxiety), where mood can be broadly either anxious, fearful or depressed.
7. **Diabetes distress,** a diabetes-specific emotional burden distinct from clinical depression, comprises the emotional burden of self-management, physician-related distress, regimen-related distress, and interpersonal distress.

Summary points

1. Depression increases the risk of developing Type 2 diabetes, while having diabetes increases the risk of depression.
2. Peripheral insulin resistance compromises insulin transport across the blood-brain barrier, impairing signaling pathways essential for neuroplasticity and neurotransmitter function.
3. The dopamine reward system in the brain loses insulin receptors, reducing D2/D3 receptor activity and contributing to anhedonia.
4. Elevated cytokines like IL-6 and TNF-α cross the blood-brain barrier to activate microglia and divert tryptophan away from serotonin synthesis toward neurotoxic kynurenine pathways.
5. Chronic stress and depression overactivate the HPA axis, resulting in elevated cortisol levels that damage the hippocampus and further worsen insulin resistance.
6. Management requires a multidisciplinary model that combines metabolic control with treatments such as CBT and pharmacotherapy to address both physiological and psychological symptoms.
7. The estimated prevalence of sexual dysfunction among diabetic patients is 61.4%, caused by a triad of vascular, neurologic, and hormonal deteriorations.
8. Chronic hyperglycemia induces endothelial dysfunction that impairs the synthesis of NO, preventing the smooth muscle relaxation required for erections and vaginal lubrication.
9. A psychological feedback loop exists where anticipatory anxiety maintains sexual dysfunction even after organic causes are treated, creating a self-sustaining cycle of distress.
10. Management protocols integrate PDE5i to enhance nitric oxide signaling with psychotherapeutic interventions like sensate focus to reduce performance anxiety.

References

Alhomoud, I. S., Wheeler, S., & Dixon, D. (2025). Repurposing incretin therapies: A narrative review of emerging indications across cardiometabolic, liver, kidney, neurological, psychiatric, and other systems. *Canadian Journal of Physiology and Pharmacology, 103*(9), 298–311.

Allen, J., Caruncho, H. J., & Kalynchuk, L. E. (2021). Severe life stress, mitochondrial dysfunction, and depressive behavior: A pathophysiological and therapeutic perspective. *Mitochondrion, 56*, 111–117.

Balhara, Y. P., Sarkar, S., & Gupta, R. (2015). Phosphodiesterase-5 inhibitors for erectile dysfunction in patients with diabetes mellitus: A systematic review and meta-analysis of randomized controlled trials. *Indian Journal of Endocrinology and Metabolism, 19*(4), 451–461.

Buchberger, B., Huppertz, H., Krabbe, L., Lux, B., Ells, L., & Glaesmer, H. (2016). Psychological factors and depression in children and adolescents with type 1 diabetes mellitus: A systematic review. *Diabetic Medicine, 33*(8), 988–999.

Cannon, A., Jacoby, C., & Hughes, A. S. (2024). Mind in metabolism - A comprehensive literature review on diabetes and its connections to obsessive compulsive disorder, schizophrenia, and bipolar disorder. *Current Diabetes Reports, 25*(1), 10.

Carmellini, P., Cuomo, A., Rescalli, M. B., & Fagiolini, A. (2025). GLP-1 receptor agonists in mood disorders: A psychiatric perspective. *Life, 15*(9), 1422.

Colberg, S. R., Sigal, R. J., Yardley, J. E., Riddell, M. C., Dunstan, D. W., Dempsey, P. C., Horton, E. S., Castorino, K., & Tate, D. F. (2016). Physical activity/exercise and diabetes: A position statement of the American Diabetes Association. *Diabetes Care, 39*(11), 2065–2079.

Coppola, A., Esposito, K., & Gazzaruso, C. (2025). Editorial: Sexual function and dysfunction in men and women with diabetes. *Frontiers in Endocrinology, 16*, 1629375.

Corona, G., Wu, F. C., Rastrelli, G., Lee, D. M., Forti, G., O'Connor, D. B., O'Neill, T. W., Pendleton, N., Bartfai, G., Boonen, S., Casanueva, F. F., Finn, J. D., Huhtaniemi, I. T., Kula, K., Punab, M., Vanderschueren, D., Rutter, M. K., Maggi, M., & EMAS Study Group. (2014). Low prolactin is associated with sexual dysfunction and psychological or metabolic disturbances in middle-aged and elderly men: the European Male Aging Study (EMAS). *The Journal of Sexual Medicine, 11*(1), 240–253.

Duan, D., Kim, L. J., Jun, J. C., & Polotsky, V. Y. (2023). Connecting insufficient sleep and insomnia with metabolic dysfunction. *Annals of the New York Academy of Sciences, 1519*(1), 94–117.

Fanelli, G., Mota, N. R., Salas-Salvadó, J., Bulló, M., Fernandez-Aranda, F., Camacho-Barcia, L., Testa, G., Jiménez-Murcia, S., Bertaina-Anglade, V., Franke, B., Poelmans, G., van Gils, V., Jansen, W. J., Vos, S. J. B., Wimberley, T., Dalsgaard, S., Barta, C., Serretti, A., Fabbri, C., & Bralten, J. (2022). The link between cognition and somatic conditions related to insulin resistance in the UK Biobank study cohort: A systematic review. *Neuroscience & Biobehavioral Reviews, 143*, 104927.

Fang, W., Zhang, J., Hong, L., Huang, W., Dai, X., Ye, Q., & Chen, X. (2020). Metformin ameliorates stress-induced depression-like behaviors via enhancing the expression of BDNF by activating AMPK/CREB-mediated histone acetylation. *Journal of Affective Disorders, 260*, 302–313.

Friedman, J. M. (2019). Leptin and the endocrine control of energy balance. *Nature Medicine, 25*(1), 131–144.

Gao, J., Zhao, L., Cheng, Y., Lei, W., Wang, Y., Liu, X., Zheng, N., Shao, L., Chen, X., Sun, Y., Ling, Z., & Xu, W. (2023). Probiotics for the treatment of depression and its comorbidities: A systemic review. *Frontiers in Cellular and Infection Microbiology, 13*, 1167116.

Gebeyehu, N. A., Gesese, M. M., Tegegne, K. D., Kebede, Y. S., Kassie, G. A., Mengstie, M. A., Zemene, M. A., Bantie, B., Feleke, S. F., Dejenie, T. A., Abebe, E. C., Anley, D. T., Dessie, A. M., Bayih, W. A., & Adela, G. A. (2023). Global prevalence of sexual dysfunction among diabetic patients from 2008 to 2022: Systematic review and meta-analysis. *Metabolism Open, 18*, 100247.

Grillo, C. A., Piroli, G. G., Lawrence, R. C., Wrighten, S. A., Green, A. J., Wilson, S. P., Sakai, Buchberger, R. R., Kelly, S. J., Wilson, M. A., Mott, D. D., & Reagan, L. P. (2015). Hippocampal insulin resistance impairs spatial learning and synaptic plasticity. *Diabetes, 64*(11), 3927–3936.

Guo, J., Zhang, Y., & Xiao, L. (2023). Metformin as a potential antidepressant: Mechanisms and clinical evidence. *Pharmacological Research, 192*, 106876.

He, Z., Yin, G., Li, Q. Q., Zeng, Q., & Duan, J. (2021). Diabetes mellitus causes male reproductive dysfunction: A review of the evidence and mechanisms. *In Vivo, 35*(5), 2503–2511.

Huang, S., Li, Z., & Santtila, P. (2024). The effectiveness of online sensate focus exercises in enhancing sexual function and intimacy among Chinese heterosexual couples: A randomized controlled trial. *Journal of Sex & Marital Therapy, 50*(6), 707–724.

Joseph, J. J., & Golden, S. H. (2017). Cortisol dysregulation: The bidirectional link between stress, depression, and type 2 diabetes mellitus. *Annals of the New York Academy of Sciences, 1391*(1), 20–34.

Kautzky-Willer, A., Harreiter, J., & Pacini, G. (2016). Sex and gender differences in risk, pathophysiology and complications of type 2 diabetes mellitus. *Endocrine Reviews, 37*(3), 278–316.

Kumari, K., Kumar, R., Memon, A., Kumari, B., Tehrim, M., Kumari, P., Shehryar, M., Islam, H., Islam, R., Khatri, M., Kumar, S., & Kumar, A. (2023). Treatment with testosterone therapy in type 2 diabetic hypogonadal adult males: A systematic review and meta-analysis. *Clinical Practice, 13*(2), 454–469.

Lach, G., Schellekens, H., Dinan, T. G., & Cryan, J. F. (2018). Anxiety, depression, and the microbiome: A role for gut peptides. *Neurotherapeutics, 15*(1), 36–59.

Lassale, C., Batty, G. D., Baghdadli, A., Jacka, F., Sánchez-Villegas, A., Kivimäki, M., & Akbaraly, T. (2019). Healthy dietary indices and risk of depressive outcomes: A systematic review and meta-analysis of observational studies. *Molecular Psychiatry, 24*(7), 965–986.

Leonard, B. E., & Wegener, G. (2020). Inflammation, insulin resistance and neuroprogression in depression. *Acta Neuropsychiatrica, 32*(1), 1–9.

Li, C., Xu, D., Hu, M., Tan, Y., Zhang, P., Li, G., & Chen, L. (2017). A systematic review and meta-analysis of randomized controlled trials of cognitive behavior therapy for patients with diabetes and depression. *Bioscience Reports, 37*(2), Article BSR20160557.

Mansour, A., Dehghan Najm Abadi, A., Etesam, F., Danaei, S., Abaszadeh, Y., Torabi, F., Murshed, A., Nasli-Esfahani, E., Vahabi, Z., & Sajjadi-Jazi, S. M. (2025). Association of antidiabetic medications with psychiatric disorders in patients with type 2 diabetes: A cross-sectional study. *European Journal of Medical Research, 30*(1), 654.

Melis, M. R., & Argiolas, A. (2021). Oxytocin, erectile function and sexual behavior: Last discoveries and possible advances. *International Journal of Molecular Sciences, 22*(19), 10376.

Mestre-Bach, G., Blycker, G. R., & Potenza, M. N. (2022). Behavioral therapies for treating female sexual dysfunctions: A state-of-the-art review. *Journal of Clinical Medicine, 11*(10), 2794.

Miedlich, S. U., & Lamberti, J. S. (2025). Connecting the dots: Understanding and addressing the metabolic impact of antipsychotic and antidepressant medications. *Annals of the New York Academy of Sciences, 1546*(1), 35–57.

Mostafa, T., & Abdel-Hamid, I. A. (2021). Ejaculatory dysfunction in men with diabetes mellitus. *World Journal of Diabetes, 12*(7), 954–974.

Moulton, C. D., Pickup, J. C., & Ismail, K. (2015). The link between depression and diabetes: The role of inflammation and HPA axis dysregulation. *Lancet Diabetes & Endocrinology, 3*(6), 461–471.

Osimo, E. F., Baxter, L. J., Lewis, G., Jones, P. B., & Khandaker, G. M. (2020). Inflammatory markers in depression: A meta-analysis of mean differences and variability in 5,166 patients and 5,083 controls. *Brain, Behavior, and Immunity, 87*, 901–912.

Prosperi, S., & Chiarelli, F. (2025). Insulin resistance, metabolic syndrome and polycystic ovaries: An intriguing conundrum. *Frontiers in Endocrinology, 16*, 1669716.

Ray, S., & Pramanik, S. (2020). Reproductive dysfunctions in males with type 2 diabetes mellitus: An updated review. *EMJ Diabet, 8*(1), 79–89.

Romero Otero, J., García Gómez, B., Medina Polo, J., García Rojo, E., García Cruz, E., Peinado, F., Sopeña Sutil, R., Bozzini, G., & Rodríguez Antolín, A. (2018). The brief sexual symptom checklist as a screening tool for sex-related problems in the primary care setting: Focus on erectile dysfunction. *Urologia Internationalis, 101*(1), 98–105.

Rosen, C., Brown, J., Heiman, S., Leiblum, C., Meston, R., Shabsigh, D., Ferguson, R., & D'Agostino, R. (2000). The female sexual function index (FSFI): A multidimensional self-report instrument for the assessment of female sexual function. *Journal of Sex & Marital Therapy, 26*(2), 191–208.

Sartorius, N. (2018). Depression and diabetes. *Dialogues in Clinical Neuroscience, 20*(1), 47–52.

Semenkovich, K., Brown, M. E., Svrakic, D. M., & Lustman, P. J. (2015). Depression in type 2 diabetes mellitus: Prevalence, impact, and treatment. *Drugs, 75*(6), 577–587.

Shapiro, M. S. (2022). Special psychosocial issues in diabetes management: Diabetes distress, disordered eating, and depression. *PrimaryCare, 49*(2), 363–374.

Shindel, A. W., & Lue, T. F. (2021). In K. R. Feingold, S. F. Ahmed, B. Anawalt, et al. (Eds.), *Sexual dysfunction in diabetes*. Endotext.

Slyepchenko, A., Maes, M., Machado-Vieira, R., Anderson, G., Solmi, M., Sanz, Y., Berk, M., Köhler, C. A., & Carvalho, A. F. (2016). Intestinal dysbiosis, gut hyperpermeability and bacterial translocation: Missing links between depression, obesity and type 2 diabetes. *Current Pharmaceutical Design, 22*(40), 6087–6106.

Tabassum, A., Badulescu, S., Singh, E., Asoro, R., McIntyre, R. S., Teopiz, K. M., Llach, C. D., Shah, H., & Mansur, R. B. (2024). Central effects of acute intranasal insulin on neuroimaging, cognitive, and behavioural outcomes: A systematic review. *Neuroscience and Biobehavioral Reviews, 167*, 105907.

Tanaka, M., Szabó, Á., Spekker, E., Polyák, H., Tóth, F., & Vécsei, L. (2022). Mitochondrial impairment: A common motif in neuropsychiatric presentation? The link to the Tryptophan-Kynurenine metabolic system. *Cells, 11*(16), 2607.

Tareen, R. S., & Tareen, K. (2017). Psychosocial aspects of diabetes management: dilemma of diabetes distress. *Translational Pediatrics, 6*(4), 383–396.

Taylor, P. N., Razvi, S., Pearce, S. H., & Dayan, C. M. (2023). Clinical review: Thyroid disease, metabolism, and mental health in diabetes. *Lancet Diabetes & Endocrinology, 11*(3), 189–200.

Xu, S., Lu, F., Gao, J., & Yuan, Y. (2024). Inflammation-mediated metabolic regulation in adipose tissue. *Obesity Reviews, 25*(6), e13724.

Yang, T., Nie, Z., Shu, H., Kuang, Y., Chen, X., Cheng, J., Yu, S., & Liu, H. (2020). The role of BDNF on neural plasticity in depression. *Frontiers in Cellular Neuroscience, 14*, 82.

Yang, Q., Zhang, J., Li, X., et al. (2025). Prevalence and risk factors of depression in patients with diabetes: A systematic review and meta-analysis. *Frontiers in Endocrinology, 16*, 1660478.

Chapter 16

Anxiety related behavior in parents of diabetic children

Ivonne Carosi Arcangeli[1], Mauro Cozzolino[1], Francesco Di Prinzio[1], Stefania Sinesi[1] and Giovanna Celia[2]
[1]*Department of Humanities, Philosophy and Education, University of Salerno, Fisciano, Italy;* [2]*eCampus University, Novedrate, Italy*

Abbreviations

ACT acceptance and commitment therapy
ACTH adrenocorticotropic hormone
BWM-T brain wave modulation technique
CBT cognitive-behavioral therapy
CGM continuous glucose monitoring
CRH corticotropin-releasing hormone
HPA hypothalamic-pituitary-adrenal axis
HPG hypothalamic-pituitary-gonadal axis
HRV heart rate variability
MBSR mindfulness-based stress reduction
MBTS mind-body therapies
SAM sympathetic-adrenomedullary system
T1D type 1 diabetes

Introduction

A child's diabetes diagnosis abruptly changes family life, posing major psychological challenges and demanding mastery of complex medical tasks. This chapter focuses on the behavioral expressions, psychological mechanisms, and biological correlates of parental anxiety in the context of pediatric diabetes. Parental anxiety manifests behaviorally through hypervigilance, risk avoidance, and excessive control, and psychologically through guilt, attachment insecurity, and intolerance of uncertainty, interacting with neuroendocrine stress systems (HPA and SAM axes). While these behaviors often aim to protect, they can reinforce stress cycles, hinder the child's self-management, and strain family relationships. Protective factors-such as social support, self-efficacy, adaptive coping, and family cohesion-can buffer these effects, allowing anxiety-driven behaviors to transform into resilience and empowerment. This chapter bridges empirical evidence and clinical practice to support multidisciplinary teams in promoting family well-being and sustainable diabetes management.

In sum (Fig. 16.1), parental anxiety in diabetes management manifests primarily as hypercontrol, avoidance, and repeated checking, with consequences for family dynamics and the child's self-management.

The psychological experience of parents caring for a child with type 1 diabetes is shaped by a constellation of affective and relational mechanisms that influence their behavior, emotional reactivity, and caregiving strategies (Fig. 16.2).

Physiologically, the hypothalamic-pituitary-adrenal (HPA) axis represents the primary response pathway (Fig. 16.3): in the presence of anxiety-provoking stimuli, the hypothalamus secretes corticotropin-releasing hormone (CRH), which stimulates the anterior pituitary to release adrenocorticotropic hormone (ACTH), leading to cortisol secretion from the adrenal cortex. In parents chronically preoccupied with their child's health, cortisol may show dysregulated patterns, such as elevated basal levels or a flattened diurnal rhythm, both associated with chronic stress and burnout (Gunnar & Quevedo, 2007).

In summary (Fig. 16.4), while caregiving for a child with diabetes is inherently demanding, the presence and reinforcement of specific protective factors, such as social support, self-efficacy, adaptive coping styles, and family cohesion, can significantly reduce anxiety-related behaviors in parents.

Parental psychological adaptation to a child's type 1 diabetes diagnosis

The onset of type 1 diabetes (T1D) is typically abrupt, leaving parents with limited opportunities to prepare for the profound changes that affect family life. The necessity to rapidly acquire clinical management skills while simultaneously coping with the emotional and organizational impact renders the diagnosis a highly destabilizing

Hormones in Neuroscience. https://doi.org/10.1016/B978-0-443-30263-3.00014-2

FIGURE 16.1 **Behavioral manifestations of parental anxiety.** Illustrates how parental anxiety manifests in diabetes management through hypervigilance, risk avoidance, compulsive monitoring, and intrusive communication-behaviors intended to protect but that may undermine child autonomy and perpetuate family anxiety cycles.

and frequently traumatic event for most parents (Whittemore et al., 2012). Parents commonly report a range of negative emotional experiences, including fear, frustration, helplessness, anger, sadness, and guilt, which are often associated with elevated levels of anxiety, depression, and stress (Carosi Arcangeli et al., 2024; Simpson et al., 2021; Vayisoglu et al., 2025). In the immediate post-discharge period, parents are required to reorient their expectations of normality in order to adjust to the new family reality imposed by the diagnosis (Jespersen et al., 2023). Parental psychological adaptation to a child's T1D diagnosis represents a complex and multifactorial process, influenced by individual, familial, and contextual variables, and constitutes a central determinant of both disease management and family psychological well-being. Over the medium- and long-term, trajectories of adaptation appear heterogeneous: while some parents develop effective coping strategies, reframing the illness experience in terms of learning and resilience, others remain entrenched in hypervigilance and excessive control, leading to heightened anxiety, depressive symptoms, and strained parent-child relationships (Jespersen et al., 2023; Simpson et al., 2021; Whittemore et al., 2012). In adaptive trajectories, the new reality is progressively integrated into everyday life, with the establishment of effective management routines and increased confidence in caregiving abilities (Jespersen et al., 2023; Whittemore et al., 2012). Conversely, maladaptive pathways are characterized by persistent psychological distress and relational difficulties. Importantly, longitudinal studies suggest that parents who

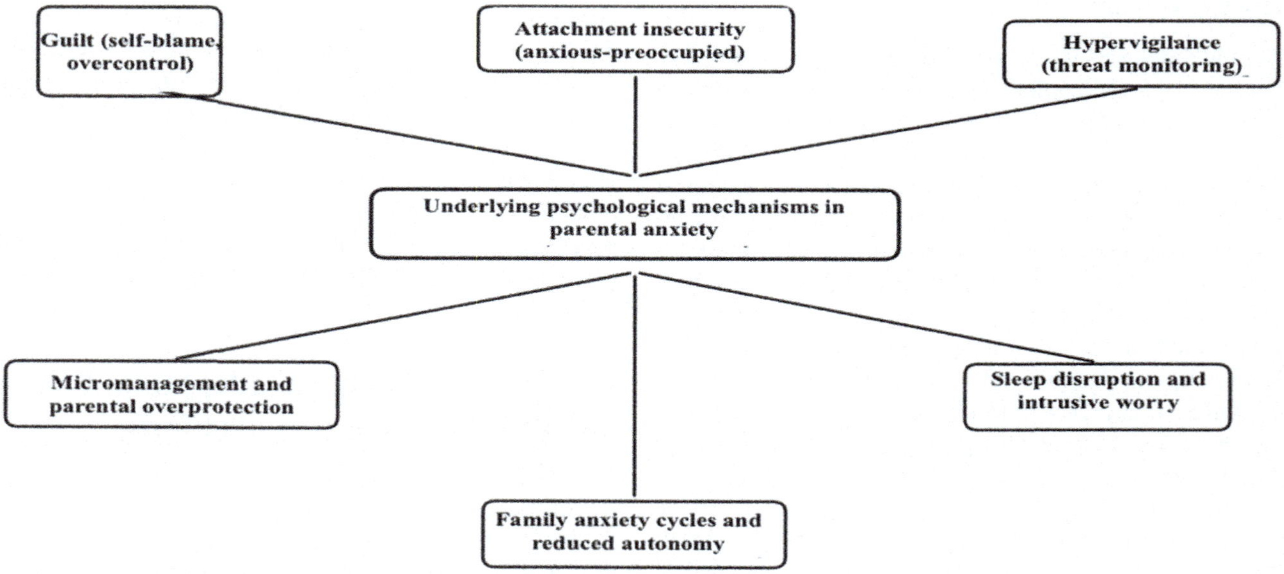

FIGURE 16.2 **Core psychological mechanisms driving parental anxiety in type 1 diabetes.** Depicts the core psychological mechanisms underlying parental anxiety in type 1 diabetes management-heightened stress reactivity, negative cognitive appraisals, and maladaptive coping that sustain worry and shape caregiving behaviors.

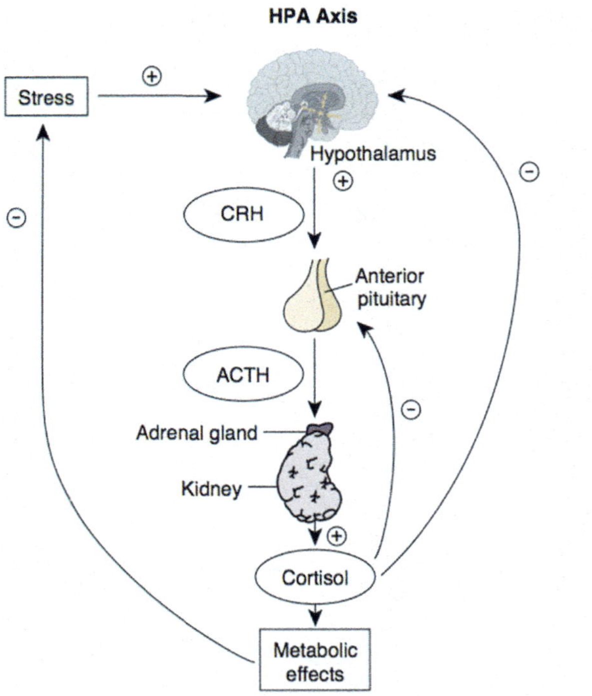

FIGURE 16.3 HPA axis diagram. Schematic of the hypothalamic-pituitary-adrenal (HPA) axis, illustrating the stress-activated cascade from hypothalamic CRH release to pituitary ACTH secretion and adrenal cortisol production. https://embryology.med.unsw.edu.au/embryology/index.php?curid=10637.

experience high levels of uncertainty at the time of diagnosis are significantly more likely to report long-term psychological distress, even 5–6 years post-diagnosis (Jespersen et al., 2023). Social support has been consistently identified as a critical factor in parental adaptation. The concept of social support encompasses the emotional, instrumental, and motivational assistance individuals receive from their interpersonal networks, underscoring the functional and structural diversity of these ties (Schwarzer et al., 2004). This encompasses professional support from healthcare providers-essential for acquiring disease-management skills and ensuring effective daily care-as well as personal support from spouses, extended family, friends, and other parents of children with T1D. Insufficient support, contradictory information, and unrealistic expectations may significantly hinder adaptation. Similarly, the school context has emerged as a source of parental concern, particularly in cases of inadequate T1D management, although the presence of trained personnel and regular communication with parents have been associated with more positive experiences. Engagement in community activities and peer interactions with other families living with T1D has also been shown to mitigate feelings of isolation, foster a sense of belonging, and instill hope (Simpson et al., 2021; Vayisoglu et al., 2025; Whittemore et al., 2012). To facilitate effective parental adaptation to their child's chronic condition and enhance the quality of life of both the child and the entire family, it is essential to provide structured support beginning immediately after diagnosis. Such support should include: psychological interventions aimed at addressing the initial trauma and emotional responses; clear and consistent medical guidance to support daily disease management; and social support that promotes effective communication within the family and across relevant social contexts, which must be both immediate and sustained, integrated longitudinally throughout the care trajectory (Villaécija et al., 2023; Wakelin et al., 2025).

Behavioral manifestations of parental anxiety in diabetes management

Among parents of children and adolescents with diabetes, anxiety is expressed not only as an emotional state but, more prominently, through a cluster of controlling and protective behaviors aimed at preventing adverse events—most notably hypoglycemia and its consequences—and at reducing daily uncertainty (Carlsund et al., 2025). A common expression is *hyper-monitoring*: excessively frequent glucose checks, repeated device verifications, and meticulous data reviews even in the absence of immediate risk signals. This "expanded vigilance" may be adaptive during therapeutic adjustment phases, but when persistent, it is often maintained by fear of hypoglycemia and catastrophic cognitions (e.g., "*if I do not check, the worst may happen*"), with potential repercussions on family stress and sleep quality (Monaghan et al., 2009; Patton et al., 2008).

Parental anxiety is also associated with avoidance behaviors, such as restricting activities perceived as risky (sports, school trips, overnight stays) or delaying the child's autonomy in managing carbohydrate intake, insulin boluses, and corrections, even when age and skills would allow it (Jaser & Whittemore, 2014; Sullivan-Bolyai et al., 2003). Although these patterns are motivated by protective intentions, they may hinder the child's acquisition of self-efficacy and foster vicious cycles of dependency and conflict.

Another frequent manifestation is compulsive double-checking (repeated verification with requests for reassurance from the partner or the child), which alleviates anxiety in the short term but maintains it in the long term through the negative reinforcement mechanisms typical of anxious coping (Hilliard et al., 2016).

At the interactional level, anxiety may translate into directive or intrusive communication—pressing requests for measurements, immediate corrections, or critical feedback on dietary "errors"- which, particularly in adolescence, are more likely to elicit oppositional behavior,

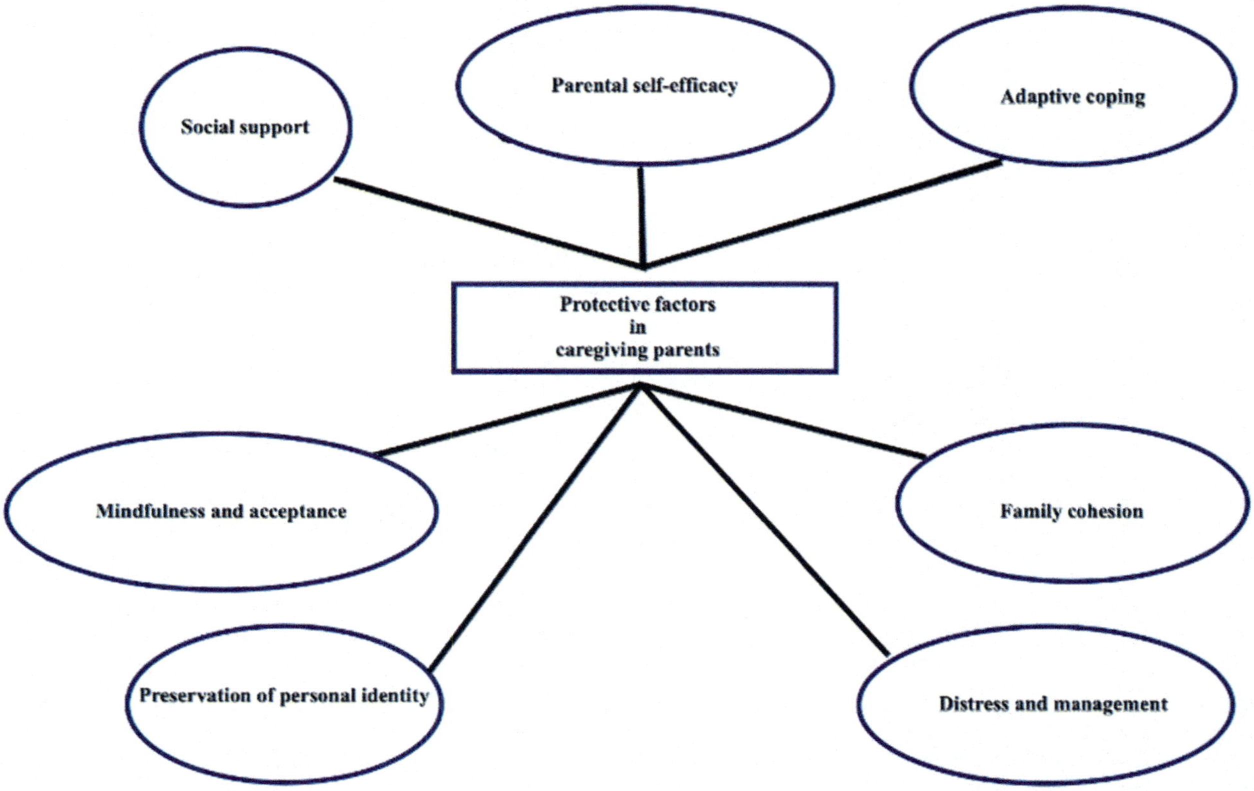

FIGURE 16.4 Protective factors in caregiving parents of children with type 1 diabetes. Highlights the main factors that buffer parental anxiety and promote adaptive coping, including social support, self-efficacy, mindfulness, family cohesion, which together foster resilience and positive caregiving practices.

skipped doses, and glycemic variability (Hilliard et al., 2016; Streisand et al., 2005). Even micro-behaviors in daily life reflect hypervigilance: frequent nocturnal awakenings for glucose checks, redundant stocks of food and glucagon, rigid meal planning, and heightened alarm at glycemic variations within physiologically expected ranges (Patton et al., 2008; Young-Hyman et al., 2016).

Guidelines emphasize that such behaviors must be contextualized within the family's heavy decision-making burden and the constant balancing of metabolic safety with "normal" life (Delamater et al., 2018; Young-Hyman et al., 2016). Psychoeducational and supportive interventions -such as training in CGM data interpretation, family problem-solving, gradual exposure to feared situations, and promotion of progressive autonomy-can reduce dysfunctional anxious behaviors and improve both psychological and glycemic outcomes (Delamater et al., 2018; Jaser & Whittemore, 2014).

In sum (Fig. 16.1), parental anxiety in diabetes management manifests primarily as hypercontrol, avoidance, and repeated checking, with consequences for family dynamics and the child's self-management. Effective clinical approaches should integrate emotional, behavioral, and community dimensions, in line with international recommendations (Delamater et al., 2018; Young-Hyman et al., 2016).

Underlying psychological mechanisms: guilt, attachment, and hypervigilance

The psychological experience of parents caring for a child with type 1 diabetes is shaped by a constellation of affective and relational mechanisms that influence their behavior, emotional reactivity, and caregiving strategies (Fig. 16.2). Prior research has highlighted how stress reactivity, anxiety, and hypervigilance intertwine with caregiving behaviors, shaping both parental adjustment and family dynamics (Hilliard, Holmes et al., 2011; Hilliard, Monaghan et al., 2011). Systematic reviews further emphasize that these affective and relational processes constitute core mechanisms underlying parental adaptation to the demands of diabetes management (Streisand & Monaghan, 2014; Whittemore et al., 2012).

Among the most consistently observed underlying psychological processes are guilt, attachment-related anxieties, and hypervigilance, each of which plays a significant role in the development and maintenance of parental anxiety.

Parental guilt frequently emerges in response to the perception of having failed to protect the child from illness or of not doing "enough" to manage their condition. This guilt may be exacerbated by the chronic and unpredictable nature of type 1 diabetes, where even minor deviations in blood glucose levels can trigger feelings of responsibility and self-blame (Streisand et al., 2005). While moderate levels of guilt can motivate attentive caregiving, persistent or unresolved guilt can lead to maladaptive behaviors such as overcontrol, micromanagement, and emotional over-involvement, ultimately undermining the child's autonomy and increasing relational tension. For many parents, especially mothers, caregiving becomes intertwined with the implicit belief that any adverse outcome is a direct reflection of personal inadequacy.

Attachment dynamics also shape the way parental anxiety is expressed and regulated. Parents with insecure attachment styles, particularly those characterized by anxious-preoccupied patterns, are more likely to exhibit heightened emotional reactivity, fear of separation, and difficulty tolerating uncertainty in their child's medical trajectory (Mikulincer & Shaver, 2007). In the context of type 1 diabetes, this can translate into overprotective parenting, resistance to delegating care tasks to the child or others, and difficulty in supporting the child's gradual move toward self-management. Moreover, anxiously attached parents may experience caregiving as a form of validation of their role and worth, further entrenching dependency loops within the parent-child dyad.

Hypervigilance, defined as a state of persistent alertness and scanning for potential threats, is both a symptom and a driver of chronic parental anxiety. In parents of children with type 1 diabetes, hypervigilance often centers around the fear of nocturnal hypoglycemia, dietary transgressions, or treatment errors. While some degree of vigilance is necessary for safe diabetes management, excessive hypervigilance is associated with sleep disturbances, intrusive thoughts, and reduced quality of life (Monaghan et al., 2009; Patton et al., 2008). Over time, it can impair the parent's emotional availability, reinforce anxious cycles, and paradoxically increase child anxiety by signaling that the world is unpredictable and dangerous.

Parental anxiety does not remain confined to individual experience but permeates family interactions, shaping the emotional climate and relational patterns (Monaghan et al., 2009).

Taken together, guilt, attachment insecurity, and hypervigilance constitute a psychological triad that can entrench anxiety-related behavior in caregiving parents. These mechanisms not only influence parental well-being but also shape family functioning and the child's capacity to internalize effective self-care habits. Integrating a developmental and psychodynamic lens into pediatric diabetes care, acknowledging these underlying processes, is essential for designing more comprehensive, psychologically attuned interventions that support both the parent and the child.

Neuroendocrine correlates of parental anxiety: hormones and stress pathways

Parental anxiety in the management of chronic conditions such as pediatric diabetes is not only expressed through hypercontrolling or avoidant behaviors but is closely intertwined with neuroendocrine alterations reflecting prolonged activation of stress systems.

Physiologically, the hypothalamic-pituitary-adrenal (HPA) axis represents the primary response pathway (Fig. 16.3): in the presence of anxiety-provoking stimuli, the hypothalamus secretes corticotropin-releasing hormone (CRH), which stimulates the anterior pituitary to release adrenocorticotropic hormone (ACTH), leading to cortisol secretion from the adrenal cortex. In parents chronically preoccupied with their child's health, cortisol may show dysregulated patterns, such as elevated basal levels or a flattened diurnal rhythm, both associated with chronic stress and burnout (Gunnar & Quevedo, 2007). Beyond its metabolic and anti-inflammatory functions, cortisol interacts with brain structures involved in threat processing-amygdala, hippocampus, and prefrontal cortex-thereby amplifying anxious reactivity and shaping memory for stressful events (Lupien et al., 2009).

Alongside the HPA system, the sympathetic-adrenomedullary (SAM) system constitutes another key circuit. Sympathetic activation triggers the release of catecholamines (adrenaline and noradrenaline) from the adrenal medulla, generating tachycardia, hypervigilance, and motor readiness typical of the anxious response. In parents under constant alert-e.g., anticipating possible nocturnal hypoglycemia-chronic sympathetic activation may foster sleep disturbances, irritability, and heightened threat perception even in neutral contexts (Chida & Hamer, 2008).

These endocrine and autonomic responses are inseparable from psychological dimensions: cognitive appraisals of risk and responsibility fuel perceived stress, which in turn reinforces biological activation through feedback loops. Evidence further shows that HPA dysregulation in anxious parents is associated with behavioral markers of overprotection and control, suggesting that family emotional dynamics share a neuroendocrine substrate (Laurent et al., 2011).

Interventions targeting emotion regulation, particularly mind-body therapies such as mindfulness, Brain Wave Modulation Technique and relaxation techniques, have been linked to reductions in salivary cortisol and improvements in heart rate variability, a marker of autonomic resilience (Borgese et al., 2025; Cozzolino et al., 2021; Creswell & Lindsay, 2014).

These findings confirm that neuroendocrine correlates should not be conceptualized merely as biological consequences of stress, but as dynamic mechanisms mediating the impact of parental anxiety on behavior and well-being.

Finally, sustained neuroendocrine activation may have intergenerational repercussions: a family climate characterized by hypervigilance and tension also modulates the child's physiological responses, contributing to dysfunctional stress co-regulation (Hostinar & Gunnar, 2015). Understanding these correlates requires integrating psychological, neurobiological, and family perspectives.

Other Hormones, Endocrine Tissues, and Pathways: Translational aspects related to the neuroendocrine and psychophysiological framework of parental anxiety.

Building upon the neuroendocrine mechanisms previously discussed, it is essential to examine how additional physiological systems interact to sustain or attenuate stress reactivity within caregiving contexts.

We highlighted that the hypothalamic-pituitary-adrenal (HPA) and sympathetic-adrenomedullary (SAM) axes represent the main biological correlates of parental anxiety. However, a broader network of hormonal and neurophysiological processes contributes to the modulation of chronic stress responses in caregivers.

The autonomic nervous system (ANS) represents a key regulatory interface between endocrine activity and emotional functioning. Through the coordinated action of its sympathetic and parasympathetic branches, the ANS governs cardiovascular, respiratory, and visceral adjustments to stress, thereby influencing both arousal and recovery processes. Autonomic imbalance-commonly reflected in reduced heart rate variability (HRV) and diminished vagal tone-has been consistently associated with sustained hyperarousal, emotional dysregulation, and vulnerability to anxiety symptoms among caregivers (Thayer & Lane, 2000). Consequently, autonomic flexibility has been proposed as a biological marker of resilience and a potential therapeutic target for interventions aimed at restoring psychophysiological homeostasis.

A crucial role is exerted by several neuropeptides, including vasopressin and oxytocin, which serve as central modulators of social bonding, caregiving motivation, and the regulation of stress-related responses. Elevated oxytocin levels have been associated with enhanced empathy and nurturing behaviors, whereas chronic stress and emotional exhaustion may blunt oxytocinergic functioning, thereby weakening stress-buffering mechanisms and emotional attunement between parent and child (Carter, 2014). In caregivers exhibiting heightened anxiety and hypervigilance, reduced oxytocin responsiveness may exacerbate difficulties in emotion regulation and sensitivity to the child's signals, reinforcing maladaptive patterns of overcontrol and intrusive caregiving. Conversely, interventions that promote emotional safety and interpersonal connection-such as partner support or mindfulness-based relational programs-may facilitate oxytocin release and foster dyadic co-regulation.

Vasopressin, frequently co-regulated with oxytocin, exerts complementary effects on stress and vigilance. Animal and human studies indicate that elevated vasopressin levels intensify defensive and protective behaviors, particularly under conditions of perceived threat to offspring. Within the parental anxiety model, dysregulated vasopressin signaling may thus contribute to persistent hyperarousal and excessive protective drive, mirroring patterns of overmonitoring and difficulty delegating care (Donaldson & Young, 2020).

Interactions between the hypothalamic-pituitary-gonadal (HPG) axis and stress-related pathways further modulate emotional reactivity. Fluctuations in sex hormones such as estrogen, progesterone, and testosterone influence stress sensitivity, affective tone, and cognitive control. In mothers, estrogen deficiency-whether stress-induced or postpartum-has been linked to anxiety and depressive symptoms via reduced serotonergic transmission and altered glucocorticoid feedback (Albert et al., 2015). Similarly, low testosterone levels in fathers have been associated with empathic distress and emotional fatigue (Gettler, 2016; Porges, 2007), potentially amplifying caregiving strain.

Emerging evidence also highlights the regulatory influence of the gut-brain axis and peripheral endocrine tissues in shaping stress responsivity. Microbiota-derived metabolites, including short-chain fatty acids and tryptophan derivatives, affect systemic inflammation, cortisol regulation, and mood (Cryan et al., 2019). Chronic anxiety and sleep disturbance-common in parents of children with chronic illness-may disrupt gut microbiota composition, influencing emotional stability and immune function. Moreover, metabolic hormones such as insulin, leptin, and ghrelin contribute to emotional regulation and energy balance. Insulin resistance, often linked to stress and sleep deprivation, impairs cognitive flexibility and mood (McEwen, 2007), while leptin and ghrelin differentially modulate anxiety, arousal, and reward-related processes.

Taken together, these interconnected systems demonstrate that parental anxiety in the context of pediatric chronic illness arises from a dynamic integration of stress, metabolic, reproductive, and affiliative mechanisms. Autonomic regulation serves as a central bridge linking neuroendocrine activation with physiological expression. Interventions that enhance vagal tone-such as mindfulness, breathing regulation, or HRV biofeedback-alongside approaches targeting sleep, nutrition, and emotional connection, may help recalibrate endocrine homeostasis and promote psychophysiological resilience.

Protective factors and adaptive coping strategies in caregiving parents

Parenting a child with type 1 diabetes presents continuous psychological, emotional, and logistical demands that can significantly impact parental mental health. Elevated levels of anxiety, hypervigilance, and even symptoms of post-traumatic stress have been frequently reported among parents of children with type 1 diabetes, particularly in the early years following diagnosis (Landolt et al., 2002; Streisand et al., 2005). However, a growing body of research highlights the presence of protective factors and adaptive coping strategies that can mitigate the psychological burden and promote resilience in caregiving parents.

Perceived social support, from family members, peers, and healthcare professionals, is one of the most robust protective factors identified in the literature. Supportive networks can buffer stress, reduce feelings of isolation, and enhance problem-solving abilities (Hilliard et al., 2012). Parental self-efficacy, defined as the belief in one's ability to manage the illness-related demands of caregiving, is another key element associated with lower anxiety levels and improved family functioning (Streisand et al., 2008; Carosi Arcangeli et al., 2025). Parents who feel competent in managing insulin administration, carbohydrate counting, and glycemic monitoring are more likely to adopt proactive and adaptive responses to daily challenges.

Adaptive coping strategies include cognitive reappraisal, planning, emotional regulation, and the integration of illness into a coherent family narrative (McDonnell et al., 2019). For example, parents who reframe diabetes as a manageable condition rather than a constant threat tend to experience lower psychological distress. Emotion-focused strategies such as mindfulness, acceptance-based approaches, and guided relaxation have been shown to reduce stress biomarkers (e.g., cortisol) and improve parental well-being (Serlachius et al., 2014). Importantly, these strategies appear more effective when they are flexible and context-sensitive, allowing parents to alternate between problem-solving and emotional regulation depending on the situation (Compas et al., 2012).

Family-level protective factors also play a significant role. High family cohesion, effective communication, and shared responsibility for diabetes care among caregivers and co-parents contribute to reduced caregiver burnout and improved glycemic control in children (Lewin et al., 2006). Furthermore, the preservation of personal identity outside the caregiving role, such as maintaining social activities, work, or personal interests, emerges as a critical buffer against caregiver fatigue and emotional depletion (Haugstvedt et al., 2011).

In summary (Fig. 16.4), while caregiving for a child with diabetes is inherently demanding, the presence and reinforcement of specific protective factors, such as social support, self-efficacy, adaptive coping styles, and family cohesion, can significantly reduce anxiety-related behaviors in parents. Promoting these resources through psychoeducational programs, family therapy, and structured parental support interventions may enhance not only parental mental health but also the overall quality of care provided to the child.

Psychological interventions targeting parental anxiety

Raising a child with type 1 diabetes exposes parents to a high risk of psychological distress, even in the long term (Simpson et al., 2021; Whittemore et al., 2012), with particular emphasis on anxiety, which is closely linked to the constant perception of threat and the need for continuous disease monitoring. The literature highlights that such a condition, when prolonged, may lead to parental burnout, characterized by chronic stress, emotional fatigue, and negative consequences for the caregivers' physical and psychological health (Carosi Arcangeli et al., 2024; Vayisoglu et al., 2025). Research on psychological interventions aimed at caregivers and families of children with type 1 diabetes remains relatively limited; however, the available evidence suggests that such programs-despite their heterogeneous approaches and methodologies-may provide clinically relevant benefits. Interventions, both individual and family-based, range from counseling to cognitive-behavioral therapy (CBT), and often include specific techniques such as relaxation, goal setting, and problem solving (Wakelin et al., 2025). Data indicate that, even when effect sizes are modest, it is possible to achieve reductions in psychological distress among parents and children, with favorable repercussions also on metabolic control. A recurring theme is the central role of parental well-being, which emerges as a key determinant of children's psychological adjustment and overall health (Wakelin et al., 2025). The DELFIN program demonstrated positive effects on parenting style, reduction of children's behavioral problems, and stability of glycemic control, while more targeted behavioral approaches promoted greater child collaboration in disease management (Lohan et al., 2015). The First STEPS project, in turn, was perceived as highly acceptable by parents, who reported improvements particularly in emotional well-being and a reduction in depressive symptoms, largely due to the perceived support and the opportunity to share experiences with other caregivers and professionals (Baudino et al., 2023). A meta-analysis confirmed the effectiveness of parent-focused interventions in reducing stress and depression among caregivers, although consistent improvements in anxiety and specific worries were not always observed. The discrepancies found appear to be related both to the assessment tools employed and to sample characteristics, such as children's age (Zhao et al., 2019). The heterogeneity of interventions reflects the complexity of everyday life with pediatric type 1 diabetes and the diverse challenges faced by each family. This diversification, while testifying to the richness of clinical practices, also complicates the identification of a uniform intervention model, hindering systematic comparisons of results and the development of shared guidelines. Positive outcomes are often not maintained consistently over time, owing both to the limited number of controlled studies and the substantial heterogeneity of techniques, which complicates the identification of the most effective approaches (Wakelin et al., 2025). In this context, the need becomes evident for programs that are both accessible and personalized: accessible, to ensure feasibility in terms of costs, time, and delivery methods; personalized, to respond to the specific needs of each family unit. This must be accompanied by a solid methodological framework, based on rigorous studies and large samples, to generate reliable evidence. Only in this way will it be possible to develop interventions that are effective, transferable to clinical practice, and truly useful in supporting both parents and children.

Applications of the material: how translate chapter's findings into research, clinical practice, and psychosocial intervention

The theoretical and empirical findings discussed throughout this chapter have direct translational implications for both clinical practice and health policy. Understanding parental anxiety as a biopsychosocial and neuroendocrine phenomenon allows for a redefinition of intervention targets-from symptom reduction to systemic regulation and empowerment (McEwen & Gianaros, 2010; Ryan & Deci, 2017).

Clinical and psychotherapeutic applications

Integrating psychophysiological markers into clinical assessment enables healthcare providers to tailor interventions to the specific stress profiles of parents. For instance, elevated basal cortisol or blunted diurnal variation may identify caregivers at risk of chronic stress or burnout, guiding referrals for stress management or psychological support (Laurent et al., 2011; Miller et al., 2007). Interventions such as CBT, Acceptance and Commitment Therapy (ACT), mind-body therapies such as Brain-Wave Modulation Technique, and mindfulness-based stress reduction (MBSR) can reduce perceived threat and normalize autonomic activity, leading to improved parental emotional regulation and more balanced caregiving behaviors (Baer, 2003; Borgese et al., 2025; Cozzolino et al., 2021; Grossman et al., 2004). Furthermore, biofeedback and HRV training allow parents to directly monitor and modulate physiological arousal, translating neurobiological insights into daily self-regulation tools (Thayer et al., 2012; Thayer & Lane, 2000).

Family and community-based interventions

Given the interdependence between parent and child emotional states, family-centered approaches are critical. Programs emphasizing supportive supervision rather than control, joint problem-solving, and structured autonomy transfer help to dismantle anxiety-driven dynamics (Anderson et al., 2009; Hilliard et al., 2016). Community-based support-through school health plans, peer groups, or telehealth platforms-can reduce parental hypervigilance and isolation while promoting confidence in the child's self-care abilities (Walker et al., 2024; Whittemore et al., 2012). Translationally, this means embedding psychoeducational modules into standard diabetes education curricula, ensuring that emotional well-being is treated as an integral component of disease management (Delamater et al., 2018).

Digital health and translational research

Advances in digital health technologies, including continuous glucose monitoring (CGM) systems and wearable biosensors, provide opportunities for integrating physiological data with psychological monitoring. Combining glucose trends with stress indicators (e. g., HRV, sleep quality) can help parents and clinicians identify the reciprocal influence between anxiety and glycemic control (Walker et al., 2024). Translational research could leverage such data to develop predictive algorithms or adaptive interventions that alert caregivers when anxiety-driven behaviors (e.g., overchecking) emerge, allowing for timely behavioral adjustments (Zhao et al., 2019).

Educational and professional development implications

Training healthcare providers in the psychoneuroendocrinology of stress and caregiving enhances empathic understanding and reduces stigma around parental anxiety (Hostinar & Gunnar, 2015). Workshops on communication, emotional regulation, and interdisciplinary collaboration empower teams to deliver more coherent and compassionate care (Young-Hyman et al., 2016). Educational materials should also incorporate translational findings about hormones such as oxytocin and cortisol to help families understand the biological underpinnings of stress reactions, fostering self-compassion and adherence (Carter, 2014; Cozzolino, Girelli et al., 2020; Cozzolino, Vivo et al., 2020).

Future directions

Translational research should continue to explore integrated interventions that are two-way, top down and bottom up—such as whether reductions in parental anxiety predict normalization of cortisol rhythms or improved child glycemic variability (Borgese et al., 2025; Hostinar & Gunnar, 2015). Longitudinal and multi-method approaches (combining qualitative narratives with physiological measures) will be essential to bridge laboratory insights and everyday experience (Simpson et al., 2021; Whittemore et al., 2012). Moreover, integrating lifestyle medicine—nutrition, exercise, and sleep optimization—into psychosocial programs could further enhance endocrine balance and resilience (Cryan et al., 2019; McEwen, 2007).

Conclusions

Parental anxiety in pediatric type 1 diabetes is a complex, multidimensional phenomenon, affecting behavioral, psychological, and physiological domains. It can create feedback loops that impact both parent well-being and child development, potentially undermining family communication, autonomy, and treatment adherence. Protective factors-such as social support, self-efficacy, flexible coping, and family cohesion-can buffer stress and foster resilience. Evidence-based interventions, including CBT, MBTs, family therapy, and psychoeducation, show promise in reducing distress and improving family functioning, though research is limited by small sample sizes and methodological heterogeneity. A bio-psycho-social, empowerment-focused approach, integrating psychosocial screening and multidisciplinary collaboration, is essential to support families. Future efforts should develop accessible, personalized interventions that enhance resilience, promote parental empowerment, and support children's autonomy and psychological well-being.

Mini-dictionary of terms

Attachment: represents the enduring emotional connection that forms between an infant (whether human or nonhuman) and a primary caregiver or parental figure.

Caregiver: any individual who offers care, supervision, and emotional or practical support to a person who cannot function independently, such as infants and individuals affected by physical or mental health conditions.

Coping Strategy: deliberate actions, behaviors, or thought processes employed to manage stressful or unpleasant situations, or to modify one's emotional response to them.

Distress Management: Distress management is a structured framework involving screening, assessment, triage, intervention, and outcome evaluation of patient distress. By leveraging patient-reported outcomes, distress management enables timely identification of distressed individuals and directs them to appropriate supportive care services for evidence-based interventions tailored to their needs.

Family Dynamics: processes through which family members shape one another's thoughts, emotions, and behaviors, as well as the transformations in family relationships that emerge from these ongoing interactions.

Hypervigilance: state of excessive alertness and sensory sensitivity, characterized by an intensified focus on potential threats or sources of danger in the environment.

Oxytocin: a peptide produced in the hypothalamus and released by the posterior pituitary, acting as a hormone and neurotransmitter; it regulates uterine contractions and lactation, supports social bonding, parental and sexual behaviors, pair-bond formation, social recognition, and modulates stress.

Psychoeducation: an educational approach aimed at providing individuals with psychological disorders and their families with information about the disorder, its symptoms, causes, medication effects, and management strategies.

Self-Efficacy: an individual's belief in their ability to succeed in specific tasks or achieve desired outcomes, considered by Bandura a key factor influencing emotions, motivation, and behavior.

Summary points

- T1D diagnosis is abrupt and emotionally challenging, requiring rapid acquisition of caregiving skills.
- Parental adaptation varies: resilience and effective coping versus persistent anxiety and hypercontrol.
- Anxiety manifests behaviorally as hypervigilance, avoidance, compulsive checking, and intrusive communication.
- Psychological drivers include guilt, attachment insecurity, and hypervigilance, affecting family dynamics.
- Neuroendocrine and physiological mechanisms (HPA, SAM, ANS, oxytocin, vasopressin) mediate stress responses and caregiving behaviors.
- Protective factors and interventions-social support, self-efficacy, adaptive coping, CBT, mindfulness, family-centered programs-promote parental well-being and child self-management.

References

Albert, K., Pruessner, J., & Newhouse, P. (2015). Estradiol levels modulate brain activity and negative affect in postmenopausal women. *Neuropsychopharmacology, 40*(6), 1390–1397.

Anderson, B. J., Holmbeck, G., Iannotti, R. J., McKay, S. V., Lochrie, A., Volkening, L. K., & Laffel, L. M. B. (2009). Dyadic measures of the parent-child relationship during the transition to adolescence and glycemic control in children with type 1 diabetes. *Families, Systems, & Health, 27*(2), 141–152.

Baer, R. A. (2003). Mindfulness training as a clinical intervention: A conceptual and empirical review. *Clinical Psychology: Science and Practice, 10*(2), 125–143.

Baudino, M. N., Inverso, H., Wang, C., Tully, C., Levy, W., Clary, L., Gallagher, K. A., Monaghan, M., Streisand, R., & Hilliard, M. E. (2023). Satisfaction with participation in the first STEPS behavioral intervention: Experiences of parents of young children with newly diagnosed type 1 diabetes. *Journal of Pediatric Psychology, 48*(7), 605–613.

Borgese, M., Tinella, L., Cozzolino, M., & Celia, G. (2025). The impact of the Brain Wave Modulation Technique (BMW-T) on heart rate variability: A single session for short-term stress reduction. *Journal of Clinical Medicine, 14*(3), 715.

Carlsund, Å., Jutengren, G., Anderzén, I., & Jansson, A. (2025). Stress and burden experienced by parents of children with type 1 diabetes. *Children, 12*(8), 984.

Carosi Arcangeli, I., Celia, G., Girelli, L., Fioretti, C., & Cozzolino, M. (2024). A study on the sense of parental competence, health locus of control and levels of anxiety, depression and somatization in parents of children with type 1 diabetes: Evidence on a possible relationship. *Journal of Clinical Medicine, 13*(8), 2259.

Carosi Arcangeli, I., Ciavatta, V., & Celia, G. (2025). Self-efficacy among parents of children and adolescents with type 1 diabetes: A systematic review. *Diabetes Research and Clinical Practice, 229*, 112939.

Carter, C. S. (2014). Oxytocin pathways and the evolution of human behavior. *Annual Review of Psychology, 65*, 17–39.

Chida, Y., & Hamer, M. (2008). Chronic psychosocial factors and acute physiological responses to laboratory-induced stress. *Biological Psychology, 77*(1), 1–16.

Compas, B. E., Jaser, S. S., Dunn, M. J., & Rodriguez, E. M. (2012). Coping with chronic illness in childhood and adolescence. *Annual Review of Clinical Psychology, 8*, 455–480.

Cozzolino, M., Celia, G., Girelli, L., & Limone, P. (2021). Effects of the Brain Wave Modulation Technique administered online on stress, anxiety, global distress, and affect during the first wave of the COVID-19 pandemic: A randomized clinical trial. *Frontiers in Psychology, 12*, 648978.

Cozzolino, M., Girelli, L., Vivo, D. R., Limone, P., & Celia, G. (2020). A mind-body intervention for stress reduction as an adjunct to an information session on stress management in university students. *Brain and Behavior, 10*(6).

Cozzolino, M., Vivo, D. R., Girelli, L., Limone, P., & Celia, G. (2020). The evaluation of a mind-body intervention (MBT-T) for stress reduction in academic settings: A pilot study. *Behavioral Sciences, 10*(8), 124.

Creswell, J. D., & Lindsay, E. K. (2014). How does mindfulness training affect health? A mindfulness stress buffering account. *Psychological Science, 25*(7), 1133–1156.

Cryan, J. F., O'Riordan, K. J., Cowan, C. S., Sandhu, K. V., Bastiaanssen, T. F., Boehme, M., ... Dinan, T. G. (2019). The microbiota-gut-brain axis. *Physiological Reviews, 99*(4), 1877–2013.

Delamater, A. M., de Wit, M., McDarby, V., Malik, J. A., Hilliard, M. E., Northam, E., & Acerini, C. L. (2018). Psychological care of children and adolescents with type 1 diabetes. ISPAD Clinical Practice Consensus Guidelines. *Pediatric Diabetes, 19*(Suppl. 27), 237–249.

Donaldson, Z. R., & Young, L. J. (2020). The neurobiology and genetics of social behavior. *Science, 369*(6506), eaax9559.

Gettler, L. T. (2016). Becoming DADS: Considering the role of testosterone in nurturing human fathers. *American Journal of Human Biology, 28*(3), 317–330.

Grossman, P., Niemann, L., Schmidt, S., & Walach, H. (2004). Mindfulness-based stress reduction and health benefits: A meta-analysis. *Journal of Psychosomatic Research, 57*(1), 35–43.

Gunnar, M. R., & Quevedo, K. (2007). The neurobiology of stress and development. *Annual Review of Psychology, 58*, 145–173.

Haugstvedt, A., Wentzel-Larsen, T., Rokne, B., & Graue, M. (2011). Perceived family burden and emotional distress: Similarities and differences between mothers and fathers of children with type 1 diabetes in a population-based study. *Pediatric Diabetes, 12*(2), 107–114.

Hilliard, et al. (2012). Diabetes resilience: A model of risk and protection in type 1 diabetes. *Current Diabetes Reports, 12*(6), 739–748. https://doi.org/10.1007/s11892-012-0314-3

Hilliard, et al. (2016). Evidence-based behavioral interventions to promote diabetes management in children, adolescents, and families. *The American Psychologist, 71*(7), 590–601. https://doi.org/10.1037/a0040359

Hilliard, M. E., Holmes, C. S., Chen, R., Maher, K., & Streisand, R. (2011). Disentangling the roles of parental monitoring and family conflict in adolescents' management of type 1 diabetes. *Health Psychology, 30*(6), 597–604.

Hilliard, M. E., Monaghan, M., Cogen, F. R., & Streisand, R. (2011). Parent stress and child behaviour among young children with type 1 diabetes. Child: Care. *Health and Development, 37*(2), 224–232.

Hostinar, C. E., & Gunnar, M. R. (2015). Social support can buffer against stress in children. *Development and Psychopathology, 27*(2), 1–15.

Jaser, S. S., & Whittemore, R. (2014). Parent stress and youth diabetes management. *Current Diabetes Reports, 14*(1), 1–8.

Jespersen, L. N., Mikkelsen, K. Z., & Grabowski, D. (2023). "It's all about calculations-but there are no definite results"-parental adaptation and coping during the first month following their child's diabetes diagnosis. *Healthcare, 11*(2), 280.

Landolt, M. A., Ribi, K., Laimbacher, J., Vollrath, M., Gnehm, H. E., & Sennhauser, F. H. (2002). Posttraumatic stress disorder in parents of children with newly diagnosed type 1 diabetes. *Journal of Pediatric Psychology, 27*(7), 647–652.

Laurent, H. K., Ablow, J. C., & Measelle, J. (2011). Risky shifts: How the timing and course of mothers' depressive symptoms shape children's cortisol functioning. *Developmental Psychobiology, 53*(2), 217–228.

Lewin, A. B., Storch, E. A., Williams, L. B., Greco, L. A., Geffken, G. R., & Silverstein, J. H. (2006). Brief report: Diabetes-related family conflict and psychological distress in pediatric type 1 diabetes. *Journal of Pediatric Psychology, 31*(10), 977–981.

Lohan, A., Morawska, A., & Mitchell, A. (2015). A systematic review of parenting interventions for parents of children with type 1 diabetes. *Child: Care, Health and Development, 41*(6), 803–817.

Lupien, S. J., McEwen, B. S., Gunnar, M. R., & Heim, C. (2009). Effects of stress throughout the lifespan on the brain, behaviour and cognition. *Nature Reviews Neuroscience, 10*(6), 434–445.

McDonnell, C. M., Northam, E. A., Donath, S. M., Werther, G. A., & Cameron, F. J. (2019). A prospective study of depressive symptoms, quality of life and psychological adjustment in adolescents with type 1 diabetes. *Pediatric Diabetes, 20*(7), 937–945.

McEwen, B. S. (2007). Physiology and neurobiology of stress and adaptation: Central role of the brain. *Physiological Reviews, 87*(3), 873–904.

McEwen, B. S., & Gianaros, P. J. (2010). Central role of the brain in stress and adaptation: Links to socioeconomic status, health, and disease. *Annals of the New York Academy of Sciences, 1186*(1), 190–222.

Mikulincer, M., & Shaver, P. R. (2007). *Attachment in adulthood: Structure, dynamics, and change*. New York, NY: Guilford Press.

Miller, G. E., Chen, E., & Zhou, E. S. (2007). Chronic stress and HPA axis regulation: A meta-analysis. *Psychological Bulletin, 133*(1), 25–45.

Monaghan, M., Hilliard, M. E., Cogen, F. R., & Streisand, R. (2009). Nighttime caregiving behaviors among parents of young children with type 1 diabetes: Associations with illness characteristics and parent functioning. *Families, Systems, & Health, 27*(1), 28–38.

Patton, S. R., Dolan, L. M., Henry, R., & Powers, S. W. (2008). Parental fear of hypoglycemia: Young children treated with continuous subcutaneous insulin infusion. *Diabetic Medicine, 25*(10), 1127–1132.

Porges, S. W. (2007). The polyvagal perspective. *Biological Psychology, 74*(2), 116–143.

Ryan, R. M., & Deci, E. L. (2017). *Self-determination theory: Basic psychological needs in motivation, development, and wellness*. New York, NY: Guilford Press.

Schwarzer, R., Knoll, N., & Rieckmann, N. (2004). Social support. In A. Kaptein, & J. Weinman (Eds.), *Health psychology* (pp. 158–181). Blackwell Publishing.

Serlachius, A., Northam, E., Frydenberg, E., Cameron, F., & Court, J. (2014). Adaptive coping strategies in adolescents with type 1 diabetes: Relationships with continuous glucose monitoring and glycemic control. *Journal of Pediatric Psychology, 39*(10), 1112–1121.

Simpson, S., Hanna, P., & Jones, C. J. (2021). A systematic review of parents' experiences of raising a child with type 1 diabetes. *British Journal of Diabetes, 21*(1), 43–58.

Streisand, R., Mackey, E. R., & Herge, W. (2008). Associations of parent coping, stress, and family functioning in pediatric type 1 diabetes. *Journal of Child and Family Studies, 17*(1), 81–90.

Streisand, R., & Monaghan, M. (2014). Young children with type 1 diabetes: Challenges, research, and future directions. *Current Diabetes Reports, 14*(9), 520.

Streisand, R., Swift, E., Wickmark, T., Chen, R., & Holmes, C. S. (2005). Pediatric parenting stress among parents of children with type 1 diabetes: The role of self-efficacy, responsibility, and fear. *Journal of Pediatric Psychology, 30*(6), 513–521.

Sullivan-Bolyai, S., Grey, M., Deatrick, J., Gruppuso, P., Giraitis, P., & Tamborlane, W. (2003). Helping other mothers and fathers of young children with type 1 diabetes: A substudy of the parent-to-parent support program. *Diabetes Educator, 29*(5), 616–622.

Thayer, J. F., Åhs, F., Fredrikson, M., Sollers, J. J., & Wager, T. D. (2012). A meta-analysis of heart rate variability and neuroimaging studies: Implications for heart rate variability as a marker of stress and health. *Neuroscience & Biobehavioral Reviews, 36*(2), 747–756.

Thayer, J. F., & Lane, R. D. (2000). A model of neurovisceral integration in emotion regulation and dysregulation. *Journal of Affective Disorders, 61*(3), 201–216.

Vayisoglu, S. K., Koc, E., & Oncu, E. (2025). Distinguishing risk factors for depression and burnout in parents managing type 1 diabetes mellitus in children. *Journal of Pediatric Nursing, 80*, 32–40.

Villaécija, J., Luque, B., Castillo-Mayén, R., Farhane-Medina, N. Z., & Tabernero, C. (2023). Influence of family social support and diabetes self-efficacy on the emotional wellbeing of children and adolescents with type 1 diabetes: A longitudinal study. *Children, 10*(7), 1196.

Wakelin, K. E., Read, R. K., Williams, A. Y., Francois-Walcott, R. R., O'Donnell, N., Satherley, R. M., Harrington, M. P., John, M., & Jones, C. J. (2025). The effectiveness of psychological interventions for families of children with type 1 diabetes on caregiver and child functioning: A systematic review and meta-analysis. *Journal of Diabetes, 17*(6), e70112.

Walker, A. F., Hood, K. K., & Monaghan, M. (2024). Integrating psychosocial and digital interventions for parents managing type 1 diabetes in children. *Pediatric Diabetes, 25*(5), 621–630.

Whittemore, R., Jaser, S., Chao, A., Jang, M., & Grey, M. (2012). Psychological experience of parents of children with type 1 diabetes: A systematic mixed-studies review. *The Diabetes Educator, 38*(4), 562–579.

Young-Hyman, D., de Groot, M., Hill-Briggs, F., Gonzalez, J. S., Hood, K., & Peyrot, M. (2016). Psychosocial care for people with diabetes: A position statement of the American Diabetes Association. *Diabetes Care, 39*(12), 2126–2140.

Zhao, X., Ai, Z., Chen, Y., Wang, J., Zou, S., & Zheng, S. (2019). The effectiveness of parenting interventions on psychosocial adjustment in parents of children and adolescents with type 1 diabetes: A meta-analysis. *Worldviews on Evidence-Based Nursing, 16*(6), 462–469.

Chapter 17

Insulin signaling and impairments in executive functions in preclinical models: A narrative approach

Edoardo Pisa and Simone Macrì
Centre for Behavioural Sciences and Mental Health, Istituto Superiore di Sanità, Rome, Italy

Abbreviations

AD Alzheimer disease
ADHD attention-deficit/hyperactivity disorders
ASST attentional set-shifting task
CNS central nervous system
HFD high-fat diet
IR insulin receptors
IRS-1 insulin receptor substrate 1
OCD obsessive-compulsive disorder
PI3K phosphatidylinositol 3-kinase
PKB protein kinase B
STZ streptozotocin
T1D type 1 diabetes
T2D type 2 diabetes
TH TALLYHO/JngJ
ZDF Zucker diabetic fatty rat

Introduction

Historical notes

It is often said that to understand the present and predict the future, it is important to know the past well. It is therefore certainly interesting to begin this chapter by including some historical notes about the main character of this review, insulin. The term "insulin" was coined in 1910 in London, thanks to Sir Edward Albert Sharpey-Schafer. He was the first to describe the most important function of the islets of Langerhans—discovered in 1869—namely that of secreting a substance capable of controlling glucose metabolism. Drawing from the Latin term "insula" (island) he coined the term "insulin", precisely because it was produced by the islets of Langerhans. However, insulin was isolated and discovered for the first time a few years later, in 1921, thanks to Frederick Grant Banting and his assistant Charles Herbert Best, who developed a methodology to isolate insulin from the dog's pancreas. Less than a year later, this hormone was used by Banting at Toronto General Hospital on a young 14-year-old diabetic patient named Leonard Thompson. In 1923, the production of insulin by pharmaceutical companies was initiated. Also in 1923, Banting was awarded the Nobel Prize for this discovery (Vecchio et al., 2018).

The canonical hypoglycemic role of insulin

Today we know that insulin, produced by β cells found in the islets of Langerhans of the pancreas, is a peptide hormone involved in the regulation of glucose metabolism. Functionally, when blood glucose concentration rises, for example after a meal, pancreatic β cells activate a mechanism leading to the release of insulin through exocytosis. The first organ to intervene in the regulation of blood insulin concentrations is the liver: hepatocytes remove a significant portion of the hormone. Only the portion of insulin that escapes hepatic degradation reaches the systemic circulation, passing through the heart which acts as a "distribution pump" toward target tissues, such as striated muscle and adipose tissue. In these districts, insulin facilitates the entry and accumulation of glucose, which can thus be used immediately or stored as an energy reserve. Insulin action begins when the hormone binds to its receptor (IR), a transmembrane glycoprotein belonging to the receptor tyrosine kinase family. This receptor comprises four polypeptide chains connected through disulfide bonds. When insulin binds to the extracellular portion of the receptor, it stimulates enhanced kinase activity within the receptor's cytoplasmic domain, triggering autophosphorylation of multiple tyrosine residues. The activated receptor subsequently recruits and phosphorylates several target proteins on tyrosine residues. One of these is insulin receptor substrate 1 (IRS-1),

Hormones in Neuroscience. https://doi.org/10.1016/B978-0-443-30263-3.00025-7

an adaptor protein that facilitates the recruitment of additional kinases. Consequently, the signaling cascade initiated by the receptor propagates throughout the cell, generating insulin's numerous metabolic and mitogenic effects. One major signal transduction pathway involves activation of the enzyme phosphatidylinositol 3-kinase (PI3K), which binds to IRS1 through its SH2 domain and, once activated, converts the membrane lipid phosphatidylinositol 2-phosphate (PIP2) to phosphatidylinositol 3-phosphate (PIP3). PIP3 then activates protein kinase B (PKB or AKT). AKT phosphorylates glycogen synthase kinase 3 (GSK3), and when phosphorylated, GSK3 becomes inactivated, stimulating GLUT-4 translocation to the cell surface—critical for glucose utilization by skeletal muscle and adipose tissue, as shown in Fig. 17.1 (Posner, 2017; Tokarz et al., 2018).

Therefore, as is well known, insulin's primary action is fundamentally hypoglycemic. However, limiting insulin to "only" this role is extremely reductive and, in this chapter, it will be described its role beyond its hypoglycemic effect, focusing on the central nervous system.

Aim of the chapter

Given the rising prevalence of metabolic disorders associated with cognitive deficits, an increasing number of preclinical studies focused on understanding how impaired brain insulin signaling compromises executive functions. This chapter aims to connect these studies to detail the mechanistic relationship between insulin signaling impairments and executive function. Executive functions comprise complex cognitive processes including planning, impulse control, working memory and cognitive flexibility, controlled by the prefrontal cortex and fronto-striatal neural circuits (Zoratto et al., 2023), and are necessary to handle complicated situations and find effective solutions.

The role of insulin as not only a metabolic hormone, but also a neuromodulator

The first study that demonstrated the presence of insulin within the rat brain, indistinguishable from that produced in the pancreas, was conducted by Havrankova et al. (1978). This pioneering work, using multiple analytical techniques (radioimmunoassay, radioreceptor assay, bioassay, gel filtration chromatography, and immunocytochemistry), demonstrated that insulin is present in the brain in concentrations much higher than in plasma, and with properties indistinguishable from pancreatic insulin. Following the discovery of insulin in the brain, research focused on determining its source: specifically, whether it crosses the blood-brain barrier or whether it is produced within the brain itself. Although the second option might

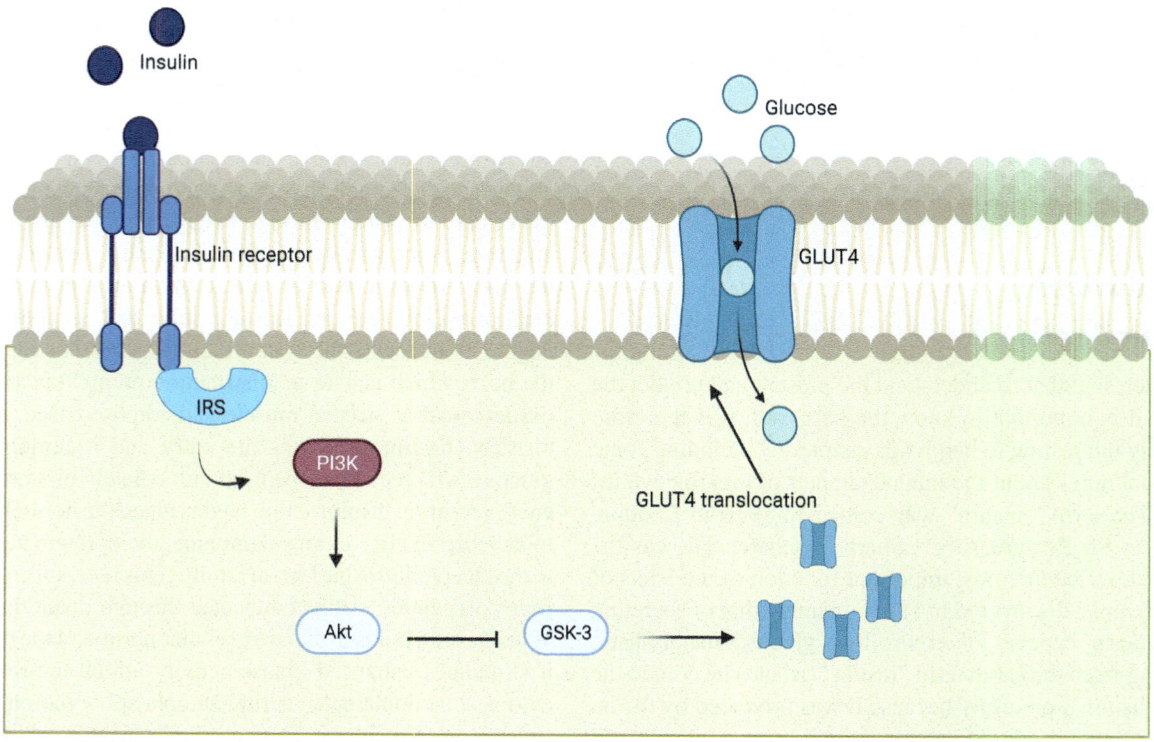

FIGURE 17.1 Schematic representation of the canonical mechanism by which insulin regulates cellular glucose homeostasis. Upon insulin binding to its receptor, a cascade involving IRS, PI3K, and Akt is activated, thus leading in the translocation of GLUT4 glucose transporters to the cell membrane. This process facilitates glucose uptake and storage.

have seemed unrealistic at the time, we now know well that both options are true.

Insulin can cross the blood-brain barrier through the capillary endothelial cells via a selective, saturable, and receptor-dependent mechanism (Banks, 2004). Once in the brain, insulin binds to the IR to form an insulin-receptor complex, which undergoes transcytosis through the brain endothelial cells. This transport process can be modulated by various factors, including high-fat diet (HFD), obesity, inflammation, and diabetes (Banks et al., 2012). All cell types in the brain express insulin receptors; however, their expression varies between different brain regions. Animal studies indicate that the highest density of IR occurs in the CA1 and CA3 regions of the hippocampus, olfactory bulbs, entorhinal cortex, and hypothalamus, as well as in the cerebral cortex, striatum, and cerebellum (Werther et al., 1987). Most receptors are localized on neurons, and are also highly expressed in presynaptic axon terminals as components of postsynaptic density. Indeed, the synapse appears to be an important site of specialized insulin signaling in the brain. The widespread distribution of IR throughout the brain supports multiple pivotal functions (Schulingkamp et al., 2000; Sędzikowska & Szablewski, 2021).

Simultaneously, studies in rodents showed that insulin mRNA existed within the central nervous system (CNS) which suggested that brain could synthesize de novo its own insulin (Marks et al., 1990; Young, 1986). Neuronal cells express insulin genes localized in olfactory and limbic system areas (Devaskar et al., 1994). Moreover, C-peptide (a metabolic product in insulin biosynthesis) is also present in some pyramidal cells of the neocortex and hippocampus (Dorn et al., 1982). Therefore, the pancreas is not the sole organ synthetizing insulin.

The fact that neurons actively synthesize insulin, rather than relying solely on circulating pancreatic insulin, demonstrates that insulin operates autonomously within the central nervous system to execute vital functions which extend beyond glucose management.

The previously discussed canonical insulin signaling cascade that controls glucose homeostasis acts as a complex regulatory system which manages both metabolic and cell growth pathways (i.e., mitogenic response). IRS-1 activation recruits both the PI3K cascade and the mitogen-activated protein kinase (MAPK) pathway, as shown in Fig. 17.2 (Rahman et al., 2021). Thus, it activates transcriptional regulators controlling gene expression for cell cycle progression, protein biosynthesis, and cellular hypertrophy (e.g., growth factor receptors and extracellular matrix, remodeling enzymes and metabolic regulators (Saltiel & Kahn, 2001; Siddle, 2011)). Then, insulin functions as CNS-modulator through receptors in pre- and postsynaptic membrane compartments (Taniguchi et al.,

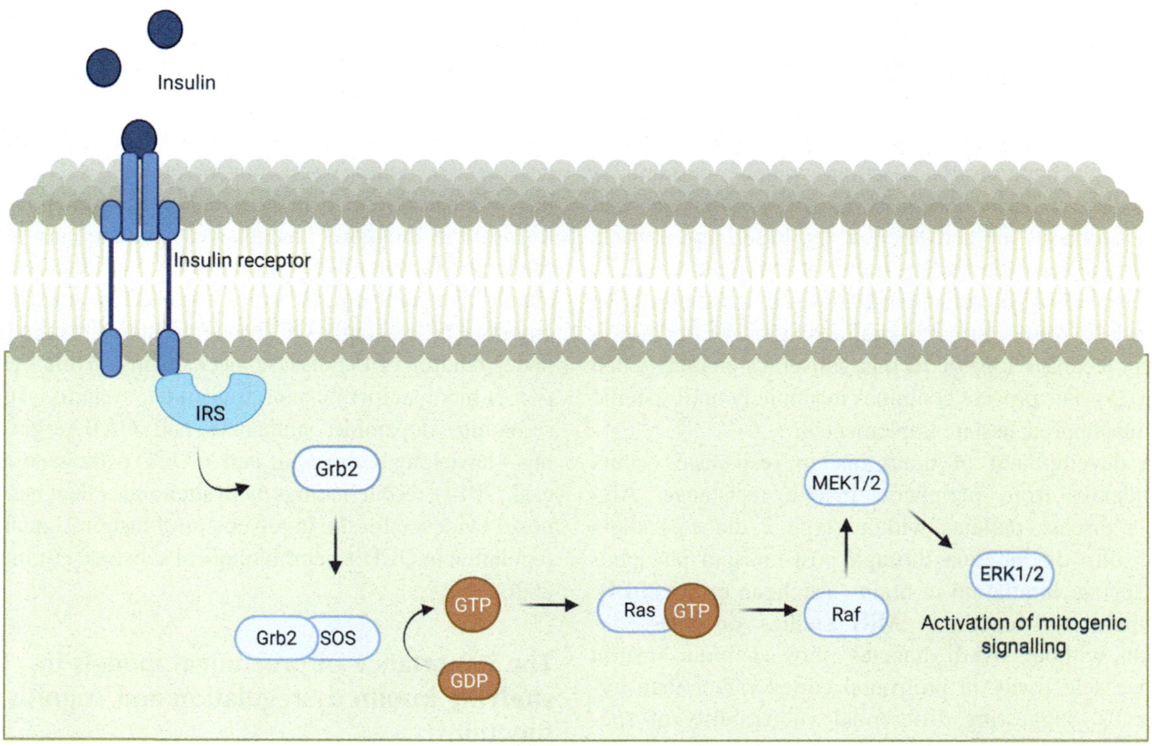

FIGURE 17.2 Schematic representation of insulin's mitogenic role. Upon insulin binding and IRS recruitment, Grb2 associates with SOS to facilitate GDP-to-GTP exchange on Ras proteins. Activated Ras recruits and phosphorylates Raf kinase, which sequentially activates MEK1/2 and ERK1/2. This pathway leads to gene expression for cell proliferation, protein synthesis, and cellular hypertrophy.

2006). Current research shows that insulin signaling influences various neuronal functions by promoting neurite growth and branching through PI3K/AKT activation which stabilizes cytoskeletal proteins including tubulin (Chiu et al., 2008). This process extends to controlling monoaminergic neurotransmitters production, packaging and ionotropic receptors trafficking (NMDA, AMPA and GABA receptor subtypes) (Beattie et al., 2000; Kleinridders et al., 2014). Insulin exerts neuroprotective effects also through survival pathway activation (Ding et al., 2006; Hooper et al., 2008), controlling GSK-3β phosphorylation to modulate tau protein phosphorylation and microtubule stability. Insulin signaling improves synaptic plasticity through NMDA receptor regulation and calcium-dependent signaling pathways linked to long-term potentiation and long-term depression (Llorens-Martín et al., 2014).

Central insulin resistance: from receptor desensitization to executive dysfunction

Brain insulin resistance develops through a gradual deterioration of insulin signaling pathways inside cells instead of being caused by insulin hormone shortages. The core molecular process involves abnormal serine phosphorylation of IRS-1 (at Ser307 in mice; Ser312 and Ser616 in humans) instead of its typical tyrosine phosphorylation sites, blocking PI3K binding and stopping signal transmission (Woo et al., 2024). The pathological process leading to this condition emerges from JNK activation which results from pro-inflammatory cytokines and endoplasmic reticulum stress (Yarchoan et al., 2014).

The inability of Akt to inhibit GSK-3β results in abnormal tau protein hyperphosphorylation which destabilizes microtubules at serine residues, creates a blockage that strongly decreases the activation of the PI3K, and disrupts fronto-striatal network synaptic transmission (Schubert et al., 2004). The lack of Akt activation leads to NF-κB suppression failure which results in microglial activation and creates a persistent feedback loop of declining insulin signaling (Chen et al., 2025). The process continues indefinitely until specific therapeutic approaches are implemented.

The development of brain insulin resistance occurs independently from peripheral insulin resistance. Alzheimer's disease patients without type 2 diabetes show brain insulin dysfunction through post-mortem phospho-IRS-1(Ser) accumulation in brain (Yarchoan et al., 2014).

Furthermore, functional MRI studies on obese individuals without overt diabetes showed brain insulin resistance selectively in prefrontal cortex rather than hypothalamus, suggesting differential vulnerability of specific circuits. This distinction is particularly relevant to executive functions, as it implies that mechanisms compromising cognitive control operate independently of peripheral metabolic defects—potentially explaining why insulin-related deficits in central executive functions may occur in the absence of hyperglycemia or peripheral metabolic dysfunction.

Insulin signaling dysfunctions in neurological and neuropsychiatric disorders

Considering the role of insulin in the CNS, it seems now clear that dysfunctions in its signaling pathways may represent a critical factor in the pathogenesis of diverse neurological and psychiatric conditions (Blázquez et al., 2014; Bosco et al., 2011; Klimova et al., 2018). This expanding body of evidence links dysregulated insulin homeostasis to the onset and progression of neurodegenerative diseases including Alzheimer's disease (AD) (Blázquez et al., 2014; Bosco et al., 2011; Klimova et al., 2018), as well as neuropsychiatric disorders such obsessive compulsive disorder (OCD) (Bralten et al., 2020; van de Vondervoort et al., 2016), mild cognitive impairment (MCI) (Biessels & Despa, 2018; Koekkoek et al., 2015), and attention deficit hyperactivity disorder (ADHD) (Chen et al., 2013). The relationship between Type 2 Diabetes (T2D) and AD has been extensively documented in contemporary literature (Blázquez et al., 2014; Bosco et al., 2011; Klimova et al., 2018). The two diseases share common pathological mechanisms: insulin resistance, inflammation, amyloid accumulation and cognitive decline. The connection between insulin resistance and cognitive decline exists across all dementia stages from mild to severe including MCI (Biessels & Despa, 2018; Koekkoek et al., 2015). Recent research has shown that insulin-dependent molecular pathways create a direct connection to obsessive-compulsive symptoms (van de Vondervoort et al., 2016). Van de Vondervoort and colleagues demonstrated that OCD pathophysiology can also stem from disrupted insulin signaling, leading to intrusive thoughts, compulsive behaviors and repetitive thinking (Goodman et al., 2014). The identification of abnormal neural circuit activity related to emotional and cognitive processing in OCD patients has not resulted in a full understanding of the disease mechanisms. While dysregulation of monoaminergic neurotransmitter systems—including serotonin, dopamine, glutamate and GABAergic signaling—have long been implicated in OCD pathogenesis (Pauls et al., 2014), recent findings by Bralten and colleagues provide novel evidence for the involvement of insulin signaling dysregulation in OCD's neurobiological substrate (Bralten et al., 2020).

The importance of preclinical models in studying insulin dysregulation and cognitive functions

The experimental use of animal models has become indispensable to help understand human body functions and disease processes and evaluate new treatments. The study of the intricate relationship between metabolic dysfunction and cognitive decline largely benefits from rodent models as useful research tools. Given T2D complexity and insulin dysregulation heterogeneity in human, animal models recapitulating key pathological features are essential for identifying the biological substrates underlying related cognitive impairment (Ottomana et al., 2023). As noted by Biessels and Despa (Biessels & Despa, 2018), "Animal models that adequately capture the heterogeneity of diabetes seen in humans are essential to uncover a pathological substrate for cognitive dysfunction and dementia in T2D." This statement establishes a basic scientific rule which states that experimental systems must be appropriate to prove the mechanistic connections between insulin signaling impairments, metabolic disturbances and cognitive impairment. The experimental findings from well-studied rodent models provided substantial information about these mechanisms. Research using APP/PS1 transgenic models which mimic Alzheimer's disease characteristics showed that also poor glycemic control may lead to cognitive deficits (Ramos-Rodriguez et al., 2016). Similarly, transgenic mice with reduced insulin receptor expression show consistent impairments in memory formation and abnormalities in the cellular mechanisms underlying synaptic plasticity (Nisticò et al., 2012). More recently, scientists have leveraged spontaneously diabetic-like mouse strains (such as TALLYHO/JngJ mice, which naturally develop late-onset hyperglycemia and insulin resistance) to establish causal relationships between disrupted insulin signaling and behavioral abnormalities including elevated compulsivity and anxiety-like responses (van de Vondervoort et al., 2019). The experimental methods deliver precise mechanistic information which enables researchers to study molecular processes like insulin receptor autophosphorylation defects and IRS-1 recruitment dysregulation and PI3K/AKT signaling irregularities that lead to specific behavioral and cognitive changes. Preclinical models allow scientists to study the effects of specific insulin pathway component manipulation through pharmacological or genetic methods which help determine if targeted insulin pathway restoration can treat or improve cognitive and emotional problems.

Rodent preclinical models used to study insulin dysfunction

The preclinical research field has leveraged multiple experimental methods to study insulin-related cognitive impairments. The most widely adopted method for pharmacological induction is streptozotocin (STZ) administration (Lenzen, 2008). STZ acts as a highly specific pancreatic β-cell cytotoxic agent which leads to β-cell death through necrosis, thereby creating a model that closely resembles type 1 diabetes (T1D) in terms of hyperglycemia and insulin deficiency. When administered in lower, multiple doses, STZ causes a gradual inflammatory process which leads to partial β-cell damage, mimicking the pathophysiology of human diabetes progression (Lenzen, 2008). Beside pharmacology, environmental manipulations are often used to induce insulin signaling impairments. HFD administration represents one of the most significant experimental methods employed (Guillemot-Legris & Muccioli, 2017). HFD exposure leads to insulin resistance through inflammation which causes fasting glucose and insulin levels to rise and glucose tolerance to decrease (Guillemot-Legris & Muccioli, 2017). This methodology successfully models the nutritional determinants of T2D. Also, genetic models are used to reproduce insulin signaling impairments. Among them, the most used are represented by the db/db mice (Zheng et al., 2017), lacking leptin receptors, the Irs2-deficient mice (Tanokashira et al., 2021) and Goto-Kakizaki rats which develop diabetes-related symptoms spontaneously (Goto et al., 1976). These models offer advantages in terms of construct validity, as they develop diabetes through mechanisms that more closely approximate human genetic predispositions to metabolic disorders. Cognitive evaluation in these models usually refers to spatial navigation tests for spatial memory (e.g., Morris Water Maze, hereafter MWM), object recognition tests for recognition memory, anxiety tests (elevated plus maze), and cognitive flexibility tests (attentional set-shifting task) that measure executive functions. The MWM measures the animal's ability to locate, in a water tank, a hidden escape box, which position was learned during an acquisition phase using visual cues. The novel object recognition memory test measures the animal's ability to discriminate a novel object from a familiar one (Hauser et al., 2021). The elevated zero-maze test measures the conflict between exploratory drive and innate fear of open spaces, through the animal's exploration of open versus closed sectors of an elevated apparatus (Hauser et al., 2021). The attentional set-shifting task evaluates individual capability to acquire a rule in order to obtain a reward, and to subsequently disregard such rule in favor of a new one, once the learning criterion has been attained by trial and error (Zoratto et al., 2023). In the following sections of the chapter, it has been described how each model replicates specific cognitive deficit profiles through these paradigms (see also Table 17.1).

Pharmacological induction: Streptozotocin-induced diabetes models

The high-dose STZ treatment (50–65 mg/kg i.p.) replicates T1D phenotype, by destroying pancreatic β-cells, thus

TABLE 17.1 Rodent models of impaired insulin signaling: metabolic, cognitive, and molecular characterization.

Model	Induction type	Metabolic impairments	Cognitive impairments	Biological features
STZ high dose	Pharmacological	T1D-like phenotype, severe hyperglycaemia, insulin deficiency, rapid ketoacidosis	Anxiety-like behavior	Pancreatic β-cells destroyed; cumulative metabolic stress
STZ low dose	Pharmacological	T2D-like phenotype, insulin deficiency, chronic hyperglycaemia	Recognition memory deficits	Slow and partial β-cell destruction
High-fat diet	Environmental	Insulin resistance, elevated fasting glucose and insulin, reduced glucose tolerance	Recognition memory and cognitive flexibility impairments, working memory deficits	Prefrontal monoamine dysregulation, mitochondrial stress, IR desensitization
db/db mice	Genetic	T2D-like phenotype, leptin receptor deficiency, obesity	Impaired cognitive flexibility (Morris WM reversal)	Alteration in energy metabolism proteins, neuroinflammation, systemic inflammation
Zucker Diabetic Fatty (ZDF) rats	Genetic	T2D-like phenotype, obesity, hyperinsulinemia, severe hypertriglyceridemia, leptin receptor mutation	Short-term and long-term memory deficits	Reduced GLUT4 expression, impaired BBB insulin transport, mitochondrial dysfunction, oxidative stress
TALLYHO/JngJ (TH) mice	Genetic	Polygenic T2D-like phenotype, spontaneous hyperglycaemia, insulin resistance	Compulsivity-like behavior (reduced spontaneous alternation), executive function deficits	White matter microstructure alterations, increased glucose in dorsomedial striatum, reduced cerebellar Igf1/increased plasma Igf1, IRS1 degradation.

resulting in severe hyperglycemia (blood glucose>250–300 mg/dL), hypoinsulinemia, and ketoacidosis. Young streptozotocin-induced diabetic rats exhibit anxiety-like behaviors (avoidance of open arms in the approach-avoidance elevated plus maze test, together with prolonged latency in the exhibition of exploratory behaviors). The development of anxiety-like behaviors in diabetic rats emerges progressively with the onset of diabetic-like phenotype, indicating that hyperglycemia and metabolic stress produce cumulative damage to amygdala-prefrontal circuits, responsible for the control of emotional responses (Rajashree et al., 2011).

The low-dose STZ treatment (typically 40–50 mg/kg administered as multiple daily injections over 5–7 consecutive days) causes pancreatic β-cell destruction at a slow rate. This approach firstly induces insulin deficiency, accompanied by chronic hyperglycemia, thereby replicating a metabolic state similar to the human T2D progression. Research using low-dose STZ protocols in mice has shown that diabetes-related cognitive decline develops steadily and directly relates to blood sugar levels. These models demonstrate that recognition memory performance – a declarative long-term memory capacity dependent on hippocampal-prefrontal circuitry – deteriorates in parallel with glycemic control. Complementarily, pancreatic tissue recovery and restoration of normal blood sugar levels lead to partial regaining of cognitive abilities, suggesting reversibility of some insulin dysregulation-induced deficits (Venuti et al., 2025).

Dietary induction: High-fat diet models

The cognitive consequences of HFD exposure are striking, particularly during prefrontal cortex maturation (e.g., adolescence). Multiple independent studies showed that adolescent mice who received HFD exhibited impairments in cognitive function such as recognition memory (Zhang et al., 2022). Exposure to HFD during adolescence also impairs executive functions. Adolescent rats exposed to HFD show significant deficits in cognitive flexibility, as observed in the MWM reversal learning task. Specifically, compared to controls, HFD rats commit more errors and require more time to reach the learning criterion when requested to switch from a previously acquired rule to a new one upon changed circumstances. These deficits emerge even after a brief HFD exposure (1–3 weeks), and in the absence of significant weight gain or overt metabolic dysfunction, suggesting that central insulin signaling dysregulation may precede substantial peripheral metabolic derangement (Cordner & Tamashiro, 2015; Labouesse et al., 2018). Moreover, the cognitive deficits that follow an HFD are not uniform but neural circuits specific. Specifically, it has been demonstrated across independent laboratories that HFD exposure impacts both hippocampus-dependent tasks—such as acquiring memory in the MWM—and prefrontal-dependent functions. This prefrontal selectivity aligns with the hypothesis that executive function circuits—especially the dorsolateral prefrontal cortex and anterior cingulate cortex—are especially vulnerable to insulin dysregulation. At a molecular level, these regions show a higher expression of IR and a marked HFD-induced desensitization of downstream signaling cascades.

The mechanistic substrate for HFD-induced prefrontal cognitive dysfunction involves monoaminergic dysregulation and mitochondrial stress. HFD exposure induces upregulation of monoamine oxidase-A and MAO-B in prefrontal cortex and striatum through reduced Akt-mediated suppression. This shift accelerates dopamine catabolism by reducing the neurotransmitter's reserve. The resulting dopamine deficit directly impairs working-memory capacity, a function that depends critically on D1 dopamine receptor signaling within the pyramidal neurons of the dorsolateral prefrontal cortex. Additionally, HFD exposure induces dysregulation of insulin-responsive transcriptional programs, leading to reduced neuroprotection and increased neuronal vulnerability to metabolic stress.

Genetic models: Leptin receptor-deficient mice (db/db) and Zucker rats

Among genetic models, db/db mice, exhibiting diabetic-like phenotype caused from an autosomal recessive point mutation in the leptin receptor gene (LepR), show impaired cognitive flexibility in the reversal learning phase of MWM. Specifically, they committed more errors during the reversal phase when the previously learned associations must be inhibited, and new stimulus-response associations need to be formed. While cognitive flexibility is compromised, their performance on spatial acquisition tasks—an index of spatial memory—remains largely intact. This dissociation suggests that insulin dysregulation selectively undermines prefrontal-mediated response flexibility and inhibitory control whereas the more hippocampal-dependent spatial memory appears comparatively resilient (Yermakov et al., 2019).

The proteomics profiling of brain tissue from db/db mice indicated a widespread alteration in energy metabolism proteins, cytoskeletal structural proteins (consistent with tau-driven microtubule disruption), and neural functioning proteins, including reduced expression of synaptic plasticity markers. Complementary plasma proteomics showed an increase in pro-inflammatory messengers, suggesting that systemic inflammation contributes to neuroinflammation that perpetuates brain insulin resistance (Ernst et al., 2013).

The Zucker diabetic fatty (ZDF) rat develops obesity, hyperinsulinemia, and hypertriglyceridemia. These metabolic impairments are caused by a missense mutation in the leptin receptor gene, leading to disrupted leptin signaling and representing a model of T2D. In hippocampal-dependent learning tasks such as the arm maze, ZDF rats show compromised performance in both short- and long-term memory (Winocur et al., 2005). At biological level, ZDF rats show reduced expression of insulin-responsive glucose transporters (GLUT4) in hippocampus, reduced blood brain barrier insulin receptor density (impairing insulin transport into brain), impaired mitochondrial function, elevated oxidative stress production, and reduced ATP content in brain tissue. When ZDF rats are treated with agents that restore insulin signaling through the IRS-1/Akt/GSK-3β pathway, both cognitive impairments and anxiety-like behaviors are mitigated, providing evidence that restoring the insulin pathway can partly mitigate cognitive function impairments (Raza et al., 2015).

Spontaneous genetic models: TALLYHO/JngJ mice

The TALLYHO/JngJ (TH) mouse represents a polygenic model of T2D that spontaneously develops hyperglycemia, hyperinsulinemia, and Langerhans' islet enlargement. Together with this T2D-phenotype, TH mice also develop compulsivity-like behavior and impaired executive functions, see Table 17.2 (Presta et al., 2024; van de Vondervoort et al., 2019).

Specifically, TH mice showed reduced spontaneous alternation in the T-maze, a measure of compulsivity-like behavior that reflects reduced exploratory flexibility and a bias toward perseverative responding. Concurrently, TH mice performed poorly in several stages of the attentional set-shifting task, aimed at measuring attentional capabilities and executive functions. Specifically, TH mice committed more errors in the: (i) compound discrimination reversal phase, which evaluates the ability to reverse learned associations; (ii) intra-dimensional shift phase, which assesses rule generalization abilities; (iii) extra-dimensional shift, which requires animals to move their attention from important features to new stimulus dimensions which were irrelevant during the task and evaluates learned irrelevance and persistent behavioral patterns. The observed impairments indicate that higher-order cognitive functions suffer from a wide-ranging disruption which affects three essential cognitive abilities: inhibitory control for appropriate motor response suppression; working memory for active information maintenance; and behavioral adaptability for environmental contingency adjustment. The wide range of executive function deficits in TH mice suggest that they have multiple prefrontal cognitive impairments instead of having a single specific executive function problem (Presta et al., 2024).

Other hormones: Incretins and GLP1-RAs

Among other important hormones, there is a category that is gaining momentum in recent years: incretins. Incretins

TABLE 17.2 Cognitive and metabolic impairments of TALLYHO/JngJ mice.

Cognitive impairment	Methods	Metabolic impairment	Methods
Compulsivity-like behavior	T-maze test: reduced spontaneous alternation	Hyperglycaemia	Fasting blood glucose measurements; glucose tolerance test
Inhibitory control deficits	Attentional set-shifting task: compound discrimination reversal phase (inability to suppress previously learned and now inappropriate responses)	Insulin resistance	Insulin sensitivity test
Working memory impairment	Attentional set-shifting task: intra-dimensional shift phase (reduced active information maintenance across task demands)	Altered lipid metabolism	Respiratory exchange ratio measurement in automatic metabolic cages
Cognitive flexibility and rule abstraction deficits	Attentional set-shifting task: intra-dimensional shift and extra-dimensional shift phases (more errors when switching from previously acquired rule to new one; impaired learned irrelevance)	Polydipsia and altered energy expenditure	Drink intake measurement; indirect calorimetry in automatic metabolic cages
Attentional deficits	Attentional set-shifting task: extra-dimensional shift phase (inability to shift attention from important stimulus dimensions to new irrelevant ones)	Moderate obesity	Body weight measurements

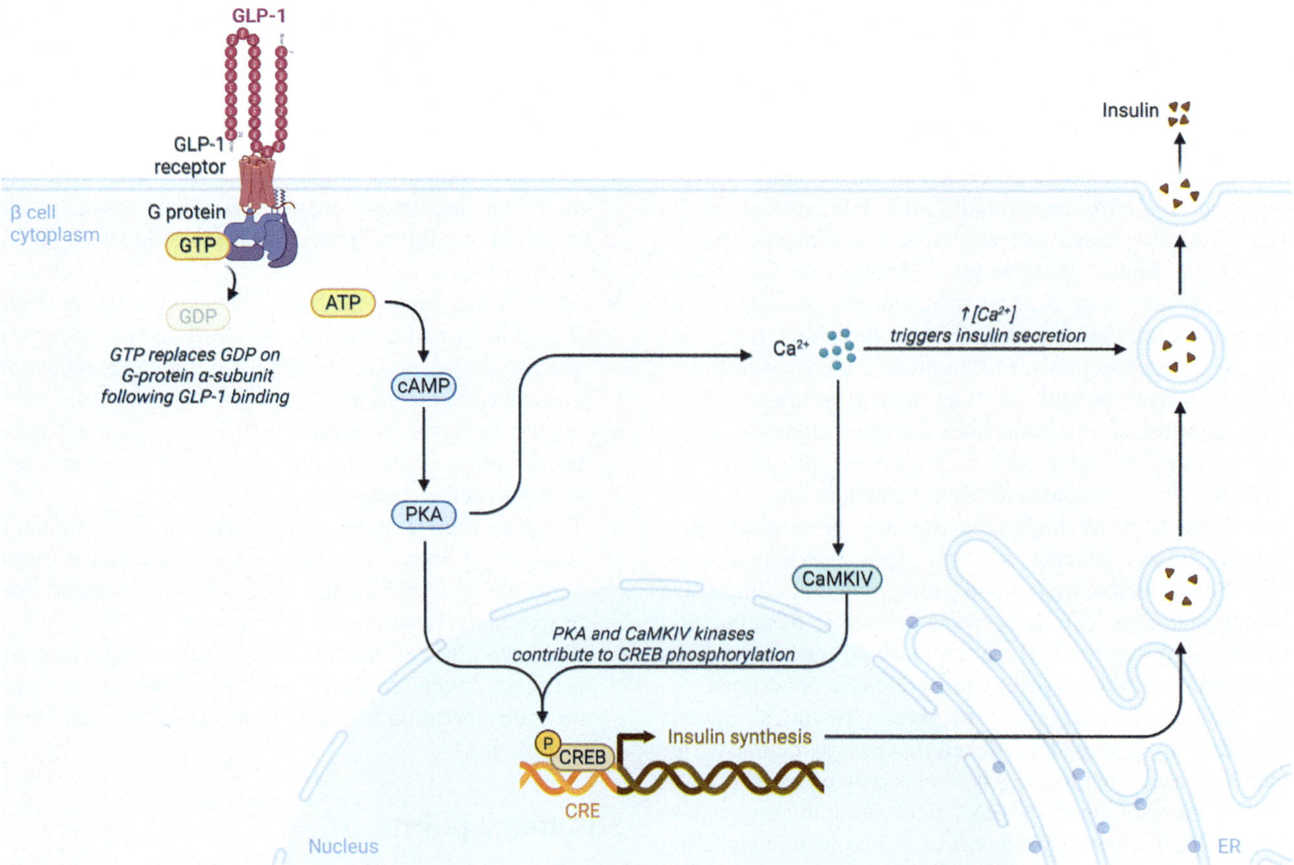

FIGURE 17.3 Schematic representation GLP-1 receptor-mediated signaling in pancreatic β-cells. Upon GLP-1 binding, the GLP-1 receptor activates associated G-proteins, triggering GDP-to-GTP exchange on the α-subunit. This leads to increased intracellular cAMP levels, activating PKA and CaMKIV kinases. Both kinases phosphorylate CREB, promoting insulin gene transcription and synthesis. Concurrently, GLP-1 signaling elevates intracellular calcium concentrations, triggering insulin secretion from the endoplasmic reticulum into the cytoplasm.

are peptide hormones produced in the intestine, exerting a key role in the regulation of metabolic homeostasis. Among incretins, glucagon-like peptide-1 (GLP-1) and glucose-dependent insulinotropic peptide (GIP) are secreted by intestinal L- and K-cells, respectively, following food ingestion. Once secreted, these hormones bind their receptors and are responsible for approximately 60%–70% of insulin secretion after glucose intake. Thus, they are key molecules in the coordination between intestine and pancreas for proper glycemic regulation (Reed et al., 2024). However, it is reductive to limit incretins' role to glucose control alone. GLP-1 receptor agonists (GLP-1RAs) can activate GLP-1 receptors, usually coupled to G-proteins. Receptor activation leads to stimulation of cellular proliferation, increased resistance to apoptosis, inhibition of glucagon's secretion, delayed gastric emptying, and reduced gastric acid secretion. Together with improved glucose-dependent insulin secretion (see Fig. 17.3), GLP-1R activation acts as a key contributor to the post-prandial homeostatic regulation of glucose metabolism (Reed et al., 2024). On the therapeutic level, the use of GLP-1RAs has extended from the treatment of T2D to weight control, with agents such as liraglutide and semaglutide widely used for these indications. Cardiovascular studies have demonstrated that GLP-1RAs treatment is able to significantly reduce adverse cardiovascular episodes and mortality in diabetic patients, by inducing pleiotropic effects including amelioration of endothelial functions and reduction of atherosclerosis (Kalayci et al., 2025). Importantly, recent studies suggest that GLP-1RAs may also exert neuroprotective effects. Epidemiological evidence indicates that diabetic patients treated with GLP-1RAs benefit from delayed progression of dementia, and clinical trials involving exenatide in Parkinson's disease have shown significant improvement in both motor and cognitive symptoms (Ghosh-Swaby et al., 2022; Monney et al., 2023). These findings pave the way for promising perspectives in the use of GLP-1RAs in neurodegenerative and potentially neuropsychiatric conditions. Nonetheless, further research is required to elucidate the underlying mechanisms and to confirm clinical efficacy in non-diabetic populations.

Applications of the material

The preclinical models described in this chapter provide important tools for understanding the link between insulin signaling disruptions and cognitive impairments. On the neurological level, these models demonstrate that insulin resistance may contribute to AD pathophysiology, via tau hyperphosphorylation, microtubule destabilization, and synaptic dysfunction. Furthermore, these preclinical models offer insight at the neuropsychiatric level, including obsessive-compulsive disorder, attention deficit hyperactivity disorder, and substance use disorders. The TH mice show compulsivity-like behaviors and perseveration which parallel OCD symptoms in humans, thus making these models suitable for studying insulin-sensitizing treatments targeting repetitive and compulsive symptomatology. Similarly, monoaminergic dysregulation in mice exposed to HFD causes working memory deficits analogous to those observed in psychiatric conditions.

The translational value of these preclinical models extends to therapeutic development, particularly for incretin-based therapies. The neuroprotective effects of GLP-1 receptor agonists (e.g., neuroinflammation reduction, synaptic plasticity enhancement and insulin signaling pathway restoration) have been confirmed through clinical studies of T2D patients and AD and PD clinical trials. The preclinical models presented in this chapter may enable researchers to develop new dual or triple agonists which target GLP-1R, GIPR, and glucagon receptors, and detail their underlying mechanisms of action before assessing their efficacy in clinical trials.

Importantly, the studies reviewed in this chapter demonstrate that early blood sugar normalization leads to the mitigation of cognitive impairments. These findings show that mitigating metabolic dysfunction at early stages can prevent permanent neuronal damage. Moreover, since different models show that peripheral and central insulin resistance develop independently, it could be important to start such interventions before people develop full metabolic syndrome.

In conclusion, this chapter offers insulin signaling as an essential target in executive functions. The evidence presented through multiple preclinical models demonstrates that brain insulin resistance is a mechanism extending beyond metabolic regulation, and that it seems to be very selective for key brain areas related to cognition (i.e., prefrontal cortex and hippocampus). The emerging therapeutic potential of incretin-based therapies underscores the urgency of early detection and treatment. Future research must continue bridging preclinical work with clinical evidence to optimize neuroprotective strategies in at-risk populations, potentially transforming our approach to cognitive preservation across the neuropsychiatric spectrum (Fig. 17.4).

Mini-dictionary of terms

- Executive functions: higher-order cognitive processes, including inhibitory control, working memory and cognitive flexibility, that control lower-level cognitive functions.
- GLP-1RAs: drugs that bind and activate the receptor for glucagon-like peptide 1, leading to improved glucose homeostasis, neuroprotection, and reduction of neuroinflammation.
- Insulin Receptor Substrate 1 (IRS-1): protein activated by IR after insulin binding, promoting metabolic and neuroprotective pathways.
- TALLYHO/JngJ: mice strain exhibiting T2D phenotype (e.g., hyperglycemia, insulin resistance, lipid metabolism impairments) together with impairments in executive functions.
- Tau: proteins of neural microtubules important for neuronal structure. Their hyperphosphorylation causes neurodegeneration and represents one of the main hallmarks of AD.

Summary points

Central insulin signaling operates as a neuromodulator controlling executive functions through PI3K/AKT/GSK-3β and MAPK cascades, with distinct molecular mechanisms independent of peripheral metabolic homeostasis.

Brain insulin resistance causes abnormal IRS-1 serine phosphorylation and subsequent tau hyperphosphorylation, microtubule destabilization, and synaptic dysfunction, replicating Alzheimer's disease pathophysiology.

The prefrontal cortex demonstrates selective vulnerability to insulin dysregulation, resulting in specific impairments in cognitive flexibility, working memory, and impulse control.

Multiple preclinical models of metabolic impairments demonstrate that insulin signaling disruption leads to executive function impairments and compulsivity-like behavior, thus mimicking human neuropsychiatric symptoms.

GLP-1 receptor agonists exert neuroprotective effects through restoration of insulin signaling pathways and reduction of neuroinflammation, with emerging clinical evidence supporting cognitive benefits in type 2 diabetes, Alzheimer's disease, and Parkinson's disease populations.

Use of AI and AI-assisted technologies

During the preparation of this work, the authors used Perplexity to improve the clarity, coherence, and

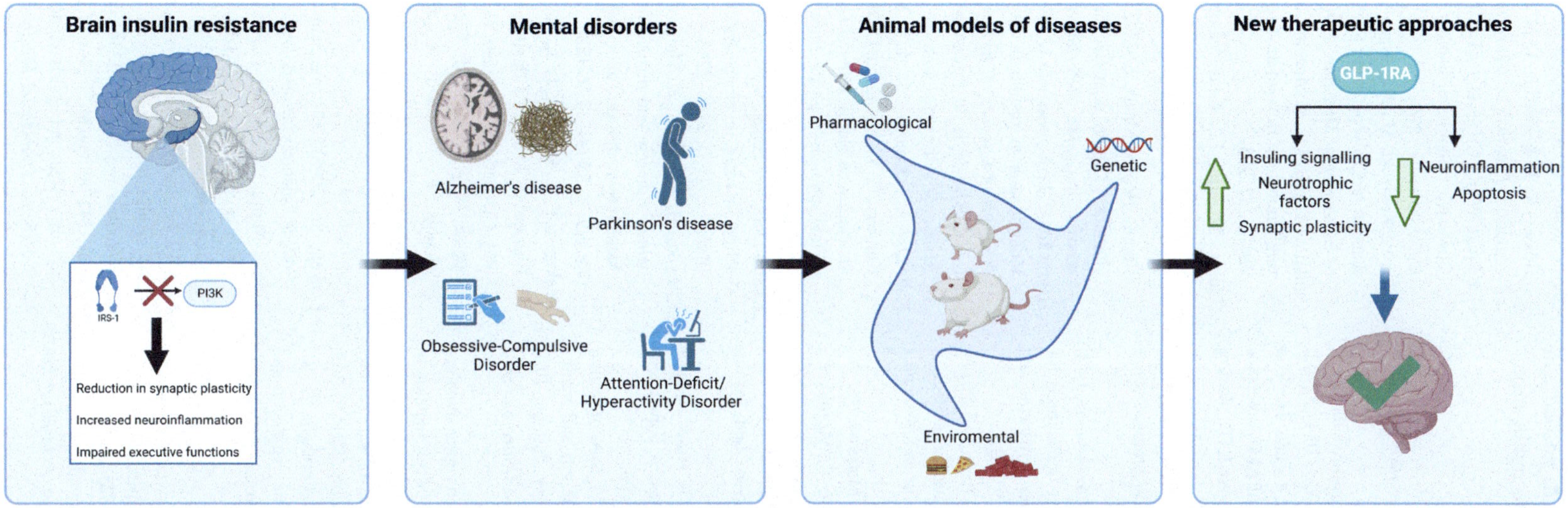

FIGURE 17.4 Brain insulin resistance represents a critical pathogenic mechanism underlying executive dysfunction and neuropsychiatric disorders. Central insulin signaling impairments, driven by aberrant IRS-1 and PI3K dysregulation, lead to reduced synaptic plasticity, increased neuroinflammation, and neuronal apoptosis. Preclinical models—encompassing pharmacological, environmental, and genetic inductions—recapitulate these molecular alterations and associated cognitive phenotypes, including deficits in executive functions and compulsivity-like behaviors observed in Alzheimer's disease, Parkinson's disease, obsessive-compulsive disorder, and attention-deficit/hyperactivity disorder. GLP-1 receptor agonists restore insulin signaling through multiple mechanisms (neurotrophic factor enhancement, neuroinflammation reduction, and apoptosis inhibition), representing a promising translational therapeutic strategy for cognitive restoration across neurological and neuropsychiatric conditions.

conciseness of some paragraphs because the authors are not English native speakers. After using this tool/service, the authors reviewed and edited the content as needed and took full responsibility for the content of the publication.

References

Banks, W. A. (2004). The source of cerebral insulin. *European Journal of Pharmacology, 490*(1–3), 5–12.

Banks, W. A., Owen, J. B., & Erickson, M. A. (2012). Insulin in the brain: There and back again. *Pharmacology and Therapeutics, 136*(1), 82–93.

Beattie, E. C., Carroll, R. C., Yu, X., Morishita, W., Yasuda, H., Von Zastrow, M., & Malenka, R. C. (2000). Regulation of AMPA receptor endocytosis by a signaling mechanism shared with LTD. *Nature Neuroscience, 3*(12), 1291–1300.

Biessels, G. J., & Despa, F. (2018). Cognitive decline and dementia in diabetes mellitus: Mechanisms and clinical implications. *Nature Reviews. Endocrinology, 14*(10), 591–604.

Blázquez, E., Velázquez, E., Hurtado-Carneiro, V., & Ruiz-Albusac, J. M. (2014). Insulin in the brain: Its pathophysiological implications for states related with central insulin resistance, type 2 diabetes and Alzheimer's disease. *Frontiers in Endocrinology, 5*(Oct).

Bosco, D., Fava, A., Plastino, M., Montalcini, T., & Pujia, A. (2011). Possible implications of insulin resistance and glucose metabolism in Alzheimer's disease pathogenesis. *Journal of Cellular and Molecular Medicine, 15*(9), 1807–1821.

Bralten, J., Widomska, J., Witte, W. De, Yu, D., Mathews, C. A., Scharf, J. M., Buitelaar, J., Crosbie, J., Schachar, R., Arnold, P., Lemire, M., Burton, C. L., Franke, B., & Poelmans, G. (2020). Shared genetic etiology between obsessive-compulsive disorder, obsessive-compulsive symptoms in the population, and insulin signaling. *Translational Psychiatry, 10*(1).

Chen, H. J., Lee, Y. J., Yeh, G. C., & Lin, H. C. (2013). Association of attention-deficit/hyperactivity disorder with diabetes: A population-based study. *Pediatric Research, 73*(4 Pt 1), 492–496.

Chen, W., Liu, X., Muñoz, V. R., & Kahn, C. R. (2025). Loss of insulin signaling in microglia impairs cellular uptake of Aβ and neuroinflammatory response exacerbating AD-like neuropathology. *Proceedings of the National Academy of Sciences of the United States of America, 122*(21). e2501527122.

Chiu, S. L., Chen, C. M., & Cline, H. T. (2008). Insulin receptor signaling regulates synapse number, dendritic plasticity, and circuit function in vivo. *Neuron, 58*(5), 708–719.

Cordner, Z. A., & Tamashiro, K. L. K. (2015). Effects of high-fat diet exposure on learning & memory. *Physiology & Behavior, 152*(Pt B), 363.

Devaskar, S. U., Giddings, S. J., Rajakumar, P. A., Carnaghi, L. R., Menon, R. K., & Zahm, D. S. (1994). Insulin gene expression and insulin synthesis in mammalian neuronal cells. *Journal of Biological Chemistry, 269*(11), 8445–8454.

Ding, Q., Vaynman, S., Akhavan, M., Ying, Z., & Gomez-Pinilla, F. (2006). Insulin-like growth factor I interfaces with brain-derived neurotrophic factor-mediated synaptic plasticity to modulate aspects of exercise-induced cognitive function. *Neuroscience, 140*(3), 823–833.

Dorn, A., Rinne, A., Hahn, H. J., Bernstein, H. G., & Ziegler, M. (1982). C-peptide immunoreactive neurons in human brain. *Acta Histochemica, 70*(2), 326–330.

Ernst, A., Sharma, A. N., Elased, K. M., Guest, P. C., Rahmoune, H., & Bahn, S. (2013). Diabetic db/db mice exhibit central nervous system and peripheral molecular alterations as seen in neurological disorders. *Translational Psychiatry, 3*(5). e263.

Ghosh-Swaby, O. R., Reichelt, A. C., Sheppard, P. A. S., Davies, J., Bussey, T. J., & Saksida, L. M. (2022). Metabolic hormones mediate cognition. *Frontiers in Neuroendocrinology, 66*, 101009.

Goodman, W. K., Grice, D. E., Lapidus, K. A. B., & Coffey, B. J. (2014). Obsessive-compulsive disorder. *Psychiatric Clinics of North America, 37*(3), 257–267.

Goto, Y., Kakizaki, M., & Masak, N. (1976). Production of spontaneous diabetic rats by repetition of selective breeding. *The Tohoku Journal of Experimental Medicine, 119*(1), 85–90.

Guillemot-Legris, O., & Muccioli, G. G. (2017). Obesity-induced neuroinflammation: Beyond the hypothalamus. *Trends in Neurosciences, 40*(4), 237–253.

Hauser, J., Pisa, E., Arias Vásquez, A., Tomasi, F., Traversa, A., Chiodi, V., Martin, F. P., Sprenger, N., Lukjancenko, O., Zollinger, A., Metairon, S., Schneider, N., Steiner, P., Martire, A., Caputo, V., & Macrì, S. (2021). Sialylated human milk oligosaccharides program cognitive development through a non-genomic transmission mode. *Molecular Psychiatry, 26*(7), 2854–2871.

Havrankova, J., Schmechel, D., Roth, J., & Brownstein, M. (1978). Identification of insulin in rat brain. *Proceedings of the National Academy of Sciences of the United States of America, 75*(11), 5737.

Hooper, C., Killick, R., & Lovestone, S. (2008). The GSK3 hypothesis of Alzheimer's disease. *Journal of Neurochemistry, 104*(6), 1433–1439.

Kalayci, A., Januzzi, J. L., Mitsunami, M., Tanboga, I. H., Karabay, C. Y., & Gibson, C. M. (2025). Clinical features modifying the cardiovascular benefits of GLP-1 receptor agonists: A systematic review and meta-analysis. *European Heart Journal—Cardiovascular Pharmacotherapy, 11*(6), 552–561.

Kleinridders, A., Ferris, H. A., Cai, W., & Kahn, C. R. (2014). Insulin action in brain regulates systemic metabolism and brain function. *Diabetes, 63*(7), 2232–2243.

Klimova, B., Kuca, K., & Maresova, P. (2018). Global view on Alzheimer's disease and diabetes mellitus: Threats, risks and treatment Alzheimer's disease and diabetes mellitus. *Current Alzheimer Research, 15*(14), 1277–1282.

Koekkoek, P. S., Kappelle, L. J., van den Berg, E., Rutten, G. E. H. M., & Biessels, G. J. (2015). Cognitive function in patients with diabetes mellitus: Guidance for daily care. *The Lancet Neurology, 14*(3), 329–340.

Labouesse, M. A., Polesel, M., Clementi, E., Müller, F., Markkanen, E., Mouttet, F., Cattaneo, A., & Richetto, J. (2018). MicroRNA expression profiling in the prefrontal cortex: Putative mechanisms for the cognitive effects of adolescent high fat feeding. *Scientific Reports, 8*(1), 8344.

Lenzen, S. (2008). The mechanisms of alloxan- and streptozotocin-induced diabetes. *Diabetologia, 51*(2), 216–226.

Llorens-Martín, M., Jurado, J., Hernández, F., & Ávila, J. (2014). GSK-3β, a pivotal kinase in Alzheimer disease. *Frontiers in Molecular Neuroscience, 7*(May).

Marks, J. L., Porte, D., Stahl, W. L., & Baskin, D. G. (1990). Localization of insulin receptor mRNA in rat brain by in situ hybridization. *Endocrinology, 127*(6), 3234–3236.

Monney, M., Jornayvaz, F. R., & Gariani, K. (2023). GLP-1 receptor agonists effect on cognitive function in patients with and without type 2 diabetes. *Diabetes & Metabolism, 49*(5), 101470.

Nisticò, R., Cavallucci, V., Piccinin, S., Macrì, S., Pignatelli, M., Mehdawy, B., Blandini, F., Laviola, G., Lauro, D., Mercuri, N. B., &

D'Amelio, M. (2012). Insulin receptor β-subunit haploinsufficiency impairs hippocampal late-phase LTP and recognition memory. *Neuromolecular Medicine, 14*(4), 262–269.

Ottomana, A. M., Presta, M., O'Leary, A., Sullivan, M., Pisa, E., Laviola, G., Glennon, J. C., Zoratto, F., Slattery, D. A., & Macrì, S. (2023). A systematic review of preclinical studies exploring the role of insulin signalling in executive function and memory. *Neuroscience & Biobehavioral Reviews, 155*, 105435.

Pauls, D. L., Abramovitch, A., Rauch, S. L., & Geller, D. A. (2014). Obsessive-compulsive disorder: An integrative genetic and neurobiological perspective. *Nature Reviews. Neuroscience, 15*(6), 410–424.

Posner, B. I. (2017). Insulin signalling: The inside story. *Canadian Journal of Diabetes, 41*(1), 108–113.

Presta, M., Zoratto, F., Mulder, D., Ottomana, A. M., Pisa, E., Arias Vásquez, A., Slattery, D. A., Glennon, J. C., & Macrì, S. (2024). Hyperglycemia and cognitive impairments anticipate the onset of an overt type 2 diabetes-like phenotype in TALLYHO/JngJ mice. *Psychoneuroendocrinology, 167*.

Rahman, M. S., Hossain, K. S., Das, S., Kundu, S., Adegoke, E. O., Rahman, M. A., Hannan, M. A., Uddin, M. J., & Pang, M. G. (2021). Role of insulin in health and disease: An update. *International Journal of Molecular Sciences, 22*(12).

Rajashree, R., Kholkute, S. D., & Goudar, S. S. (2011). Effects of duration of diabetes on behavioural and cognitive parameters in Streptozotocin-induced juvenile diabetic rats. *The Malaysian Journal of Medical Sciences, 18*(4), 26.

Ramos-Rodriguez, J. J., Infante-Garcia, C., Galindo-Gonzalez, L., Garcia-Molina, Y., Lechuga-Sancho, A., & Garcia-Alloza, M. (2016). Increased spontaneous central bleeding and cognition impairment in APP/PS1 mice with poorly controlled diabetes mellitus. *Molecular Neurobiology, 53*(4), 2685–2697.

Raza, H., John, A., & Howarth, F. C. (2015). Increased oxidative stress and mitochondrial dysfunction in Zucker diabetic rat liver and brain. *Cellular Physiology and Biochemistry : International Journal of Experimental Cellular Physiology, Biochemistry, and Pharmacology, 35*(3), 1241–1251.

Reed, J., Bain, S. C., & Kanamarlapudi, V. (2024). The regulation of metabolic homeostasis by incretins and the metabolic hormones produced by pancreatic islets. *Diabetes, Metabolic Syndrome and Obesity, 17*, 2419.

Sędzikowska, A., & Szablewski, L. (2021). Insulin and insulin resistance in Alzheimer's disease. *International Journal of Molecular Sciences, 22*(18).

Saltiel, A. R., & Kahn, C. R. (2001). Insulin signalling and the regulation of glucose and lipid metabolism. *Nature, 414*(6865), 799–806.

Schubert, M., Gautam, D., Surjo, D., Ueki, K., Baudler, S., Schubert, D., Kondo, T., Alber, J., Galldiks, N., Küstermann, E., Arndt, S., Jacobs, A. H., Krone, W., Ronald Kahn, C., & Brüning, J. C. (2004). Role for neuronal insulin resistance in neurodegenerative diseases. *Proceedings of the National Academy of Sciences of the United States of America, 101*(9), 3100–3105.

Schulingkamp, R. J., Pagano, T. C., Hung, D., & Raffa, R. B. (2000). Insulin receptors and insulin action in the brain: Review and clinical implications. *Neuroscience and Biobehavioral Reviews, 24*(8), 855–872.

Siddle, K. (2011). Signalling by insulin and IGF receptors: Supporting acts and new players. *Journal of Molecular Endocrinology, 47*(1).

Taniguchi, C. M., Emanuelli, B., & Kahn, C. R. (2006). Critical nodes in signalling pathways: Insights into insulin action. *Nature Reviews. Molecular Cell Biology, 7*(2), 85–96.

Tanokashira, D., Wang, W., Maruyama, M., Kuroiwa, C., White, M. F., & Taguchi, A. (2021). Irs2 deficiency alters hippocampus-associated behaviors during young adulthood. *Biochemical and Biophysical Research Communications, 559*, 148–154.

Tokarz, V. L., MacDonald, P. E., & Klip, A. (2018). The cell biology of systemic insulin function. *The Journal of Cell Biology, 217*(7), 2273–2289.

van de Vondervoort, I., Amiri, H., Bruchhage, M. M. K., Oomen, C. A., Rustogi, N., Cooper, J. D., van Asten, J. J. A., Heerschap, A., Bahn, S., Williams, S. C. R., Buitelaar, J. K., Poelmans, G., & Glennon, J. C. (2019). Converging evidence points towards a role of insulin signaling in regulating compulsive behavior. *Translational Psychiatry, 9*(1).

van de Vondervoort, I., Poelmans, G., Aschrafi, A., Pauls, D. L., Buitelaar, J. K., Glennon, J. C., & Franke, B. (2016). An integrated molecular landscape implicates the regulation of dendritic spine formation through insulin-related signalling in obsessive-compulsive disorder. *Journal of Psychiatry & Neuroscience, 41*(4), 280–285.

Vecchio, I., Tornali, C., Bragazzi, N. L., & Martini, M. (2018). The discovery of insulin: An important milestone in the history of medicine. *Frontiers in Endocrinology, 9*.

Venuti, M. T., Roda, E., Brandalise, F., Sarkar, M., Cappelletti, M., Speciani, A. F., Soffientini, I., Priori, E. C., Giammello, F., Ratto, D., Locatelli, C. A., & Rossi, P. (2025). A pathophysiological intersection between metabolic biomarkers and memory: A longitudinal study in the STZ-induced diabetic mouse model. *Frontiers in Physiology, 16*, 1455434.

Werther, G. A., Hogg, A., Oldfield, B. J., Mc Kinley, M. J., Figdor, R., Allen, A. M., & Mendelsohn, F. A. O. (1987). Localization and characterization of insulin receptors in rat brain and pituitary gland using in vitro autoradiography and computerized densitometry. *Endocrinology, 121*(4), 1562–1570.

Winocur, G., Greenwood, C. E., Piroli, G. G., Grillo, C. A., Reznikov, L. R., Reagan, L. P., & McEwen, B. S. (2005). Memory impairment in obese Zucker rats: An investigation of cognitive function in an animal model of insulin resistance and obesity. *Behavioral Neuroscience, 119*(5), 1389–1395.

Woo, J. R., Bae, S. H., Wales, T. E., Engen, J. R., Lee, J., Jang, H., & Park, S. Y. (2024). The serine phosphorylations in the IRS-1 PIR domain abrogate IRS-1 and IR interaction. *Proceedings of the National Academy of Sciences of the United States of America, 121*(17). e2401716121.

Yarchoan, M., Toledo, J. B., Lee, E. B., Arvanitakis, Z., Kazi, H., Han, L. Y., Louneva, N., Lee, V. M. Y., Kim, S. F., Trojanowski, J. Q., & Arnold, S. E. (2014). Abnormal serine phosphorylation of insulin receptor substrate 1 is associated with tau pathology in Alzheimer's disease and tauopathies. *Acta Neuropathologica, 128*(5), 679.

Yermakov, L. M., Griggs, R. B., Drouet, D. E., Sugimoto, C., Williams, M. T., Vorhees, C. V., & Susuki, K. (2019). Impairment of cognitive flexibility in type 2 diabetic db/db mice. *Behavioural Brain Research, 371*, 111978.

Young, W. S. (1986). Periventricular hypothalamic cells in the rat brain contain insulin mRNA. *Neuropeptides, 8*(2), 93–97.

Zhang, Q., Jin, K., Chen, B., Liu, R., Cheng, S., Zhang, Y., & Lu, J. (2022). Overnutrition induced cognitive impairment: Insulin resistance, gut-brain axis, and neuroinflammation. *Frontiers in Neuroscience, 16*, 884579.

Zheng, H., Zheng, Y., Zhao, L., Chen, M., Bai, G., Hu, Y., Hu, W., Yan, Z., & Gao, H. (2017). Cognitive decline in type 2 diabetic db/db mice may be associated with brain region-specific metabolic disorders. *Biochimica et Biophysica Acta—Molecular Basis of Disease, 1863*(1), 266–273.

Zoratto, F., Pisa, E., Soldati, C., Barezzi, C., Ottomana, A. M., Presta, M., Santangelo, V., & Macrì, S. (2023). Automation at the service of the study of executive functions in preclinical models. *Scientific Reports, 13*(1), 16890.

Section 4

Treatment regimens modulating pancreatic hormones status and impact on neurological tissues and behavior

Chapter 18

Drugs used in treating diabetes and insulin resistance: Neurological and behavioral effects

Tanu Dixit[2], Soumya Verma[2], Riya Kumari[2], Tanvi Desai[1] and Selvan Ravindran[2]

[1]*Symbiosis School of Biological Sciences, Symbiosis International (Deemed University), Pune, Maharashtra, India;* [2]*Symbiosis School of Biological Sciences, Faculty of Medical and Health Sciences, Symbiosis International (Deemed University), Pune, Maharashtra, India*

Abbreviations

AD Alzheimer's disease
BBB blood–brain barrier
CNS central nervous system
DPP-4 dipeptidyl peptidase-4
GLP-1 glucagon-like peptide-1
IR insulin resistance
MRI magnetic resonance imaging
PET positron emission tomography
PPAR-γ peroxisome proliferator-activated receptor gamma
ROS reactive oxygen species
SGLT-2 sodium–glucose cotransporter 2
T2DM type 2 diabetes mellitus
WHO World Health Organization

Introduction

Diabetes mellitus, notably the type 2 form, has attained epidemic status worldwide, with hundreds of millions being affected, as estimated by the WHO (World Health Organization). Its effects on cerebral functioning are now becoming clearer, though its effects on the metabolic system, such as nephropathy and cardiovascular disease, have been well documented. Of particular focus is the type II diabetes insulin resistance, which is imposed on peripheral tissues and neuronal circuits. Uncoordinated brain insulin signaling has an effect on neurotransmission and can cause neuroplasticity in brain circuits that control cognitive and mood homeostasis (Barber et al., 2021). With the global chronicle of diabetic patients growing daily, particularly among the elderly, there is a corresponding rise in the rate of cognitive disruption. Neurological impairment of diabetics has ceased to be a peripheral phenomenon but has become a direct consequence of the inability to conduct metabolic signaling in the brain. It requires a shift of paradigms of treatment, including metabolic restoration and neuroprotection (Arnold et al., 2018).

Recent research has shown that insulin has a leading role to play in neural integrity. Besides peripheral glycemic control, insulin is also active in the modulation of synaptic activity, memory, and neuronal protection. The neurodegenerative traits of amyloid beta (Aβ) deposition and tau pathological mechanisms are linked to dysfunction of insulin receptor activity in the brain (Benedict & Grillo, 2018). These results promoted the idea of a type 3 diabetes, that is, Alzheimer's. It is likely that the insulin resistance in the neurons, along with the oxidative stress, low-grade inflammation, and glial dysfunction, are nourished by neural insulin resistance and engaged in neuronal injury (Arnold et al., 2018). This neuro-pathological cascade has been monitored in both research models and in clinical groups. The next common question is whether or not brain degeneration can be stopped or even reversed as soon as the systemic metabolic derangement is eradicated (Blázquez et al., 2022). The shared double pathology of neurodegenerative and metabolic disease suggests shared biological substrates, therefore opening the way to the study of antidiabetic medication to achieve cognitive and behavioral benefits (Boccardi et al., 2019).

Patients with type 2 diabetes generally report forgetfulness and decreased mental sharpness even in early phases of the disease. These indications are now well-supported by an epidemiological study, which demonstrated that diabetics have a significantly greater risk of mild cognitive impairment and progressive dementias.

Hormones in Neuroscience. https://doi.org/10.1016/B978-0-443-30263-3.00034-8

According to structural brain imaging, the diabetic brain exhibits frontal cortical atrophy and significant hippocampus volume reduction, which is consistent with the observed impairment in memory and executive function (Batista et al., 2018). Toward this is the dysregulation of adiponectin and other metabolic hormones that take part in regulating synaptic plasticity and brain inflammation (Bloemer et al., 2018). The psychiatric burden of diabetic populations entails elevated levels of depression and anxiety disorders that also complicate the self-care of the disease. These overlapping cognitive and emotional disturbances emphasize the need to examine whether already available antidiabetic medications have the potential to affect neurobehavioral results.

Metformin, a first-line medication for type 2 diabetes, has also been shown to have antioxidative and anti-inflammatory effects in the brain, though concerns still exist for its effects on neurotrophic signaling with long-term administration (Allard et al., 2016). Similarly, incretin-based treatments like GLP-1 receptor agonists can cross the blood–brain barrier and may ameliorate synaptic deficits as well as amyloid burden (Bassil et al., 2017). To be more precise, GLP-1 analogs were found to restore cognitive impairment in models of Alzheimer's disease without even altering peripheral glucose levels. The extensive distribution of both the GLP-1 and the insulin receptors in the hippocampus is in favor of the fact that there exists a robust physiological rationale for the central action of these drugs (Athauda & Foltynie, 2016). This body of evidence substantiates the rationale for screening such compounds not just for metabolic gain but also for the ability to protect or restore neurological function in those with insulin resistance.

This chapter will discuss the pathophysiological link between diabetes and brain dysfunction, and will also evaluate the neurological and behavioral effects of antidiabetic medications that are routinely utilized in the management of type 2 diabetes and insulin resistance. It will also attempt to discuss the mechanisms and translational potential of these drug agents in preventing neurodegenerative and neuropsychiatric complications that are common in diabetic patients.

Pathophysiological link between diabetes and brain dysfunction

Insulin signaling in the brain

Insulin is traditionally recognized for its peripheral role in glucose uptake, but it also plays pivotal neuromodulatory roles in the central nervous system (CNS). In the brain, insulin regulates synaptic plasticity, promotes neuronal growth through neurotrophic signaling, and facilitates overall neuronal survival. These effects have specific significance in regions of the brain that control learning and memory, such as the hippocampus. When insulin signaling is disrupted, as it is in individuals with insulin resistance, these neurobiological events are interfered with. Such interference is one contributor to the diminishment of cognitive functions, such as the creation of new memories and executive function, and is one potential mechanism for the onset of Alzheimer's disease (Diehl et al., 2017). Neuronal insulin resistance may hinder intracellular cascades such as PI3K/Akt and MAPK pathways, which are necessary for cell survival and synaptic regulation. De la Monte (2017) noted that insulin resistance in the brain is a significant aspect of the pathogenesis of neurodegenerative disease, implicating it in neuronal apoptosis and tau phosphorylation. Of interest is treatment with drugs such as metformin, which is widely used in the management of type 2 diabetes, and has been shown to revive elements of insulin signaling within the CNS. By acting on AMPK pathways and mitochondrial augmentation, metformin may reduce oxidative stress and restore neural insulin signaling, thereby conferring a hypothetical neuroprotective action in diabetic patients and individuals at risk of cognitive decline (Cao et al., 2022). This suggests that insulin's fundamental activity extends far beyond metabolism, positioning it as a key orchestrator of neurological health in metabolic disease.

Oxidative stress and neuroinflammation

One of the characteristic features of diabetic neuropathology is the presence of oxidative stress, which results from a chronic imbalance between the production of reactive oxygen species (ROS) and the brain's capacity to detoxify these reactive intermediates. Hyperglycemia and insulin resistance augment mitochondrial dysfunction, allowing for excess ROS accumulation and causing cellular damage. Neurons are particularly vulnerable since they possess a high oxygen demand and low antioxidant protection. Such an oxidative environment not only paralyzes cellular metabolism but also promotes the accumulation of aggregated misfolded proteins and lipid peroxidation, both of which are associated with neurodegenerative pathology (Dewanjee et al., 2022). Moreover, chronic oxidative stress is a stimulus for neuro-inflammation, facilitating the activation of microglia and astrocytes. The immune activation encourages the damage of neurons and degeneration of synapses, thereby inhibiting the plasticity of neurons and cognitive function (De la Monte, 2017). The interplay between mitochondrial dysfunction and chronic inflammation creates a vicious cycle that accelerates the progression of diseases like Alzheimer's in diabetics. Cao et al. (2022) suggest that targeting mitochondrial integrity and enhancing insulin sensitivity are ways to interrupt this pathologic cascade.

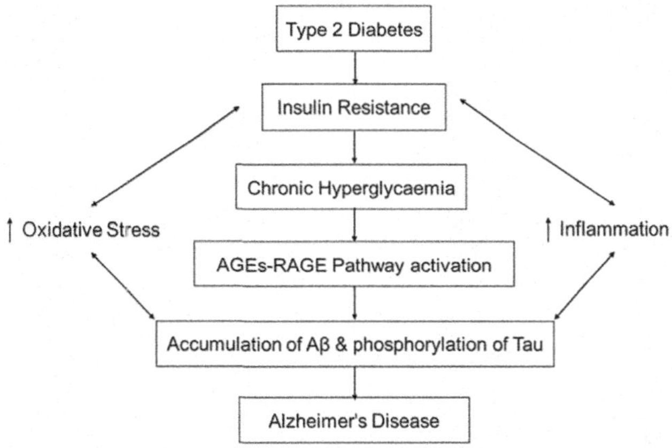

FIGURE 18.1 Pathophysiological mechanisms linking type 2 diabetes to Alzheimer's disease development.

For example, metformin not only enhances peripheral glucose use but also modulates mitochondrial respiration and oxidative burden in the CNS. Such findings point to the control of metabolic stress at the cellular level as a strategy for preserving cognitive function in diabetic patients. Overall, oxidative stress and neuroinflammation represent important mechanistic bridges between insulin resistance and progressive neurological dysfunction, as shown in Fig. 18.1.

Cognitive and behavioral manifestations

Cognitive impairment is now recognized as a complication of diabetes, particularly in insulin-resistant patients. Not just memory but also attention, information processing speed, executive function, and mood regulation are affected. Preclinical studies have also demonstrated that insulin signaling is important for the maintenance of synaptic integrity and neuronal communication, especially in the hippocampus and prefrontal cortex. Cognitive behaviors, including decreased spatial learning and working memory, have been disrupted in rodent models of diet-induced insulin resistance. However, these deficits are apparently reversible. Chunchai et al. (2016) found that vagus nerve stimulation in insulin-resistant rats significantly improved memory and learning outcomes via modulation of central inflammation and insulin sensitivity. In the clinical setting, cognitive impairment is often accompanied by behavioral symptoms of depression and anxiety, both of which can worsen diabetes control. De la Monte (2017) implicated deranged neurotransmitter homeostasis in the causation of these symptoms, as insulin modulates dopamine and serotonin pathways. Diehl et al. (2017) also suggested the link between defective insulin receptor activity and impaired synaptic plasticity, and hence reduced cognitive flexibility and affective stability. Such behavioral actions support the argument that resistance to insulin in the brain is not a consequence of metabolic disease but an essential player in the development of neuropsychiatric symptoms. This relationship is what makes the effort of integrating neuropsychological assessment with diabetes care worthwhile and valid, and offers the prospect of using pharmacologic intervention wherein both metabolic and cognitive activities are not dormant.

Central nervous effects of antidiabetic medications

Metformin

The most commonly used type 2 diabetes medication, metformin, has beneficial neurologic impacts beyond glycemic control. Its ability to penetrate the BBB (Blood Brain Barrier) and to modulate the major processes, such as neurogenesis, synaptic plasticity, and mTOR signaling, has been proven. Most importantly, metformin has potential in the treatment of fragile X syndrome and cognitive impairment by regulating protein synthesis pathways associated with brain activity (Gantois et al., 2019). When it comes to cognitive impairment in diabetes, the role of this hormone in increasing the insulin sensitivity of the hippocampus cannot be underrated (Grillo et al., 2019). Furthermore, it has been demonstrated that metformin enhances the brain insulin action, therefore, stimulating the neurotransmitter balancing and neuron survival mechanisms that contribute to the inhibition of neurodegeneration (Gantois et al., 2019). Although the drug has been linked to the normalization of cognition and lower neuroinflammation, the long-term effect of the drug on mood, motivation, and reward pathway is inadequately reported (Fiory et al., 2019). In spite of such restrictions, metabolic homeostasis restoration in the CNS by metformin presents its value as a neurotherapeutic agent in metabolic and neurodegenerative pathology. It can also be demonstrated that insulin resistance-related neurotoxicity can be normalized by chronic treatment, which further implies its neuroprotective effect on the CNS (Kamat et al., 2016). They need, however, to be clinically tested to establish their cognitive effect on aging and diabetic patients. Collectively, the central effects of metformin develop new opportunities for treatment interventions that target metabolism regulation and brain functioning. It is also characterized by a good safety profile, which highlights its translational application in neuroendocrinology and psychopharmacology.

Thiazolidinediones

The operating mechanism of TZDs (Thiazolidinedione) is activated PPAR-gamma (peroxisome proliferator-activated receptor gamma) in an attempt to control insulin sensitivity. This form of activation was demonstrated to minimize inflammation and oxidative stress, which are two significant factors that may cause diabetes associated brain neurodegeneration and cognitive deficiency (Hamed, 2017). Pioglitazone is capable of enhancing the performance of the hippocampal and synaptic plasticity and, hence, can be used to enhance memory and cognitive performance under insulin-resistant states. Also, the pioglitazone has an impact on the dopaminergic transit, which is significant insofar as motivation and mood regulation are concerned. This infers that pioglitazone may exert an antidepressant-like effect in diabetic patients who also have comorbid psychiatric conditions (Fiory et al., 2019). Most notably, its primary mechanism of action extends to the diminishment of glial cell activation and a decrease in neuroinflammatory markers, particularly in Alzheimer's disease and metabolic encephalopathy models. Although promising, this is countered by clinical use being limited by side effects such as fluid retention and possible congestive heart failure, particularly in older patients who have pre-existing cardiovascular or renal comorbidities. Accordingly, besides mechanistic proof of its neuroprotective action, long-term therapy with TZDs for overlapping metabolic and cognitive disease ought to be administered with trepidation. Optimization of dosage regimens or redesign of PPAR-γ agonists with less peripheral side effects could be future avenues of study. However, pioglitazone demonstrates the promise of repurposing insulin-sensitizing agents for the correction of central nervous system homeostasis in neuropsychiatric syndromes of diabetes.

GLP-1 receptor agonists

GLP-1 receptor agonists such as liraglutide have attracted increasing attention for their neuroprotective effect in addition to glucose-lowering action. The medications can cross the blood-brain barrier and activate GLP-1 receptors directly in regions involved in cognition, hunger control, and emotional processing, such as the hippocampus and hypothalamus (Ferreira et al., 2018). In animal models of Alzheimer's disease, liraglutide was able to decrease amyloid-beta deposition and tau hyperphosphorylation, two features of neurodegeneration. It has also been shown to enhance neurogenesis and synaptic plasticity, providing a mechanistic rationale for its cognitive-improving activity (Hölscher, 2020). Characteristically, the drug alters hypothalamic and mesolimbic circuit activity that tunes reward behavior and mood, something that can be used to treat diabetes-related emotional dysregulation and binge eating (Fulton et al., 2022). Even though animal studies confirm liraglutide's efficacy in improving spatial memory and learning, human trials are few but show promise for cognitive effects, particularly in individuals with mild cognitive impairment and type 2 diabetes (Kamat et al., 2016). The safety profile and dual mechanism render liraglutide a promising anti-neuroendocrine medication. Yet its long-term efficacy on psychiatric measures like depression or anxiety remains to be determined with certainty. Its efficacy in broader neurological applications necessitates clinical trials for evaluation. In general, liraglutide stands out among antidiabetic medications for its novel neurotropic activity and as an interface medication between metabolic and cognitive medications.

DPP-4 inhibitors

Dipeptidyl peptidase-4 (DPP-4) inhibitors, a class of antidiabetic drugs, inhibit the degradation of incretin hormones such as GLP-1. Besides contributing to the enhancement of glycemic control, it is possible that they can also affect central insulin signaling. Despite the fact that DPP-4 inhibitors sitagliptin and saxagliptin have poor penetration across the blood-brain barrier, they also have systemic GLP-1-level modulating effects, which can indirectly produce neuroprotection. According to the experience with animal models, hippocampal functioning, learning, and memory are facilitated by the chronic inhibition of DPP-4, but this can be due to increased synaptic plasticity and abrogation of neuroinflammation (Ferreira et al., 2018). The outcomes are especially pronounced when it comes to cognitive decline related to diabetes and brain insulin resistance, which have both been linked to neurodegenerative disorders, such as Alzheimer's disease (Hamer et al., 2019). Besides that, DPP-4 inhibitors have the potential to indirectly affect dopaminergic and serotonergic systems, which provides a possible therapeutic window for psychiatric symptoms, anhedonia, and motivational impairment (Fiory et al., 2019). Human studies are still at an early stage, and the central effects of these medications are less robust than for GLP-1 receptor agonists. However, their favorable safety profile and ease of delivery via oral administration make them a compelling candidate for adjunctive therapy addressing the cognitive and mood symptoms in diabetic patients. Further exploration of their neuromodulatory effects, especially in extended randomized controlled trials, is necessary to determine their translational benefit. In the meantime, DPP-4 inhibitors are a quiet but compelling addition to the neuroendocrine management of metabolic and cognitive dysfunction.

SGLT-2 inhibitors

Sodium-glucose cotransporter 2 (SGLT-2) inhibitors were first imagined as glucose-lowering drugs by promoting the excretion of glucose in the urine, but there are new findings indicating a broader spectrum of benefits, e.g., neuroprotection. They enhance systemic insulin sensitivity and reduce circulating inflammatory markers, both effects that indirectly offer the potential for improved brain performance (Drummen et al., 2019). Their ability to limit oxidative stress and abate low-grade chronic inflammation is particularly relevant to neurodegenerative conditions associated with diabetes, including Alzheimer's disease and vascular dementia. While SGLT-2 inhibitors do not permeate the blood-brain barrier readily, their metabolic effects, such as augmenting ketone body availability and optimizing cerebrovascular function, have been shown to preserve neuronal health and cognitive function in preclinical models. Furthermore, SGLT-2 inhibitor-treated patients usually report heightened satiety, reduced craving for food, and better emotional control, implying potential implications for brain reward systems (Fulton et al., 2022). These behavioral impacts are supported by evidence connecting insulin resistance with impaired brain response to food cues and emotional processing. Even though current clinical trials focus primarily on cardiovascular and renal outcomes, ancillary data regarding mood, cognition, and quality of life are now increasingly being reported, as shown in Table 18.1. The absence of noteworthy CNS side effects and the secondary cognitive effects position SGLT-2 inhibitors as interesting drugs for diabetes treatment in a holistic context. Yet, conclusive long-term neurological trials' data are required to establish their role in neurodegenerative or psychiatric disorders. The summary of the mechanisms of these drugs discussed above is shown in Fig. 18.2.

Translational and clinical implications

Cognitive disorders and Alzheimer's disease

Brain insulin resistance has come to be considered a principal pathophysiologic mechanism in Alzheimer's disease (AD). Insulin action within the central nervous system (CNS) involves regulation of neuronal survival, synaptic plasticity, and amyloid-beta clearance. Without adequate insulin signaling, there is heightened amyloid aggregation, hyperphosphorylation of tau, and decreased synaptic resilience markers of AD pathology (Kellar & Craft, 2020). Neuroimaging examinations such as PET and MRI scans always reveal structural and functional changes in the brain of lower insulin-sensitive patients even before clinical presentation of cognitive deficiency (Kullmann et al., 2016). Moreover, interventions aimed at enhancing CNS insulin signaling, i.e., intranasal insulin and GLP-1 receptor agonists, have also been shown to reverse early cognitive deficits and reduce neurodegenerative progression (Nowell et al., 2023). The mechanistic explanation can involve restoration of insulin receptor sensitivity, inhibition of neuroinflammation, and recapturing of neurotransmitter systems responsible for memory formation. These are particularly effective therapeutic approaches in patients with comorbid metabolic and cognitive dysfunctions. Since AD currently has no treatment in the pipeline, repurposing of antidiabetic drugs for neurodegenerative indications is a major translational success. It also underlines the necessity for early metabolic screening for dementia in high-risk populations. Future clinical trials will necessitate the inclusion of metabolic and cognitive endpoints to facilitate more efficient patient stratification and drug targeting. These convergence-based approaches have the potential to transform the treatment model of AD from symptomatic to mechanistically preventive, especially in the elderly with type 2 diabetes mellitus (T2DM) and insulin resistance.

Psychiatric comorbidities

Psychiatric illnesses such as depression and anxiety tend to co-occur with metabolic syndromes and insulin resistance, which hints at a shared neuropathological origin. Insulin signaling in brain regions like the prefrontal cortex, limbic system, and hypothalamus has direct effects on mood control, cognitive evaluation, and reward processing (Kleinridders & Pothos, 2019). Ablation of these circuits through impaired insulin sensitivity contributes to affective disturbances, emotional dysregulation, and anhedonia. This is evident in clinical presentations in which T2DM patients are characterized by higher rates of major depressive disorder (MDD) and generalized anxiety disorder. Additionally, antipsychotic medications, particularly those of the second-generation, have the potential to exacerbate insulin resistance, therefore undermining both psychiatric symptoms and glucose metabolism (Nicol et al., 2018). Two of the main symptoms of depression, anhedonia and cognitive blunting, have been linked with brain insulin resistance (Pomytkin et al., 2018). These findings are reflective of a bidirectional feedback loop between metabolic and psychiatric conditions that would govern therapeutic approaches based on both at the same time. For example, GLP-1 receptor agonists and intranasal insulin are assessed for their psychotropic effects, presenting novel options against conventional antidepressants. The evidence is building up that insulin sensitivity marker screening would be useful in psychiatric patients, particularly in those who are nonresponsive to SSRIs or mood stabilizers. Approaches like exercise and diet that are known to raise insulin sensitivity are perhaps another potential adjunctive treatment of mood disorders. By including endocrinological elements into diagnostic and therapeutic procedures, the neuroendocrine axis has the capability to revolutionize psychiatric care.

TABLE 18.1 CNS effects of antidiabetic drugs.

Drug class	Example	Blood-brain barrier penetration	Primary CNS effects	Neuroprotective mechanisms	Behavioral impact	References
Biguanides	Metformin	Moderate	Enhances cognition, synaptic plasticity	Activates AMPK, reduces oxidative stress	May support mood, improve memory	Gantois et al. (2019) and Hölscher (2020)
Thiazolidinediones	Pioglitazone	Moderate	Reduces neuroinflammation, supports memory	PPAR-γ activation decreases glial activity	May aid mood, improve motivation	Hamed (2017) and Fiory et al. (2019)
GLP-1 receptor agonists	Liraglutide	High	Decreases amyloid, improves synaptic plasticity	Promotes neurogenesis, anti-amyloid action	Enhances reward processing, mood regulation	Ferreira et al. (2018) and Hölscher (2020)
DPP-4 inhibitors	Sitagliptin	Low	Enhances hippocampal learning, memory	Indirect GLP-1 increase, anti-inflammatory	Potential benefit in mood stability	Hamer et al. (2019)
SGLT-2 inhibitors	Empagliflozin	Low	Improves metabolic regulation, cognitive resilience	Reduces inflammation, optimizes ketone usage	Affects satiety, improves emotional control	Drummen et al. (2019) and Fulton et al. (2022)
Common mechanisms	—	—	Modulate CNS insulin signaling	Anti-inflammatory, antioxidant, neurotrophic	Improve cognition, stabilize affect	Hölscher (2020)
Clinical relevance	—	—	Treat diabetes-associated cognitive symptoms	Target brain-metabolism interaction	Potential therapy for neurodegeneration	Fiory et al. (2019) and Kamat et al. (2016)

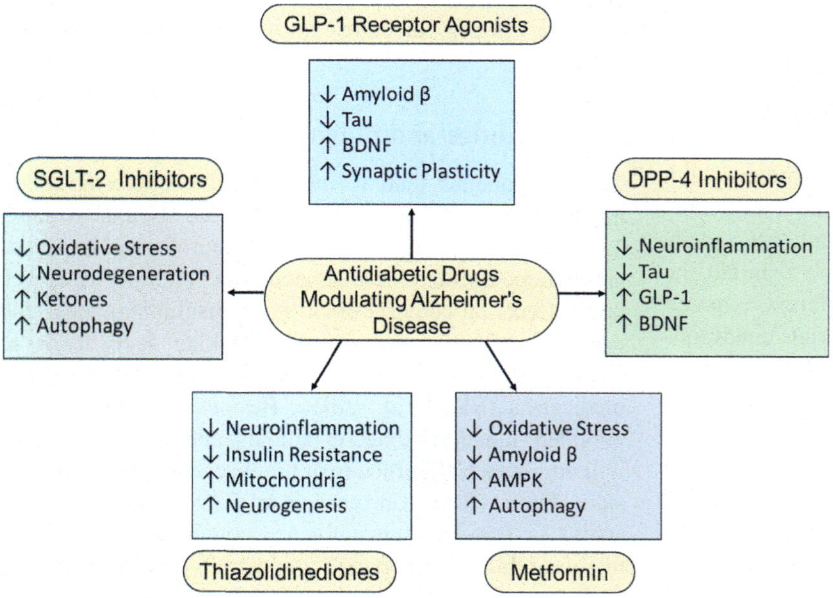

FIGURE 18.2 Mechanistic effects of antidiabetic drug classes in Alzheimer's disease.

Precision medicine perspective

Precision medicine is transforming the industry in the war on insulin resistance and the associated neurological impact. It is an approach that personalizes therapies based on the genetic, metabolic, and neuroimaging profile of each patient rather than a standardized therapy approach. The differences in insulin resistance in the brain of individuals influence the predisposition of an individual to mood and cognitive disorders and the response to medication treatment (Kullmann et al., 2020). Techniques such as fMRI, PET scans, and CSF biomarker testing allow the physician to detect the first changes in the brain associated with insulin signaling dysfunctions (Milstein & Ferris, 2021). These methods can be used to cluster patients based on their central insulin sensitivity, which enhances the treatment response predictions. Additionally, genetic variants that impact insulin receptors or signaling pathways can be identified by genomic screening, again improving therapy selection (Li et al., 2022). As an example, centrally acting insulin-resistant drugs, such as GLP-1 analogs, would maximize those with hippocampal insulin resistance. Integrating this precision with digital health platforms and longitudinal follow-up could revolutionize neuroendocrinology. Integrative models linking metabolic biomarkers with neuropsychological tests would enable clinicians to more reliably predict neurodegenerative risk among metabolic patients. Repurposing antidiabetic treatments in these stratified subgroups maximizes their utility without inducing unwanted side effects. Furthermore, combining pharmacologic interventions with lifestyle and behavioral changes may enhance synergy. With continued advancements in brain-metabolism interaction knowledge, precision medicine will be a cornerstone in treating metabolic and neuropsychiatric disorders (Labandeira et al., 2022).

Challenges and limitations

Limited CNS entry of antidiabetic drugs

Despite growing evidence for the role of insulin in brain function, the majority of antidiabetic drugs poorly penetrate the CNS. The blood-brain barrier (BBB) is an effective barrier to the delivery of therapeutic drugs to brain tissue. For example, while both metformin and GLP-1 receptor agonists exert some CNS effects, their concentrations in cerebral regions are subtherapeutic in most models (Simsir et al., 2018). This limitation is a significant hindrance to utilizing metabolic medications for cognitive protection or enhancement. Transporters of drugs like P-glycoprotein also transport therapeutic drugs out of the brain, reducing their intracerebral efficacy (Rad et al., 2018). Further, while intranasal administration of insulin has been encouraging in bypassing the BBB, the concerns of standardizing doses and long-term safety remain to be addressed (Shpakov et al., 2023). Without CNS bioavailability, the potential for treating Alzheimer's or depression with insulin-sensitizing medications remains speculative. Ultimately, until future studies can determine means of enhancing BBB permeability or designing CNS-directed analogs, the therapeutic potential of available antidiabetic drugs for neurological ailments will be limited (Sacramento et al., 2018).

Preclinical-clinical inconsistencies

Preclinical studies of the neurologic activity of antidiabetic drugs routinely yield positive outcomes, but in clinical translation, it has been wanting. Rodent models of Alzheimer's and diabetes typically respond positively to GLP-1 analogs, metformin, and other insulin sensitizers, with cognitive enhancement and reduced amyloid load (Song et al., 2021). But huge clinical trials do not necessarily replicate these findings. Such disagreement might be attributed to species-specific metabolic differences, exaggerated disease models, or controlled trial conditions failing to reflect human heterogeneity (Soto et al., 2018). Additionally, preclinical models use young or genetically uniform animals, whereas human populations consist of diverse genetic, lifestyle, and comorbid backgrounds (Stenvers et al., 2019). In addition, differences in drug metabolism, CNS penetrability, and interaction along the gut-brain axis can also limit extrapolation. Lack of availability of biomarker validation in humans also prevents clinical progress, as enhancements identified in animal cognition are not easily measurable in clinical cognitive assessments (Simsir et al., 2018). Closing this translational gap includes matching trial designs to real disease conditions and maximizing human trial designs so that they better replicate the mechanisms established in animal experiments.

Long-term neurocognitive data inadequacy

While short-term studies have provided some promise for the neurocognitive benefits of antidiabetic drugs, robust long-term data are scarce. The majority of trials only last several months to a few years, which is not enough time to track progressive neurodegenerative changes or residual cognitive effects (Rojas-Carranza et al., 2018). Neurodegenerative diseases such as Alzheimer's disease take decades to evolve, and an evaluation of the preventative impact of metabolic interventions requires long-term follow-up. Furthermore, most studies concentrate on glycemic control or cardiovascular outcomes, making neurocognitive testing a secondary outcome measure (Sędzikowska & Szablewski, 2021). Additionally, neurocognitive function is further influenced by a variety of factors, including sleep, circadian disturbance, and chronic inflammation, making it difficult to disentangle drug-specific effects without longitudinal follow-up (Stenvers et al., 2019). Without longitudinal imaging, electrophysiological data, or serial cognitive assessments, it is difficult to determine whether the benefits seen are persistent or dissipate with time. Registries or inclusion of cognitive endpoints in diabetes care protocols could improve the accrual of such data. While CNS diabetic complications remain underdiagnosed, long-term studies must shift their emphasis toward overall patient outcomes, especially in diabetics' aging diabetic group who are at risk for dementia.

Comorbidities and polypharmacy impact

Type 2 diabetics tend to have multiple comorbid states hypertension, dyslipidemia, and cardiovascular disease, adding complexity that obscures the neurological impact of antidiabetic agents. These concomitant conditions can each have effects on cerebral blood flow, insulin signaling, and neuroinflammation individually, making it difficult to attribute cognitive outcomes solely to glucose-lowering medication (Rad et al., 2018). Besides, polypharmacy is prevalent in elderly diabetic patients, and drug interactions will attenuate CNS efficacy or amplify antidiabetic therapy side effects (Rojas-Carranza et al., 2018). For example, statins or beta-blockers influence cognition or mood and therefore bias the apparent impact of interventions like GLP-1 agonists (Ruud et al., 2017). Moreover, pharmacogenomic variability, such as polymorphisms in drug metabolism, influences the therapeutic responses. Disruption of circadian rhythm and sleep, and dietary intake also place confounding effects on cognitive function and insulin resistance (Stenvers et al., 2019). Unless controlled for these variables, clinical studies can produce false-positive efficacy or safety findings, as shown in Table 18.2. Full management regimens with drug reconciliation, control of comorbid conditions, and pharmacogenomic profiling are required to accurately determine the CNS benefits of antidiabetic drugs.

Future directions

CNS-targeted antidiabetic therapy development

CNS-targeted antidiabetic therapy development holds promise in the treatment of cognitive impairment secondary to insulin resistance. Emerging evidence supports that brain insulin action is central to the pathogenesis of neurodegenerative diseases such as Alzheimer's disease (Tumminia et al., 2018). Therapies avoiding peripheral metabolism and directly enhancing brain insulin sensitivity are desperately needed. Intranasal insulin and drugs like liraglutide hold potential to affect neuroplasticity, synaptic function, and neuroinflammation (Watson et al., 2019). These CNS-permeating agents are especially crucial in the high-risk population, for which peripheral therapies cannot be effective in reversing cognitive symptoms. Targeting pathways such as PI3K/AKT and mTOR in neurons has been reported to enhance memory deficits and attenuate neurodegeneration (Vinuesa et al., 2021). Also, therapeutic delivery systems with the ability to effectively cross the blood-brain barrier are a key design target. Nanoparticles

TABLE 18.2 Mechanistic pathways linking insulin resistance to neuronal cell death in neurodegenerative disorders.

Challenge area	Description	Cause	Example drug	Clinical impact	Proposed solution	References
Limited CNS entry	Poor penetration of drugs across the blood-brain barrier (BBB)	Efflux transporters, tight junctions	Metformin, GLP-1 RA	Limited neurotherapeutic utility	BBB-permeable analogs or intranasal delivery	Simsir et al. (2018) and Shpakov et al. (2023)
Preclinical-clinical inconsistencies	Positive animal results have not been replicated in human studies	Species differences, artificial models	GLP-1 RA, Metformin	Uncertain clinical translation	Align trial models with real-world variability	Soto et al. (2018) and Stenvers et al. (2019)
Data inadequacy (long-term)	Lack of extended studies for neurocognitive outcomes	Short trial duration, secondary outcome focus	All classes	Unverified long-term brain effects	Include neurocognitive endpoints in trials	Rojas-Carranza et al. (2018)
Polypharmacy impact	Drug-drug interactions and comorbidity effects on CNS activity	Multiple medications and chronic conditions	GLP-1 RA, DPP-4 inhibitors	Masked or biased CNS effects	Comprehensive medication review and stratified analyses	Ruud et al. (2017)
Pharmacogenomic variability	Genetic differences affect drug metabolism and CNS action	CYP polymorphisms, transporter variation	Pioglitazone, Metformin	Variable CNS response and side effects	Personalized medicine using genomic profiling	Rojas-Carranza et al. (2018)
Non-standardized dosing	Variability in CNS dose delivery (esp. intranasal insulin)	Technique differences, poor standardization	Intranasal insulin	Inconsistent cognitive outcomes	Controlled dosing systems, pharmacokinetic validation	Shpakov et al. (2023)
Confounding lifestyle factors	Diet, sleep, and circadian rhythm alter insulin sensitivity and cognition	Lifestyle heterogeneity	All	Reduced reliability of CNS impact studies	Control lifestyle variables in study design	Stenvers et al. (2019)

or conjugated molecules can be added to future drugs to enhance CNS selectivity. In general, the development of new drugs should be guided by a mechanistic knowledge of central insulin signaling derangements.

Synergy of neuroimaging and biomarker-based monitoring

In diabetic neurology, neuroimaging and biomarkers work together to track the disease's course and response to treatment. The biomarkers of fMRI, positron emission tomography (PET), and cerebral spinal fluid (CSF) evidence the impact of antidiabetic drugs on the connection and ABs deposition in the brain (Watson et al., 2019). An example is that liraglutide has been observed to regulate the hippocampus in conjunction with the prefrontal cortex, brain regions of memory and emotional processing, using fMRI. Moreover, the insulin resistance indicators can be used in the tests of treatment responsiveness and customized treatment (HOMA-IR, BDNF and neurofilament light chain) (Tumminia et al., 2018). In particular, standardized sets of biomarkers would permit disease classification and preventive intervention in the preclinical populations of Alzheimer's. Imaging and molecular diagnostics will have to be incorporated regularly in future clinical trials so that a better appreciation of the multidimensional effects of antidiabetic medicines on the CNS is achieved.

Personalized medicine approaches in diabetic neurology

Personalized medicine in the field of diabetic neurology is a new paradigm, and its goal is to individualize the treatment on the basis of personalized genetic, metabolic, and behavioral characteristics. Insulin resistance is a heterogeneous condition, and its neuropsychiatric (among other) symptoms require targeted treatment (Watson et al., 2018). They include age, APOE genotype, status of inflammation, and level of sex hormone, all of which have a significant influence on the action of drugs in the CNS as well as the susceptibility to cognitive impairment (Vinuesa et al., 2021). Metformin, as an example, demonstrated anxiolytic and antidepressant activity in insulin-resistant patients, yet the behavioral response pattern is dependent on the level of branched-chain amino acids and microbiome (Zemdegs et al., 2019). Individualized programs could combine genomic profiles with neural imaging outcomes and peripheral markers to make informed decisions on pharmacology. Mood testing and psychometry can also be useful in selecting patients who are most likely to respond positively to CNS-selective treatments. The future directions are to develop artificial intelligence-based systems of decision support in predicting therapeutic responses and side-effect profiles.

Longitudinal and multimodal clinical trials expansion

Robust longitudinal and multimodal clinical trials are indicated to validate long-term cognitive efficacy and safety of antidiabetic drugs in neurologically at-risk populations. The study has a tendency to focus on short-term or solitary metabolic endpoints, with disregard for neurobehavioral course. Nevertheless, conditions such as Alzheimer's and depression share mechanisms with diabetes, including neuroinflammation and disrupted insulin signaling (Vinuesa et al., 2021). Longitudinal trials that combine cognitive testing, neuroimaging, and biomarker profiling can uncover modest longitudinal changes in brain function and structure (Watson et al., 2018). Essentially, for results to be generalizable, trials should include diverse groups in terms of age, sex, and metabolic profile. For example, metformin's effects on affective symptoms and brain plasticity in insulin-resistant mice highlight the need for months-long or even years-long human studies (Zemdegs et al., 2019). To explain the entire spectrum of central effects and interactions between neuroendocrine, genetic, and psychosocial factors, multimodal designs will be utilized.

Applications of the material

Considering the translational significance of antidiabetic medication treatment in lowering neurological consequences, the study examines the relationship between metabolic disease and brain dysfunction. With the increasing prevalence of type II diabetes across the globe, particularly among the elderly, insulin resistance has been the central cause of metabolic imbalance, the etiology of Alzheimer's disease, and other impairments of cognitive functioning. The regulation of synapse function, neuron survival, and neurotransmitter balance is among the functions of insulin in the brain. These central effects function as the foundation for the neurological repurposing of antidiabetic medications like metformin and GLP-1 receptor agonists. In addition to its metabolic advantages, metformin affects brain function by modifying the AMPK and mTOR pathways, which are linked to decreased neuroinflammation and oxidative stress. Because GLP-1 analogs, like liraglutide, can pass the blood-brain barrier, alter synaptic signaling, and reduce the production of amyloid plaque, they have attracted attention in neurodegenerative models. These neurological consequences are now being evaluated in real time using innovative diagnostic imaging modalities like functional MRI and PET scans. Following antidiabetic medication, the imaging methods enable clinicians to measure changes in brain structure and cognitive functions. Direct food delivery to brain insulin receptors due to bypassing of systemic metabolism is also attained with insulin delivery into the nasal passages, giving another route of administration with reduced peripheral side effects. The same can be said about psychiatric comorbidities that demonstrate the potential of translation. The metabolic dysregulation is associated with depression and anxiety, as insulin signaling is interlinked with dopaminergic and serotonergic neurotransmission. Considering that mood disorders are negatively impactful on glycemic control, therapeutic interventions that affect metabolic as well as psychological health are required. Drugs acting on insulin signaling in the brain may thus confer dual therapeutic benefit. Lastly, this chapter redefines antidiabetic drugs as multimodal agents with the capacity to treat metabolic, cognitive, and affective symptoms. This paradigm encourages longitudinal clinical trial designs involving neurocognitive measures, biomarker studies, and behavioral rating scales to establish and optimize the effectiveness of treatment.

Conclusion

The complex interaction between diabetes, insulin resistance, and the central nervous system function has been recently recognized, and antidiabetic agents have shown both effective and heterogeneous neurological and behavioral effects. There are indications that drugs like metformin, GLP-1 receptor agonists, and pioglitazone not only impact glycemic control but may also have an impact on neuroinflammation, cognitive processes, and mood control. However, such therapeutic advantages are often restricted by the lack of central nervous system penetration, inconsistent clinical effect, and a shortage of long-term neurocognitive data. Also, pathologies and polypharmacy issues make treatment difficult in diabetic subjects. The research in the future should be aimed at the development of brain-based antidiabetic treatments, biomarker combinations with neuroimaging, and personalized approaches to the neuroimaging strategies to derive maximum therapeutic effects. Longitudinal, multimodal clinical trials will be the key to concluding the long-term effects of the drugs on the brain. To sum up, current antidiabetic agents have the potential to improve CNS pathways, but their medical application to treat or prevent neurodegenerative and mood disorders associated with diabetes is still to be fully validated. This research on insulin signaling and neuroinflammation, as well as the gut-brain axis, is relevant to the identification of novel methods of therapeutic intervention that can enhance the metabolic and neurological health of diabetic patients.

Mini-dictionary of terms

1. **Adiponectin:** Protein hormone that enhances insulin sensitivity and regulates lipid metabolism, is generally lower in T2DM.
2. **Amyloid-beta (Aβ):** Neurotoxic peptide that accumulates in Alzheimer's disease, linked with dysfunctional insulin signaling.
3. **Blood–brain barrier (BBB):** Protective brain barrier that restricts molecular exchange, affecting drug delivery to the CNS.
4. **Cognitive Dysfunction:** Condition marked by defects in memory, attention, or decision-making, commonly linked with insulin resistance.
5. **GLP-1:** Intestinal-released hormone that lowers blood glucose and exhibits neuroprotective effects on neurons.
6. **Hippocampus:** Brain region important for memory and learning, frequently disturbed in T2DM-induced neurodegeneration.
7. **Insulin Resistance (IR):** Physiological state in which cells exhibit decreased sensitivity toward insulin, leading to hyperglycemia and neuroinflammation.
8. **Metformin:** Antidiabetic medication with glycemic control and new cognition advantages through central action.
9. **Neuroinflammation:** Progressive inflammation in the brain, responsible for cognitive impairment in metabolic disorders.
10. **Oxidative Stress:** Harmful cellular state caused by an overabundance of reactive oxygen species, associated with T2DM as well as neurodegeneration.
11. **PPAR-γ:** Nuclear receptor playing a role in lipid metabolism and inflammation, the target of thiazolidinediones.
12. **SGLT-2 Inhibitors:** Medications that reduce blood glucose by enhancing urinary excretion of glucose, which may prove useful in the context of brain metabolism.
13. **Synaptic Plasticity:** Brain's capacity to change synaptic strength, compromised by insulin resistance.

Summary points

1. Brain insulin resistance is an important cause of cognitive and mood disturbances in T2DM patients.
2. Metformin enhances insulin signaling in neurons and may be able to reverse memory deficits by influencing intracellular pathways such as mTOR.

3. GLP-1 receptor agonists can cross the BBB and have neuroprotective effects against amyloid-induced neuronal damage.
4. CNS penetration of medications remains a significant hurdle to most current antidiabetic therapies.
5. Neuroimaging techniques like fMRI and PET allow assessment of the cognitive impact of metabolic treatment.
6. Individualized treatment by metabolic and genetic analysis may enhance treatment outcomes in T2DM patients with neurological complications.
7. Depression and cognitive impairment present in T2DM may be alleviated by drugs targeting the central insulin pathway.
8. Future studies need long-term trials that follow metabolic and neurological markers for an extended duration.
9. Intranasal delivery of insulin is a non-invasive approach to direct activation of insulin receptors within the brain.
10. Antidiabetic drugs are promising beyond their glycemic effects, especially in inhibiting or preventing neurodegenerative diseases.

References

Allard, J. S., Perez, E. J., Fukui, K., Carpenter, P., Ingram, D. K., & de Cabo, R. (2016). Prolonged metformin treatment leads to reduced transcription of Nrf2 and neurotrophic factors without cognitive impairment in older C57BL/6J mice. *Behavioural Brain Research, 301*, 1–9.

Arnold, S. E., Arvanitakis, Z., Macauley-Rambach, S. L., Koenig, A. M., Wang, H. Y., Ahima, R. S., ... Nathan, D. M. (2018). Brain insulin resistance in type 2 diabetes and Alzheimer disease: Concepts and conundrums. *Nature Reviews Neurology, 14*(3), 168–181.

Athauda, D., & Foltynie, T. (2016). The glucagon-like peptide 1 (GLP) receptor as a therapeutic target in Parkinson's disease: Mechanisms of action. *Drug Discovery Today, 21*(5), 802–818.

Barber, T. M., Kyrou, I., Randeva, H. S., & Weickert, M. O. (2021). Mechanisms of insulin resistance at the crossroad of obesity with associated metabolic abnormalities and cognitive dysfunction. *International Journal of Molecular Sciences, 22*(2), 546.

Bassil, F., Canron, M. H., Vital, A., Bezard, E., Li, Y., Greig, N. H., ... Meissner, W. G. (2017). Insulin resistance and exendin-4 treatment for multiple system atrophy. *Brain, 140*(5), 1420–1436.

Batista, A. F., Forny-Germano, L., Clarke, J. R., Lyra e Silva, N. M., Brito-Moreira, J., Boehnke, S. E., ... De Felice, F. G. (2018). The diabetes drug liraglutide reverses cognitive impairment in mice and attenuates insulin receptor and synaptic pathology in a non-human primate model of Alzheimer's disease. *The Journal of Pathology, 245*(1), 85–100.

Benedict, C., & Grillo, C. A. (2018). Insulin resistance as a therapeutic target in the treatment of Alzheimer's disease: A state-of-the-art review. *Frontiers in Neuroscience, 12*, 215.

Blázquez, E., Hurtado-Carneiro, V., LeBaut-Ayuso, Y., Velázquez, E., García-García, L., Gómez-Oliver, F., ... Pozo, M.Á. (2022). Significance of brain glucose hypometabolism, altered insulin signal transduction, and insulin resistance in several neurological diseases. *Frontiers in Endocrinology, 13*, 873301.

Bloemer, J., Pinky, P. D., Govindarajulu, M., Hong, H., Judd, R., Amin, R. H., ... Suppiramaniam, V. (2018). Role of adiponectin in central nervous system disorders. *Neural Plasticity, 2018*(1), 4593530.

Boccardi, V., Murasecco, I., & Mecocci, P. (2019). Diabetes drugs in the fight against Alzheimer's disease. *Ageing Research Reviews, 54*, 100936.

Cao, G., Gong, T., Du, Y., Wang, Y., Ge, T., & Liu, J. (2022). Mechanism of metformin regulation in the central nervous system: Progression and future perspectives. *Biomedicine & Pharmacotherapy, 156*, 113686.

Chunchai, T., Samniang, B., Sripetchwandee, J., Pintana, H., Pongkan, W., Kumfu, S., ... Chattipakorn, S. C. (2016). Vagus nerve stimulation exerts the neuroprotective effects in obese-insulin resistant rats, leading to the improvement of cognitive function. *Scientific Reports, 6*(1), 26866.

De la Monte, S. M. (2017). Insulin resistance and neurodegeneration: Progress towards the development of new therapeutics for Alzheimer's disease. *Drugs, 77*, 47–65.

Dewanjee, S., Chakraborty, P., Bhattacharya, H., Chacko, L., Singh, B., Chaudhary, A., ... Kandimalla, R. (2022). Altered glucose metabolism in Alzheimer's disease: Role of mitochondrial dysfunction and oxidative stress. *Free Radical Biology and Medicine, 193*, 134–157.

Diehl, T., Mullins, R., & Kapogiannis, D. (2017). Insulin resistance in Alzheimer's disease. *Translational Research, 183*, 26–40.

Drummen, M., Dorenbos, E., Vreugdenhil, A. C., Raben, A., Westerterp-Plantenga, M. S., & Adam, T. C. (2019). Insulin resistance, weight, and behavioral variables as determinants of brain reactivity to food cues: A prevention of diabetes through lifestyle intervention and population studies in Europe and around the World–a PREVIEW study. *The American Journal of Clinical Nutrition, 109*(2), 315–321.

Ferreira, L. S., Fernandes, C. S., Vieira, M. N., & De Felice, F. G. (2018). Insulin resistance in Alzheimer's disease. *Frontiers in Neuroscience, 12*, 830.

Fiory, F., Perruolo, G., Cimmino, I., Cabaro, S., Pignalosa, F. C., Miele, C., ... Oriente, F. (2019). The relevance of insulin action in the dopaminergic system. *Frontiers in Neuroscience, 13*, 868.

Fulton, S., Décarie-Spain, L., Fioramonti, X., Guiard, B., & Nakajima, S. (2022). The menace of obesity to depression and anxiety prevalence. *Trends in Endocrinology & Metabolism, 33*(1), 18–35.

Gantois, I., Popic, J., Khoutorsky, A., & Sonenberg, N. (2019). Metformin for treatment of fragile X syndrome and other neurological disorders. *Annual Review of Medicine, 70*(1), 167–181.

Grillo, C. A., Woodruff, J. L., Macht, V. A., & Reagan, L. P. (2019). Insulin resistance and hippocampal dysfunction: Disentangling peripheral and brain causes from consequences. *Experimental Neurology, 318*, 71–77.

Hölscher, C. (2020). Brain insulin resistance: Role in neurodegenerative disease and potential for targeting. *Expert Opinion on Investigational Drugs, 29*(4), 333–348.

Hamed, S. A. (2017). Brain injury with diabetes mellitus: Evidence, mechanisms and treatment implications. *Expert Review of Clinical Pharmacology, 10*(4), 409–428.

Hamer, J. A., Testani, D., Mansur, R. B., Lee, Y., Subramaniapillai, M., & McIntyre, R. S. (2019). Brain insulin resistance: A treatment target for cognitive impairment and anhedonia in depression. *Experimental Neurology, 315*, 1–8.

Kamat, P. K., Kalani, A., Rai, S., Tota, S. K., Kumar, A., & Ahmad, A. S. (2016). Streptozotocin intracerebroventricular-induced neurotoxicity and brain insulin resistance: A therapeutic intervention for treatment of sporadic Alzheimer's disease (sAD)-like pathology. *Molecular Neurobiology, 53*, 4548–4562.

Kellar, D., & Craft, S. (2020). Brain insulin resistance in Alzheimer's disease and related disorders: Mechanisms and therapeutic approaches. *The Lancet Neurology, 19*(9), 758–766.

Kleinridders, A., & Pothos, E. N. (2019). Impact of brain insulin signaling on dopamine function, food intake, reward, and emotional behavior. *Current Nutrition Reports, 8*, 83–91.

Kullmann, S., Heni, M., Hallschmid, M., Fritsche, A., Preissl, H., & Häring, H. U. (2016). Brain insulin resistance at the crossroads of metabolic and cognitive disorders in humans. *Physiological Reviews, 96*(4), 1169–1209.

Kullmann, S., Kleinridders, A., Small, D. M., Fritsche, A., Häring, H. U., Preissl, H., & Heni, M. (2020). Central nervous pathways of insulin action in the control of metabolism and food intake. *The Lancet Diabetes & Endocrinology, 8*(6), 524–534.

Labandeira, C. M., Fraga-Bau, A., Ron, D. A., Alvarez-Rodriguez, E., Vicente-Alba, P., Lago-Garma, J., & Rodriguez-Perez, A. I. (2022). Parkinson's disease and diabetes mellitus: Common mechanisms and treatment repurposing. *Neural Regeneration Research, 17*(8), 1652–1658.

Li, M., Chi, X., Wang, Y., Setrerrahmane, S., Xie, W., & Xu, H. (2022). Trends in insulin resistance: Insights into mechanisms and therapeutic strategy. *Signal Transduction and Targeted Therapy, 7*(1), 216.

Milstein, J. L., & Ferris, H. A. (2021). The brain as an insulin-sensitive metabolic organ. *Molecular Metabolism, 52*, 101234.

Nicol, G. E., Yingling, M. D., Flavin, K. S., Schweiger, J. A., Patterson, B. W., Schechtman, K. B., & Newcomer, J. W. (2018). Metabolic effects of antipsychotics on adiposity and insulin sensitivity in youths: A randomized clinical trial. *JAMA Psychiatry, 75*(8), 788–796.

Nowell, J., Blunt, E., & Edison, P. (2023). Incretin and insulin signaling as novel therapeutic targets for Alzheimer's and Parkinson's disease. *Molecular Psychiatry, 28*(1), 217–229.

Pomytkin, I., Costa-Nunes, J. P., Kasatkin, V., Veniaminova, E., Demchenko, A., Lyundup, A., ... Strekalova, T. (2018). Insulin receptor in the brain: Mechanisms of activation and the role in the CNS pathology and treatment. *CNS Neuroscience & Therapeutics, 24*(9), 763–774.

Rad, S. K., Arya, A., Karimian, H., Madhavan, P., Rizwan, F., Koshy, S., & Prabhu, G. (2018). Mechanism involved in insulin resistance via accumulation of β-amyloid and neurofibrillary tangles: Link between type 2 diabetes and Alzheimer's disease. *Drug Design, Development and Therapy*, 3999–4021.

Rojas-Carranza, C. A., Bustos-Cruz, R. H., Pino-Pinzón, C. J., Ariza-Marquez, Y. V., Gómez-Bello, R. M., & Canadas-Garre, M. (2018). Diabetes-related neurological implications and pharmacogenomics. *Current Pharmaceutical Design, 24*(15), 1695–1710.

Ruud, J., Steculorum, S. M., & Brüning, J. C. (2017). Neuronal control of peripheral insulin sensitivity and glucose metabolism. *Nature Communications, 8*(1), 15259.

Sędzikowska, A., & Szablewski, L. (2021). Insulin and insulin resistance in Alzheimer's disease. *International Journal of Molecular Sciences, 22*(18), 9987.

Sacramento, J. F., Chew, D. J., Melo, B. F., Donegá, M., Dopson, W., Guarino, M. P., ... Conde, S. V. (2018). Bioelectronic modulation of carotid sinus nerve activity in the rat: A potential therapeutic approach for type 2 diabetes. *Diabetologia, 61*, 700–710.

Shpakov, A. O., Zorina, I. I., & Derkach, K. V. (2023). Hot spots for the use of intranasal insulin: Cerebral ischemia, brain injury, diabetes mellitus, endocrine disorders and postoperative delirium. *International Journal of Molecular Sciences, 24*(4), 3278.

Simsir, I. Y., Soyaltin, U. E., & Cetinkalp, S. (2018). Glucagon-like peptide-1 (GLP-1) is associated with Alzheimer's disease. *Diabetes & Metabolic Syndrome: Clinical Research & Reviews, 12*(3), 469–475.

Song, X., Tan, L., Wang, M., Ren, C., Guo, C., Yang, B., ... Pei, J. (2021). Myricetin: A review of the most recent research. *Biomedicine & Pharmacotherapy, 134*, 111017.

Soto, M., Herzog, C., Pacheco, J. A., Fujisaka, S., Bullock, K., Clish, C. B., & Kahn, C. R. (2018). Gut microbiota modulates neurobehavior through changes in brain insulin sensitivity and metabolism. *Molecular Psychiatry, 23*(12), 2287–2301.

Stenvers, D. J., Scheer, F. A., Schrauwen, P., la Fleur, S. E., & Kalsbeek, A. (2019). Circadian clocks and insulin resistance. *Nature Reviews Endocrinology, 15*(2), 75–89.

Tumminia, A., Vinciguerra, F., Parisi, M., & Frittitta, L. (2018). Type 2 diabetes mellitus and Alzheimer's disease: Role of insulin signalling and therapeutic implications. *International Journal of Molecular Sciences, 19*(11), 3306.

Vinuesa, A., Pomilio, C., Gregosa, A., Bentivegna, M., Presa, J., Bellotto, M., ... Beauquis, J. (2021). Inflammation and insulin resistance as risk factors and potential therapeutic targets for Alzheimer's disease. *Frontiers in Neuroscience, 15*, 653651.

Watson, K., Nasca, C., Aasly, L., McEwen, B., & Rasgon, N. (2018). Insulin resistance, an unmasked culprit in depressive disorders: Promises for interventions. *Neuropharmacology, 136*, 327–334.

Watson, K. T., Wroolie, T. E., Tong, G., Foland-Ross, L. C., Frangou, S., Singh, M., ... Rasgon, N. L. (2019). Neural correlates of liraglutide effects in persons at risk for Alzheimer's disease. *Behavioural Brain Research, 356*, 271–278.

Zemdegs, J., Martin, H., Pintana, H., Bullich, S., Manta, S., Marqués, M. A., ... Guiard, B. P. (2019). Metformin promotes anxiolytic and antidepressant-like responses in insulin-resistant mice by decreasing circulating branched-chain amino acids. *Journal of Neuroscience, 39*(30), 5935–5948.

Further readings

Chatterjee, S., & Mudher, A. (2018). Alzheimer's disease and type 2 diabetes: A critical assessment of the shared pathological traits. *Frontiers in Neuroscience, 12*. https://doi.org/10.3389/fnins.2018.00383

Min, S., Joon, H., & Mook-Jung, I. (2011). Insulin resistance and Alzheimer's disease. In *InTech eBooks*. https://doi.org/10.5772/23409

Chapter 19

Insulin treatments and effects on neurotransmitters

Omkar Hulawale[1], Mudra Dwivedi[1] and Girdhari Lal Gupta[2]
[1]School of Pharmacy & Technology Management, SVKM'S NMIMS Deemed-to-be University, Shirpur, Maharashtra, India; [2]Department of Pharmacology, School of Pharmacy & Technology Management, SVKM'S NMIMS Deemed-to-be University, Shirpur, Maharashtra, India

Abbreviations

fMRI functional magnetic resonance imaging
GABA gamma-aminobutyric acid
IDE insulin-degrading enzyme
NPY neuropeptide Y
PFK-1 phosphofructokinase-1

Introduction

Insulin is a peptide hormone synthesized by the pancreatic beta cells and plays an important role in maintaining glucose homeostasis in the body. The unique hormone decreases blood sugar and lies at the core of controlling carbohydrate metabolism. Endogenous insulin has a molecular mass of 5807.69 Da. It is first produced as proinsulin, which is subsequently cleaved by the preprotein convertases 1 and 2 (PC1 and PC2) within the rough endoplasmic reticulum. This is cleaved to release the C peptide, thereby giving mature insulin with two polypeptide chains (A and B chains) bridged by three disulfide bridges, stabilizing its functional form (Rahman et al., 2021) (Tables 19.1 and 19.2).

Insulin Production within the pancreas cells is closely regulated, and it is kept in nearly zinc-stabilized hexameric form in the secretory bucket. Insulin is produced in the form of a monomer and released into the circulatory system in response to stimuli such as high blood glucose levels. On secretion, insulin binds to insulin receptors (IR) on cells of diverse tissues, leading to subsequent metabolic signaling and actions. Upon completion of its functions, insulin is broken down mainly by insulin-degrading enzyme (IDE), a metalloprotease that requires zinc. IDE has specific binding sites on insulin and begins degrading in a complicated manner. This incorporates an attack of a carbonyl group of the insulin substrate with a hydroxide ion bound to zinc, resulting in a reactive intermediate. One of the most important residues, Glu111, helps rearrange the bonds due to its formation of hydrogen bonds and proton transfer. Actual breakdown of the peptide by cleavage of the peptide bond leads to inactivating insulin to prevent any extended action in the final stage of degradation (Wang et al., 2025).

Insulin stimulates the synthesis of glucose into glycogen, especially in the liver (hepatic glycogen) and muscles (myoglycogen), to be stored. This starts by phosphorylating the glucose to glucose-6-phosphate (G-6-P) with ATP. The initial step entails the conversion of G-6-P into glucose-1-phosphate (G-1-P) by the action of phosphoglucose mutase enzyme, followed by the conversion initiated by UDP-glucose pyrophosphorylase enzyme to UDP-glucose. The polymerization process is catalyzed by the glycogen synthase that forms alpha-1,4 glycosidic bonds, then converts UDP-glucose to glycogen. The resulting structure is the branched structure produced by the glycogen branching enzyme (GBE), which provides 1,6-linkages. A glycogen primer and initiator is glycogenin (Lv et al., 2025).

In the process of glycolysis, insulin converts glucose into pyruvate to obtain ATP to be used as energy. It controls major enzymes in the process. To begin with, the insulin activates hexokinase, which converts glucose to G-6-P through phosphorylation of glucose. Activating phosphofructokinase-1 (PFK-1) enhances the activity by escalating the concentration of fructose-2,6-bisphosphate to fructose-1,6-bisphosphate, which is then one of the rate-limiting steps. Finally, pyruvate kinase is also activated, catalyzing the last stage of the reaction that leads to the conversion of phosphoenolpyruvate (PEP) into pyruvate. These activities guarantee proper glucose digestion and ATP production, promoting cell energy supply (Rojas-Pirela et al., 2025).

Hormones in Neuroscience. https://doi.org/10.1016/B978-0-443-30263-3.00012-9

TABLE 19.1 Hormones acting on the pancreas.

Hormone	Source (cell type)	Stimulus for secretion	Effect on the pancreas	Type of effect	Regulation
Secretin	Duodenum (S cells)	Acidic chyme in the duodenum	Stimulates bicarbonate secretion from pancreatic ductal cells	Stimulatory (exocrine)	Negative feedback via normalization of duodenal pH
Cholecystokinin (CCK)	Duodenum (I cells)	Presence of fatty acids and proteins in the duodenum	Stimulates enzyme secretion by pancreatic acinar cells	Stimulatory (exocrine)	Feedback inhibition
Gastrin	Stomach (G cells)	Peptides in the stomach	Mild stimulation of insulin secretion	Stimulatory (endocrine)	Inhibited by somatostatin and low gastric pH
Somatostatin	Hypothalamus and GI tract	Elevated nutrient levels	Inhibits secretion of insulin, glucagon, and exocrine enzymes	Inhibitory	Negative feedback

TABLE 19.2 Hormones secreted by the pancreas.

Hormone	Source (cell type)	Stimulus for secretion	Target organs/ tissues	Major functions	Signaling pathway	Regulation
Insulin	β-cells	Elevated blood glucose	Liver, skeletal muscle, adipose tissue	Lowers blood glucose by promoting glucose uptake and storage	Tyrosine kinase pathway (insulin receptor–mediated)	Negative feedback from reduced blood glucose levels
Glucagon	α-cells	Reduced blood glucose	Liver	Increases glucose production via glycogenolysis and gluconeogenesis	Gs–cAMP–PKA pathway	Negative feedback from elevated blood glucose levels
Somatostatin	δ-cells	Elevated glucose and amino acid levels	Pancreatic cells, gastrointestinal tract	Inhibits secretion of insulin, glucagon, and growth hormone	Gi pathway → decreased intracellular cAMP	Local inhibition via negative feedback
Pancreatic polypeptide	PP (F) cells	Fasting, protein ingestion	Pancreas, liver, gastrointestinal tract	Modulates pancreatic exocrine secretion and regulates hepatic glycogen levels	Likely Gi or Gq pathway	Not well-defined; subject to nutrient status cues

Mechanism involved in insulin crossing the brain barrier

Insulin absorption into the central nervous system (CNS) is carried along a plethora of complex pathways, mostly by receptor-mediated transcytosis. Insulin should cross the blood-brain barrier to impact the CNS and affect metabolism regulation, cognition, and others. Transport of insulin via the blood-brain barrier is affected by different factors such as diet, exercise, and disease. The main processes and variables involved in insulin transport across the blood-brain barrier (Ho et al., 2025) (Fig. 19.1).

Receptor-mediated transcytosis

Overall translocation of insulin through the brain endothelial cells (BECs) occurs mainly by receptor-mediated export through insulin receptors (IR). This process involves insulin being bound to IRs and internalised into vesicles to be transported across the BECs, and the mechanism is known as vesicle-mediated transport (Yang et al., 2025).

Endocytosis methods

Two endocytosis pathways thought to be involved in insulin transport are caveolin-mediated endocytosis, particularly in the hypothalamus, and clathrin-mediated endocytosis of insulin binding to the cell surface. BEC Insulin absorption is also associated with lipid raft-mediated endocytosis, which is independent of the effects of the insulin on either the PI3K pathway or the MEK pathway (Zhu et al., 2025).

Factors influencing insulin transport

Although receptor-mediated transcytosis is the primary route for insulin to cross the blood–brain barrier, its efficiency can be modified by various physiological and pathological factors. A high-fat diet (HFD) alters signaling pathways—such as increasing NF-κB transcription factor activity—which does not change insulin receptor (IR) expression but reduces insulin uptake by brain endothelial cells (BECs). Nitric oxide inhibits insulin transcytosis,

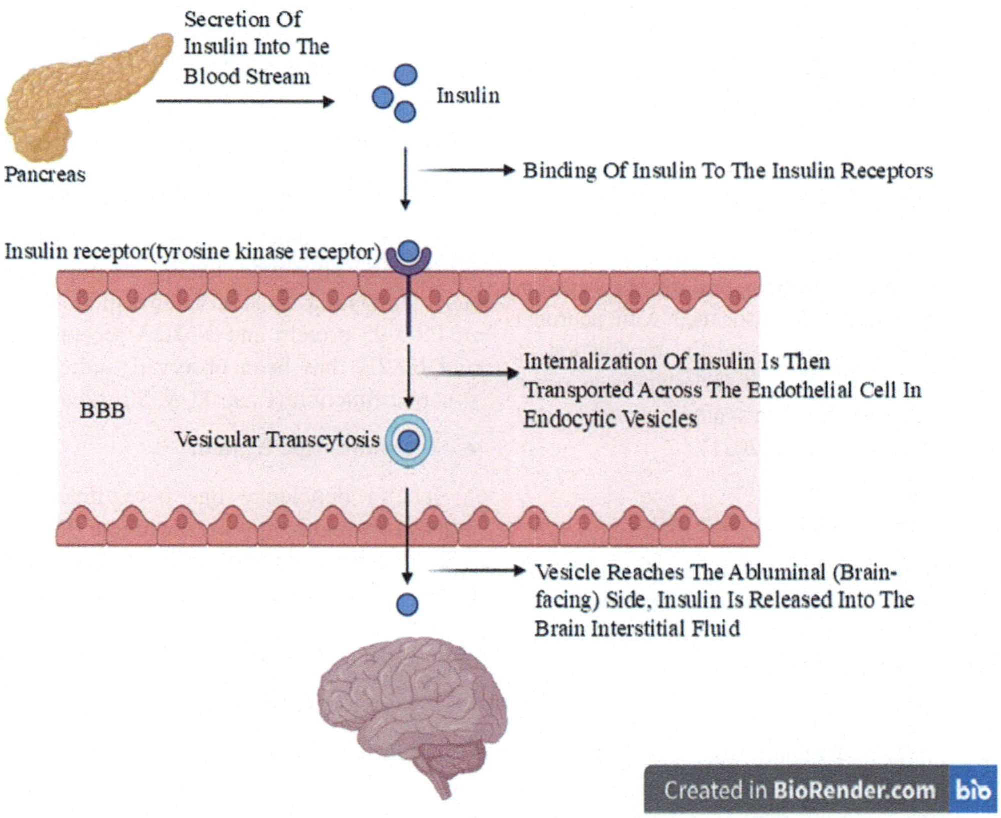

FIGURE 19.1 Mechanism of insulin transport across the blood–brain barrier (BBB). Insulin is released from pancreatic β-cells and enters the bloodstream. It then binds to tyrosine kinase receptors on brain endothelial cells. This begins receptor-mediated endocytosis, which makes vesicles that use vesicular transcytosis to move insulin through the endothelial cell. When the vesicle reaches the abluminal (brain-facing) side of the blood-brain barrier, it releases insulin into the brain's interstitial fluid. It can change how neurons send signals and how they work.

whereas astrocyte activation, for example via L-glutamate stimulation, enhances it.

In vivo studies further reveal that insulin can mediate the transvascular clearance of other molecules, such as opioid peptides, across the blood–brain barrier, suggesting additional transport mechanisms independent of receptor-mediated endocytosis (Misra et al., 2003).

Astrocyte activity, diet composition, and physical exercise influence insulin transport and its functional effects within the central nervous system. Understanding these modulatory pathways is essential for developing targeted therapeutic strategies that exploit CNS insulin signaling to address metabolic and cognitive disorders (Banks et al., 2022).

Localization of insulin receptors in key brain regions

The spatial distribution of insulin receptors (InsR) in critical brain areas—notably the cortex, hypothalamus, and hippocampus—provides essential insights into their roles in regulating metabolism, cognition, and neuronal signaling. Beyond their mere presence, these receptors participate in synaptic plasticity, neuroendocrine regulation, and higher-order cognitive functions. The following subsections detail the localization of InsR in these regions and their functional implications (Belsham & Dalvi, 2021).

Hippocampal distribution of insulin receptors

The hippocampus contains a high density of insulin receptors, underscoring its importance in learning and memory. Insulin signaling in this region has been associated with neuronal survival, functional neural circuitry, and the modulation of synaptic plasticity. The localization of InsR within synaptic compartments suggests a direct role in synapse regulation and information processing (Yang et al., 2021).

Cortical distribution of insulin receptors

In the cerebral cortex, insulin receptors are abundant, particularly in the deeper cortical layers. This distribution pattern supports the involvement of insulin in advanced cognitive processes, including executive function, attention, and decision-making. By influencing neuronal communication within cortical networks, InsR may regulate complex behaviors and higher-order brain functions (Rahman et al., 2021).

Hypothalamic localization of insulin receptors

Within the hypothalamus, insulin receptors play a pivotal role in maintaining metabolic homeostasis and regulating appetite. The highest receptor expression is observed in nuclei critical for energy balance and feeding behavior, such as the arcuate and paraventricular nuclei. Their strategic localization enables the integration of metabolic and neuroendocrine signals, coordinating peripheral and central responses (De la Cruz-Concepción et al., 2023).

The widespread distribution of insulin receptors across key brain regions highlights their dual role in metabolic regulation and cognitive function. Disruptions in insulin signaling have been linked to various neurological disorders, emphasizing the complex interplay between metabolic pathways and brain health (Agrawal et al., 2021).

Modulation of neurotransmitters by insulin

Insulin plays a crucial role in metabolic regulation and modulating key neurotransmitter systems that influence emotional and cognitive functions. Evidence suggests that insulin interacts with glutamatergic, dopaminergic, GABAergic, and serotonergic pathways—systems integral to neurobiological processes and implicated in numerous psychiatric disorders. The interplay between insulin signaling and neurotransmitter function underscores its importance in maintaining metabolic balance and mental health (Chen et al., 2022).

Neurotransmitter systems influenced by insulin

- **GABAergic and Glutamatergic Systems**

Insulin resistance is associated with reduced release of GABA and glutamate, leading to impaired synaptic plasticity. In insulin-resistant animal models, lower expression of PSD-95 protein and NMDA receptor subunits (NR2A and NR2B) has been observed, indicating compromised synaptic function (Czapski & Strosznajder, 2021).

- **Dopaminergic System**

Insulin deficiency has been linked to vasopressin-induced upregulation of dopamine D2 receptor mRNA and alterations in dopamine transport. These changes suggest that insulin modulates dopaminergic signaling, critical for mood regulation and reward processing (Szczepanska-Sadowska et al., 2024).

- **Serotonergic System**

Insulin directly affects serotonergic neurons, particularly in the dorsal raphe nucleus, by modulating 5-HT1A receptor–mediated feedback mechanisms. Insulin resistance, as observed in type 2 diabetes (T2D), can impair serotonergic transmission, contributing to an increased prevalence of depressive symptoms (Georgescu et al., 2021) (Fig. 19.2).

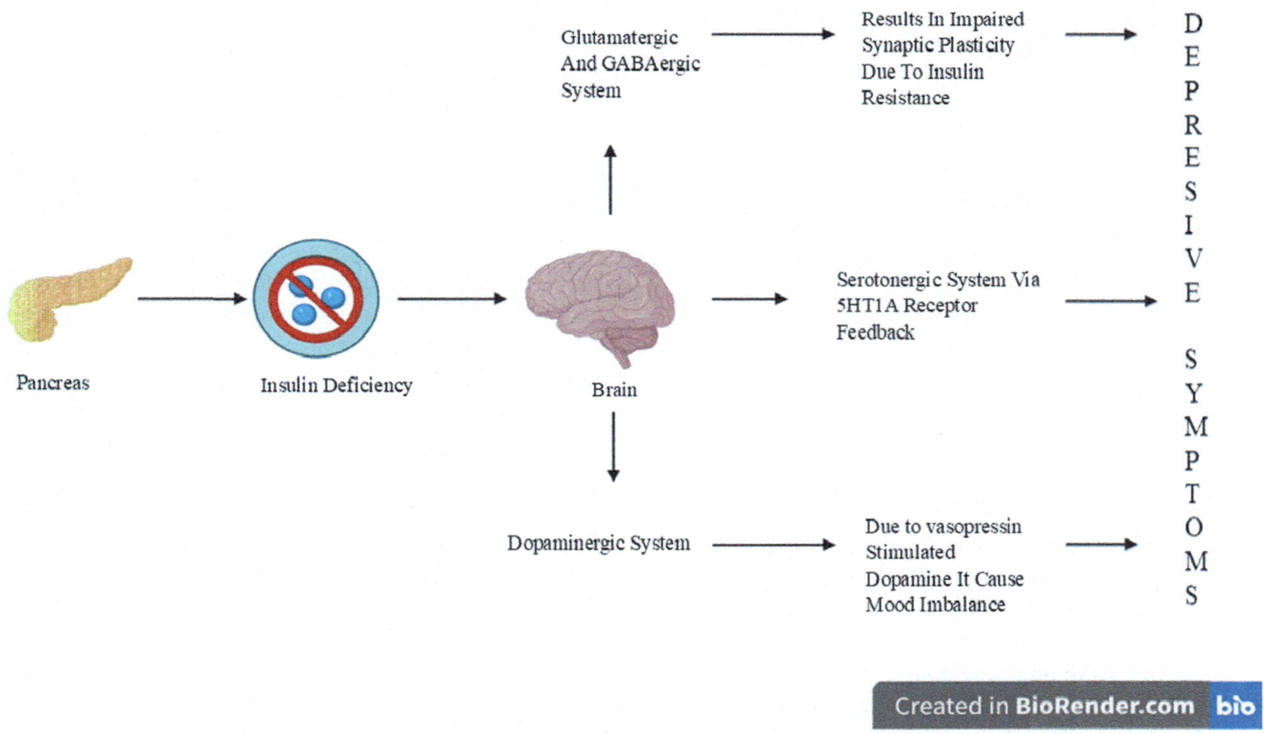

FIGURE 19.2 **Impact of insulin deficiency on neurotransmitter systems leading to depressive symptoms.** Insulin deficiency, stemming from diminished pancreatic secretion, impacts various neurotransmitter pathways in the brain. Insulin resistance diminishes synaptic plasticity by disrupting glutamatergic and GABAergic signaling. Changes in the serotonergic system, especially through feedback from the 5-HT1A receptor, lead to mood dysregulation. Also, dopamine activity in the dopaminergic system that is caused by vasopressin can make moods unstable. These problems all together make depressive symptoms worse.

Implications for mood disorders

- The comorbidity between T2D and depression highlights the importance of the insulin–serotonin relationship. Insulin resistance within serotonergic pathways has been identified as a contributing factor to severe depressive symptoms in patients with T2D (Martin et al., 2021).
- Certain antidiabetic agents demonstrate antidepressant-like properties, suggesting that targeting insulin signaling may offer novel treatment avenues for mood disorders with underlying metabolic disturbances (Martin et al., 2021).

Bi-directional regulation of insulin and neurotransmitters

Notably, neurotransmitters can also regulate insulin secretion from pancreatic β-cells, indicating a bidirectional relationship between these systems. Understanding the dynamics of neurotransmitter–insulin interactions is essential for elucidating their combined effects on metabolic and mental health (Martinez-Sanchez et al., 2022).

Insulin influences neurotransmitter systems in ways vital for emotional stability and metabolic homeostasis. Disruption of this regulatory network—whether through insulin resistance or neurotransmitter imbalance—can contribute to the development of severe psychiatric and metabolic disorders (Blázquez et al., 2022).

Neuropeptide and hormonal crosstalk in metabolic and neuropsychiatric regulation

Neuropeptides, particularly neuropeptide Y (NPY) and orexins, are pivotal in regulating hunger, energy homeostasis, and interactions with hormonal signals. Their crosstalk involves multiple signaling pathways influencing feeding behavior, metabolic processes, and emotional regulation (Mavanji et al., 2022).

Neuropeptide-endocrine regulation of appetite

Orexins and NPY act as potent orexigenic agents, stimulating appetite and food intake. NPY, a primary regulator within the hypothalamus, works synergistically with orexins to enhance feeding behavior.

Hormonal Modulation: Hormones such as cholecystokinin (CCK) and leptin modulate these neuropeptides.

CCK reduces appetite by influencing NPY and orexin signaling, while leptin suppresses NPY expression (Jászberényi et al., 2024).

Mechanistic insights

Signaling Pathways: Orexins activate OX1R and OX2R receptors, triggering intracellular calcium signaling in NPY neurons—counteracting leptin-mediated inhibition of NPY activity.

Cytokine Effects: Pro-inflammatory cytokines, such as interleukin-6 (IL-6), suppress NPY production via STAT3 signaling, linking inflammation to appetite regulation (Affinati & Myers Jr, 2021).

Metabolic and psychiatric implications

Obesity and Dysregulation: NPY–orexin balance disruption contributes to obesity and related metabolic disorders. Conversely, excessive cytokine activity reducing NPY signaling can lead to anorexia and weight loss.

Mood Disorders and Suicidality: Alterations in neuropeptide levels have been associated with depression and suicidal behavior, suggesting a role in emotional regulation (Xie et al., 2025).

Proper NPY–orexin–hormone signaling is essential for maintaining energy balance and healthy appetite regulation. Dysregulation of these pathways may result in adverse metabolic and mental health outcomes (Gebrye et al., 2023).

Insulin resistance, mood disorders, and neurotransmitter dysregulation

Insulin resistance is increasingly recognized as a contributor to mood disorders, including depression and bipolar disorder. This section examines the biological mechanisms, neurotransmitter alterations, inflammatory processes, and therapeutic strategies related to IR in mental health (Miola et al., 2023).

Biological mechanisms

Insulin resistance impairs brain regions involved in mood regulation and can cause blood–brain barrier (BBB) leakage, facilitating neuroprogression in bipolar disorder. It also disrupts mitochondrial function, elevates oxidative stress, and damages neuronal integrity (Calkin et al., 2021).

Neurotransmitter dysregulation

Insulin resistance reduces tryptophan uptake into the brain, impairing serotonin synthesis and contributing to depressive symptoms. In bipolar disorder, IR increases dopamine turnover, destabilizing mood (Zou et al., 2020).

Inflammation and oxidative stress

Chronic IR activates pro-inflammatory cytokine pathways and exacerbates oxidative stress, further impairing synaptic plasticity and worsening mood disturbances.

Therapeutic approaches

Therapeutic strategies for managing insulin resistance in mood disorders include pharmacological interventions such as metformin, which enhances insulin sensitivity, restores blood–brain barrier integrity, boosts serotonergic activity, and alleviates depressive symptoms; pioglitazone and GLP-1 agonists, which improve both metabolic control and mood; and certain antidepressants or mood stabilizers, such as lithium, that also influence glucose metabolism. Lifestyle measures, including diets low in branched-chain amino acids and regular physical activity, further support insulin sensitivity and emotional well-being. Given that sex and ancestry can affect IR susceptibility, personalized treatment approaches are essential. Since IR impairs brain function through neurotransmitter imbalance, inflammation, and oxidative stress, targeting insulin sensitivity offers a promising avenue for improving mood disorders.

Antipsychotic-induced insulin resistance in schizophrenia

Both typical and atypical antipsychotics are linked to metabolic side effects, including insulin resistance, a concern heightened by the fact that metabolic dysfunction often exists in schizophrenia even before treatment begins. Shared genetic predispositions may increase susceptibility to IR, and elevated cortisol levels in drug-naïve patients suggest a role for stress-related hormonal dysregulation. Among antipsychotics, olanzapine shows the strongest association with elevated IR, fasting blood glucose, and fasting insulin levels, while agents such as risperidone and ziprasidone also contribute to varying degrees. Effective management requires regular metabolic monitoring and pharmacological strategies, such as samidorphan or empagliflozin, alongside lifestyle interventions to reduce IR risk. While antipsychotic therapy remains indispensable for controlling schizophrenia, proactive measures are essential to minimize its metabolic consequences.

Intranasal insulin: an emerging CNS-targeted therapy

Intranasal insulin provides a non-invasive delivery method that bypasses the gastrointestinal tract and avoids first-pass metabolism, offering therapeutic potential for diabetes and neurodegenerative disorders. Absorption through the nasal mucosa enables rapid systemic and central nervous system availability, more closely replicating physiological insulin release than subcutaneous administration. Early clinical findings suggest it can effectively manage blood glucose in type 1 and type 2 diabetes, with pharmacokinetic properties that may improve clinical outcomes. While generally well-tolerated, long-term pulmonary safety—particularly.

Precision medicine approaches for metabolic–neurological comorbidities

The incorporation of precision medicine in managing metabolic–neurological comorbidities, especially those related to insulin dysregulation, signifies a revolutionary strategy in personalized healthcare. By using cutting-edge technologies and patient-specific profiles, interventions can be customized to improve therapeutic outcomes in conditions like type 2 diabetes and related neurological disorders. The main goal is to improve the effectiveness of treatment while reducing side effects by using personalized approaches.

Precision neuronutrition

Personalized Nutrient Therapies: Precision nutrition focuses on finding and meeting the specific nutrient needs of each patient. This method is especially useful for reducing neuroinflammation and oxidative stress, two major causes of neurological disorders caused by metabolism.

Important Vitamins and Minerals: Nutrients like omega-3 fatty acids, vitamin B12, and magnesium may help keep the brain healthy and stop neurological problems linked to insulin.

Metabolomics in diabetes management

- Biomarker Identification: Metabolomic profiling enables the discovery of biomarkers to guide individualized diabetes management, particularly in understanding how insulin dysregulation influences neurological function.
- Precision Interventions: By analyzing metabolic patterns, clinicians can tailor glucose-lowering therapies to improve glycemic control and neurological outcomes in patients with diabetes-related comorbidities.

Insulin signaling and neurological conditions

- Comorbidity Insights: Research has highlighted strong links between insulin dysregulation and neurological disorders such as Alzheimer's disease and ADHD. Targeting metabolic pathways can help mitigate the progression or severity of these conditions.
- Holistic Solutions: Precision medicine offers the potential to integrate metabolic and neurological care into a unified strategy. However, challenges remain in standardizing treatment protocols across diverse populations, necessitating further research to ensure equitable and effective therapeutic outcomes.

Other hormones, endocrine tissues, and regulatory pathways

Hormones acting on the pancreas

Several gastrointestinal hormones modulate pancreatic endocrine and exocrine activity:

- Secretin—Released from S cells in the duodenum in response to acidic chyme, stimulating bicarbonate secretion by pancreatic ductal cells (exocrine stimulation). This effect is regulated through negative feedback as pH levels normalize.
- Cholecystokinin (CCK)—Secreted by I cells in the duodenum upon detecting fatty acids and proteins, CCK stimulates acinar cells to release digestive enzymes (exocrine stimulation), regulated by feedback inhibition.
- Gastrin—Produced by G cells in the stomach in response to peptides, it mildly stimulates insulin secretion (endocrine effect). Gastrin release is inhibited by somatostatin and low gastric pH.
- Somatostatin—Secreted by the hypothalamus and gastrointestinal tract in response to elevated nutrient levels, somatostatin inhibits insulin, glucagon, and exocrine pancreatic enzyme secretion, acting as a broad suppressor of pancreatic function.

Pancreatic hormones

The pancreas produces several key hormones essential for glucose and metabolic regulation:

- Insulin—Synthesized by β-cells and released in response to elevated blood glucose. Insulin acts on the liver, muscle, and adipose tissue to promote glucose uptake, thereby reducing blood glucose levels. It signals via the tyrosine kinase pathway and is downregulated when glucose levels fall (negative feedback).
- Glucagon—Secreted by α-cells during hypoglycemia, glucagon stimulates hepatic glycogenolysis and gluconeogenesis via the Gs–cAMP–PKA pathway, increasing blood glucose. Its release is inhibited when glucose levels are restored.
- Somatostatin—Produced by δ-cells in response to elevated glucose and amino acid levels, somatostatin inhibits the secretion of insulin, glucagon, and growth hormone through Gi protein–mediated pathways that reduce intracellular cAMP.
- Pancreatic Polypeptide (PP)—Released by PP/F cells during fasting or after protein ingestion, PP influences pancreatic secretion, hepatic glycogen regulation, and

gastrointestinal motility. Although its signaling mechanism is not fully understood, it is thought to involve Gi or Gq pathways.

Intranasal insulin in neurodegeneration and mental health

Insulin-based therapies are being increasingly explored for their potential role in mental health disorders and neurodegenerative diseases, particularly Parkinson's disease (PD) and Alzheimer's disease (AD). Among these, intranasal insulin delivery has emerged as a promising approach, attracting significant research interest due to its ability to enhance cognitive performance and reduce central insulin resistance. This delivery route offers the advantage of bypassing the blood–brain barrier while minimizing systemic side effects, making it a viable therapeutic option for neurological conditions linked to metabolic dysfunction.

Intranasal insulin has shown notable promise in improving metabolic integrity and memory function in individuals with AD and mild cognitive impairment (MCI). Its mechanisms of action are thought to involve the counteraction of amyloid-β toxicity, enhancement of synaptic plasticity, and increased glucose uptake in brain tissue. These effects collectively support neuronal health and cognitive processes. Clinical studies to date indicate that intranasal insulin is generally safe and well-tolerated. However, more research is needed to determine optimal dosing regimens and to understand factors such as sex-specific responses that may influence treatment outcomes.

Neuroimaging techniques have been widely applied to better understand how insulin impacts brain function. Structural imaging studies have revealed associations between central insulin resistance, brain volume loss, and hippocampal atrophy, which are significant contributors to the cognitive decline seen in AD. Functional imaging methods, including functional magnetic resonance imaging (fMRI), have been used to track changes in neural activity following intranasal insulin administration, providing valuable insights into its influence on cognitive processing and neural network connectivity.

The therapeutic relevance of insulin signaling extends beyond Alzheimer's disease. In Parkinson's disease, similar mechanisms of insulin resistance have been observed, suggesting that disrupted insulin signaling may contribute to its pathogenesis. Moreover, the link between insulin resistance and certain psychiatric conditions, such as depression and schizophrenia, raises the possibility that therapies targeting insulin signaling could yield cognitive benefits across a wider spectrum of mental health disorders.

Looking ahead, several areas warrant further investigation. Long-term clinical trials are essential to evaluate whether intranasal insulin can slow or halt cognitive decline over extended periods. Additionally, there is growing interest in combination therapies that target multiple neurodegenerative pathways, including insulin and insulin-like growth factor (IGF) signaling, to produce synergistic effects.

Intranasal insulin therapy represents a compelling and innovative avenue in precision neuropharmacology. Its potential to address both metabolic and neurological aspects of disease highlights the importance of integrated treatment approaches. While further research is needed to refine its application and confirm its long-term benefits, the growing body of evidence supports its promise as a valuable tool in managing Alzheimer's disease, Parkinson's disease, and other cognitive and psychiatric disorders.

Conclusion and future directions

Insulin, which has long been known to help with peripheral glucose metabolism, has now been shown to be an important neuromodulator that affects mood, cognition, appetite control, and neurotransmitter balance. The fact that it can cross the blood-brain barrier and interact with important neural circuits shows how important it is for metabolic and neuropsychiatric health. Disruptions in insulin signaling—stemming from peripheral resistance, antipsychotic treatment, or neurodegenerative conditions—are increasingly associated with mood disorders, cognitive deterioration, and altered neurotransmitter dynamics. Moreover, the interaction of insulin with neuropeptides like neuropeptide Y and orexins underscores a more extensive neuroendocrine network that synthesizes energy homeostasis with emotional regulation.

Intranasal insulin delivery, precision nutrition, and metabolomics-guided interventions are some therapeutic strategies that show promise for treating metabolic–neurological comorbidities. Initial evidence indicates their efficacy in conditions including Alzheimer's disease, Parkinson's disease, depression, bipolar disorder, and schizophrenia. However, difficulties persist in refining dosing protocols, reducing long-term risks, and accommodating population-specific responses.

Future research directions ought to concentrate on:

- Long-term clinical trials to assess enduring cognitive and metabolic advantages of intranasal insulin and other CNS-targeted interventions.

  Mechanistic studies clarifying differences in central insulin signaling and therapeutic responsiveness based on sex, age, and ancestry.
- Combination therapies that combine insulin modulation with drugs that fight neuroinflammation, oxidative stress, and protein buildup in diseases that cause nerve damage.

- Advanced neuroimaging and biomarker profiling to monitor treatment responses, enhance patient selection, and facilitate precision medicine methodologies.
- Investigation of gut–brain–pancreas axis interactions, specifically the influence of microbiota on insulin sensitivity and neuropsychiatric outcomes.
- Preventive strategies that include changes to lifestyle, targeted nutrition, and early metabolic monitoring in groups that are at risk.

To sum up, insulin is at the crossroads of metabolism and neuroscience. Utilizing its dual function via integrated therapeutic strategies could revolutionize the treatment of intricate disorders characterized by the intersection of metabolic and neurological pathways. It will be important to keep working together across disciplines to turn these ideas into safe, effective, and personalized treatments.

Summary points

1. Insulin penetrates the brain via the mechanism of receptor-mediated transcytosis and acts on cognition, mood, and neurotransmission.
2. It has several neurotransmitter systems, such as glutamate, dopamine, serotonin, and GABA, which it governs.
3. We associate mood disorders, including depression and bipolar disorder, with insulin resistance and impaired messages to the brain.
4. Insulin pathways are associated with neuropeptides such as neuropeptide Y and orexins and regulate hunger and energy balance.
5. Non-invasive methods of cognitive deterioration and neurodegenerative diseases, such as intranasal insulin, have good prospects as a candidate therapy.
6. There is insulin resistance in schizophrenia patients who are treated with antipsychotic drugs, particularly olanzapine.
7. Neuronutrition and precision medicine present a special solution in treating metabolic-neurological comorbidities.

Mini-dictionary

1. Astrocytes: Star-shaped glial cells in the central nervous system that support neuronal function, regulate neurotransmitter levels, and maintain the blood-brain barrier.
2. Caveolin-Mediated Endocytosis: Cells pull in things like insulin through small bubble-like openings.
3. Neuropeptide Y (NPY): A brain chemical that makes you hungry and controls energy use.
4. Orexins: Chemicals in the brain that help control sleep, hunger, and energy levels.
5. Receptor-Mediated Transcytosis: Substances move across cells by attaching to specific receptors.

References

Affinati, A. H., & Myers Jr, M. G. (2021). *Neuroendocrine control of body energy homeostasis* (Endotext [Internet]).

Agrawal, R., Reno, C. M., Sharma, S., Christensen, C., Huang, Y., & Fisher, S. J. (2021). Insulin action in the brain regulates central and peripheral functions. *American Journal of Physiology-Endocrinology and Metabolism, 321*(1), E156–E163.

Banks, W. A., Noonan, C., & Rhea, E. M. (2022). Evidence for an alternative insulin transporter at the blood-brain barrier. *Aging Pathobiology and Therapeutics, 4*(4), 100.

Belsham, D. D., & Dalvi, P. S. (2021). Insulin signalling in hypothalamic neurones. *Journal of Neuroendocrinology, 33*(4), e12919.

Blázquez, E., Hurtado-Carneiro, V., LeBaut-Ayuso, Y., Velázquez, E., García-García, L., Gómez-Oliver, F., … Pozo, M.Á. (2022). Significance of brain glucose hypometabolism, altered insulin signal transduction, and insulin resistance in several neurological diseases. *Frontiers in Endocrinology, 13*, 873301.

Calkin, C., McClelland, C., Cairns, K., Kamintsky, L., & Friedman, A. (2021). Insulin resistance and blood-brain barrier dysfunction underlie neuroprogression in bipolar disorder. *Front Psychiatry, 12*, 636174. https://doi.org/10.3389/fpsyt.2021.636174

Chen, W., Cai, W., Hoover, B., & Kahn, C. R. (2022). Insulin action in the brain: Cell types, circuits, and diseases. *Trends in Neurosciences, 45*(5), 384–400.

Czapski, G. A., & Strosznajder, J. B. (2021). Glutamate and GABA in microglia-neuron cross-talk in Alzheimer's disease. *International Journal of Molecular Sciences, 22*(21), 11677.

De la Cruz-Concepción, B., Flores-Cortez, Y. A., Barragán-Bonilla, M. I., Mendoza-Bello, J. M., & Espinoza-Rojo, M. (2023). Insulin: A connection between pancreatic β cells and the hypothalamus. *World Journal of Diabetes, 14*(2), 76.

Gebrye, A. T., BPT, D. S., & Kumar, S. (2023). Hormones, peptides, and neurotransmitters, effects on appetite regulation and their relationship to obesity: Systematic review. *Journal of Chemical Health Risks, 14* (1), 317–329.

Georgescu, T., Lyons, D., & Heisler, L. K. (2021). Role of serotonin in body weight, insulin secretion, and glycaemic control. *Journal of Neuroendocrinology, 33*(4), e12960.

Ho, S., Narayan, S. I., & Tropea, D. (2025). Insulin-like growth factor-1 as brain-body connector. *Brain-Body Connections: Bidirectional Communication Between the Brain and Body Systems*, 83–104.

Jászberényi, M., Thurzó, B., Bagosi, Z., Vécsei, L., & Tanaka, M. (2024). *Master regulators of vigilance and stress response: The intricate involvement of orexin/hypocretin neuropeptides in physiology and behavior.*

Lv, S., Zhu, Z., & Xiao, H. (2025). Flavonoids and their metal complexes as potential agents for diabetes mellitus with future perspectives. *Critical Reviews in Food Science and Nutrition*, 1–31.

Martin, H., Bullich, S., Guiard, B. P., & Fioramonti, X. (2021). The impact of insulin on the serotonergic system and consequences on diabetes-associated mood disorders. *Journal of Neuroendocrinology, 33*(4), e12928.

Martinez-Sanchez, N., Sweeney, O., Sidarta-Oliveira, D., Caron, A., Stanley, S. A., & Domingos, A. I. (2022). The sympathetic nervous system in the 21st century: Neuroimmune interactions in metabolic homeostasis and obesity. *Neuron, 110*(21), 3597–3626.

Mavanji, V., Pomonis, B., & Kotz, C. M. (2022). Orexin, serotonin, and energy balance. *WIREs Mechanisms of Disease, 14*(1), e1536.

Miola, A., Alvarez-Villalobos, N. A., Ruiz-Hernandez, F. G., De Filippis, E., Veldic, M., Prieto, M. L., ... Resendez, M. G. (2023). Insulin resistance in bipolar disorder: A systematic review of illness course and clinical correlates. *Journal of Affective Disorders, 334*, 1–11.

Misra, A., Ganesh, S., Shahiwala, A., & Shah, S. P. (2003). Drug delivery to the central nervous system: A review. *Journal of Pharmacy & Pharmaceutical Sciences, 6*(2), 252–273.

Rahman, M. S., Hossain, K. S., Das, S., Kundu, S., Adegoke, E. O., Rahman, M. A., ... Pang, M.-G. (2021). Role of insulin in health and disease: An update. *International Journal of Molecular Sciences, 22* (12), 6403.

Rojas-Pirela, M., Andrade-Alviárez, D., Rojas, V., Marcos, M., Salete-Granado, D., Chacón-Arnaude, M., ... Michels, P. A. (2025). Exploring glycolytic enzymes in disease: Potential biomarkers and therapeutic targets in neurodegeneration, cancer and parasitic infections. *Open Biology, 15*(2), 240239.

Szczepanska-Sadowska, E., Cudnoch-Jędrzejewska, A., & Żera, T. (2024). Molecular interaction between vasopressin and insulin in regulation of metabolism: Impact on cardiovascular and metabolic diseases. *International Journal of Molecular Sciences, 25*(24), 13307.

Wang, S., Wang, X., Luo, Y., & Liang, Y. (2025). A comprehensive review of conventional and stimuli-responsive delivery systems for bioactive peptides: From food to biomedical applications. *Advanced Composites and Hybrid Materials, 8*(1), 12.

Xie, L.-Q., Hu, B., Lu, R.-B., Cheng, Y.-L., Chen, X., Wen, J., ... Dai, Y. (2025). Raptin, a sleep-induced hypothalamic hormone, suppresses appetite and obesity. *Cell Research, 35*(3), 165–185.

Yang, K., Li, Q., Ruan, Y., Xia, Y., & Fang, Z. (2025). Caveolae-mediated transcytosis and its role in neurological disorders. *Biomolecules, 15*(4), 456.

Yang, H., Tang, L., Qu, Z., Lei, S.-H., Li, W., & Wang, Y.-H. (2021). Hippocampal insulin resistance and the sirtuin 1 signaling pathway in diabetes-induced cognitive dysfunction. *Neural Regeneration Research, 16*(12), 2465–2474.

Zhu, Y., Verkhratsky, A., Chen, H., & Yi, C. (2025). Understanding glucose metabolism and insulin action at the blood–brain barrier: Implications for brain health and neurodegenerative diseases. *Acta Physiologica, 241*(2), e14283.

Zou, X. H., Sun, L. H., Yang, W., Li, B. J., & Cui, R. J. (2020). Potential role of insulin in the pathogenesis of depression. *Cell Proliferation, 53*(5), e12806. https://doi.org/10.1111/cpr.12806

Chapter 20

Pancreas and pancreatic hormones: Insulin dosing behavior and "insulin metrics"

Jie Xue and Xinyue Chang
Eli Lilly and Company, Indianapolis, IN, United States

Abbreviations

ADA American Diabetes Association
BGM blood glucose meter
CGM continuous glucose monitoring
CV coefficient of variation
DIA duration of insulin action
ICR insulin-to-carbohydrate ratio
IFM insulin fault metric
IOB insulin on board
ISF insulin sensitivity factor
PIE plasma insulin estimate
PK pharmacokinetics
SMBG self monitoring of blood glucose
T1D type 1 diabetes
T2D type 2 diabetes
TDB total daily basal
TDD total daily doses
TIR time in range

Introduction

Effective insulin therapy depends on both insulin dosing behaviors and the metrics used to evaluate them. Dosing behaviors refer to how insulin is initiated, adjusted, and maintained over time, while insulin metrics provide standardized measures to quantify these behaviors and make them comparable across individuals and populations. Practices vary widely by country, health system, and culture. For example, some settings emphasize carbohydrate exchanges, others include fat protein units, and local eating habits and resources shape what is practical for patients and clinicians. Because approaches differ, the terms used here are intended as commonly used examples rather than universal standards.

The relationship between behaviors and metrics is useful but it is not sufficient by itself. Metrics translate patterns into data that can be monitored and analyzed, while the person living with diabetes provides context that numbers alone cannot capture. Lived experience, preferences, daily routines, the burden of dose decisions, fear of hypoglycemia, and affordability all influence what is safe and sustainable for an individual. Considering both dimensions helps clinicians and researchers keep therapy goal directed and responsive to individual needs while avoiding one size fits all conclusions (Danne et al., 2017).

In clinical practice, insulin metrics can help assess dosing appropriateness, guide treatment adjustments, compare the performance of different titration strategies, and monitor progress toward glycemic targets. In research, they can support the evaluation of new therapies and inform guideline development (Battelino et al., 2023; Davies et al., 2018). The growing use of connected insulin delivery devices and mobile applications now makes it possible, in some settings, to integrate patient-reported context with dosing and glucose data in near real time, potentially enabling more individualized care where such technology is available (Ekberg et al., 2024).

Insulin dosing behavior

Insulin dosing behavior can be characterized by the timing, dose, and frequency of insulin administration within a defined observation window (e.g., per meal, per day, per week, depending on the analytic objective). The impact of these dosing decisions is reflected in blood-glucose patterns (Rodbard, 2007; Toschi et al., 2025) which can be obtained by self-monitoring of blood glucose (SMBG) using a blood glucose meter (BGM) or by continuous

Hormones in Neuroscience. https://doi.org/10.1016/B978-0-443-30263-3.00006-3

glucose monitoring (CGM). SMBG provides discrete, event-based measurements, whereas modern CGM systems provide near-continuous streams, typically one reading about every 5 minutes (e.g., Dexcom G7), allowing finer temporal resolution of dosing effects and postprandial excursions (Cappon et al., 2019).

Because dosing behavior is closely linked to meal timing and glucose monitoring modality, the following sections describe SMBG-based and CGM-based analyses, respectively, to illustrate common analytic frameworks rather than prescriptive standards.

BGM based insulin dosing behavior

SMBG is recommended whenever results are likely to change therapy or enhance safety (ADA Standards of Care in Diabetes 2024), recommended practical times include (ADA Professional Practice Committee, 2024): fasting (pre-breakfast), before meals and snacks, 1–2 hours after meals (to gauge postprandial control), at bedtime and overnight if indicated, etc. These SMBG time points can be mapped to insulin-action domains to guide dose adjustments. For example, fasting, bedtime and overnight readings are used to evaluate glucose control and then determine basal insulin dose increments (Riddle et al., 2003). Also, pre-meal and 1–2 hour post-meal readings are used to adjust prandial insulin dose and timing relative to carbohydrate intake, aiming to minimize postprandial excursions while avoiding late hypoglycemia (Rodbard, 2007; Toschi et al., 2025). From a dosing time perspective, common bolus dosing patterns can be categorized into pre-meal, delayed, missed, and correction bolus based on SMBG reading time points.

A pre-meal bolus is usually given about 10–30 minutes before eating. Evidence supports dosing roughly 15–20 minutes before the first bite to reduce postprandial excursions and improve time in range (TIR) without increasing hypoglycemia. Randomized crossover studies in type 1 diabetes (T1D) and a clinical review both showed benefits with rapid acting insulin 15 minutes before meals and with doses 20 minutes before meals compared with at-meal or post-meal dosing (Cobry et al., 2010; Luijf et al., 2010). Ultra rapid acting insulin can be dosed immediately before meals and significantly reduce post prandial glucose levels compared with rapid acting insulin (e.g., Humalog), as demonstrated in PRONTO-T1D and PRONTO-T2D (Leohr et al., 2021).

A delayed bolus is dosing after the meal has begun, typically within the first 2 hours. Some analyses extend observation to 4 hours or until the next meal to capture late dosing effects. Delayed dosing is associated with higher post-meal peaks and greater glycemic exposure compared with timely pre-meal dosing. Evaluating delayed doses by comparing the 1—2 hour post-meal reading with the pre-meal baseline is appropriate and aligns with structured BGM approaches (Cobry et al., 2010).

A missed bolus is operationally defined as no dose recorded in a set window around a meal, such as 15 minutes before to 60 minutes after, or 2 hours before to 4 hours after. In practice, a 2-hour post-meal glucose above 180 mg/dL or a marked rise from the pre-meal baseline is often used to confirm impact. Connected-pen studies link missed dosing with worse glycemic outcomes, while smart pen was associated with fewer missed boluses and higher TIR (Adolfsson et al., 2020; Ekberg et al., 2024; Parkin et al., 2011).

A correction bolus is given when a BGM reading exceeds the user's correction threshold. An insulin regimen with basal, nutritional, and correction bolus is the preferred glycemic treatment strategy in patients with oral nutritional intake (Diabetes Care, 2025). To limit insulin stacking, many clinical programs advise spacing correction doses by roughly 3—4 hours or using "insulin-on-board" calculations when available (ADA Professional Practice Committee, 2024). There are also insulin boluses that add meal and correction boluses together.

CGM based insulin dosing behavior

Compared with BGM, continuous glucose monitoring (CGM) captures glucose dynamics at higher temporal resolution, allowing glucose excursions to be identified and quantified more reliably. An excursion is typically defined as a rapid rise in interstitial glucose within a specified window (Fig. 20.1). For example, Norlander et al. (2018) operationalized an excursion as a $\geq$70 mg/dL increase within 2 hours in the absence of preceding hypoglycaemia.

Building on CGM-detected excursions, insulin dosing behaviors can be classified by the relationship of a bolus to the onset of the excursion and, where relevant, the recent rate of change in sensor glucose. Wolpert et al. (2025) proposed the following criteria in Table 20.1.

Using these definitions, clinicians and researchers can derive metrics, e.g., time with missed boluses, distribution of excursion onsets, and response profiles (magnitude, duration, area under the excursion)—and relate them to outcomes such as TIR, hypoglycemia burden, and HbA1c. Fig. 20.2 illustrates how dosing-behavior categories can be used to analyze the mean change in CGM glucose after excursion onset in T1D and type 2 diabetes (T2D) cohorts and to compare results between masked and unmasked CGM periods. In practice, accurate classification requires precise timestamp alignment between CGM data and insulin delivery records, along with consistent handling of rate-of-change calculations and exclusion of periods with preceding hypoglycemia when applying excursion criteria (Edwards et al., 2022; Norlander et al., 2018; Wolpert et al., 2025).

Besides bolus patterns classification, insulin behavior based on CGM has similar impact on blood glucose and

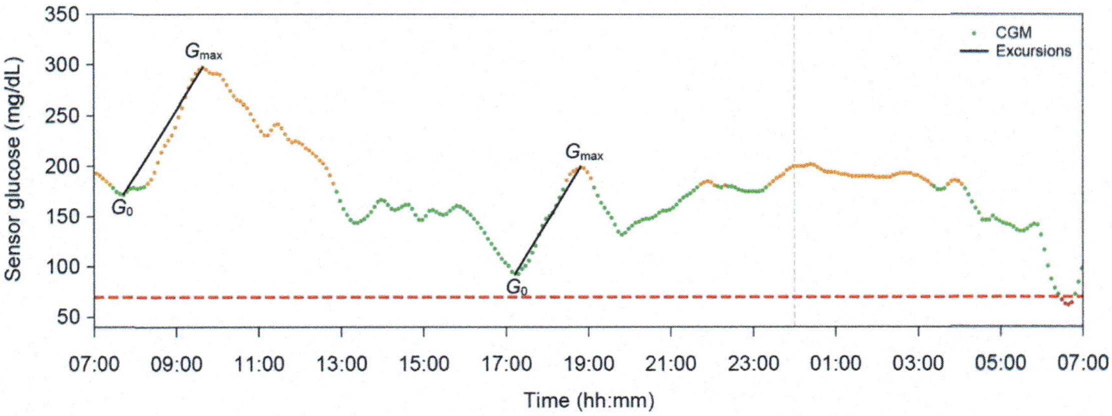

FIGURE 20.1 Glucose excursion example based on modified MAGE+ (Wolpert et al., 2025). Individual sensor measurements are shown as points: red for glucose <70 mg/dL, orange for >180 mg/dL, and green for 70–180 mg/dL. The dashed red line marks 70 mg/dL.

glycemic control. Analogous to discussion in Section 2.1, delayed mealtime boluses are common in T1D and are associated with worse glycemic control (Annuzzi et al., 2024). In T2D, delaying the bolus by approximately 59 minutes reduced 4-hour post-meal TIR from 48% (pre-meal dosing) to 24% (delayed dosing), indicating a marked deterioration in immediate postprandial control (Alexanian et al., 2025). In a NovoPen 6 study of individuals with T1D, differences of roughly two missed meal boluses per week were associated with an ≈8.5% difference in TIR, which corresponds to an ≈0.4–0.7% difference in HbA1c (Adolfsson et al., 2020). Finally, repeated correction bolus within a short insulin-action window can lead to insulin stacking, elevating hypoglycemia risk for both pen and pump users. Appropriate settings for insulin action duration and tracking of bolus insulin on board are therefore essential to minimize unsafe stacking (Heise & Meneghini, 2014; Walsh et al., 2018).

Insulin metrics

As introduced in the previous section, several insulin dosing behaviors were defined. To quantify dosing patterns and evaluate the adequacy, effectiveness, and safety of both rapid-acting and basal insulin, we require insulin metrics. These metrics extend beyond insulin data alone and integrate complementary information, including SMBG and CGM readings, meal timing and composition, insulin dose-and-time records, carbohydrate logs, insulin action duration, and relevant patient characteristics.

Total insulin exposure summary metrics

Total daily doses (TDD) is defined as the sum of all insulin units a person receives in 24 hours, including basal insulin and prandial insulin. TDD can be estimated based on body weight, with typical requirements ranging from 0.4–1 unit

TABLE 20.1 Criteria to characterize glucose excursions and associated insulin boluses based on timing of insulin bolus relative to onset of excursion (Wolpert et al., 2025).

Bolus type	Time from insulin injection to onset of detectable excursion (minimum to maximum, hours)	Criteria: Bolus time	Criteria: Rate of change in sensor glucose within past 20 minutes
Pre-meal	−2 to 0	Bolus within 2 hours before onset of excursion	<1 mg/dL/min
Delayed	0 to 4	Bolus after onset of excursion	≥1 mg/dL/min
Missed	−2 to 4	No recorded bolus between 2 hours before onset of excursion to 4 hours after onset of excursion	NA
Correction	0 to 4	Bolus after peak of the excursion and up to 4 hours after the onset of excursion	NA

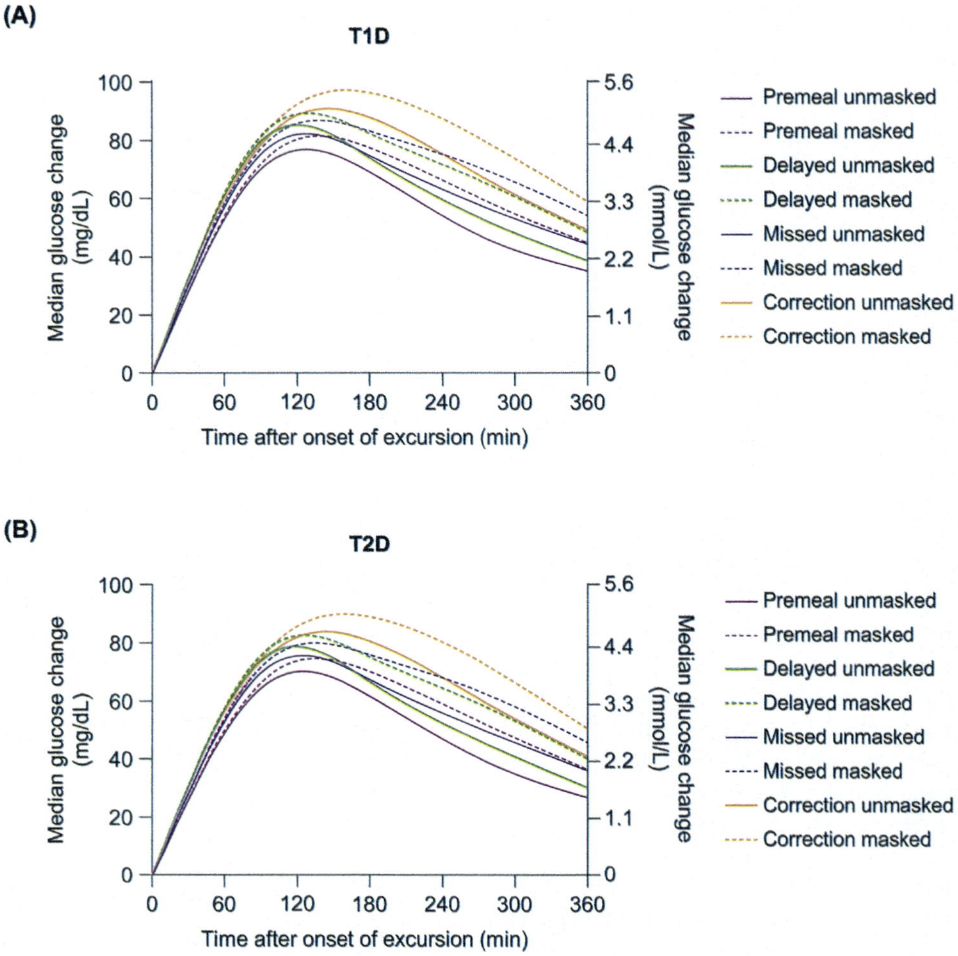

FIGURE 20.2 Glucose excursion profiles for different insulin bolus dose types during the masked and unmasked CGM periods in participants with (A) T1D and (B) T2D. T1D, type 1 diabetes; T2D, type 2 diabetes (Wolpert et al., 2025).

per kilogram per day. On the other hand, the total daily dose unit per kilogram (U/kg) calculated by TDD/weight is an essential insulin metric for initial dose setup and titration process. For example, a common starting dose is approximately 0.5 U/kg per day for metabolically stable adults with T1D. For insulin-naïve adults with T2D, an initial dose is commonly 0.1–0.2 U/kg per day, or a fixed dose of 10 units daily, followed by regular basal insulin titration (ADA Professional Practice Committee, 2024; ADA Professional Practice Committee, 2025). Day-to-day variability in insulin dose, expressed as the coefficient of variation (CV%) over a 7-to 14-day period, can be used to assess the stability of daily insulin requirements, meaning that higher CV indicates more variability due to diet, exercise, adherence, or other factors.

Rapid-acting insulin metrics

One approach to understanding the metrics of rapid-acting insulin is to begin with the standard bolus calculator equation,

$$\text{Dose} = \frac{\text{BG} - \text{BG}_{\text{target}}}{\text{ISF}} + \frac{\text{CHO}}{\text{ICR}} - \text{IOB},$$

which is widely implemented in insulin pumps and insulin pen applications (Schmidt & Nørgaard, 2014; Zisser et al., 2008). The three important insulin metrics in this equation, the insulin sensitivity factor (ISF), the insulin-to-carbohydrate ratio (ICR) and insulin on board (IOB), will be introduced in the following sections, together with other related measures.

ISF and ICR

The ISF expressed in mg/dL per unit., sometimes referred to as the correction factor (CF), represents the expected decrease in blood glucose produced by 1 unit of rapid-acting insulin. The rule of 1800 is commonly recommended to approximate the ISF using the total daily dose (TDD) of insulin (Donner & Sarkar, 2015), i.e.,

$$\text{ISF} = \frac{1800}{\text{TDD}}.$$

Alternative formulas have been proposed, including ISF = 1700/TDD or ISF = (1076/TDD) + 12 (King & Armstrong, 2007). In clinical practice, lower constants (e.g., 1600 or 1400) are also sometimes used to account for individual variability and different insulin sensitivities.

The ICR reflects the amount of carbohydrate in grams that is covered by 1 unit of rapid-acting insulin, effectively offsetting its glucose-lowering effect (Schmidt & Nørgaard, 2014). Similarly, the rule of 500 can be applied to estimate the insulin-to-carbohydrate ratio (Donner & Sarkar, 2015), i.e.,

$$\mathrm{ICR} = \frac{500}{\mathrm{TDD}}.$$

As the 500 rule or its variants may overestimate the ICR in some cases, a modified approach of using 350/TDD for breakfast and 400/TDD for lunch and dinner can yield improved postprandial glycemic outcomes (Alcántara-Aragón et al., 2015).

In practice, additional factors such as dietary protein and fat can also influence postprandial glycemia, and some individuals may benefit from using split or extended bolus strategies, particularly for meals rich in fat or for grazing patterns (Bell et al., 2015; Cobry et al., 2010).

However, due to intra-day (within-day) variability and inter-day (day-to-day) variability in insulin sensitivity, using a fixed ISF is challenging when aiming for tight glucose control (Hinshaw et al., 2013; Moberg et al., 1995; Walsh et al., 2018). Research on ICR has suggested that different values may be optimal for breakfast, lunch, and dinner (Alcántara-Aragón et al., 2015), and further studies have explored automatic adjustment of ICR and ISF to optimize health outcomes in both insulin pen and insulin pump platforms. For example, Herrero et al. (2015) conducted an in silico evaluation of an algorithm that simultaneously adjusted both parameters, assuming a fixed relationship between them. Similarly, Sun et al. (2019) developed a reinforcement learning–based adaptive insulin advisor which updated basal insulin and ICR values daily based on summary glucose metrics. Most of the work in this area remains simulation-based, with few algorithms progressing to clinical trials. One example is the study by Avari et al. (2021), which evaluated the Predictive Personalized Decision Support (PEPPER) algorithm for an adaptive bolus advisor where ICR and/or ISF are adjusted based on postprandial glucose.

IOB

Multiple methods exist for calculating IOB, all of which depend on the key parameter known as the duration of insulin action (DIA) (León-Vargas et al., 2013; Walsh et al., 2014). The simplest approach is the linear decay model, in which the residual insulin effect decreases at a constant rate from the moment of delivery until complete clearance,

$$\mathrm{IOB}(t) = \begin{cases} \text{bolus amount} \times \left(1 - \frac{t}{\mathrm{DIA}}\right), 0 < t < \mathrm{DIA} \\ 0, t \geq \mathrm{DIA} \end{cases}.$$

While this method is straightforward, it assumes a fixed decline in action that does not reflect the actual pharmacokinetic profile. The exponential decay models, such as the two-compartment dynamical model, describe insulin kinetics more accurately,

$$\frac{dC(t)}{dt} = \begin{bmatrix} -K_{\mathrm{DIA}} & 0 \\ K_{\mathrm{DIA}} & -K_{\mathrm{DIA}} \end{bmatrix} C(t) + \begin{bmatrix} 1 \\ 0 \end{bmatrix} u(t),$$

$$\mathrm{IOB}(t) = [1 \quad 1]C(t), C(t) = [C_1(t) \quad C_2(t)]^T,$$

where C_1 and C_2 represent two compartments, $u(t)$ is the delivered insulin dose, and K_{DIA} is a rate constant related to the DIA. This formulation captures the gradual ramp-up to peak action followed by a prolonged tail-off, providing a more realistic representation of insulin action (León-Vargas et al., 2013). Community-driven projects, such as OpenAPS, have compared linear and exponential curves, reporting that exponential models align more closely with measured insulin activity and reduce the risk of stacking in closed-loop systems (OpenAPS, n.d.) (Fig. 20.3).

DIA configuration is a critical factor. A short DIA underestimates IOB, causing excessive bolus doses and insulin stacking. A long DIA overestimates IOB, leading to conservative dosing and hyperglycemia. Extensions to standard IOB methodologies have been proposed to account for physiological variability. Notably, Toffanin et al. (2013) integrated circadian variation in insulin sensitivity into the IOB decay function, enabling dynamic adjustment of residual insulin activity according to the time of day.

Insulin metric based on pharmacokinetics

A Pharmacokinetics (PK) model can be identified using measured drug concentration in animal experiments or clinical trials, resulting in the estimated pharmacokinetics profile that describes drug concentration changing over time. A typical PK curve for an insulin analogue starts 5–10 minutes after injection, achieves peak in 30–90 minutes and eliminates in a longer time than time to peak (Home, 2012). Commonly used insulin metrics derived from pharmacokinetics are summarized as following.

- $C_{\max}$: peak plasma insulin after injection, which indicates the peak exposure and informs potential hypoglycemia risk.
- $T_{\max}$: time needed to reach the peak concentration, which implies absorption rate and how earlier postprandial insulin taken before a meal.

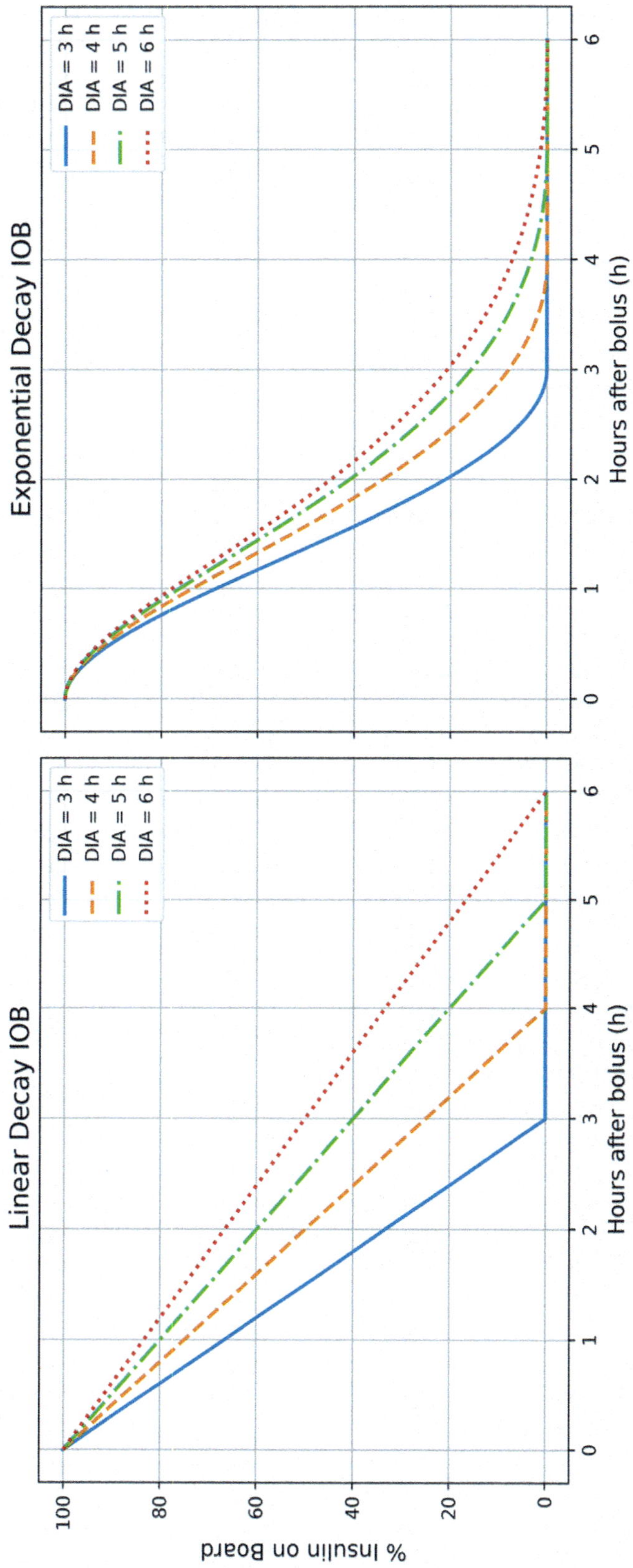

FIGURE 20.3 IOB decline trajectory against time under different DIA options.

- Area under curve (AUC): area under the insulin concentration–time curve that measures overall systemic exposure, from time zero to the last measurements or infinity.
- $t_{1/2}$: time for drug concentration to fall from the peak to half in the elimination phase, which also reflects the duration of insulin presence.
- Partial AUCs: area under curve from zero to 2h, 4h, 6h, etc, which can be used to evaluate early, mid, or late insulin exposure phase.

For insulin analogs formulation comparisons, these metrics are central to demonstrating bioequivalence, where $C_{\max}$ and AUC are two fundamental metrics that quantify exposure and absorption. In Shiramoto et al. (2022) where a study compares two insulin lispro in Japanese patients, the primary PK endpoints were $C_{\max}$ and AUC from time zero to the last quantifiable concentration. Similarly, in an insulin apart PK/PD comparison, Hövelmann et al. (2021) used partial AUC (from time zero to 12h) and $C_{\max}$ as primary PK endpoints.

PIE and IFM

Researchers have proposed additional insulin metrics to address specific needs in automated insulin delivery. For example, Howsmon et al. (2017) introduced the plasma insulin estimate (PIE) as the basis for a novel insulin fault metric (IFM).

The PIE represents an estimate of the active plasma insulin concentration, rather than the total insulin administered. It is obtained from a linear pharmacokinetic system and can be calculated recursively,

$$\begin{bmatrix} x_{k+1}^1 \\ x_{k+1}^2 \end{bmatrix} = \begin{bmatrix} 0.98 & 0.02 \\ 0 & 0.98 \end{bmatrix} \begin{bmatrix} x_k^1 \\ x_k^2 \end{bmatrix} + \begin{bmatrix} 0 \\ 1 \end{bmatrix} \text{Insulin}_k \quad \text{PIE}_k = \begin{bmatrix} 1 & 0 \end{bmatrix} \begin{bmatrix} x_k^1 \\ x_k^2 \end{bmatrix},$$

assuming that peak of absorption is after 50 minutes. Building on this, the IFM is designed to detect meaningful deviations in insulin delivery over short time scales,

$$\text{IFM}_k = \frac{\overline{\text{PIE}}_{k|k-SW} - \overline{\text{PIE}}_{k|k-LW}}{\overline{\text{PIE}}_{k|k-LW}}.$$

At each time step, the PIE's are computed from the combined bolus and basal contributions, then the short-term average is compared to a longer-term baseline, with the ratio defined as IFM. A higher ratio indicates increased insulin delivery, while a lower ratio suggests a reduction.

Long-acting insulin metrics

Total daily dose

One core long-acting insulin metric used in titration is the total daily basal (TDB). The TDB is defined analogously to the total daily dose (TDD) but includes only basal insulin (e.g., Glargine, Degludec, or Detemir). When considering once-weekly basal insulin, such as icodec and efsitora alfa, the dose can be converted to an equivalent daily amount by dividing by seven. It is important to note that a "high" or "low" TDB does not automatically indicate good or poor therapy. Its interpretation must be made in the context of glycemic outcomes, particularly fasting glucose, nocturnal TIR, and the frequency of hypoglycemic episodes. Furthermore, because different basal insulins have distinct pharmacokinetic and pharmacodynamic characteristics, formulations with flatter profiles, such as glargine U-300, may require modestly higher final TDB to achieve comparable glycemic outcomes (Bolli et al., 2015; Heise et al., 2015; Philis-Tsimikas et al., 2020). In addition, the weight-normalized basal dose, expressed in units per kilogram per day (U/kg/day) adjusts the total daily basal amount for body weight, allowing for fair comparisons across individuals from different populations.

Insulin metrics in titration

Basal insulin titration is the process of regularly adjusting insulin doses in small steps, based on fasting blood glucose, to reach personal glucose goals while reducing hypoglycemia risk (ADA Professional Practice Committee, 2025; Riddle et al., 2003). Traditionally, titration has been carried out during scheduled clinic visits or through remote guidance, with patients self-adjusting their dose according to a written algorithm. More recently, the availability of connected devices has enabled research into potential digitized titration, in which dosing tables are integrated into mobile applications or partly replaced by algorithm-driven adjustment systems (Katz et al., 2025; Luo et al., 2023; Martens et al., 2025).

Given the necessity of titration in long-acting insulin therapy, standardized metrics have been established to objectively assess whether dosing is appropriate and to compare the effectiveness of different titration strategies (Chun et al., 2019; Riddle et al., 2003). Regardless of the titration approach, its overall appropriateness should be considered in relation to multiple elements, including the starting dose, the magnitude of each adjustment step, the frequency of dose adjustments, and the criteria for reducing the dose for safety considerations.

The starting dose is the initial amount of basal insulin prescribed when therapy begins. It can be set as a fixed dose or calculated based on body weight. In some clinical situations, a loading dose, which is a higher initial amount than the usual starting dose, may be used to more rapidly establish adequate basal coverage (Trevisan et al., 2024). For example, a one-time starting dose, three times the weekly dose, enables efsitora concentrations to reach efficacious levels more quickly (Heise et al., 2024).

The magnitude of each adjustment step, referred to as titration step size, and the dose adjustment frequency are interdependent and both strongly influence the speed of achieving glycemic targets, the stability of glucose control, and hypoglycemia risk. In clinical trials, basal insulin is often increased by 1–8 units at intervals ranging from daily to weekly when fasting glucose remains above target. For daily basal insulins, less frequent adjustments can accelerate attainment of target glycemia but may increase variability and hypoglycemia risk. In contrast, smaller, more frequent changes provide finer control and may enhance safety, albeit with slower progress toward goals. Different titration protocols reflect varying balances between step size and frequency. Examples include:

- AT.LANTUS: adjustments every 3 days by +2 units (Davies et al., 2005).
- Treat to Target: weekly adjustments with increments of up to +8 units (Riddle et al., 2003).

Dose reduction criteria for safety are primarily based on the prevention of hypoglycemia. For example, in some clinical trials, a fasting glucose reading below 3.9 mmol/L (70 mg/dL) suggests a reduction of 1–2 units (Davies et al., 2005; Riddle et al., 2003). Some protocols recommend dose reduction when glucose readings approach the hypoglycemia threshold (e.g., 80 mg/dL), even if the target range has not been reached (Gerstein et al., 2006). Safety criteria for once-weekly basal insulins, such as icodec and efsitora, were more detailed, specifying dose reduction if there were at least three readings at or below 70 mg/dL, at least one reading below 54 mg/dL, or any episode of severe hypoglycemia (Bergenstal et al., 2023; Trevisan et al., 2024).

Application of materials

In a fully closed-loop study in T2D, CV (Section 3.1) was evaluated across specific time intervals on morning, afternoon, evening and overnight respectively to characterize challenges in glucose control (Boughton et al., 2021). A closed-loop insulin delivery study in T1D further demonstrated that comparing nighttime and daytime CVs can help explain the commonly observed variability in morning glycemia (Ruan et al., 2016).

As introduced in Section 3.2.1, the insulin-to-carbohydrate ratio (ICR) and insulin sensitivity factor (ISF) are essential part in bolus calculation equation, both contributing directly to rapid acting insulin dosing. Insulin on board (IOB) estimation (Section 3.2.2) is essential for modern insulin dosing strategies, designed to prevent insulin stacking that may cause hypoglycemia. Clinically, the insulin pumps or dosing software perform the IOB calculation in the backend.

In Section 3.2.3, the PK estimate curve infers directly on insulin formulation and kinetics characteristics, but also further indicates on clinical consequences and practice guidance (Haahr & Heise, 2020). The faster absorption and quicker onset of action in pharmacokinetics is believed to lead significantly lower postprandial glucose excursions (Home, 2012). So $C_{\max}$, $T_{\max}$ and $t_{1/2}$ together can help compare efficacy and hypoglycemia risk for different kinds of rapid-acting insulins and dosing regimens, which guides clinical decisions on optimal dosing behavior and hypoglycemia risk mitigation.

The PIE metric (Section 3.2.4) was incorporated into the LISA algorithm (Losses in Infusion Set Actuation) to detect infusion set failures in artificial pancreas systems. The IFM (Section 3.2.4), a normalized comparison metric, helps identify abrupt changes in dosing patterns and supports timely fault detection in closed-loop control (Cameron et al., 2012; Steil et al., 2011).

For long-acting insulin metrics, it is common to use weight normalized TDB (Section 3.3.1) in clinical trials, observational studies, and guideline recommendations to describe typical dosing ranges, identify patterns of insulin sensitivity or resistance, and guide intensification strategies. For example, in type 2 diabetes, a basal requirement persistently above approximately 0.5 U/kg/day without achieving glycemic targets is often cited as a threshold for considering prandial insulin initiation rather than further basal escalation, as typically higher doses than the threshold increase the risk of hypoglycemia (Rosenstock & Del Prato, 2022; Galindo et al., 2023).

Regarding insulin metrics in titration (Section 3.3.2), a comprehensive assessment should integrate starting dose, titration step size, dose adjustments frequency, and dose reduction criteria together within the broader clinical context, patient characteristics, and treatment objectives. This multi-dimensional perspective ensures that titration strategies are both effective in achieving glycemic targets and aligned with patient safety and quality of care.

Limitations and considerations

While insulin dosing behavior and insulin metrics provide valuable insight into therapy patterns, it is essential to recognize that the person living with diabetes remains at the center of interpretation and care. Metrics can help organize and quantify patterns, but real understanding comes from talking with the patient about meals, timing, activity, stress, and other daily factors that shape those numbers.

First, insulin metrics and dosing behavior should be evaluated together with glucose metrics obtained from SMBG or CGM. A given dosing pattern, whether described by total daily basal, weight normalized basal dose, bolus timing, or adjustment frequency, can only be meaningfully interpreted in relation to the resulting glycemic outcomes such as time in range, fasting glucose, and the frequency of hypoglycemia (Battelino et al., 2023; Bergenstal et al., 2024). Without this combined assessment, there is a risk of misinterpretation, for example labeling a dose as high or low without considering whether it meets individualized glucose targets while maintaining safety (Bolli et al., 2015; Philis-Tsimikas et al., 2020).

Second, the practical use of these metrics depends on both the completeness of recorded data and the conversations that bring those data to life. In many real-world settings, self-reported insulin doses, meal information, and glucose readings may be incomplete or inaccurately timed, reducing the reliability of calculated metrics (Kuritzky et al., 2019). Advanced technologies such as connected insulin pens, smart pumps, and integrated continuous glucose monitoring systems can help address these limitations by automatically capturing insulin delivery, glucose values, and sometimes contextual information such as meal timing (Adolfsson et al., 2020; Ekberg et al., 2024; Luo et al., 2023). Yet even with automation, understanding therapy patterns still depends on engaging with the person behind the data, discussing routines, meal timing, stress, or lifestyle factors that influence dosing and glucose responses. Data completeness supports accuracy, but dialog supports insight, and both are essential for effective clinical care (Bee et al., 2016).

Third, the therapeutic landscape is changing with the introduction of new delivery approaches such as once-weekly basal insulins (Russell-Jones et al., 2023; Trevisan et al., 2024) and formulations with extended or flatter pharmacokinetic profiles (Heise et al., 2015). These therapies can modify established titration practices and may alter the relationship between dosing behaviors, insulin metrics, and glucose outcomes. For instance, weekly dosing requires different definitions of step size, adjustment frequency, and safety criteria (Bergenstal et al., 2024). In addition, changes in day-to-day glycemic variability for weekly compared with daily regimens may provide useful insight for research, even if their relevance to routine clinical care remains to be determined. As these approaches are adopted, it will be important to adjust the existing metrics such that it reflects weekly dosing behavior, insulin action, and treatment response to support safe and effective care (Galindo et al., 2023; Rosenstock & Del Prato, 2022).

In summary, the analysis of insulin metrics and dosing behavior should be placed within a broader clinical framework that incorporates glycemic outcomes, recognizes the challenges of data capture, and adapts to emerging therapies. This integrated approach will help clinicians and researchers transform raw dosing data into meaningful insights that can improve patient care.

Other hormones, endocrine tissues, and pathways

The pancreas regulates metabolism through a coordinated network of hormones secreted by specialized islet cells. While β-cells and insulin are crucial in glucose management, α-, δ-, PP-, and ε-cells secrete complementary hormones that influences energy balance, counter-regulation, and appetite. Although this chapter focuses on insulin-related metrics and dosing behaviors, this section highlights additional pancreatic hormones that underpin recent therapeutic advances in diabetes.

Glucagon

Glucagon, secreted by α-cells, stimulates hepatic glycogenolysis and gluconeogenesis, opposing insulin's anabolic effects. Clinically, understanding glucagon dynamics is critical when adjusting basal insulin to avoid over-correction and nocturnal hypoglycemia. Therapeutically, recombinant glucagon provides emergency rescue for severe hypoglycemia. Emerging dual agonists—such as tirzepatide (glucose-dependent insulinotropic polypeptide (GIP) and glucagon-like peptide-1 (GLP-1) receptor), demonstrate how modulating glucagon pathways can enhance glucose control while promoting weight loss.

Amylin

Amylin is co-released with insulin from β-cells and slows gastric emptying, suppresses postprandial glucagon, and promotes satiety. In type 1 diabetes and in advanced stages of type 2 diabetes, amylin deficiency contributes to faster gastric emptying and exaggerated postprandial glucose excursions. The amylin analog pramlintide restores these physiologic effects, which blunts postprandial glucose excursions and typically permits reductions in prandial insulin dosing.

Incretin

Incretins, notably GLP-1 (glucagon-like peptide-1) and GIP (glucose-dependent insulinotropic peptide), amplify glucose-stimulated insulin secretion. The GLP-1 receptor agonist class has transformed type 2 diabetes care, improving glycemia, reducing insulin dose needs, and facilitating weight loss. Newer dual agonists combines GIP and GLP-1 activity to synergistically enhance β-cell responsiveness. The incretin system exemplifies how multi-hormonal modulation complements exogenous insulin therapy by targeting both pancreatic secretion and peripheral metabolism.

Dosing behavior and metrics

These pancreatic hormones influence metabolic regulation through mechanisms that differ from insulin. In terms of dosing behavior mentioned in Section 2, defined based on its impacts on blood-glucose patterns, the hormone-based therapies beyond insulin should have different definitions and clinical focus. Although similar quantitative measures might be applied to assess their therapeutic impact (e.g., total daily dose, IOB, PIE, and pharmacokinetic-based metrics), metrics related to insulin titration or bolus calculators are not directly transferable to non-insulin based therapies. In clinical practice, physicians will have the data and frontier knowledge to determine which metrics are informative to support patients for optimizing the process of achieving individual glycemic target.

Summary points

- Insulin dosing behavior is defined by the timing, dose, and frequency of administration, which can be analyzed using SMBG or CGM data.
- SMBG-based analysis categorizes bolus dosing into patterns such as pre-meal, delayed, missed, and correction boluses, each impacting glycemic control differently.
- CGM-based analysis allows for more precise classification of dosing patterns and their effects on postprandial glucose excursions, time in range, and hypoglycemia risk.
- Total daily insulin dose (TDD) and its weight-based calculation are foundational metrics for initial dosing and ongoing titration in both type 1 and type 2 diabetes.
- Key rapid-acting insulin metrics include the insulin sensitivity factor (ISF), insulin-to-carbohydrate ratio (ICR), and insulin on board (IOB), with various equations and models used to individualize and optimize dosing.
- Pharmacokinetic metrics such as C_{max}, T_{max}, area under the curve (AUC), and $t_{1/2}$ provide insights into insulin absorption, action, and duration, guiding clinical comparisons and risk assessment for different insulin formulations.
- The plasma insulin estimate (PIE) and insulin fault metric (IFM) are advanced measures developed to enhance automated insulin delivery monitoring and detect dosing anomalies or device failures.
- Long-acting insulin metrics, including total daily basal (TDB), weight-normalized dosing, and titration strategies (starting dose, step size, adjustment frequency, and safety criteria), are essential for individualized therapy adjustment and evaluating treatment effectiveness and safety.

Mini-dictionary of terms

- **Hyperglycemia.** Elevated blood glucose, mostly defined as >180 mg/dL (10 mmol/L).
- **Hypoglycemia.** Abnormally low blood glucose, mostly defined as <70 mg/dL (3.9 mmol/L).
- **Postprandial excursion.** The rise in blood glucose concentrations following a meal and the subsequent return toward baseline.
- **Rapid-acting insulin.** Rapid-acting insulin (e.g., aspart, lispro, glulisine), usually taken before meal, have a fast onset of action, peak within 1–2 hours, and primarily control postprandial glucose.
- **Long-acting insulin.** Long-acting (basal) insulins (e.g., glargine, detemir, degludec), usually taken daily or weekly, provide a slow, steady release of insulin to control blood glucose between meals and overnight.
- **Insulin sensitivity.** Insulin sensitivity refers to how effectively the physiological mechanism responds to insulin, with increased insulin sensitivity indicating a greater ability to utilize insulin for glucose uptake.
- **Insulin action.** The process by which insulin binds to receptors and triggers a signaling cascade that regulates glucose metabolism, hence stimulating glucose uptake.

References

Adolfsson, P., Hartvig, N. V., Kaas, A., Møller, J. B., & Hellman, J. (2020). Increased time in range and fewer missed bolus injections after introduction of a smart connected insulin pen. *Diabetes Technology & Therapeutics, 22*(10), 709–718.

Alcántara-Aragón, V., Gonzalez, C., Corcoy, R., Ubeda, J., & Chico, A. (2015). Carbohydrate-to-insulin ratio in a Mediterranean population of type 1 diabetic patients on continuous subcutaneous insulin infusion therapy. *Journal of Diabetes Science and Technology, 9*(3), 588–592.

Alexanian, S. M., Cheney, M. C., Bello Ramos, J. C., Spartano, N. L., Wolpert, H. A., & Steenkamp, D. W. (2025). Impact of meal insulin bolus timing and bedtime snacking on continuous glucose monitoring–derived glycemic metrics in hospitalized inpatients. *Diabetes Technology & Therapeutics, 27*(7), 511–516.

American Diabetes Association Professional Practice Committee. (2024). Pharmacologic approaches to glycemic treatment: Standards of care in diabetes—2024. *Diabetes Care, 47*(Suppl. 1), S158–S178.

American Diabetes Association Professional Practice Committee. (2025). Pharmacologic approaches to glycemic treatment: Standards of care in diabetes—2025. *Diabetes Care, 48*(Suppl. 1), S181–S206.

Annuzzi, G., Triggiani, R., De Angelis, R., Rainone, C., Corrado, A., Scida, G., ... Bozzetto, L. (2024). Delayed prandial insulin boluses are an important determinant of blood glucose control and relate to fear of hypoglycemia in people with type 1 diabetes on advanced technologies. *Journal of Diabetes and Its Complications, 38*(2), 108689.

Avari, P., Leal, Y., Herrero, P., Wos, M., Jugnee, N., Arnoriaga-Rodríguez, M., ... Reddy, M. (2021). Safety and feasibility of the PEPPER adaptive bolus advisor and safety system: A randomized control study. *Diabetes Technology & Therapeutics, 23*(3), 175–186.

Battelino, T., Alexander, C. M., Amiel, S. A., Arreaza-Rubin, G., Beck, R. W., Bergenstal, R. M., ... Phillip, M. (2023). Continuous glucose monitoring and metrics for clinical trials: An international consensus statement. *Lancet Diabetes & Endocrinology, 11*(1), 42–57.

Bee, Y. M., Batcagan-Abueg, A. P. M., Chei, C. L., Do, Y. K., Haaland, B., Goh, S. Y., ... Matchar, D. B. (2016). A smartphone application to deliver a treat-to-target insulin titration algorithm in insulin-naive patients with type 2 diabetes: A pilot randomized controlled trial. *Diabetes Care, 39*(10), e174–e176.

Bell, K. J., Smart, C. E., Steil, G. M., Brand-Miller, J. C., King, B., & Wolpert, H. A. (2015). Impact of fat, protein, and glycemic index on postprandial glucose control in type 1 diabetes: Implications for intensive diabetes management in the continuous glucose monitoring era. *Diabetes Care, 38*(6), 1008–1015.

Bergenstal, R. M., Philis-Tsimikas, A., Wysham, C., Carr, M. C., Bue-Valleskey, J. M., Botros, F. T., Blevins, T., Rosenstock, J., et al. (2024). Once-weekly insulin efsitora alfa: Design and rationale for the QWINT phase 3 clinical development programme. *Diabetes, Obesity and Metabolism, 26*(8), 3020–3030.

Bergenstal, R. M., Hachmann-Nielsen, E., Kvist, K., Peters, A. L., Tarp, J. M., Buse, J. B., et al. (2023). Increased derived time in range is associated with reduced risk of major adverse cardiovascular events, severe hypoglycemia, and microvascular events in type 2 diabetes: A post hoc analysis of DEVOTE. *Diabetes Technology & Therapeutics, 25*(6), 378–383.

Bolli, G. B., Riddle, M. C., Bergenstal, R. M., Ziemen, M., Sestakauskas, K., Goyeau, H., ... EDITION 3 Study Investigators. (2015). New insulin glargine 300 U/ml compared with glargine 100 U/ml in insulin-naive people with type 2 diabetes on oral glucose-lowering drugs: A randomized controlled trial (EDITION 3). *Diabetes, Obesity and Metabolism, 17*(4), 386–394.

Boughton, C. K., Daly, A., Thabit, H., Hartnell, S., Herzig, D., Vogt, A., ... Hovorka, R. (2021). Day-to-day variability of insulin requirements in the inpatient setting: Observations during fully closed-loop insulin delivery. *Diabetes, Obesity and Metabolism, 23*(8), 1978–1982.

Cameron, F., Niemeyer, G., & Bequette, B. W. (2012). Extended multiple model prediction with application to blood glucose regulation. *Journal of Process Control, 22*(8), 1422–1432.

Cappon, G., Vettoretti, M., Sparacino, G., & Facchinetti, A. (2019). Continuous glucose monitoring sensors for diabetes management: A review of technologies and applications. *Diabetes & Metabolism Journal, 43*(4), 383.

Chun, J., Strong, J., & Urquhart, S. (2019). Insulin initiation and titration in patients with type 2 diabetes. *Diabetes Spectrum, 32*(2), 104–111.

Cobry, E., McFann, K., Messer, L., Gage, V., VanderWel, B., Horton, L., & Chase, H. P. (2010). Timing of meal insulin boluses to achieve optimal postprandial glycemic control in patients with type 1 diabetes. *Diabetes Technology & Therapeutics, 12*(3), 173–177.

Danne, T., Nimri, R., Battelino, T., Bergenstal, R. M., Close, K. L., DeVries, J. H., ... Phillip, M. (2017). International consensus on use of continuous glucose monitoring. *Diabetes Care, 40*(12), 1631–1640.

Davies, M. J., D'Alessio, D. A., Fradkin, J., Kernan, W., Mathieu, C., Mingrone, G., Rossing, P., Tsapas, A., Wexler, D. J., Buse, J. B., et al. (2018). Management of hyperglycemia in type 2 diabetes, 2018. A consensus report. *Diabetes Care, 41*(12), 2669–2701.

Davies, M., Storms, F., Shutler, S., Bianchi-Biscay, M., Gomis, R., & At. Lantus Study Group. (2005). Improvement of glycemic control in subjects with poorly controlled type 2 diabetes: Comparison of two treatment algorithms using insulin glargine. *Diabetes Care, 28*(6), 1282–1288.

Diabetes care in the hospital: Standards of care in diabetes—2025. *Diabetes Care, 48*(Suppl. 1), (2025), S321–S334.

Donner, T., & Sarkar, S. (2015). *Insulin–pharmacology, therapeutic regimens, and principles of intensive insulin therapy.*

Edwards, S., He, X., Wang, W., Poon, J. L., Meadows, E., Price, D., ... Polonsky, W. (2022). Use of connected pen as a diagnostic tool to evaluate missed bolus dosing behavior in people with type 1 and type 2 diabetes. *Diabetes Technology & Therapeutics, 24*(1), 61–66.

Ekberg, N. R., Hartvig, N. V., Kaas, A., Møller, J. B., & Adolfsson, P. (2024). Smart pen exposes missed basal insulin injections and reveals the impact on glycemic control in adults with type 1 diabetes. *Journal of Diabetes Science and Technology, 18*(1), 66–73.

Galindo, R. J., Trujillo, J. M., Wang, C. C. L., & McCoy, R. G. (2023). Advances in the management of type 2 diabetes in adults. *BMJ Medicine, 2*(1), e000372.

Gerstein, H. C., Yale, J. F., Harris, S. B., Issa, M., Stewart, J. A., & Dempsey, E. (2006). A randomized trial of adding insulin glargine vs. avoidance of insulin in people with type 2 diabetes on either no oral glucose-lowering agents or submaximal doses of metformin and/or sulphonylureas. The Canadian INSIGHT (implementing new strategies with insulin glargine for hyperglycaemia treatment) study. *Diabetic Medicine, 23*(7), 736–742.

Haahr, H., & Heise, T. (2020). Fast-acting insulin aspart: A review of its pharmacokinetic and pharmacodynamic properties and the clinical consequences. *Clinical Pharmacokinetics, 59*(2), 155–172.

Heise, T., Andersen, G., Pratt, E. J., Leohr, J., Bergenstal, R. M., Philis-Tsimikas, A., Wysham, C., Carr, M. C., Bue-Valleskey, J. M., Botros, F. T., ... Rosenstock, J. (2024). Once-weekly insulin efsitora alfa: Design and rationale for the QWINT phase 3 clinical development programme. *Diabetes, Obesity and Metabolism, 26*(8), 3020–3030.

Heise, T., & Meneghini, L. F. (2014). Insulin stacking versus therapeutic accumulation: Understanding the differences. *Endocrine Practice, 20*(1), 75–83.

Heise, T., Hoevelmann, U., Nosek, L., Hermanski, L., Bøttcher, S. G., & Haahr, H. (2015). Comparison of the pharmacokinetic and pharmacodynamic profiles of insulin degludec and insulin glargine. *Expert Opinion on Drug Metabolism and Toxicology, 11*(8), 1193–1201.

Herrero, P., Pesl, P., Bondia, J., Reddy, M., Oliver, N., Georgiou, P., & Toumazou, C. (2015). Method for automatic adjustment of an insulin bolus calculator: In silico robustness evaluation under intra-day variability. *Computer Methods and Programs in Biomedicine, 119*(1), 1–8.

Hinshaw, L., Dalla Man, C., Nandy, D. K., Saad, A., Bharucha, A. E., Levine, J. A., ... Basu, A. (2013). Diurnal pattern of insulin action in type 1 diabetes: Implications for a closed-loop system. *Diabetes, 62*(7), 2223–2229.

Home, P. D. (2012). The pharmacokinetics and pharmacodynamics of rapid-acting insulin analogues and their clinical consequences. *Diabetes, Obesity and Metabolism, 14*(9), 780–788.

Hövelmann, U., Raiter, Y., Chullikana, A., Liu, M., Donnelly, C., Lawrence, T., ... Barve, A. (2021). Pharmacokinetic and pharmacodynamic bioequivalence of biosimilar MYL-1601D with US and European insulin aspart in healthy volunteers: A randomized, double-blind, crossover, euglycaemic glucose clamp study. *Diabetes, Obesity and Metabolism, 23*(12), 2670–2678.

Howsmon, D. P., Cameron, F., Baysal, N., Ly, T. T., Forlenza, G. P., Maahs, D. M., ... Bequette, B. W. (2017). Continuous glucose monitoring enables the detection of losses in infusion set actuation (LISAs). *Sensors, 17*(1), 161.

Katz, M., Kazda, C., Xue, J., Frías, J., Brazg, R., Mitra, J., ... Dassau, E. (2025). Safety and functionality of a novel clinical decision-support algorithm with insulin efsitora alfa in adults with type 2 diabetes:

Early feasibility study. *Diabetes Technology & Therapeutics, 27*(9), 737–746.

King, A. B., & Armstrong, D. U. (2007). A prospective evaluation of insulin dosing recommendations in patients with type 1 diabetes at near normal glucose control: Bolus dosing. *Journal of Diabetes Science and Technology, 1*(1), 42–46.

Kuritzky, L., Reid, T. S., & Wysham, C. H. (2019). Practical guidance on effective basal insulin titration for primary care providers. *Clinical Diabetes, 37*(4), 368–376.

León-Vargas, F., Garelli, F., De Battista, H., & Vehí, J. (2013). Postprandial blood glucose control using a hybrid adaptive PD controller with insulin-on-board limitation. *Biomedical Signal Processing and Control, 8*(6), 724–732.

Leohr, J., Dellva, M. A., Carter, K., LaBell, E., & Linnebjerg, H. (2021). Ultra rapid lispro (URLi) accelerates insulin lispro absorption and insulin action vs Humalog® consistently across study populations: A pooled analysis of pharmacokinetic and glucodynamic data. *Clinical Pharmacokinetics, 60*(11), 1423–1434.

Luijf, Y. M., van Bon, A. C., Hoekstra, J. B., & DeVries, J. H. (2010). Premeal injection of rapid-acting insulin reduces postprandial glycemic excursions in type 1 diabetes. *Diabetes Care, 33*(10), 2152–2155.

Luo, Y., Chang, Y., Zhao, Z., Xia, J., Xu, C., Bee, Y. M., … Ji, L. (2023). Device-supported automated basal insulin titration in adults with type 2 diabetes: A systematic review and meta-analysis of randomized controlled trials. *Lancet Regional Health. Western Pacific, 35*.

Martens, T. W., Johnson, J., Davidson, M. B., Schneck, K., Xue, J., Chang, X., & Dassau, E. (2025). Continuous glucose monitoring versus fasting blood glucose basal insulin titration: A retrospective analysis. *Diabetes & Metabolic Syndrome: Clinical Research Reviews*, 103266.

Moberg, E., Kollind, M., Lins, P. E., & Adamson, U. (1995). Day-to-day variation of insulin sensitivity in patients with type 1 diabetes: Role of gender and menstrual cycle. *Diabetic Medicine, 12*(3), 224–228.

Norlander, L. M., Anderson, S., Levy, C. J., Ekhlaspour, L., Lam, D. W., Hsu, L., … Buckingham, B. (2018). Late and missed meal boluses with multiple daily insulin injections. *Diabetes, 67*(Suppl. 1). 992-P.

OpenAPS. (n.d.). Understanding insulin on board calculations. In: OpenAPS documentation. Retrieved August 13, 2025, from https://openaps.readthedocs.io/en/latest/docs/While%20You%20Wait%20For%20Gear/understanding-insulin-on-board-calculations.html.

Parkin, C. G., Hinnen, D. A., & Tetrick, D. L. (2011). Effective use of structured self-management of blood glucose in type 2 diabetes: Lessons from the STeP study. *Clinical Diabetes, 29*(4), 131–138.

Philis-Tsimikas, A., Klonoff, D. C., Khunti, K., Bajaj, H. S., Leiter, L. A., Hansen, M. V., … CONCLUDE Study Group. (2020). Risk of hypoglycaemia with insulin degludec versus insulin glargine U300 in insulin-treated patients with type 2 diabetes: The randomised, head-to-head CONCLUDE trial. *Diabetologia, 63*(4), 698–710.

Riddle, M. C., Rosenstock, J., Gerich, J., & Insulin Glargine 4002 Study Investigators. (2003). The treat-to-target trial: Randomized addition of glargine or human NPH insulin to oral therapy of type 2 diabetic patients. *Diabetes Care, 26*(11), 3080–3086.

Rodbard, D. (2007). Optimizing display, analysis, interpretation and utility of self-monitoring of blood glucose (SMBG) data for management of patients with diabetes. *Journal of Diabetes Science and Technology, 1*(1), 62–71.

Rosenstock, J., & Del Prato, S. (2022). Basal weekly insulins: The way of the future! *Metabolism, 126*, 154924.

Ruan, Y., Thabit, H., Leelarathna, L., Hartnell, S., Willinska, M. E., Dellweg, S., … Hovorka, R. (2016). Variability of insulin requirements over 12 weeks of closed-loop insulin delivery in adults with type 1 diabetes. *Diabetes Care, 39*(5), 830–832.

Russell-Jones, D., Babazono, T., Cailleteau, R., Engberg, S., Irace, C., Kjaersgaard, M. I. S., … Klonoff, D. C. (2023). Once-weekly insulin icodec versus once-daily insulin degludec as part of a basal-bolus regimen in individuals with type 1 diabetes (ONWARDS 6): A phase 3a, randomised, open-label, treat-to-target trial. *The Lancet, 402*(10413), 1636–1647.

Schmidt, S., & Nørgaard, K. (2014). Bolus calculators. *Journal of Diabetes Science and Technology, 8*(5), 1035–1041.

Shiramoto, M., Yoshihara, T., Schmider, W., Takahashi, Y., Nowotny, I., Kajiwara, M., & Muto, H. (2022). Similar pharmacokinetics and pharmacodynamics of biosimilar SAR342434 insulin lispro and japan-approved humalog insulin lispro in healthy Japanese subjects. *Clinical Pharmacology in Drug Development, 11*(6), 754–760.

Steil, G. M., Palerm, C. C., Kurtz, N., Voskanyan, G., Roy, A., Paz, S., & Kandeel, F. R. (2011). The effect of insulin feedback on closed loop glucose control. *The Journal of Clinical Endocrinology & Metabolism, 96*(5), 1402–1408.

Sun, Q., Jankovic, M. V., & Mougiakakou, S. G. (July 2019). Reinforcement learning-based adaptive insulin advisor for individuals with type 1 diabetes patients under multiple daily injections therapy. In *2019 41st annual international conference of the IEEE engineering in medicine and biology society (EMBC)* (pp. 3609–3612). IEEE.

Toffanin, C., Zisser, H., Doyle III, F. J., & Dassau, E. (2013). Dynamic insulin on board: Incorporation of circadian insulin sensitivity variation. *Journal of Diabetes Science and Technology, 7*(4), 928–940.

Toschi, E., Edwards, S., Kao, C. Y., Xue, J., Atakov-Castillo, A., Wang, W., … Wolpert, H. (2025). What really matters?: How insulin dose, timing, and distribution relate to meal composition in free-living people with type 1 diabetes. *Diabetes Technology & Therapeutics, 27*(1), 66–71.

Trevisan, R., Conti, M., & Ciardullo, S. (2024). Once-weekly insulins: A promising approach to reduce the treatment burden in people with diabetes. *Diabetologia, 67*(8), 1480–1492.

Walsh, J., Roberts, R., Bailey, T. S., & Heinemann, L. (2018). Bolus advisors: Sources of error, targets for improvement. *Journal of Diabetes Science and Technology, 12*(1), 190–198.

Walsh, J., Roberts, R., & Heinemann, L. (2014). Confusion regarding duration of insulin action: A potential source for major insulin dose errors by bolus calculators. *Journal of Diabetes Science and Technology, 8*(1), 170–178.

Wolpert, H., Rodbard, D., Xue, J., Johnson, J., & Dassau, E. (2025). Characterizing insulin dosing behaviour and glycaemic excursions: Development of metrics using connected insulin pen and continuous glucose monitoring. *Diabetes, Obesity and Metabolism, 27*(5), 2507–2514.

Zisser, H., Robinson, L., Bevier, W., Dassau, E., Ellingsen, C., Doyle III, F. J., & Jovanovic, L. (2008). Bolus calculator: A review of four "smart" insulin pumps. *Diabetes Technology & Therapeutics, 10*(6), 441–444.

Chapter 21

Metformin treatment and effects on psychiatric behaviors: Measures of depression

Riya Kumari, Tanu Dixit, Soumya Verma, Mansi Sawardekar and Selvan Ravindran

Symbiosis School of Biological Sciences, Faculty of Medical and Health Sciences, Symbiosis International (Deemed University), Pune, Maharashtra, India

List of abbreviations

AMPK activated protein kinase
BDNF brain-derived neurotrophic factor
FST forced swim test
HAM-D hamilton depression rating scale
IL-1β interleukin
IL-6 interleukin
MDD major depressive disorder
NF-κB nuclear factor kappa B
PCOS polycystic ovary syndrome
PHQ-9 patient health questionnaire
RCT randomized controlled trials
ROS reactive oxygen species
SPT sucrose preference test
T2DM type-2 diabetes mellitus
TNF-α tumor necrosis factor
TST tail suspension test

Introduction

Depression is a common and disabling psychiatric illness, a major contributor to the global disease burden, and responsible for a significant level of disability and morbidity globally. Prevalent in more than 280 million people worldwide, it appears as an interaction between the biological, psychological, and sociological determinants that merge to form its origins and development. Despite extensive application of pharmacotherapy, such as antidepressants, and diverse psychotherapeutic therapies, these interventions commonly fail to effect full remission or to halt relapse in most patients. This reflects the importance of current treatments, which are largely based on monoaminergic mechanisms like serotonin and dopamine modulation. The ongoing rates of partial response and relapse indicate the necessity for new therapeutic strategies that target essential biological processes involved in depression pathophysiology. Targeting processes outside traditional neurotransmitter paradigms, including metabolic derangement, impaired insulin signaling, neuroinflammation, and neuroplasticity, is promising for creating more powerful and longer-lasting treatments for this ubiquitous mental illness (Tabara, 2025).

Metabolic dysregulation link

Evidence accumulates to elucidate the involvement of metabolic dysfunctions, including impaired glucose metabolism and insulin resistance, in the pathogenesis and maintenance of depressive disorders. Metabolic disorders such as Type 2 Diabetes Mellitus (T2DM) and obesity have been shown to have an extremely elevated prevalence of depressive symptoms, suggesting an underlying pathophysiologic mechanism. Systemic and neuroinflammation due to chronic hyperglycemia is also responsible, along with mitochondrial dysfunction and increased oxidative stress that further contributes to neuronal damage and interfering with neurotransmission. Together, all these create a bidirectional relationship in which the metabolic disturbances impact brain function, and depression can impair metabolic health in turn. The intricate interplay between central nervous system and peripheral metabolic dysregulation shows that it is possible to use therapeutic approaches that interfere with metabolic processes as a way of treating depressive symptoms. This offers new prospects for the use of metabolic modulators, such as insulin-sensitizing medication, in managing depression by

Hormones in Neuroscience. https://doi.org/10.1016/B978-0-443-30263-3.00017-8

addressing its biological basis. The integration of metabolic management into psychiatric treatment can lead to more effective and integrated treatment protocols for individuals who experience both mood and metabolic issues (Li et al., 2024).

Metformin's emerging role

Metformin, a first-line time-honored oral hypoglycemic drug in widespread practice for the management of Type 2 Diabetes Mellitus (T2DM), has been increasingly on the spotlight due to its postulated therapeutic effects beyond glycemic regulation. With an established safety profile and widespread clinical practice, metformin is famously known to possess pleiotropic effects, such as the modulation of inflammatory pathways, endothelial function improvement, and the activation of AMP-activated protein kinase (AMPK). AMPK activation in itself has been linked with improved neurogenesis, improved synaptic plasticity, and mood stabilization, all pivotal processes often disrupted in depression (Drzewoski & Hanefeld, 2021; Garcia-Guzman et al., 2024). Preclinical and some clinical trials have all yielded findings supporting the postulation that metformin has positive effects on psychiatric symptoms and signs, particularly in those with comorbid mood and metabolic illness. These observations indicate that metformin's mechanism of action is not limited to regulation of metabolism but may have a direct role in brain wellness and mood stabilization. The collective findings show promising opportunities to explore metformin as a repurposed medication in the treatment of depression, targeting the metabolic and neuropsychiatric features of the disease (Fanelli et al., 2025; Wilson et al., 2023).

This chapter is intended to thoroughly integrate present evidence regarding the impact of metformin on depression and psychiatric behaviors. Through a convergence of understanding from mechanism studies, animal models, and human clinical research, this study intends to clarify underlying biological mechanisms, assess preclinical and clinical results, and emphasize the therapeutic potential of metformin in the treatment of depressive illness. The chapter also points out present gaps in knowledge and suggests directions for future work to inform investigative endeavors in this nascent interdisciplinary research area.

Mechanistic insights: Metformin's pathways in depression

Molecular and cellular mechanisms

Chronic neuroinflammation is now being identified as an important causative factor for depression, characterized by high levels of pro-inflammatory cytokines IL-1β, IL-6, and TNF-α. These cytokines facilitate neurodegenerative processes and impair the subtle balance in neurotransmitter signaling that results in mood dysregulation and worsening of depressive symptoms. The sustained inflammatory response may impair synaptic plasticity and functional connectivity within mood-modulating neural circuits, thereby increasing the magnitude and duration of depression. Metformin has exhibited potent anti-inflammatory effects primarily through inhibition of NF-κB activation, the key transcription factor regulating the release of pro-inflammatory cytokines. By inhibiting microglial activation and release of inflammatory mediators, metformin has been shown in preclinical rodent models to significantly reduce neuroinflammation (Kouba et al., 2024). Reduced neuroinflammation is correlated with improved behavioral performance on tests reporting decreased depressive-like behavior. In addition to its anti-inflammatory properties, metformin also targets cellular energy pathways, specifically by AMPK activation, thus offering a further key mechanism for mood control.

AMPK is an important cellular energy sensor and regulator with a central role in neuronal health, synaptic plasticity, and neurogenesis all of which are essential processes engaged in mood stabilization and cognition. Metformin, a well-documented AMPK activator, has been shown in preclinical models to promote hippocampal neurogenesis as well as enhance long-term potentiation, processes linked with improved learning and memory. Activation of AMPK results not only in enhanced insulin sensitivity and consequently, general metabolic health but also in activation of Brain-Derived Neurotrophic Factor (BDNF) signaling, an important molecule required for synaptic plasticity and stress-induced depressive behavior resistance. By activation of AMPK pathways, metformin has the ability to reverse neuronal abnormalities that have been characteristically present in depression (Garcia & Shaw, 2017). This approach illustrates metformin's therapeutic potential to modulate both the metabolic and neuropsychiatric domains, rendering it an innovative and intriguing treatment for mood disorders, especially in co-morbid metabolic abnormality patients. Coupled with these neurogenic activities, metformin's insulin-sensitizing property further reinforces its psychiatric potential, especially in brain areas that are important for mood and cognition.

The insulin receptors are also highly concentrated in the regions of the brain that play the most crucial roles in mood and cognitive functioning such as hippocampus and prefrontal cortex. These regions have been linked to subtle alterations in insulin signaling that cause impaired synaptic flexibility, disrupted neurotransmitter homeostasis and increased susceptibility to depression. Metformin, with insulin-sensitizing effect, has potential to re-establish insulin signaling in the central nervous system and enhance glucose metabolism, and efficiency and connectivity of synapses. Those metabolic changes result in typical interactions between neurotransmitters and encourage neuroplasticity, which is needed to sustain emotional stability and mental resilience. Regulation of depressive symptoms could potentially be achieved with the use of the standard drug metformin as it has a positive impact on the brain insulin resistance of individuals with metabolic comorbidities, including diabetes or obesity (Liu et al., 2022).

These outcomes highlight the possibility of using metformin as an integrative therapy that would span the conceptual gap between the treatment of metabolic and psychiatric disease to achieve overall mental health outcomes. In addition to mitigating the loss of insulin signaling, metformin has protective effects, including on mitochondrial health and oxidative stress, which further contributes to neuronal resilience.

Mitochondrial function and oxidative stress

Mitochondrial dysfunction and oxidative stress are currently acknowledged to be major contributors to the emergence of the depressive disorder as well as its maintenance. Overproduction of Reactive Oxygen Species (ROS) can cause oxidative damage resulting in cellular integrity loss and impaired neurotransmission. It was found that Metformin suppressed the production of ROS and did not affect the structure or activity of mitochondria, thus, preserving the integrity of neurons. Higher mitochondrial performance helps sustain adequate energy to promote synaptic plasticity and neurotransmitter signal, both of which serve in mood stability. Mitochondrial stabilization with metformin could prevent the neuronal apoptosis and increase the stress resistance to the neurodegeneration processes. The protective effect proves that metformin perhaps provides the chance to rectify the ultimate biological deficit of the depression and provides an additional example of how moods could be manipulated and improved the thinking procedure. In the treatment of depression, metformin can also positively interact with other drugs because, in addition to their effect on the level of oxidative stress, it improves the functioning of the mitochondrial complex and its composition is disturbed, which in turn reduces the metabolic activity of the patient (Zhang et al., 2023). To add another dimension to its therapeutic profile, complementing these cellular effects, metformin also acts on systemic physiology via the gut-brain axis.

Gut–brain axis

The intestinal microbiome also plays a role in regulating brain function and behavior via complex interactions with immune, neural and endocrine systems. The development and persistence of depressive symptoms have been linked to a disruption in the microbial ecosystem or dysbiosis. Metformin has been reported to influence the gut microbiota composition by expanding the number of beneficial, short-chain fatty acid-producing bacteria. This gut microbial shift plays a role in restoring gut-brain axis homeostasis, which is essential for sustained balanced neurochemical signaling. Through the diminution of systemic inflammation and the augmentation of neurotransmitter communication, metformin's gut-modulatory actions could foster mood stabilization and alleviate depressive symptoms (Mitrea et al., 2022). These observations highlight metformin's value as a therapeutic compound acting on both metabolic and neuropsychiatric axes via the gut-brain axis.

Figure 21.1 illustrate the multifactorial processes by which metformin could have antidepressant activity. The most prominent pathways involve modulation of

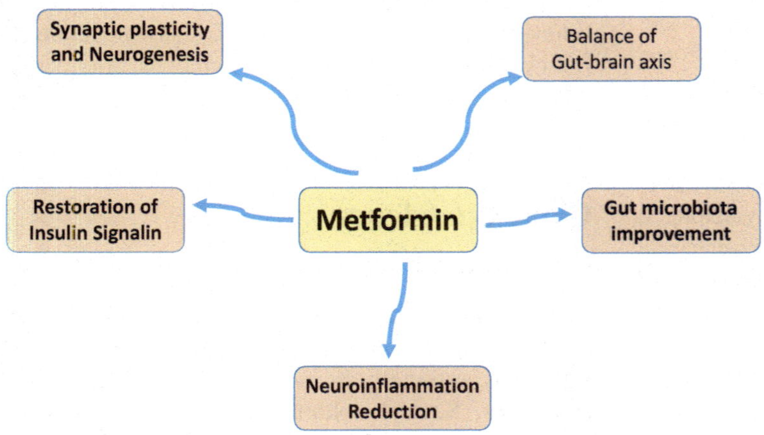

FIGURE 21.1 Metformin's mechanisms in depression relief.

neuroinflammation, AMPK activation, enhanced insulin signaling, and modulation of the gut-brain axis, all converging upon symptomatic relief from depression. These convergent pathways highlight the promise of metformin as a metabolic-psychiatric treatment.

Preclinical evidence

Animal models and behavioral outcomes

Smaller preclinical studies on the potential antidepressant effects of metformin have been primarily founded on valid animal models of depression. They are Forced Swim Test (FST), Tail Suspension Test (TST) and Sucrose Preference Test (SPT) that provides behavioral evaluation of depressive states in rodents. In the FST and the TST, increased immobility is measured over time to reflect behavioral despair, which is a marker of unmotivation and hopelessness typically observed in human depression. The main symptom of depression is anhedonia; SPT measures it by measuring the decreased consumption of the sucrose solution, which means that the ability to experience pleasure is decreased. The models fail to fully capture the complexity of human depressive disorders, but they nonetheless have a lot to say about the behavioral and neurobiological reaction to interventions such as metformin (Ai et al., 2020; Kessing et al., 2020). With these models, a researcher is able to determine the effectiveness of metformin in managing mood behavior and consequently will have a baseline of knowing the therapeutic use of metformin in depression. It is based upon these models that preclinical studies have observed that metformin can exert measurable antidepressant-like effects across behavioral paradigms with a number of studies also indicating that insulin signaling modulation may be one of the primary mediators of these effects in stress- and diet-induced paradigms.

A number of preclinical investigations have indicated that metformin therapy is associated with a substantive decrease in depressive-like behavior in rodent models of chronic stress or metabolic disruption, and in rodent models of high-fat diet-induced stress. In particular, the metformin-treated animals showed less immobility time in the Forced Swim Test (FST), and Tail Suspension Test (TST), which are measuring antidepressant-like behavior and decreased behavioral despair. Furthermore, metformin also raised the preference of sucrose in Sucrose Preference Test (SPT), which indicates the relief of one of the major symptoms of depression, anhedonia. The process by which such enhanced behaviors were mediated was attributed to increased hippocampal neurogenesis, and metformin promoted proliferation and differentiation of the dentate gyrus neural progenitor cells. Furthermore, metformin has been found to exhibit strong anti-inflammatory effects, including blocking the activation of microglia and inhibition of pro-inflammatory cytokines like interleukin-1β (IL-1β) and tumor necrosis factor-alpha (TNF-α) within the central nervous system, thereby eliminating neuroinflammatory events that play a pathogenic role in causing depression. Besides, metformin has also been shown to enhance the expression of Brain-Derived Neurotrophic Factor (BDNF) in brain regions most important for mood regulation, enhancing synaptic plasticity and stress resilience (Chen et al., 2020). These results, collectively, highlight the pleiotropic effect of metformin to reduce depressive behaviors by neurogenic, anti-inflammatory, and neurotrophic mechanisms. These beneficial effects, however, are strongly influenced by both dosage and duration of treatment.

Dose–response and duration of treatment

The antidepressant-like effect of metformin described in preclinical models has been said to be dose-dependent and a function of treatment duration. Numerous doses between 50 and 250 mg/kg/day have been employed, administered either orally or intraperitoneally, with treatment durations ranging from at least 1 week up to several months. Generally, higher doses and longer treatment periods are associated with stronger behavioral improvement, e.g., reduced immobility in depression-like tests, and stronger neuroprotection. It is critical to recognize that very high doses of metformin can cause negative results in animal models, like hypoglycemia and gastrointestinal upset, highlighting the importance of optimal titration of doses to balance safety with efficacy. Also, the treatment duration is instrumental, with chronic treatment being required to induce persistent antidepressant and neural actions (Pecikoza et al., 2022). These observations highlight the need to identify optimal dosing schedules and treatment durations to maximize therapeutic benefit and limit side effects. Standardization of these values will be instrumental in effectively translating preclinical results to useful and safe clinical utility in metformin use in treating depression.

Figure 21.2 shows how the interaction between metformin dosage and duration of treatment influences the outcome of behavior in depression. Although maximal benefits are gained from increased doses with longer durations, brief periods of high dosing can cause side effects, mandating personalized, well-managed therapeutic approaches.

Clinical evidence

Observational and randomized trials

Observational studies have offered important data on the association between metformin therapy and depressive symptoms, especially in groups with metabolic comorbid conditions. Several cohort and cross-sectional research have reported that patients with T2DM using metformin have reduced levels of depressive symptoms as compared

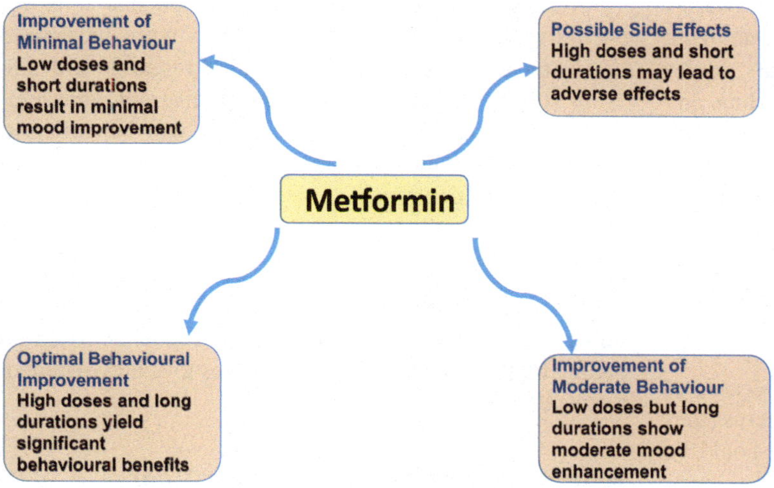

FIGURE 21.2 Metformin's antidepressant effects in animal models.

to those not on the drug. For example, large-scale studies using national health databases have reported an inverse correlation between metformin exposure and diagnoses of Major Depressive Disorder (MDD). Outside diabetic patients, new evidence indicates that the mood-modifying effects of metformin could potentially be applied to patients with polycystic ovary syndrome (PCOS) and obesity, both diseases featuring insulin resistance (Infante et al., 2021). Given that insulin resistance is a shared feature across these disorders, improvements in mood may partly reflect metformin's ability to restore central insulin signaling in addition to improving glycemic control. These studies, however, are limited by confounding influences, such as unmeasured behavioral factors and concomitant treatments. While these observational findings suggest a possible association, randomized controlled trials provide stronger evidence of causality.

Observational studies indicate a possible connection, whereas RCTs yield stronger causality evidence. Various RCTs have compared the impact of metformin on depressive symptoms among heterogeneous populations with T2DM, PCOS, and metabolic syndrome. Findings from these trials report modest improvement in standardized depression rating scales like the Hamilton Depression Rating Scale (HAM-D) and the Patient Health Questionnaire (PHQ-9) after treatment with metformin. Interestingly, a double-blind RCT in patients with T2DM with comorbid depression showed a marked decrease in depressive symptoms at 12 weeks of metformin adjunctive treatment versus placebo. The degree of improvement is heterogeneous between studies, with some studies showing little or no effect (Abdalla et al., 2022). Study design variability, sample size, metformin dose, and duration of treatment account for the heterogeneity of results. Given the variability across individual trials, meta-analyses and systematic reviews have sought to integrate these findings.

Meta-analyses, limitations, and confounders

To integrate the heterogeneous clinical evidence, recent systematic reviews and meta-analyses have been performed. These meta-analyses provide evidence of a statistically significant but clinically modest depression effect of metformin when administered in patients with metabolic disorders. Overall evidence quality is moderate on account of diversity in study designs, small sample sizes, and short follow-up duration (Chele et al., 2024). These analyses highlight the requirement for large, well-conducted trials to definitively determine metformin's antidepressant activity and identify the populations that will most benefit. Despite these encouraging results, interpretation must account for potential confounding factors.

Careful attention to potential confounding factors must be given when interpreting the clinical data. Most studies are conducted in the setting of T2DM or PCOS, in which symptom improvement in depression could be due to improved glycemic control or loss of weight instead of direct neuropsychiatric activity. Simultaneous use of psychotropic drugs and differences in lifestyle variables can impact outcomes. Future research needs to include stratification by initial metabolic condition, thorough adjustment for confounding variables, and standardized depression measures to reveal metformin's real antidepressant potential (Guixue et al., 2024).

Population-specific factor considerations

- **Sex Differences:** There is emerging evidence that sex differences could affect the antidepressant effectiveness of metformin. Biological variations in sex hormones, fat distribution, and insulin sensitivity give rise to differing metabolic and neuropsychiatric reactions. Female rodents have shown increased sensitivity to the mood-stabilizing action of metformin in preclinical experiments due to estrogen-mediated regulation of AMPK signaling and neuroinflammation. Clinically, women with PCOS, which is characterized by insulin resistance and hyperandrogenism, have been noted to experience both metabolic and mood symptom improvement after treatment with metformin. This twofold benefit underscores the therapeutic promise of metformin in treating the entwined metabolic and psychological issues specific to PCOS women. In contrast, other evidence indicates men might have a less robust antidepressant response to metformin because of differences in fat distribution and hormonal patterns (Ilias et al., 2022). Such sex-specific mechanisms should be investigated further in future research, such as the action of sex hormones and brain insulin sensitivity, to maximize the use of metformin in depression treatment in both sexes.

- **Age Differences:** Variations in metformin's effects on depression by age are also of interest due to the different neurobiological environments throughout the life course. In the younger age groups, such as adolescents and young adults, the neuroprotective and anti-inflammatory actions of metformin might confer special advantages in reducing depressive symptoms, especially in the context of early metabolic dysfunction. The information in this age group is limited. In contrast, older individuals tend to have a greater comorbidity burden, polypharmacy, and mitochondrial dysfunction related to age, which can impact metformin's safety and efficacy profile. Early findings indicate that metformin can have neurocognitive advantages in older adults with insulin resistance, but its effects on mood in this group are not yet understood (Wang et al., 2017). Knowledge of age-related pharmacokinetics and pharmacodynamics, such as dose modification and monitoring, is needed to individualize metformin treatments to best target both metabolic and psychiatric effects throughout life.
- **Comorbidities:** The existence of comorbid diseases like obesity, PCOS, and metabolic syndrome has a dramatic impact on both the risk of depression and treatment response. The majority of these metabolic illnesses result in increased inflammation, oxidative stress, and disrupted insulin signaling, all of which have an additive effect on depressive symptomatology. Metformin's proven effects in enhancing insulin sensitivity, lowering systemic inflammation, and correcting adipokine profiles imply a promising intervention in targeting the metabolic roots of depression within these groups. Interestingly, metformin has been demonstrated to have beneficial effects on weight loss and hormonal regulation in PCOS, potentially extending to better mood benefits. Heterogeneity within comorbid presentations and compounded treatments calls for individualized treatment strategies with metformin (Chen & Pang, 2021). Subsequent research will aim to stratify patients by metabolic profile and comorbidities to determine which subgroups will benefit most from metformin's dual action to modify both metabolism and depression.

Table 21.1 demonstrates the subtle influence of sex, age, and comorbidities on the therapeutic effectiveness of metformin in the treatment of depression. It emphasizes the importance of individualized treatment regimens based on unique biological and metabolic profiles to achieve optimal results and reap maximum benefits.

Safety, tolerability, and practical implications

Adverse effects and safety considerations

Metformin is generally regarded as a safe and well-tolerated medication, with an extensive history of use in the management of type 2 diabetes mellitus and other metabolic disorders. Its application in psychiatric populations requires careful consideration of potential adverse effects. The most commonly reported side effects are gastrointestinal disturbances, including nausea, diarrhea, abdominal discomfort, and bloating. These are usually dose-dependent and transient, resolving in conjunction with gradual dose titration or reduction. Rare but serious adverse events, such as lactic acidosis, have been documented more frequently in patients with significant renal impairment or hepatic dysfunction, with careful patient selection and renal function monitoring being critical. Additionally, limited data exist regarding potential interactions between metformin and commonly prescribed psychotropic medications, such as Selective Serotonin Reuptake Inhibitors (SSRIs) or atypical antipsychotics (Drzewoski & Hanefeld, 2021). While no clinically significant pharmacokinetic interactions have been established, clinicians should remain vigilant for possible synergistic or antagonistic effects on mood regulation and metabolic parameters, particularly in complex patient profiles. In light of these safety concerns, practical considerations for patient selection and monitoring become essential.

Practical use and adjunctive potential

Practically, the addition of metformin to depression treatment requires careful patient selection and monitoring. Those patients with concurrent metabolic dysfunction, including insulin resistance, obesity, or polycystic ovary syndrome, might be the best candidates for treatment with metformin, as it has the potential to influence both

TABLE 21.1 Population-specific factors influencing metformin's effectiveness and safety in depression management, summarizing sex, age, and comorbidity considerations.

Factor	Details	Implications for metformin use
Sex differences	Biological differences in sex hormones, fat distribution, and insulin sensitivity may alter metformin's antidepressant effectiveness. Female rodents show heightened sensitivity due to estrogen-mediated AMPK signaling. PCOS women benefit metabolically and psychologically. Men may respond less robustly due to hormonal and fat distribution differences.	Future research should investigate sex-specific mechanisms, including hormonal modulation and brain insulin sensitivity, to optimize metformin therapy.
Age differences	Younger individuals may benefit from metformin's neuroprotective and anti-inflammatory properties, but data are limited. Older adults face higher comorbidity, polypharmacy, and mitochondrial dysfunction, which may affect metformin's safety and efficacy. Early data suggest possible neurocognitive benefits in insulin-resistant elderly.	Dosing and monitoring should be individualized based on age-related pharmacokinetics and pharmacodynamics. Further research is needed to clarify mood effects in elderly populations.
Comorbidities	Conditions like obesity, PCOS, and metabolic syndrome increase inflammation, oxidative stress, and disrupted insulin signaling, exacerbating depression. Metformin improves insulin sensitivity, reduces systemic inflammation, and modulates adipokine profiles. Particularly effective in PCOS with added weight loss and hormonal regulation benefits.	Individualized strategies are essential, considering metabolic profiles and comorbidities. Further stratification research is required to identify patients who would benefit most from metformin's dual actions.

TABLE 21.2 Summary of metformin's safety, tolerability, and practical considerations in depression management.

Aspect	Details	Clinical implications
Adverse effects	Common: GI disturbances (nausea, diarrhea, bloating); dose-dependent and transient. Rare: Lactic acidosis in patients with renal/hepatic impairment. Limited data on interactions with psychotropics like SSRIs and antipsychotics.	Careful patient selection and monitoring are required; gradual dose titration is advised; vigilance for synergistic or antagonistic effects with psychotropics.
Practical considerations	Best suited for patients with metabolic dysfunction (e.g., insulin resistance, obesity, PCOS). Baseline renal and liver assessments needed. Start with low doses (e.g., 500 mg/day) with titration. Monitor adherence, side effects, and lifestyle factors.	Individualized dosing and monitoring are essential; patient education on lifestyle optimization enhances treatment success.
Potential for adjunctive therapy	Metformin may enhance antidepressant efficacy by addressing metabolic dysfunction and neuroinflammation. Promising for treatment-resistant depression and comorbid metabolic conditions.	Requires robust RCTs to establish optimal regimens; potential to integrate metabolic and psychiatric care.

metabolic and psychiatric functions. Baseline evaluation should encompass renal and liver function tests to detect contraindications and prevent adverse effects. Low doses of initiating treatment (such as 500 mg/day) and titration help to improve tolerability and reduce gastrointestinal side effects. Follow-up visits are required to monitor adherence, symptom improvement, and the development of side effects. Exercise and diet represent lifestyle factors which could also be key determinants of the effects of depressive symptoms and metformin effectiveness in psychiatric patients (Doan et al., 2023). Patient education on these issues can lead to improved treatment and compliance over time. In addition to its use as a first-line therapy, metformin also has a potential to be used as an adjunctive therapy to improve the efficacy of antidepressants currently in use.

Another area of research with a lot of interest is the potential therapeutic uses of metformin as a co-agent to conventional antidepressant treatments. Preliminary evidence suggests that metformin may augment the efficacy of selective serotonin reuptake inhibitors and other groups of antidepressants by rectifying the underlying metabolic dysregulation and neuroinflammation, both of which have become emerging pathophysiologic contributors to depression. The effects of a concomitant metformin and antidepressant may be synergistic, especially in patients with comorbid metabolic disease or in cases of treatment-resistant depression. Large randomized controlled trials, however, need to be performed to establish optimal dosing regimens, safety profiles, and long-term outcomes of such combination therapy (Zhou et al., 2020). If confirmed, adjunctive metformin treatment would be a novel, low-cost approach to enhance outcomes of depression treatment, a crossroad between psychiatric and metabolic treatment.

Table 21.2 shows the safety, tolerability, and clinical utility of metformin in depression treatment, with an emphasis on individualized dosing, careful monitoring, and possible adjunctive use with antidepressants. It reinforces metformin's potential role in bridging metabolic and psychiatric care, pending strong evidence to support its efficacy and safety.

Summary points

1. Metabolic disorders such as type-2 diabetes and obesity have been shown to have an elevated prevalence of depressive symptoms
2. The interplay between central nervous system and peripheral metabolic dysregulation shows that it is possible to use therapeutics that interfere with metabolic processes to treat depressive symptoms
3. Metformin in practice for management of Type-2 diabetes, is known to modulate inflammatory pathways, activation of AMP-activated protein kinase and thus improve synaptic plasticity and mood stabilization
4. Metformin as a repurposed medication in the treatment of depression, targeting the metabolic and neuropsychiatric features of the disease
5. Metformin's therapeutic potential to modulate both the metabolic and neuropsychiatric domains, rendering it an intriguing treatment for mood disorders in co-morbid metabolic abnormality patients
6. Use of metformin has a positive impact on the brain insulin resistance of individual with metabolic comorbidities, including diabetes or obesity
7. Metformin influence the gut microbiota composition by increasing the beneficial short chain fatty acid

producing bacteria which is essential for sustained balanced neurochemical signaling

8. Preclinical investigations have indicated that metformin therapy is associated with a substantive decrease in depressive-like behavior in rodent models of chronic stress or metabolic disruption, and in rodent models of high-fat diet-induced stress.
9. The antidepressant-like effect of metformin described in preclinical models has been found to be dose-dependent and a function of treatment duration.
10. Sex differences, Age differences and comorbidities to be considered as population specific factors that can influence metformin's effectiveness and safety in depression management
11. Metformin to depression treatment requires careful patient selection and monitoring. Those patients with concurrent metabolic dysfunction, including insulin resistance, obesity, or polycystic ovary syndrome, might be the best candidates for treatment with metformin, as it has the potential to influence both metabolic and psychiatric functions.

Conclusion

This review identifies the emerging promise of metformin as a new therapeutic approach in the treatment of depressive symptoms and concomitant psychiatric behaviors. Depression, being a multifaceted psychiatric illness, tends to overlap with metabolic abnormalities such as insulin resistance and persistent inflammation. Metformin's well-established safety profile and pleiotropic actions, involving AMPK activation, restoration of insulin signaling, reduction of neuroinflammation, and modulation of the gut–brain axis, render it a potential agent in treating the interlinked pathophysiology of depression and metabolic disturbances. Preclinical studies exhibit metformin's antidepressant-like properties in animal models through indications of behavioral improvement, increased hippocampal neurogenesis, and a decrease in neuroinflammatory markers. Although clinical evidence remains limited and heterogeneous at present, it points toward a moderate yet meaningful link between metformin exposure and lower depressive symptoms, specifically among groups of individuals with comorbid metabolic disorders like type 2 diabetes and polycystic ovary syndrome. Yet, confounding variables such as lifestyle modification, concomitant therapies, and study design heterogeneity weaken the present evidence. The tolerability and safety of metformin, added to its accessibility and cost-effectiveness, provide a useful benefit for the potential repurposing of metformin in psychiatric treatment. Rigorously designed randomized controlled trials are necessary to resolve its immediate impact on depressive symptoms, best-dosing regimens, and long-term consequences. Additionally, particular attention to population-related considerations such as sex, age, and comorbidities will be critical in the adaptation of metformin interventions across various clinical settings.

Acknowledgment

Authors thank Symbiosis International (Deemed University) for facility and support.

References

Abdalla, M. A., Shah, N., Deshmukh, H., Sahebkar, A., Östlundh, L., Al-Rifai, R. H., Atkin, S. L., & Sathyapalan, T. (2022). Impact of metformin on the clinical and metabolic parameters of women with polycystic ovary syndrome: A systematic review and meta-analysis of randomised controlled trials. *Therapeutic Advances in Endocrinology and Metabolism, 13*. https://doi.org/10.1177/20420188221127142, 20420188221127142.

Ai, H., Fang, W., Hu, H., Hu, X., & Lu, W. (2020). Antidiabetic drug metformin ameliorates depressive-like behavior in mice with chronic restraint stress via activation of AMP-activated protein kinase. *Aging and Disease, 11*(1), 31–43. https://doi.org/10.14336/AD.2019.0403

Chele, D., Sirbu, C., Mitrica, M., Toma, M., Vasiliu, O., Sirbu, A., Authier, F. J., Mischianu, D., & Munteanu, A. E. (2024). Metformin's effects on cognitive function from a biovariance perspective: A narrative review. *International Journal of Molecular Sciences, 26*(4), 1783. https://doi.org/10.3390/ijms26041783

Chen, W., & Pang, Y. (2021). Metabolic syndrome and PCOS: Pathogenesis and the role of metabolites. *Metabolites, 11*(12), 869. https://doi.org/10.3390/metabo11120869

Chen, J., Zhou, T., Guo, A. M., Chen, W. B., Lin, D., Liu, Z. Y., & Fei, E. K. (2020). Metformin ameliorates lipopolysaccharide-induced depressive-like behaviors and abnormal glutamatergic transmission. *Biology, 9*(11), 359. https://doi.org/10.3390/biology9110359

Doan, S. N., Patel, S. K., Xie, B., Nelson, R. A., & Yee, L. D. (2023). Disrupting the mood and obesity cycle: The potential role of metformin. *Obesities, 3*(1), 59–75. https://doi.org/10.3390/obesities3010006

Drzewoski, J., & Hanefeld, M. (2021). The current and potential therapeutic use of metformin—The good old drug. *Pharmaceuticals, 14*(2), 122. https://doi.org/10.3390/ph14020122

Fanelli, G., Raschi, E., Hafez, G., Matura, S., Schiweck, C., Poluzzi, E., & Lunghi, C. (2025). The interface of depression and diabetes: Treatment considerations. *Translational Psychiatry, 15*(1), 22. https://doi.org/10.1038/s41398-025-03234-5

Garcia, D., & Shaw, R. J. (2017). AMPK: Mechanisms of cellular energy sensing and restoration of metabolic balance. *Molecular Cell, 66*(6), 789–800. https://doi.org/10.1016/j.molcel.2017.05.032

Garcia-Guzman, P., Schifter-Aceves, L., L, O. A, Romero-Canta, P. (2024). Taguchi design for development of lipid-polymer effervescent floating tablets for metformin prolonged release. *International Journal of Drug Delivery Technology, 14*(01), 38–44. https://doi.org/10.25258/ijddt.14.1.07

Guixue, G., Yifu, P., Xiaofeng, T., Qian, S., Yuan, G., Wen, Y., Conghui, H., & Zuobin, Z. (2024). Investigating the causal impact of polycystic ovary syndrome on gestational diabetes mellitus: A two-sample Mendelian randomization study. *Frontiers in Endocrinology, 15*, 1337562. https://doi.org/10.3389/fendo.2024.1337562

Ilias, I., Rizzo, M., & Zabuliene, L. (2022). Metformin: Sex/gender differences in its uses and effects-narrative review. *Medicine (Kaunas, Lithuania), 58*(3), 430. https://doi.org/10.3390/medicina58030430

Infante, M., Leoni, M., Caprio, M., & Fabbri, A. (2021). Long-term metformin therapy and vitamin B12 deficiency: An association to bear in mind. *World Journal of Diabetes, 12*(7), 916–931. https://doi.org/10.4239/wjd.v12.i7.916

Kessing, L. V., Rytgaard, H. C., Ekstrøm, C. T., Knop, F. K., Berk, M., & Gerds, T. A. (2020). Antidiabetes agents and incident depression: A nationwide population-based study. *Diabetes Care, 43*(12), 3050–3060. https://doi.org/10.2337/dc20-1561

Kouba, R., de Araujo Borba, L., Borges de Souza, P., Gil-Mohapel, J., & Rodrigues, A. L. S. (2024). Role of inflammatory mechanisms in major depressive disorder: From etiology to potential pharmacological targets. *Cells, 13*(5), 423. https://doi.org/10.3390/cells13050423

Li, S., Yang, D., Zhou, X., Chen, L., Liu, L., Lin, R., Li, X., Liu, Y., Qiu, H., Cao, H., Liu, J., & Cheng, Q. (2024). Neurological and metabolic related pathophysiologies and treatment of comorbid diabetes with depression. *CNS Neuroscience and Therapeutics, 30*(4), e14497. https://doi.org/10.1111/cns.14497

Liu, Q., Wang, Z., Cao, J., Dong, Y., & Chen, Y. (2022). The role of insulin signaling in hippocampal-related diseases: A focus on Alzheimer's disease. *International Journal of Molecular Sciences, 23*(22), 14417. https://doi.org/10.3390/ijms232214417

Mitrea, L., Nemeş, S. A., Szabo, K., Teleky, B. E., & Vodnar, D. C. (2022). Guts imbalance imbalances the brain: A review of gut microbiota association with neurological and psychiatric disorders. *Frontiers of Medicine, 9*, 813204. https://doi.org/10.3389/fmed.2022.813204

Pecikoza, U., Tomić, M., Nastić, K., Micov, A., & Stepanović-Petrović, R. (2022). Synergism between metformin and analgesics/vitamin B12 in a model of painful diabetic neuropathy. *Biomedicine and Pharmacotherapy, 153*, 113441. https://doi.org/10.1016/j.biopha.2022.113441

Tabara, M. F. (2025). Addressing social function in the treatment of depression: A call for holistic approaches. *World Journal of Psychiatry, 15*(2), 102658. https://doi.org/10.5498/wjp.v15.i2.102658

Wang, C. P., Lorenzo, C., Habib, S. L., Jo, B., & Espinoza, S. E. (2017). Differential effects of metformin on age-related comorbidities in older men with type 2 diabetes. *Journal of Diabetes and its Complications, 31*(4), 679–686. https://doi.org/10.1016/j.jdiacomp.2017.01.013

Wilson, C., Carpenter, J. S., Nichles, A., Zmicerevska, N., Song, Y. J. C., McHugh, C., Hamilton, B., Hockey, S., Crouse, J., Koethe, D., Scott, E. M., & Hickie, I. B. (2023). Double-blind, randomised placebo-controlled clinical trial of metformin as an adjunct to a sleep-wake, activity and metabolically focused behavioural intervention to improve cardiometabolic outcomes and mood symptoms in youth with major mood syndromes: Study protocol. *BMJ Open, 13*(2), e064682. https://doi.org/10.1136/bmjopen-2022-064682

Zhang, Z., Huang, Q., Zhao, D., Lian, F., Li, X., & Qi, W. (2023). The impact of oxidative stress-induced mitochondrial dysfunction on diabetic microvascular complications. *Frontiers in Endocrinology, 14*, 1112363. https://doi.org/10.3389/fendo.2023.1112363

Zhou, C., Kong, D., Xue, R., Chen, M., Li, G., Xu, Y., Liu, S., Tian, H., & Zhuo, C. (2020). Metformin enhances antidepressant/antipsychotic combination therapy of schizophrenia with comorbid depression in a murine model. *Frontiers in Neuroscience, 14*, 544600. https://doi.org/10.3389/fnins.2020.00517

Chapter 22

Psychotherapeutic interventions in diabetes care: A narrative review and conceptual framework of CBT and beyond

Kenni Wojujutari Ajele and Erhabor Sunday Idemudia
Faculty of Humanities, North-West University, Mafikeng, South Africa

Introduction

Type 2 diabetes mellitus (T2DM) involves both metabolic and psychological processes. Chronic stress contributes to its development and progression by activating the HPA axis and the sympathetic nervous system. This leads to elevated cortisol and catecholamines, which impair insulin sensitivity, increase glucose production, and promote the accumulation of visceral fat. Depression and psychosocial stress further disrupt cortisol rhythms and increase allostatic load, contributing to insulin resistance and poor glycemic control. Stress also activates inflammatory pathways, thereby worsening metabolic dysfunction. Brain regions critical for emotion and self-care, such as the hippocampus and prefrontal cortex, are affected by prolonged exposure to cortisol. Emerging psychometabolic perspectives, including the concept of psychoglycemia, further emphasize the bidirectional relationship between psychological distress and glycemic dysregulation through behavioral, neuroendocrine, and inflammatory pathways. This chapter reviews psychotherapeutic interventions for type 2 diabetes mellitus (T2DM), with a focus on Cognitive Behavioral Therapy (CBT) and related approaches. It introduces the Integrated CBT Model, which maps strategies across cognitive, emotional, behavioral, and temporal domains to support both psychological health and diabetes management.

Psychological stress and diabetes: a complex interaction

Type 2 diabetes mellitus (T2DM) is a multifactorial condition in which metabolic dysregulation interacts with behavioral and psychological processes. Psychological stress is increasingly recognised not just because of living with diabetes but also as a contributing factor in its pathophysiology (Joseph & Golden, 2017). Chronic stress activates the hypothalamic-pituitary-adrenal (HPA) axis and the sympathetic nervous system (SNS), leading to increased secretion of glucocorticoids and catecholamines (Merabet et al., 2022; Yaribeygi et al., 2022). These hormonal shifts elevate hepatic glucose output, impair insulin signaling, and promote visceral adiposity, ultimately worsening glycemic control (Merabet et al., 2022; Yajurvedi, 2018; Yaribeygi et al., 2022).

Moreover, psychosocial stress and depression are linked to the dysregulation of the HPA axis, as evidenced by the flattened diurnal cortisol rhythms and heightened allostatic load. Such patterns are associated with insulin resistance and increased risk of T2DM (Bursać et al., 2018; Joseph & Golden, 2017; Xiao et al., 2020). In individuals with diabetes, higher morning cortisol levels are associated with poorer glycemic control, as measured by time-in-range metrics (Lundqvist et al., 2024; Yaribeygi et al., 2022).

Neuroendocrine and immune pathways

The neuroendocrine stress response is also pro-inflammatory. Activation of the SNS increases the release of epinephrine and norepinephrine, which elevate glucose levels and stimulate inflammatory cytokines like interleukin-6 and TNF-alpha (Janssen, 2022; Kumari et al., 2023; Liang et al., 2018; Liu et al., 2025; Perrelli et al., 2024; Thorens, 2024). This creates a feedback loop that reinforces both neuroendocrine and immune dysfunction,

Hormones in Neuroscience. https://doi.org/10.1016/B978-0-443-30263-3.00016-6

further impairing insulin action (Wilson et al., 2023; Xu & Chen, 2025).

Allostatic load, defined as the cumulative burden of chronic stress on the body, is a key mediator in the relationship between psychological factors and diabetes. Persistent elevation of cortisol not only disrupts metabolic balance but also damages neural substrates such as the hippocampus and prefrontal cortex, which are essential for executive functioning and emotional regulation (Almby et al., 2021; De Alcubierre et al., 2023). These brain regions play a critical role in the cognitive control required for diabetes self-care behaviors.

Depression and anxiety in diabetes

Depression and anxiety are among the most prevalent comorbidities in individuals with diabetes, and both are associated with worse glycemic outcomes and increased risk of complications. Depression has been linked to alterations in HPA axis function, pro-inflammatory cytokine production, and impaired insulin sensitivity (Iob et al., 2020; Joseph & Golden, 2017; Lei et al., 2025; Xu & Chen, 2025). Anxiety disorders, particularly generalized anxiety disorder (GAD), contribute to increased sympathetic nervous system activation, elevated cortisol, and disrupted sleep, all of which are metabolic risk factors (Atasoy et al., 2021; López-Cano et al., 2019; Stahl-Pehe et al., 2022).

The presence of depression or anxiety not only amplifies biological stress responses but also undermines diabetes self-management by impairing motivation, planning, and routine maintenance of health behaviors. Bidirectional relationships are common, as diabetes can increase the likelihood of depression through physiological and psychosocial mechanisms, while depression can impair glycemic control and increase the risk of diabetes onset (Chilimoniuk et al., 2022; Iob et al., 2020; Joseph & Golden, 2017; Lei et al., 2025; Watanabe et al., 2024; Xu & Chen, 2025).

Behavioral and self-care consequences

Psychological distress can impair adherence to diabetes self-management behaviors such as medication compliance, dietary regulation, physical activity, and glucose monitoring (Abdelghaffar et al., 2020; Alzubaidi et al., 2022; Carpenter et al., 2019). These behavioral disruptions may have additive or even synergistic effects with biological stress pathways, compounding glycemic dysregulation. Resilience factors, such as secure attachment and social support, modulate HPA activity and are associated with improved metabolic outcomes (Gunnar & Hostinar, 2015; Hostinar et al., 2014; Thomas et al., 2018). Social buffering mechanisms, particularly those mediated by oxytocin, mitigate the stress response and may offer protective metabolic effects (Wu, 2021). These findings underscore the importance of integrating psychological and social dimensions into diabetes care.

Cognitive behavioral therapy in diabetes care

Cognitive Behavioral Therapy (CBT) is a structured, time-limited psychotherapeutic approach developed in the early 1960s. It aims to reduce psychological distress by helping individuals identify, challenge, and replace distorted cognitions with more realistic and adaptive thoughts (Kora, 2024). In the context of diabetes care, CBT targets both emotional challenges, such as depression and anxiety, as well as diabetes-specific distress and behaviors affecting self-management, including medication adherence, blood glucose monitoring, and lifestyle changes (Abbas et al., 2023; Saha et al., 2024).

The core components of CBT include psychoeducation, cognitive restructuring, behavioral activation, problem-solving and skills training (Abbas et al., 2023; Saha et al., 2024). These sessions are often goal-oriented and include homework assignments that reinforce learning between meetings. For individuals with diabetes, CBT usually includes modules focused on dietary adherence, glucose monitoring, and coping with the stress of chronic illness (Alshehri et al., 2021; Saha et al., 2024). The aim is to build motivation, increase problem-solving skills, and promote long-term disease management (Abbas et al., 2023; Mansour et al., 2022; Saha et al., 2024).

CBT can be delivered in various formats, including individual therapy, group therapy, and digital platforms. Group CBT has demonstrated efficacy in reducing diabetes-related distress and improving glycemic control within a relatively short time frame (Tunsuchart et al., 2020). Digital CBT programmes are also gaining popularity due to their accessibility and cost-effectiveness. For example, a mobile-based CBT program has shown promise in improving emotional well-being and glucose regulation in individuals with type 2 diabetes (Abbas et al., 2023; Vlachou et al., 2022).

Tailored versions of the CBT

Tailored CBT approaches have improved accessibility and effectiveness across diverse patient groups. CBT for insomnia (CBT-I) has been adapted for people with diabetes and shown to enhance sleep quality and glycemic control (Alshehri et al., 2021; Groeneveld et al., 2024). Group CBT (G-CBT) formats and telephone-delivered CBT (T-CBT) have also demonstrated positive outcomes. G-CBT, involving 597 participants, yielded significant psychological improvements (Xu et al., 2021). In contrast, a diabetes-tailored intervention (T-CBT) was associated with reductions in HbA1c and distress scores (Cummings et al., 2019).

Culturally adapted CBT has proven beneficial across diverse populations. One program targeting Asian and Pacific Islander adults improved depressive symptoms and patient attitudes, although the HbA1c benefits were inconsistent (Inouye et al., 2015). A large randomised controlled trial showed a significant decrease in HbA1c (Li et al., 2023). Enhanced CBT (eCBT), incorporating motivational interviewing or emotion regulation, further reduced distress and depression (Newby et al., 2017; Van Bastelaar et al., 2011). Mobile-based CBT apps have also been shown to improve emotional functioning and glucose regulation (Canonico et al., 2022).

Mapping CBT intervention components to key dimensions

Fig. 22.1 and Table 22.1 illustrate the Integrated CBT Model for Diabetes and Depression Comorbidity Management. This framework organises intervention strategies into three core domains: Thoughts, Emotions, and Behaviors, providing a multidimensional approach to psychosocial care. Each domain includes specific CBT components that address the psychological and behavioral challenges commonly encountered in diabetes management.

In the cognitive (thoughts) domain, interventions such as cognitive restructuring help patients identify and challenge negative or irrational thoughts related to diabetes. Psychoeducation supports the understanding of the connection between mental and physical health. Goal setting and self-efficacy techniques build motivation and confidence in managing both conditions. The emotional domain focuses on mood monitoring, stress management strategies such as mindfulness and relaxation, and fostering a strong therapist-patient relationship to enhance emotional safety and engagement. The behavioral domain includes behavioral activation to encourage mood-enhancing activities, as well as problem-solving skills to manage obstacles in self-care and diabetes self-care routines, such as medication adherence, diet, exercise and glucose monitoring. A temporal structure, including scheduled follow-up, reinforces sustained behavior change.

This model builds upon earlier frameworks such as Snoek's biopsychosocial model (Snoek et al., 2008), which highlighted the psychological impact of diabetes, and Glasgow's self-management model (Glasgow et al., 2002), which emphasized health behavior change. The Integrated

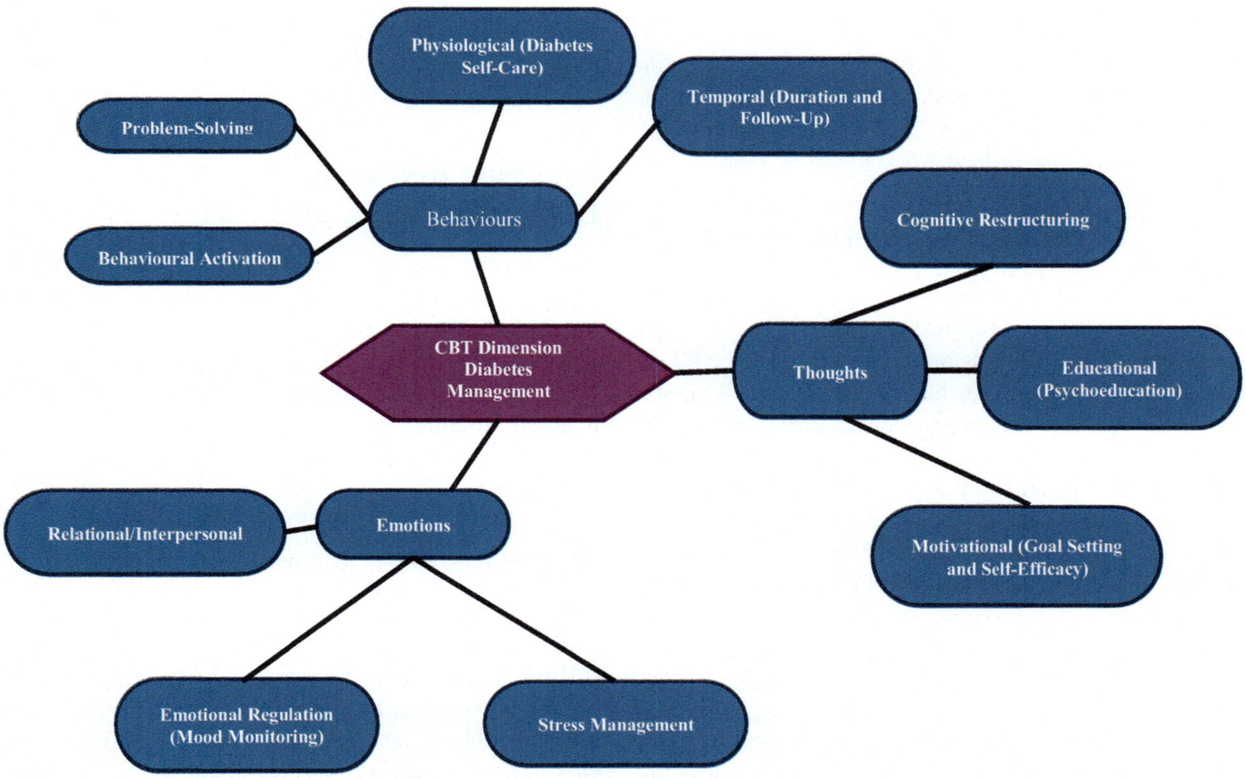

FIGURE 22.1 Mapping of CBT intervention components to core therapeutic dimensions in diabetes management. The model categorizes CBT strategies into three main domains: Thoughts, emotions, and behaviors. Each domain included specific intervention components, such as cognitive restructuring, emotional regulation, problem-solving, and diabetes self-care adherence. This structured framework supports the integration of psychological and behavioral therapies in chronic disease care.

TABLE 22.1 Mapping of the CBT intervention components to dimensions.

Category	CBT dimension	Description
Thoughts	Cognitive restructuring	Identifying and challenging negative or irrational thoughts related to diabetes care and replacing them with more realistic ones.
	Educational (psychoeducation)	Educating patients on how diabetes and depression interact, empowering them to manage both conditions effectively
	Motivational (goal setting and self-efficacy)	Helping patients set realistic goals and build confidence in managing their health, fostering a positive mindset about self-care
Emotions	Emotional regulation (mood monitoring)	Tracking emotional states and mood changes over time, especially in relation to diabetes management
	Stress management	Teaching relaxation techniques (e.g., mindfulness, breathing exercises) to help reduce stress and its negative impact on health
	Relational/Interpersonal	Building a strong therapist-patient relationship that fosters emotional safety and engagement in therapy
Behaviors	Behavioral activation	Encouraging participation in activities that improve mood and promote positive health behaviors like exercise
	Problem-solving	Teaching practical strategies to overcome obstacles in managing diabetes and depressive symptoms
	Physiological (diabetes self-care)	Focusing on adherence to self-care routines such as medication, diet, exercise, and blood glucose monitoring
	Temporal (duration and follow-up)	Structured intervention timelines and follow-up sessions to ensure continued progress in behavior

CBT Model adds clarity by linking specific therapeutic strategies to practical challenges in diabetes management. This design supports modular, personalized interventions that address both mental health and glycemic control.

Evidence from reviews, meta-analyses, and RCTs

Multiple randomised controlled trials (RCTs) and meta-analyses have demonstrated the efficacy of CBT in diabetes care. Meta-analyses have found that CBT significantly improves HbA1c levels, reduces depressive symptoms (Ajele & Deacon, 2025; Li et al., 2022; Yang et al., 2020), and is especially effective when combined with stress management and homework components (Yang et al., 2020). In addition, meta-analyses focusing on individuals with diabetes and comorbid depression found that CBT improved depression, anxiety, quality of life, and fasting glucose levels (Li et al., 2022; Yu et al., 2024). A systematic review highlighted its effectiveness in reducing emotional distress but noted inconsistent improvements in long-term glycemic outcomes (Uchendu & Blake, 2017; Yang et al., 2020). CBT is a well-established and adaptable approach for managing the psychological and behavioral aspects of diabetes. It offers clear benefits in both emotional well-being and self-care although ongoing research is necessary to refine its long-term effectiveness and broaden its applicability.

Emerging psychotherapy in diabetes care: MBIs, ACT, and DBT

As diabetes care increasingly incorporates psychosocial support, third-wave behavioral therapies such as Mindfulness-Based Interventions (MBIs), Acceptance and Commitment Therapy (ACT), and Dialectical Behavior Therapy (DBT) are gaining prominence. These approaches integrate mindfulness, acceptance, and values-based strategies to support emotional regulation and diabetes self-management.

Mindfulness-based interventions (MBIs)

MBIs, including Mindfulness-Based Stress Reduction (MBSR) and Mindfulness-Based Cognitive Therapy (MBCT), use structured practices like breathing exercises, body scans, and meditation to reduce distress and enhance self-care. These interventions, typically delivered in 8-week group sessions, have been adapted for individual, digital, and culturally specific contexts. Studies have shown that MBIs lead to improvements in psychological outcomes and, in some cases, glycemic control (Chen et al., 2021; Sasikumar & Padmapriya, 2022). MBSR reportedly reduced both mental health outcomes and HbA1c (Xia et al., 2022). Similarly, reductions in distress and improved emotional functioning have been found across MBI formats (DiNardo et al., 2022; Nikkhah Ravari et al., 2020). Although glycemic effects vary, psychological gains are consistent and clinically meaningful.

Acceptance and commitment therapy (ACT)

ACT promotes psychological flexibility by encouraging individuals to accept uncomfortable experiences while pursuing actions aligned with their values. Its six core components are often applied through group, digital, or brief formats. Research supports ACT's benefits in improving self-care behavior, reducing anxiety and stabilizing HbA1c (Hajati et al., 2024; Ngan et al., 2023). It is particularly beneficial for patients facing emotional avoidance, motivational deficits, or diabetes burnout.

Dialectical behavior therapy (DBT)

DBT integrates cognitive-behavioral methods with skills in mindfulness, emotion regulation, distress tolerance and interpersonal effectiveness. Initially developed for emotional dysregulation, it has been adapted for diabetes patients with comorbid psychological issues. Although fewer studies exist, DBT has shown utility in improving emotional regulation and diabetes self-care (Aghili & Farhang, 2022; Lotfi et al., 2024). It may be particularly effective for individuals with high emotional reactivity that undermines adherence and metabolic stability.

Integrated care in diabetes

Integrated care combines medical, psychological, and behavioral services to improve outcomes in people with diabetes, especially those facing emotional distress or self-care difficulties. Routine screening using tools like the PHQ-9, GAD-7, and Diabetes Distress Scale (DDS) identifies psychological barriers to management. These tools can be administered during visits or digitally via apps and patient portals, enabling real-time scoring and automated referrals.

Embedding screening in electronic records ensures that mental health is monitored alongside clinical metrics. Early detection supports timely intervention with therapies such as CBT, MBIs, or ACT. Evidence shows that integrated models improve outcomes. Depression and self-care improved with mental health integration (Coventry et al., 2015); structured care enhanced blood pressure and lipid control (Schouten et al., 2010). Hemoglobin A1c reductions were greater in safety-net clinics using collaborative care and were cost-effective (Chwastiak et al., 2017; Schouten et al., 2010). Coordinated, timely care supports sustainable diabetes self-management.

Applications of the material

The Integrated CBT Model provides a structured approach to improving diabetes care through psychological interventions. It aligns key cognitive, emotional, and behavioral strategies with the challenges of type 2 diabetes. In clinical settings, the model helps mental health and diabetes care providers coordinate more personalized treatments. Techniques such as cognitive restructuring, mood monitoring, and problem-solving can be used to improve both emotional well-being and glycemic outcomes. Structured follow-up supports long-term engagement. The model can also guide digital delivery. Online programmes and mobile apps may incorporate their core components, such as stress reduction and goal setting, to extend therapy access in underserved areas.

Cultural adaptation is supported by the model's flexible design. Interventions can be tailored to align with local beliefs, languages, and family dynamics, thereby making them more relevant to diverse populations. For training, the model can be used to educate health professionals on integrating CBT into diabetes care. This may enhance the psychological support in routine practice. Finally, the model offers a framework for research. It can be used to test which CBT components are most effective and how they interact to improve outcomes in people with diabetes.

Research and future directions

While CBT shows promise in diabetes care, the long-term effects on glycemic outcomes remain unclear. Most studies emphasize short-term psychological benefits, with limited evidence on the metabolic impact or diabetes complications. More longitudinal trials are needed to assess the sustained effects of this approach. Digital CBT tools and mobile apps offer scalable options but require testing across diverse populations. Future research should explore how age, gender, and cultural factors affect treatment outcomes. Few studies have investigated the biological mechanisms underlying the relationship between psychological interventions and metabolic changes. Including markers such as cortisol, inflammation, and neural activation in future trials may clarify how CBT influences diabetes physiology.

Conclusion

Psychological stress and depression worsen diabetes outcomes and complicate self-care. CBT and related therapies offer structured approaches to address these challenges. The Integrated CBT Model aligns therapeutic strategies with the core domains of diabetes management. Embedding psychological care in routine diabetes treatment can improve both mental and metabolic health. Further research will help refine and expand these interventions for broader clinical use.

Summary points

- Psychological stress contributes to the development and progression of type 2 diabetes through neuroendocrine and behavioral pathways.
- Diabetes distress and depression are highly prevalent and negatively affect glycemic control and treatment adherence.
- Cognitive behavioral therapy (CBT) is one of the most effective psychotherapeutic approaches for managing psychological and behavioral barriers in diabetes care.
- The Integrated CBT Model organises interventions into cognitive, emotional, and behavioral domains that align with diabetes-specific challenges.
- The key components included cognitive restructuring, mood monitoring, stress management, behavioral activation, and structured follow-up.
- The model incorporates psychoeducation, goal setting, and cultural adaptation to enhance patient engagement and relevance.
- Digital platforms such as eCoaching and mobile health tools can deliver model-based interventions at scale.
- The model supports personalized care planning and can be used in primary, specialist or community settings.
- It provides a framework for clinician training and interdisciplinary collaboration in integrated care.
- Future research can apply this model to test the effectiveness of specific intervention domains and optimize delivery strategies.

Mini-dictionary of terms

- **Allostatic Load:** The cumulative physiological burden of chronic stress that disrupts the hormonal and metabolic systems, increasing disease vulnerability.
- **Behavioral Activation:** A CBT technique that encourages patients to engage in enjoyable or goal-directed activities to improve mood and reduce avoidance behaviors.
- **Cognitive Restructuring:** A process in CBT that involves identifying, challenging, and modifying unhelpful or distorted thoughts.
- **Diabetes Distress:** Emotional strain related to the ongoing management of diabetes, often distinct from clinical depression but with a significant impact on self-care.
- **Emotion Regulation:** Strategies used to monitor, evaluate, and modify emotional reactions in ways that support adaptive functioning.
- **eCoaching:** A digital or remote health coaching method using electronic platforms to support behavior change and chronic disease management.
- **Goal Setting:** A motivational technique that helps individuals establish and work toward realistic, achievable objectives, particularly in behavior change.
- **Mindfulness:** A practice of focused, non-judgmental awareness of this moment, often used in therapy to reduce stress and enhance emotional regulation.
- **Psychoeducation:** Providing individuals with knowledge about the connection between mental and physical health to empower self-management.
- **Self-Efficacy:** One's belief in their ability to successfully perform actions needed to manage health or achieve behavioral goals.
- **Cognitive Distortion:** Inaccurate or exaggerated thought patterns that contribute to emotional distress and maladaptive behavior, commonly addressed in CBT.
- **Diabetes Distress:** Emotional strain and frustration are associated explicitly with managing diabetes, including worries about complications, self-care burden, and perceived failure.
- **Experiential Avoidance:** A psychological tendency to avoid thoughts, feelings, or bodily sensations, often at the expense of effective behavior, central to ACT frameworks.
- **Glycemic Variability:** Fluctuations in blood glucose levels over time may contribute to both physiological complications and psychological distress.
- **Hypothalamic-Pituitary-Adrenal (HPA) Axis:** A neuroendocrine system involved in the stress response. Chronic activation is associated with elevated cortisol levels and metabolic disturbances.
- **Psychological Flexibility:** The capacity to adapt to situational demands with openness and persistence, particularly in the presence of negative internal experiences.

Psychoglycemia: A proposed psychometabolic framework describing the measurable and bidirectional interaction between psychological distress and glycemic dysregulation through behavioral, neuroendocrine, inflammatory, and psychosocial pathways.

Psychometabolic: Relating to the interaction between psychological processes and metabolic functioning.

Psychosocial Stress: Stress arising from psychological and social factors that may negatively affect mental and physical health.

References

Abbas, Q., Latif, S. M. W., Habib, H. A., Shahzad, S., Sarwar, U., Shahzadi, M., Ramzan, Z., & Washdev, W. (2023). Cognitive behaviour therapy for diabetes distress, depression, health anxiety, quality of life and treatment adherence among patients with type-II diabetes mellitus: A randomised controlled trial. *BMC Psychiatry, 23*(1). https://doi.org/10.1186/s12888-023-04546-w

Abdelghaffar, W., Amiri, H., & Zouari, B. (2020). Self-care activities adherence in Tunisian patients with type 2 diabetes: Role of diabetes-related distress and self-efficacy. *Acta Diabetologica, 57*(5), 631–633. https://doi.org/10.1007/s00592-020-01476-x

Aghili, S. M., & Farhang, A. (2022). Investigating the effectiveness of dialectical behaviour therapy on resilience, psychological flexibility, and glycemic control of diabetic patients. *Quarterly of Horizon of Medical Sciences, 28*(4), 514–529. https://doi.org/10.32598/hms.28.4.3866.2

Ajele, K. W., & Deacon, E. (2025). Targeting depression and diabetes comorbidity: A generalisation meta-analysis of randomised controlled trials on cognitive-behavioural therapy efficacy. *Primary Care Diabetes, 19*(2), 93–102. https://doi.org/10.1016/j.pcd.2025.01.004

Almby, K. E., Katsogiannos, P., Pereira, M. J., Karlsson, F. A., Sundbom, M., Wiklund, U., Kamble, P. G., & Eriksson, J. W. (2021). Time course of metabolic, neuroendocrine, and adipose effects during 2 years of follow-up after gastric bypass in patients with type 2 diabetes. *Journal of Clinical Endocrinology and Metabolism, 106* (10), e4049–e4061. https://doi.org/10.1210/clinem/dgab398

Alshehri, M. M., Alenazi, A. M., Alothman, S. A., Rucker, J. L., Phadnis, M. A., Miles, J. M., Siengsukon, C. F., & Kluding, P. M. (2021). Using cognitive behavioural therapy for insomnia in people with type 2 diabetes, pilot RCT part I: Sleep and concomitant symptom. *Behavioural Sleep Medicine, 19*(5), 652–671. https://doi.org/10.1080/15402002.2020.1831501

Alzubaidi, H., Sulieman, H., Namara, K. M., Samorinha, C., & Browning, C. (2022). Relationship between diabetes distress, medication taking, glycaemic control and self-management. *International Journal of Clinical Pharmacy, 44*(1), 127–137. https://doi.org/10.1007/s11096-021-01322-2

Atasoy, S., Johar, H., Herder, C., Rathmann, W., Koenig, W., Roden, M., Peters, A., Kruse, J., & Ladwig, K.-H. (2021). Generalised anxiety disorder symptoms and type 2 diabetes onset: Findings from the prospective cooperative health research in the region of Augsburg f4 and ff4 studies. *Journal of Psychosomatic Research, 145*(110480), 110480. https://doi.org/10.1016/j.jpsychores.2021.110480

Bursać, B., Djordjevic, A., Veličković, N., Milutinović, D. V., Petrović, S., Teofilović, A., Gligorovska, L., Preitner, F., Tappy, L., & Matić, G. (2018). Involvement of glucocorticoid receptor metabolism and signalling in rat visceral adipose tissue lipid metabolism after chronic stress combined with a high-fructose diet. *Molecular and Cellular Endocrinology, 476*, 110–118. https://doi.org/10.1016/j.mce.2018.04.015

Canonico, M. E., Hsia, J., Guthrie, N. L., Simmons, M., Mehta, P., Lupinacci, P., Edwards, K., Mosesso, K., Gearhart, M., Skuban, A., Bonaca, M. P., & Berman, M. A. (2022). Cognitive behavioural therapy delivered via digital mobile application for the treatment of type 2 diabetes: Rationale, design, and baseline characteristics of a randomised, controlled trial. *Clinical Cardiology, 45*(8), 850–856. https://doi.org/10.1002/clc.23853

Carpenter, R., DiChiacchio, T., & Barker, K. (2019). Interventions for self-management of type 2 diabetes: An integrative review. *International Journal of Nursing Science, 6*(1), 70–91. https://doi.org/10.1016/j.ijnss.2018.12.002

Chen, Q., Yang, G., Lin, S., Li, M., Liu, Z., Fu, Y., & Chen, Y. (2021). Effects of mindfulness-based stress reduction therapy combined with intensive education on the effectiveness of the care and the awareness rate in patients with arthritis and diabetes. *American Journal of Translational Research, 13*(4), 3190–3197.

Chilimoniuk, Z., Dobosz, M., Chałupnik, A., Borkowska, A., & Mizerski, G. (2022). The bidirectional relationship between depression and diabetes. *Journal of Education, Health and Sport, 12* (3), 67–77. https://doi.org/10.12775/jehs.2022.12.03.006

Chwastiak, L. A., Jackson, S. L., Russo, J., DeKeyser, P., Kiefer, M., Belyeu, B., Mertens, K., Chew, L., & Lin, E. (2017). A collaborative care team to integrate behavioural health care and treatment of poorly-controlled type 2 diabetes in an urban safety net primary care clinic. *General Hospital Psychiatry, 44*, 10–15. https://doi.org/10.1016/j.genhosppsych.2016.10.005

Coventry, P., Lovell, K., Dickens, C., Bower, P., Chew-Graham, C., McElvenny, D., Hann, M., Cherrington, A., Garrett, C., Gibbons, C. J., Baguley, C., Roughley, K., Adeyemi, I., Reeves, D., Waheed, W., & Gask, L. (2015). Integrated primary care for patients with mental and physical multimorbidity: Cluster randomised controlled trial of collaborative care for patients with depression comorbid with diabetes or cardiovascular disease. *BMJ, 350*, h638. https://doi.org/10.1136/bmj.h638

Cummings, D. M., Lutes, L. D., Littlewood, K., Solar, C., Carraway, M., Kirian, K., Patil, S., Adams, A., Ciszewski, S., Edwards, S., Gatlin, P., & Hambidge, B. (2019). Randomised trial of a tailored cognitive behavioural intervention in type 2 diabetes with comorbid depressive and regimen-related distress symptoms: 12-month outcomes from COMRADE. *Diabetes Care, 42*(5), 841–848.

De Alcubierre, D., Ferrari, D., Mauro, G., Isidori, A. M., Tomlinson, J. W., & Pofi, R. (2023). Glucocorticoids and cognitive function: A walkthrough in endogenous and exogenous alterations. *Journal of Endocrinological Investigation, 46*(10), 1961–1982. https://doi.org/10.1007/s40618-023-02091-7

DiNardo, M. M., Greco, C., Phares, A. D., Beyer, N. M., Youk, A. O., Obrosky, D. S., Morone, N. E., Owen, J. E., Saba, S. K., Suss, S. J., & Siminerio, L. (2022). Effects of an integrated mindfulness intervention for veterans with diabetes distress: A randomised controlled trial. *BMJ Open Diabetes Research & Care, 10*(2), e002631.

Glasgow, R. E., Funnell, M. M., Bonomi, A. E., Davis, C., Beckham, V., & Wagner, E. H. (2002). Self-management aspects of the improving chronic illness care breakthrough series: Implementation with diabetes and heart failure teams. *Annals of Behavioral Medicine, 24*(2), 80–87. https://doi.org/10.1207/s15324796abm2402_04

Groeneveld, L., Beulens, J. W. J., Blom, M. T., van Straten, A., van der Zweerde, T., Elders, P. J. M., & Rutters, F. (2024). The effect of cognitive behavioural therapy for insomnia on sleep and glycemic outcomes in people with type 2 diabetes: A randomised controlled trial. *Sleep Medicine, 120*, 44–52. https://doi.org/10.1016/j.sleep.2024.05.029

Gunnar, M. R., & Hostinar, C. E. (2015). The social buffering of the hypothalamic-pituitary-adrenocortical axis in humans: Developmental and experiential determinants. *Social Neuroscience, 10*(5), 479–488. https://doi.org/10.1080/17470919.2015.1070747

Hajati, E., Gharraee, B., Fathali Lavasani, F., Farahani, H., & Rajab, A. (2024). Comparing the effectiveness of acceptance-based emotion regulation therapy and acceptance and commitment therapy on haemoglobin glycosylated and self-care in patients with type II diabetes: A randomised controlled trial. *Journal of Behavioral Medicine, 47*(5), 874–885. https://doi.org/10.1007/s10865-024-00507-0

Hostinar, C. E., Sullivan, R. M., & Gunnar, M. R. (2014). Psychobiological mechanisms underlying the social buffering of the hypothalamic-pituitary-adrenocortical axis: A review of animal models and human studies across development. *Psychological Bulletin, 140*(1), 256–282. https://doi.org/10.1037/a0032671

Inouye, J., Li, D., Davis, J., & Arakaki, R. (2015). Psychosocial and clinical outcomes of cognitive behavioural therapy for Asians and Pacific islanders with type 2 diabetes: A randomised clinical trial. *Hawai'i Journal of Medicine & Public Health: A Journal of Asia Pacific Medicine & Public Health, 74*(11), 360–368.

Iob, E., Kirschbaum, C., & Steptoe, A. (2020). Persistent depressive symptoms, HPA-axis hyperactivity, and inflammation: The role of cognitive-affective and somatic symptoms. *Molecular Psychiatry, 25* (5), 1130–1140. https://doi.org/10.1038/s41380-019-0501-6

Janssen, J. A. M. J. L. (2022). New insights into the role of insulin and hypothalamic-pituitary-adrenal (HPA) axis in the metabolic syndrome. *International Journal of Molecular Sciences, 23*(15), 8178. https://doi.org/10.3390/ijms23158178

Joseph, J. J., & Golden, S. H. (2017). Cortisol dysregulation: The bidirectional link between stress, depression, and type 2 diabetes mellitus. *Annals of the New York Academy of Sciences, 1391*(1), 20–34. https://doi.org/10.1111/nyas.13217

Kora, M. (2024). *The foundational principles behind cognitive-behavioural therapies*. Ceres Yayınları.

Kumari, S., Dhapola, R., Sharma, P., Singh, S. K., & Reddy, D. H. (2023). Implicative role of cytokines in neuroinflammation-mediated AD and associated signalling pathways: Current progress in molecular signalling and therapeutics. *Ageing Research Reviews, 92* (102098), 102098. https://doi.org/10.1016/j.arr.2023.102098

López-Cano, C., Gutiérrez-Carrasquilla, L., Sánchez, E., González, J., Yeramian, A., Martí, R., Hernández, M., Cao, G., Ribelles, M., Gómez, X., Barril, S., Barbé, F., Hernández, C., Simó, R., & Lecube, A. (2019). Sympathetic hyperactivity and sleep disorders in individuals with type 2 diabetes. *Frontiers in Endocrinology, 10*, 752. https://doi.org/10.3389/fendo.2019.00752

Lei, A. A., Phang, V. W. X., Lee, Y. Z., Kow, A. S. F., Tham, C. L., Ho, Y.-C., & Lee, M. T. (2025). Chronic stress-associated depressive disorders: The impact of HPA axis dysregulation and neuroinflammation on the hippocampus- A mini review. *International Journal of Molecular Sciences, 26*(7), 2940. https://doi.org/10.3390/ijms26072940

Li, Y., Buys, N., Ferguson, S., Li, Z., Shi, Y.-C., Li, L., & Sun, J. (2023). The evaluation of cognitive-behavioural therapy-based intervention on type 2 diabetes patients with comorbid metabolic syndrome: A randomised controlled trial. *Diabetology and Metabolic Syndrome, 15*(Issue 1), 158. https://doi.org/10.1186/s13098-023-01100-2

Li, Y., Storch, E. A., Ferguson, S., Li, L., Buys, N., & Sun, J. (2022). The efficacy of cognitive behavioural therapy-based intervention in patients with diabetes: A meta-analysis. *Diabetes Research and Clinical Practice, 189*, 109965. https://doi.org/10.1016/j.diabres.2022.109965

Liang, Y.-Z., Dong, J., Zhang, J., Wang, S., He, Y., & Yan, Y.-X. (2018). Identification of neuroendocrine stress response-related circulating microRNAs as biomarkers for type 2 diabetes mellitus and insulin resistance. *Frontiers in Endocrinology, 9*. https://doi.org/10.3389/fendo.2018.00132

Liu, Z., Lei, M., & Bai, Y. (2025). Chronic stress mediates alterations in inflammatory cytokines and their role in tumorigenesis. *Journal of Inflammation Research, 18*, 1067–1090. https://doi.org/10.2147/JIR.S485159

Lotfi, Z., Sheikhi, M., Damavandi, M., Mirzaei, F., & Towsyfyan, N. (2024). Comparing the effectiveness of solution-focused therapy and dialectical behaviour therapy on self-care and quality of life in patients with type 2 diabetes. *Health Nexus, 2*(2), 86–94. https://doi.org/10.61838/kman.hn.2.2.10

Lundqvist, M. H., Pereira, M. J., Almby, K., Hetty, S., & Eriksson, J. W. (2024). Regulation of the cortisol axis, glucagon, and growth hormone by glucose is altered in prediabetes and type 2 diabetes. *Journal of Clinical Endocrinology and Metabolism, 109*(2), e675–e688. https://doi.org/10.1210/clinem/dgad549

Mansour, N., Labib, N., Khalil, M., & Esmat, S. (2022). Brief cognitive behavioural therapy for patients with comorbid depression and type 2 diabetes in an urban primary care facility: Randomised controlled trial. *Open Access Macedonian Journal of Medical Sciences, 10*, 60–67. https://doi.org/10.3889/oamjms.2022.7883

Merabet, N., Lucassen, P. J., Crielaard, L., Stronks, K., Quax, R., Sloot, P. M. A., la Fleur, S. E., & Nicolaou, M. (2022). How exposure to chronic stress contributes to the development of type 2 diabetes: A complexity science approach. *Frontiers in Neuroendocrinology, 65*, 100972. https://doi.org/10.1016/j.yfrne.2021.100972

Newby, J., Robins, L., Wilhelm, K., Smith, J., Fletcher, T., Gillis, I., Ma, T., Finch, A., Campbell, L., & Andrews, G. (2017). Web-based cognitive behaviour therapy for depression in people with diabetes mellitus: A randomised controlled trial. *Journal of Medical Internet Research, 19*(5). https://doi.org/10.2196/jmir.7274

Ngan, H. Y., Chong, Y. Y., Loo, K. M., & Chien, W. T. (2023). Preliminary efficacy of an acceptance-based diabetes education (ACT-DE) programme for people with type 2 diabetes on diabetes distress and self-care behaviours: A pilot randomised controlled trial. *Journal of Contextual Behavioural Science, 30*, 50–60. https://doi.org/10.1016/j.jcbs.2023.09.002

Nikkhah Ravari, O., Mousavi, S. Z., & Babak, A. (2020). Evaluation of the effects of 12 weeks of mindfulness-based stress reduction on glycemic control and mental health indices in women with diabetes mellitus type 2. *Advanced Biomedical Research, 9*, 61.

Perrelli, M., Goparaju, P., Postolache, T. T., Del Bosque-Plata, L., & Gragnoli, C. (2024). Stress and the CRH system, norepinephrine, depression, and type 2 diabetes. *Biomedicines, 12*(6), 1187. https://doi.org/10.3390/biomedicines12061187

Saha, C. K., Shubrook, J. H., Guyton Hornsby, W., Yang, Z., Pillay, Y., Mather, K. J., & de Groot, M. (2024). Programme ACTIVE II: 6- and 12-month outcomes of a treatment approach for major depressive disorder in adults with type 2 diabetes. *Journal of Diabetes and Its Complications, 38*(Issue 2). https://doi.org/10.1016/j.jdiacomp.2023.108666

Sasikumar, S., & Padmapriya, S. (2022). Beneficial effects of mindfulness-based stress reduction (MBSR) on the bio-physiological and psychological parameters among people with type 2 diabetes. *Indian Journal of Traditional Knowledge, 21*(4), 766–773.

Schouten, L. M. T., Niessen, L. W., van de Pas, J. W. A. M., Grol, R. P. T. M., & Hulscher, M. E. J. L. (2010). Cost-effectiveness of a quality improvement collaborative focusing on patients with diabetes. *Medical Care, 48*(10), 884–891. https://doi.org/10.1097/MLR.0b013e3181eb318f

Snoek, F. J., van der Ven, N. C. W., Twisk, J. W. R., Hogenelst, M. H. E., Tromp-Wever, A. M. E., van der Ploeg, H. M., & Heine, R. J. (2008). Cognitive behavioural therapy (CBT) compared with blood glucose awareness training (BGAT) in poorly controlled type 1 diabetic patients: Long-term effects on HbA moderated by depression. A randomised controlled trial. *Diabetic Medicine: A Journal of the British Diabetic Association, 25*(11), 1337–1342.

Stahl-Pehe, A., Selinski, S., Bächle, C., Castillo, K., Lange, K., Holl, R. W., & Rosenbauer, J. (2022). Screening for generalised anxiety disorder (GAD) and associated factors in adolescents and young adults with type 1 diabetes: Cross-sectional results of a Germany-wide population-based study. *Diabetes Research and Clinical Practice, 184*(109197), 109197. https://doi.org/10.1016/j.diabres.2022.109197

Thomas, J. C., Letourneau, N., Campbell, T. S., Giesbrecht, G. F., & Apron Study Team. (2018). Social buffering of the maternal and infant HPA axes: Mediation and moderation in the intergenerational transmission of adverse childhood experiences. *Development and Psychopathology, 30*(3), 921–939. https://doi.org/10.1017/S0954579418000512

Thorens, B. (2024). Neuronal glucose sensing mechanisms and circuits in the control of insulin and glucagon secretion. *Physiological Reviews, 104*(4), 1461–1486. https://doi.org/10.1152/physrev.00038.2023

Tunsuchart, K., Lerttrakarnnon, P., Srithanaviboonchai, K., Likhitsathian, S., & Skulphan, S. (2020). Benefits of brief group cognitive behavioural therapy in reducing diabetes-related distress and HbA1c in uncontrolled type 2 diabetes mellitus patients in Thailand. *International Journal of Environmental Research and Public Health, 17*(15), 5564. https://doi.org/10.3390/ijerph17155564

Uchendu, C., & Blake, H. (2017). Effectiveness of cognitive–behavioural therapy on glycaemic control and psychological outcomes in adults with diabetes mellitus: A systematic review and meta-analysis of randomised controlled trials. *Diabetic Medicine, 34*(3), 328–339. https://doi.org/10.1111/dme.13195

Van Bastelaar, K. M. P., Pouwer, F., Cuijpers, P., Riper, H., & Snoek, F. J. (2011). Web-based depression treatment for type 1 and type 2 diabetic patients: A randomised, controlled trial. *Diabetes Care, 34*(2), 320–325. https://doi.org/10.2337/dc10-1248

Vlachou, E., Ntikoudi, A., Owens, D. A., Nikolakopoulou, M., Chalimourdas, T., & Cauli, O. (2022). Effectiveness of cognitive behavioural therapy-based interventions on psychological symptoms in adults with type 2 diabetes mellitus: An updated review of randomised controlled trials. *Journal of Diabetes and its Complications, 36*(5), 108185. https://doi.org/10.1016/j.jdiacomp.2022.108185

Watanabe, M., Reynolds, E. L., Banerjee, M., Charles, M., Mizokami-Stout, K., Albright, D., Ang, L., Lee, J. M., Pop-Busui, R., & Feldman, E. L. (2024). Bidirectional associations between mental health disorders and chronic diabetic complications in individuals with type 1 or type 2 diabetes. *Diabetes Care, 47*(9), 1638–1646.

Wilson, J. B., Epstein, M., Lopez, B., Brown, A. K., Lutfy, K., & Friedman, T. C. (2023). The role of neurochemicals, stress hormones and the immune system in the positive feedback loops between diabetes, obesity and depression. *Frontiers in Endocrinology, 14*, 1224612. https://doi.org/10.3389/fendo.2023.1224612

Wu, A. (2021). Social buffering of stress–physiological and ethological perspectives. *Applied Animal Behaviour Science, 239*, 105325. https://doi.org/10.1016/j.applanim.2021.105325

Xia, T., Lopes, S., Chen, L., Roth, R., Zinzow, H., Jones, K., Zhang, L., Shi, L., & Jindal, M. (2022). A feasibility study on low-dose mindfulness-based stress reduction intervention among prediabetes and diabetes patients. *Complementary Therapies in Medicine, 65*, 102810. https://doi.org/10.1016/j.ctim.2022.102810

Xiao, Y., Liu, D., Cline, M. A., & Gilbert, E. R. (2020). Chronic stress, epigenetics, and adipose tissue metabolism in the obese state. *Nutrition and Metabolism, 17*(1), 88. https://doi.org/10.1186/s12986-020-00513-4

Xu, H., & Chen, Q. (2025). The bidirectional influence between type 2 diabetes mellitus and the state of depression and anxiety. *Journal of Affective Disorders, 386*, 119467. https://doi.org/10.1016/j.jad.2025.119467

Xu, C., Dong, Z., Zhang, P., Chang, G., Xiang, Q., Zhang, M., Zhou, J., Qiao, C., Yang, Q., Qin, Y., & Lou, P. (2021). Effect of group cognitive behavioural therapy on psychological stress and blood glucose levels in people with type 2 diabetes mellitus: A community-based cluster randomised controlled trial in China. *Diabetic Medicine, 38*(2). https://doi.org/10.1111/dme.14491

Yajurvedi, H. N. (2018). Stress and glucose metabolism: A review. *Imaging Journal of Clinical and Medical Sciences, 5*(1), 008–012. https://doi.org/10.17352/2455-8702.000037

Yang, X., Li, Z., & Sun, J. (2020). Effects of cognitive behavioural therapy–based intervention on improving glycaemic, psychological and physiological outcomes in adult patients with diabetes mellitus: A meta-analysis of randomised controlled trials. *Frontiers in Psychiatry, 11*. https://doi.org/10.3389/fpsyt.2020.00711

Yaribeygi, H., Maleki, M., Butler, A. E., Jamialahmadi, T., & Sahebkar, A. (2022). Molecular mechanisms linking stress and insulin resistance. *EXCLI Journal, 21*, 317–334. https://doi.org/10.17179/excli2021-4382

Yu, Z., An, Q., Hawkins, J., & Zhang, A. (2024). Cognitive behavioural therapy for depression among Chinese patients with diabetes: A systematic review and meta-analysis. *Research on Social Work Practice, 34*(3), 277–292.

Chapter 23

Modulation of oxidative stress in the management of pancreatitis

Samata Pradhan and Chandraiah Godugu

Department of Biological Sciences (Regulatory Toxicology), National Institute of Pharmaceutical Education & Research (NIPER), Hyderabad, Telangana, India

List of abbreviations

AP Acute pancreatitis
CP Chronic pancreatitis
Drp1 Dynamic related protein 1
ECM Extracellular matrix
GSH Glutathione
IL-1β Interleukin 1β
IL-6 Interleukin 6
MDA Malondialdehyde
NADPH Nicotinamide adenine dinucleotide phosphate
NF-κB Nuclear factor kappa B
Nrf2 Nuclear factor erythroid 2-related factor
NSAIDs Nonsteroidal antiinflammatory drugs
PDAC Pancreatic ductal adenocarcinoma
PERT Pancreatic enzyme replacement therapy
PSC Pancreatic stellate cells
RNS Reactive nitrogen species
ROS Reactive oxygen species
SOD Superoxide dismutase
TNF-α Tumor necrosis factor-alpha

Introduction

A pathological condition like inflamed pancreas is called pancreatitis. A mild to moderate damage to the gastrointestinal tract (GIT) leads to nausea, vomiting and abdominal pain due to inflammation in the pancreas, a temporary, reversible and less life-threatening pancreatic disease like acute pancreatitis (AP). In contrast, chronic pancreatitis (CP), a severe inflammation, fibrosis, irreversible, nonmalignant and progressive medical conditions defined by persistent pain and dysfunction of endocrine and exocrine glands. Altogether, pancreatitis continues to be linked to substantial morbidity and death, particularly in severe forms, despite advancements in supportive therapy. The global incidence of this inflammatory disease of pancreatitis has been increasing with the notable increase throughout the previous decades. Acute pancreatitis affects 5 to 80 people out of every 100,000 people worldwide each year and it is on the rise in many areas, mainly due to gallstones and uncontrolled consumption of alcohol (Iannuzzi et al., 2022). Acute pancreatitis is becoming more common internationally. The incidence of acute pancreatitis has been significantly increased in the United States, with an increased mortality rates with age. Despite this, alcohol-induced pancreatitis is more prevalent in younger persons (less than 56 years of age), a median age of onset is about 50 years. According to estimates, there are 34–42 instances of CP for every 100,000 people worldwide, with rates of incidence varying greatly according to factors including alcohol consumption, gallstones, genetics, and geographic risk factors, raising the possibility of pancreatic ductal adenocarcinoma (PDAC). There are several etiological risk factors linked to CP, including ductal blockage, genetic alterations, environmental variables (such as alcohol and nicotine), diet and autoimmune factors (Wang, Zou, et al., 2023). Progressive fibrotic deterioration of the pancreas is a hallmark of CP. Inflammation, duct blockage, necrosis, apoptosis, and other mechanisms damage the secretory parenchyma, leading to the development of a fibrotic pancreas. A growing body of research suggests that pancreatic stellate cells (PSC) are the primary mediators of fibrosis, leading to the production of extracellular matrix (ECM) in the interstitial spaces as well as in regions where duct cells are damaged or acinar cells disappear (Klöppel et al., 2004).

One of the significant factors in the onset and progression of pancreatitis is oxidative stress, characterized by an imbalance between the generation of oxidants i.e., reactive oxygen species (ROS) and body's inability to neutralize the oxidants. In addition to damaging cellular

Hormones in Neuroscience. https://doi.org/10.1016/B978-0-443-30263-3.00015-4

structures, ROS intensify inflammatory cascades, inflammatory cells infiltration exacerbating tissue damage, necrosis and fibrosis (Gossart et al., 1996). The higher concentration release of inflammatory cytokines such as interleukins, indicating local and systemic inflammation (Mittal et al., 2014). Lipid peroxidation and mitochondrial dysfunction in pancreatic acinar cells are accelerated by the overproduction of ROS, causing trypsinogen activation and auto-digestion, which ends with inflammation, necrosis, fibrosis and finally pancreatic dysfunction. Therefore, an effective treatment option for pancreatitis is to focus on oxidative stress. Improving the treatment possible by comprehending how oxidative stress contributes to pancreatitis. This chapter aims to explore common approaches of pancreatitis treatment, how oxidative stress influences the severity of the condition, and novel strategies for minimizing oxidative stress as an adjunct therapy.

Role of oxidative stress in the pathogenesis of pancreatitis

The precise molecular etiology of pancreatitis is still up for debate. Oxidative stress is believed to be a critical contributor to the pathogenic pathway of several diseases, especially acute and chronic pancreatitis. Over the past decade, several preclinical and clinical experiments have been used to assess the involvement of oxidative stress in the cellular damage and inflammation in the pancreas (Fig. 23.1). Overproduction of ROS and improper function of body's antioxidant system combine to result in mitochondrial dysfunction, lipid peroxidation, inflammation and acinar cell damage (Rothe & Kolb, 1999). The extremely unstable intermediate oxygen metabolites that are generated during oxygen metabolism are mainly contribute to the production of ROS in the body. Despite of having many physiological functions like immune defense, high concentration of ROS results in many pathological conditions such as lipid peroxidation, cellular injury, inflammation, DNA methylation, necrosis leading to multi-organ damage, failure and death (Robertson & Harmon, 2006). However, in pancreatitis, a number of mechanisms combine to produce oxidative stress which are listed below (Table 23.1).

Mitochondrial dysfunction

It is commonly recognized that mitochondria play a critical role in many biological processes such as cell signaling, homeostasis, ROS generation, apoptosis, and inflammation. Nevertheless, altered mitochondrial dynamics can lead to many pathophysiological changes including oxidative stress and inflammatory cascade, which subsequently trigger overproduction of ROS (Brookes et al., 2004). Additionally, several complicated inflammatory disorders are closely associated with the improper function of mitochondria, encompassing acute and chronic pancreatitis as well. Malondialdehyde (MDA), glutathione peroxidase (GSH-Px) and superoxide dismutase (SOD) etc are the major measurable indicators of oxidative stress. Extremely reactive ROS generated by mitochondrial dysfunction accumulate in the pancreas, stimulating various pro-inflammatory cytokine cascades and inflammatory mediators, leading to pancreatic acinar cell damage and cell death (Fig. 23.1). Moreover, the severity of pancreatitis is widely correlated with an improper balance between oxidants and antioxidants in the pre-clinical and clinical models of pancreatitis (Verlaan et al., 2006).

Premature activation of digestive enzymes

Premature enzyme activation is one of the most important initial stages in the pathogenesis of pancreatitis. However, many pathological mediators such as alcohol consumption, gall stones, genetic mutations trigger inappropriate activation of proteolytic enzymes such as trypsinogen resulting auto-digestion and inflammatory response amplification. In pancreatitis mitochondrial dysfunction leads to generation of ROS, which acts as intracellular calcium elevator that effectively cause trypsinogen activation (Petersen & Tepikin, 2008). Moreover, excessive ROS reduce the plasma level of antioxidative thiols such as GSH and cysteine in chronic pancreatitis patients. A clinical study conducted by Mariette Verlaan et al. demonstrated that the prolonged consumption of alcohol is one of the major risk factors for chronic pancreatitis (Verlaan et al., 2006). Additionally, improper metabolism of ethanol may result in generation of ROS and development of pancreatitis which consequently stimulates the auto-digestion of pancreatic enzymes resulting in pancreatic acinar cell death (Verlaan et al., 2006). Oxidative stress and improper antioxidative defense mechanism in the body directly cause premature activation of inactive pancreatic enzymes leading to pancreatic dysfunction and multi-organ failure. It is well reported that activation of digestive enzymes is highly associated with the pathogenesis of pancreatitis. Furthermore, pancreatic acinar cell injury is noticeably absent in the caerulein induced animals lacking trypsinogen and cathepsin B indicating an adequately clear role of premature activation of trypsinogen in the development of pancreatitis (Dawra et al., 2011).

Endocrine hormones and oxidative stress in the pancreatitis

Endocrine hormones actively and simultaneously modulate immunological responses, oxidative stress and cellular repair procedures in pancreatic tissue, playing a significant impact on the both pancreatic disease development and

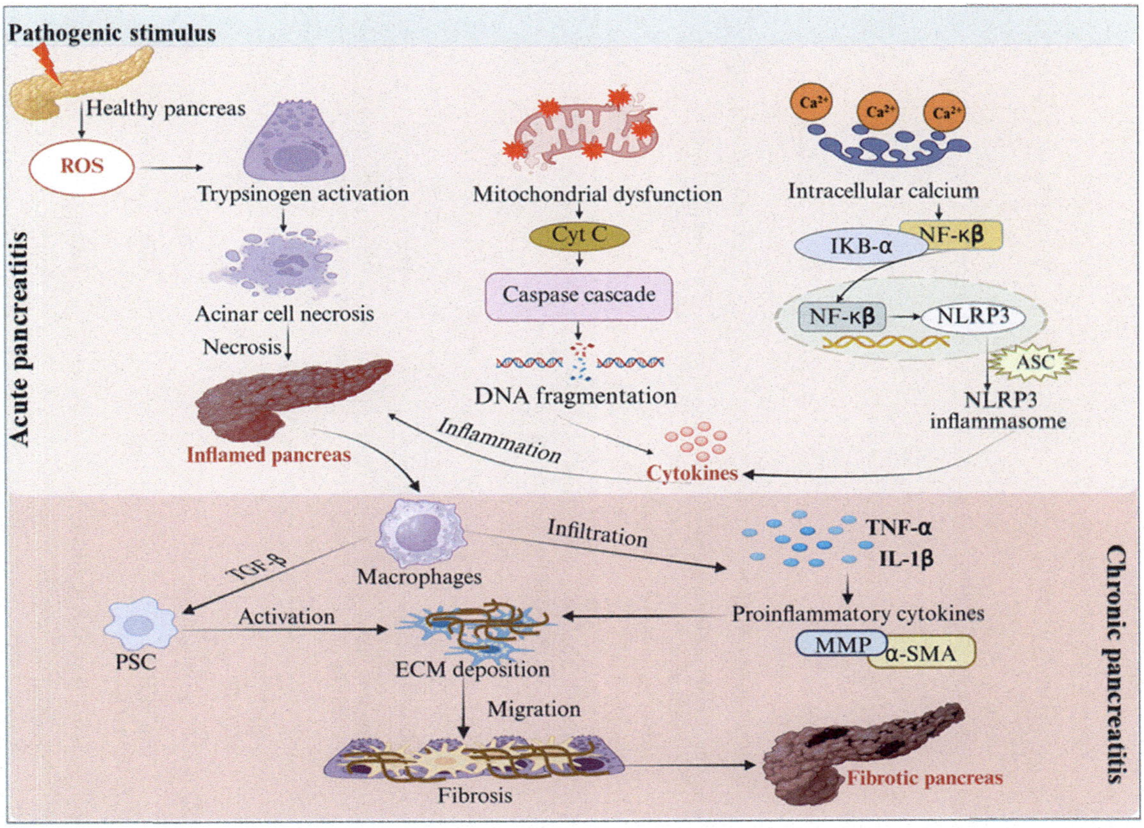

FIGURE 23.1 The figure demonstrates the physiological distinctions and commonalities in the development of acute and chronic pancreatitits emphasizing on the transition of irreversible chronic progressive fibrotic pancreas from acute reversible pancreatic injury. In acute pancreatitis, premature pancreatic enzyme activation, oxidaitive stress, mitochondrial dysfunction, intracellular calcium overload result in activation of inflamamatory cascades and acinar cell injury contributing to the tissue necrosis and systemic inflamamtion. Persistsnt inflamamtion, acivated PSCs, deposition of ECMs leading to progression of fibrosis in chronic pancreatitis.

oxidative stress control However, pancreatitis majorly affects the exocrine portion such as acinar and ductal cells of the pancreas while, hormones specially produced by the pancreas and other endocrine glands, plays significant role in the regulation of inflammation and oxidative damage. An imbalance may either worsen or lessen the pancreatic damage. Melatonin deficiency can activate signaling cascades like NF-κB, p38 MAPK pathways leading to inflammation, cellular apoptosis, necrosis and tissue destruction. Inadequate production of insulin causes hyperglycaemia aggravating acinar cell damage and inflammation in the pancreas. According to certain research, individuals with acute pancreatitis are susceptible to elevated level of a hormone secreted by the pancreatic alpha cells which further contributes to the pancreatitis associated hyperglycaemia. Furthermore, chronic stress and elevated cortisol level revealed a substantial potential link with severe acute pancreatitis contributing to the generation of mitochondrial ROS and pancreatic tissue injury. A study by Jiang et al. highlighted a severe complication between catecholamine neurotransmitters like dopamine and severe acute pancreatitis (Jiang et al., 2014). The endocrine system plays an essential role in pancreatitis particularly in development of chronic pancreatitis, affecting its hormonal activity that can result diabetes mellitus like several metabolic disorders by destroying pancreatic beta cells alongside exocrine tissue in the pancreas. In contrast, Hormones including glucagon, insulin, melatonin, leptin and ghrelin regulates inflammatory responses in the pancreas.

Immune cells infiltration

One of the key pathological mechanisms in pancreatitis is inflammation. Increased levels of cytokines like, IL-6 and TNF-α, which stimulate the activation of pancreatic enzymes, resulting in pancreatic tissue necrosis and systemic inflammation. Inhibition of IL-6 signaling pathway significantly reduced the severity of acute pancreatitis in mice models (Chao et al., 2006; Lima et al., 2013). Free radicals produced due to improper oxygen metabolism, causes infiltration of inflammatory cells, especially macrophages and neutrophils, ultimately resulting in lipid peroxidation, DNA damage and apoptosis. Inhibition ROS

TABLE 23.1 A detailed outline of pancreatitis management strategies focusing on antioxidative defense.

A summary of protective effect of vitamins in pancreatitis					
S No.	**Therapeutic approach**	**Dose of the drug**	**Evidence in AP/CP**	**Outcomes**	**References**
1	α-tocopherol	600 mg/kg	CP	↓lipid peroxidation ↓profibrogenic activity Restores pancreatic morphology	Gómez et al. (2004)
2		60 mg/kg	AP	↓amylase, LDH, MPO ↓TNF-α ↑IL-10 ↓leukocyte infiltration	Al-Hashem et al. (2023)
3	Ascorbic acid	100–1000 mg/kg	AP	↓amylase, lipase ↓IL-6, IL-10 ↓Bax, ↑Bcl-2 ↑Nrf2, HO-1	Xu et al. (2020)
4		500 mg/kg	SAP	↓amylase, lipase ↓IL-6, TNF-α	Gui et al. (2023)
5		200 mg/kg	SAP	↓CRP, fluid retention	Zhao et al. (2025)
6	Pyridoxin	100 mg/kg	AP	Inhibited caspase-3 signaling pathway ↓proinflammatory cytokines ↓reduced neutrophil infiltration	Xu et al. (2024)
Phytoconstituents comprise a potentially effective treatment options for pancreatitis					
7	Quercetin	100, 150, 200 mg/kg	AP	↓MPO, IL-1β, TNFα, IL-6, NF-κB ↓IRE1α, sXBP1, C/EBPα and C/EBPβ	Zheng et al. (2016)
8		50 mg/kg	AP	↓IL-1β, TNFα, IL-6, MPO ↓TLR4 ↓amylase	Junyuan et al. (2018)
9		40 mg/kg	AP	Inhibited p38 and MAPK pathway	Sheng et al. (2021)
10	Visnagin	10, 30, 60 mg/kg	AP	↓amylase, lipase ↓cytokines	Pasari et al. (2019)
11	Resveratrol	20 mg/kg	AP	↓apoptosis, necrosis	Liu et al. (2020)
12	Isorhamnetin	10, 30 mg/kg	SAP	↓HtrA2 ↑KDM5B	Li et al. (2024)
13	Stigmasterol	50, 100 mg/kg	AP	↓MPO, TNFα, IL-1β, IL-6	Zhao, Li, et al. (2024)
14	Curcumin	100 mg/kg	AP	↓iNOS TNFα, IL-6 ↓NF-κB,	Gulcubuk et al. (2013)
15		100 mg/kg	SAP	↓TNFα, IL-6, TLR-4, NF-κB	Zhong (2015)
16		50 mg/kg	AP	↓amylase ↓NF-κB, MPO	Siriviriyakul et al. (2019)
17	Grape seed proanthocyanidins	25, 50, 100 mg/kg	AP	↓TNF-α, IL-1β, NF-κB, IFN-γ ↑Nrf-2 and HO-1 ↓NO, OH–	Bashir et al. (2016)

18	Hesperidin	100 mg/kg	Sub-acute	↓iNOS, NF-κB, IL-6 and TNF-α ↓amylase, lipase	Aja et al. (2020)
19	Emodin	0.5 mg/kg	SAP	↓IL-1β and IL-18 ↓caspase −1	Ning et al. (2017)
20		6 mg/mL	SAP	↓TNF-α, IL-18 ↓amylase, lipase ↓ROS ↓NLRP3	Xia et al. (2019)
11	Borneol	100, 300 mg/kg	AP	↑Nrf2, SOD ↓NF-α, IL-1β, IL-6, ↓NF-κB	Bansod et al. (2021)
22	Nimbolide	0.3, 1 mg/kg	CP	↓amylase, lipase ↓MDA, NO, ↑GSH Inhibited TGF-β/Smad signaling	Bansod et al. (2020)
23		0.3, 1 mg/kg	AP	↓ROS ↓TNF-α, IL-6, iNOS, IL-1β, NF-κB, ↓MDA, NO,	Bansod and Godugu (2021)
24	Berberine	30, 100 mg/kg	AP	↑AMPKα, Nrf2 ↓NOX2	Bansod et al. (2025)
An overview of preventive role of hormone in pancreatitis					
25	Melatonin	25 mg/kg	AP	↓lipid peroxidation ↑glutathione peroxidase activity	Carrasco et al. (2014)
26		50 mg/kg	AP	↓JNK/NF-κB ↓TNFα, IL-6, Bax, Caspase-3 ↑Bcl2	Zhao et al. (2018)
27		50 mg/kg	AP	↓iNOS TNFα, IL-1β, IL-6 ↓NF-κB	Uluşık et al. (2024)
28		20 mg/kg	SAP	↓LDH, ↑Nrf2 ↓Keap1, p- NF-κB	Zhao, Fang, et al. (2024)
Potential effects of trace elements in the management of pancreatitis					
29	Zinc	5 mg/kg	SAP	↑GSH, SOD ↓MDA	Tang et al. (2014)
30	Selenium	1 mg/kg	SAP	↑Nrf2, ↓MAPK, NF-κB, STAT-3.	Swetha et al. (2025)
31	Selenium sod. selenite	1 mg/kg	AP	Improved mitophagy and lipophagy	Han et al. (2025)

SAP: severe acute pancreatitis, MDA: malondialdehyde, GSH: glutathione, SOD: superoxide dismutase, LDH: lactate Dehydrogenase, NF-κB: nuclear factor kappa B, NO: nitric oxide, IL-6: interleukin 6, IL-1β: interleukin-1β, Nrf2: nuclear factor erythroid 2-related factor 2, TNF-α: tumor necrosis factor, IFN-γ: interferon gamma, MPO: myeloperoxidase, HtrA2: HtrA serin peptidase 2, CRP: C reactive protein, TLR4: toll like receptor 4, MAPK: Mitogen-activated protein kinase, JNK1/2: c-Jun N-terminal kinase 1/2, KDM5B: Lysine demethylase 5B, AMPK: AMP-activated protein kinase.

improved the levels of antioxidative enzyme (SOD) in the mice with AP. Meanwhile, myeloperoxidase (MPO), an enzyme highly expressed in inflammatory conditions has been reported with a high concentration in caerulein induced mice (Bansod et al., 2021; Lima et al., 2013). Jun He et al. revealed that the increased ROS levels involved in the inactivation of a protein with antioxidative properties named sulfiredoxin-1 (Srxn1) by promoting the infiltration of inflammatory cells and inducing apoptosis (He et al., 2021). A transcription factor called Nrf2 activates protective gene Heme oxygenase (HO-1) contributing to protection against oxidative stress, apoptosis and inflammation, causing protection against pathological conditions including pancreatitis. Moreover, significant decrease in the protein levels of Nrf2/HO-1in L-arginine induced mice indicates its role to protect from oxidative stress in the pathogenesis of pancreatitis. Mitogen-activated protein kinase (MAPK), a strong regulator of many biological functions in the body such as cell signaling, apoptosis and inflammation. Recent studies have reported that impaired regulation of MAPK increases the degree of apoptosis by enhancing cytokine production which includes IL-6 and TNF-α (Li et al., 2021).

Pancreatic stellate cells (PSCs) activation and oxidative stress

Pancreatic stellate cells (PSCs), myofibroblast like cells present in the pancreas, can be proliferated by the ROS and pro-inflammatory cytokines and trigger to the development of pancreatic fibrosis (Zhang et al., 2022). Phosphorylation of the components of the MAPK family (p38, ERK1/2 and JNK1/2) contribute significantly to the pancreatic fibrosis by stimulating the activation of PSCs and TGF-β1 signaling pathways (Fig. 23.1). It is confirmed that inhibition of phosphorylation of p38 and other MAPKs provide a protective mechanism against chronic pancreatitis. Dynamic related protein (Drp1) associated with the mitochondrial fission, which regulates various biological processes such as apoptosis, inflammation Etc. Intracellular ROS promote the activation of PSCs by up-regulation of Drp1, which is essential in developing pancreatic fibrosis (Yuan et al., 2022). An enzyme named Sphingosine kinase 1 (SPHK1) catalyzes to generate sphingosine-1-phosphate (S1P) which is involved in various physiological and pathophysiological activities including inflammation and apoptosis. Moreover, scientists have reported the role of over expression of SPHK1/S1P in activation of PSCs, which plays vital role in the development of CP (Wang, Han, et al., 2023).

Current management of pancreatitis

Consistent epigastric pain, nausea and vomiting are the observable clinical hallmarks of acute pancreatitis, while these manifestations become broader in the patients with chronic pancreatitis. An integrative approach that incorporates medical support, dietary plans, pharmaceutical treatments, operative alternatives and prevention of consequences is necessary for the mitigation of acute pancreatitis, whereas pancreatic enzyme replacement therapy (PERT), pain management, antioxidative therapy are necessary for the treatment of chronic pancreatitis, which involves irreversible progressive fibrosis.

Supportive care in the management of pancreatitis

The lives of patients with SAP and chronic pancreatitis are really at risk and this is becoming a challenging portion in the management of pancreatitis. Supportive care which primarily includes fluid resuscitation, pain management, respiratory, renal and cardiac support. Lactate ringer's solution widely known for its ability to reduce systemic inflammation, has been reported to increase the rate of survival of the individuals suffering from pancreatitis. In order to successfully medicate pancreatitis pain management is still crucial. Circulatory collapse due to persistent pain might have serious consequences. However, opioids continue to be the first choice of analgesics in pancreatitis, whereas NSAIDs can be utilized in moderate conditions. Given that most instances of pancreatitis do not benefit from the administration of antibiotics but they are still essential for treating bacterial infection and especially necrotic pancreas. A better blood circulation in the pancreas or epidural analgesia which increases the clinical effectiveness, reduces inflammation and severity of the pancreatitis. Hyperbaric oxygen therapy (HBOT), an improved breathing technique helps to heal from many chronic diseases, inflammation and microbial infections. Moreover, investigation in experimental animals suggest that HBOT may minimize pancreatic necrosis and mortality of the disease (Nikfarjam et al., 2007). Nutritional therapy provides specialized diet to the patients suffering from acute and chronic pancreatitis as well. It is not required to introduce the oral diet gradually, starting with clear liquids, because a high-calorie, nutritious meal may be started safely. Despite the fact that previous investigations have employed a low-fat diet (less than 30% of the entire calories). Nasogastric and/or nasojejunal routes are also recommended when patients are not able to receive oral dose of nutritional support.

Pancreatic enzyme replacement therapy

Pancreatic enzyme replacement therapy (PERT) which improves digestion, absorption of nutrients, steatorrhea and abdominal pain can be implemented after pancreatic surgery, exocrine dysfunction and pancreatic necrosis. This therapy uses pancreatic enzymes such as amylase and lipase as supplements (Fig. 23.2). The particular dose and specific enzyme may differ in patients. Consequently, the recommended initial dosage of 30–40,000 IU with main meal and 15–20,000 with IU with a snack (Imrie et al., 2010; Keller & Layer, 2005). Overdose of PRET can cause side effects like GIT disturbances, diarrhea, nausea, vomiting.

Surgical care for treating pancreatitis

Surgical intervention is less targeted and a delayed approach in case of pancreatitis treatment. Mild to moderate AP is often reversible with supportive care, while surgical care may help in the complications like infected and necrotic pancreas and reduces organ failure and rate of mortality. Open necrosectomy, laparoscopic necrosectomy and video-assisted retroperitoneal debridement (VARD) are commonly preferred surgical techniques implemented mostly when medical care fails. When severe complications like pancreatic necrosis fails to go away with non-surgical approaches such as medications are the reason to seek surgical procedures in acute pancreatitis. In chronic pancreatitis, persistent inflammation or chronic pain that is unresponsive to alternative therapies, ductal blockage, ongoing inflammation, or severe or recurring consequences such as strictures or pseudocysts may demand surgery. Due to inadequate identification of necrotic pancreas, early surgery (less than 2 weeks) is linked to a greater fatality rate. Endoscopy and video-assisted retroperitoneal techniques are increasingly adopted owing to less mortality and morbidity relative to open surgery. Because of its increased risk of negative outcomes, open necrosectomy is currently only performed on individuals who are unresponsive to less invasive methods. The upgraded method is now accepted for the infected pancreas and surgery for pancreatitis should be carried out in hospitals with competence in extensive pancreatic surgery. Supportive care, nutritional diet and antioxidant therapy may enhance the results even further.

Alcohol cessation

Almost 80% of the pathology of pancreatitis is contributed by prolonged use of alcohol. Therefore, a fundamental approach in prevention of pancreatitis is alcohol cessation. It has been reported that those who abstain have fewer serious incidents like pain intensity than those who

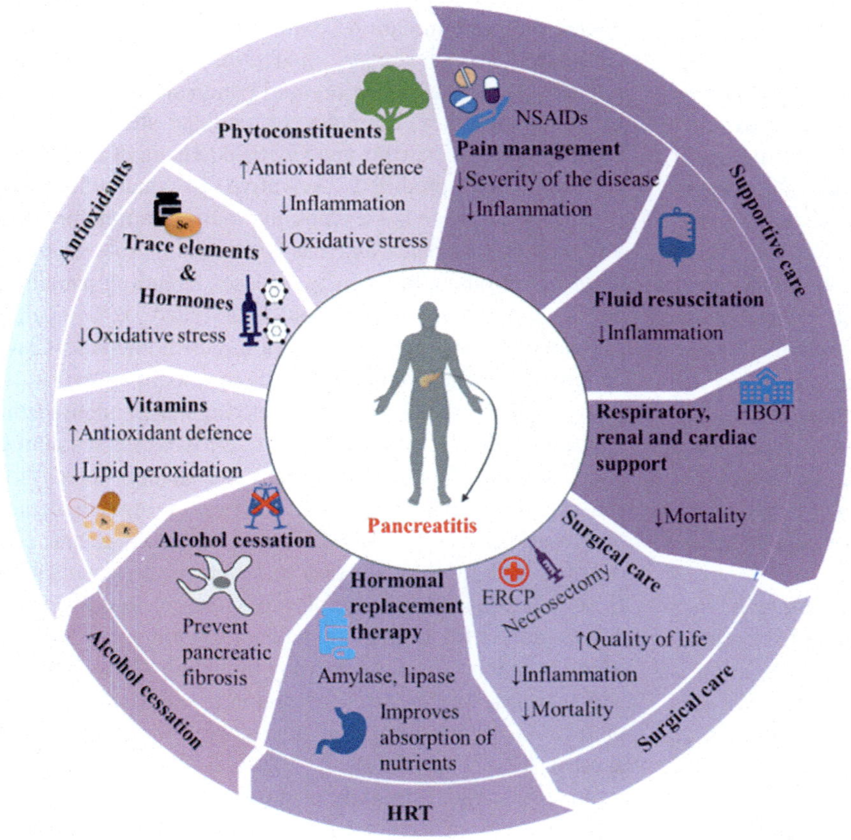

FIGURE 23.2 Integrated approach for the management of acute and chronic pancreatitis focusing on supportive care, surgical procedure, pain maangement, alcohol cessation and particularly antioxidant therapy. PERT: pancreatic enzyme replacement therapy, HBOT: hyperbaric oxygen therapy, ERCP: endoscopic retrograde cholangiopancreatography.

continue to drink. Alcohol cessation reduces the risk of developing pancreatic fibrosis and pancreatic ductal adenocarcinoma.

Role of antioxidative therapy in pancreatitis

A key pathogenic process in both acute pancreatitis and chronic pancreatitis is oxidative stress. In pancreatitis, excessive production of ROS and reactive nitrogen species (RNS) alter the body's natural antioxidant defense mechanism resulting in trypsinogen activation, necrosis, inflammation and fibrosis. Therefore, antioxidative treatment that targets minimizing ROS production has been suggested as a possible supplement to the supportive care for pancreatitis. The impact of antioxidants in alleviating oxidative stress in experimental models has been extensively studied.

Vitamins

Ascorbic acid (vitamin C) is one of the most popular vitamins with scavenging properties present in plasma with adequate role in antioxidative defense and immune modulation. A rapid decline in serum levels of vitamin C has been reported in the patients with pancreatitis. Mortality has been observed in many disease conditions such as pancreatitis, sepsis patients with lower concentration of vitamin C in plasma. In acute pancreatitis, exogenous and endogenous toxins induced high concentration of ROS reduce the levels of antioxidant enzymes. Ascorbic acid minimizes the activation of oxidative enzymes by preventing formation of highly reactive superoxides due to its direct scavenging properties of superoxide and hydroxy radicals. Additionally, vitamin C reduces cytokine (IL-6, IL-10) production by inhibiting the NF-κB pathway, suggesting a strong antioxidant activity (Xu et al., 2020). Findings from a study by Bing Zhao et al. demonstrated the beneficial effect of intravenous administration of vitamin C on prevention of AP, where patients have identified with lesser inflammation (Zhao et al., 2025). Previous studies have indicated that intravenous dose of vitamin C reduces risk of organ failure and improves inflammatory conditions. Furthermore, alcohol reduces the circulation of vitamin C and contribute to the oxidative stress and progression of fibrosis (Fig. 23.2).

A fat-soluble antioxidant having cell membrane protective ability is vitamin E, it reduces lipid peroxidation, a hallmark of pancreatitis, slowing down the activation of PSCs and inhibiting the activation of cytokines cascades. Vitamin E supplementation successfully minimized the pancreatic enzyme levels, necrosis and provided an ultimate protection from the ROS mediated apoptosis. A water-soluble vitamin pyridoxin (vitamin B6) improves the severe complications of pancreatitis and reduces oxidative stress and inflammation, inhibiting caspase-3 signaling pathway (Xu et al., 2024). Although, there have been few clinical investigations which typically indicate a combinational approach of vitamin E with other antioxidants as a potential oxidant reductor. Thus, a combinational therapy of Vitamin E, vitamin C, selenium, some phytoconstituents help in improving the progressive acinar cell damage and fibrosis in the pancreatitis patients.

Phytomedicines

Bioactive components known as phytoconstituents can be extracted from different part of plants and have antioxidant and antifibrotic properties. Phytochemicals have become promising alternative treatments in pancreatitis by controlling the oxidative stress, fibrosis and inflammation, the key aspects in the pathogenesis of pancreatitis. Many phytochemicals enhance oxidative defense by activating Nrf2 signaling, inhibiting NF-κB signaling and inflammatory cytokines thereby, controlling PSCs activation, ECM deposition and fibrosis, acting as a potent ROS scavenger. During pancreatitis, *Curcuma longa* or curcumin reduced neutrophil and macrophages infiltration, fibrosis, serum levels of pancreatic enzymes by slowing down fibroblast formation, upregulation of Nrf2 and downregulation of NF-κB. Similarly, an antioxidant and NF-κB inhibitor, resveratrol minimized the risk of pancreatic injury by attenuating inflammation, oxidative stress and fibrosis. Another ROS scavenger, quercetin, a plant flavonoid reduced the acinar cell injury and pancreatic inflammation by supressing key regulator of cellular stress such as IRE1α, sXBP1, C/EBPα and C/EBPβ (Zheng et al., 2016). A chemically modified derivative of quercetin, Isorhamnetin likewise has exceptional anti-inflammatory and antioxidant properties and has a promising role in treatment of various inflammatory diseases including pancreas. High temperature requirement A2 (HtrA2), a central regulator of mitochondrial function, cell homeostasis and apoptosis, which eventually result in apoptosis by facilitating caspase dependent and caspase independent pathways. In order to restore HtrA2 activity, while sustaining mitochondrial function in pancreatic acinar cells, Isorhamnetin directly inhibit the activity of histone demethylation activity of KDM5B and reduced acinar cell necrosis and ROS production in severe acute pancreatitis (Li et al., 2024). Stigmasterol, a phytosterol, is considered as a precursor for antioxidant and antiinflammatory activity in inflammatory diseases, thereby preventing ROS induced acinar cell death through suppression of the ERK signaling pathways, stigmasterol has a notable beneficial effect against AP (Zhao, Li, et al., 2024). A flurocoumarin derivative, visnagin originally originated from *Ammi visnaga* is widely recognised for having anti-inflammatory actions and other possible therapeutic benefits. Visnagin was shown to be useful in lowering the levels of plasma levels of lipase and amylase, inflammatory cytokines indicating a potential anti-oxidant activity against pancreatitis (Pasari et al., 2019).

Hormones

Some hormones and its analogs have been investigated as alternative treatments to improve pancreatic functions and supress oxidative stress and inflammation. A meta-analysis carried out in randomized controlled trials revealed that a combinational therapy with somatostatin and ulinastatin decreased the complications in patients suffering from acute pancreatitis (Horvath et al., 2022). Melatonin, a neurohormone widely recognised for its ability to regulate sleep-wake cycle. It has been reported that melatonin as a potential antioxidant and antiinflammatory agent, reducing lipid peroxidation, mitochondrial dysfunctions and cellular apoptosis (Fig. 23.2). Furthermore, melatonin stimulates the production of antioxidant enzymes, inhibit pro-inflammatory signaling and thus reduces oxidative stress in the cells and minimizes the serum levels of amylase, lipase and cytokines in acute pancreatitis (Sidhu et al., 2010).

Trace elements

Trace elements widely recognised as micronutrients, act as a cofactor of enzymes highly required by the body at a very low concentration and play a very important role in many physiological conditions like immune system, metabolism and oxidative stress balance. Changes in some trace elements levels (Se, Zn) in serum are strongly correlated with the severity of pancreatitis and consequently being important for the both therapy and prognosis. Interestingly, the lower concentration of certain trace elements has been reported with several pathological conditions like oxidative stress, immune cell infiltrations, contributing to the pancreatic cell injury (Bowrey et al., 1999). Besides medications, some of the trace elements become important nutritional support in both acute and chronic pancreatitis. Pre-treatment with selenium at a dose of 1 mg/kg shown clinically relevant to the mice treated with caerulein and LPS, stimulating Nrf2 pathway and reducing pancreatic enzymes and inflammatory marker levels (Swetha et al., 2025). Both acute and chronic pancreatitis are frequently accompanied by zinc deficiency, linking to increased severity and high mortality rate, whereas supplementation with zinc may improve the exocrine function and diminish oxidative stress induced cellular damage (Tang et al., 2014).

Targeting several molecular pathogeneses including oxidative stress, inflammation and fibrosis, antioxidants may become appropriate therapy in the treatment of pancreatitis.

Application to the material

Addressing regulation of oxidative stress in the therapeutic management of pancreatitis is a potential approach for pathological understanding and therapeutic innovation. Investigating the formation of ROS in the pancreas and their impact on the pancreas may provide the significant insights into various pancreatic diseases development specifically pancreatitis and potential intervention options. Considering its intricate etiology, fluctuating clinical manifestation, and possibility of serious systemic consequences, pancreatitis continues to be a challenging clinical issue. Despite therapeutic advances like sophisticated approaches emphasizing on early diagnosis, supportive care and surgical procedures, mortality rate is still high in both acute and chronic pancreatitis. According to increasing findings, the accumulation ROS encounter a pathogenic role in acute and chronic pancreatitis. However, acinar cell necrosis, inflammation, fibrotic remodeling all have been associated with the excess levels of intracellular ROS. Antioxidative therapies comprising vitamins C and E, certain phytochemicals, trace elements (Se) and hormones (melatonin) may lessen the inflammation and improve the activity of antioxidant enzymes and minimize the pancreatic damage. From a clinical perspective this strategy encourages the use of modulating agents like hormones and antioxidants as supplements to traditional therapies, which may enhance the patient outcomes and reduce complications and development of pancreatitis. Furthermore, knowing the biomarkers for oxidative stress may help with the disease evaluation and early diagnosis and specific treatment. In general, investigation into oxidative stress regulation and hormonal control in pancreatitis is important for advancement of medication and preventive care in order to maintain the overall functioning of the pancreas. The best opportunity of improving patient outcomes is with an integrated approach that includes targeted antioxidative therapies and combination therapies to improve pancreatitis care and reduce disease burden.

Mini dictionary of terms

- **Pancreatic stellate cells**. Myofibroblast-like cells are key mediators of pancreatic fibrosis and produce excess extracellular matrix proteins and trigger inflammatory cascade.
- **Pancreatic ductal adenocarcinoma**. Most prevalent form of cancers that arises from the epithelial cells of the pancreatic duct.
- **Hyperbaric oxygen therapy**. A medical procedure provides breathing pure oxygen in several life-threatening conditions by using a hyperbaric chamber.

- **Pancreatic enzyme replacement therapy**. A safe and effective therapy for an underlying condition in which pancreas fails to generate digestive enzymes.
- **Video-assisted retroperitoneal debridement**. An invasive surgical technique to remove necrotic and infected pancreatic tissue.

Summary points

- Exploration of how ROS contribute to the acinar cell injury, inflammation and fibrosis in the pancreas.
- Pancreatic necrosis and systemic inflammation become worse due to excess ROS produced by mitochondria of inflammatory cells.
- Imbalance of endocrine hormones such as insulin, glucagon and melatonin may prone to the dysregulation of antioxidant defense system, which may result in tissue damage and inflammation in the pancreas.
- Antioxidants such as Vitamin C, hormones, trace elements and phytochemicals that manage oxidative stress are essential for improving pancreatic tissue recovery.
- Treatment strategies that target both oxidative and endocrine pathways such as hormonal control and antioxidant treatment may provide effective defense against pancreatitis.

References

Aja, P. M., Izekwe, F. I., Famurewa, A. C., Ekpono, E. U., Nwite, F. E., Igwenyi, I. O., … Obasi, N. A. (2020). Hesperidin protects against cadmium-induced pancreatitis by modulating insulin secretion, redox imbalance and iNOS/NF-κB signaling in rats. *Life Sciences, 259*, 118268.

Al-Hashem, F., Abd Ellatif, M., ShamsEldeen, A. M., Kamar, S. S., Al-Ani, B., & Haidara, M. A. (2023). Vitamin E protects against the modulation of TNF-α-AMPK axis and inhibits pancreas injury in a rat model of L-arginine-induced acute necrotising pancreatitis. *Archives of Physiology and Biochemistry, 129*(1), 148–156.

Bansod, S., Chilvery, S., Saifi, M. A., Das, T. J., Tag, H., & Godugu, C. (2021). Borneol protects against cerulein-induced oxidative stress and inflammation in acute pancreatitis mice model. *Environmental Toxicology, 36*(4), 530–539.

Bansod, S., & Godugu, C. (2021). Nimbolide ameliorates pancreatic inflammation and apoptosis by modulating NF-κB/SIRT1 and apoptosis signaling in acute pancreatitis model. *International Immunopharmacology, 90*, 107246.

Bansod, S. P., Saifi, M. A., Chilvery, S., Doijad, N., & Godugu, C. (2025). Berberine attenuates cerulein-induced acute pancreatitis by modulating Nrf2/NOX2 signaling pathway via AMPK activation. *Environmental Toxicology, 40*(5), 764–773.

Bansod, S., Saifi, M. A., Khurana, A., & Godugu, C. (2020). Nimbolide abrogates cerulein-induced chronic pancreatitis by modulating β-catenin/Smad in a sirtuin-dependent way. *Pharmacological Research, 156*, 104756.

Bashir, N., Manoharan, V., & Miltonprabu, S. (2016). Grape seed proanthocyanidins protects against cadmium induced oxidative pancreatitis in rats by attenuating oxidative stress, inflammation and apoptosis via Nrf-2/HO-1 signaling. *The Journal of Nutritional Biochemistry, 32*, 128–141.

Bowrey, D. J., Morris-Stiff, G. J., & Puntis, M. C. (1999). Selenium deficiency and chronic pancreatitis: Disease mechanism and potential for therapy. *HPB Surgery, 11*(4), 207–216.

Brookes, P. S., Yoon, Y., Robotham, J. L., Anders, M. W., & Sheu, S. S. (2004). Calcium, ATP, and ROS: A mitochondrial love-hate triangle. *American Journal of Physiology: Cell Physiology, 287*(4), C817–833.

Carrasco, C., Rodríguez, A. B., & Pariente, J. A. (2014). Effects of melatonin on the oxidative damage and pancreatic antioxidant defenses in cerulein-induced acute pancreatitis in rats. *Hepatobiliary and Pancreatic Diseases International, 13*(4), 442–446.

Chao, K. C., Chao, K. F., Chuang, C. C., & Liu, S. H. (2006). Blockade of interleukin 6 accelerates acinar cell apoptosis and attenuates experimental acute pancreatitis in vivo. *British Journal of Surgery, 93*(3), 332–338.

Dawra, R., Sah, R. P., Dudeja, V., Rishi, L., Talukdar, R., Garg, P., & Saluja, A. K. (2011). Intra-acinar trypsinogen activation mediates early stages of pancreatic injury but not inflammation in mice with acute pancreatitis. *Gastroenterology, 141*(6), 2210-2217. e2212.

Gómez, J. A., Molero, X., Vaquero, E., Alonso, A., Salas, A., & Malagelada, J. R. (2004). Vitamin E attenuates biochemical and morphological features associated with development of chronic pancreatitis. *American Journal of Physiology - Gastrointestinal and Liver Physiology, 287*(1), G162–169.

Gossart, S., Cambon, C., Orfila, C., Séguélas, M.-H., Lepert, J.-C., Rami, J., … Pipy, B. (1996). Reactive oxygen intermediates as regulators of TNF-α production in rat lung inflammation induced by silica. *The Journal of Immunology, 156*(4), 1540–1548.

Gui, M., Huang, J., Sheng, H., Chen, Y., Yang, Z., Ma, L., … Liu, J. (2023). High-dose vitamin C alleviates pancreatic necrosis by inhibiting platelet activation through the CXCL12/CXCR4 pathway in severe acute pancreatitis. *Journal of Inflammation Research*, 2865–2877.

Gulcubuk, A., Haktanir, D., Cakiris, A., Ustek, D., Guzel, O., Erturk, M., … Altunatmaz, K. (2013). Effects of curcumin on proinflammatory cytokines and tissue injury in the early and late phases of experimental acute pancreatitis. *Pancreatology, 13*(4), 347–354.

Han, P., Ma, X., Ge, Y., Ren, H., & Wu, Y. (2025). Selenium corrects dysregulated mitophagy and lipophagy in obesity-related acute pancreatitis through the GPX1-Mfn2-PLIN2 axis. *Biochemical and Biophysical Research Communications*, 151916.

He, J., Ma, M., Li, D., Wang, K., Wang, Q., Li, Q., … Yang, L. (2021). Sulfiredoxin-1 attenuates injury and inflammation in acute pancreatitis through the ROS/ER stress/cathepsin B axis. *Cell Death & Disease, 12*(7), 626.

Horvath, I. L., Bunduc, S., Fehervari, P., Vancsa, S., Nagy, R., Garmaa, G., … Csupor, D. (2022). The combination of ulinastatin and somatostatin reduces complication rates in acute pancreatitis: A systematic review and meta-analysis of randomized controlled trials. *Scientific Reports, 12*(1), 17979.

Iannuzzi, J. P., King, J. A., Leong, J. H., Quan, J., Windsor, J. W., Tanyingoh, D., … Kaplan, G. G. (2022). Global incidence of acute pancreatitis is increasing over time: A systematic review and meta-analysis. *Gastroenterology, 162*(1), 122–134.

Imrie, C., Connett, G., Hall, R., & Charnley, R. (2010). Enzyme supplementation in cystic fibrosis, chronic pancreatitis, pancreatic and periampullary cancer. *Alimentary Pharmacology and Therapeutics, 32*, 1–25.

Jiang, H., Li, F., Liu, S., Sun, H., Cui, Y., & Wu, Y. (2014). Acute pancreatitis affects the metabolism of catecholamine neurotransmitters in rats. *Neuroscience, 268*, 112–117.

Junyuan, Z., Hui, X., Chunlan, H., Junjie, F., Qixiang, M., Yingying, L., ... Yue, Z. (2018). Quercetin protects against intestinal barrier disruption and inflammation in acute necrotizing pancreatitis through TLR4/MyD88/p38 MAPK and ERS inhibition. *Pancreatology, 18*(7), 742–752.

Keller, J., & Layer, P. (2005). Human pancreatic exocrine response to nutrients in health and disease. *Gut, 54*(Suppl. 6), 1–28.

Klöppel, G., Detlefsen, S., & Feyerabend, B. (2004). Fibrosis of the pancreas: The initial tissue damage and the resulting pattern. *Virchows Archiv, 445*(1), 1–8.

Li, C., Cui, L., Zhang, L., Yang, L., Zhuo, Y., Cui, J., ... Zhang, S. (2021). Saikosaponin D attenuates pancreatic injury through suppressing the apoptosis of acinar cell via modulation of the MAPK signaling pathway. *Frontiers in Pharmacology, 12*, 735079.

Li, X., Wang, T., Zhou, Q., Li, F., Liu, T., Zhang, K., ... Huang, W. (2024). Isorhamnetin alleviates mitochondrial injury in severe acute pancreatitis via modulation of KDM5B/HtrA2 signaling pathway. *International Journal of Molecular Sciences, 25*(7).

Lima, P. R., de Melo, T. S., Carvalho, K. M., de Oliveira Í, B., Arruda, B. R., de Castro Brito, G. A., ... Santos, F. A. (2013). 1,8-cineole (eucalyptol) ameliorates cerulein-induced acute pancreatitis via modulation of cytokines, oxidative stress and NF-κB activity in mice. *Life Sciences, 92*(24–26), 1195–1201.

Liu, D., Song, G., Ma, Z., Geng, X., Dai, Y., Yang, T., ... Song, Z. (2020). Resveratrol improves the therapeutic efficacy of bone marrow-derived mesenchymal stem cells in rats with severe acute pancreatitis. *International Immunopharmacology, 80*, 106128.

Mittal, M., Siddiqui, M. R., Tran, K., Reddy, S. P., & Malik, A. B. (2014). Reactive oxygen species in inflammation and tissue injury. *Antioxidants and Redox Signaling, 20*(7), 1126–1167.

Nikfarjam, M., Cuthbertson, C. M., Malcontenti-Wilson, C., Muralidharan, V., Millar, I., & Christophi, C. (2007). Hyperbaric oxygen therapy reduces severity and improves survival in severe acute pancreatitis. *Journal of Gastrointestinal Surgery, 11*(8), 1008–1015.

Ning, J.-W., Zhang, Y., Yu, M.-S., Gu, M.-L., Xu, J., Usman, A., & Ji, F. (2017). Emodin alleviates intestinal mucosal injury in rats with severe acute pancreatitis via the caspase-1 inhibition. *Hepatobiliary and Pancreatic Diseases International, 16*(4), 431–436.

Pasari, L. P., Khurana, A., Anchi, P., Aslam Saifi, M., Annaldas, S., & Godugu, C. (2019). Visnagin attenuates acute pancreatitis via Nrf2/NFκB pathway and abrogates associated multiple organ dysfunction. *Biomedicine & Pharmacotherapy, 112*, 108629.

Petersen, O. H., & Tepikin, A. V. (2008). Polarized calcium signaling in exocrine gland cells. *Annual Review of Physiology, 70*, 273–299.

Robertson, R. P., & Harmon, J. S. (2006). Diabetes, glucose toxicity, and oxidative stress: A case of double jeopardy for the pancreatic islet β cell. *Free Radical Biology and Medicine, 41*(2), 177–184.

Rothe, H., & Kolb, H. (1999). Strategies of protection from nitric oxide toxicity in islet inflammation. *Journal of Molecular Medicine (Berlin), 77*(1), 40–44.

Sheng, B., Zhao, L., Zang, X., Zhen, J., Liu, Y., Bian, W., & Chen, W. (2021). Quercetin inhibits caerulein-induced acute pancreatitis through regulating miR-216b by targeting MAP2K6 and NEAT1. *Inflammopharmacology, 29*(2), 549–559.

Sidhu, S., Pandhi, P., Malhotra, S., Vaiphei, K., & Khanduja, K. (2010). Melatonin treatment is beneficial in pancreatic repair process after experimental acute pancreatitis. *European Journal of Pharmacology, 628*(1–3), 282–289.

Siriviriyakul, P., Chingchit, T., Klaikeaw, N., Chayanupatkul, M., & Werawatganon, D. (2019). Effects of curcumin on oxidative stress, inflammation and apoptosis in L-arginine induced acute pancreatitis in mice. *Heliyon, 5*(8).

Swetha, K., Indumathi, M. C., Kishan, R., Siddappa, S., Chen, C.-H., & Marathe, G. K. (2025). Selenium mitigates caerulein and LPS-induced severe acute pancreatitis by inhibiting MAPK, NF-κB, and STAT3 signaling via the Nrf2/HO-1 pathway. *Biological Trace Element Research*, 1–23.

Tang, Q.-q., Su, S.-y., & Fang, M.-y. (2014). Zinc supplement modulates oxidative stress and antioxidant values in rats with severe acute pancreatitis. *Biological Trace Element Research, 159*(1), 320–324.

Uluışık, D., Keskin, E., Özaydın, T., & Öznurlu, Y. (2024). Ameliorative effects of the melatonin on some cytokine levels, NF-κB immunoreactivity, and apoptosis in rats with cerulein-induced acute pancreatitis. *Iranian Journal of Basic Medical Sciences, 27*(3), 279–285.

Verlaan, M., Roelofs, H. M., van-Schaik, A., Wanten, G. J., Jansen, J. B., Peters, W. H., & Drenth, J. P. (2006). Assessment of oxidative stress in chronic pancreatitis patients. *World Journal of Gastroenterology, 12*(35), 5705–5710.

Wang, D., Han, S., Lv, G., Hu, Y., Zhuo, W., Zeng, Z., ... Wang, J. (2023). Pancreatic acinar cells-derived sphingosine-1-phosphate contributes to fibrosis of chronic pancreatitis via inducing autophagy and activation of pancreatic stellate cells. *Gastroenterology, 165*(6), 1488-1504. e1420.

Wang, Y. C., Zou, W. B., Tang, D. H., Wang, L., Hu, L. H., Qian, Y. Y., ... Liao, Z. (2023). High clinical and genetic similarity between chronic pancreatitis associated with light-to-moderate alcohol consumption and classical alcoholic chronic pancreatitis. *Gastro Hep Advances, 2*(2), 186–195.

Xia, S., Ni, Y., Zhou, Q., Liu, H., Xiang, H., Sui, H., & Shang, D. (2019). Emodin attenuates severe acute pancreatitis via antioxidant and anti-inflammatory activity. *Inflammation, 42*(6), 2129–2138.

Xu, H., Yue, H., Ge, H., & Wang, F. (2024). Vitamin B6 ameliorates acute pancreatitis by suppressing the caspase3 signaling pathway. *BMC Gastroenterology, 24*(1), 151.

Xu, L. L., Zhao, B., Sun, S. L., Yu, S. F., Wang, Y. M., Ji, R., ... Mao, E. Q. (2020). High-dose vitamin C alleviates pancreatic injury via the NRF2/NQO1/HO-1 pathway in a rat model of severe acute pancreatitis. *Annals of Translational Medicine, 8*(14), 852.

Yuan, Y., Li, Z., Li, M., Jin, T., Zhang, X., Liu, X., & Hao, J. (2022). Mitochondria oxidative stress mediated nicotine-promoted activation of pancreatic stellate cells by regulating mitochondrial dynamics. *Toxicology in Vitro, 84*, 105436.

Zhang, J., Bai, J., Zhou, Q., Hu, Y., Wang, Q., Yang, L., ... Wang, Y. (2022). Glutathione prevents high glucose-induced pancreatic fibrosis by suppressing pancreatic stellate cell activation via the ROS/TGFβ/SMAD pathway. *Cell Death and Disease, 13*(5), 440.

Zhao, T., Fang, R., Ding, J., Liu, Y., Cheng, M., Zhou, F., ... Zhou, L. (2024). Melatonin ameliorates multiorgan injuries induced by severe acute pancreatitis in mice by regulating the Nrf2 signaling pathway. *European Journal of Pharmacology, 975*, 176646.

Zhao, X., Li, F., Wen, A., Yu, X., Xu, X., Wan, C., … Huang, W. (2024). Elucidating the mechanism of stigmasterol in acute pancreatitis treatment: Insights from network pharmacology and in vitro/in vivo experiments. *Frontiers in Pharmacology, 15*, 1485915.

Zhao, B., Sun, W., Wang, Y., Ma, L., Gui, M., Li, J., … Sun, S. (2025). High-dose intravenous vitamin C reduce C-reactive protein levels, fluid retention, and APACHE II scores in patients with moderately severe acute pancreatitis: A prospective, randomized, double-blinded, placebo-controlled study. *Annals of Intensive Care, 15*(1), 30.

Zhao, Q., Tang, X., Huang, J., Li, J., Chen, Q., Sun, Y., & Wu, J. (2018). Melatonin attenuates endoplasmic reticulum stress in acute pancreatitis. *Pancreas, 47*(7), 884–891.

Zheng, J., Wu, J., Chen, J., Liu, J., Lu, Y., Huang, C., … Zeng, Y. (2016). Therapeutic effects of quercetin on early inflammation in hypertriglyceridemia-related acute pancreatitis and its mechanism. *Pancreatology, 16*(2), 200–210.

Zhong, K. (2015). Curcumin mediates a protective effect via TLR-4/NF-κB signaling pathway in rat model of severe acute pancreatitis. *Cell Biochemistry and Biophysics, 73*(1), 175–180.

Chapter 24

Diabetic complications with emphasis on neuropathy: Mechanistic insights and therapeutic role of aldose reductase inhibitors

Preety Kumari, Rupali Kohal, Navneesh Kumar and Sant Kumar Verma
Department of Pharmaceutical Chemistry, ISF College of Pharmacy, Moga, Punjab, India

List of abbreviations

ACE angiotensin-converting enzyme
ADCT aldose reductase inhibitor–diabetes complications trial
ALR1 aldehyde reductase 1
ALR2 aldose reductase 2
AR aldose reductase
ARB angiotensin receptor blocker
ARI/ARIs aldose reductase inhibitor(s)
GI gastrointestinal
GIP glucose-dependent insulinotropic polypeptide
GLP-1 glucagon-like peptide-1
HPA axis hypothalamic–pituitary–adrenal axis
IGF-1 insulin-like growth factor-1
MNCV median motor nerve conduction velocity
NAD^+ nicotinamide adenine dinucleotide (oxidized form)
NADH nicotinamide adenine dinucleotide (reduced form)
$NADP^+$ nicotinamide adenine dinucleotide phosphate (oxidized form)
NADPH nicotinamide adenine dinucleotide phosphate (reduced form)
RAAS renin–angiotensin–aldosterone system
RCTs randomized controlled trials

Introduction

Diabetic neuropathy is one of the common causes of morbidity and death among patients with diabetes, generating a substantial economic burden (Edmonds et al., 2021). Apart from tight glycemic control, no other evidence-based treatments are known to ameliorate or prevent neuropathy (Rowin, 2019). Diabetic neuropathy affected an estimated 206 million people globally as of 2021, representing more than a threefold increase since 1990 (Liao et al., 2025; Yang et al., 2025). Researchers have now recognized that the condition does not arise from a single defect, but rather from the cumulative impact of multiple disrupted metabolic pathways (Steinmetz et al., 2024). Together, these changes accelerate both the onset and progression of nerve damage. Among these mechanisms, the polyol pathway has received particular attention as a promising therapeutic target (Giri et al., 2018; Lin et al., 2023). In this pathway, the enzyme aldose reductase converts glucose into sorbitol, which accumulates within cells and triggers osmotic imbalance, oxidative stress, and metabolic disruption, ultimately injuring nerves (Singh et al., 2021). This mechanistic insight positioned aldose reductase as an attractive target for drug development, giving rise to aldose reductase inhibitors (ARIs) (El-Kabbani et al., 2004; Maccari & Ottanà, 2015). Over the last 5 decades, a wide range of ARIs have been designed and evaluated in clinical trials, with outcomes ranging from promising to limited (Mitchell et al., 2021; Sanyal et al., 2011). This chapter reviews the current understanding of diabetic complications, focusing on neuropathy, and provides a critical appraisal of the therapeutic promise and challenges associated with ARIs.

Diabetic complications: Emphasis on diabetic neuropathy

Persistent high blood sugar in diabetes triggers a series of biochemical and cellular imbalances, one of the most critical being the excessive build-up of sorbitol in insulin-insensitive cells (Galicia-Garcia et al., 2020; Malone et al., 1984). This occurs through the polyol pathway, where

Hormones in Neuroscience. https://doi.org/10.1016/B978-0-443-30263-3.00020-8

glucose is converted to sorbitol, leading to its accumulation inside cells (Srikanth & Orrick, 2022). The excess sorbitol disrupts cellular balance by causing osmotic stress and interfering with normal cellular functions, eventually resulting in tissue damage (Godisela et al., 2020; Zhu et al., 2024). Because of these mechanisms, diabetic complications are generally grouped into two major categories: microvascular and macrovascular complications (Armengol et al., 2021; Islam et al., 2025) (Fig. 24.1).

Microvascular complications arise from structural and functional damage to small blood vessels and include diabetic neuropathy, nephropathy, and retinopathy (Li et al., 2023; Vithian & Hurel, 2010). On the other hand, macrovascular complications involve large blood vessels, manifesting as cardiovascular disease, cerebrovascular disease, and peripheral arterial disease (Fowler, 2008; Zakir et al., 2023). These complications significantly increase the risk of premature mortality among individuals with diabetes (Cade, 2008).

Among these, diabetic neuropathy presents the highest prevalence and clinically diverse manifestations of chronic diabetes (Bodman et al., 2024; Bondar & POPA, 2018). It is a progressive nerve disorder resulting from metabolic and vascular disturbances, which lead to impaired nerve conduction, axonal degeneration, and loss of neurotrophic support (Eid et al., 2023; Feldman et al., 2019). The most common form is distal symmetric sensorimotor polyneuropathy, characterized by numbness, tingling, burning pain, and sensory loss in a "glove and stocking" pattern (Jensen et al., 2021; Lee & Won, 2025). In contrast, Autonomic neuropathy affects the cardiovascular, gastrointestinal, and genitourinary systems, while acute sensory, focal, and multifocal neuropathies add further complexity to its presentation (Verrotti et al., 2014; Yang et al., 2025). Diabetic neuropathy is strongly associated with foot ulcers, recurrent infections, and amputations, making it a leading cause of morbidity in diabetes (Edmonds et al., 2021; Malone et al., 1984). Its economic burden is substantial, with global healthcare costs running into billions of dollars annually due to hospitalizations and long-term care (Maity et al., 2024). Early recognition and management remain crucial to reducing both patient suffering and the impact on healthcare.

The polyol concept from hyperglycemia to nerve damage

When blood sugar levels rise abnormally, as seen in diabetes, the body begins to handle glucose differently (Balaji

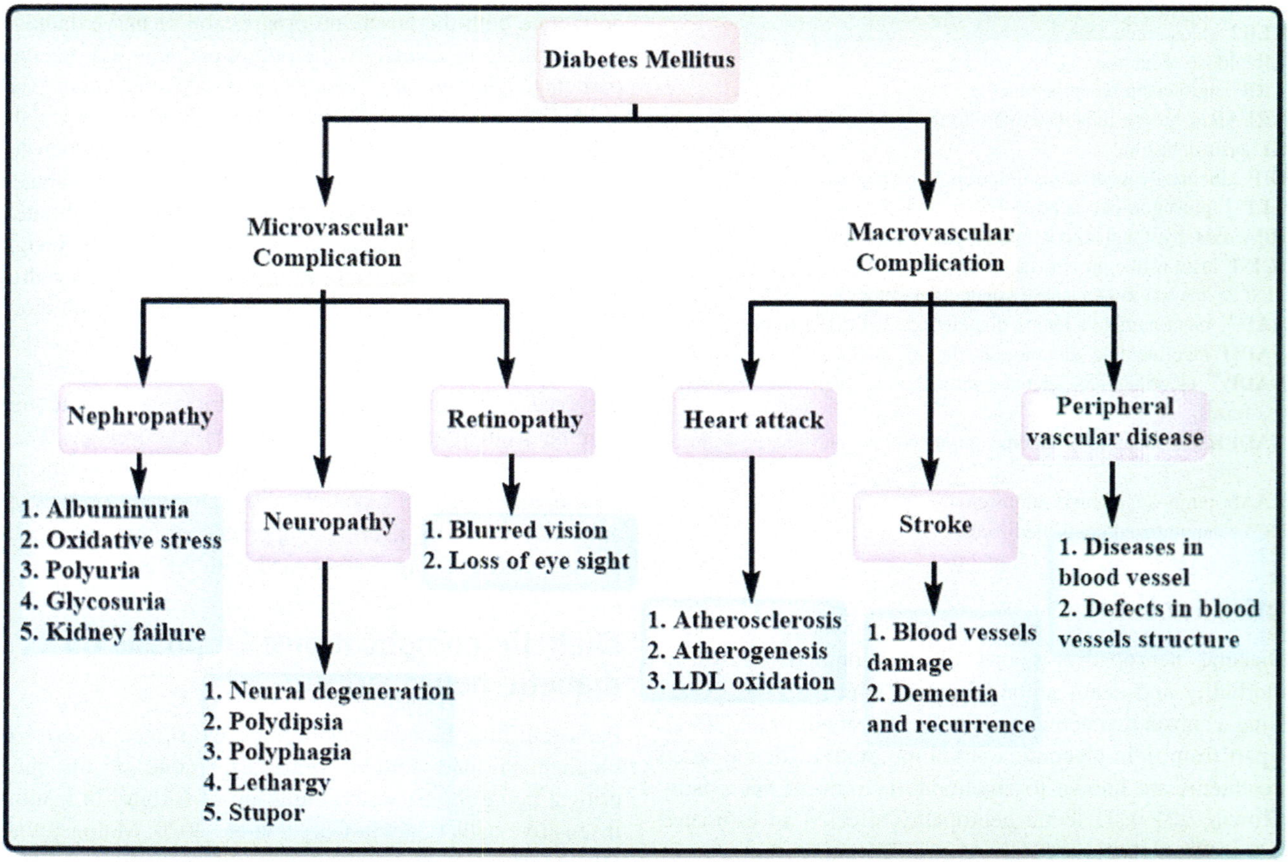

FIGURE 24.1 Classification of diabetic complications.

et al., 2019). Instead of taking the usual energy-producing pathways, a considerable portion of glucose is shunted into the polyol pathway (Guthrie & Guthrie, 2004). Under normal conditions, this route is almost negligible; only about 3% of glucose is metabolized here (Wood, 1955). However, during diabetes, more than 30% of circulating glucose may be diverted into this pathway, placing an unusual burden on the system (Bano, 2013).

The polyol pathway unfolds in two sequential steps. In the first step, aldose reductase (AR) reduces glucose into sorbitol (consuming NADPH) (Dunlop, 2000). Subsequently, sorbitol dehydrogenase converts sorbitol into fructose (while NAD^+ is simultaneously reduced to NADH) (Jeffery & Jörnvall, 1983) as shown in Fig. 24.2. Although seemingly harmless, this shift in metabolism sets off a chain of harmful events.

The problem begins because sorbitol cannot easily cross cell membranes. As it builds up inside cells, it draws water with it, creating osmotic stress that injures tissues (Chandler-Brown et al., 2015). At the same time, the pathway drains vital cofactors like NADPH and disturbs the delicate balance of NAD^+/NADH (Martinière et al., 2019). This redox imbalance mimics a state of "*pseudohypoxia*"-as if cells were starved of oxygen even when oxygen is plentiful. The result is a rise in oxidative stress, metabolic disruption, and tissue injury, all of which contribute to the chronic complications of diabetes (Kumari et al., 2024).

The nervous system, particularly the peripheral nerves, is one of the most brutal hits. AR is most active in Schwann cells, which are essential for insulating and nourishing nerve fibers (Fallon & Tadi, 2019). Excessive sorbitol here interferes with Na^+, K^+-ATPase activity, slows down axonal transport, and disrupts the myelination process. This gradually undermines nerve conduction and forms the foundation of diabetic neuropathy (Greene et al., 1988; Hao et al., 2015) (Fig. 24.3). Encouragingly, research has shown that aldose reductase inhibitors (ARIs) can block this harmful pathway at its very first step. By preventing the accumulation of sorbitol, these inhibitors restore the nerve conduction velocity, improve blood flow, and protect the nerve structure in experimental diabetic models (Dănilă et al., 2024). Strong ARIs can nearly shut down the enzyme at therapeutic doses, effectively normalizing glucose metabolism in affected tissues. This makes them a highly promising strategy to not just slow, but potentially reverse, the course of diabetic neuropathy (Greene et al., 1999).

Clinical evidence from major trials

The clinical development of ARIs for diabetic neuropathy has spanned more than 3 decades. A comprehensive meta-analysis of 32 randomized controlled trials (RCTs) involving 4970 participants revealed that ARIs produced measurable improvements in nerve conduction and neurophysiological parameters, although overall clinical benefits remained modest and heterogeneous (Chalk et al., 2007; Nguyen et al., 2025). The pooled analysis reported no statistically significant overall difference between ARI therapy and placebo in neurological function (standardized mean difference: 0.25; 95% CI, −0.56 to 0.05) (Ran et al., 2024). However, subgroup analyses indicated potential benefits for specific ARIs and selected patient populations. The most compelling clinical evidence emerged from the Aldose Reductase Inhibitor–Diabetes Complications Trial (ADCT), a 3-year multicenter study conducted in Japan with 594 patients with mild diabetic neuropathy (Cumbie & Hermayer, 2007). In this trial, epalrestat (150 mg/day) effectively prevented deterioration of median motor nerve conduction velocity (MNCV), achieving a between-group

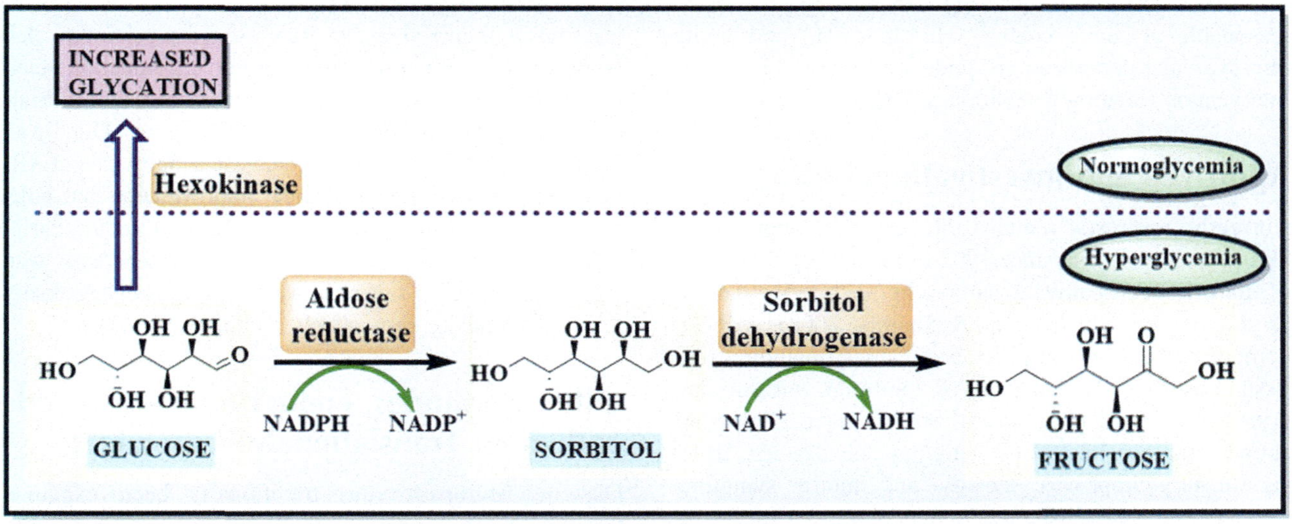

FIGURE 24.2 Polyol pathway.

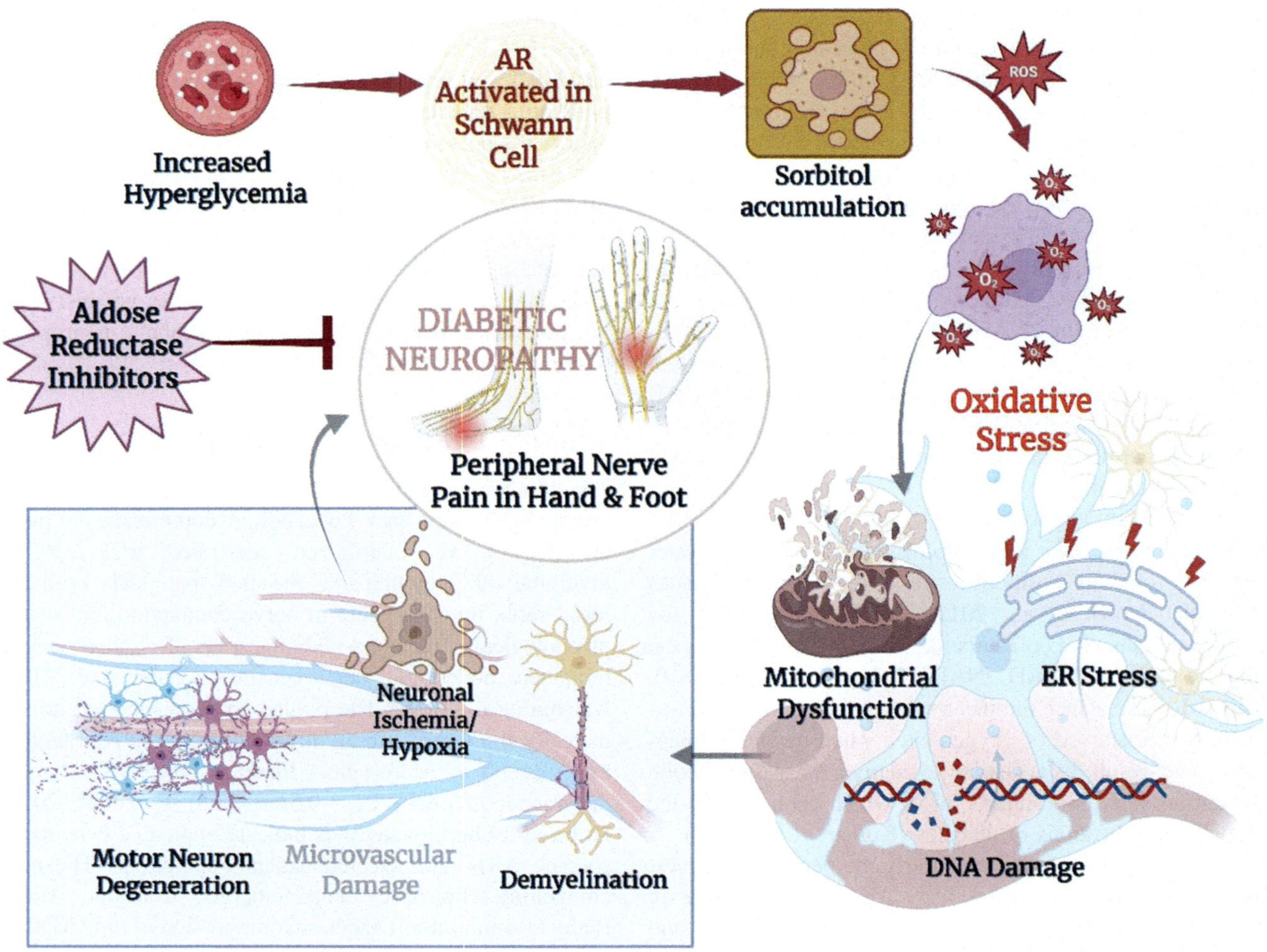

FIGURE 24.3 Polyol pathway leading to chronic diabetic neuropathy.

difference of 1.6 m/s ($P < .001$) (Hotta et al., 2006). Epalrestat also improved F-wave latency, vibration perception thresholds, and patient-reported symptoms such as numbness, abnormal sensations, and muscle cramps. Importantly, therapeutic benefit was most significant among patients with reasonable glycemic control (HbA1c < 7%) and limited microvascular complications, underscoring the role of early intervention (Sharma & Sharma, 2008).

Approved and investigational ARIs

Currently, epalrestat remains the only ARI approved for clinical use, available since 1992 in Japan for the treatment of diabetic neuropathy. A carboxylic acid derivative with an IC_{50} of ≈98 nM (Sharma & Sharma, 2008), epalrestat demonstrates favorable pharmacokinetics, including rapid tissue absorption and potent AR inhibition with minimal systemic side effects. Clinical studies have consistently shown that epalrestat (150 mg/day in divided doses) significantly improves sensory and motor symptoms, electrophysiological parameters, and quality of life, with a strong safety record. Post-marketing surveillance across 2190 patients reported adverse effects in only 2.5%, mainly mild gastrointestinal discomfort or reversible liver enzyme elevations (Kawai et al., 2010; Steele et al., 1993).

Other investigational ARIs have advanced through clinical trials with mixed results. Ranirestat, a succinimide-class inhibitor, showed improvements in nerve conduction velocity (particularly sensory nerves) during Phase II and III trials, though symptomatic benefits were less consistent (Dutta et al., 2022; Sekiguchi et al., 2019). Fidarestat, a highly potent ARI, demonstrated electrophysiological improvements but failed to meet primary efficacy endpoints in Phase III studies. Earlier compounds such as sorbinil, tolrestat, and zenarestat were discontinued due to severe hypersensitivity, hepatotoxicity, and nephrotoxicity, respectively (Hotta et al., 2001).

Other hormones, endocrine tissues, and pathways: Translational perspectives

Diabetic neuropathy has traditionally been explained through hyperglycemia-driven biochemical pathways, especially the polyol pathway and aldose reductase activation. However, recent research makes it clear that the condition does not evolve in isolation. It is shaped by a much broader hormonal environment that constantly communicates with metabolic organs and peripheral nerves. Hormones released from the pancreas, thyroid, adipose tissue, adrenal axis, and even the gut help regulate glucose use, oxidative balance, vascular tone, and neuronal repair mechanisms. When these endocrine signals become disrupted, the vulnerability of nerves increases sharply. Recognizing this interconnected landscape is essential for translating basic mechanisms into more effective therapeutic strategies.

Insulin-IGF-1 axis

Insulin is not only a regulator of blood glucose but also a key neurotrophic signal. Neurons and Schwann cells rely on insulin and IGF-1 to maintain axonal transport, mitochondrial activity, and cellular survival pathways. In chronic hyperglycemia or insulin resistance, these signals weaken, reducing Akt pathway activation and impairing axonal regeneration. IGF-1, in particular, has been shown to enhance nerve repair, raising interest in therapies that restore or support this pathway in conjunction with aldose reductase inhibitors (Aghanoori et al., 2019).

Thyroid hormones

Thyroid hormones regulate metabolic rate and mitochondrial efficiency-two factors essential for healthy nerve function. Even mild hypothyroidism, frequently seen in long-standing diabetes, is associated with slower nerve conduction, impaired microvascular flow, and increased oxidative stress. Because these changes can intensify sorbitol pathway activation, identifying and correcting thyroid imbalance may help stabilize neuronal energy metabolism (Mullur et al., 2014).

Cortisol and HPA axis

Chronic stress and persistent activation of the hypothalamic–pituitary–adrenal (HPA) axis raises cortisol levels. Elevated cortisol promotes gluconeogenesis, worsens insulin resistance, and increases inflammatory signaling. These changes heighten oxidative stress and can make neurons more vulnerable to hyperglycemic injury. Strategies that normalize sleep, reduce stress, or modulate cortisol rhythms may therefore offer indirect neuroprotective benefits (Knezevic et al., 2023).

Adipokines from adipose tissue

Adipose tissue acts as an endocrine organ, releasing hormones and cytokines that regulate inflammation and insulin sensitivity. Imbalances-especially low adiponectin and high leptin-promote endothelial dysfunction, chronic inflammation, and mitochondrial stress. Low adiponectin has been linked with worsened nerve conduction and higher pain sensitivity. Improving adipokine balance through weight management, lifestyle interventions, or pharmacologic modulation may enhance the response to ARI-based therapy (Ahima, 2006).

Renin-angiotensin-aldosterone system (RAAS)

Activation of the RAAS pathway contributes to microvascular injury in diabetes. Angiotensin II reduces nitric oxide availability and promotes vasoconstriction, directly impairing nerve perfusion. Clinical use of ACE inhibitors or ARBs has shown improvements in nerve blood flow and modest slowing of neuropathy progression, suggesting that RAAS modulation and ARIs may offer complementary effects (Hsueh & Wyne, 2011).

Gut hormones and microbiota

Hormones derived from the gut-such as GLP-1 and GIP-regulate glucose metabolism, inflammation, and neuronal survival. GLP-1 receptor agonists, now widely used in diabetes management, also demonstrate neuroprotective effects by enhancing mitochondrial stability and promoting axonal repair. Gut dysbiosis can further disturb hormone release and enhance oxidative stress, indicating a role for microbiota-targeted therapies as supportive interventions (Urkon et al., 2025).

Pancreatic hormones and glucose-neural homeostasis

Insulin, glucagon, and somatostatin collectively determine glucose availability for neurons. Insulin supports synaptic activity and axonal maintenance, while glucagon ensures glucose supply during fasting. Somatostatin fine-tunes both pancreatic secretion and intestinal absorption. Disturbances in these hormones alter neuronal energy balance and increase susceptibility to oxidative injury (Felig et al., 1976).

Integrative translational perspective

Viewing neuropathy through an endocrine systems lens highlights why single-target strategies often fall short. A combined approach-pairing aldose reductase inhibition with hormonal modulation, improved insulin sensitivity, thyroid correction, adipokine balancing, stress reduction, or gut-endocrine therapies-holds promise for early and more effective neuroprotection. This integrative model aligns with modern efforts toward personalized, multimodal management of diabetic neuropathy.

Safety profile and adverse effects

The clinical development of aldose reductase inhibitors (ARIs) has been shaped as much by their safety concerns as by their therapeutic promise. While early agents initially appeared effective, their use was soon limited by severe toxicities (Krans, 1992). For instance, sorbinil was discontinued after reports of life-threatening hypersensitivity reactions such as Stevens–Johnson syndrome (Ahmad et al., 2024). Tolrestat, despite its efficacy in alleviating neuropathic symptoms, was globally withdrawn in 1996 due to fatal hepatotoxicity (Chalk et al., 2007), and Zenarestat faced a similar fate owing to kidney-related adverse effects (Shimoshige et al., 2000). These setbacks underscored the importance of drug safety in the development of ARIs. Encouragingly, newer agents have demonstrated a more favorable profile. Epalrestat, currently the only widely approved ARI, has been evaluated in thousands of patients for up to 3 years, showing predominantly mild and reversible side effects such as transient gastrointestinal discomfort or occasional elevations in liver enzymes, with no evidence of severe organ damage (Shende et al., 2018). This progress highlights a significant transition, suggesting that ARIs can be developed as safer, long-term therapeutic options for diabetic complications.

Combination therapy approaches

Diabetic neuropathy involves multiple pathological mechanisms, making single-drug approaches often insufficient. Combination therapies that target these diverse pathways are emerging as a promising strategy (Javed et al., 2015). For example, pairing epalrestat, an aldose reductase inhibitor, with α-lipoic acid, a strong antioxidant, has shown superior improvement in nerve conduction, vibration sensitivity, and symptom relief compared to either agent alone (Wang et al., 2018; Zhao et al., 2018). Other combinations-such as ARIs with pregabalin for pain relief or methyl cobalamin for nerve repair-aim to address metabolic, oxidative, and neurodegenerative processes concurrently (Gupta et al., 2024). These synergistic approaches offer a scientifically grounded pathway toward more effective management of diabetic neuropathy.

Patient stratification and predictive factors

Clinical trials have highlighted the importance of **patient selection** in maximizing ARI efficacy. The ADCT study confirmed that ARIs are most effective in patients with:

- Good glycemic control (HbA1c $< 7\%$)
- Mild neuropathy
- Minimal microvascular complications

Patients with advanced retinopathy, nephropathy, or poorly controlled diabetes showed limited benefit, suggesting that ARIs should be initiated early, before irreversible nerve damage develops. This stratification underscores the importance of early diagnosis and timely therapeutic intervention (Buse et al., 2020; Davies et al., 2022).

Global market access and regional variations

Epalrestat has been approved and widely used in Japan and several Asian countries, including China, India, and South Korea, where it is considered a standard therapy for diabetic neuropathy (Shi et al., 2023). However, it remains unavailable in the United States and the European Union, primarily due to historical regulatory challenges and the difficulty of conducting large-scale trials in regions where alternative treatments are already established (Hotta et al., 1996, 2006; Jain & Singh, 2021). These disparities highlight global inequities in patient access to ARI therapy. With the rising prevalence of diabetes worldwide, newer ARIs under development may achieve broader approval and expand the international market in the future.

Applications and translational relevance of aldose reductase inhibitors in diabetic neuropathy

Aldose reductase inhibitors (ARIs) have progressed beyond their earlier reputation as experimental agents and now represent a meaningful link between biochemical understanding and clinical management of diabetic neuropathy. Their relevance spans clinical practice, drug-development programs, mechanistic research, and the pharmaceutical market, making them an important translational outcome of polyol-pathway science.

Clinical applications

In clinical settings, ARIs have emerged as one of the very few therapies capable of modifying the underlying course of diabetic neuropathy rather than merely easing symptoms. By blocking the first and rate-limiting step of the polyol pathway, ARIs help reduce intracellular sorbitol accumulation, restore redox balance, and improve nerve conduction properties. Epalrestat, the only ARI in routine clinical use, has repeatedly shown benefits in patients with early or mild neuropathy-improving vibration perception, reducing paresthesia, and slowing the decline in nerve conduction velocity. Because its therapeutic impact is strongest when glycemic control is reasonable, clinicians increasingly view ARIs as an early-intervention strategy that complements glucose management rather than as a last-line therapy (Sharma & Sharma, 2008). In addition, ARIs support biomarker-based monitoring through measurable parameters such as erythrocyte sorbitol levels and oxidative stress indicators, enabling more precise assessment of treatment response. These features make ARIs a valuable component of modern, personalized diabetes care, with emerging data also suggesting supportive roles in retinopathy and early nephropathy.

Pharmaceutical and medicinal chemistry applications

Medicinal chemistry has played a crucial role in addressing the limitations of earlier ARIs, especially toxicity and poor selectivity. Structural insights into the aldose reductase active site have enabled chemists to design derivatives with improved affinity toward ALR2 while minimizing off-target interactions with ALR1. A significant advancement in this direction is the work by Pasala et al. (2021), who developed a series of coumarin-thiazolidinedione hybrids by merging functional elements from both epalrestat and sorbinil. Their 14 synthesized molecules demonstrated how finely tuned substitutions influence ALR2 selectivity. In particular, compound 1, featuring a bromine atom at the sixth position of the coumarin ring and a cyclopentyl group on the thiazolidinedione moiety-showed exceptional potency, inhibiting ALR2 at 0.012 μM with more than 300-fold selectivity over ALR1. Compound 2, with chlorine substitution in place of bromine, displayed similarly impressive ALR2 inhibition and strong isoform preference (Fig. 24.4) (Pasala et al., 2021). In parallel, Imran et al. (2022) explored coumarin-based thiosemicarbazones and identified compound 3 as the most promising ALR2 inhibitor (IC_{50} = 3.035 μM), showing moderate ALR1 suppression (Fig. 24.5) (Imran et al., 2022). These two independent SAR studies highlight the coumarin scaffold as a versatile structural platform, providing clear direction for next-generation ARIs with improved safety, potency, and drug-like behavior.

Research applications

ARIs have also become reliable tools in experimental neurobiology. By selectively shutting down the polyol pathway, researchers can dissect its contribution to oxidative stress, Schwann-cell dysfunction, mitochondrial imbalance, and pseudohypoxic injury. ARIs are widely used to explore nerve-repair pathways, test antioxidant combinations, and study mechanisms of axonal degeneration. Their use has expanded into regenerative research, where they help investigate interactions between metabolic stress, inflammatory responses, and myelin maintenance.

These research applications have clarified the complementary benefits of combining ARIs with α-lipoic acid, neurotrophic agents, or mitochondrial stabilizers-providing a scientific basis for multimodal therapy.

Future prospects

The path forward for aldose reductase inhibitors (ARIs) in diabetic neuropathy is both promising and transformative. Next-generation ARIs are being designed with precision, enhancing selectivity, improving tissue penetration, and extending drug action, while minimizing off-target effects that limited earlier candidates. Advances in medicinal

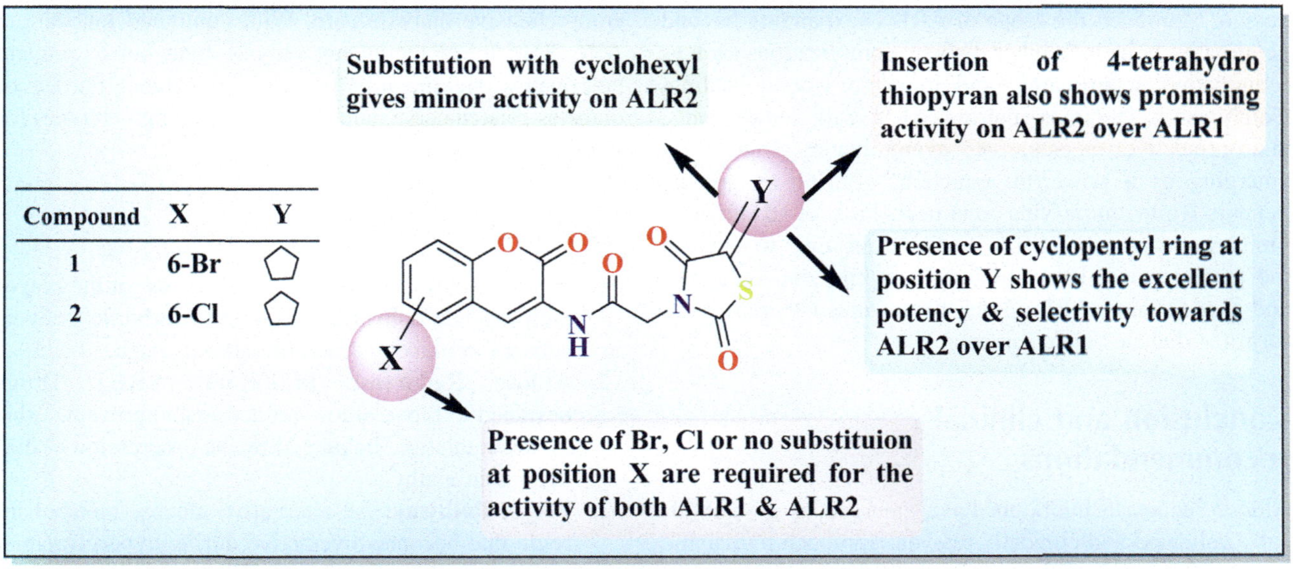

FIGURE 24.4 SAR of coumarin thiazolidinedione hybrids.

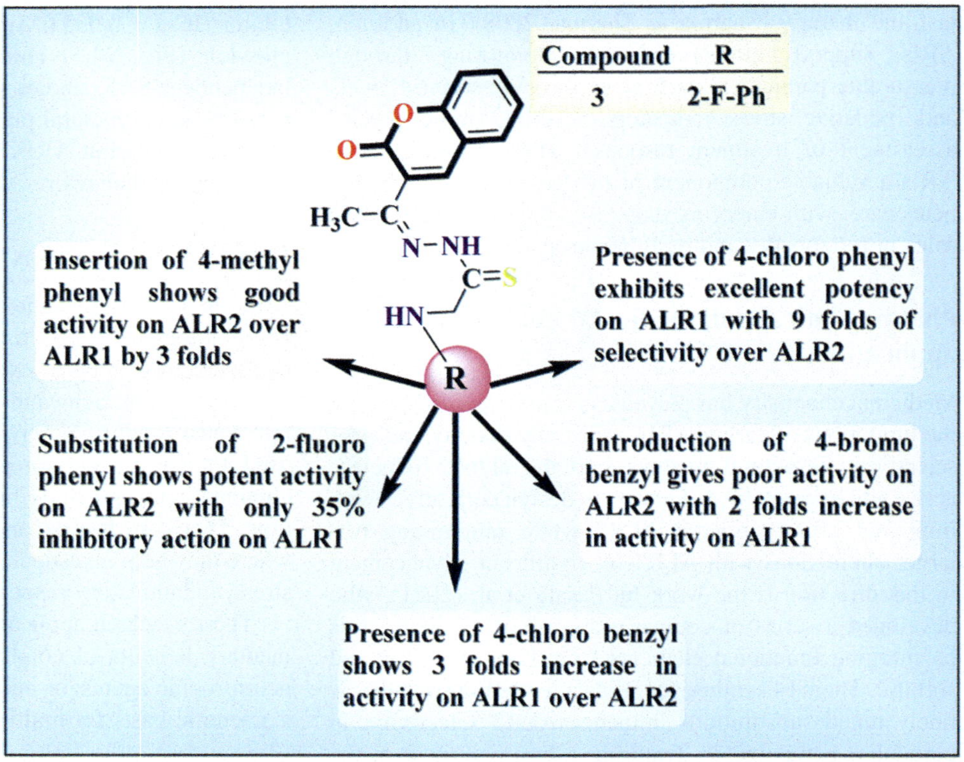

FIGURE 24.5 SAR of coumarin-based thiosemicarbazone derivatives.

chemistry and structure-activity relationship studies are paving the way for compounds that not only inhibit aldose reductase more effectively but also integrate seamlessly into modern therapeutic regimens. A significant shift in research focus is toward prevention rather than cure. Evidence suggests ARIs are most effective in the early stages of neuropathy, particularly in patients with reasonable glycemic control, opening the door to proactive strategies that could halt disease before irreversible nerve damage occurs. Moreover, the scope of ARIs is expanding beyond neuropathy to include other diabetic complications such as retinopathy, nephropathy, and impaired wound healing. Combination therapies, pairing ARIs with antioxidants, neuroprotective agents, or symptomatic drugs, are emerging as a powerful approach, offering synergistic benefits while simplifying treatment. Parallel advances in biomarker-driven precision medicine promise to identify the patients most likely to benefit, optimizing outcomes and transforming ARIs into a personalized therapy platform for diabetes management.

Conclusion and clinical recommendations

Aldose reductase inhibitors have emerged as a scientifically validated and clinically relevant approach to diabetic neuropathy, bridging the gap between mechanistic insight and therapeutic innovation. Epalrestat, the only widely approved ARI to date, exemplifies the potential of this class to slow disease progression, restore nerve function, and improve patient quality of life when applied early in the disease course. The future of ARI therapy lies in integration-combining early intervention, combination strategies, and precision medicine to address the multifactorial nature of diabetic neuropathy. This vision transforms ARIs from isolated agents into a cornerstone of comprehensive diabetic care. With continued innovation, ARIs hold the potential not only to treat but to prevent neuropathy, significantly reducing the global burden of diabetes-related nerve damage and improving outcomes for millions worldwide.

Mini-dictionary of key terms

1. **Aldose Reductase (AR)**: A key enzyme of the polyol pathway that converts glucose into sorbitol; its overactivity in diabetes leads to nerve damage.
2. **Aldose Reductase Inhibitors (ARIs)**: Drugs designed to block aldose reductase and prevent sorbitol accumulation, helping slow the progression of diabetic neuropathy.
3. **Polyol Pathway**: An alternative glucose-metabolism route that becomes overactive during hyperglycemia, contributing to oxidative stress and tissue injury.

4. **Sorbitol Accumulation**: Build-up of sorbitol inside cells due to excess aldose reductase activity, causing osmotic stress and nerve dysfunction.
5. **Oxidative Stress**: An imbalance between free radicals and antioxidant defenses, leading to cellular damage—commonly seen in diabetic nerves.
6. **Diabetic Neuropathy**: A nerve disorder caused by long-standing hyperglycemia, leading to numbness, burning pain, and loss of sensory function.
7. **Motor Nerve Conduction Velocity (MNCV)**: A test that measures how fast electrical signals travel in nerves; slowing of this speed indicates nerve injury.
8. **Schwann Cells**: Cells that form the myelin sheath around peripheral nerves; they are highly vulnerable to sorbitol-induced stress in diabetes.
9. **Pseudohypoxia**: A metabolic state where cells behave as if oxygen is low because of altered NAD+/NADH balance, despite a normal oxygen supply.
10. **Adipokines**: Hormones released from fat tissue (like leptin and adiponectin) that influence inflammation, insulin sensitivity, and nerve health.
11. **IGF-1 (Insulin-Like Growth Factor-1)**: A growth factor that supports nerve repair and axonal maintenance; its levels often fall in insulin-resistant states.
12. **Retinopathy**: A microvascular complication of diabetes affecting the retina; AR activation and oxidative stress play major roles in its progression.
13. **Glycemic Control**: Maintaining blood glucose within target levels; essential for improving ARI effectiveness and slowing nerve damage.
14. **Combination Therapy**: Using ARIs with antioxidants or neuroprotective agents to target multiple mechanisms of neuropathy for better outcomes.
15. **Selectivity (ALR1 versus ALR2)**: The ability of an ARI to inhibit aldose reductase (ALR2) without affecting similar enzymes like ALR1, ensuring better safety and efficacy.

Summary points

- Diabetic neuropathy remains one of the most widespread and disabling complications of diabetes, arising from a mix of metabolic, vascular, and oxidative disturbances rather than a single mechanism.
- The polyol pathway plays a central role in nerve injury, with excessive aldose reductase activity driving sorbitol accumulation, osmotic stress, redox imbalance, and eventual impairment of nerve conduction.
- Aldose reductase inhibitors (ARIs) act at the first step of this pathway, helping restore metabolic balance, reduce oxidative stress, and slow or stabilize early-stage neuropathy when used alongside good glycemic control.
- Epalrestat is the only ARI approved for regular clinical use, and long-term studies consistently show that it can preserve nerve conduction velocity, reduce sensory symptoms, and demonstrate an excellent safety profile.
- Earlier ARIs were limited by toxicity, but modern medicinal chemistry-supported by recent SAR studies including coumarin-based hybrids-has opened the way for safer, more selective, and more potent next-generation ARIs.
- Combination therapies (e.g., epalrestat with α-lipoic acid, pregabalin, or methylcobalamin) target multiple pathways at once, offering more effective symptom control and metabolic protection than single-agent treatment.
- Hormones beyond insulin-such as thyroid hormones, cortisol, adipokines, and gut-derived peptides-also influence nerve health, showing that neuropathy is an endocrine-metabolic condition requiring a broader therapeutic lens.
- Early diagnosis and proper patient selection greatly increase ARI effectiveness, especially in individuals with mild neuropathy, minimal microvascular complications, and stable glucose levels.
- Market access for ARIs varies globally, with epalrestat widely available in Asia but still absent in the US and EU, reflecting regulatory, historical, and trial-based challenges.
- Future progress lies in integrating ARIs with precision medicine, biomarker-guided monitoring, and preventive strategies, ultimately aiming not only to treat but also to avert diabetes-related nerve damage.

Acknowledgment

The authors are thankful to the ISF College of Pharmacy, Moga, Punjab, India, for providing the necessary facilities.

Consent for publication

All authors have reviewed and approved the final manuscript and provided their final consent for its publication as a book chapter.

Conflict of interest

There is no conflict of interest.

References

Aghanoori, M.-R., Smith, D. R., Shariati-Ievari, S., Ajisebutu, A., Nguyen, A., Desmond, F., Jesus, C. H., Zhou, X., Calcutt, N. A., & Aliani, M. (2019). Insulin-like growth factor-1 activates AMPK to augment mitochondrial function and correct neuronal metabolism in sensory neurons in type 1 diabetes. *Molecular Metabolism, 20*, 149–165.

Ahima, R. S. (2006). Adipose tissue as an endocrine organ. *Obesity, 14* (S8), 242S–249S.

Ahmad, S., Ahmad, M. F. A., Khan, S., Alouffi, S., Khan, M., Prakash, C., Khan, M. W. A., & Ansari, I. A. (2024). Exploring aldose reductase inhibitors as promising therapeutic targets for diabetes-linked disabilities. *International Journal of Biological Macromolecules, 280*, 135761.

Armengol, G. D., Hayfron-Benjamin, C. F., Van den Born, B.-J. H., Galenkamp, H., & Agyemang, C. (2021). Microvascular and macrovascular complications in type 2 diabetes in a multi-ethnic population based in Amsterdam. The HELIUS study. *Primary Care Diabetes, 15*(3), 528–534.

Balaji, R., Duraisamy, R., & Kumar, M. (2019). Complications of diabetes mellitus: A review. *Drug Invention Today, 12*(1).

Bano, G. (2013). Glucose homeostasis, obesity and diabetes. *Best Practice and Research Clinical Obstetrics and Gynaecology, 27*(5), 715–726.

Bodman, M. A., Dreyer, M. A., & Varacallo, M. A. (2024). Diabetic peripheral neuropathy. In *StatPearls [Internet]*. StatPearls Publishing.

Bondar, A. C., & Popa, A. R. (2018). Diabetic neuropathy prevalence and its associated risk factors in two representative groups of type 1 and type 2 diabetes mellitus patients from Bihor county. *Maedica, 13*(3), 229.

Buse, J. B., Wexler, D. J., Tsapas, A., Rossing, P., Mingrone, G., Mathieu, C., D'Alessio, D. A., & Davies, M. J. (2020). 2019 update to: Management of hyperglycemia in type 2 diabetes, 2018. A consensus report by the American Diabetes Association (ADA) and the European Association for the Study of Diabetes (EASD). *Diabetes Care, 43*(2), 487–493.

Cade, W. T. (2008). Diabetes-related microvascular and macrovascular diseases in the physical therapy setting. *Physical Therapy, 88*(11), 1322–1335.

Chalk, C., Benstead, T. J., & Moore, F. (2007). Aldose reductase inhibitors for the treatment of diabetic polyneuropathy. *Cochrane Database of Systematic Reviews*, (4).

Chandler-Brown, D., Choi, H., Park, S., Ocampo, B. R., Chen, S., Le, A., Sutphin, G. L., Shamieh, L. S., Smith, E. D., & Kaeberlein, M. (2015). Sorbitol treatment extends lifespan and induces the osmotic stress response in *Caenorhabditis elegans*. *Frontiers in Genetics, 6*, 316.

Cumbie, B. C., & Hermayer, K. L. (2007). Current concepts in targeted therapies for the pathophysiology of diabetic microvascular complications. *Vascular Health and Risk Management, 3*(6), 823–832. https://doi.org/10.2147/vhrm.s12187421

Dănilă, A.-I., Ghenciu, L. A., Stoicescu, E. R., Bolintineanu, S. L., Iacob, R., Săndesc, M.-A., & Faur, A. C. (2024). Aldose reductase as a key target in the prevention and treatment of diabetic retinopathy: A comprehensive review. *Biomedicines, 12*(4), 747.

Davies, M. J., Aroda, V. R., Collins, B. S., Gabbay, R. A., Green, J., Maruthur, N. M., Rosas, S. E., Del Prato, S., Mathieu, C., & Mingrone, G. (2022). Management of hyperglycaemia in type 2 diabetes, 2022. A consensus report by the American Diabetes Association (ADA) and the European Association for the Study of Diabetes (EASD). *Diabetologia, 65*(12), 1925–1966.

Dunlop, M. (2000). Aldose reductase and the role of the polyol pathway in diabetic nephropathy. *Kidney International, 58*, S3–S12.

Dutta, D., Mohindra, R., Kumar, M., Kumar, A., & Sharma, M. (2022). Ranirestat improves electrophysiologic but not clinical measures of diabetic polyneuropathy: A meta-analysis. *Indian Journal of Endocrinology and Metabolism, 26*(5), 399–406.

Edmonds, M., Manu, C., & Vas, P. (2021). The current burden of diabetic foot disease. *Journal of Clinical Orthopaedics and Trauma, 17*, 88–93.

Eid, S. A., Rumora, A. E., Beirowski, B., Bennett, D. L., Hur, J., Savelieff, M. G., & Feldman, E. L. (2023). New perspectives in diabetic neuropathy. *Neuron, 111*(17), 2623–2641.

El-Kabbani, O., Ruiz, F., Darmanin, C., & Chung, R.-T. (2004). Aldose reductase structures: Implications for mechanism and inhibition. *Cellular and Molecular Life Sciences CMLS, 61*(7), 750–762.

Fallon, M., & Tadi, P. (2019). *Histology, Schwann cells.*

Feldman, E. L., Callaghan, B. C., Pop-Busui, R., Zochodne, D. W., Wright, D. E., Bennett, D. L., Bril, V., Russell, J. W., & Viswanathan, V. (2019). Diabetic neuropathy. *Nature Reviews Disease Primers, 5*(1), 41. https://doi.org/10.1038/s41572-019-0092-1

Felig, P., Wahren, J., Sherwin, R., & Hendler, R. (1976). The lilly lecture 1976: Insulin, glucagon, and somatostatin in normal physiology and diabetes mellitus. *Diabetes, 25*(12), 1091–1099.

Fowler, M. J. (2008). Microvascular and macrovascular complications of diabetes. *Clin Diabetes, 26*(2), 77–82.

Galicia-Garcia, U., Benito-Vicente, A., Jebari, S., Larrea-Sebal, A., Siddiqi, H., Uribe, K. B., Ostolaza, H., & Martín, C. (2020). Pathophysiology of type 2 diabetes mellitus. *International Journal of Molecular Sciences, 21*(17), 6275.

Giri, B., Dey, S., Das, T., Sarkar, M., Banerjee, J., & Dash, S. K. (2018). Chronic hyperglycemia mediated physiological alteration and metabolic distortion leads to organ dysfunction, infection, cancer progression and other pathophysiological consequences: An update on glucose toxicity. *Biomedicine and Pharmacotherapy, 107*, 306–328.

Godisela, K. K., Reddy, S. S., Reddy, P. Y., Kumar, C. U., Reddy, V. S., Ayyagari, R., & Reddy, G. B. (2020). Role of sorbitol-mediated cellular stress response in obesity-associated retinal degeneration. *Archives of Biochemistry and Biophysics, 679*, 108207.

Greene, D. A., Arezzo, J., Brown, M., & Group, Z. S. (1999). Effect of aldose reductase inhibition on nerve conduction and morphometry in diabetic neuropathy. *Neurology, 53*(3), 580.

Greene, D. A., Lattimer, S. A., & Sima, A. A. (1988). Are disturbances of sorbitol, phosphoinositide, and Na+-K+-ATPase regulation involved in pathogenesis of diabetic neuropathy? *Diabetes, 37*(6), 688–693.

Gupta, P. K., Roy, R., Jha, R. K., Paudel, M., Dewasy, B., & Gupta, P. (2024). Efficacy of pregabalin with methylcobalamin in neuropathic pain: A study in the Nepalese population of Eastern region. *Journal of Nobel Medical College, 13*(2), 77–82.

Guthrie, R. A., & Guthrie, D. W. (2004). Pathophysiology of diabetes mellitus. *Critical Care Nursing Quarterly, 27*(2), 113–125.

Hao, W., Tashiro, S., Hasegawa, T., Sato, Y., Kobayashi, T., Tando, T., Katsuyama, E., Fujie, A., Watanabe, R., & Morita, M. (2015). Hyperglycemia promotes Schwann cell de-differentiation and demyelination via sorbitol accumulation and Igf1 protein downregulation. *Journal of Biological Chemistry, 290*(28), 17106–17115.

Hotta, N., Akanuma, Y., Kawamori, R., Matsuoka, K., Oka, Y., Shichiri, M., Toyota, T., Nakashima, M., Yoshimura, I., & Sakamoto, N. (2006). Long-term clinical effects of epalrestat, an aldose reductase inhibitor, on diabetic peripheral neuropathy: The 3-

year, multicenter, comparative aldose reductase inhibitor-diabetes complications trial. *Diabetes Care, 29*(7), 1538–1544.

Hotta, N., Sakamoto, N., Shigeta, Y., Kikkawa, R., Goto, Y., & Japan, D. N. S. G. (1996). Clinical investigation of epalrestat, an aldose reductase inhibitor, on diabetic neuropathy in Japan: Multicenter study. *Journal of Diabetes and Its Complications, 10*(3), 168–172.

Hotta, N., Toyota, T., Matsuoka, K., Shigeta, Y., Kikkawa, R., Kaneko, T., Takahashi, A., Sugimura, K., Koike, Y., & Ishii, J. (2001). Clinical efficacy of fidarestat, a novel aldose reductase inhibitor, for diabetic peripheral neuropathy: A 52-week multicenter placebo-controlled double-blind parallel group study. *Diabetes Care, 24*(10), 1776–1782.

Hsueh, W. A., & Wyne, K. (2011). Renin-angiotensin-aldosterone system in diabetes and hypertension. *The Journal of Clinical Hypertension, 13*(4), 224–237.

Imran, A., Shehzad, M. T., Shah, S. J. A., Al Adhami, T., Laws, M., Rahman, K. M., Alharthy, R. D., Khan, I. A., Shafiq, Z., & Iqbal, J. (2022). Development and exploration of novel substituted thiosemicarbazones as inhibitors of aldose reductase via in vitro analysis and computational study. *Scientific Reports, 12*(1), 5734.

Islam, K., Islam, R., Nguyen, I., Malik, H., Pirzadah, H., Shrestha, B., Lentz, I. B., Shekoohi, S., & Kaye, A. D. (2025). Diabetes mellitus and associated vascular disease: Pathogenesis, complications, and evolving treatments. *Advances in Therapy*, 1–20.

Jain, R., & Singh, S. K. (2021). Study of epalrestat in diabetic distal symmetric polyneuropathy and correlation of its therapeutic efficacy with erythrocyte sorbitol levels: A step towards precision medicine. *Clinical Diabetology, 10*(4), 354–358.

Javed, S., Alam, U., & Malik, R. A. (2015). Treating diabetic neuropathy: Present strategies and emerging solutions. *The Review of Diabetic Studies: RDS, 12*(1–2), 63.

Jeffery, J., & Jörnvall, H. (1983). Enzyme relationships in a sorbitol pathway that bypasses glycolysis and pentose phosphates in glucose metabolism. *Proceedings of the National Academy of Sciences, 80*(4), 901–905.

Jensen, T. S., Karlsson, P., Gylfadottir, S. S., Andersen, S. T., Bennett, D. L., Tankisi, H., Finnerup, N. B., Terkelsen, A. J., Khan, K., & Themistocleous, A. C. (2021). Painful and non-painful diabetic neuropathy, diagnostic challenges and implications for future management. *Brain, 144*(6), 1632–1645.

Kawai, T., Takei, I., Tokui, M., Funae, O., Miyamoto, K., Tabata, M., Hirata, T., Saruta, T., Shimada, A., & Itoh, H. (2010). Effects of epalrestat, an aldose reductase inhibitor, on diabetic peripheral neuropathy in patients with type 2 diabetes, in relation to suppression of Nε-carboxymethyl lysine. *Journal of Diabetes and Its Complications, 24*(6), 424–432.

Knezevic, E., Nenic, K., Milanovic, V., & Knezevic, N. N. (2023). The role of cortisol in chronic stress, neurodegenerative diseases, and psychological disorders. *Cells, 12*(23), 2726.

Krans, H. (1992). Recent clinical experience with aldose reductase inhibitors. *Journal of Diabetes and Its Complications, 6*(1), 39–44.

Kumari, P., Kohal, R., Gupta, G. D., & Verma, S. K. (2024). Selectivity challenges for aldose reductase inhibitors: A review on comparative SAR and interaction studies. *Journal of Molecular Structure, 1318*, 139207.

Lee, J.-E., & Won, J. C. (2025). Clinical phenotypes of diabetic peripheral neuropathy: Implications for phenotypic-based therapeutics strategies. *Diabetes and Metabolism Journal, 49*(4), 542–564.

Li, Y., Liu, Y., Liu, S., Gao, M., Wang, W., Chen, K., Huang, L., & Liu, Y. (2023). Diabetic vascular diseases: Molecular mechanisms and therapeutic strategies. *Signal Transduction and Targeted Therapy, 8*(1), 152.

Liao, X., Zhang, Y., Xu, J., Yin, J., Li, S., Dong, K., Shi, X., Xu, W., Ma, D., & Chen, X. (2025). A narrative review on cognitive impairment in type 2 diabetes: Global trends and diagnostic approaches. *Biomedicines, 13*(2), 473.

Lin, Q., Li, K., Chen, Y., Xie, J., Wu, C., Cui, C., & Deng, B. (2023). Oxidative stress in diabetic peripheral neuropathy: Pathway and mechanism-based treatment. *Molecular Neurobiology, 60*(8), 4574–4594.

Maccari, R., & Ottanà, R. (2015). Targeting aldose reductase for the treatment of diabetes complications and inflammatory diseases: New insights and future directions. *Journal of Medicinal Chemistry, 58*(5), 2047–2067.

Maity, S., Leton, N., Nayak, N., Jha, A., Anand, N., Thompson, K., Boothe, D., Cromer, A., Garcia, Y., & Al-Islam, A. (2024). A systematic review of diabetic foot infections: Pathogenesis, diagnosis, and management strategies. *Frontiers in Clinical Diabetes and Healthcare, 5*, 1393309.

Malone, J., Knox, G., & Harvey, C. (1984). Sorbitol accumulation is altered in type 1 (insulin-dependent) diabetes mellitus. *Diabetologia, 27*(5), 509–513.

Martinière, A., Fiche, J. B., Smokvarska, M., Mari, S., Alcon, C., Dumont, X., Hematy, K., Jaillais, Y., Nollmann, M., & Maurel, C. (2019). Osmotic stress activates two reactive oxygen species pathways with distinct effects on protein nanodomains and diffusion. *Plant Physiology, 179*(4), 1581–1593.

Mitchell, P., Holz, F. G., Hykin, P., Midena, E., Souied, E., Allmeier, H., Lambrou, G., Schmelter, T., Wolf, S., & Investigators, A. S. (2021). Efficacy and safety of intravitreal aflibercept using a treat-and-extend regimen for neovascular age-related macular degeneration: The ARIES study: A randomized clinical trial. *Retina, 41*(9), 1911–1920.

Mullur, R., Liu, Y.-Y., & Brent, G. A. (2014). Thyroid hormone regulation of metabolism. *Physiological Reviews, 94*(2), 355–382.

Nguyen, V., Dinh, Q., Yu, F., Jia, S., & Wang, X. (2025). Interventional effects of exercise on neuropathy in patients with diabetes: A systematic review with meta-analysis. *BMC Sports Science, Medicine and Rehabilitation, 17*(1), 82.

Pasala, V. K., Gudipudi, G., Sankeshi, V., Basude, M., Gundla, R., singh Jadav, S., Srinivas, B., Goud, E. Y., & Nareshkumar, D. (2021). Design, synthesis and biological evaluation of selective hybrid coumarin-thiazolidinedione aldose reductase-II inhibitors as potential antidiabetics. *Bioorganic Chemistry, 114*, 104970.

Ran, G.-L., Li, Y.-P., Lu, L.-C., & Lan, S.-H. (2024). Disease-modifying therapies for diabetic peripheral neuropathy: A systematic review and meta-analysis of randomized controlled trials. *Journal of Diabetes and Its Complications, 38*(2), 108691.

Rowin, J. (2019). Integrative neuromuscular medicine: Neuropathy and neuropathic pain: Consider the alternatives. *Muscle and Nerve, 60*(2), 124–136.

Sanyal, A. J., Brunt, E. M., Kleiner, D. E., Kowdley, K. V., Chalasani, N., Lavine, J. E., Ratziu, V., & McCullough, A. (2011). *Endpoints and clinical trial design for nonalcoholic steatohepatitis*. Wiley Online Library.

Sekiguchi, K., Kohara, N., Baba, M., Komori, T., Naito, Y., Imai, T., Satoh, J., Yamaguchi, Y., Hamatani, T., & Group, R. (2019). Aldose reductase inhibitor ranirestat significantly improves nerve conduction velocity in diabetic polyneuropathy: A randomized double-blind placebo-controlled study in Japan. *Journal of Diabetes Investigation, 10*(2), 466–474.

Sharma, S., & Sharma, N. (2008). Epalrestat, an aldose reductase inhibitor, in diabetic neuropathy: An Indian perspective. *Annals of Indian Academy of Neurology, 11*(4), 231–235.

Shende, S. D., Baig, M. S., & Doifode, S. (2018). Evaluation of efficacy and safety of epalrestat (150 mg) compared to epalrestat (50 mg) in patients suffering from diabetic peripheral neuropathy. *Journal of Clinical and Diagnostic Research, 12*(4).

Shi, W., Xu, G., Gao, Y., Zhao, J., Liu, T., Zhao, J., Yang, H., Wei, Z., Li, H., & Xu, A.-L. (2023). Novel role for epalrestat: Protecting against NLRP3 inflammasome-driven NASH by targeting aldose reductase. *Journal of Translational Medicine, 21*(1), 700.

Shimoshige, Y., Ikuma, K., Yamamoto, T., Takakura, S., Kawamura, I., Seki, J., Mutoh, S., & Goto, T. (2000). The effects of zenarestat, an aldose reductase inhibitor, on peripheral neuropathy in Zucker diabetic fatty rats. *Metabolism-Clinical and Experimental, 49*(11), 1395–1399.

Singh, M., Kapoor, A., & Bhatnagar, A. (2021). Physiological and pathological roles of aldose reductase. *Metabolites, 11*(10), 655.

Srikanth, K. K., & Orrick, J. A. (2022). *Biochemistry, polyol or sorbitol pathways.*

Steele, J. W., Faulds, D., & Goa, K. L. (1993). Epalrestat: A review of its pharmacology, and therapeutic potential in late-onset complications of diabetes mellitus. *Drugs and Aging, 3*(6), 532–555.

Steinmetz, J. D., Seeher, K. M., Schiess, N., Nichols, E., Cao, B., Servili, C., Cavallera, V., Cousin, E., Hagins, H., & Moberg, M. E. (2024). Global, regional, and national burden of disorders affecting the nervous system, 1990–2021: A systematic analysis for the global Burden of disease study 2021. *The Lancet Neurology, 23*(4), 344–381.

Urkon, M., Ferencz, E., Szász, J. A., Szabo, M. I. M., Orbán-Kis, K., Szatmári, S., & Nagy, E. E. (2025). Antidiabetic GLP-1 receptor agonists have neuroprotective properties in experimental animal models of Alzheimer's disease. *Pharmaceuticals, 18*(5), 614.

Verrotti, A., Prezioso, G., Scattoni, R., & Chiarelli, F. (2014). Autonomic neuropathy in diabetes mellitus. *Frontiers in Endocrinology, 5*, 205.

Vithian, K., & Hurel, S. (2010). Microvascular complications: Pathophysiology and management. *Clinical Medicine, 10*(5), 505–509.

Wang, X., Lin, H., Xu, S., Jin, Y., & Zhang, R. (2018). Alpha lipoic acid combined with epalrestat: A therapeutic option for patients with diabetic peripheral neuropathy. *Drug Design, Development and Therapy*, 2827–2840.

Wood, H. G. (1955). Significance of alternate pathways in the metabolism of glucose. *Physiological Reviews, 35*(4), 841–859.

Yang, Y., Zhao, B., Wang, Y., Lan, H., Liu, X., Hu, Y., & Cao, P. (2025). Diabetic neuropathy: Cutting-edge research and future directions. *Signal Transduction and Targeted Therapy, 10*(1), 132.

Zakir, M., Ahuja, N., Surksha, M. A., Sachdev, R., Kalariya, Y., Nasir, M., Kashif, M., Shahzeen, F., Tayyab, A., & moazzam Khan, M. S. (2023). Cardiovascular complications of diabetes: From microvascular to macrovascular pathways. *Cureus, 15*(9).

Zhao, M., Chen, J.-Y., Chu, Y.-D., Zhu, Y.-B., Luo, L., & Bu, S.-Z. (2018). Efficacy of epalrestat plus α-lipoic acid combination therapy versus monotherapy in patients with diabetic peripheral neuropathy: A meta-analysis of 20 randomized controlled trials. *Neural Regeneration Research, 13*(6), 1087–1095.

Zhu, J., Hu, Z., Luo, Y., Liu, Y., Luo, W., Du, X., Luo, Z., Hu, J., & Peng, S. (2024). Diabetic peripheral neuropathy: Pathogenetic mechanisms and treatment. *Frontiers in Endocrinology, 14*, 1265372.

Chapter 25

Phytomedicine in diabetes care: Molecular pathways and therapeutic mechanisms

Halimat Yusuf Lukman, Christina Peter, Aaliyah Khan and Saheed Sabiu
Department of Biotechnology and Food Science, Faculty of Applied Sciences, Durban University of Technology, Durban, South Africa

List of abbreviations

AA alpha amylase
AG alpha glucosidase
AGE-RAGE advanced glycation end-products and its receptor
AMPK 5′ adenosine monophosphate-activated protein kinase
BMI body mass index
BW body weight
DC diabetic complications
DPP4 dipeptidyl peptidase-IV
FOXO1 forkhead box protein O1
GK glucokinase
GLP-1 glucagon-like peptide-1
GLUT 2/4 glucose-transporter 2/4
GSK3 glycogen synthase kinase-3
HbA1c glycated hemoglobin
HOMO-IR homeostasis model assessment-estimated insulin resistance
IR insulin resistance
IRS insulin receptor substrates
JNK/SAPK Jun N-terminal kinase/stress-activated protein kinase
MM9 matrix metalloproteinase-9
NF-kB nuclear factor-kappa B
p38MAPK p38 mitogen-activated protein kinases
PFK-1 phosphofructokinase-1
PI3K phosphoinositide-3 kinase
PK pyruvate kinase
PPAR peroxisome proliferator-activated receptors
PRKCD protein kinase C delta type
ROS reactive oxygen species
SREBP1c sterol regulatory element binding protein-1c
T2D type 2 diabetes
TC total cholesterol
TG triglycerides

Introduction

Attention toward the exploration of phytomedicine to support the advancement of effective strategies against global diabetes burden is increasingly growing (Governa et al., 2018). For centuries, human civilizations have utilized plants for medical and therapeutic purposes (Srivastava et al., 2019). Today, many of these plant-derived bioactive compounds serve as the basis for modern pharmaceuticals (Chihomvu et al., 2024). These phytomedicines offer potential to simultaneously target a range of metabolic pathways, are costeffective, and culturally acceptable among diverse patients. However, the practical implementation of phytomedicines into diabetes care has yet to reach its potential, attributable to the lack of understanding regarding the underlying molecular pathways and the variations in formulations and delivery methods.

Studies have revealed that some notable mechanisms of action associated with antidiabetic agents include the prevention of oxidative stress, stimulating insulin secretion, attenuation of insulin degradation, improvement in insulin sensitivity, delaying carbohydrate metabolism, and regulating glucose metabolizing enzymes. However, a significant portion of these phytomedicines possess poorly elucidated mechanisms of action and are believed to act through non-specific or multifactorial pathways (Okoduwa et al., 2024). Therefore, mechanistic insights into the antidiabetic activity of these phytochemicals are fundamental to validate their therapeutic efficacy and enhance their integration and augmentation within clinical practice.

Hormones in Neuroscience. https://doi.org/10.1016/B978-0-443-30263-3.00008-7

It is noteworthy that beyond pharmacological action, the success of phytomedicines as clinical drugs is heavily influenced by factors such as their formulation strategies and delivery methods, which in turn, determine their bioavailability and target specificity. Sustainability and biodiversity of plants, standardization, safety and regulatory concerns are part of the challenges associated with the use of phytomedicines in diabetes care. Thus, this chapter provides an overview of the therapeutic potential of phytomedicines in diabetes management and their targeted molecular pathways, formulation strategies, as well as setbacks and future perspectives of phytomedicines in diabetes intervention.

Hormones, endocrine tissues, and molecular pathways in diabetes pathogenesis

Deciphering the underlying molecular pathways associated with diabetes pathogenesis is imperative for prevention, diagnosis, and precise treatment. Hormonal signaling and molecular pathways such as insulin signaling dysregulation, oxidative stress, and inflammatory have been linked to diabetes (Pizzino et al., 2017) (Fig. 25.1). By mapping these pathways, clinicians and researchers can identify therapeutic targets and develop mechanism-based diabetes interventions. Moreover, phytomedicines and nutraceuticals with multi-target effects can be rationally integrated into care when their molecular mechanisms are elucidated. Furthermore, such mechanistic insights can inform biomarker discovery, enabling early detection of therapeutic response and complication risk.

Insulin signaling pathway

Insulin is an endocrine polypeptide hormone, secreted within the Islets of Langerhans by pancreatic β-cells. The hormone controls metabolic fluxes in all animals, thereby regulating blood glucose levels and facilitating storage of glucose in hepatic, adipose and muscle tissue (Rahman et al., 2021). Insulin binds to its surface cell receptor, triggering a molecular cascade of events involving carbohydrate, lipid and protein metabolism. Under physiological conditions, the binding of insulin to its receptor, a transmembrane tyrosine kinase, results in the

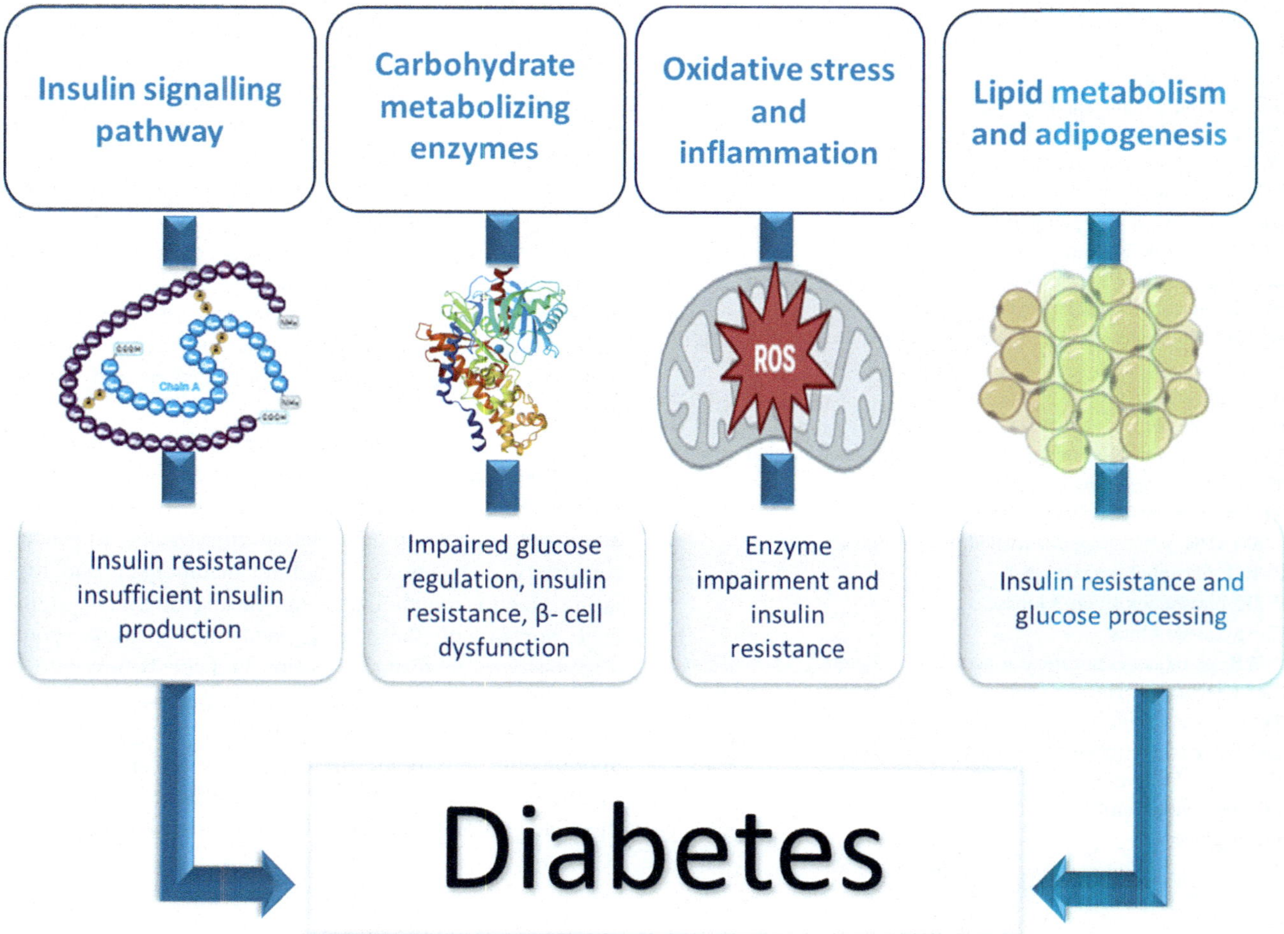

FIGURE 25.1 **Key molecular pathways and their features in diabetes.** Hormonal signaling and molecular pathways such as insulin signaling dysregulation, oxidative stress and inflammatory have been linked to diabetes.

autophosphorylation and recruitment of insulin receptor substrates (IRS) (Lu et al., 2024).

This activates downstream cascades, including the phosphoinositide-3 kinase (PI3K)/Akt pathway that promotes glucose uptake via GLUT4 translocation, glycogen synthesis through glycogen synthase kinase-3 (GSK-3) inhibition, and suppression of hepatic gluconeogenesis through forkhead box protein O1 (FOXO1) inactivation (He, 2021). Disruption in this sequence of events results in type-1 diabetes mellitus (T1DM), characterized by insufficient insulin production or type-2 diabetes mellitus (T2DM) leading to insulin resistance. This phenomenon, therefore, marks the insulin signaling pathway as a key focus in diabetes management.

Carbohydrate metabolizing enzymes

As a disorder of carbohydrate metabolism, diabetes, particularly T2D, disrupts the balance between glucose utilization, storage and production via pathways such as glycolysis, gluconeogenesis, and glycogenesis which are the hallmark in maintaining glucose homeostasis (Lu et al., 2024). Damage to the integral enzymes and pathways in carbohydrate metabolism results in impaired glucose regulation, insulin resistance, β-cell dysfunction, and glucagon dysregulation. For example, glycolytic pathway that primarily metabolize glucose is regulated by the allosteric activity of three main enzymes of the pathway: glucokinase (GK), phosphofructokinase-1 (PFK-1), and pyruvate kinase (PK) (Rojas-Pirela et al., 2025) are implicated in hyperglycaemia.

Oxidative stress and inflammation

Oxidative stress occurs when a biological system is unable to maintain a balance between reactive oxygen species (ROS) production and detoxification, this in turn exacerbating diabetes pathogenesis by dysregulating integral signaling and regulatory pathways associated with diabetes. During electron transport chain in the mitochondria, energy is produced, and a few molecular oxygen is reduced to water, whilst the remainder are transformed to a superoxide anion, a primary ROS. Additional ROS, including hydrogen peroxide and hydroxyl radical, are also generated through processes such as enzymatic glycation of proteins, heightened lipid peroxidation, and glucose oxidation, which results in disrupted cellular homeostasis (González et al., 2023). Consequently, these events ultimately lead to cellular and enzyme impairment, and increased insulin resistance.

Other oxidative stress biomarkers, such as reactive nitrogen intermediates, such as nitric oxide, peroxynitrite, nitrogen dioxide, and nitrous acid, further contribute to metabolic dysregulation, including diabetes (Aranda-Rivera et al., 2022). For instance, the presence of excess ROS triggers the development of sequential stress pathways, like NF-kB, JNK/SAPK, and p38MAPK. These pathways involve various serine/threonine kinases, which exhibit a detrimental effect on insulin signaling.

Lipid metabolism and adipogenesis

Lipids are involved in the production of energy, certain vitamins, hormones and bile salts, and contribute to the regulation of some cellular signaling (Chandel, 2021). However, prolonged consumption of lipid-rich diets leads to the buildup of lipid intermediates, such as diacylglycerols, ceramides, and free fatty acids, which lead to fat accumulation and contribute to the development of insulin resistance, and thus T2D. When excess free fatty acids and dietary lipids accumulate in non-adipose tissue like the liver, pancreas, muscles, they are stored as ectopic fat and contribute toward lipotoxicity. The abnormal lipid levels and their uncontrolled deposition in tissues forms a strong link to disorders such as T2D (Zhu et al., 2024). In addition to its role in lipid storage and mobilization, adipose tissues also function as a major endocrine organ by releasing cytokines and other proinflammatory mediators that impair insulin action and promote inflammation, collectively contributing to insulin resistance and its associated metabolic implications.

Application of phytomedicines in diabetes care

Since ancient times, folkloric medicines from Indian, Korean and Chinese cultures have utilized plants for their curative benefits (Alam et al., 2022). Medicinal plants or herbal medicines with antidiabetic activity usually stimulate pancreatic islets to enhance insulin secretion or prevent intestinal glucose retention (Okoduwa et al., 2024). Numerous reviews have harnessed medicinal plants that have been studied using *in vitro, in vivo* or *in silico* models. For example, the review of Salehi et al. (2019) highlighted more than 500 antidiabetic plant species that spread across about 140 genera, these include *Gymnema sylvestre, Momordica charantia, Vernonia amygdalina, Pterocarpus marsupium, Trigonella foenum-graecum, Cinnamomum verum, Viscum album,* and *Allium sativum* (Fig. 25.2). Phytoconstituents such as flavonoids, phenolics, terpenoids, stilbene, alkaloids, and tannins have also been documented as antidiabetic (Rahman et al., 2022). A few widely studied antidiabetic plants and the target molecular pathway in diabetes care are presented in this chapter.

Gymnema sylvestre

Gymnema sylvestre also known as Gur-Mar, is medicinal plant belonging to the Asclepiadaceae family (Khan et al., 2019). Native to India and tropical Africa, *G. syvestre* leaves comprise chlorophyll, resins, carbohydrates, albumins, butyric acid, tartaric acid, formic acid and anthraquinone derivatives (Shariati et al., 2021). The leaves also

FIGURE 25.2 Examples of widely studied antidiabetic plants. Representative phytomedicines with well-established antidiabetic activities.

contain gymnemic acid which exerts its antidiabetic effects by preventing glucose absorption and weight gain (Garg, 2016). Gymnemic acid molecules share similar atomic arrangement with sugar molecules, as such, they prevent glucose absorption in the intestine by filling receptor cavities in the intestinal absorptive external layers. Other mechanisms include upregulation of enzymes involved in glucose metabolism and phosphorylation and downregulation of sorbitol dehydrogenase and gluconeogenesis enzymes (Kashif et al., 2023).

Momordica charantia

Momordica charantia commonly referred to as karela, bitter gourd or bitter melon forms part of *Cucurbitaceae* family. It is mainly grown and consumed in east India, and other tropical, subtropical and temperate regions (Liu, Gong, et al., 2021). *M. charantia* is a nutritionally rich plant containing several vitamins and minerals, antioxidants and phytochemicals such as flavonoids, phenolic compounds, cucurbitane-type triterpenoids and triterpenes glycosides, oleanne-type triterpenes, saponins and insulin-like peptides responsible for their bioactive properties (Pahlavani et al., 2019). Mechanistically, *M. charantia* stimulates insulin secretion, increases glucose uptake in adipose and muscle tissues and prevents intestinal glucose absorption and hepatic glucose production, while maintaining fewer adverse effects (Richter et al., 2023).

Trigonella foenum-graecum

Trigonella foenum-graecum also known as fenugreek, is a member of the *Fabaceae* family consumed as food and medicine (Singh et al., 2022). Fenugreek's phytochemical profile includes protodioscin, diosgenin, 4-hydroxyisoleucine, trigoneoside, galactomannans, yamogenin and trigonelline (Mukherjee et al., 2022). Most of fenugreek's bioactive properties including antidiabetic, can be attributed to trigonelline. The antidiabetic activity of fenugreek has been linked to protection of pancreatic beta cells, enhanced insulin secretion, hexokinase activity and GLUT4 translocation as well as decreased fructose-1,6-bisphosphatase and glucose-6-phosphatase activity and DPP4, alpha amylase and maltase inhibition (Sarker et al., 2024).

Cinnamomum verum

Cinnamomum verum is the most harvested of the 250 *cinnamomum* species that belong to the *Lauraceae* family, utilized as food and medicine (Pundarikakshudu et al., 2024). It is commonly cultivated in tropical rainforests and is native to southern India and Sri Lanka. Therapeutically, *C. verum* has been used as interventive care for diabetes and other metabolic conditions, with its aqueous extract known to upregulate glycogen synthase and insulin receptor activity by various mechanisms (Qais et al., 2019). These bioactive properties can be attributed to *C. verum's* vast phytochemical profile comprising cinnamic acid, cinnamate, cinnamaldehyde and polyphenols such as chlorogenic acid, quercetin, protocatechuic acid, esculetin, catechin and icariin (Batiha et al., 2020).

Several computational, pre-clinical, and clinical studies have been conducted to establish the antidiabetic potential of these plants (Tables 25.1–25.4).

Several phytomedicines contained in nutraceuticals also elicit antidiabetic effects via the identified molecular pathways and even produce greater effect on the gut microbiome toward diabetes care e.g., berberine, cinnamon, and catechins. Notably, phytomedicine can be formulated using a single or combination of plant/plant constituents. Polyherbal formulation made from a mixture of single medicinal entities contains many bioactive compounds and often presents better and multiple medicinal effects. For example, *Musa paradisiaca* L., *Tribulus terrestris* L., *Phyllanthus emblica* L., *Salacia reticulata* Wight, *Santalum album* L. and *Sida rhombifolia* subsp. *alnifolia* (L.) Ugbor., decoction is used in Ayurvedic medicine to treat T2D (Kuriyan et al., 2008) while *Coccinia grandis* (L.) extracts which reportedly contain Cucurbitacins B, and D, Cephalandrol, Cephalandrin A and B and associated analogs can induce beta cell regeneration (Wasana et al., 2021).

Delivery of phytomedicine for improved antidiabetic efficacy

With the shift and identification of several antidiabetic plants, efforts are ongoing to address limitations such as poor solubility, stability, absorption, permeability, and bioavailability for improved therapeutic effectiveness by employing methods such as nanotechnology (Abdulsalam et al., 2025). With nanotechnology—which entails the fabrication of materials at 1–100 nm, the biological, physical, and chemical properties of drugs can be enhanced allowing for better biological functions, surface area, shape, controlled release, and conductivity, overall promoting the delivery of antidiabetic drugs to the targeted site. Based on these benefits, antidiabetic phytochemicals have been produced in safe, unreactive, and biocompatible nano-systems such as liposomes, niosomes, solid lipid nanoparticles, metallic nanoparticles, nanostructured lipid carrier, and nanoemulsions for targeted delivery, controlled release, and improved antidiabetic properties (Afzal et al., 2022). For instance, the incorporation of betanin into lecithin to produce a nanoliposome enhanced its antidiabetic activity.

Safety concerns and pharmacokinetics of phytomedicine use in diabetes care

Generally perceived as safe, eco-friendly, and consumed without prescription, the use of phytomedicine in disease treatment, including diabetes requires pharmacovigilance study. This helps to detect, assess, understand, and prevent their side effects. Unlike adverse drug-drug or herb-drug interactions associated with synthetic antidiabetic drugs, phytomedicines can be used concurrently or synergistically to produce therapeutic effects, reduced drug dependency, with little or reduced herb–herb interactions and side effects (Gamil et al., 2025). Despite these benefits, pharmacokinetic interactions, quality control of phytomedicines, and regulatory approval should not be undermined because while they offer promising antidiabetic properties at acceptable therapeutic dosages, they have been reported to produce adverse effects such as hypoglycaemia, organ toxicity (e.g., hepatotoxicity, nephrotoxicity, and allergic reactions beyond their lethal dose as well as other risks from contaminants (Mugale et al., 2024).

Therapeutic or unwanted/unexpected adverse effects of phytomedicines, including antidiabetic medications depend on the frequency, site of action, and quantity of drug present in systemic circulation as phytoconstituents elicit varying degrees of pharmacokinetics profiles. Pre-clinical evaluation in the drug discovery pipeline is crucial in elucidating the site/mechanism of action, potency and toxicity levels of antidiabetic phytochemicals (Rombolà et al., 2020).

Challenges and future directions of phytomedicine in diabetes care

Despite the promising and multifaceted approaches of phytomedicine in diabetes care, challenges such as standardization, bioavailability, reliable clinical evidence, sustainability and diversity concerns, ethical and regulatory frameworks hinder their full application in diabetes care. While numerous plants offer antidiabetic potential preclinically, limited translational/clinical trials are still a challenge in the use of phytomedicine as commercial antidiabetic agents (Willcox et al., 2021). For example, preclinical evaluation shows curcumin's capability to affect FBS levels, HOMO-IR, HbA1c, and insulin levels; however, randomized controlled trials are lacking to advance curcumin into the next stage of antidiabetic drug development (Zheng et al., 2024).

Standardization of phytomedicine as therapeutic agents is quite challenging due to inconsistency associated with plant constituents due to factor such as varying climate conditions, preparation methods, and validations. Therefore, it becomes imperative that harmonized specifications and Good agricultural and collection practices for starting materials of herbal origin and good manufacturing practices for herbal medicinal products are implemented globally for wider acceptance of phytomedicine in diabetes care. Standardization of medicinal plants will ensure qualitative, safe, and effective antidiabetic drug

TABLE 25.1 Computational antidiabetic study of selected medicinal plants.

Plant specie	Plant extract/metabolite	Mechanism of action	References
Gymnema sylvestre	Gymnemic acid	Regulates the EGFR tyrosine kinase inhibitor resistance, AGE-RAGE signaling pathway in diabetic complications (DC) and Lipid and atherosclerosis pathways through modulation of AKT1	Mayyas et al. (2025)
	Gymnemic acid I and II Gymnemoside A and B	Binds to the insulin receptor	Hossain et al. (2016)
	Orcinol 3-Hydroxy-3′- methoxyflavone Curcumin Glyinflanin A	Binds to alpha amylase and alpha glucosidase	Abdullahi et al. (2022)
Momordica charantia	Cucurbitane-type triterpenoids	Regulates lipid and atherosclerosis, AGE-RAGE signaling pathway in DC and the PI3K-Akt signaling pathway through modulation of AKT1, SRC and IL6 genes	Soliman et al. (2020)
	Momordicoside C Momorcharaside B	Exerts effects on T2D by modulating IR through association the with PRKCD gene	Yadav et al. (2025)
	Saponins (Cucurbitacin, Charantin and Momordicoside-D)	Stimulates insulin secretion by binding to TGR5, GLP-1 and DPP4	Elekofehinti et al. (2018)
Trigonella foenum-graecum	Diosgenin, Luteolin and quercetin	Exerts potential anti-oxidative stress, anti-inflammatory and β-cell protective effects through modulation of the AGE-RAGE and NF-κB signaling pathways.	Luo et al. (2023)
	Trigonelline	Potential therapeutic effect against diabetic nephropathy by targeting the MMP9 gene	Wang et al. (2025)
Cinnamomum verum	Gallic acid quercetin Vanillic acid Ferulic acid	Binds to PPAR- γ and GLUT4	Mnge et al. (2025)

In silico mechanistic antidiabetic studies on selected medicinal plants.

TABLE 25.2 *In vitro* **antidiabetic molecular mechanisms of selected medicinal plants.**

Plant specie	Plant extract/metabolite	Mechanism of action	References
Gymnema sylvestre	Methanolic leaf extract	Increased glucose uptake by upregulating GLUT4 and PPAR-γ Reduced insulin resistance by increasing adiponectin and leptin	Kumar et al. (2016)
	GS extract OSA®	Reduces cytokine-induced apoptosis of β-cells by upregulating cell survival pathways and scavenge superoxide dismutase and catalase	Al-Romaiyan et al. (2020)
	Gymnemosides ND7-9, isolated from Vietnamese cultivar	Increased glucose uptake	Pham et al. (2018)
	Encapsulated triterpene glycosides	Increased glucose uptake by upregulating GLUT4	Shenoy et al. (2025)
	Gymnemagenin	Upregulated TGs hydrolyzing, PPAR- γ, and anti-inflammatory genes	DasNandy et al. (2022)
	Methanolic leaf extract	Inhibits AA and AG activity	Abdullahi et al. (2022)
	Gymnemagenin	Inhibits AA and AG activity	Kamran and Showkat (2023)
Momordica charantia	Methanolic extracts and Charantin	Pro-angiogenic signaling through deactivation of RAGE for alleviation of diabetes-impaired wound healing	Aljohi et al. (2018)
	3β,7β,25-trihydroxycucurbita-5,23(E)-dien-19-al Charantal Charantoside XI 5ξ-isopropenylchole-5, 6-ene-3-O-D-glucopyranoside	Inhibits AA and AG activity Downregulate inflammatory genes implicated in DC	Shivanagoudra et al. (2019)
	Saponins	Increases insulin secretion through modulation of the PI3K/Akt/FoxO1 signaling pathway.	Liu, Mu, et al. (2021)
	Kuguaovin H Kuguaovin J	Stimulates GLP-1 secretion	Liaw et al. (2022)
Trigonella foenum-graecum	Fenugreek seed extract	Increases phosphorylation of IR-β, Akt, GSK-3α/β, and GLUT2. Increases glucose uptake, by regulating downstream insulin signaling, GS, GK, SREBP1c, and GLUT2 expression	Naicker et al. (2016)
	Ethanolic fenugreek seed extract	Enhanced GLP-1 signaling	King et al. (2015)
Diosgenin, Luteolin and quercetin	Improves glucose consumption in insulin resistant liver cells	Luo et al. (2023)	
Trigonelline	Alleviated high glucose induced kidney injury by targeting MMP9	Wang et al. (2025)	
Water fenugreek leaf extract	Improved glucose uptake	Hajra and Paul (2018)	
Cinnamomum verum	Bark ethanolic extract	Inhibits AA and AG activity and promote adipocyte glucose uptake	Mnge et al. (2025)

Cell line and enzyme studies elucidating the antidiabetic activities of selected medicinal plants.

TABLE 25.3 ***In vivo* studies on the mechanism of antidiabetic activity of selected medicinal plants.**

Plant	Plant part/metabolite	Antidiabetic effect	References
Gymnema sylvestre	Ethanolic leaf extract	Decreased fasting blood sugar (FBS), HbA1c and increased serum insulin levels	Choudhary and Tahir (2023)
	Gymnemagenin	Decreased FBS levels	Kamran and Showkat (2023)
	Leaf extract	Decreased FBS, TC, TG, HbA1c and increased serum insulin, body weight and improved pancreatic function	Kamble et al. (2016)
Momordica charantia	*M. charantia* formulated into rats feed (5, 10, 20%)	Increased GLP-1 levels and via reduced DPP4 activity	Elekofehinti et al. (2018)
	Methanolic leaf extract	Decreased FBS, HbA1c and increased serum insulin levels	Hussain et al. (2022)
Trigonella foenum-graecum	Freeze-dried fenugreek seeds	Decreased glucose and lipid metabolism, hepatic oxidative stress and inflammatory responses, and modulated liver function and intestinal injury	Chi et al. (2025)
	Ethanolic extract fenugreek capsules	Improves glucose uptake	Jyothi et al. (2017)

Animal model studies of the mechanism of antidiabetic activity of selected medicinal plants.

TABLE 25.4 Clinical studies on the mechanism of antidiabetic activity of selected medicinal plants.

Plant	Plant part/metabolite	Antidiabetic effect	References
Gymnema sylvestre	Swanson Premium *G. sylvestre* leaf	Decreased body weight, BMI and very low-density lipoprotein	Zuniga et al. (2017)
	Oral capsules of *G. sylvestre*	Improved glucose metabolism by decreasing HbA1c and blood glucose levels and increasing insulin sensitivity	Gaytan Martinez et al. (2021)
Momordica charantia	*Momordica charantia* in combination with a conventional therapeutic	Decreased BGL and alleviate IR	Kumari et al. (2018)
	500 mg of *M. charantia* fruit pulp powder	Decreased HbA1c, blood glucose levels, body weight, BMI and insulin secretion	Cortez-Navarrete et al. (2018)
	Ethanolic extract	Decreased BGL by reducing glucagon levels in prediabetic individuals	Kim et al. (2023)
Trigonella foenum-graecum	Fenugreek dried seed extract capsules (335 mg)	Ameliorated diabetes-induced dyslipidaemia	Chehregosha et al. (2024)
	Galactomannan	Decreased FBS and HbA1c and ameliorate diabetes-induced dyslipidaemia	Rashid et al. (2019)
Cinnamomum verum	500 mg of cinnamon bark powder capsules	Improved BM1, BW, visceral fat Decreased IR, BGL, and HbA1C, and ameliorated dyslipidaemia	Zare et al. (2019)

Representative outcomes of clinical studies conducted with selected medicinal plants and their bioactive constituents.

development that will further promote global trade, cultural acceptance, research and development (Wang et al., 2023).

Limited standardized regulatory framework greatly affects/delay the development of new plant products' approval into commercial antidiabetic products (Gatt et al., 2024). Factors contributing to having substantive regulatory frameworks in the use of phytomedicine for diabetes care include lack of harmonized regulatory frameworks across countries, limited guidelines by stakeholders and regulatory agencies on development, lack of distinct classification of medicinal plants (as drugs or foods resulting in confusion and varying regulations), and quality control. Diverse regional frameworks such as the United States Food and Drug Administration botanical pathway, European Medicines Agency and Committee on Herbal Medicinal Products worsen global development through standardization, stability, and evidence level requirements. Sustainable sourcing, cultivation strategies, and fair and ethical benefit-sharing models are critical to safeguard biodiversity and respect traditional knowledge.

Multidisciplinary approaches and stakeholder involvement are cornerstones in the use and acceptance of phytomedicine in diabetes care. Integration of modern analytical tools and techniques like Omics techniques (e.g., network pharmacology, systems biology, and metabolomics) and computer-aided drug designs have shown mechanistic elucidation of diabetes molecular pathway towards evidence-based personalized and integrative diabetic care with medicinal plants (Akoonjee et al., 2025). Phytocompounds such as berberine, curcumin, and essential oils with promising antidiabetic properties have low bioavailability and/or poor pharmacokinetics profiles warrant for advanced extraction, formulation, and optimization strategies through interdisciplinary approaches.

While nanotechnology holds a promise in tackling phytomedicine bioavailability and delivery to targeted site for antidiabetic effects, rigorous clinical evaluations and validations, addressing regulatory, sustainable and biodiversity issues, and community engagement can greatly promote the wide acceptance of phytomedicine in diabetes care. To standardize phytomedicine for diabetes intervention, having plant products with little or no compositional variability, reproducible and appropriate extract preparation methods, and validated targeted molecular pathway is critical (Wang et al., 2023). Additionally, to mitigate the challenges of regulatory frameworks, international collaborations, promotion of research and product development towards standard, effective, and safe antidiabetic drug design are important.

Conclusion

This chapter harnesses the role of phytomedicines in diabetes care focusing on their interaction with critical molecular targets. Molecular pathway targeting has demonstrated better outcomes in therapeutic efficacies of various diseases including diabetes. Crude Plant extracts and plant compounds have demonstrated the ability to activate/modulate insulin signaling pathways and diabetes related biomarkers and improve pancreatic β-cells and insulin sensitivity. Advances in formulation technologies, such as nanoencapsulation and polyherbal synergies, that will further enhance the therapeutic potential of phytomedicines were put forth while challenges, possible solutions, and prospects of phytomedicine in diabetes care were identified.

Mini-dictionary of terms

β-cell dysfunction: Inability of pancreatic β -cells to produce function and produce enough insulin.

Drug delivery: Process in which administered therapeutic compounds travel to specific molecular sites to bring about desired effects in the body.

Endocrine: Tissue or glands that produce and release hormones into the bloodstream.

Gluconeogenesis: Metabolic process through which glucose is produced from non-carbohydrate sources.

Glycogenesis: Metabolic process whereby excess glucose is stored in the liver and muscles as glycogen.

Mechanism: Series of combined units that work together to create a specific outcome.

Molecular pathways: Series of molecular events that occur in cellular networks that lead to a specific function.

Pharmacokinetics: Analysis of the ways in which pharmaceutical compounds interact with the body upon administration.

Phytomedicine: Therapeutic compounds derived directly from plant sources.

Signaling: Process whereby molecules convey information to other molecules, which ultimately directs the behavior of related biological processes.

Standardization of formulations: Stringent evaluation of novel therapeutic compounds to assess its behavior, efficacy and safety.

Summary points

- Molecular pathways such as insulin signaling, carbohydrate metabolism, oxidative stress, and inflammation have been linked to diabetes.
- Insulin, an endocrine polypeptide hormone produced by pancreatic β-cells regulates blood glucose levels.
- Phytomedicines offer significant potential to target molecular pathways involved in diabetes.
- Phytomedicines are formulated as pharmaceutical or nutraceutical agents in diabetes care
- Pharmacokinetics and safety evaluations are critical in phytomedicine use in diabetes care.

- Standardization, regulatory frameworks could enhance research and development and acceptance of phytomedicine for diabetes care.

References

Abdullahi, Z., Magaji, Y., Vantsawa, P. A., Sheshe, S. M., & Alhaji, J. A. (2022). Anti-diabetic potential of gymnema sylvestre: In vitro and in silico analysis. *International journal of research in pharmaceutical and biomedical sciences, 2*, 233–248.

Abdulsalam, R. A., Lukman, H. Y., & Sabiu, S. (2025). Phenolics and terpenoids in nanostructures for biomedicine. In *Natural products: Phytochemistry, botany, metabolism of alkaloids, phenolics and terpenes* (pp. 1–22). Springer.

Afzal, O., Altamimi, A. S., Nadeem, M. S., Alzarea, S. I., Almalki, W. H., Tariq, A., Mubeen, B., Murtaza, B. N., Iftikhar, S., & Riaz, N. (2022). Nanoparticles in drug delivery: From history to therapeutic applications. *Nanomaterials, 12*(24), 4494.

Akoonjee, A., Lukman, H. Y., Ajao, A. A., Uthman, T. O., & Sabiu, S. (2025). A network pharmacology-and molecular dynamics simulation-based bioprospection of Khaya grandifoliola C. DC. for diabetes care. *Journal of Biomolecular Structure and Dynamics, 43*(13), 7155–7174.

Alam, S., Sarker, M. M. R., Sultana, T. N., Chowdhury, M. N. R., Rashid, M. A., Chaity, N. I., Zhao, C., Xiao, J., Hafez, E. E., & Khan, S. A. (2022). Antidiabetic phytochemicals from medicinal plants: Prospective candidates for new drug discovery and development. *Frontiers in Endocrinology, 13*, 800714.

Aljohi, A., Matou-Nasri, S., Liu, D., Al-Khafaji, N., Slevin, M., & Ahmed, N. (2018). Momordica charantia extracts protect against inhibition of endothelial angiogenesis by advanced glycation endproducts in vitro. *Food & Function, 9*(11), 5728–5739.

Al-Romaiyan, A., Liu, B., Persaud, S., & Jones, P. (2020). A novel Gymnema sylvestre extract protects pancreatic beta-cells from cytokine-induced apoptosis. *Phytotherapy Research, 34*(1), 161–172.

Aranda-Rivera, A. K., Cruz-Gregorio, A., Arancibia-Hernández, Y. L., Hernández-Cruz, E. Y., & Pedraza-Chaverri, J. (2022). RONS and oxidative stress: An overview of basic concepts. *Oxygen, 2*(4), 437–478.

Batiha, G. E.-S., Beshbishy, A. M., Guswanto, A., Nugraha, A., Munkhjargal, T., M. Abdel-Daim, M., Mosqueda, J., & Igarashi, I. (2020). Phytochemical characterization and chemotherapeutic potential of Cinnamomum verum extracts on the multiplication of protozoan parasites in vitro and in vivo. *Molecules, 25*(4), 996.

Chandel, N. S. (2021). Lipid metabolism. *Cold Spring Harbor Perspectives in Biology, 13*(9). https://doi.org/10.1101/cshperspect.a040576

Chehregosha, F., Maghsoumi-Norouzabad, L., Mobasseri, M., Fakhr, L., & Tarighat-Esfanjani, A. (2024). The effect of fenugreek seed dry extract supplement on glycemic indices, lipid profile, and prooxidant-antioxidant balance in patients with type 2 diabetes: A double-blind randomized clinical trial. *Journal of Cardiovascular and Thoracic Research, 16*(3), 184.

Chi, L., Niu, H., Niu, Y., Yao, R., Shi, D., Lu, B., & Pang, Z. (2025). Trigonella foenum-graecum L. ameliorates metabolism-associated fatty liver disease in type 2 diabetic mice: A multi-omics mechanism analysis. *Journal of Ethnopharmacology*, 119862.

Chihomvu, P., Ganesan, A., Gibbons, S., Woollard, K., & Hayes, M. A. (2024). Phytochemicals in drug discovery—A confluence of tradition and innovation. *International Journal of Molecular Sciences, 25*(16), 8792.

Choudhary, A. N., & Tahir, F. (2023). The therapeutic effect of gymnema sylvestre extract against hyperglycemia: In Vivo study. *Agrobiological Records, 14*, 50–58.

Cortez-Navarrete, M., Martínez-Abundis, E., Pérez-Rubio, K. G., González-Ortiz, M., & Méndez-del Villar, M. (2018). Momordica charantia administration improves insulin secretion in type 2 diabetes mellitus. *Journal of Medicinal Food, 21*(7), 672–677.

DasNandy, A., Patil, V. S., Hegde, H. V., Harish, D. R., & Roy, S. (2022). Elucidating type 2 diabetes mellitus risk factor by promoting lipid metabolism with gymnemagenin: An in vitro and in silico approach. *Frontiers in Pharmacology, 13*, 1074342.

Elekofehinti, O. O., Ariyo, E. O., Akinjiyan, M. O., Olayeriju, O. S., Lawal, A. O., Adanlawo, I. G., & Rocha, J. B. T. (2018). Potential use of bitter melon (Momordica charantia) derived compounds as antidiabetics: In silico and in vivo studies. *Pathophysiology, 25*(4), 327–333.

Gamil, N. M., Elsayed, H. A., Hamed, R. M., Salah, E. T., Ahmed, A. M., Mostafa, H. A., El-Shiekh, R. A., & Abou-Hussein, D. (2025). Insights from herb interactions studies: A foundational report for integrative medicine. *Future Journal of Pharmaceutical Sciences, 11*(1), 46.

Garg, R. C. (2016). Nutraceuticals in glucose balance and diabetes. In *Nutraceuticals* (pp. 145–160). Elsevier.

Gatt, A. R., Vella Bonanno, P., & Zammit, R. (2024). Ethical considerations in the regulation and use of herbal medicines in the European Union. *Frontiers in Medical Technology, 6*, 1358956.

Gaytan Martinez, L. A., Sánchez-Ruiz, L. A., Zuñiga, L. Y., González-Ortiz, M., & Martínez-Abundis, E. (2021). Effect of Gymnema sylvestre administration on glycemic control, insulin secretion, and insulin sensitivity in patients with impaired glucose tolerance. *Journal of Medicinal Food, 24*(1), 28–32.

González, P., Lozano, P., Ros, G., & Solano, F. (2023). Hyperglycemia and oxidative stress: An integral, updated and critical overview of their metabolic interconnections. *International Journal of Molecular Sciences, 24*(11), 9352.

Governa, P., Baini, G., Borgonetti, V., Cettolin, G., Giachetti, D., Magnano, A. R., Miraldi, E., & Biagi, M. (2018). Phytotherapy in the management of diabetes: A review. *Molecules, 23*(1), 105.

Hajra, D., & Paul, S. (2018). Study of glucose uptake enhancing potential of fenugreek (Trigonella foenum graecum) leaves extract on 3T3 L1 cells line and evaluation of its antioxidant potential. *Pharmacognosy Research, 10*(4).

He, L. (2021). Alterations of gut microbiota by overnutrition impact gluconeogenic gene expression and insulin signaling. *International Journal of Molecular Sciences, 22*(4), 2121.

Hossain, M. U., Khan, M. A., Rakib-Uz-Zaman, S., Ali, M. T., Islam, M. S., Keya, C. A., & Salimullah, M. (2016). Treating diabetes mellitus: Pharmacophore based designing of potential drugs from Gymnema sylvestre against insulin receptor protein. *BioMed Research International, 2016*(1), 3187647.

Hussain, F., Hafeez, J., Khalifa, A. S., Naeem, M., Ali, T., & Eed, E. M. (2022). In vitro and in vivo study of inhibitory potentials of α-glucosidase and acetylcholinesterase and biochemical profiling of M. charantia in alloxan-induced diabetic rat models. *American Journal of Translational Research, 14*(6), 3824.

Jyothi, D., Koland, M., Priya, S., & James, J. P. (2017). Formulation of herbal capsule containing Trigonella foenum-graecum seed extract for the treatment of diabetes. *Journal of Young Pharmacists, 9*(3), 352.

Kamble, B., Gupta, A., Moothedath, I., Khatal, L., Janrao, S., Jadhav, A., & Duraiswamy, B. (2016). Effects of Gymnema sylvestre extract on the pharmacokinetics and pharmacodynamics of glimepiride in streptozotocin induced diabetic rats. *Chemico-Biological Interactions, 245*, 30–38.

Kamran, J., & Showkat, R. (2023). Gymnemagenin a promising drug candidate for management of hyperglycemia: In-vitro and In-Vivo study. *Integrative Biomedical Research, 7*(1), 1–6.

Kashif, M., Nasir, A., Gulzaman, Rafique, M. K., Abbas, M., ur Rehman, A., Riaz, M., Rasool, G., & Mtewa, A. G. (2023). Unlocking the anti-diabetic potential of Gymnema sylvestre, Trigonella foenum-graecum, and their combination thereof: An in-vivo evaluation. *Food Science and Nutrition, 11*(12), 7664–7672.

Khan, F., Sarker, M. M. R., Ming, L. C., Mohamed, I. N., Zhao, C., Sheikh, B. Y., Tsong, H. F., & Rashid, M. A. (2019). Comprehensive review on phytochemicals, pharmacological and clinical potentials of Gymnema sylvestre. *Frontiers in Pharmacology, 10*, 1223.

Kim, B., Lee, H. S., Kim, H.-J., Lee, H., Lee, I.-y., Ock, S., Kwon, S., Kang, S.-S., & Choi, Y. (2023). Momordica charantia (bitter melon) efficacy and safety on glucose metabolism in Korean prediabetes participants: A 12-week, randomized clinical study. *Food Science and Biotechnology, 32*(5), 697–704.

King, K., Lin, N.-P., Cheng, Y.-H., Chen, G.-H., & Chein, R.-J. (2015). Isolation of positive modulator of glucagon-like peptide-1 signaling from Trigonella foenum-graecum (fenugreek) seed. *Journal of Biological Chemistry, 290*(43), 26235–26248.

Kumar, P. M., Venkataranganna, M. V., Manjunath, K., Viswanatha, G. L., & Ashok, G. (2016). Methanolic leaf extract of Gymnema sylvestre augments glucose uptake and ameliorates insulin resistance by upregulating glucose transporter-4, peroxisome proliferator-activated receptor-gamma, adiponectin, and leptin levels in vitro. *Journal of intercultural ethnopharmacology, 5*(2), 146.

Kumari, S., Dash, I., & Behera, K. K. (2018). Therapeutic effect of Momordica charantia on blood glucose, lipid profile and oxidative stress in type 2 diabetes mellitus patients: A randomised controlled trial. *Journal of Clinical and Diagnostic Research, 12*(9).

Kuriyan, R., Rajendran, R., Bantwal, G., & Kurpad, A. V. (2008). Effect of supplementation of Coccinia cordifolia extract on newly detected diabetic patients. *Diabetes Care, 31*(2), 216–220.

Liaw, C.-C., Huang, H.-T., Liu, H.-K., Lin, Y.-C., Zhang, L.-J., Wei, W.-C., Shen, C.-C., Wu, C.-L., Huang, C.-Y., & Kuo, Y.-H. (2022). Cucurbitane-type triterpenoids from the vines of Momordica charantia and their anti-inflammatory, cytotoxic, and antidiabetic activity. *Phytochemistry, 195*, 113026.

Liu, Z., Gong, J., Huang, W., Lu, F., & Dong, H. (2021). The effect of Momordica charantia in the treatment of diabetes mellitus: A review. *Evidence-Based Complementary and Alternative Medicine, 2021*(1), 3796265.

Liu, Y., Mu, S., Chen, W., Liu, S., Cong, Y., Liu, J., & Jia, N. (2021). Saponins of Momordica charantia increase insulin secretion in INS-1 pancreatic β-cells via the PI3K/Akt/FoxO1 signaling pathway. *Endocrinología, Diabetes y Nutrición (English ed.), 68*(5), 329–337.

Lu, X., Xie, Q., Pan, X., Zhang, R., Zhang, X., Peng, G., Zhang, Y., Shen, S., & Tong, N. (2024). Type 2 diabetes mellitus in adults: Pathogenesis, prevention and therapy. *Signal Transduction and Targeted Therapy, 9*(1), 262.

Luo, W., Deng, J., He, J., Yin, L., You, R., Zhang, L., Shen, J., Han, Z., Xie, F., & He, J. (2023). Integration of molecular docking, molecular dynamics and network pharmacology to explore the multi-target pharmacology of fenugreek against diabetes. *Journal of Cellular and Molecular Medicine, 27*(14), 1959–1974.

Mayyas, A., Al-Samydai, A., Oraibi, A. I., Debbabi, N., Hassan, S. S., Al-Hussainy, H. A., Salamatullah, A. M., Dauelbait, M., Bourhia, M., & Almaary, K. S. (2025). Deciphering the anti-diabetic potential of Gymnema sylvestre using integrated computer-aided drug design and network pharmacology. *Journal of Cellular and Molecular Medicine, 29*(1), e70349.

Mnge, U. L., Ngnameko, C. R., Salau, V. F., Olofinsan, K. A., Mishra, A. P., & Matsabisa, M. G. (2025). Cinnamomum verum (Syn. C. zeylanicum) bark ethanolic extract inhibits carbohydrate digestive enzymes and enhances glucose uptake in 3T3-adipocytes: Insights from in vitro and computational perspectives. *Scientific African, 27*, e02539.

Mugale, M. N., Dev, K., More, B. S., Mishra, V. S., Washimkar, K. R., Singh, K., Maurya, R., Rath, S. K., Chattopadhyay, D., & Chattopadhyay, N. (2024). A comprehensive review on preclinical safety and toxicity of medicinal plants. *Clinical Complementary Medicine and Pharmacology, 4*(1), 100129.

Mukherjee, P. K., Banerjee, S., Gupta, B. D., & Kar, A. (2022). Evidence-based validation of herbal medicine: Translational approach. In *Evidence-based validation of herbal medicine* (pp. 1–41). Elsevier.

Naicker, N., Nagiah, S., Phulukdaree, A., & Chuturgoon, A. (2016). Trigonella foenum-graecum seed extract, 4-hydroxyisoleucine, and metformin stimulate proximal insulin signaling and increase expression of glycogenic enzymes and GLUT2 in HepG2 cells. *Metabolic Syndrome and Related Disorders, 14*(2), 114–120.

Okoduwa, S. I., Mhya, D. H., Abdulwaliyu, I., Igiri, B. E., Okoduwa, U. J., Arthur, D. E., Laleye, A. O., Osang, G. J., Onaleye, O. L., & Nathyns-Pepple, E. (2024). Phytomedicine approach for management of diabetes mellitus: An overview of scientifically confirmed medicinal plants with hypoglycaemic properties and their probable mechanism of action. *Phytochemistry Reviews*, 1–61.

Pahlavani, N., Roudi, F., Zakerian, M., Ferns, G. A., Navashenaq, J. G., Mashkouri, A., Ghayour-Mobarhan, M., & Rahimi, H. (2019). Possible molecular mechanisms of glucose-lowering activities of Momordica charantia (Karela) in diabetes. *Journal of Cellular Biochemistry, 120*(7), 10921–10929.

Pham, H. T. T., Hoang, M. C., Ha, T. K. Q., Dang, L. H., Tran, V. O., Nguyen, T. B. T., Lee, C. H., & Oh, W. K. (2018). Discrimination of different geographic varieties of Gymnema sylvestre, an anti-sweet plant used for the treatment of type 2 diabetes. *Phytochemistry, 150*, 12–22.

Pizzino, G., Irrera, N., Cucinotta, M., Pallio, G., Mannino, F., Arcoraci, V., Squadrito, F., Altavilla, D., & Bitto, A. (2017). Oxidative stress: Harms and benefits for human health. *Oxidative Medicine and Cellular Longevity, 2017*(1), 8416763.

Pundarikakshudu, K., Patel, M. G., & Shah, P. A. (2024). An overview of some Indian vegetables, fruits, and spices effective in diabetes and metabolic disorders: Current status and future scenarios. *Antidiabetic Medicinal Plants*, 75–139.

Qais, F. A., Khan, M. S., Althubiani, A. S., Al-Ghamdi, S. B., & Ahmad, I. (2019). Understanding biochemical and molecular mechanism of complications of glycation and its management by herbal medicine. In *New look to phytomedicine* (pp. 331–366). Elsevier.

Rahman, M. M., Dhar, P. S., Anika, F., Ahmed, L., Islam, M. R., Sultana, N. A., Cavalu, S., Pop, O., & Rauf, A. (2022). Exploring the plant-derived bioactive substances as antidiabetic agent: An extensive review. *Biomedicine and Pharmacotherapy, 152*, 113217.

Rahman, M. S., Hossain, K. S., Das, S., Kundu, S., Adegoke, E. O., Rahman, M. A., Hannan, M. A., Uddin, M. J., & Pang, M.-G. (2021). Role of insulin in health and disease: An update. *International Journal of Molecular Sciences, 22*(12), 6403.

Rashid, R., Ahmad, H., Ahmed, Z., Rashid, F., & Khalid, N. (2019). Clinical investigation to modulate the effect of fenugreek polysaccharides on type-2 diabetes. *Bioactive Carbohydrates and Dietary Fibre, 19*, 100194.

Richter, E., Geetha, T., Burnett, D., Broderick, T. L., & Babu, J. R. (2023). The effects of Momordica charantia on type 2 diabetes mellitus and Alzheimer's disease. *International Journal of Molecular Sciences, 24*(5), 4643.

Rojas-Pirela, M., Andrade-Alviárez, D., Rojas, V., Marcos, M., Salete-Granado, D., Chacón-Arnaude, M., Pérez-Nieto, M.Á., Kemmerling, U., Concepción, J. L., Michels, P. A. M., & Quiñones, W. (2025). Exploring glycolytic enzymes in disease: Potential biomarkers and therapeutic targets in neurodegeneration, cancer and parasitic infections. *Open Biology, 15*(2), 240239.

Rombolà, L., Scuteri, D., Marilisa, S., Watanabe, C., Morrone, L. A., Bagetta, G., & Corasaniti, M. T. (2020). Pharmacokinetic interactions between herbal medicines and drugs: Their mechanisms and clinical relevance. *Life, 10*(7), 106.

Salehi, B., Ata, A., V. Anil Kumar, N., Sharopov, F., Ramirez-Alarcon, K., Ruiz-Ortega, A., Abdulmajid Ayatollahi, S., Valere Tsouh Fokou, P., Kobarfard, F., & Amiruddin Zakaria, Z. (2019). Antidiabetic potential of medicinal plants and their active components. *Biomolecules, 9*(10), 551.

Sarker, D. K., Ray, P., Dutta, A. K., Rouf, R., & Uddin, S. J. (2024). Antidiabetic potential of fenugreek (Trigonella foenum-graecum): A magic herb for diabetes mellitus. *Food Science and Nutrition, 12*(10), 7108–7136.

Shariati, M. A., Khan, M. U., Ntsomboh-Ntsefong, G., Hussain, M. B., Akram, M., Mishra, A. P., & Egbuna, C. (2021). Herbal drug technology. In *Preparation of phytopharmaceuticals for the management of disorders* (pp. 77–87). Elsevier.

Shenoy, R. S., Skariyachan, S., Bhat, S., Prashanth, K. H., Kumar, K. M., Manonmani, H. K., Kumar, P. S., & Niranjan, V. (2025). Elucidation of the encapsulated triterpene glycosides of Gymnema sylvestre on GLUT4 expression and integrative in silico modeling of potential α-glucosidase inhibitors. *Journal of Molecular Liquids*, 128024.

Shivanagoudra, S. R., Perera, W. H., Perez, J. L., Athrey, G., Sun, Y., Jayaprakasha, G., & Patil, B. S. (2019). Cucurbitane-type compounds from Momordica charantia: Isolation, in vitro antidiabetic, anti-inflammatory activities and in silico modeling approaches. *Bioorganic Chemistry, 87*, 31–42.

Singh, N., Yadav, S. S., Kumar, S., & Narashiman, B. (2022). Ethnopharmacological, phytochemical and clinical studies on fenugreek (Trigonella foenum-graecum L.). *Food Bioscience, 46*, 101546.

Soliman, G. A., Abdel-Rahman, R. F., Ogaly, H. A., Althurwi, H. N., Abd-Elsalam, R. M., Albaqami, F. F., & Abdel-Kader, M. S. (2020). Momordica charantia extract protects against diabetes-related spermatogenic dysfunction in male rats: Molecular and biochemical study. *Molecules, 25*(22), 5255.

Srivastava, A., Srivastava, P., Pandey, A., Khanna, V., & Pant, A. (2019). Phytomedicine: A potential alternative medicine in controlling neurological disorders. In *New look to phytomedicine* (pp. 625–655). Elsevier.

Wang, H., Chen, Y., Wang, L., Liu, Q., Yang, S., & Wang, C. (2023). Advancing herbal medicine: Enhancing product quality and safety through robust quality control practices. *Frontiers in Pharmacology, 14*, 1265178.

Wang, J., Zhang, P., Bao, N., Du, P., Duan, B., & Ding, T. (2025). Network pharmacology prediction and molecular docking-based strategy to explore the potential mechanism of Trigonelline against diabetic nephropathy via targeting MMP9. *Letters in Drug Design and Discovery*, 100095.

Wasana, K. G. P., Attanayake, A. P., Weerarathna, T. P., & Jayatilaka, K. A. P. W. (2021). Efficacy and safety of a herbal drug of Coccinia grandis (Linn.) Voigt in patients with type 2 diabetes mellitus: A double blind randomized placebo controlled clinical trial. *Phytomedicine, 81*, 153431.

Willcox, M. L., Elugbaju, C., Al-Anbaki, M., Lown, M., & Graz, B. (2021). Effectiveness of medicinal plants for glycaemic control in type 2 diabetes: An overview of meta-analyses of clinical trials. *Frontiers in Pharmacology, 12*, 777561.

Yadav, R., Nambiar, N., Shah, M., & Patel, B. D. (2025). Exploring the therapeutic potential of Momordica charantia in targeting protein kinase C delta (PRKCD) for type 2 diabetes mellitus: Insights from network pharmacology, molecular docking, and molecular dynamics simulations. *Silico Pharmacology, 13*(2), 99.

Zare, R., Nadjarzadeh, A., Zarshenas, M. M., Shams, M., & Heydari, M. (2019). Efficacy of cinnamon in patients with type II diabetes mellitus: A randomized controlled clinical trial. *Clinical Nutrition, 38*(2), 549–556.

Zheng, X., Zhu, J., Haedi, A. R., & Zhou, M. (2024). The effect of curcumin supplementation on glycemic indices in adults: A meta-analysis of meta-analyses. *Prostaglandins and Other Lipid Mediators, 175*, 106908.

Zhu, Y., Wan, F., Liu, J., Jia, Z., & Song, T. (2024). In *The critical role of lipid metabolism in health and diseases* (Vol 16, p. 4414). MDPI.

Zuniga, L. Y., González-Ortiz, M., & Martinez-Abundis, E. (2017). Effect of Gymnema sylvestre administration on metabolic syndrome, insulin sensitivity, and insulin secretion. *Journal of Medicinal Food, 20*(8), 750–754.

Chapter 26

Hypothalamic neurons and treatment with metformin and propionate

Natrus Larysa[1], Osadchuk Yuliia[2] and Klys Yuliia[3]

[1]*Department of Modern Medical Diagnostic and Treatment Technology, Bogomolets National Medical University, Kyiv, Ukraine;* [2]*University Clinic, Bogomolets National Medical University, Kyiv, Ukraine;* [3]*Research Institute of Experimental and Clinical Medicine, Bogomolets National Medical University, Kyiv, Ukraine*

List of abbreviations

ATF6 activating transcription factor 6
BBB blood-brain barrier
ER endoplasmic reticulum
ERAD ER-associated degradation
FAs fatty acids
GRP78 glucose-regulated protein 78
HFD high-fat diet
IR insulin resistance
IRE-1 inositol-requiring enzyme-1
mRNA messenger ribonucleic acid
PERK PKR-like endoplasmic reticulum kinase
SCFAs short-chain fatty acids
STZ streptozotocin
T2DM type 2 diabetes mellitus
UPR unfolded protein response
VMH ventromedial nucleus of the hypothalamus

Introduction

During life, the body constantly experiences fluctuations in homeostatic constants within physiological limits, but at the same time, the maintenance of homeostasis ensures the free and independent existence of the organism from external conditions. A persistent deviation of any homeostatic indicator leads to the development of disease. In most cases, even the most modern therapeutic approaches and treatment regimens aimed at restoring the altered homeostatic parameters do not lead to the restoration of physiological balance. Therefore, the problem of treating a wide range of diseases, primarily hypothalamic and metabolic syndromes, autonomic disorders accompanied by persistent and prolonged deviation of homeostatic constants, is very relevant.

It has been established that obesity, metabolic syndrome and T2DM are based on a combined disorder of carbohydrate and lipid metabolism, which is formed by a complex of related pathophysiological mechanisms, but primarily by the development of insulin resistance (IR). In 1992, Haffner et al. proposed the term "insulin resistance syndrome," which means a decrease in the response of insulin-sensitive tissues to insulin at sufficient concentrations, leading to chronic hyperinsulinemia (Haffner et al., 1992). A high fasting insulin concentration correlated with hypertension, decreased high-density lipoprotein cholesterol concentration, increased triglycerides, and the development of T2DM and was not associated with increased low-density lipoprotein or total cholesterol concentration (Haffner et al., 1992).

The term "dyslipidemia" in the context of IR should be understood not as a direct increase in large-molecular plasma lipids, but as a redistribution of fatty acids (FAs) of phospholipids in membranes and intracellular inclusions (Natrus et al., 2024). Glucose restriction forces the body to slow down energy expenditure and switch to alternative sources (fats, ketone bodies). From the point of view of the rational use of energy "fuel" in the body, according to the "thrifty gene" hypothesis, this mechanism is entirely justified (Neel, 1962). Adipocytes absorb and store FAs when food is available, and during fasting, FAs are released and used by other tissues as energy fuel. However, if the level of FAs in plasma rises for more than a few hours, FAs-induced IR occurs, which eventually becomes counterproductive. Switching to the beta-oxidation pathway of FAs for energy (ATP) is accompanied by a number of complications, including the toxic effects of FAs and acyl-CoA, damage to membranes and mitochondria, increased lipid peroxidation, ketoacidosis, oxidative-nitrosative stress, etc (Natrus et al., 2024).

Hormones in Neuroscience. https://doi.org/10.1016/B978-0-443-30263-3.00029-4

Brain tissue is particularly sensitive to complications of beta-oxidation of FAs. Neurons are highly protected from the use of FAs as an energy substrate due to the low activity of β-oxidation enzymes (acyl-CoA dehydrogenases, thiolases) and the presence of the blood-brain barrier (BBB), which almost completely blocks FAs. However, astrocytes have a high capacity for beta-oxidation. They use this pathway both to maintain energy homeostasis and to detoxify excess FAs. Although some of the energy obtained can be indirectly transferred to neurons (via lactate or ketone bodies), damage to neurons from oxidative stress, glial activation, and low-intensity inflammation is very significant. Since these pathogenic mechanisms affect brain neurons involved in the central regulation of carbohydrate and lipid metabolism, this prevents the restoration of homeostatic constants when they are disrupted and creates a vicious cycle.

Functional uniqueness of neurons in the mediobasal hypothalamus

The hypothalamus is the center of regulation of the body's metabolism. Many researchers have studied the structures of the hypothalamus as substrates for maintaining important homeostatic constants. Early classical works and historical reviews (Kennedy, 1953; Lévêque, 1974) and more recent works (Anthony & Yeo, 2013; Sakurai et al., 1998; Williams et al., 2001) provide extensive data on the relationship between thermogenesis and energy expenditure, and the role of hormones and neurotransmitters as important regulators of metabolism. Tran et al. (2022) summarize current data on the mechanisms through which the hypothalamus influences metabolism, energy expenditure, and adaptation to changes in nutrition and ambient temperature.

It is now recognized that the neuroregulatory network that controls eating behavior, peripheral insulin sensitivity, glucose metabolism, etc., extends beyond the mediobasal hypothalamus. It is joined by other hypothalamic nuclei, parts of the posterior hypothalamus, and autonomic, parasympathetic, and sympathetic preganglionic neurons of the brain. However, the main structures of central regulation of glucose and lipid levels, ensuring energy balance according to the body's needs are neurons of the ventromedial (VMH), lateral, and arcuate nuclei (Fig. 26.1). Dysregulation of hypothalamic nucleus function may be one of the main mechanisms of glucose metabolism disorders, low-grade inflammation, and the development of T2DM (Ruud et al., 2017).

One of the reasons for the disruption of the central mechanisms that maintain homeostasis is a change in the

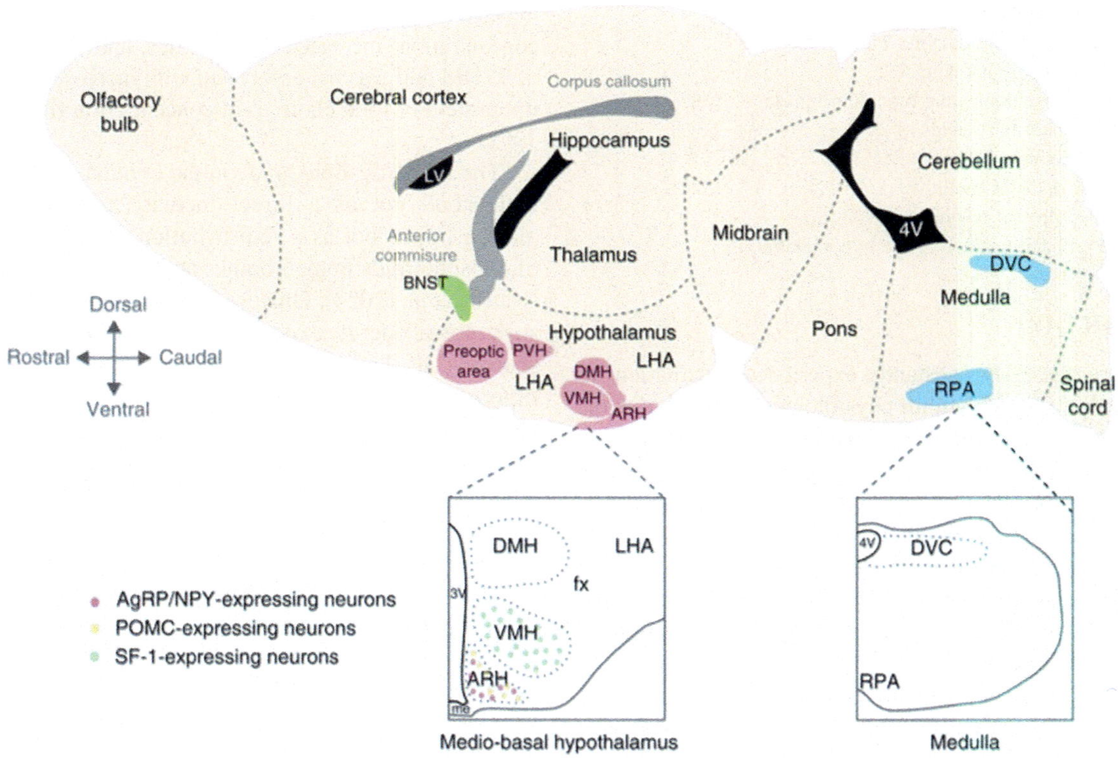

FIGURE 26.1 This figure shows schematic representing a sagittal section of a mouse brain in which critical brain regions controlling glucose homeostasis and peripheral insulin sensitivity. *From Ruud, J., Steculorum, S. M., & Brüning, J. C. (2017). Neuronal control of peripheral insulin sensitivity and glucose metabolism. Nature Communications, 8, 15259. https://doi.org/10.1038/ncomms15259 with permission.*

functional properties of neurons in the corresponding parts of the hypothalamus. Using an electrophysiological approach (Kazakov & Natrus, 2005; Natrus, 2006), the response of individual neurons in the anterior hypothalamus and preoptic area to thermo-, press-, osmo-, and glucostimulation was analyzed and it was found that only 19% of neurons in the studied structures are monosensory. The majority (81%) of neurons are polysensory, i.e., sensitive to fluctuations in combined visceral stimuli and their combinations. The existence of polysensory neurons is explained by the absence of an isolated change in any single homeostatic parameter in the body. Fluctuations in any homeostasis constant are necessarily associated with changes in other homeostatic parameters (Kazakov & Natrus, 2005). However, this complex functional organization of the neurons of the anterior hypothalamus does not correspond to the morphological homogeneity of the neurons that form the regulatory centers (Fig. 26.2).

Therefore, it can be assumed that all regulatory mechanisms for restoring combined homeostatic shifts are provided in the neurons of the hypothalamus at the molecular level, including regulatory signaling pathways and cascades that maintain intracellular homeostasis.

When studying functional disorders of hypothalamic neurons, considerable attention is paid to the study of signaling neuropeptides. The hormone leptin, which is produced by adipose tissue, acts as an afferent signal to the brain, in particular to the hypothalamus. It coordinates energy homeostasis by modulating food intake and energy expenditure (Friedman & Halaas, 1998; Liu et al., 2023; Morris & Rui, 2009). One of the pathogenic causes of obesity and related metabolic diseases is the development of resistance of hypothalamic neurons to leptin. Morris and Rui (2009) suggested that leptin resistance is the result of a disruption in leptin-receptor signaling, which is particularly dependent on ER stress and/or disruption of energy balance neural circuits (Morris & Rui, 2009; Ramírez & Claret, 2015).

Endoplasmic reticulum stress as a sign of functional tension in neurons

The ER is a dynamic organelle, in the granular part of which synthesis and folding occur—the assembly of secretory and transmembrane proteins into secondary and tertiary structures. It is here that they undergo various posttranslational modifications, including glycosylation and the formation of disulfide bonds (Braakman & Hebert, 2013; Mao et al., 2019; Ramírez & Claret, 2015). These

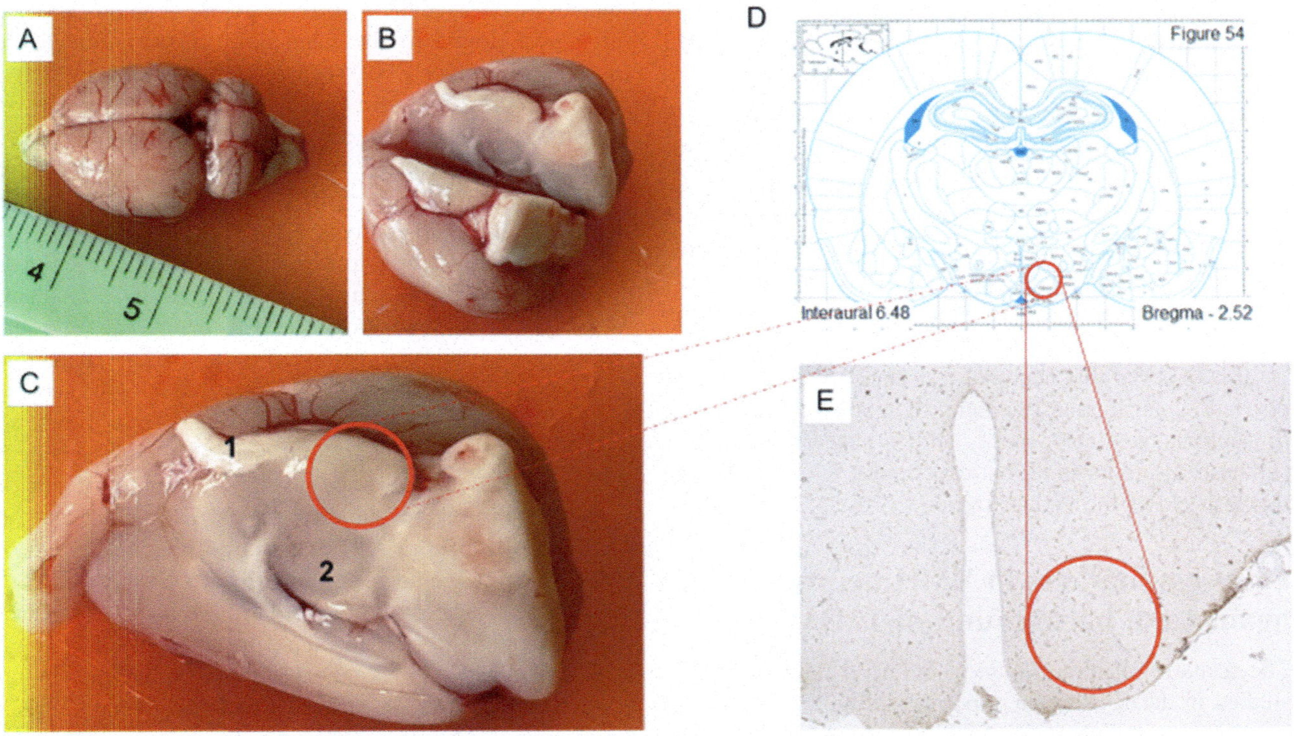

FIGURE 26.2 Morphological homogeneity of the rat medio-basal hypothalamus neurons, demonstrated through an experimental approach to the area. (A) top view of the whole brain; (B) bottom view of the sagittally divided brain; (C) brain preparation in sagittal section (1—chiasma opticus; 2—thalamus); (D) scheme of the rat atlas Paxinos and Watson "the rat brain in stereotaxic coordinates," Bregma −2.52 mm; E – zone of the ventromedial nucleus of the hypothalamus, immunohistochemical staining Iba-1.

processes in the ER are regulated (catalyzed) by important complexes of signaling molecules and pathways, in particular a network of chaperones, glycosylating enzymes, and oxidoreductases, which provide a perfect mechanism for the folding and modification of proteins needed by the cell (Urra et al., 2013). Compared to the cytosol, the ER has a much higher calcium concentration and high redox potential. The ionic and electronic environment of the ER is optimal for reproducing the regulatory system and mechanisms for rapid and efficient folding of proteins into final structures. The stability and maturation of neuropeptides is ensured by key specialized enzymes, chaperones (Braakman & Hebert, 2013), and "quality control" mechanisms, thanks to which deformed proteins in the cytosol undergo proteasomal degradation, or ER-associated degradation (ERAD) (Ruggiano et al., 2014).

Despite this, significant and prolonged exogenous influences or stress on the cell can alter ER homeostasis, leading to the accumulation of misfolded proteins that become potentially toxic to the cell. This state of functional overload is characterized as ER stress (Xu et al., 2024).

Under these conditions, the cell activates a set of phylogenetically conserved signaling pathways that respond to stress, known as the Unfolded Protein Response (UPR) system (Wei & Fang, 2021). This is the cell's response to misfolding, the main task of which is to eliminate the shift in intracellular homeostasis and return the cell to a state of adequate protein production, thus adapting it to the changed conditions (Bravo et al., 2012; Walter & Ron, 2011). UPR is mediated by three main ER transmembrane mediators: IRE-1 (inositol-requiring enzyme-1), PERK (PKR-like endoplasmic reticulum kinase), and ATF6 (activating transcription factor 6), which are found in large quantities in the ER and, under non-stressful conditions, are bound to the chaperone BiP/GRP78 (Glucose-Regulated Protein 78) (Gomez & Rutkowski, 2016; Pincus et al., 2010).

It is believed that neuronal resistance to leptin, glucose restriction, and forced switching to another energy source in neurons inevitably causes and/or exacerbates protein synthesis disorders and ER stress (Jung & Kim, 2013). And this also closes the vicious cycle.

ER stress markers in VMH neurons against the background of T2DM

The scenario of intracellular events in VMH neurons was studied in detail using an experimental model of rats that had been on a high-fat diet (HFD) for a long time: 43% pork fat + 20% fructose and were additionally induced with streptozotocin (STZ) (25 mg/kg). The study of functional changes in neurons is described through the measurement of UPR marker protein content, assessment of gene expression of each protein by determining the amount of messenger ribonucleic acid (mRNA) using real-time PCR, and ER morphometry in electron microscopic studies (Natrus et al., 2022a, 2022b, 2024) (Fig. 26.3).

The described ultrastructural changes in neurons caused by T2DM were as follows: ER tubules and cisternae were enlarged in most neurons; clusters of fragmented mitochondria with signs of cristae destruction and matrix swelling were detected; nuclei were pyknotically altered with signs of chromatin condensation, deep invaginations of the nuclear membrane were detected. Most neurons showed signs of apoptosis and loss of dendrites. Morphopathological changes in glial cells were detected, including cytoplasmic swelling and a decrease in the number of organelles. Morphometry of ER parameters revealed a significant 2.25-fold increase in the relative area of cisterns and a 44% expansion of the perinuclear space (Natrus et al., 2022a).

A functional sign of ER stress in VMH neurons in T2DM was a twofold decrease in GRP78 content and an increase of transmembrane proteins PERK, ATF6, and IRE1. However, despite significant fluctuations in UPR system proteins in the cytoplasm of neurons, their gene expression was only 30%–40% higher than in the control group. It is likely that cytoplasmic processes (protein accumulation, destruction, depletion, etc.) are more reactive processes and are significantly altered by exogenous factors, such as hyperglycemia, hyperlipidemia, and the effects of drugs. At the same time, transcription in the nucleus of certain DNA segments is a genetically determined process. The mRNA content shows greater conservatism. Therefore, when compensating for ER stress, more pronounced fluctuations in PERK, ATF6, and IRE1 are observed in the cytoplasm of VMH neurons than in their nuclear transcription activity.

The data obtained formed a view of the ER stress scenario in VMH neurons. In T2D, the BiP/GRP78 chaperone is sequestered from the ER membrane into the cisternae and binds to misfolded proteins. The chaperone inactivates damaged proteins, thereby reducing proteotoxicity in the cell. As BiP/GRP78 detaches from the membrane, transmembrane UPR proteins are also released, transported to the nucleus, and activate transcription of specific gene regions in the nucleus. IRE1, through the active splice form of XBP-1, initiates the transcription of BiP/GRP78 in the nucleus to increase its production (Park et al., 2021). ATF6 is transported to the nucleus to regulate the expression of genes encoding chaperones (Wei & Fang, 2021). Activated PERK phosphorylates eukaryotic initiation factor 2 α (eIF2 α) and suppresses overall mRNA translation. Although a temporary pause in protein translation is beneficial and reduces functional load, prolonged translation block due to a sustained PERK signal is incompatible with survival (Urra et al., 2013). Moreover, PERK hyperactivation can activate the transcription factor

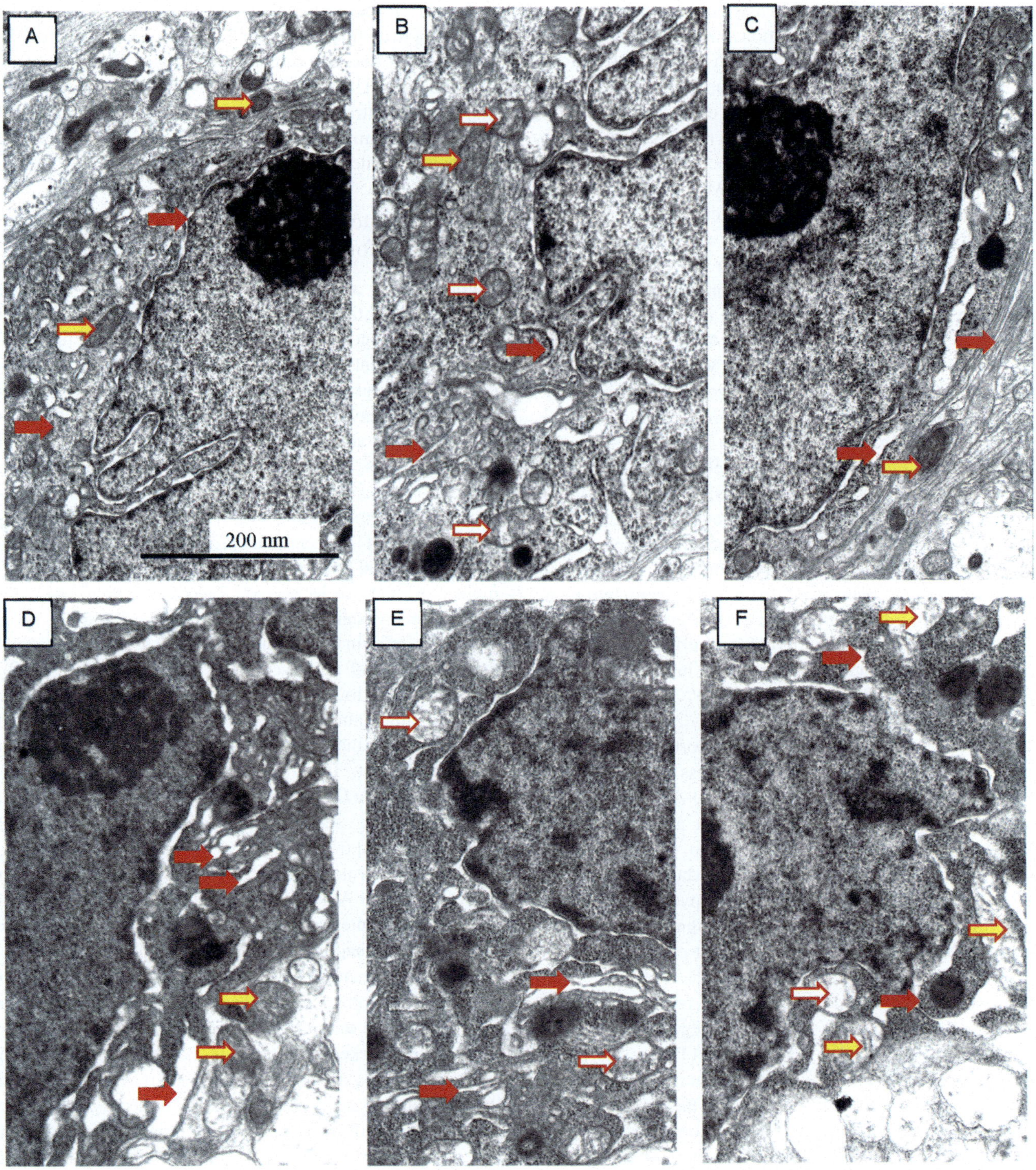

FIGURE 26.3 Ultrastructural changes in VMH neurons were studied by electron microscopy: (A–C) intact rats; (D–F) rats of the T2DM group. Red arrow indicate ER, yellow arrow—mitochondria, white arrows—autophagosome. Scale bar: 200 nm.

CHOP/GADD153, which suppresses the expression of the gene encoding the anti-apoptotic Bcl-2, thereby enhancing the expression of pro-apoptotic proteins and oxidative stress proteins (Rozpedek et al., 2016).

Apoptosis as one of the outcomes of ER stress

It should be noted that the UPR system can produce paradoxical results: on the one hand, it has cytoprotective properties that restore homeostasis, and on the other hand, it has destructive cellular functions, in particular apoptosis. This process of "switching" between cytoprotective action and proapoptotic activation continues to be studied, but there is evidence that this course of events depends on the activity of individual UPR pathways (Natrus et al., 2022b).

The study of the quantitative content of regulatory proteins proves the fact that in T2DM, apoptosis processes in VMH neurons reach maximum activity: Caspase-3 and BAX significantly increase, while anti-apoptotic Bcl-xL decreases. BAX is a key regulator of apoptosis. Under physiological conditions, BAX monomer is an inactive molecule that is mainly found in the cytoplasm of cells. In the presence of apoptotic stimuli, BAX translocates to the outer mitochondrial membrane, where it undergoes conformational changes, oligomerizes, and converts to a BAX dimer, which promotes the formation of pores in the outer mitochondrial membrane. This leads to the release of cytochrome *c* into the cytosol and cell destruction through mitoptosis (Walensky & Gavathiotis, 2011). Thus, the increase in BAX dimer in the VMH neurons and endothelial cells proves that, against the background of T2DM, apoptosis is realized through mitoptosis, which is the pathogenetic mechanism of endothelial dysfunction and neurodegeneration (Natrus et al., 2022b).

It is believed that cytoprotective effects will prevail over proapoptotic ones as long as the half-life of mRNA and proteins of factors such as BiP/GRP78 continues (Chen & Brandizzi, 2013; Wang et al., 2009). This phase provides cells with a "window of opportunity" to correct the ER status in order to cope with stress. If these steps do not restore homeostasis, then IRE1 and ATF6 signaling is weakened, creating an imbalance that activates the proapoptotic pathway, leading the cell to death (Arensdorf et al., 2013; Hetz & Papa, 2018). The stage of irreversible stress, precisely through the UPR pathway, activates proinflammatory and pro-apoptotic signals to quickly destroy the cell. In this case, the result of UPR activation leading to apoptosis is considered protective, since in the context of chronic disease, the lack of regulation of these pathways will inevitably lead to cell degeneration.

The study of IRE1α, PERK, and ATF6 opens up opportunities for researching these molecular regulators in the pathogenesis of many diseases. In the future, research will likely focus on finding ways to control UPR regulators, as well as the scenario and sequence of their activation. Therefore, the study and control of these signaling events are considered promising directions for therapeutic strategies (He et al., 2024; Park & Ozcan, 2013; Yuan et al., 2024). However, in addition to activating the UPR system, quality control in the ER is regulated by the ERAD system. Thanks to it, misfolded proteins are recognized and delivered to the cytosol for ubiquitination. At the same time, the ERAD process is ATP-dependent, so the cell makes a "cautious" decision about the induction of degradation components, since ATP is also used for the primary synthesis of proteins (Neal et al., 2019).

Thus, the UPR system is the main mechanism for maintaining intracellular homeostasis and relieving ER stress by reducing the intensity of protein synthesis and enhancing the degradation of misfolded proteins.

Neuronal autophagy is an important adaptive mechanism for preventing neuronal death

Autophagy is an important mechanism that regulates cellular homeostasis, as it ensures selective degradation of cellular components and effective destruction of functionally incompetent proteins, organelles, and lipids. For hypothalamic neurons, which actively produce regulatory proteins and may undergo proteotoxicity under conditions of misfolding, autophagosome formation is an important rescue process. Autophagosomes undergo a stepwise maturation process. Initially, phagosomes fuse with endosomal vesicles to combine with lysosomal enzymes. At this stage, most of the complex components of the phagosome's internal contents are processed, broken down into molecules, and can be reused (Zhao & Zhang, 2019).

Autophagy mainly plays an adaptive role in "programmed cell survival" and is activated to protect cells during periods of increased stress (Kroemer et al., 2010). Morton (2007) found that in the mediolateral region, the normally high autophagic activity of the hypothalamus decreases with the development of chronic obesity on a high-calorie diet. Thus, the self-cleaning process in neurons is protective (Morton, 2007). However, paradoxically, in some cases, self-cannibalistic mechanisms can be harmful and lead to cell death (Gozuacik & Kimchi, 2007).

The LC3 protein, which is a ubiquitin-like protein (encoded by the Atg8 gene), is a reliable marker of autophagy, as its content positively correlates with the number of active autophagosomes. During autophagy, LC3 synthesis increases and cytoplasmic LC3-I is translocated to the autophagosome, where the protein is isolated in the autophagosomal membrane (Kabeya et al., 2000).

Another marker of autophagy is the Beclin-1 protein (encoded by the Atg6 gene), which also plays a key role in the initiation of autophagy. Beclin-1 is necessary for the normal course of mitophagy (mitochondrial autophagy). Since autophagy and apoptosis are interrelated, it is believed that the fate of a cell during stress adaptation is determined by the "switching" of the main autophagy regulator Beclin-1 and its interaction with Bcl-2 and other apoptotic cell death pathways. Moreover, depletion of Beclin-1 triggers caspase-dependent programmed cell death. Interestingly, caspases can cleave Beclin-1 during apoptosis, thereby destroying its pro-autophagic activity. There is evidence that caspase-3 initiates the cleavage of

Beclin-1, causing cells to lose their ability to undergo autophagy (Kang et al., 2011; Zhu et al., 2010). Thus, the ratio of regulatory proteins found in cells determines which path the cell takes under conditions of functional load and ER stress.

Searching for ways to effectively pharmacologically correct neuroinflammation in T2DM

In experimental models of HFD-induced obesity, inflammation in hypothalamic neurons is mainly caused by excess saturated fatty acids (SFAs), which reach the nervous tissue mainly through areas devoid of the BBB, where there is fenestrated vascular endothelium. Perivascular macrophages in the brain also respond to excess free FAs circulating in the blood and contribute to even greater BBB permeability. Glial cells, such as astrocytes and microglia, respond rapidly to the presence of FAs in the hypothalamic parenchyma and release pro-inflammatory cytokines, chemokines, and reactive oxygen species, products of nitric oxide. If the stimulus persists for a long time, oxidative-nitrosative stress occurs in the neurons of the hypothalamus, leading to neuroinflammation (Natrus et al., 2024).

A clear understanding that the development and progression of T2DM is associated with inflammation in the hypothalamic nuclei encourages the search for effective therapeutic regimens aimed at the described pathogenetic links.

Why metformin treatment does not prevent neuroinflammation

Metformin is considered the "gold standard" for treating type 2 diabetes and is a well-known hypoglycemic drug. It is often used as an antidiabetic monotherapy, which lowers plasma glucose levels through several mechanisms and has a positive effect on serum lipid profiles and the process of carbohydrate and lipid homeostasis. However, very little is known about the effect of metformin treatment on the state of hypothalamic neurons. Nevertheless, the inadequacy of metformin monotherapy in the treatment of T2DM to compensate for nervous system disorders is actively discussed.

Moreover, it has been shown in rats with T2DM that when metformin is used in VMH neurons, lipofuscin granules are found in the cytoplasm. It is believed that lipofuscin accumulation is a sign of cell "aging" suppression of the ubiquitin-proteasome system, stress response, and increased oxidation of unsaturated fatty acids, leading to neuron degeneration. Changes characterized as reactive astrogliosis in the form of astrocyte hypertrophy and proliferation have also been described. Myelin fibers show axoplasmic edema, accumulation of autophagosomes, and damage to the myelin sheath in the form of enlarged sequestra (Natrus et al., 2022a). Therefore, despite the traditional prescription of metformin for T2DM, metformin monotherapy is insufficient because it does not prevent the development of diabetic encephalopathy.

Sodium propionate as a potential neuroprotective agent in T2DM

Today, there is growing evidence of the positive effect of short-chain fatty acids (SCFAs) (SCFAs with unbranched chains, C2:0-C5:0), which are absorbed in the intestine and are the main class of signaling molecules produced as a result of microbial fermentation of dietary carbohydrates. SCFAs directly affect the host's digestive tract through phenotypic changes in colon epithelial cells and act as the main energy substrates. They are modulators of the intestinal neuroendocrine system. There is already sufficient successful clinical experience with the use of propionic acid preparations and its salt (sodium propionate) as an additional component in the treatment of patients with neurodegenerative (autoimmune) diseases (Duscha et al., 2020). It is believed that propionate, when entering the intestine, stimulates immune cells, in particular T-lymphocytes of the intestinal wall, causes the synthesis of neurotransmitters, cytokines, and hormones that are accepted by neuroglia, which inhibits demyelination and indirectly affects the properties of neurons.

Research into the structural and functional properties of VMH neurons in a rat model of experimental T2DM induced by HFD + STZ demonstrated the positive effect of combined administration of metformin and propionate (Natrus et al., 2022a, 2022b, 2024).

Moreover, administration of propionate alone to rats caused improvement in ER architecture, with an even greater effect than that observed with metformin monotherapy. The relative area of ER membranes and cisternae, as well as the total proportion of ER in cells, decreased significantly. In addition, ER cisternae and tubules were well visualized, with clear ribosomes on the surface, which were practically indistinguishable from the control group. The administration of propionate contributed to a decrease in the number of apoptotic neurons with pyknotic nuclei. However, the main difference in the effect of propionate was an increase in autophagosomes in the cytoplasm of neurons.

According to electron microscopy data, the combined administration of metformin and propionate did not reveal the expected most effective changes in nerve tissue compared to the separate administration of the drugs. Nevertheless, after combined administration, an increase in the number of microglial cells, preapoptotic and apoptotic neurons were detected, which in turn can activate microglia. It should be noted that the combined administration of metformin with propionate had a positive effect on the ER: a significant decrease in the relative area of membranes and cisterns, in the ratio of cistern/membrane area, and in the ratio of perinuclear space/nucleus perimeter length, which were almost restored to the values of intact rats. An interesting observation with propionate administration was a significant decrease in the relative area of ER membranes, which is associated with an increase in autophagosomes located in the ER zone, an increase in the content of autophagy markers LC3 and Beclin-1 in nervous tissue, and a decrease in the activity of neuronal apoptosis, which was increased in T2DM. These data provide grounds for hypothesizing that microvesicles are formed in the VMH neurons of rats with T2DM from the ER membrane as a mechanism for the implementation of the ERAD system for the effective utilization of damaged proteins within the ER.

Changes in ER stress markers against the background of combined administration of metformin and propionate

A series of studies on the components of the UPR system, the content of GRP78 as a recognized main regulator of ER stress, and its transmembrane partner regulators provides grounds for imagining a scenario of maintaining intracellular homeostasis in neurons against the background of T2DM as an important mechanism of cell survival. It also reveals the deep intracellular mechanisms of pharmacological correction of the condition.

The decrease in GRP78 levels in the VMH can be explained by the structural rearrangement of the protein during its functional activity. Under conditions of ER stress, the chaperone is sequestered into the cistern and binds to misfolded proteins to inactivate them. These "chaperone + unstructured protein" complexes are effectively utilized by the ERAD system. Transmembrane partners of the chaperone immediately respond to the decrease in protein content, and their content increases significantly. This is especially true for ATF6 and IRE1, sensors that stimulate transcription of the GRP78 gene in the nucleus. Incidentally, there is even a tendency for increased transcription of the corresponding GRP78 gene (as measured by mRNA).

The combined administration of propionate and metformin causes a decrease in ATF6 and IRE1 content in the cytoplasm of neurons as GRP78 increases, indicating successful pharmacological correction. The PERK marker increases rapidly with the development of T2DM. This is natural, since the cell must suspend folding, which is ineffective. Against the background of correction with metformin and propionate separately, an additional increase in PERK is observed. However, the combination of drugs reduces PERK, although not to control values, but most effectively. This is probably a beneficial effect, since prolonged translation block due to a persistent PERK signal is incompatible with survival.

The obtained data are presented in Fig. 26.4, which demonstrates the interrelationship between the components of the UPR system of VMH neurons at rest, under conditions of TD2M and correction, and forms a specific concept.

Under T2DM conditions, ER stress occurs in VMH neurons: the content of GRP78 chaperone decreases by half in the cytoplasm, as it is sequestered in the ER cistern and combines with misfolded proteins for their inactivation. The content of transmembrane UPR sensors ATF6, IRE1, PERK significantly increases to stimulate transcription in the nucleus of the GRP78 gene and inhibit folding. The administration of drugs increases the transcription of the corresponding GRP78 gene in the nucleus (according to mRNA data), and as GRP78 increases in the cytoplasm, it reduces ATF6 and IRE1 levels. However, the administration of metformin against the background of T2DM is accompanied by the accumulation of lipofuscin, which is a sign of cell "aging," suppression of the ubiquitin-proteasome system, and neuronal degeneration. The administration of propionate improves the ultrastructure of VMH, induces autophagy due to ER structures, which, against the background of a decrease in apoptotic changes in neurons, demonstrates an effective mechanism of ER stress compensation by the ER degradation system, while accompanied by maximum activation of glial structures. Combined administration of drugs most effectively establishes the level of UPR markers for ER functioning and promotes the continuation of protein synthesis by reducing PERK.

Switching the path of apoptotic death to autophagy mechanisms

Since autophagy and apoptosis are interrelated processes, it is believed that the fate of a cell adapting to stress is determined by the "switching" of apoptosis processes to autophagy by the main autophagy regulator Beclin-1 and its interaction with Bcl-2 and other apoptotic cell death pathways.

In models of various options for correcting T2DM, a high correlation ($R = 0.779$, $P < .01$) was found between Caspase-3 and the BAX dimer, indicating the implementation of VMH neuron apoptosis due to mitoptosis (Natrus et al., 2022b, 2024). These assumptions are confirmed by edema and destruction of mitochondria against the background of T2DM, which were detected by ultramicroscopic studies.

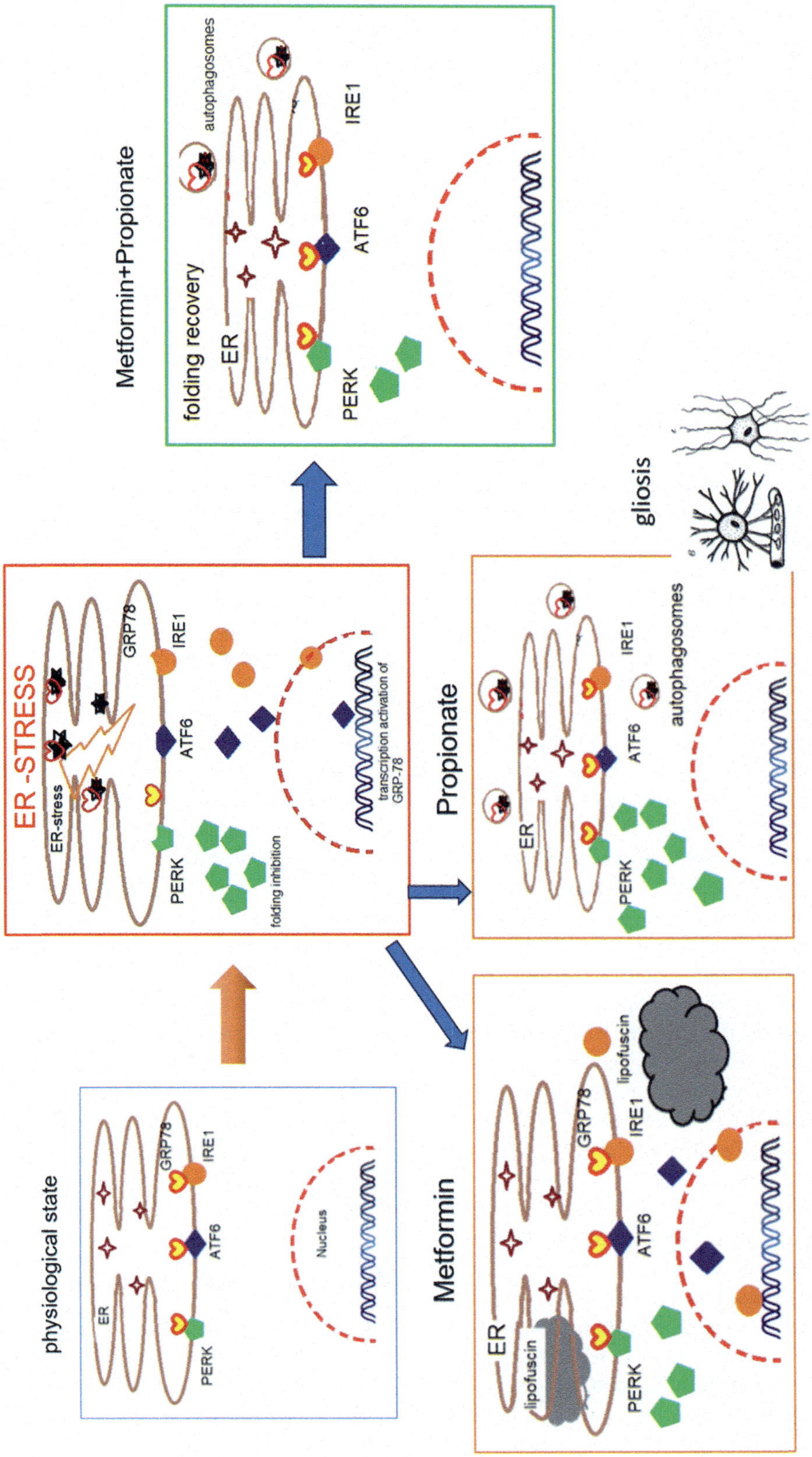

FIGURE 26.4 Schematic representation of changes in a neuron in the state of ER-stress and various pharmacological corrections.

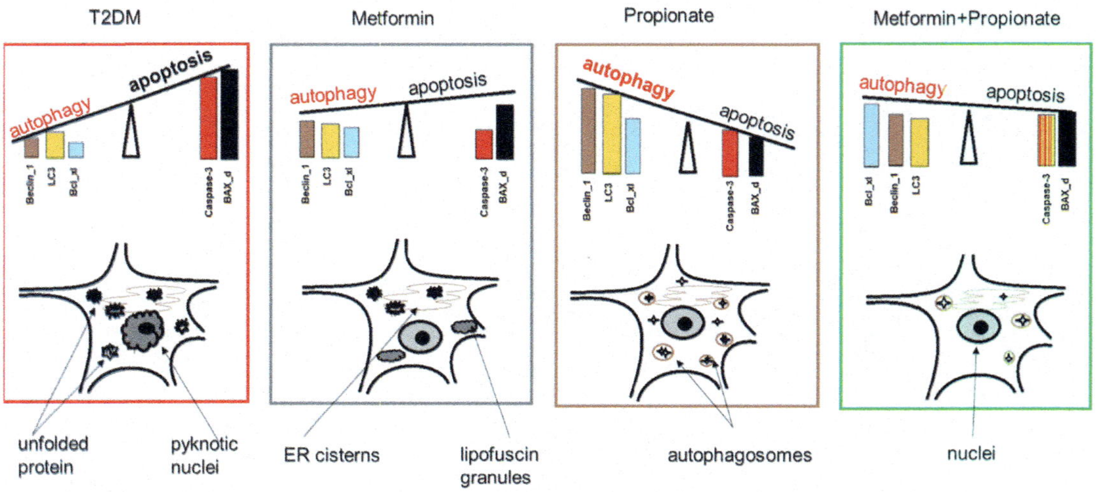

FIGURE 26.5 The scheme of apoptosis/autophagy switch by propionic acid in VMH of rats with T2DM (Natrus et al., 2022b).

The high correlation between the autophagy markers Beclin-1 and LC3 and the anti-apoptotic protein Bcl-xL indicates mechanisms of apoptosis counteraction and adaptation of neurons to ER stress conditions precisely due to the formation of autophagosomes.

Therefore, the process diagram (Fig. 26.5) appears logical, demonstrating the "switch" from the path of potential apoptotic death of neurons, the threat of which arises against the background of uncorrected hyperglycemia, to the mechanisms of enhanced autophagy, protein utilization by the ERAD system, and, as a result, the preservation of neurons, a certain adaptation to shifts in homeostasis, and compensation for ER stress.

In T2DM, VMH neurons show a maximum increase in the content of apoptosis markers Caspase-3 and BAX dimer, indicating neuronal death, mainly through mitoptosis. The administration of metformin to rats with T2DM leads to the most effective reduction of Caspase-3 levels, but does not provide an effective system for the removal of protein degradation products. The administration of propionate maximally enhances autophagy and demonstrates a "switch" from the apoptotic death pathway to autophagy mechanisms. The best balance of ER stress compensation mechanisms and neuron adaptation is observed with combined administration of drugs.

Thus, the development of T2DM in VMH neurons is characterized by significant ultrastructural changes in the form of enlargement of cisterns, perinuclear space, desquamation of ribosomes, damage due to mitochondrial edema and a large number of neurons with pyknotic nuclei. Such morphological changes, together with biochemical markers of the UPR system, fully explain the mechanism of neuronal damage against the background of HFD due to the development of ER stress. At the same time, the anti-apoptotic preservation system is maximally suppressed in neurons and apoptosis is activated.

The administration of drugs improves the condition of neurons to a certain extent. However, against the background of metformin monotherapy, only one marker, Caspase-3, is effectively reduced. At the same time, a feature of metformin administration is the detection of cells with a significant amount of lipofuscin, which signals the suppression of the ubiquitin-proteasome system, increased oxidation of unsaturated fatty acids, and is a sign of "aging" and degeneration of neurons. Therefore, metformin monotherapy does not prevent the development of diabetic encephalopathy and is insufficient in T2DM. The administration of propionate in the context of T2DM maximally activates autophagy in neurons, which can be regarded as an adaptive response to ER stress, but, due to its effect on the "gut-glia-neuron" axis, it leads to neurogliosis (Natrus et al., 2022b, 2024).

The combined administration of metformin and propionate demonstrates an effective effect on neurons in most indicators, given the implementation of UPR system signals, anti- and pro-apoptosis markers in the cell, which causes the greatest adaptation of neurons to survival under conditions of ER stress. This may be the basis for the further prospect of using these drugs and developing neuroprotection schemes.

Mini-dictionary of terms

Mitoptosis—a controlled process of mitochondrial destruction that is triggered in response to damage or dysfunction of these organelles and is aimed at maintaining cellular homeostasis; it is part of the mitochondrial quality control system.

ERAD system (Endoplasmic Reticulum-Associated Degradation) —a protein degradation mechanism that ensures the recognition, retrotranslocation, and destruction of misfolded or defective endoplasmic reticulum proteins via the ubiquitin-proteasome pathway.

Autophagy—an evolutionarily conserved process in which a cell degrades its own damaged organelles, proteins, or cytoplasmic components in autophagosomes, followed by fusion with lysosomes; it is important for maintaining cellular homeostasis and neuronal survival.

Chaperone—protein that promotes the correct folding of newly synthesized polypeptides, prevents protein aggregation, and participates in their renaturation or degradation under conditions of cellular stress.

Protein folding—the process by which a polypeptide chain acquires the specific three-dimensional structure necessary for its biological activity; folding disorders lead to proteotoxic stress and neurodegeneration.

The perinuclear space—the narrow space between the inner and outer membranes of the nuclear envelope, which continuously transitions into the lumen of the endoplasmic reticulum and participates in the transport of ions and proteins.

Summary points

- Persistent deviations in homeostatic constants in metabolic diseases are accompanied by dysregulation of the central mechanisms controlling energy balance, in particular the neurons of the mediolateral hypothalamus.
- Insulin resistance and associated dyslipidemia lead to impaired fatty acid metabolism, causing toxic effects on neurons and astrocytes through excessive beta-oxidation.
- In neurons of the ventromedial nucleus of the hypothalamus (VMH), type 2 diabetes mellitus causes endoplasmic reticulum stress, which is characterized by dilation of ER cisternae and accumulation of damaged proteins in them.
- Activation of the PERK–eIF2α–CHOP pathway promotes the transition from a cytoprotective to a proapoptotic UPR scenario, leading to neuronal loss and disruption of central metabolic regulation.
- ER stress, apoptosis, and autophagy dysfunction are key links in neurodegenerative changes in T2DM, opening up prospects for targeted neuroprotective therapy.
- Combined administration of metformin and propionate reduces signs of stress and enhances autophagy in neurons, which plays an adaptive role by ensuring the degradation of damaged proteins and organelles.

References

Anthony, P. C., & Yeo, G. S. H. (2013). The hypothalamus and metabolism: Integrating signals to control energy and glucose homeostasis. *Current Opinion in Pharmacology, 13*(6), 970–976.

Arensdorf, A. M., Diedrichs, D., & Rutkowski, D. T. (2013). Regulation of the transcriptome by ER stress: Non-canonical mechanisms and physiological consequences. *Frontiers in Genetics, 4*, 256. https://doi.org/10.3389/fgene.2013.00256

Braakman, I., & Hebert, D. N. (2013). Protein folding in the endoplasmic reticulum. *Cold Spring Harbor Perspectives in Biology, 5*(5), a013201. https://doi.org/10.1101/cshperspect.a013201

Bravo, R., Gutierrez, T., Paredes, F., Gatica, D., Rodriguez, A. E., Pedrozo, Z., Chiong, M., Parra, V., Quest, A. F., Rothermel, B. A., & Lavandero, S. (2012). Endoplasmic reticulum: ER stress regulates mitochondrial bioenergetics. *International Journal of Biochemistry and Cell Biology, 44*(1), 16–20. https://doi.org/10.1016/j.biocel.2011.10.012

Chen, Y., & Brandizzi, F. (2013). IRE1: ER stress sensor and cell fate executor. *Trends in Cell Biology, 23*(11), 547–555. https://doi.org/10.1016/j.tcb.2013.06.005

Duscha, A., Gisevius, B., Hirschberg, S., Yissachar, N., Stangl, G. I., Dawin, E., Bader, V., Haase, S., Kaisler, J., David, C., Schneider, R., Troisi, R., Zent, D., Hegelmaier, T., Dokalis, N., Gerstein, S., Del Mare-Roumani, S., Amidror, S., Staszewski, O., … Haghikia, A. (2020). Propionic acid shapes the multiple sclerosis disease course by an immunomodulatory mechanism. *Cell, 180*(6), 1067–1080.e16. https://doi.org/10.1016/j.cell.2020.02.035

Friedman, J. M., & Halaas, J. L. (1998). Leptin and the regulation of body weight in mammals. *Nature, 395*(6704), 763–770. https://doi.org/10.1038/27376

Gomez, J. A., & Rutkowski, D. T. (2016). Experimental reconstitution of chronic ER stress in the liver reveals feedback suppression of BiP mRNA expression. *eLife, 5*, e20390. https://doi.org/10.7554/eLife.20390

Gozuacik, D., & Kimchi, A. (2007). Autophagy and cell death. *Current Topics in Developmental Biology, 78*, 217–245. https://doi.org/10.1016/S0070-2153(06)78006-1

Haffner, S. M., Valdez, R. A., Hazuda, H. P., Mitchell, B. D., Morales, P. A., & Stern, M. P. (1992). Prospective analysis of the insulin-resistance syndrome (syndrome X). *Diabetes, 41*(6), 715–722. https://doi.org/10.2337/diab.41.6.715

He, J., Zhou, Y., & Sun, L. (2024). Emerging mechanisms of the unfolded protein response in therapeutic resistance: From chemotherapy to immunotherapy. *Cell Communication and Signaling, 22*, 89. https://doi.org/10.1186/s12964-023-01438-0

Hetz, C., & Papa, F. R. (2018). The unfolded protein response and cell fate control. *Molecular Cell, 69*(2), 169–181. https://doi.org/10.1016/j.molcel.2017.06.017

Jung, C. H., & Kim, M. S. (2013). Molecular mechanisms of central leptin resistance in obesity. *Archives of Pharmacal Research, 36*(2), 201–207. https://doi.org/10.1007/s12272-013-0020-y

Kabeya, Y., Mizushima, N., Ueno, T., Yamamoto, A., Kirisako, T., Noda, T., … Yoshimori, T. (2000). LC3, a mammalian homologue of yeast Apg8p, is localized in autophagosome membranes after processing. *Journal of Cell Biology, 151*(2), 263–276. https://doi.org/10.1083/jcb.151.2.263

Kang, R., Zeh, H., Lotze, M., & Kroemer, G. (2011). The Beclin 1 network regulates autophagy and apoptosis. *Cell Death and Differentiation, 18*(4), 571–580. https://doi.org/10.1038/cdd.2010.191

Kazakov, V. N., & Natrus, L. V. (2005). Modulation of neuronal impulse activity of the anterior hypothalamus as a functional basis of the mechanisms underlying hypothalamic control. *Neurophysiology, 37* (5), 407–417. https://doi.org/10.1007/s11062-006-0018-x

Kennedy, G. C. (1953). The role of depot fat in the hypothalamic control of food intake in the rat. *Proceedings of the Royal Society of London. Series B, Biological Sciences, 140*(901), 578–592. https://doi.org/10.1098/rspb.1953.0009

Kroemer, G., Mariño, G., & Levine, B. (2010). Autophagy and the integrated stress response. *Molecular Cell, 40*(2), 280–293. https://doi.org/10.1016/j.molcel.2010.09.023

Lévêque, T. F. (1974). The endocrine hypothalamus: An historical review. *Canadian Journal of Neurological Sciences, 1*(1), 24–28. https://doi.org/10.1017/S0317167100019545

Liu, Z., Xiao, T., & Liu, H. (2023). Leptin signaling and its central role in energy homeostasis. *Frontiers in Neuroscience, 17*, 1238528. https://doi.org/10.3389/fnins.2023.1238528

Mao, J., Hu, Y., Ruan, L., Ji, Y., & Lou, Z. (2019). Role of endoplasmic reticulum stress in depression. *Molecular Medicine Reports, 20*(5), 4774–4780. https://doi.org/10.3892/mmr.2019.10789 (Review).

Morris, D. L., & Rui, L. (2009). Recent advances in understanding leptin signaling and leptin resistance. *American Journal of Physiology-Endocrinology and Metabolism, 297*(6), E1247–E1259. https://doi.org/10.1152/ajpendo.00274.2009

Morton, G. J. (2007). Hypothalamic leptin regulation of energy homeostasis and glucose metabolism. *Journal of Physiology, 583*(2), 437–443. https://doi.org/10.1113/jphysiol.2007.135590

Natrus, L. V. (2006). Changes in activity of the hypothalamic neurons in response to physiological fluctuations of the homeostasis constants. *Fiziologicheskii Zhurnal, 52*(3), 57–63.

Natrus, L. V., Osadchuk, Y. S., Lisakovska, O. O., Labudzinskyi, D. O., Klys, Y. G., & Chaikovsky, Y. B. (2022a). Effect of propionic acid on diabetes-induced impairment of unfolded protein response signaling and astrocyte/microglia crosstalk in rat ventromedial nucleus of the hypothalamus. *Neural Plasticity*. , Article 6404964. https://doi.org/10.1155/2022/6404964

Natrus, L., Osadchuk, Y., Lisakovska, O., Labudzinskyi, D., Klys, Y., & Chaikovsky, Y. (2022a). Effect of propionic acid on diabetes-induced impairment of unfolded protein response signaling and astrocyte/microglia crosstalk in rat ventromedial nucleus of the hypothalamus. *Neural Plasticity, 2022*. https://doi.org/10.1155/2022/6404964. Article 6404964.

Natrus, L., Osadchuk, Y., Lisakovska, O., Labudzynskyi, D., Klys, Y., & Chaikovsky, Y. (2022b). Regulation of the apoptosis/autophagy switch by propionic acid in ventromedial hypothalamus of rats with type 2 diabetes mellitus. *Heliyon, 8*(11), e11529. https://doi.org/10.1016/j.heliyon.2022.e11529

Neal, S., Duttke, S. H., & Hampton, R. Y. (2019). Assays for protein retrotranslocation in ERAD. In M. Hochstrasser (Ed.), *Methods in enzymology* (Vol 619, pp. 1–26). Academic Press. https://doi.org/10.1016/bs.mie.2019.01.002

Neel, J. V. (1962). Diabetes mellitus: A "thrifty" genotype rendered detrimental by "progress". *American Journal of Human Genetics, 14* (4), 353–362. https://doi.org/10.1097/00000441-196212000-00005

Park, S. M., Kang, T. I., & So, J. S. (2021). Roles of XBP1s in transcriptional regulation of target genes. *Biomedicines, 9*(7), 791. https://doi.org/10.3390/biomedicines9070791

Park, S. W., & Ozcan, U. (2013). Potential for therapeutic manipulation of the UPR in disease. *Seminars in Immunopathology, 35*(3), 351–373. https://doi.org/10.1007/s00281-013-0370-z

Pincus, D., Chevalier, M. W., Aragón, T., van Anken, E., Vidal, S. E., El-Samad, H., & Walter, P. (2010). BiP binding to the ER-stress sensor Ire1 tunes the homeostatic behavior of the unfolded protein response. *PLoS Biology, 8*(7), e1000415. https://doi.org/10.1371/journal.pbio.1000415

Ramírez, S., & Claret, M. (2015). Hypothalamic ER stress: A bridge between leptin resistance and obesity. *FEBS Letters, 589*(14), 1678–1687. https://doi.org/10.1016/j.febslet.2015.04.025

Rozpedek, W., Pytel, D., Mucha, B., Leszczynska, H., Diehl, J. A., & Majsterek, I. (2016). The role of the PERK/eIF2α/ATF4/CHOP signaling pathway in tumor progression during endoplasmic reticulum stress. *Current Molecular Medicine, 16*(6), 533–544. https://doi.org/10.2174/1566524016666160523143937

Ruggiano, A., Foresti, O., & Carvalho, P. (2014). Quality control: ER-associated degradation—Protein quality control and beyond. *Journal of Cell Biology, 204*(6), 869–879. https://doi.org/10.1083/jcb.201312042

Ruud, J., Steculorum, S. M., & Brüning, J. C. (2017). Neuronal control of peripheral insulin sensitivity and glucose metabolism. *Nature Communications, 8*, 15259. https://doi.org/10.1038/ncomms15259

Sakurai, T., Amemiya, A., Ishii, M., Matsuzaki, I., Chemelli, R. M., Tanaka, H., … Yanagisawa, M. (1998). Orexins and orexin receptors: A family of hypothalamic neuropeptides and G protein-coupled receptors that regulate feeding behavior. *Cell, 92*(4), 573–585. https://doi.org/10.1016/s0092-8674(00)80949-6

Tran, L. T., Park, S., & Kim, S. K. (2022). Hypothalamic control of energy expenditure and thermogenesis. *Experimental and Molecular Medicine, 54*, 358–369. https://doi.org/10.1038/s12276-022-00741-z

Urra, H., Dufey, E., Lisbona, F., Rojas-Rivera, D., & Hetz, C. (2013). When ER stress reaches a dead end. *Biochimica et Biophysica Acta, 1833*(12), 3507–3517. https://doi.org/10.1016/j.bbamcr.2013.07.024

Walensky, L. D., & Gavathiotis, E. (2011). BAX unleashed: The biochemical transformation of an inactive cytosolic monomer into a toxic mitochondrial pore. *Trends in Biochemical Sciences, 36*(12), 642–652. https://doi.org/10.1016/j.tibs.2011.08.009

Walter, P., & Ron, D. (2011). The unfolded protein response: From stress pathway to homeostatic regulation. *Science, 334*(6059), 1081–1086. https://doi.org/10.1126/science.1209038

Wang, M., Wey, S., Zhang, Y., Ye, R., & Lee, A. S. (2009). Role of the unfolded protein response regulator GRP78/BiP in development, cancer, and neurological disorders. *Antioxidants and Redox Signaling, 11*(9), 2307–2316. https://doi.org/10.1089/ars.2009.2485

Wei, J., & Fang, D. (2021). Endoplasmic reticulum stress signaling and the pathogenesis of hepatocarcinoma. *International Journal of Molecular Sciences, 22*(4), 1799. https://doi.org/10.3390/ijms22041799

Williams, G., Bing, C., Cai, X. J., Harrold, J. A., King, P. J., & Liu, X. H. (2001). The hypothalamus and the control of energy homeostasis: Different circuits, different purposes. *Physiology and Behavior, 74* (4–5), 683–701. https://doi.org/10.1016/s0031-9384(01)00612-6

Xu, F., et al. (2024). Deciphering ER stress–unfolded protein response relationship by visualizing unfolded proteins in the ER. *Cell Reports, 43*(6), 114358. https://doi.org/10.1016/j.celrep.2024.114358

Yuan, S., She, D., & Jiang, S. (2024). Endoplasmic reticulum stress and therapeutic strategies in metabolic, neurodegenerative diseases and cancer. *Molecular Medicine, 30*, 40. https://doi.org/10.1186/s10020-024-00808-9

Zhao, Y. G., & Zhang, H. (2019). Autophagosome maturation: An epic journey from the ER to lysosomes. *Journal of Cell Biology, 218*(3), 757–770. https://doi.org/10.1083/jcb.201810099

Zhu, Y., Zhao, L., Liu, L., Gao, P., Tian, W., Wang, X., Jin, H., Xu, H., & Chen, Q. (2010). Beclin 1 cleavage by caspase-3 inactivates autophagy and promotes apoptosis. *Protein and Cell, 1*(5), 468–477. https://doi.org/10.1007/s13238-010-0048-4

Section 5

Nonpancreatic treatment regimens and effects on neuroscientific domains

Chapter 27

Statins and insulin resistance

Yulia Yusrini Djabir[1] and Aryadi Arsyad[2]
[1]*Laboratory of Clinical Pharmacy, Faculty of Pharmacy, Hasanuddin University, Makassar, Indonesia;* [2]*Department of Physiology, Faculty of Medicine, Hasanuddin University, Makassar, Indonesia*

List of abbreviations

Akt protein kinase B
AMPK AMP-activated protein kinase
ASCVD atherosclerotic cardiovascular disease
ATP adenosine triphosphate
CK creatine kinase
DAG diacylglycerol
EGR early growth response protein
ERK extracellular signal-regulated kinase
FPP farnesyl pyrophosphate
GGPP geranylgeranyl pyrophosphate
GIP glucose-dependent insulinotropic peptide
GLP-1 glucagon-like peptide-1
GLUT2 glucose transporter type 2
GLUT4 glucose transporter type 4
Grb2 growth factor receptor–bound protein 2
GSK-3 glycogen synthase kinase-3
HDL-C high-density lipoprotein cholesterol
HMG-CoA 3-hydroxy-3-methylglutaryl-coenzyme A
IKKβ inhibitor of kappa B kinase beta
IL-6 interleukin-6
IRS insulin receptor substrate
JNK c-Jun N-terminal kinase
LDL-C low-density lipoprotein cholesterol
MAPK mitogen-activated protein kinase
MEK MAPK/ERK kinase
NF-κB nuclear factor kappa-light-chain-enhancer of activated B cells
NOD new-onset diabetes
PDK1 phosphoinositide-dependent kinase-1
PI3K phosphatidylinositol 3-kinase
PIP2 phosphatidylinositol-4,5-bisphosphate
PIP3 phosphatidylinositol-3,4,5-trisphosphate
PKC protein kinase C
PPARγ peroxisome proliferator-activated receptor gamma
ROS reactive oxygen species
SOCS suppressor of cytokine signaling
SOS son of sevenless
T2DM type 2 diabetes mellitus
TNF-α tumor necrosis factor-alpha
VLDL very low-density lipoprotein

Introduction

Statins have become the foundational therapy for the management of dyslipidemia and the prevention of atherosclerotic cardiovascular disease (ASCVD) (Grundy et al., 2019). Statins are among the most commonly prescribed medications worldwide (Blais et al., 2021). The global popularity of statins reflects not only their broad acceptance in clinical practice but also their favorable safety, tolerability, and efficacy in lowering cardiovascular morbidity and mortality (Collins et al., 2016).

The primary mechanism of action of statins is the inhibition of 3-hydroxy-3-methylglutaryl-CoA (HMG-CoA) reductase, the rate-limiting enzyme of the mevalonate pathway. This blockade diminishes hepatic cholesterol synthesis, which in turn triggers compensatory upregulation of low-density lipoprotein (LDL) receptors on hepatocyte membranes. The result is enhanced clearance of circulating LDL, very-low-density lipoprotein (VLDL), and chylomicron remnants, leading to a robust reduction in atherogenic lipoproteins (Ahmadi et al., 2017). These lipid-lowering effects translate directly into a reduction in major adverse cardiovascular events and deaths. Importantly, delayed initiation of statin therapy in high-risk individuals, such as those with diabetes, has been linked to worse cardiovascular outcomes (Shah et al., 2025). Consequently, international guidelines consistently endorse statins as the first-line pharmacological therapy for individuals at elevated ASCVD risk (Grundy et al., 2019).

Over the past decade, however, increasing attention has been directed toward the metabolic consequences of long-term statin therapy, particularly its association with the development of insulin resistance. Insulin resistance refers to a reduced biological response of tissues such as skeletal muscle, adipose tissue, and the liver to circulating insulin. Accumulating evidence has consistently demonstrated a modest but significant increase in the incidence of type 2

Hormones in Neuroscience. https://doi.org/10.1016/B978-0-443-30263-3.00001-4

diabetes (T2DM) among individuals receiving statins (Lembo et al., 2025; Reith et al., 2024).

As statins continue to play a central role in cardiovascular prevention, understanding their potential metabolic effects has become an important area of clinical and research interest. This chapter provides an overview of statin pharmacology and insulin action and intracellular signaling to comprehend their roles in metabolic regulation and the development of insulin resistance in response to statin therapy.

Statins

Pharmacology

Statins act mainly by competitively blocking HMG-CoA reductase, the key regulatory enzyme of the mevalonate pathway in cholesterol synthesis (Fig. 27.1). Under normal conditions, HMG-CoA reductase catalyzes the conversion of HMG-CoA to mevalonic acid, a precursor of cholesterol and other isoprenoid intermediates. By blocking this step, statins reduce the intracellular pool of cholesterol within hepatocytes (Schmidt et al., 2025). A reduction in intracellular cholesterol activates sterol regulatory element–binding proteins (SREBPs), which in turn upregulate the transcription of LDL receptor genes. As a result, the number of LDL receptors on the hepatocyte surface increases, enhancing the clearance of circulating LDL particles from plasma. This mechanism accounts for the potent LDL-lowering effect of statins, often achieving a reduction of 30%–50% depending on drug potency and dosage (Stone et al., 2014).

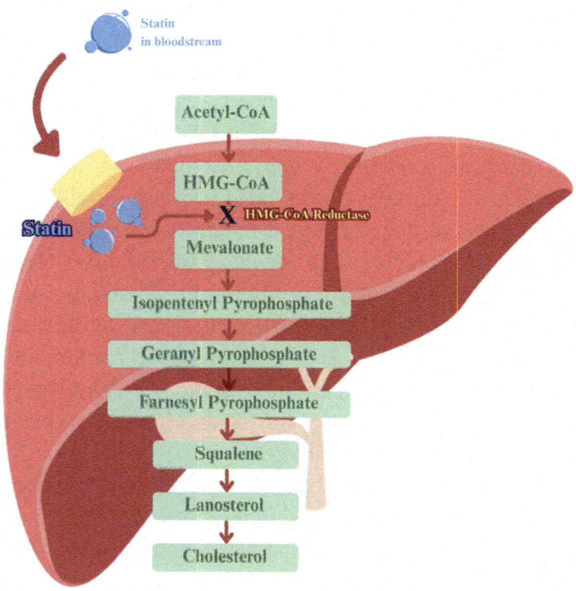

FIGURE 27.1 Statin inhibition of the mevalonate pathway in hepatic cholesterol biosynthesis. Statins act as competitive inhibitors of HMG-CoA, 3-hydroxy-3-methylglutaryl coenzyme A (HMG-CoA) reductase, the rate-limiting enzyme in the mevalonate pathway. By blocking the conversion of HMG-CoA to mevalonate, statins reduce the downstream synthesis, including isopentenyl pyrophosphate, geranyl pyrophosphate, farnesyl pyrophosphate, squalene, lanosterol, and ultimately cholesterol.

Beyond lowering LDL cholesterol, inhibition of the mevalonate pathway also reduces the synthesis of isoprenoid intermediates like farnesyl pyrophosphate (FPP) and geranylgeranyl pyrophosphate (GGPP). These molecules are critical for the posttranslational prenylation of small GTP-binding proteins, including Ras, Rho, and Rac. By inhibiting RhoA, statin increases endothelial nitric oxide synthase expression, boosting nitric oxide production that important for vasodilation, preventing platelet aggregation, and improving endothelial function. Meanwhile, Rac1 inhibition reduces activation of NADPH oxidase, lowering production of reactive oxygen species (ROS). Additionally, Ras inhibition blunts proliferative and inflammatory signaling in vascular smooth muscle cells, preventing plaque growth and instability. This secondary action contributes to the so-called pleiotropic effects of statins (Liao & Laufs, 2005).

Additionally, statins have been shown to increase HDL-C levels by approximately 4%–10% in patients with hypercholesterolemia, with a more pronounced effect observed in individuals who present with low baseline HDL levels or elevated triglycerides (McTaggart & Jones, 2008). The underlying mechanisms are thought to involve two principal pathways: (1) suppression of CETP-mediated transfer of cholesteryl esters from HDL to apoB-containing lipoproteins, and (2) upregulation of apoA-I synthesis in hepatocytes. Notably, this HDL-raising effect appears to be independent of LDL-C reduction, although its magnitude may vary according to the specific statin administered (Barter et al., 2010).

Classification

Statins can be categorized by origin, chemical structure, and LDL-C–lowering intensity.

- By origin, statins are classified as natural, semisynthetic, or synthetic. The naturally derived statins (lovastatin, pravastatin, and mevastatin) are produced by fungal fermentation (Endo, 1979; Sadowska et al., 2023). Semisynthetic statins, such as simvastatin, are chemically modified from natural precursors to improve potency and pharmacokinetics. Synthetic statins, including atorvastatin, rosuvastatin, fluvastatin, and pitavastatin, are designed through chemical synthesis and generally offer greater potency and longer half-lives (Murphy et al., 2020).
- By chemical structure, statins are broadly divided into lipophilic and hydrophilic agents (Fig. 27.2). Lipophilic statins (simvastatin, lovastatin, fluvastatin, atorvastatin, pitavastatin) readily diffuse across membranes, distribute widely into extrahepatic tissues, and are mainly metabolized by cytochrome P450 enzymes such as CYP3A4 or CYP2C9 (Hirota et al., 2020).

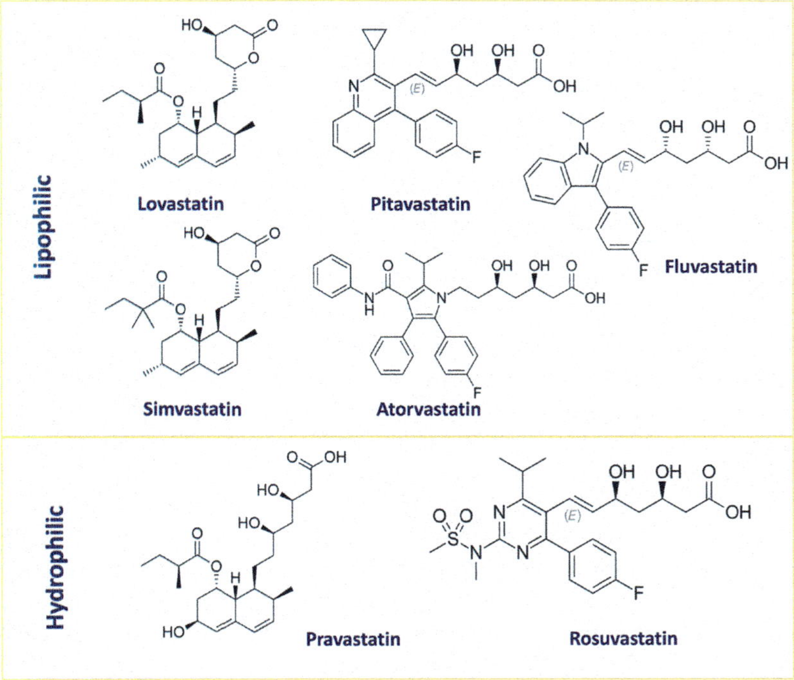

FIGURE 27.2 Chemical structures of statins. Lipophilic statins include simvastatin, lovastatin, atorvastatin, fluvastatin, and pitavastatin, while hydrophilic statins include pravastatin and rosuvastatin.

Their broad tissue penetration explains the higher association with muscle-related adverse effects such as myalgia (Climent et al., 2021). In contrast, hydrophilic statins (rosuvastatin, pravastatin) depend on active hepatic uptake via organic anion–transporting polypeptides (OATPs), display greater hepatoselectivity, and have a lower propensity for muscle side effects and drug–drug interactions (Shi & Han, 2025; Svec & Adameova, 2023).

- By LDL-C–lowering intensity, statins are classified as high-, moderate-, or low-intensity according to the average percentage reduction in LDL-C achieved. High-intensity statins (atorvastatin 40–80 mg and rosuvastatin 20–40 mg) lower LDL-C by ≥50%. Moderate-intensity (atorvastatin 10–20 mg, rosuvastatin 5–10 mg, or simvastatin 20–40 mg) achieves reductions of about 30%–49%. Low-intensity regimens (simvastatin 10 mg, pravastatin 10–20 mg, or lovastatin 20 mg) generally reduce LDL-C by less than 30% (Grundy et al., 2019).

Toxicity

The most clinically relevant adverse effects of statins involve skeletal muscle, with presentations ranging from mild myalgia to myopathy with elevated creatine kinase (CK) levels, and in rare cases, rhabdomyolysis leading to myoglobinuria and acute kidney injury. Severe rhabdomyolysis is exceedingly uncommon, occurring in only a few cases per 100,000 patient-years, yet its potential severity necessitates careful monitoring (Backes et al., 2017; Law & Rudnicka, 2006).

Statin-induced muscle toxicity may result from reduced coenzyme Q10 synthesis impairing mitochondrial function, altered prenylation of small GTP-binding proteins disrupting cellular signaling, and increased oxidative stress promoting protein degradation. Risk factors include older age, hypothyroidism, renal impairment, high-dose therapy, polypharmacy, and genetic variants such as SLCO1B1, which reduce hepatic statin uptake and predispose to myopathy (Ward et al., 2019).

Statin-associated liver toxicity is relatively uncommon in clinical practice. The most frequent manifestation is a mild, asymptomatic rise in aminotransferase levels, observed in approximately 0.2%–3% of patients, typically within the first few months of therapy (Averbukh et al., 2022). Importantly, the majority of these cases resolve spontaneously even when treatment is continued, while persistent elevations exceeding three times the upper limit of normal are rare.

Beyond musculoskeletal and hepatic adverse effects, statins have also been linked to an increased risk of new-onset T2DM. The diabetogenic effect of statins is mediated through the development of insulin resistance, providing a mechanistic link between statin therapy and altered glucose homeostasis (Reith et al., 2024).

Insulin and insulin resistance

Insulin physiological function

Insulin is the principal anabolic hormone in human metabolism, orchestrating the coordinated regulation of glucose, lipid, and protein homeostasis. It ensures that nutrients derived from food intake are efficiently utilized and stored, thereby maintaining energy balance at both the cellular and systemic levels (Tokarz et al., 2018).

The secretion of insulin originates from the pancreatic β-cells located within the islets of Langerhans. These specialized endocrine cells are uniquely equipped to sense changes in circulating glucose concentrations. Following a rise in plasma glucose—typically after a meal—β-cells rapidly increase insulin secretion in a biphasic manner: an initial burst reflecting the release of preformed insulin granules, followed by a sustained phase driven by new granule mobilization and synthesis (Henquin et al., 2002). This glucose-stimulated insulin secretion from pancreatic cells is tightly coupled to cellular metabolism (Fig. 27.3). Begin with glucose uptake via GLUT2 transporters, phosphorylation by glucokinase, and subsequent ATP generation through glycolysis and oxidative phosphorylation. The rise in the ATP/ADP ratio leads to the closure of K^+ channels sensitive to ATP, causing β-cell membrane depolarization that subsequently triggers Ca^{2+} influx, and stimulates insulin granule exocytosis (Röder et al., 2016).

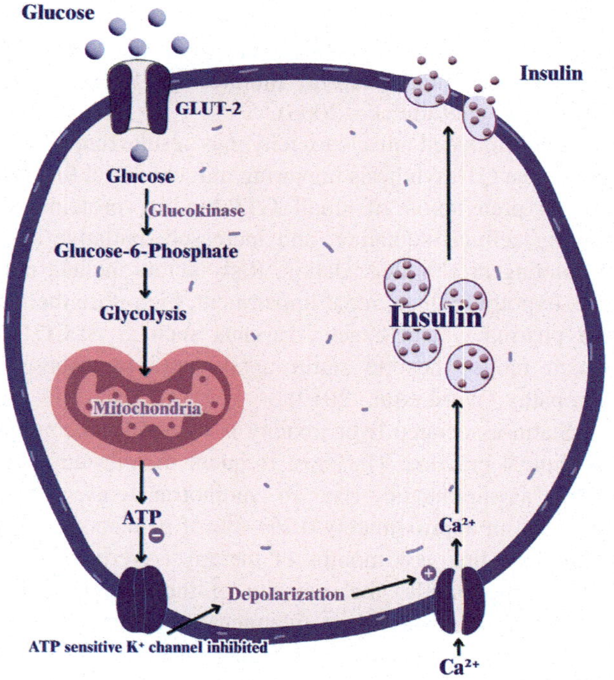

FIGURE 27.3 Insulin secretion from pancreatic β-cells. Inside the cell, glucose is phosphorylated by glucokinase to glucose-6-phosphate, initiating glycolysis and subsequent mitochondrial oxidative phosphorylation, which increases ATP production. The resulting rise in the ATP/ADP ratio causes closure of ATP-sensitive potassium (K_ATP) channels, leading to membrane depolarization. This depolarization opens voltage-gated calcium channels, allowing Ca^{2+} influx. The increase in intracellular calcium concentration triggers the exocytosis of insulin-containing secretory granules, releasing insulin into the bloodstream.

Beyond glucose, several other factors modulate insulin release. Amino acids, fatty acids, incretin hormones such as glucagon-like peptide-1 (GLP-1) and glucose-dependent insulinotropic peptide (GIP), and parasympathetic neural inputs enhance secretion. Conversely, sympathetic adrenergic signaling and somatostatin inhibit insulin release. This finely tuned regulation ensures that insulin secretion matches metabolic demands under varying physiological conditions (Aronoff et al., 2004). Once in circulation, insulin acts systemically on skeletal muscle, adipose tissue, liver, and the central nervous system, driving processes such as glucose uptake, glycogen synthesis, lipogenesis, and protein synthesis. In doing so, insulin not only prevents excessive postprandial hyperglycemia but also provides the foundation for long-term metabolic stability. Disruption of insulin secretion or action undermines this balance, contributing to pathological states such as insulin resistance, metabolic syndrome, and T2DM.

Molecular basis of insulin receptor signaling

The insulin receptor (IR) is a membrane-bound tyrosine kinase that consists of two extracellular α subunits paired with two β subunits that span the cell membrane, forming a heterotetrameric structure. When insulin binds to the α-subunit of its receptor, it activates the β-subunit's tyrosine kinase, leading to autophosphorylation at specific tyrosine residues and creating docking sites for downstream signaling proteins (Haeusler et al., 2018).

Two major signaling pathways are subsequently activated: the metabolic pathway and the mitogenic pathway (Fig. 27.4). The metabolic pathway is initiated by insulin receptor substrate (IRS), which upon phosphorylation activates phosphoinositide 3-kinase (PI3K). PI3K converts phosphatidylinositol-4,5-bisphosphate (PIP2) into phosphatidylinositol-3,4,5-trisphosphate (PIP3), leading to activation of phosphoinositide-dependent kinase-1 (PDK1) and subsequent phosphorylation of protein kinase B (Akt). Akt plays a central role in mediating insulin's metabolic actions, including translocation of glucose transporter type 4 (GLUT4) to the plasma membrane in skeletal muscle and adipose tissue, suppression of hepatic gluconeogenesis, and stimulation of glycogen synthesis via inhibition of glycogen synthase kinase-3 (GSK-3) (Saltiel, 2021).

The mitogenic pathway, in contrast, involves the recruitment of growth factor receptor–bound protein 2 (Grb2) and son of sevenless (SOS) to phosphorylated IRS or Shc proteins, leading to activation of the Ras–Raf–MEK–ERK (MAPK) cascade. ERK activation leads to translocation of

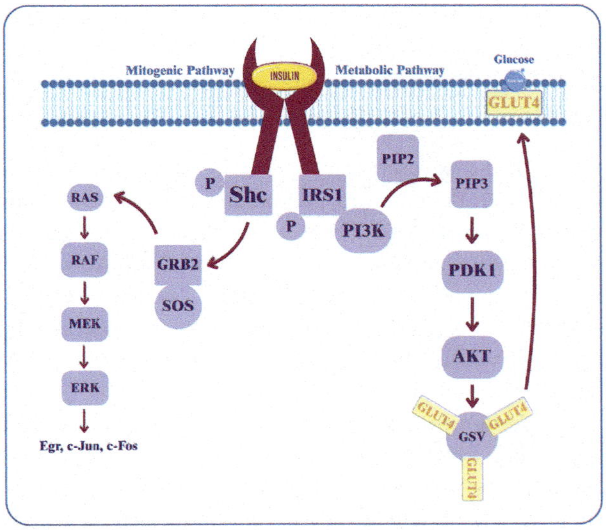

FIGURE 27.4 Insulin receptor signaling pathways. Upon insulin binding, the insulin receptor activates two major downstream arms: (1) the metabolic pathway, mediated by IRS, PI3K, and Akt, promotes GLUT4 translocation, glycogen synthesis, and suppression of hepatic gluconeogenesis; and (2) the mitogenic pathway, involving Grb2, SOS, Ras, Raf, MEK, and ERK. Activated ERK stimulates transcription factors such as EGR, c-Jun, and c-Fos, which regulate growth, differentiation, and gene expression. *IRS*, insulin receptor substrate; *PI3K*, phosphoinositide 3-kinase; *Akt*, protein kinase B; *GLUT4*, glucose transporter type 4; *Grb2*, growth factor receptor–bound protein 2; *SOS*, son of sevenless; *MEK*, mitogen-activated protein kinase kinase; *ERK*, extracellular signal–regulated kinase; *EGR*, early growth response protein.

the kinase to the nucleus and activates transcription factors such as EGR, c-Jun, and c-Fos. This pathway predominantly regulates cell growth, differentiation, and gene expression, rather than acute metabolic processes (Le et al., 2023). The balance between these two pathways is critical for maintaining normal insulin function.

Pathophysiology of insulin resistance

Insulin resistance occurs when key insulin-responsive tissues—including skeletal muscle, adipose tissue, and the liver—do not respond effectively to physiological levels of circulating insulin. Unlike the physiologic signaling cascade that ensures efficient glucose uptake and metabolic regulation, insulin resistance reflects defects at multiple levels, ranging from impaired receptor signaling to alterations in intracellular metabolism and systemic inflammatory responses.

At the postreceptor level, one of the earliest and most consistent defects is the impairment of IRS function. In contrast to tyrosine phosphorylation, which propagates normal insulin signaling, serine/threonine (Ser/Thr) phosphorylation of IRS, particularly IRS1, acts as a negative regulatory mechanism. Under conditions of metabolic stress, various kinases such as JNK, IKKβ, ERK, mTOR/S6K1, and PKC target Ser/Thr residues on IRS proteins. This modification interferes with the ability of IRS to undergo tyrosine phosphorylation and alters its three-dimensional conformation (Hoehn et al., 2008).

Many conditions have been associated with increased Ser/Thr phosphorylation of IRS. Elevated free fatty acids can increase Ser/Thr phosphorylation of IRS, which hinders PI3K activation. This reduces Akt signaling, limiting GLUT4 translocation in muscle and weakening insulin's ability to suppress liver glucose production. Such alterations shift glucose handling toward hyperglycemia and compensatory hyperinsulinemia (Petersen & Shulman, 2018).

Inflammation and adipokine dysregulation also play central roles. In obesity, hypertrophied adipocytes and infiltrating immune cells release pro-inflammatory cytokines such as TNF-α and IL-6, which activate serine kinases including JNK and IKKβ. These kinases phosphorylate IRS proteins on inhibitory residues, thereby attenuating insulin signaling. At the same time, reduced secretion of adiponectin diminishes AMP-activated protein kinase (AMPK) activation, further impairing insulin sensitivity in muscle and liver. This inflammatory–adipokine axis represents a key link between excess adiposity and systemic insulin resistance (Hotamisligil, 2017).

Furthermore, mitochondrial dysfunction and ectopic lipid accumulation can exacerbate the problem. In skeletal muscle, reduced mitochondrial oxidative capacity leads to excess diacylglycerol and ceramide act as signaling molecules that trigger novel PKC isoforms, which directly inhibit IRS–PI3K signaling (Sergi et al., 2019). In the liver, similar mechanisms impede insulin's ability to suppress hepatic glucose production, while simultaneously driving lipogenesis (Shannon et al., 2021).

Clinical manifestation of insulin resistance

The clinical presentation of insulin resistance is often silent for years, eventually surfacing through a constellation of signs and conditions that share pathophysiological roots. The hallmark metabolic defect of insulin resistance is the inability of key tissues to respond to insulin. Clinically, this progressive dysregulation is identified as impaired fasting glucose (IFG) and impaired glucose tolerance (IGT), together termed prediabetes. This state represents a critical window where intervention can alter the disease course. Ultimately, the combination of chronic hyperglycemia and increased metabolic demand places profound stress on pancreatic β-cells. The eventual decline in β-cell function and insulin secretion marks the transition from prediabetes to overt T2DM.

Insulin resistance is also central to the metabolic syndrome, a cluster of interrelated risk factors including abdominal obesity, dyslipidemia (elevated triglycerides and reduced HDL-C), hypertension, and glucose intolerance. These features reflect the interplay between adipose tissue dysfunction, ectopic fat deposition, and systemic inflammation. The presence of metabolic syndrome not only increases the likelihood of developing T2DM but also markedly elevates cardiovascular risk (James et al., 2020).

Beyond glucose and lipid abnormalities, insulin resistance may also manifests in conditions such as polycystic ovary syndrome (PCOS) and non-alcoholic fatty liver disease (NAFLD). In NAFLD, impaired hepatic insulin signaling promotes hepatic steatosis and inflammation, while in PCOS, insulin resistance exacerbates hyperandrogenism and ovulatory dysfunction (Zhao et al., 2023). At the vascular level, insulin resistance promotes endothelial dysfunction, proinflammatory signaling, and atherogenesis, linking it directly to increased risk of myocardial infarction and stroke (Ke et al., 2023).

Statin-induced insulin resistance

Evidence linking statins and insulin resistance

Over the past decade, a growing body of clinical evidence has established a modest but consistent link between statin therapy and increased insulin resistance, as well as a higher incidence of new-onset type 2 diabetes (NOD). For example, a meta-analysis of 13 randomized trials found that statin use was associated with a 9% relative increase in the risk of developing diabetes. Despite being modest in scale, this risk corresponds to nearly one additional case of diabetes for every 255 patients receiving statin therapy over 4 years (Sattar et al., 2010).

Findings from individual large-scale trials have been consistent with this pattern. In the JUPITER trial, rosuvastatin was tested for primary prevention among patients with high C-reactive protein. The study found small but significant rises in HbA1c and in the incidence of physician-diagnosed diabetes. The excess risk was most pronounced in older participants and those with metabolic risk factors. Yet, significant reductions in cardiovascular events were also observed, stressing the necessity of balancing risks with therapeutic benefits (Ridker, 2009).

Evidence indicates that statins differ in their metabolic effects, with the risk of diabetes influenced by the specific agent, dosage, intensity, and overall treatment duration. Lipophilic, high-potency statins (including atorvastatin and rosuvastatin) are more commonly associated with worsening glycemic control, while hydrophilic statins (pravastatin and pitavastatin) show little to no adverse impact (Preiss et al., 2011; Rees-Milton et al., 2020). Results from meta-analysis showed that high-intensity regimens increased diabetes risk by approximately 12% more than moderate doses (Preiss et al., 2011). Moreover, longer duration and higher cumulative exposure of statin correlate with greater likelihood of impaired glucose metabolism and NOD (Ko et al., 2019).

Mechanisms of statin-induced insulin resistance

Although the development of T2DM in statin-treated patients is well documented, the precise mechanisms remain only partly understood. Current evidence suggests that both on-target effects, which is related to inhibition of the mevalonate pathway, and off-target effects on cellular metabolism contribute to this phenomenon. Understanding these mechanisms may help explain the increase in T2DM risk observed with statin therapy.

Disruption in insulin receptor signaling. When statins inhibit the mevalonate pathway, they not only reduce cholesterol synthesis but also decrease the generation of isoprenoid intermediates. FPP and GGPP are critical isoprenoid intermediates required for protein prenylation. Adequate prenylation ensures that these G-proteins are properly localized and capable of transmitting intracellular signals with precision. When levels of FPP or GGPP are reduced—as may occur with statin therapy—prenylation is impaired, leading to mislocalization of Rab and Rho family proteins. This misregulation can trigger aberrant activation of stress-sensitive kinases, including JNK and ERK. Once activated, these kinases phosphorylate IRS at serine residues, thereby interfering with its normal tyrosine phosphorylation mediated by the insulin receptor. As a result, the efficiency of insulin signal propagation through the PI3K–PDK1–Akt pathway is reduced (Fig. 27.5). This attenuation impairs GLUT4 translocation to the plasma membrane in skeletal muscle and adipose tissue, diminishing glucose uptake, while also weakening insulin's ability to suppress hepatic gluconeogenesis (Chen et al., 2024; Wang et al., 2022). In pancreatic β-cells, defective prenylation interferes with insulin granule exocytosis, diminishing insulin secretion capacity (Henriksbo et al., 2020).

Mitochondrial dysfunction. Statins have also been implicated in the disruption of mitochondrial function, primarily through the depletion of coenzyme Q10 (ubiquinone), a critical component of the electron transport chain. A deficiency of coenzyme Q10 disrupts oxidative phosphorylation, resulting in diminished ATP synthesis and elevated reactive oxygen species (ROS) formation. The excess ROS activate redox-sensitive kinases such as JNK and p38 MAPK, which phosphorylate IRS on inhibitory serine residues, thereby attenuating downstream insulin signaling. In skeletal muscle, statin-induced mitochondrial dysfunction diminishes fatty acid oxidation, resulting in the accumulation of lipid intermediates such as DAG and ceramides. These metabolites activate PKC isoforms, which further disrupt insulin receptor signaling pathways (Fig. 27.6). Thereby, these mitochondrial

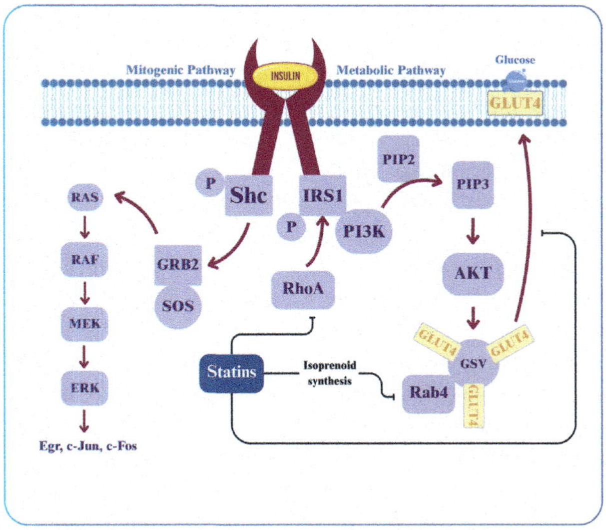

FIGURE 27.5 Statin-induced disruption of insulin signaling. Statins inhibit HMG-CoA reductase and deplete isoprenoid intermediates, impairing prenylation of small GTP-binding proteins such as RhoA and Rab4. This leads to activation of stress kinases that promote IRS serine phosphorylation, attenuate PI3K–Akt signaling, and reduce GLUT4 translocation. The combined effect is diminished glucose uptake in skeletal muscle and adipose tissue and impaired suppression of hepatic gluconeogenesis. *HMG-CoA*, 3-hydroxy-3-methylglutaryl–coenzyme A; *RhoA*, Ras homolog family member A; *Rab4*, Ras-related protein Rab-4; *IRS*, insulin receptor substrate; *PI3K*, phosphoinositide 3-kinase; *Akt*, protein kinase B; *GLUT4*, glucose transporter type 4.

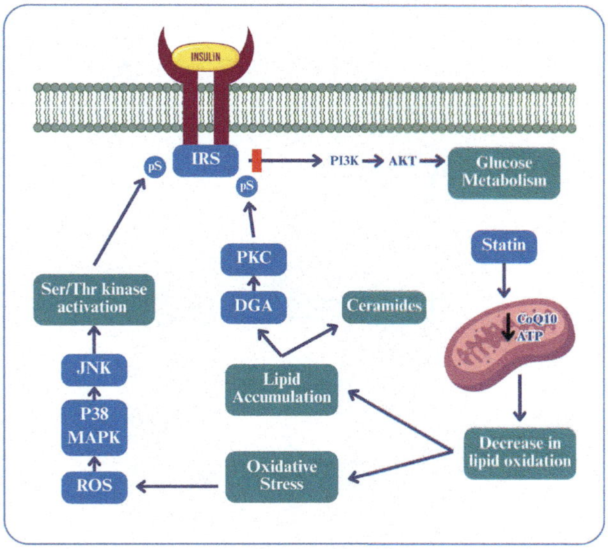

FIGURE 27.6 Statin-induced mitochondrial dysfunction. Statins deplete CoQ10, impairing oxidative phosphorylation and increasing ROD production. ROS activate stress kinases such as JNK and p38 MAPK, which in turn inhibit IRS signaling. In parallel, reduced fatty acid oxidation in muscle leads to the accumulation of lipotoxic intermediates such as DAG and ceramides. These metabolites activate PKC isoforms, promoting serine phosphorylation of IRS. This modification prevents effective recruitment of PI3K, thereby attenuating downstream insulin signaling. *CoQ10*, coenzyme Q10; *ROS*, reactive oxygen species; *JNK*, c-Jun N-terminal kinase; *p38 MAPK*, p38 mitogen-activated protein kinase; *IRS*, insulin receptor substrate; *DAG*, diacylglycerol; *PKC*, protein kinase C.

perturbations contribute to insulin resistance by coupling energy deficit with lipotoxic stress. Supplementation with coenzyme Q10 has been proposed as a potential strategy to alleviate the diabetogenic effects of statins, although evidence remains preliminary (Han et al., 2025).

Altered adipose tissue and adipokine regulation. Statins can impair adipocyte differentiation by inhibiting the mevalonate pathway, which reduces the expression of key adipogenic transcription factors such as PPARγ (Nakata et al., 2006). As a result, adipocytes remain smaller and less efficient at storing lipids, leading to impaired adipose tissue expandability. This promotes ectopic lipid deposition in nonadipose tissues such as liver and skeletal muscle, where it interferes with insulin signaling and contributes to systemic insulin resistance. Statins have also been shown to alter adipose tissue signaling, including circulating adiponectin levels, an anti-inflammatory and insulin-sensitizing hormone (Megawaty et al., 2023).

In addition, statin use has been associated with elevated levels of pro-inflammatory cytokines (IL-6, TNF-α), leptin, and saturated free fatty acids, which activate inhibitory molecules such as suppressor of cytokine signaling (SOCS) proteins and JNK (Fig. 27.7). These, in turn, promote serine phosphorylation of IRS, which diminishes its ability to recruit PI3K. As a result, downstream activation of the PDK1–AKT signaling pathway is reduced, impairing key metabolic processes of insulin (Smith & Kahn, 2016).

Chronic low-grade inflammation. Although statins are widely recognized for their antiinflammatory effects, under certain conditions they can paradoxically enhance inflammatory signaling, contributing to metabolic dysfunction. Statins inhibition on the mevalonate pathway may lead to a reduction in isoprenoid intermediates. Under normal conditions, isoprenoids are required for the prenylation and functional regulation of small GTPases, such as Rho proteins, which exert inhibitory control over inflammasome activation. Depletion of these intermediates removes this regulatory brake, thereby facilitating activation of inflammasomes, particularly the NLRP3 inflammasome. In parallel, statin-induced alterations in innate immune signaling can enhance toll-like receptor (TLR) activity, further amplifying inflammatory responses. Once activated, inflammasomes and TLRs promote the production of proinflammatory cytokines, including IL-1β (Koushki et al., 2021). This inflammatory activation reinforces defects in insulin receptor signaling and accelerates metabolic deterioration, particularly in patients with obesity or preexisting metabolic syndrome.

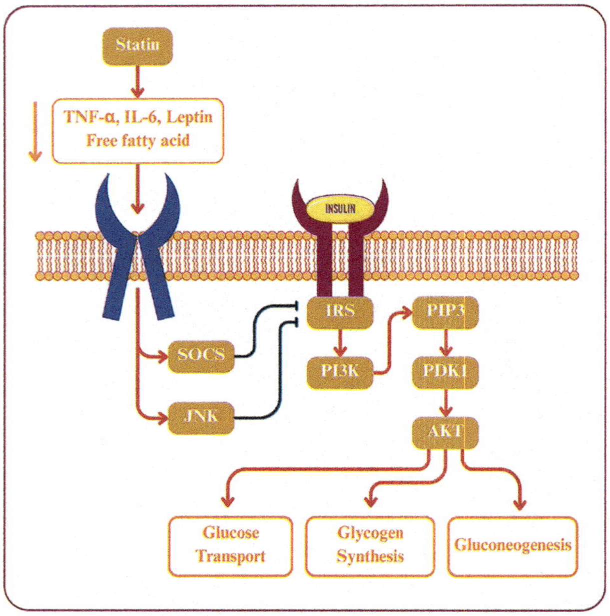

FIGURE 27.7 Statin effects on inflammatory adipokines. Pro-inflammatory signals, including TNF-α, IL-6, leptin, and saturated free fatty acids, activate inhibitory molecules such as SOCS and JNK. This promotes serine phosphorylation of IRS, impairing PI3K recruitment and thereby reducing activation of the downstream PDK1–AKT pathway. The PI3K–AKT signaling cascade normally regulates key metabolic processes, including glucose uptake, glycogen synthesis, and suppression of gluconeogenesis. *TNF-α*, tumor necrosis factor-α; *IL-6*, interleukin-6; *SOCS*, suppressor of cytokine signaling; *JNK*ç-Jun N-terminal kinase; *IRS*, insulin receptor substrate; *PI3K*, phosphoinositide 3-kinase; *PDK1*, 3-phosphoinositide-dependent protein kinase-1; *AKT*, protein kinase B.

Other hormones

Even though this chapter focuses on statins and insulin resistance, it is important to recognize that insulin does not act in isolation. Statin-induced insulin resistance illustrates how disruptions in one pathway can ripple across the broader endocrine network. Insulin's regulation of glucose uptake, hepatic glucose production, and lipid metabolism is tightly coordinated with other hormones, including glucagon, incretins, catecholamines, cortisol, and thyroid hormones. When insulin signaling in muscle, liver, or adipose tissue is impaired, these interconnected pathways are affected, leading to alterations in energy homeostasis, appetite regulation, and systemic metabolism.

At the cellular level, insulin resistance limits glucose uptake in skeletal muscle and adipose tissue, reducing substrate availability for glycolysis and the tricarboxylic acid (TCA) cycle. This results in decreased ATP production, the cell's primary energy currency. Mitochondrial function is further compromised, with reduced biogenesis, impaired electron transport chain activity, and increased generation of ROS. These changes not only lower ATP output but also promote oxidative stress, contributing to cellular dysfunction and reinforcing insulin resistance in a vicious cycle that disrupts energy balance.

Insulin resistance also influences appetite regulation through interactions with leptin, ghrelin, and gut-derived incretins. Normally, insulin signals the hypothalamus to reduce food intake and promote satiety. When this signaling is impaired, these effects are blunted, potentially increasing caloric intake despite sufficient energy stores. Altered leptin sensitivity and disrupted gut hormone responses further exacerbate postprandial glucose handling, reinforcing systemic energy imbalance.

In addition to its role in glucose regulation, insulin resistance exerts widespread systemic effects. Persistent elevations in circulating glucose and free fatty acids, coupled with chronic low-grade inflammation and disrupted hormonal signaling, create a milieu that affects multiple organ systems. Cardiovascular function is compromised through endothelial dysfunction and accelerated atherogenesis, hepatic metabolism is altered by unchecked gluconeogenesis and steatosis, and renal physiology is influenced by increased sodium retention and glomerular stress. This highlights how a defect in one hormone's action can reshape whole-body metabolic homeostasis.

Applications of the material

This chapter has discussed the key mechanisms that underlie statin-induced dysglycemia, emphasizing how perturbations in insulin signaling, lipid handling, mitochondrial function, and inflammatory pathways converge to impair glucose homeostasis. Understanding these molecular disruptions provides more than just mechanistic insight; it creates a foundation for therapeutic innovation. By identifying the nodes at which HMG-CoA reductase inhibition exerts unintended effects—whether through altered protein prenylation, stress kinase activation, or changes in adipose tissue signaling—researchers and clinicians can begin to explore pharmacological strategies that specifically counteract these processes. Such approaches may include adjunctive therapies that preserve mitochondrial function, enhance adipokine signaling, or dampen inflammatory responses, thereby mitigating the diabetogenic potential of statins while maintaining their cardiovascular benefits.

For example, supplementation with coenzyme Q10 has been investigated as a way to restore mitochondrial bioenergetics impaired by statin-induced depletion. Targeting stress kinase pathways, such as JNK or p38 MAPK, could help preserve insulin receptor signaling by preventing inhibitory phosphorylation of IRS proteins. Likewise, agents that modulate adipokines (e.g., adiponectin enhancers, PPARγ agonists, or selective modulators of adipose tissue differentiation) may counteract statin-related alterations in adipose function. Anti-inflammatory interventions, whether through IL-1β antagonism, NF-κB inhibition, or lifestyle-based strategies such as diet and exercise, could also dampen the chronic low-grade inflammation that exacerbates insulin resistance in susceptible individuals. In this way, mechanistic knowledge serves as a bridge from molecular observation to clinical translation, guiding the development of more balanced interventions in cardiometabolic care.

Mini dictionary of terms

Adipokines—Bioactive molecules secreted by adipose tissue, including leptin and adiponectin, that modulate insulin sensitivity and inflammation.

Glucose Transporter Type 4 (GLUT4)—An insulin-sensitive glucose transporter that mediates cellular glucose uptake, especially in muscle and adipose tissue.

HMG-CoA Reductase—Regulatory enzyme of the mevalonate pathway responsible for catalyzing the transformation of HMG-CoA into mevalonate during cholesterol synthesis.

Hydrophilic Statins—Statins with polar chemical substituents that limit passive membrane diffusion.

Lipophilic Statins—Statins with nonpolar chemical structures that readily diffuse across cell membranes.

Insulin Resistance—A metabolic condition in which peripheral tissues such as the liver, skeletal muscle, and adipose tissue respond inadequately to insulin.

Insulin Receptor Substrate (IRS)—Adaptor proteins phosphorylated by the insulin receptor, which mediate downstream activation of the PI3K–Akt pathway.

New-Onset Diabetes (NOD)—The development of diabetes during statin therapy, thought to result from impaired insulin signaling and β-cell dysfunction.

Summary points

- Statins are first line therapies for lowering LDL-C and reducing cardiovascular morbidity and mortality, but it is associated with impaired glucose tolerance and increased risk of new-onset diabetes.
- Statins inhibit HMG-CoA reductase, suppressing not only cholesterol biosynthesis but also isoprenoid production, which in turn activates stress-sensitive kinases.
- Statins can reduce insulin sensitivity by disrupting insulin signaling through PI3K–PDK1-Akt pathway, which leading to impaired hepatic, adipose, and skeletal muscle metabolic function.
- The diabetogenic risk appears to be influenced by statin type, dosage, treatment duration, and patient characteristics.
- Understanding the molecular and metabolic pathways of statins is crucial to preserve their cardiovascular benefits while minimizing the adverse metabolic consequences.

Declaration of AI and AI-assisted technologies in the writing process

During the preparation of this work the author used ChatGPT to improve readability and language polishing. After using this tool/service, the author reviewed and edited the content as needed and takes full responsibility for the content of the publication.

References

Ahmadi, Y., Ghorbanihaghjo, A., & Argani, H. (2017). The effect of statins on the organs: Similar or contradictory? *Journal of Cardiovascular and Thoracic Research, 9*(2), 64–70.

Aronoff, S. L., Berkowitz, K., Shreiner, B., & Want, L. (2004). Glucose metabolism and regulation: Beyond insulin and glucagon. *Diabetes Spectrum, 17*(3), 183–190.

Averbukh, L. D., Turshudzhyan, A., Wu, D. C., & Wu, G. Y. (2022). Statin-induced liver injury patterns: A clinical review. *Journal of Clinical and Translational Hepatology, 10*(3), 543–552.

Backes, J. M., Ruisinger, J. F., Gibson, C. A., & Moriarty, P. M. (2017). Statin-associated muscle symptoms-managing the highly intolerant. *Journal of Clinical Lipidology, 11*(1), 24–33.

Barter, P. J., Brandrup-Wognsen, G., Palmer, M. K., & Nicholls, S. J. (2010). Effect of statins on HDL-C: A complex process unrelated to changes in LDL-C: Analysis of the VOYAGER database. *Journal of Lipid Research, 51*(6), 1546–1553.

Blais, J. E., Wei, Y., Yap, K. K. W., Alwafi, H., Ma, T.-T., Brauer, R., et al. (2021). Trends in lipid-modifying agent use in 83 countries. *Atherosclerosis, 328*, 44–51.

Chen, J., Zha, L., Xu, J., & Chen, G. (2024). Prolonged treatment of compactin, atorvastatin, lovastatin or simvastatin decreased glucose consumption and expressions of proteins for glucose metabolism and insulin signaling in L6 skeletal muscle cells. *Food, Nutrition and Health, 1*(1), 5.

Climent, E., Benaiges, D., & Pedro-Botet, J. (2021). Hydrophilic or lipophilic statins? *Frontiers in Cardiovascular Medicine, 8*, 687585.

Collins, R., Reith, C., Emberson, J., Armitage, J., Baigent, C., Blackwell, L., et al. (2016). Interpretation of the evidence for the efficacy and safety of statin therapy. *The Lancet, 388*(10059), 2532–2561.

Endo, A. (1979). Monacolin K, a new hypocholesterolemic agent produced by a Monascus species. *Journal of Antibiotics (Tokyo), 32*(8), 852–854.

Grundy, S. M., Stone, N. J., Bailey, A. L., Beam, C., Birtcher, K. K., Blumenthal, R. S., et al. (2019). 2018 AHA/ACC/AACVPR/AAPA/ABC/ACPM/ADA/AGS/APhA/ASPC/NL/PCNA guideline on the management of blood cholesterol: A report of the American college of cardiology/american heart association task force on clinical practice guidelines. *Circulation, 139*(25), e1082–e1143.

Haeusler, R. A., McGraw, T. E., & Accili, D. (2018). Biochemical and cellular properties of insulin receptor signalling. *Nature Reviews Molecular Cell Biology, 19*(1), 31–44.

Han, X., Liu, J., Gu, Y., Li, Y., Zhang, W., Lv, N., et al. (2025). Diabetes risks of statin therapy-coenzyme Q10 May help. *Reviews in Cardiovascular Medicine, 26*(2), 26437.

Henquin, J. C., Ishiyama, N., Nenquin, M., Ravier, M. A., & Jonas, J. C. (2002). Signals and pools underlying biphasic insulin secretion. *Diabetes, 51*(S1), S60–S67.

Henriksbo, B. D., Tamrakar, A. K., Phulka, J. S., Barra, N. G., & Schertzer, J. D. (2020). Statins activate the NLRP3 inflammasome and impair insulin signaling via p38 and mTOR. *American Journal of Physiology-Endocrinology and Metabolism, 319*(1), E110–E116.

Hirota, T., Fujita, Y., & Ieiri, I. (2020). An updated review of pharmacokinetic drug interactions and pharmacogenetics of statins. *Expert Opinion on Drug Metabolism and Toxicology, 16*(9), 809–822.

Hoehn, K. L., Hohnen-Behrens, C., Cederberg, A., Wu, L. E., Turner, N., Yuasa, T., et al. (2008). IRS1-independent defects define major nodes of insulin resistance. *Cell Metabolism, 7*(5), 421–433.

Hotamisligil, G. S. (2017). Foundations of immunometabolism and implications for metabolic health and disease. *Immunity, 47*(3), 406–420.

James, M., Varghese, T. P., Sharma, R., & Chand, S. (2020). Association between metabolic syndrome and diabetes mellitus according to international diabetic Federation and national cholesterol education program adult treatment panel III criteria: A cross-sectional study. *Journal of Diabetes and Metabolic Disorders, 19*(1), 437–443.

Ke, Z., Huang, R., Xu, X., Liu, W., Wang, S., Zhang, X., et al. (2023). Long-term high level of insulin resistance is associated with an increased prevalence of coronary artery calcification: The CARDIA study. *Journal of the American Heart Association, 12*(11), e028985.

Ko, M. J., Jo, A. J., Kim, Y. J., Kang, S. H., Cho, S., Jo, S. H., et al. (2019). Time- and dose-dependent association of statin use with risk of clinically relevant new-onset diabetes mellitus in primary prevention: A nationwide observational cohort study. *Journal of the American Heart Association, 8*(8), e011320.

Koushki, K., Shahbaz, S. K., Mashayekhi, K., Sadeghi, M., Zayeri, Z. D., Taba, M. Y., et al. (2021). Anti-inflammatory action of statins in cardiovascular disease: The role of inflammasome and toll-like receptor pathways. *Clinical Review in Allergy and Immunology, 60*(2), 175–199.

Law, M., & Rudnicka, A. R. (2006). Statin safety: A systematic review. *American Journal of Cardiology, 97*(8a), 52c–60c.

Le, T. K. C., Dao, X. D., Nguyen, D. V., Luu, D. H., Bui, T. M. H., Le, T. H., et al. (2023). Insulin signaling and its application. *Frontiers in Endocrinology, 14*, 1226655.

Lembo, M., Trimarco, V., Izzo, R., Pacella, D., Jankauskas, S. S., Gallo, P., et al. (2025). Statin-induced risk of diabetes does not reduce cardiovascular benefits in primary prevention: A 6-year propensity-score matched study in a large population. *Cardiovascular Diabetology, 24*(1), 233.

Liao, J. K., & Laufs, U. (2005). Pleiotropic effects of statins. *Annual Review of Pharmacology and Toxicology, 45*, 89–118.

McTaggart, F., & Jones, P. (2008). Effects of statins on high-density lipoproteins: A potential contribution to cardiovascular benefit. *Cardiovascular Drugs Therapy, 22*(4), 321–338.

Megawaty, M., Muzayyidah, M., Djabir, Y. Y., Mappangara, I., & Palinggi, B. (2023). Comparison of atorvastatin and simvastatin modulation on adiponectin and insulin resistance in non-diabetic dyslipidemic patients. *Eastern Journal of Medicine, 28*(2), 237–243.

Murphy, C., Deplazes, E., Cranfield, C. G., & Garcia, A. (2020). The role of structure and biophysical properties in the pleiotropic effects of statins. *International Journal of Molecular Sciences, 21*(22), 8745.

Nakata, M., Nagasaka, S., Kusaka, I., Matsuoka, H., Ishibashi, S., & Yada, T. (2006). Effects of statins on the adipocyte maturation and expression of glucose transporter 4 (SLC2A4): Implications in glycaemic control. *Diabetologia, 49*(8), 1881–1892.

Petersen, M. C., & Shulman, G. I. (2018). Mechanisms of insulin action and insulin resistance. *Physiological Reviews, 98*(4), 2133–2223.

Preiss, D., Seshasai, S. R. K., Welsh, P., Murphy, S. A., Ho, J. E., Waters, D. D., et al. (2011). Risk of incident diabetes with intensive-dose compared with moderate-dose Statin therapy: A meta-analysis. *Journal of the American Medical Association, 305*(24), 2556–2564.

Rees-Milton, K. J., Norman, P., Babiolakis, C., Hulbert, M., Turner, M. E., Berger, C., et al. (2020). Statin use is associated with insulin resistance in participants of the Canadian multicentre osteoporosis study. *Journal of the Endocrine Society, 4*(8), bvaa057.

Reith, C., Preiss, D., Blackwell, L., Emberson, J., Spata, E., Davies, K., et al. (2024). Effects of statin therapy on diagnoses of new-onset diabetes and worsening glycaemia in large-scale randomised blinded statin trials: An individual participant data meta-analysis. *Lancet Diabetes Endocrinol, 12*(5), 306–319.

Ridker, P. M. (2009). The JUPITER trial. *Circulation: Cardiovascular Quality and Outcomes, 2*(3), 279–285.

Röder, P. V., Wu, B., Liu, Y., & Han, W. (2016). Pancreatic regulation of glucose homeostasis. *Experimental and Molecular Medicine, 48*(3), e219.

Sadowska, A., Osiński, P., Roztocka, A., Kaczmarz-Chojnacka, K., Zapora, E., Sawicka, D., et al. (2023). Statins-from fungi to pharmacy. *International Journal of Molecular Sciences, 25*(1), 466.

Saltiel, A. R. (2021). Insulin signaling in health and disease. *Journal of Clinical Investigation, 131*(1), e142241.

Sattar, N., Preiss, D., Murray, H. M., Welsh, P., Buckley, B. M., de Craen, A. J., et al. (2010). Statins and risk of incident diabetes: A collaborative meta-analysis of randomised statin trials. *Lancet, 375*(9716), 735–742.

Schmidt, H. M., Jarrett, K. E., de Aguiar Vallim, T. Q., & Tarling, E. J. (2025). Pathways and molecular mechanisms governing LDL receptor regulation. *Circulation Research, 136*(8), 902–919.

Sergi, D., Naumovski, N., Heilbronn, L. K., Abeywardena, M., O'Callaghan, N., Lionetti, L., et al. (2019). Mitochondrial (dys) function and insulin resistance: From pathophysiological molecular mechanisms to the impact of diet. *Frontiers in Physiology, 10*, 532.

Shah, N., Lan, Z., Brown, C. J., Martin, S. S., & Turchin, A. (2025). Impact of statin nonacceptance on cardiovascular outcomes in patients with diabetes. *Journal of the American Heart Association, 14*(11), e040464.

Shannon, C. E., Ragavan, M., Palavicini, J. P., Fourcaudot, M., Bakewell, T. M., Valdez, I. A., et al. (2021). Insulin resistance is mechanistically linked to hepatic mitochondrial remodeling in non-alcoholic fatty liver disease. *Molecular Metabolism, 45*, 101154.

Shi, Z., & Han, S. (2025). Personalized statin therapy: Targeting metabolic processes to modulate the therapeutic and adverse effects of statins. *Heliyon, 11*(1), e41629.

Smith, U., & Kahn, B. B. (2016). Adipose tissue regulates insulin sensitivity: role of adipogenesis, de novo lipogenesis and novel lipids. *Journal of Internal Medicine, 280*(5), 465–475.

Stone, N. J., Robinson, J. G., Lichtenstein, A. H., Merz, C. N. B., Blum, C. B., Eckel, R. H., et al. (2014). 2013 ACC/AHA guideline on the treatment of blood cholesterol to reduce atherosclerotic cardiovascular risk in adults. *Journal of the American College of Cardiology, 63*(25), 2889–2934.

Svec, A., & Adameova, A. (2023). Facts and ideas on statins with respect to their lipophilicity: A focus on skeletal muscle cells and bone besides known cardioprotection. *Molecular and Cellular Biochemistry, 478*(8), 1661–1667.

Tokarz, V. L., MacDonald, P. E., & Klip, A. (2018). The cell biology of systemic insulin function. *Journal of Cell Biology, 217*(7), 2273–2289.

Wang, L., Zhu, L., Zheng, Z., Meng, L., Liu, H., Wang, K., et al. (2022). Mevalonate pathway orchestrates insulin signaling via RAB14 geranylgeranylation-mediated phosphorylation of AKT to regulate hepatic glucose metabolism. *Metabolism, 128*, 155120.

Ward, N. C., Watts, G. F., & Eckel, R. H. (2019). Statin toxicity. *Circulation Research, 124*(2), 328–350.

Zhao, H., Zhang, J., Cheng, X., Nie, X., & He, B. (2023). Insulin resistance in polycystic ovary syndrome across various tissues: An updated review of pathogenesis, evaluation, and treatment. *Journal of Ovarian Research, 16*(1), 9.

Chapter 28

Linking mitochondrial integrity and insulin resistance in response to pharmacological agents

Thobeka Madide[1, 2] and Ntethelelo Sibiya[2]
[1]*Department of Human Biology and Integrated Pathology, Medical School, Nelson Mandela University, Gqeberha, South Africa;* [2]*Division of Pharmacology, Faculty of Pharmacy, Rhodes University, Makhanda, South Africa*

List of abbreviations

ACTH Adrenocorticotropic hormone
AGES Advanced glycation end products
ALT Alanine transaminase
AMP Adenosine monophosphate
AST Aspartate transaminase
ATP Adenosine triphosphate
Cbl Casitas-B-lineage lymphoma
CRH Corticotropin-releasing hormone
CTLA-4 Cytotoxic T-lymphocyte-associated protein 4
ER Endoplasmic reticulum
ERK Extracellular signal-regulated kinase
ETC Electron transport chain
$FADH_2$ Flavin adenine dinucleotide
FDA Food and drug administration
FFA Free fatty acid
FOXO Forkhead box, sub-group O
GLUT-4 Glucose transporter type 4
GnRH Gonadotropin-releasing hormone
Gpx Glutathione peroxidase
GSH Glutathione
GSK3 Glycogen synthase kinase-3
GSSG Oxidized glutathione
H_2O_2 Hydrogen peroxide
HBV Hepatitis B virus
HFD High fat diet
HIV-1 Human immunodeficiency virus type-1
ICP Immune checkpoint inhibitors
IR Insulin receptor
IRS Insulin receptor substrate
MEK Methyl ethyl ketone
mtDNA Mitochondrial DNA
mTOR Mammalian target of rapamycin
NADH Nicotinaminde adenine dinucleotide + hydrogen
NADPH Nicotinaminde adenine dinucleotide phosphate
NFkB Nuclear factor kappa-light-chain-enhancer of activated B cells
NOS Nitric oxide synthase
NRTI's Nucleoside reverse transcriptase inhibitors
O_2 Superoxide anion
OXPHOS Oxidative phosphorylation
PARP Poly [ADP-ribose] polymerase
PD-1 Programmed cell death protein 1
PD-L1 Programmed death-ligand 1
PDK Pyruvate dehydrogenase kinase
PI3K Phosphatidylinositol 3-kinase
PIP2 Phosphatidylinositol 4,5-biphosphate
PIP3 Phosphatidylinositol (3,4,5)-triphosphate
PreP Pre-exposure prophylaxis
RAF Rapidly activated fibrosarcoma
RAS Rat sarcoma
ROS Reactive oxygen species
SHC Src homology and collagen protein
SOD Superoxide dismutase
SREB Sterol regulatory element-binding protein
T2DM Type II diabetes mellitus
TCA Tricarboxylic acid
TDF Tenofovir disoproxil fumarate
TKI Tyrosine kinase inhibitor
TNF-alpha Tumor necrosis factor alpha
TSH Thyroid stimulating hormone

Introduction

Type 2 Diabetes Mellitus (T2DM) is associated with reduced insulin sensitivity and insulin resistance in peripheral tissues (Sibiya et al., 2023). Increasing evidence suggests that T2DM accounts for a majority of cases, making it the most predominant type. In this condition, adipose tissue promotes insulin resistance through various inflammatory mechanisms, including increased

Hormones in Neuroscience. https://doi.org/10.1016/B978-0-443-30263-3.00026-9

free fatty acid (FFA) release and adipokine dysregulation. Furthermore, the elevated blood glucose concentrations increase the production of reactive oxygen species (ROS), which in turn may alter a vast array of downstream pathways (Cole & Florez, 2020). Emerging evidence strongly links mitochondrial impairments with the development of insulin resistance (Montgomery & Turner, 2015).

Mitochondria are responsible for energy production in cells via biochemical reactions in glucose and lipid metabolism, by coupling the tricarboxylic acid (TCA) cycle with oxidative phosphorylation (OXPHOS). Therefore, mitochondrial functioning is imperative to metabolic health and cellular fate. Mitochondrial dysfunction has been discovered to play a significant role in the pathogenesis of various diseases such as T2DM, kidney disease, inflammation, and cardiomyopathy (Sibiya et al., 2023). Mitochondrial impairments have been implicated in the development of insulin resistance. Several commonly utilized pharmacological agents impair mitochondrial function in the skeletal muscle, liver, central nervous system and kidneys, which results in adverse effects and toxicities. Mito toxic agents can act directly on the mitochondria, by inhibiting mitochondrial DNA (mtDNA), electron transport chain (ETC), ATP synthesis and transcription. Drugs can also exert their toxicity by elevating ROS generation, resulting in mitochondrial membrane depolarization, decreased antioxidant production and mtDNA mutations.

Nucleoside reverse transcriptase inhibitors (NRTI's), lipid regulators, antibiotics, antidepressants and anticancer drugs are amongst the most widely used pharmaceuticals, globally (Jacob et al., 2020). Previous research has demonstrated the adverse effects of these drug classes. This review will focus on simvastatin, tenofovir disoproxil fumarate (TDF), tramadol, doxorubicin, amitriptyline, and valproic acid. These drugs have been shown to induce mitochondrial dysfunction. Recent efforts have focused on targeting the mitochondria to attenuate mitochondrial impairment and researchers suggest that this trajectory could have great therapeutic potential.

With the increased burden of insulin resistance prevalence globally, simultaneously with the rise of mitochondrial toxicity—it is important to establish how drug-induced mitochondrial dysfunction may compromise insulin sensitivity. This could further provide insights on the role of frequently prescribed pharmacotherapeutics on the risk of developing T2DM. The reviewed drugs are used in the management of various diseases, which affect a large population. Understanding these interactions can possibly also identify other potential metabolic defects, that can arise during the use of these drugs. Through a heightened understanding, potential agents aiming to prevent or mitigate such defects could be pursued, with a goal to avert mitochondrial dysfunction, ultimately improving insulin sensitivity. Therefore, we aim to review the selected drugs which exert adverse effect on the mitochondria, the mechanisms of action involved. Most importantly, the goal is directed at identifying potential links between drug-induced mitochondrial toxicity and insulin resistance development.

Insulin signaling pathway

Insulin is an anabolic peptide hormone produced by beta cells of the Islets of Langerhans in the pancreas, comprised of A and B chains which are stabilized by disulphide bonds. Although the main primary role of insulin is regulating blood glucose concentrations, the maintenance of other metabolic processes is also a responsibility of this hormone. Insulin regulates blood glucose concentrations by regulating the whole-body carbohydrate, protein and lipid metabolism. Following nutrient uptake, blood glucose concentrations rise, stimulating insulin secretion and release by the B-cells. These cells have the capacity to sense circulating glucose levels and secrete the relevant amounts of insulin (Lee et al., 2022). Insulin triggers glucose uptake, glycogenesis, lipogenesis and skeletal muscle protein synthesis.

Insulin binds to the insulin receptor (IR) expressed by the cell membrane, primarily in the skeletal muscle, adipose tissue and liver, and many various other tissues (Fig. 28.1). Through binding on its receptor, insulin activates multidirectional impacts on the metabolism, proliferation and survival of target cells. The insulin receptor is a dimeric protein that has 2 alpha and beta subunits that are connected by a disulphide bond. The alpha subunits are located extracellularly and the beta subunits inside the cell. Binding of insulin initiates a conformational change in its receptor that results in the autophosphorylation of its tyrosine residues, which amplifies the kinase activity of the receptor. Thereafter, the phosphorylation of the insulin receptor substrate (IRS) and Casitas-B-lineage lymphoma (Cbl) and its associated proteins, enables binding to the intracellular receptor sites (Baghaie et al., 2023).

Once the IR1 is phosphorylated, it recruits signaling molecules to generate the insulin signaling response and promotes the activation of Phosphoinositide-3-kinases (PI3K). PI3K further phosphorylates Akt which results in the mobilization of GLUT-4 containing vesicles. The fusion of these vesicles with the with the cell membrane result in the translocation of GLUT-4, which facilitate insulin-mediated glucose uptake particularly in the skeletal muscle and adipose tissue.

Insulin resistance

Insulin resistance results from numerous mechanisms involving a decreased IR expression and alteration of

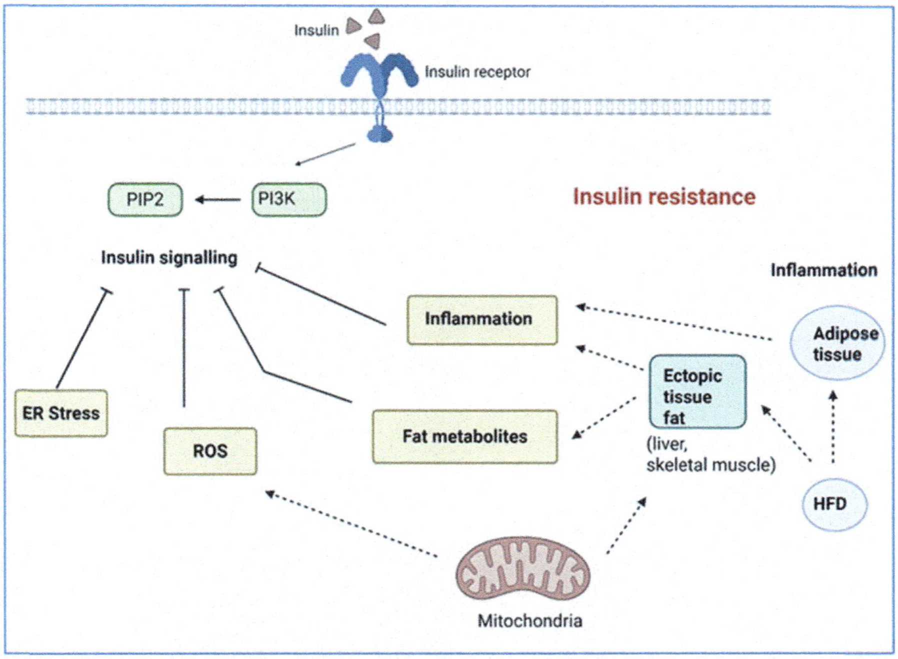

FIGURE 28.1 Potential mechanisms for insulin resistance. Endoplasmic reticulum (ER) stress, ROS and inflammation have been proposed to explain the mechanism responsible for obesity-induced insulin resistance. PIP2, phosphatidylinositol-4,5-bisphosphate; PIP3, phosphatidylinositol-3,4,5-trisphosphate; HFD, high-fat diet. *Adapted from Lee, S.-H., Park, S.-Y., & Choi, C. S. (2022). Insulin resistance: From mechanisms to therapeutic strategies.* Diabetes & Metabolism Journal, *46(1), 15–37.*

intracellular signaling cascades (Fig. 28.1). Insulin resistance disrupts the uptake of glucose in skeletal muscles, induces gluconeogenesis in the liver and promotes lipolysis in adipose tissue. Many cytokines and inflammatory mediators such as Tumor necrosis factor-α (TNF-α) and interleukins, are upregulated in insulin resistance. TNF-α over expression induces lipolysis which allows FFA release which in turn activates the Nuclear factor kappa-light-chain-enhancer of activated B cells (NFkB) pathways, which further increases TNF-α expression and other pro-inflammatory cytokines (de Mello et al., 2018). FFA's are also excessively released in this state and they act on metabolic tissues therefore contributing to metabolic syndrome (Gutiérrez-Rodelo et al., 2017). This results in chronic systematic inflammation which is considered to mediate insulin resistance. Oxidative stress has also been demonstrated to central to insulin resistance development, with increased free radicals and reduced antioxidants in this environment. ROS such as the oxygen radical and hydrogen peroxide (H_2O_2) are reactive intermediates of molecular oxygen that are physiologically formed in cells as by-products of cellular metabolism or as toxic molecules. ROS can alter and damage many intracellular molecules including DNA, lipids and proteins. Studies have highlighted reduced mitochondrial function in the pathophysiology of insulin resistance in part through increased generation of ROS.

Mitochondrial dysfunction and insulin resistance

The mitochondria have remained one of the most essential organelles in maintaining cellular homeostasis. Their biological roles in cells such as the generation of energy, proliferation and metabolic adaptations are required for the functioning of all cells. Mitochondria generate energy via OXPHOS by using glycolysis-derived pyruvate, fatty acids and amino acids. Thereafter, electrons are passed from complex I to IV in the ETC, and in turn, generating ATP. Defects and dysregulation of the mitochondria have emerged as having key roles in cytopathological mechanisms in diseases such as cancer, diabetes, and metabolic syndromes (Sorrentino et al., 2018).

Although mitochondrial dysfunction usually refers to the inability of the mitochondria to produce sufficient ATP levels, the term is also used to define the maladaptive physiological responses of the mitochondria that undergo certain deviations (de Mello et al., 2018). These include calcium buffering, mutations in mtDNA, defective iron transports, changes in mitochondrial size and morphology, ROS overproduction and apoptosis. Mitochondrial dysfunction has been implicated in the pathology of various diseases. The role of this organelle in the maintenance of cellular homeostasis and energy generation, is altered in this state. Metabolic derangements of nutrient

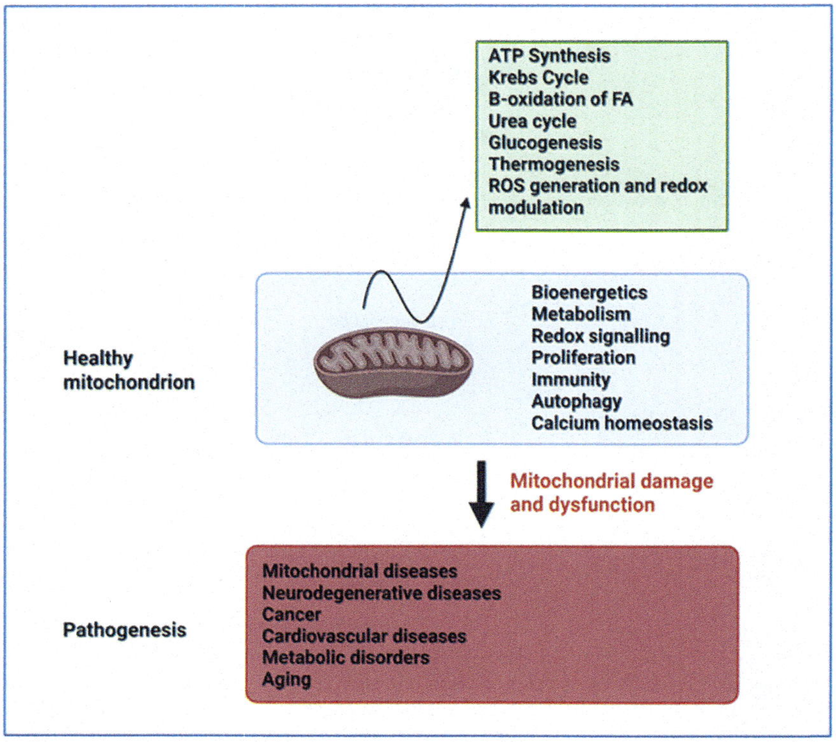

FIGURE 28.2 Representation of the involvement of mitochondria in health and disease. *Adapted from Apostolova, N., & Victor, V. M. (2015). Molecular strategies for targeting antioxidants to mitochondria: Therapeutic implications.* Antioxidants & Redox Signaling, *22(8), 686–729.*

signal input, energy production and or oxidative respiration results in mitochondrial dysfunction (Fig. 28.2) (Prasun, 2020).

High substrate levels overwhelm the Krebs cycle and the mitochondrial respiratory chain, and in turn induces mitochondrial dysfunction. Chronic excessive energy conditions also induce ROS production, oxidative stress and inflammation, which may further lead to mitochondrial dysfunction. Compromised mitochondrial function leads to decreased mitochondrial biogenesis and beta-oxidation, which compromises insulin sensitivity. Studies have also shown that mitochondrial fission and fusion proteins are downregulated in insulin resistance, leading to fragmentation of the mitochondrial architecture. Mitochondrial impairments have also been implicated in detect in metabolic control, thus leading to the development of insulin resistance (Bhatia et al., 2020; Montgomery & Turner, 2015).

Drug induced-mitochondrial dysfunction and insulin resistance

Although drugs are used for their therapeutic effects, it is imperative to recognize their adverse effects. These unfavorable effects are the reason why some drugs are removed from the market after they have been approved by the Food and Drug Administration (FDA). The mitochondria have been shown to be either a primary or secondary target by various drugs, and its impairment plays a role in drug-induced adverse effects. Therefore, it is crucial to decipher the mechanisms, in which drugs impair the mitochondria, as this defect precedes many pathologies including insulin resistance. Drug-induced mitochondrial dysfunction may be a consequence of elevated production of ROS, impaired mitochondrial respiration, inhibition of beta-oxidation of fatty acids, mtDNA damage or altered mitochondrial permeability transition (Fig. 28.3) (Fromenty, 2020). These interactions may manifest clinically as hepatoxicity, skeletal muscle injury, cardiotoxicity, renal toxicity and central nervous system injury (Sibiya et al., 2023). A decrease in beta oxidation can result to the accumulation of fatty acids which can be converted to toxic metabolites that can impair insulin sensitivity (Sibiya et al., 2023). A substantial mitochondrial toxicity database has been recently created to recognize the effects of a drug or a compound on mitochondrial functioning and its association with clinically relevant toxicity (Lin et al., 2021).

Therefore, there are sufficient grounds to establish the role of drugs in inducing mitochondrial impairment coupling with insulin resistance development. Various drugs classes such as anti-depressants, lipid regulators, anticancer agents, NRTIs in antiretroviral therapy and, are among the most prescribed pharmaceuticals and have been distinguished to induce mitochondrial toxicity. Table 28.1

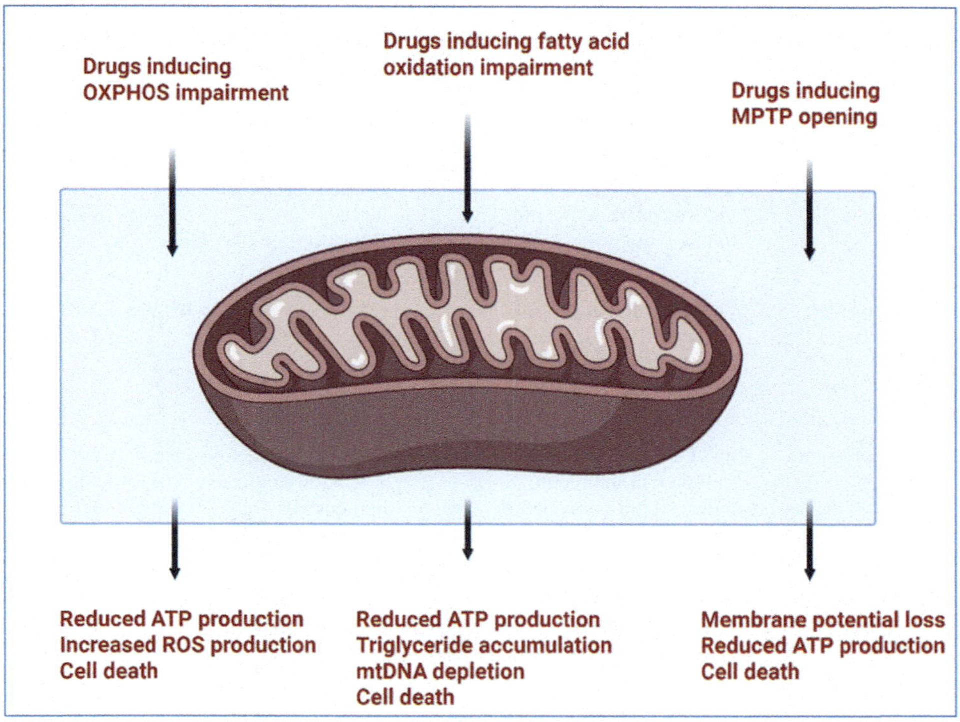

FIGURE 28.3 Schematic representation of the main mechanisms and consequences of drug-induced mitochondrial dysfunction. *Adapted from Mihajlovic, M., & Vinken, M. (2022). Mitochondria as the target of hepatotoxicity and drug-induced liver injury: Molecular mechanisms and detection methods.* International Journal of Molecular Sciences, *23(6), 3315.*

depicts numerous examples of how different drugs are implicated in mitochondrial dysfunction and insulin sensitivity. In the following sections, we will explore drugs individually with a goal to highlight link between drug-induced mitochondrial toxicity and insulin sensitivity or glucose handling impairments.

Tenofovir disoproxil fumarate

TDF is a NRT1 that is utilized in combination with other antiviral treatment used for the management of human immunodeficiency virus type-1 (HIV-1) and for hepatitis B virus (HBV) as a monotherapy. This drug is an acyclic nucleotide analog of adenosine monophosphate which inhibits viral DNA synthesis (Wassner et al., 2020).

Mitochondrial toxicity in the kidney has been shown to be an adverse effect of the drug in an in vitro kidney cell line model (Milián et al., 2017). Increased levels of oxidative stress and alteration of kidney function markers, after exposure to TDF was observed in Wister rats (Adaramoye et al., 2012). Peter et al., established that mild therapeutic doses of TDF can lead to hepatic damage, using an animal model (Peter et al., 2018). Another study showed that the mitochondrial apoptotic pathway and poly [ADP-ribose] polymerase (PARP) overactivation contribute to tenofovir-induced renal damage in rats (Ramamoorthy et al., 2019). The presence of mtDNA mutations in patients treated with TDF, shows that TDF-associated renal damage is partially mediated by the mitochondria (Samuels et al., 2017).

Antiretroviral treatment (ART) is correlated with metabolic derangements and the occurrence of T2DM. NRTI's have been examined to induce glucose impairments and insulin resistance. A study showed an increase in fasting glucose levels and insulin resistance in HIV patients treated with Tenofovir alafenamide (TAF) (Cheng et al., 2024). Research suggests that elevated fasting glucose, lipids, and increased risk of myocardial infarction, metabolic syndrome and diabetes are all associated with TAF (Wassner et al., 2020). Increased weight gain has been associated with people living with HIV treated with TDF, which could lead to an obese state, which is a major risk factor for T2DM (Sarkar & Brown, 2023). In patients treated with a combination of TDF and elvitegravir (EVG), elevated glucose levels and alteration of adipokines and GLUT-4 was observed, implicating the direct effect of these drugs on glucose metabolism (Squillace et al., 2020).

Simvastatin

Statins are a class of drugs that are utilized to lower cholesterol production and decrease dyslipidaemia associated diseases. They are the most widely prescribed and

TABLE 28.1 Commonly used pharmacological agents and their implication on mitochondrial toxicity and insulin sensitivity.

Drug	Use	Effect on mitochondrial functioning	Effect on insulin sensitivity	References
TDF	NRTI	- Increases oxidative stress - Induces mtDNA mutations - Induces apoptosis	- Elevates fasting glucose levels - Alterations of adipokines and Glut-4	Milián et al. (2017) Adaramoye et al. (2012) Samuels et al. (2017)
Simvastatin	Lipid lowering agent	- Reduced membrane potential - Decreased ATP levels - Alteration of calcium homeostasis - Complex III inhibition	- Reduces insulin sensitivity	Sanvee et al. (2019) Tian and Zhang (2025)
Amitriptyline	Anti-depressant	- Induces oxidative stress - Complex III inhibition - Induces lipid peroxidation	- Impairs glucose tolerance - Alteration in glucose homeostasis Decreased insulin secretion	Ali and Raja (2019) Miranda-Díaz and Rodríguez-Lara (2018) Abad-Segura et al. (2020)
Doxorubicin	Anticancer	- Increased ROS - Induces inflammation - Elevates mitochondrial SOD levels - Alters membrane potential	- Induces hyperglycaemia - Induces insulin resistance	Kuzu et al. (2019) Alves de Lima et al. (2014)
Valproic acid	Anticonvulsant	- Impairs mitochondrial function - Decreased mitochondrial membrane potential - Decrease SOD2 expression	- Reduces plasma glucose levels - Improvement of glucose homeostasis	Jang et al. (2021) Rakitin et al. (2015)
Tramadol	Opiod pain medication	- ROS overproduction - Membrane potential loss	- Induction of hyperglycemia - Rise in blood glucose levels	Mohammadnejad et al. (2021) Kara et al. (2013)

evidence-based lipid-lowering for lowering low-density lipoprotein (LDL) and reducing cardiovascular morbidity and mortality, both in primary and secondary prevention (Ward et al., 2019). Statins have been shown to have adverse effects and induce muscle-related disorders, renal and hepatoxicity, and T2DM (Sadowska et al., 2023). Oxidative stress may also be a result of statin-induced adverse reactions. Adverse musculoskeletal effects of simvastatin results in an increase in creatine phosphokinase levels and rhabdomyolysis (Somagutta et al., 2022).

Reduced membrane integrity and ATP content, impaired Akt phosphorylation and an increase in apoptosis via cytochrome *c* release, was observed in simvastatin-treated C2C12 myotubules (Sanvee et al., 2019). In another study, simvastatin impaired the phosphorylation of the IR, impaired oxygen consumption and elevated superoxide dismutase production in an in vitro model (Sanvee et al., 2021). Loss of 65% of metabolic activity measured by the MTT assay and increased caspase activity was observed in H9c2 cells after being treated with simvastatin (Attalah nee Rezkallah et al., 2020). Simvastatin was observed to be cytotoxic, disrupts glycolysis, impairs glucose utilisation and mitochondrial function in neuroblastoma cells (Kuzyk et al., 2020). Altered calcium homeostasis and mitochondrial complex III inhibition were observed as mechanisms of simvastatin-induced myopathy in patients (Schirris et al., 2015). Skeletal muscles are key organs in insulin sensitivity therefore their damage, in part through mitochondrial dysfunction, could precipitate insulin sensitivity impairments. Indeed, simvastatin has been associated with potential alterations in glucose homeostasis, including reduced insulin sensitivity and increased fasting glucose levels, however the effects and mechanisms are still being investigated (Tian & Zhang, 2025).

Amitriptyline

Amitriptyline is a tricyclic antidepressant drug and acts by regulating monoamine neurotransmitters, to relieve depressive symptoms (Thour & Marwaha, 2023). It is included in the WHO's list of essential medicines, providing a broad medicinal profile despite of only being an FDA approved drug for Major depressive disorder (MDD) (McClure & Daniels, 2021). Serious adverse reactions of amitriptyline have been reported, with cardiovascular complications and metabolic syndrome being the most common (Thour & Marwaha, 2023).

Oxidative stress and lipid peroxidation was identified in subjects treated with amitriptyline in a study conducted by researchers (Ali & Raja, 2019). Amitriptyline has also been reported to induce mitochondrial dysfunction via inhibition of complex III activity and induces oxidative stress in HepG2 cells (Miranda-Díaz & Rodríguez-Lara, 2018). The effect of energy metabolism was investigated and significant inhibition of complex I and II was observed after pig brain mitochondria were treated with amitriptyline (Cikánková et al., 2020). Mitochondrial dysfunction and oxidative stress have been suggested to be major mechanisms of pathophysiology of tricyclic antidepressants like amitriptyline, therefore the drug can induce mitochondrial toxicity (Lee et al., 2015).

Impaired glucose tolerance has been observed in depressed patients, and there is an increased risk of these patients to develop diabetes mellitus (Selenius et al., 2020). Chadwick et al. (2007) observed that amitriptyline had significant alterations in glucose homeostasis, as well as noteworthy changes associated with insulin secretion in male Wister rats (Chadwick et al., 2007).

Doxorubicin

Doxorubicin is an antibiotic that has been widely used since the 1960's, as a chemotherapeutic agent. It is part of the anthracycline group of chemotherapeutic agents, commonly used in the treatment of solid tumors in adult and pediatric patients (Johnson-Arbor & Dubey, 2017). Although doxorubicin is an anticancer drug, studies have shown potent toxicity in anticancer tissues. ROS over production, is a significant mechanism of mitochondrial dysfunction induced by the drug (Guven et al., 2018).

Doxorubicin has been investigated to be the leading anticancer drug that induces hyperglycaemia-related adverse drug reactions (Yang et al., 2019). Various studies have examined that doxorubicin-induced cardiotoxicity is driven by a vicious feedback loop of dysregulation of lipid and glucose metabolism, in cell cultures and rat models (Kitakata et al., 2022). A study observed that doxorubicin-mediated down-regulation of PPAR-gamma inhibited blood glucose and lipid clearance, therefore promoting hyperlipidemia and hyperglycaemia, resulting in insulin resistance, inflammation and glucotoxicity in rodents (Arunachalam et al., 2013). In Wister rats, doxorubicin induced systemic insulin resistance and modifies the expression of key genes in insulin signaling (Alves de Lima et al., 2014).

Valproic acid

Valproic acid is widely used as a broad-spectrum antiepileptic drug that has been evidenced to interfere with mitochondrial function. Chemically known as 2-propyl pentanoic or dipropyl acetic acid, this drug is able to increase the level of gamma-aminobutyric acid (GABA) transmission in the brain and blocks voltage-dependent sodium and potassium channels (Özbek et al., 2021). This drug has been studied to inhibit mitochondrial OXPHOS and, beta oxidation which may precede mitochondrial dysfunction (Komulainen et al., 2015). Valproic acid has also been investigated for its ability to induce liver injury by inhibition of fatty acid-beta oxidation, initiating mtDNA depletion, and promotion of fatty liver via inhibition of carnitine palmitoyltransferase I (Ezhilarasan & Mani, 2022). In rat mitochondria exposed to valproic acid, induction of oxidative stress by increasing mtROS, mitochondrial membrane collapse, lipid peroxidation and mitochondrial swelling, was observed (Jafarian et al., 2013). Mitochondrial MMP, SDA, ATP, and GSH were all decreased in valproic acid-treated animals and kidney injury was observed (Heidari et al., 2018).

Studies have also evaluated the effect of valproic acid on insulin sensitivity and while numerous studies observed reduced insulin resistance, others show improvement of insulin sensitivity. In diabetic rats, treatment with valproic acid resulted in reduced plasma glucose levels and insulin resistance (Khan et al., 2016). In patients with newly diagnosed epilepsy, acute glucose lowering effect of treatment with valproic acid, during the oral glucose tolerance test (OGTT) (Rakitin et al., 2015). Another study observed improvement of glucose homeostasis via increasing beta-cell proliferation and function, in juvenile diabetic rats following valproic acid administration. These variations of data suggest needed to ascertain the link between drug induced mitochondrial toxicity and insulin sensitivity.

Tramadol

Tramadol is an analgesic acting on the central nervous system (CVS) to treat moderate to severe pain. A few studies have reported the implication of the mitochondria in tramadol usage. ROS overproduction induced mitochondrial dysfunction via complex II and membrane permeability transition pores disorders, mitochondrial swelling and MMP collapse, was observed in rats following tramadol treatment (Mohammadnejad et al., 2021). In another study, induction of mitochondrial dysfunction in endometrial cancer cells, via loss of membrane potential and decreased oxygen consumption rate was observed after exposure to tramadol (Liu et al., 2022). Mehdizadeh et al., discovered that chronic administration of tramadol impaired mitochondrial function in the brain via cytochrome *c* release, increased ROS levels and increased mitochondrial swelling in rats (Mehdizadeh et al., 2017). These studies suggest that tramadol may implicate mitochondrial functioning. Evidence has also observed the effect of tramadol in insulin handling.

Tramadol initiated the activation of u-poid receptors and enhanced the use of glucose and decreased hepatic gluconeogenesis, in diabetic rats (Cheng et al., 2001). Nasouhi et al., observed similar results in tramadol overdose patients where hypoglycemia was induced by the drug (Nasouhi et al., 2015). However, some patients showed opposite results, with the induction of hyperglycemia after exposure to the drug. Another study examined tramadol administration in rats and a significant rise in blood glucose levels was observed (Kara et al., 2013).

From the above, we have provided evidence using selected drugs that drug-induced mitochondrial toxicity could be associated with insulin resistance, although there are conflicting observations. Nevertheless, this makes it still a worthwhile exploration to pursue, to further understand how drugs alter mitochondrial function, that could instigate insulin resistance. In doing so, novel pharmacological targets toward attenuating insulin resistance could be uncovered. Furthermore, strategies aiming at accelerating the screening and awareness toward drug-induced mitochondrial dysfunction and risk of developing insulin resistance could be realised.

Other hormones, endocrine tissues, and pathways

Although this chapter focuses on mitochondrial integrity, and its implication on insulin resistance, in response to pharmacological agents, various hormones are also compromised following exposure to pharmaceuticals. Endocrine hormones and mitochondria are two vital constituents that regulate cellular homeostasis (Al-Suhaimi et al., 2024). Mitochondria contain P450scc, ferredoxin and ferredoxin reductase, which are cholesterol side-chain cleavage enzymes (Al-Suhaimi et al., 2024). This characteristic provides an essential site for steroid hormone biosynthesis. Mitochondria modulate the release and production of various hormones which include glucocorticoids, mineralocorticoids, androgens, progestines, estrogens and calciferols (Miller, 2013).

Therefore, pharmacological agents that alter mitochondrial functioning, may result in hormonal dysregulation at a wider scale. Drug-induced hormonal dysregulation can occur via various mechanisms including changes in the regulation of hormonal axis, direct alteration of hormone production, effects on hormonal transport, binding and signaling (Diamanti-Kandarakis et al., 2019). Popular drug-induced metabolic and endocrine disorders involve alterations in carbohydrate metabolism, thyroid and gonadal disorders, electrolytes and calcium abnormalities (Ma et al., 2007). Anti-hypertensive, antipsychotic, anti-cancer, antiepileptic, and antidiabetic drugs are amongst the leading non-hormonal endocrine disruptors (Barroso-Sousa et al., 2018). Mitochondrial dysfunction has been examined to be the primary mechanism of action of these drugs, which then leads to hormonal dysregulation.

Immune checkpoint inhibitors (ICP) are anticancer agents that target specific proteins that are found in immune and cancer cells. Ipilimumab was the first antibody blocking immune checkpoint (CTLA4) that was authorized, followed by pembrolizumab and atezolizumab targeting programmed cell death-protein 1 (PD1) and PDl1 respectively (Robert, 2020). Although the immune response regulation has had promising results, various adverse effects have been observed. These involve alterations in the endocrine system in which the most common involve hypophysis and thyroid dysfunction, adrenalin insufficiency and hypoparathyroidism (Diamanti-Kandarakis et al., 2019). Activation of T lymphocytes occurs by the use of CTLA4, PD1 and PDL1 antibodies, which detect the endocrine glands and result in thyroid dysfunction, hypophysis and adrenalitis (Barroso-Sousa et al., 2018).

Besides immunotherapy, there is a plausible amount of drugs that supress the thyroid stimulating hormone (TSH) at the level of the hypothalamus and pituitary gland (Barroso-Sousa et al., 2018). Dopamine agonists, somatostatin analogs, glucocorticoids and rexinoids suppress TSH but do not result in significant hypothyroidism. Hyperlactatemia is also a result of certain medications. This condition results in increased levels of prolactin in the blood and results in menstrual irregularities, benign breast disease and fertility problems (Capozzi et al., 2015). The most common medications that result in hyperleptinemia are risperidone and its analogs and antipsychotics (Cloutier et al., 2018).

Opioids have been utilized as a primary method of relieving severe pain, especially in the management of cancer (Seyfried & Hester, 2012). Various studies suggest that the endocrine system is implicated by opioid use. Their mechanisms of action involve diminishing the release or disruption of the normal pulsatility of the gonadotropin-releasing hormone (GnRH) section from the hypothalamus, as well as reduction of the release of gonadotrophins form the pituitary gland of estradiol or testosterone from the gonads (Fountas et al., 2018). In patients of both sexes on opioids, hypogonadism has also been observed (Fountas et al., 2018). Clinical presentations involve infertility, erectile dysfunction, depression, decreased libido and hot flushes (Barroso-Sousa et al.,

2018). Evidence has observed that even single administration of different opioids in healthy subjects suppress ACTH and glucocorticoid secretion, resulting in decreased cortisol pituitary-adrenal response to CRH and diminished cortisol response to psychosocial stress (Bershad et al., 2015).

Applications of the material

As alluded above, the adverse effects of some drugs include altering mitochondrial functioning, which may precede many pathologies including insulin sensitivity impairments. It is therefore imperative to assess mitochondrial function as a biological and clinically relevant biomarker for drug-induced mitochondrial dysfunction. Drug-induced mitochondrial dysfunction can be assessed by various methods both at cellular and tissue levels including examining levels of mtDNA, levels of mitochondrial ROS, assessment of cellular bioenergetics, ATP generation, OXPHOS complex activities and mitochondrial membrane potential (Sharma et al., 2025).

Live cell imaging with mitochondria-selective fluorescent and phosphorescent dyes has significantly facilitated the assessment of mitochondrial biogenesis, morphology and mitochondrial function (Zhang et al., 2017). This allows for monitoring of drug effects over a period. This method has also enabled the simultaneous development of multiple biomarkers utilizing numerous probes. Mitochondrial structural damage has been assessed in cardiomyocytes utilizing fluorescent probes to detect drug-induced mitochondrial dysfunction (Tang et al., 2022). The Seahorse extracellular flux analysis also uses fluorescence and has been utilized for screening of cellular oxygen consumption rates (OCR), and is generally applied to assess mitochondrial activity in vivo and ex vivo (Yoo et al., 2024). Quantifying respiratory rates provides information about multiple mitochondrial functional parameters.

The acute metabolic switch assay is also used in screening for drug-induced mitochondrial toxicity. This assay is based on the principle that cells switch from glucose to galactose when the mitochondrial ETC is impaired (Bin Dayel et al., 2023). This assay has been used in L6 muscle cells and further validated in human skeletal muscle-derived cells to examine mitochondrial toxicants (Bin Dayel et al., 2023). Skeletal muscle biopsy has been known as the gold standard and is another diagnostic method that can be used to assess drug-induced mitochondrial dysfunction (Gharti et al., 2023). Since there are crucial morphological alterations in mitochondrial dysfunction that can be detected by staining muscle cryosections, reactive blood vessels, succinate dehydrogenase and red ragged fibers (Cai et al., 2019). However, obtaining skeletal muscle biopsies involves many processes making it a less attainable solution for research studies. Therefore, an alternative to using skeletal muscle biopsies includes the use of circulating white blood cells and platelets for the study of mitochondrial bioenergetics (Acin-Perez et al., 2021).

The screening of biological samples such as blood and urine for key metabolites, has also been utilized in assessing mitochondrial dysfunction (Muraresku et al., 2018). These biomarkers investigate the presence of certain metabolites or molecules, and lactate is a commonly used biomarker. Other biomarkers include amino acids, carnitines, creatine-kinase, pyruvate, oxidative stress parameters and circulating cytokines fibroblast growth factor-21 (FGF-21) (Finsterer & Zarrouk-Mahjoub, 2018).

It is crucial to note that there is a significant gap in standardized, validated assessment protocols to examine drug-induced mitochondrial dysfunction. Translating in vitro/vivo results to humans has posed as a significant challenge due to differences in methodologies and the absence of validated biomarkers for drug-induced mitochondrial dysfunction in humans (Tang et al., 2022). Therefore, ongoing research can aid in working toward developing models for assessing humans. Understanding the relationship between these interactions can not only assist in the therapy of these drugs, but possibly also identify potential defects that can arise during the use of these drugs. Furthermore, potential agents that can be used in conjunction with these drugs to attenuate the adverse effects, can be identified.

Mini-dictionary of terms

- **Insulin resistance.** The state in which the body cannot respond to insulin.
- **Drug-induced mitochondrial dysfunction.** An abnormal function of the mitochondria as a result to drug exposure or administration
- **Mitochondrial dysfunction.** Impairment of the mitochondria which alters the functioning of this organelle.
- **Oxidative stress.** The imbalance between oxidants and antioxidants.
- **Pharmacological agents.** Medicines used in the management and treatment of diseases and disorders.

Summary points

- Insulin resistance continues to be a global issue and the hallmark of T2DM and is characterized by elevated ROS levels, inflammation and mitochondrial dysfunction.
- The adverse effects of various widely used pharmacological agents include impairment of the mitochondria function.

- Drug induced-mitochondrial dysfunction can occur in many mechanisms including oxidative stress induction, alteration of mitochondrial respiration and inhibition of FA oxidation.
- Understanding the relationship between drug-induced mitochondrial dysfunction and insulin resistance can aid in developing preventative and curative therapy.
- The endocrine system as a whole is affected by the adverse effects of pharmacological agents.
- There is a gap in validated assessment protocols for drug-induced mitochondrial dysfunction in humans.

References

Abad-Segura, E., González-Zamar, M.-D., Infante-Moro, J. C., & Ruipérez García, G. (2020). Sustainable management of digital transformation in higher education: Global research trends. *Sustainability, 12*(5), 2107.

Acin-Perez, R., Benincá, C., Shabane, B., Shirihai, O. S., & Stiles, L. (2021). Utilization of human samples for assessment of mitochondrial bioenergetics: Gold standards, limitations, and future perspectives. *Life, 11*(9), 949.

Adaramoye, O. A., Adewumi, O. M., Adesanoye, O. A., Faokunla, O. O., & Farombi, E. O. (2012). Effect of tenofovir, an antiretroviral drug, on hepatic and renal functional indices of Wistar rats: Protective role of vitamin E. *Journal of Basic and Clinical Physiology and Pharmacology, 23*(2), 69–75.

Al-Suhaimi, E., AlQuwaie, R., AlSaqabi, R., Winarni, D., Dewi, F. R. P., AlRubaish, A. A., Shehzad, A., & Elaissari, A. (2024). Hormonal orchestra: Mastering mitochondria's role in health and disease. *Endocrine, 86*(3), 903–929.

Ali, S. H. K., & Raja, W. A. (2019). The effect of antioxidants in acute amitriptyline poisoning. *Toxicology Reports, 6*, 380–388.

Alves de Lima, E., Oliveira de Souza, C., Abílio de Souza Teixeira, A., Batatinha, H. A., de Santos Lira, F., & Neto, J. C. R. (2014). Doxorubicin leads to impaired insulin signaling in skeletal muscle. *Cancer & Metabolism, 2*(Suppl. 1), P2.

Arunachalam, S., Pichiah, P. T., & Achiraman, S. (2013). Doxorubicin treatment inhibits PPARγ and may induce lipotoxicity by mimicking a type 2 diabetes-like condition in rodent models. *FEBS Letters, 587*(2), 105–110.

Attalah nee Rezkallah, C., Thongkum, A., Zhu, C., & Chen, Q. M. (2020). Resveratrol for protection against statin toxicity in C2C12 and H9c2 cells. *Journal of Biochemical and Molecular Toxicology, 34*(6), e22484.

Baghaie, L., Bunsick, D. A., & Szewczuk, M. R. (2023). Insulin receptor signaling in health and disease. *Biomolecules, 13*, 807.

Barroso-Sousa, R., Barry, W. T., Garrido-Castro, A. C., Hodi, F. S., Min, L., Krop, I. E., & Tolaney, S. M. (2018). Incidence of endocrine dysfunction following the use of different immune checkpoint inhibitor regimens: A systematic review and meta-analysis. *JAMA Oncology, 4*(2), 173–182.

Bershad, A. K., Jaffe, J. H., Childs, E., & de Wit, H. (2015). Opioid partial agonist buprenorphine dampens responses to psychosocial stress in humans. *Psychoneuroendocrinology, 52*, 281–288.

Bhatia, D., Capili, A., & Choi, M. E. (2020). Mitochondrial dysfunction in kidney injury, inflammation, and disease: Potential therapeutic approaches. *Kidney Research and Clinical Practice, 39*(3), 244.

Bin Dayel, F. F., Alfirevic, A., & Chadwick, A. E. (2023). Developing in vitro models to define the role of direct mitochondrial toxicity in frequently reported drug-induced rhabdomyolysis. *Biomedicines, 11*(5), 1485.

Cai, C., Anthony, D. C., & Pytel, P. (2019). A pattern-based approach to the interpretation of skeletal muscle biopsies. *Modern Pathology, 32*(4), 462–483.

Capozzi, A., Scambia, G., Pontecorvi, A., & Lello, S. (2015). Hyperprolactinemia: Pathophysiology and therapeutic approach. *Gynecological Endocrinology, 31*(7), 506–510.

Chadwick, W., Wilson, G., Van De Venter, M., Oelofsen, W., & Roux, S. (2007). Shifts in metabolic parameters surrounding glucose homoeostasis resulting from tricyclic antidepressant therapy: Implications of insulin resistance? *Journal of Pharmacy and Pharmacology, 59*(1), 95–103.

Cheng, P. N., Feng, I. C., Chen, J. J., Kuo, H. T., Lee, P. L., Yu, M. L., Chiu, Y. C., Chiu, H. C., Chien, S. C., & Chen, P. J. (2024). Body weight increase and metabolic derangements after tenofovir disoproxil fumarate switch to tenofovir alafenamide in patients with chronic hepatitis B. *Alimentary Pharmacology & Therapeutics, 59*(2), 230–238.

Cheng, J.-T., Liu, I.-M., Chi, T.-C., Tzeng, T.-F., Lu, F.-H., & Chang, C. J. (2001). Plasma glucose–lowering effect of tramadol in streptozotocin-induced diabetic rats. *Diabetes, 50*(12), 2815–2821.

Cikánková, T., Fišar, Z., & Hroudová, J. (2020). In vitro effects of antidepressants and mood-stabilizing drugs on cell energy metabolism. *Naunyn-Schmiedeberg's Archives of Pharmacology, 393*(5), 797–811.

Cloutier, M., Greene, M., Touya, M., Gagnon-Sanschagrin, P., & Guerin, A. (2018). A real-world analysis of healthcare costs and relative risk of hyperprolactinemia associated with antipsychotic treatments in the United States. *Journal of Medical Economics, 21*(12), 1183–1190.

Cole, J. B., & Florez, J. C. (2020). Genetics of diabetes mellitus and diabetes complications. *Nature Reviews Nephrology, 16*(7), 377–390.

de Mello, A. H., Costa, A. B., Engel, J. D. G., & Rezin, G. T. (2018). Mitochondrial dysfunction in obesity. *Life Sciences, 192*, 26–32.

Diamanti-Kandarakis, E., Duntas, L., Kanakis, G., Kandaraki, E., Karavitaki, N., Kassi, E., Livadas, S., Mastorakos, G., Migdalis, I., & Miras, A. (2019). Diagnosis of endocrine disease: Drug-induced endocrinopathies and diabetes: A combo-endocrinology overview. *European Journal of Endocrinology, 181*(2), R73–R105.

Ezhilarasan, D., & Mani, U. (2022). Valproic acid induced liver injury: An insight into molecular toxicological mechanism. *Environmental Toxicology and Pharmacology, 95*, 103967.

Finsterer, J., & Zarrouk-Mahjoub, S. (2018). Biomarkers for detecting mitochondrial disorders. *Journal of Clinical Medicine, 7*(2), 16.

Fountas, A., Chai, S. T., Kourkouti, C., & Karavitaki, N. (2018). Mechanisms of endocrinology: Endocrinology of opioids. *European Journal of Endocrinology, 179*(4), R183–R196.

Fromenty, B. (2020). Alteration of mitochondrial DNA homeostasis in drug-induced liver injury. *Food and Chemical Toxicology, 135*, 110916.

Gharti, P., Fletcher, J. F., & Chapman, K. E. (2023). Evaluation of endpoints for the study and diagnosis of mitochondrial toxicity and disease: A narrative review. *Mutagenesis, 38*(3), 132–138.

Gutiérrez-Rodelo, C., Roura-Guiberna, A., & Olivares-Reyes, J. A. (2017). Molecular mechanisms of insulin resistance: An update. *Gaceta Médica de México, 153*(2), 214–228.

Guven, C., Sevgiler, Y., & Taskin, E. (2018). Mitochondrial dysfunction associated with doxorubicin. *Mitochondrial Diseases, 323*, 80284.

Heidari, R., Jafari, F., Khodaei, F., Shirazi Yeganeh, B., & Niknahad, H. (2018). Mechanism of valproic acid-induced Fanconi syndrome involves mitochondrial dysfunction and oxidative stress in rat kidney. *Nephrology, 23*(4), 351–361.

Jacob, R. S., de Souza Santos, L. V., d'Auriol, M., Lebron, Y. A. R., Moreira, V. R., & Lange, L. C. (2020). Diazepam, metformin, omeprazole and simvastatin: A full discussion of individual and mixture acute toxicity. *Ecotoxicology, 29*(7), 1062–1071.

Jafarian, I., Eskandari, M. R., Mashayekhi, V., Ahadpour, M., & Hosseini, M.-J. (2013). Toxicity of valproic acid in isolated rat liver mitochondria. *Toxicology Mechanisms and Methods, 23*(8), 617–623.

Jang, E.-H., Lee, J.-H., & Kim, S.-A. (2021). Acute valproate exposure induces mitochondrial biogenesis and autophagy with FOXO3a modulation in SH-SY5Y cells. *Cells, 10*(10), 2522.

Johnson-Arbor, K., & Dubey, R. (2017). In *Doxorubicin*. StatPearls Publishing.

Kara, H., Karci, A., Gümüştekin, M., Örmen, M., Taşdöğen, A., & Elar, Z. (2013). The effect of tramadol on blood glucose levels in rats. *Open Journal of Anesthesiology, 3*(1), 18–22.

Khan, S., Kumar, S., & Jena, G. (2016). Valproic acid reduces insulin-resistance, fat deposition and FOXO1-mediated gluconeogenesis in type-2 diabetic rat. *Biochimie, 125*, 42–52.

Kitakata, H., Endo, J., Ikura, H., Moriyama, H., Shirakawa, K., Katsumata, Y., & Sano, M. (2022). Therapeutic targets for DOX-induced cardiomyopathy: Role of apoptosis vs. ferroptosis. *International Journal of Molecular Sciences, 23*(3), 1414.

Komulainen, T., Lodge, T., Hinttala, R., Bolszak, M., Pietilä, M., Koivunen, P., Hakkola, J., Poulton, J., Morten, K. J., & Uusimaa, J. (2015). Sodium valproate induces mitochondrial respiration dysfunction in HepG2 in vitro cell model. *Toxicology, 331*, 47–56.

Kuzu, M., Yıldırım, S., Kandemir, F. M., Küçükler, S., Çağlayan, C., Türk, E., & Dörtbudak, M. B. (2019). Protective effect of morin on doxorubicin-induced hepatorenal toxicity in rats. *Chemico-Biological Interactions, 308*, 89–100.

Kuzyk, C. L., Anderson, C. C., & Roede, J. R. (2020). Simvastatin induces delayed apoptosis through disruption of glycolysis and mitochondrial impairment in neuroblastoma cells. *Clinical and Translational Science, 13*(3), 563–572.

Lee, M.-y., Hong, S., Kim, N., Shin, K. S., & Kang, S. J. (2015). Tricyclic antidepressants amitriptyline and desipramine induced neurotoxicity associated with Parkinson's disease. *Molecules and Cells, 38*(8), 732–738.

Lee, S.-H., Park, S.-Y., & Choi, C. S. (2022). Insulin resistance: From mechanisms to therapeutic strategies. *Diabetes & Metabolism Journal, 46*(1), 15–37.

Lin, Y.-T., Lin, K.-H., Huang, C.-J., & Wei, A.-C. (2021). MitoTox: A comprehensive mitochondrial toxicity database. *BMC Bioinformatics, 22*(10), 1–14.

Liu, L.-C., Wu, Z.-S., Chen, J.-L., Wu, Z.-F., Lai, H.-C., & Huang, Y.-H. (2022). Mitochondrial dysfunction involved in the cytotoxicity of tramadol in human endometrial carcinoma cells. *International Journal of Molecular Sciences, 24*(1), 99.

Ma, R. C., Kong, A. P., Chan, N., Tong, P. C., & Chan, J. C. (2007). Drug-induced endocrine and metabolic disorders. *Drug Safety, 30*(3), 215–245.

McClure, E. W., & Daniels, R. N. (2021). Classics in chemical neuroscience: Amitriptyline. *ACS Chemical Neuroscience, 12*(3), 354–362.

Mehdizadeh, H., Pourahmad, J., Taghizadeh, G., Vousooghi, N., Yoonessi, A., Naserzadeh, P., Behzadfar, L., Rouini, M. R., & Sharifzadeh, M. (2017). Mitochondrial impairments contribute to spatial learning and memory dysfunction induced by chronic tramadol administration in rat: Protective effect of physical exercise. *Progress in Neuro-Psychopharmacology and Biological Psychiatry, 79*, 426–433.

Milián, L., Peris, J. E., Gandía, P., Andújar, I., Pallardó, L., Górriz, J. L., & Blas-García, A. (2017). Tenofovir-induced toxicity in renal proximal tubular epithelial cells: Involvement of mitochondria. *AIDS, 31*(12), 1679–1684.

Miller, W. L. (2013). Steroid hormone synthesis in mitochondria. *Molecular and Cellular Endocrinology, 379*(1–2), 62–73.

Miranda-Díaz, A. G., & Rodríguez-Lara, S. Q. (2018). The role of oxidants/antioxidants, mitochondrial dysfunction, and autophagy in fibromyalgia. *Discussions of Unusual Topics in Fibromyalgia, 1*, 13–33.

Mohammadnejad, L., Soltaninejad, K., Seyedabadi, M., Ghasem Pouri, S. K., Shokrzadeh, M., & Mohammadi, H. (2021). Evaluation of mitochondrial dysfunction due to oxidative stress in therapeutic, toxic and lethal concentrations of tramadol. *Toxicology Research, 10*(6), 1162–1170.

Montgomery, M. K., & Turner, N. (2015). Mitochondrial dysfunction and insulin resistance: An update. *Endocrine Connections, 4*(1), R1–R15.

Muraresku, C. C., McCormick, E. M., & Falk, M. J. (2018). Mitochondrial disease: Advances in clinical diagnosis, management, therapeutic development, and preventative strategies. *Current Genetic Medicine Reports, 6*(2), 62–72.

Nasouhi, S., Talaie, H., Pajoumand, A., Aghapour, S., Rahimi, M., Khorasani, A. G., Mashayekhian, M., Aghabiklooei, A., Razi, P., & Mahdavinejad, A. (2015). Hypo and hyperglycemia among tramadol overdose patients in Loghman Hakim Hospital, Tehran, Iran. *Pakistan Journal of Pharmaceutical Sciences, 28*(6), 1959–1963.

Özbek, O., Isildak, Ö., & Isildak, I. (2021). A potentiometric biosensor for the determination of valproic acid: Human blood–based study of an anti–epileptic drug. *Biochemical Engineering Journal, 176*, 108181.

Peter, A., Naidu, E., Akang, E., Ogedengbe, O., Offor, U., Rambharose, S., Kalhapure, R., Chuturgoon, A., Govender, T., & Azu, O. (2018). Investigating organ toxicity profile of tenofovir and tenofovir nanoparticle on the liver and kidney: Experimental animal study. *Toxicological Research, 34*, 221–229. https://doi.org/10.5487/TR.2018.34.3.221

Prasun, P. (2020). Mitochondrial dysfunction in metabolic syndrome. *Biochimica et Biophysica Acta - Molecular Basis of Disease, 1866*(10), 165838.

Rakitin, A., Kõks, S., & Haldre, S. (2015). Valproate modulates glucose metabolism in patients with epilepsy after first exposure. *Epilepsia, 56*(11), e172–e175.

Ramamoorthy, H., Abraham, P., Isaac, B., & Selvakumar, D. (2019). Mitochondrial pathway of apoptosis and necrosis contribute to tenofovir disoproxil fumarate–induced renal damage in rats. *Human & Experimental Toxicology, 38*(3), 288–302.

Robert, C. (2020). A decade of immune-checkpoint inhibitors in cancer therapy. *Nature Communications, 11*(1), 3801.

Sadowska, A., Osiński, P., Roztocka, A., Kaczmarz-Chojnacka, K., Zapora, E., Sawicka, D., & Car, H. (2023). Statins—From fungi to pharmacy. *International Journal of Molecular Sciences, 25*(1), 466.

Samuels, R., Bayerri, C. R., Sayer, J. A., Price, D. A., & Payne, B. A. (2017). Tenofovir disoproxil fumarate-associated renal tubular dysfunction: Noninvasive assessment of mitochondrial injury. *AIDS, 31*(9), 1297–1301.

Sanvee, G. M., Bouitbir, J., & Krähenbühl, S. (2019). Insulin prevents and reverts simvastatin-induced toxicity in C2C12 skeletal muscle cells. *Scientific Reports, 9*(1), 7409.

Sanvee, G. M., Bouitbir, J., & Krähenbühl, S. (2021). C2C12 myoblasts are more sensitive to the toxic effects of simvastatin than myotubes and show impaired proliferation and myotube formation. *Biochemical Pharmacology, 190*, 114649.

Sarkar, S., & Brown, T. T. (2023). *Diabetes in people living with HIV*. Endotext.

Schirris, T. J., Renkema, G. H., Ritschel, T., Voermans, N. C., Bilos, A., Van Engelen, B. G., Brandt, U., Koopman, W. J., Beyrath, J. D., & Rodenburg, R. J. (2015). Statin-induced myopathy is associated with mitochondrial complex III inhibition. *Cell Metabolism, 22*(3), 399–407.

Selenius, J. S., Wasenius, N. S., Kautiainen, H., Salonen, M., von Bonsdorff, M., & Eriksson, J. G. (2020). Impaired glucose regulation, depressive symptoms, and health-related quality of life. *BMJ Open Diabetes Research and Care, 8*(1), e001568.

Seyfried, O., & Hester, J. (2012). Opioids and endocrine dysfunction. *British Journal of Pain, 6*(1), 17–24.

Sharma, E., Abadi, L. F., Kombe, J. A. K., Kandala, M., Parker, J., Winicki, N., & Kelesidis, T. (2025). Overview of methods that determine mitochondrial function in human disease. *Metabolism, 170*, 156300.

Sibiya, N., Mzimela, N., Mbatha, B., Ngubane, P., & Khathi, A. (2023). The insights on why diabetes prevalence may increase amid or post COVID-19 pandemic. *Current Diabetes Reviews, 19*(4), 37–45.

Somagutta, M. R., Shama, N., Pormento, M. L., Jagani, R. P., Ngardig, N. N., Ghazarian, K., Mahmutaj, G., El-Faramawy, K., Mahadevaiah, A., & Jain, M. S. (2022). Statin-induced necrotizing autoimmune myopathy: A systematic review. *Reumatologia/Rheumatology, 60*(1), 63–69.

Sorrentino, V., Menzies, K. J., & Auwerx, J. (2018). Repairing mitochondrial dysfunction in disease. *Annual Review of Pharmacology and Toxicology, 58*, 353–389.

Squillace, N., Ricci, E., Menzaghi, B., De Socio, G. V., Passerini, S., Martinelli, C., Mameli, M. S., Maggi, P., Falasca, K., & Cordier, L. (2020). The effect of switching from tenofovir disoproxil fumarate (TDF) to tenofovir alafenamide (TAF) on liver enzymes, glucose, and lipid profile. *Drug Design, Development and Therapy, 14*, 5515–5520.

Tang, X., Wang, Z., Hu, S., & Zhou, B. (2022). Assessing drug-induced mitochondrial toxicity in cardiomyocytes: Implications for preclinical cardiac safety evaluation. *Pharmaceutics, 14*(7), 1313.

Thour, A., & Marwaha, R. (2023). Amitriptyline. In *StatPearls*. StatPearls Publishing.

Tian, X., & Zhang, P. (2025). In *TLR4 modulates Simvastatin's impact on HDL cholesterol and glycemic control*.

Ward, N. C., Watts, G. F., & Eckel, R. H. (2019). Statin toxicity: Mechanistic insights and clinical implications. *Circulation Research, 124*(2), 328–350.

Wassner, C., Bradley, N., & Lee, Y. (2020). A review and clinical understanding of tenofovir: Tenofovir disoproxil fumarate versus tenofovir alafenamide. *Journal of the International Association of Physicians in AIDS Care, 19*, Article 2325958220919231.

Yang, J., Wang, Y., Liu, K., Yang, W., & Zhang, J. (2019). Risk factors for doxorubicin-induced serious hyperglycaemia-related adverse drug reactions. *Diabetes Therapy, 10*, 1949–1957.

Yoo, I., Ahn, I., Lee, J., & Lee, N. (2024). Extracellular flux assay (Seahorse assay): Diverse applications in metabolic research across biological disciplines. *Molecules and Cells, 47*(8), 100095.

Zhang, H., Wang, Y., Xuan, X., Wang, G., Guo, H., & Fan, J. (2017). A dynamic invertible intramolecular charge-transfer fluorescence probe: Real-time monitoring of mitochondrial ATPase activity. *Chemical Communications, 53*(40), 5535–5538.

Chapter 29

Insulin secretion and effects of beta-adrenergic blockers

Naoya Murao[1, 2], Yusuke Seino[1], Yuichiro Yamada[3] and Atsushi Suzuki[1]

[1]*Departments of Endocrinology, Diabetes and Metabolism, School of Medicine, Fujita Health University, Toyoake, Aichi, Japan;* [2]*Institute for Diabetes and Organoid Technology, Helmholtz Diabetes Center, Helmholtz Zentrum München, Deutsches Forschungszentrum für Gesundheit und Umwelt (GmbH), Neuherberg, Germany;* [3]*Yutaka Seino Distinguished Center for Diabetes Research, Kansai Electric Power Medical Research Institute, Kyoto, Japan*

List of abbreviations

AC adenylyl cyclase
ATP adenosine triphosphate
cAMP cyclic adenosine monophosphate
CICR Ca^{2+}-induced Ca^{2+} release
DAG diacylglycerol
DPP-4 dipeptidyl peptidase-4
EPAC2 exchange protein directly activated by cAMP 2
ER endoplasmic reticulum
FFA free fatty acid
GIP glucose-dependent insulinotropic polypeptide
GLP-1 glucagon-like peptide 1
GPCR G protein-coupled receptor
IP3 inositol 1,4,5-trisphosphate
OXPHOS oxidative phosphorylation
PACAP pituitary adenylate cyclase-activating polypeptide
PKA protein kinase A
PKC protein kinase C
PLCβ phospholipase Cβ
RNA-seq RNA sequencing
RT-qPCR reverse transcription-quantitative polymerase chain reaction
RYR2 ryanodine receptor 2
T2DM type 2 diabetes mellitus
TCA cycle tricarboxylic acid cycle

Introduction

Diabetes mellitus is on the rise globally, currently affecting 589 million adults, with projections reaching 853 million by 2050 (International Diabetes Federation, 2025). Given the prevalence of cardiovascular conditions in T2DM, many patients are prescribed cardiovascular medications, including β-adrenoceptor antagonists (β-blockers). These β-blockers work by competitively inhibiting epinephrine and norepinephrine at β-adrenoceptors. They are indicated for ischemic heart disease, post–myocardial infarction care, heart failure with reduced ejection fraction, arrhythmias, hypertension, hyperthyroidism, essential tremor, glaucoma, and migraine prophylaxis (Farzam & Jan, 2023). Type 2 diabetes mellitus (T2DM) is characterized by insufficient insulin secretion from pancreatic β-cells relative to insulin resistance. However, the mechanisms by which these agents act on non-cardiovascular tissues, including pancreatic islets, remain unclear. Therefore, elucidating their effects on insulin secretion and β-cell function is clinically important.

This chapter delves into the impact of β-blockers on insulin secretion by exploring β-cell biology, with a particular focus on intracellular signaling pathways. We begin by introducing canonical intracellular signaling pathways, which form the foundation for understanding the adrenergic modulation of insulin secretion. Next, we review the evidence regarding the roles of α_2- and β-adrenoceptors in insulin secretion. Finally, we discuss the effects of β-blockers on β-cells across various experimental contexts, both in vitro and in vivo, in human and animal studies, building on the preceding discussion.

Mechanisms of physiological insulin secretion

Pancreatic β-cells sense various physiological stimuli and secrete insulin to maintain blood glucose levels within a narrow range. This process is facilitated by the coordination of complex intracellular signaling pathways. Fig. 29.1 illustrates the consensus model for insulin secretion mechanisms in response to glucose, hormones, and nutrients.

Hormones in Neuroscience. https://doi.org/10.1016/B978-0-443-30263-3.00027-0

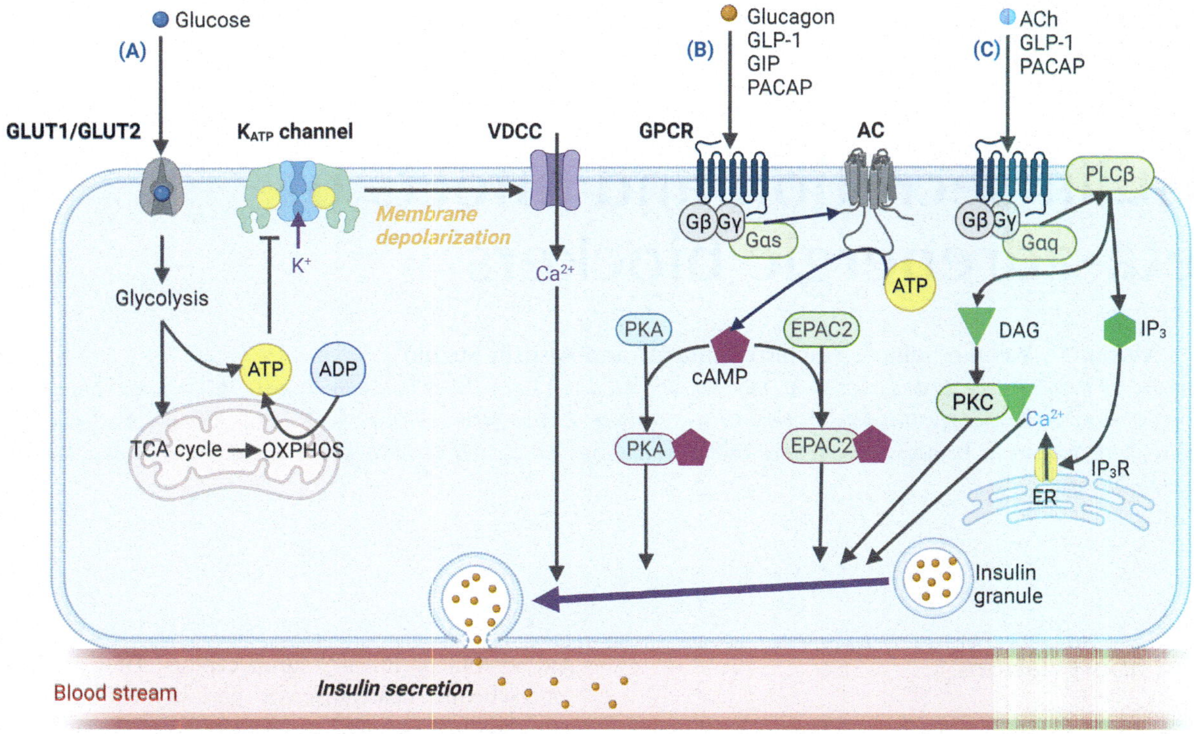

FIGURE 29.1 Molecular mechanisms of insulin secretion from pancreatic β-cells. (A) Glucose-induced insulin secretion. (B) Amplification of insulin secretion by hormones through Gαs-associated GPCRs. (C) Amplification of insulin secretion by hormones through Gαq-associated GPCRs. Refer to the main text for a detailed explanation of each pathway. AC: adenylyl cyclase; ACh: acetylcholine; cAMP: cyclic adenosine monophosphate; DAG: diacylglycerol; EPAC2: exchange protein directly activated by cAMP 2; ER: endoplasmic reticulum; FFA: free fatty acid; GCG: glucagon; GLP-1: glucagon-like peptide 1; GLUT1: glucose transporter 1; GLUT2: glucose transporter 2; IP3: inositol 1,4,5-trisphosphate; GIP: glucose-dependent insulinotropic polypeptide; OXPHOS: oxidative phosphorylation; PACAP: pituitary adenylate cyclase-activating polypeptide; PKA: protein kinase A; PKC: protein kinase C; PLCβ: phospholipase Cβ; VDCC: voltage-dependent Ca^{2+} channels. Figure drawn using BioRender.com.

(A) Glucose plays a crucial role in physiological insulin secretion. In humans, glucose enters cells via glucose transporter 1 (GLUT1), whereas in rodents, it uses glucose transporter 2 (GLUT2). Once inside, glucose undergoes glycolysis and the TCA cycle, fueling oxidative phosphorylation (OXPHOS) to produce ATP. ATP then closes ATP-sensitive K^+ (K_{ATP}) channels, leading to membrane depolarization, opening of voltage-dependent Ca^{2+} channels (VDCCs), and a subsequent influx of Ca^{2+} that triggers the exocytosis of insulin granules (Henquin 2000, 2009). In addition to this primary pathway, glucose-induced insulin secretion (GIIS) is influenced by various glucose-derived metabolites, as detailed in another review (Campbell & Newgard, 2021).

Hormones and neurotransmitters exert their effects through G protein-coupled receptors (GPCRs). While the activation of GPCRs alone is insufficient to initiate insulin secretion without elevated glucose levels (Henquin, 2009), they play a crucial role in tuning insulin secretion via distinct intracellular signals, which vary depending on the associated Gα subunit, as discussed below.

(B) Gαs-associated GPCRs include receptors for glucagon, glucagon-like peptide 1 (GLP-1), glucose-dependent insulinotropic polypeptide (GIP), and pituitary adenylate cyclase-activating polypeptide (PACAP). When a ligand binds, Gαs activates adenylyl cyclase (AC), an enzyme responsible for producing cyclic adenosine monophosphate (cAMP). cAMP then activates protein kinase A (PKA) and exchange protein directly activated by cAMP 2 (Epac2), which in turn enhances insulin exocytosis (Seino & Shibasaki, 2005).

(C) Gαq-associated GPCRs include receptors for acetylcholine (muscarinic M3), GLP-1, PACAP, and free fatty acids (FFA) (Oduori et al., 2020). Upon ligand binding, Gαq activates phospholipase Cβ (PLCβ), which transforms membranous phospholipids into diacylglycerol (DAG) and inositol 1,4,5-trisphosphate (IP3). DAG then activates protein kinase C (PKC), whereas IP3 facilitates the release of Ca^{2+} from the endoplasmic reticulum (ER) (Rorsman & Ashcroft, 2018).

Effects of catecholamines on insulin secretion

It has long been known that catecholamines, such as adrenaline and noradrenaline, inhibit insulin secretion both in vitro and in vivo (Henquin, 2021), as detailed in Table 29.1.

TABLE 29.1 Effects of catecholamines on insulin secretion.

Catecholamine	Test condition	Method	Species		Effect on insulin secretion	Drug interactions	References
In vitro							
Adr	100 nM	Static	Rat	Sorted α/β-cell suspension	Decreased; glucagon secretion increased	Effect counteracted by Phe and Yoh	Schuit and Pipeleers (1986)
Adr	100 nM at 20 mM Glc	Static	Rat	Islet	Decreased	–	Ullrich and Wollheim (1984)
Adr	10 nM	Peri	Human	Islet	Initial increase followed by sustained decrease	Initial increase counteracted by Pro; sustained decrease counteracted by Yoh.	Lacey et al. (1993)
Adr	100 nM at 8–20 mM Glc	Peri	Mouse	Pancreas slice	Decreased	–	Sluga et al. (2022)
In vivo					**Effect on plasma insulin levels**		
Adr	6 μg/min	IV infusion	Human		Decreased	Effect counteracted by Phe	Porte (1967)
Adr	6 μg/min	IV infusion	Human		Decreased acute insulin response to Glc (5 g)	–	Lerner and Porte (1971)
NAd	6 μg/min	IV infusion	Human		Decreased acute insulin response to Glc (300 mg/min)	–	Porte and Williams (1966)
NAd	6 μg/min	IV infusion	Human		Decreased	Effect counteracted by Phe	Robertson and Porte (1973)
NAd	6 μg/min	IV infusion	Human		Decreased	Effect counteracted by Phe	Meguid et al. (1975)

Catecholamine: Adr, adrenaline; NAd, noradrenaline. Test condition: Concentration of catecholamines used in each experiment are indicated. Glc, glucose. Method: Static, static incubation; Peri, perifusion; IV, intravenous. Drug interactions: Pro, propranolol; Phe, phentolamine ($\alpha_{1/2}$ -blocker); Yoh, yohimbine ($\alpha_{1/2}$ -blocker).

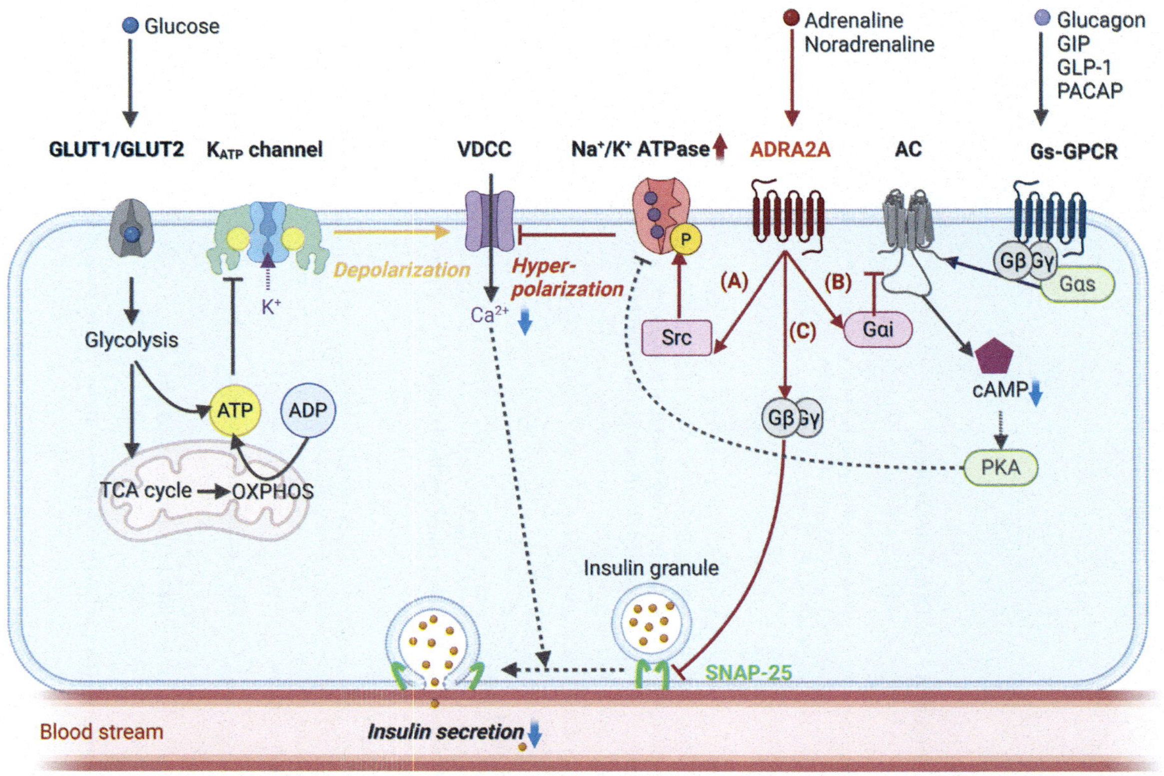

FIGURE 29.2 Effects of catecholamines on insulin secretory pathways. (A) Src-dependent phosphorylation and activation of Na^+/K^+-ATPase. (B) Downregulation of cAMP-dependent signalling pathways by Gαi/o. (C) Inhibition of synaptosomal-associated protein-25 (SNAP-25) by Gβγ. Refer to the main text for a detailed explanation of each pathway. Abbreviations are the same as those in Figure 1. Figure drawn using BioRender.com.

The infusion of catecholamines in human subjects has consistently been shown to suppress basal insulin levels (Meguid et al., 1975; Porte, 1967; Robertson & Porte, 1973) and the insulin response to glucose (Lerner & Porte, 1971; Porte & Williams, 1966). Similar effects have been observed in isolated islets and β-cells stimulated with adrenaline in vitro (Lacey et al., 1993; Schuit & Pipeleers, 1986; Sluga et al., 2022; Ullrich & Wollheim, 1984), suggesting the involvement of adrenergic receptors (adrenoceptors) in β-cells.

Adrenoceptors, a class of GPCRs, consist of nine subtypes: α_1 (α_{1A}, α_{1B}, α_{1D}), α2 (α_{2A}, α_{2B}, α_{2C}), and β (β_1, β_2, β_3) (Varney & Benovic, 2024). β-cells express the α_{2A} adrenoceptor (ADRA2A). Initially identified through RT-qPCR of islets and in situ hybridization (Lacey et al., 1996), the predominant expression of ADRA2A in human β-cells, as compared to other adrenoceptor isoforms and islet cell types, was further corroborated by recent single-cell RNA sequencing (RNA-seq) analyses (Baron et al., 2016; Segerstolpe et al., 2016). ADRA2A is regarded as the primary mediator of catecholamine-induced suppression of insulin secretion. This perspective is supported by the finding that the effect of catecholamines on insulin secretion is counteracted by α-blockers both in vitro and in vivo, as documented in multiple studies (Table 29.1). β-Cell ADRA2A has gained attention as a therapeutic target because its increased expression due to polymorphism has been linked to impaired insulin secretion and an increased risk of T2DM in humans (Rosengren et al., 2010).

Mechanistically, ADRA2A couples to Gαi and Gαo, inhibiting insulin secretion via diverse signaling pathways, as indicated in Fig. 29.2.

(A) Gαi/o-coupled GPCRs hyperpolarize β-cells by facilitating Src-dependent phosphorylation and activation of Na^+/K^+-ATPase. This process generates an outward current that counteracts depolarization and Ca^{2+} influx induced by KATP channels (Dickerson et al., 2022).

(B) Gαi/o inhibits AC and reduces intracellular cAMP, resulting in the downregulation of various PKA/EPAC2-mediated exocytotic pathways (Straub & Sharp, 2012) (Fig. 29.2). Notably, this pathway partially synergizes with pathway (A); the reduction in PKA activity alleviates the tonic inhibition of Na^+/K^+-ATPase by PKA (Dickerson et al., 2022). While this pathway may play a minor role in suppressing glucose-induced insulin secretion (Ullrich & Wollheim, 1984), it also significantly suppresses GLP-1- and GIP-amplified insulin secretion (Ito et al., 2017; Pan et al., 2011).

(C) Gβγ directly inhibits the exocytosis of insulin granules (Fig. 29.2). By binding to the C-terminus of Synaptosomal-associated protein-25 (SNAP-25),

Gβγ prevents fusion of insulin granules with the plasma membrane, even when intracellular Ca^{2+} levels are elevated (Straub & Sharp, 2012).

In summary, catecholamines interact with β-cell ADRA2A, hindering excitability and exocytosis, thereby suppressing insulin secretion in response to various physiological stimuli, including glucose and incretins.

Roles of β-adrenoceptors in insulin secretion

β-Adrenoceptors are not as well characterized in terms of their presence and functional significance in β-cells as ADRA2A. In human β-cells, β_2-adrenoceptors (ADRB2) were initially identified using in situ hybridization (Lacey et al., 1996). RNA-seq analysis has also confirmed the presence of β_1- (ADRB1) and β_2- (ADRB2) adrenoceptors, although at significantly lower levels than ADRA2A (Rorsman & Ashcroft, 2018). In mouse β-cells, the ADRB2 transcript has been reported only in neonates (Ceasrine et al., 2018). The other subtypes (ADRB1 or ADRB3) were not detected in RNA-seq analysis (Rorsman & Ashcroft, 2018), nor have they been experimentally confirmed in any other studies. Interestingly, although β-adrenoceptors in β-cells exhibit sluggish expression patterns, β-agonists significantly increase plasma insulin levels in both humans and rodents, as demonstrated in vitro and in vivo (Table 29.2).

Insulinotropic effects have been observed following the administration of a $\beta_{1/2}$-nonselective agonist (isoproterenol) or β_2-selective agonists (salbutamol, terbutaline, and clenbuterol) (Ahrén et al., 1981; Cotton et al., 1981; Harms et al., 1978; Imura et al., 1971; Jansson et al., 1989; Loubatières et al., 1971; Meister et al., 2022; Palmer et al., 1979). Similar effects were noted in vitro, where β-agonists enhanced insulin secretion from isolated islets (Lacey et al., 1990, 1993; Malaisse et al., 1967; Meister et al., 2022; Zhang et al., 1989; Zielmann et al., 1985). Interestingly, the action of β-agonists on insulin secretion in vitro appears to be concentration-dependent, increasing secretion at low concentrations, but suppressing secretion at higher concentrations (Zielmann et al., 1985).

Importantly, the insulinotropic effect of β-agonists was inhibited by a $\beta_{1/2}$-nonselective blocker (propranolol) or a β_2-blocker (ICI-118,551) (Ahrén et al., 1981), but not by a β_1-blocker (practolol) (Loubatières et al., 1971), suggesting the involvement of ADRB2. ADRB2 is primarily coupled to Gαs, which stimulates cAMP production. However, it remains unclear whether insulin secretion is directly stimulated by β-cell ADRB2 or indirectly enhanced by hormones or factors from other cell types that express ADRB2. Fig. 29.3 illustrates the potential pathways (A)–(C).

(A) α-Cell ADRB2 stimulates glucagon secretion, which subsequently enhances insulin secretion via a paracrine mechanism. This hypothesis is supported by the finding that α-cells express ADRB2 and secrete glucagon in response to adrenaline (Lorza-Gil et al., 2019; Schuit & Pipeleers, 1986) and isoproterenol (Zielmann et al., 1985).

(B) β-Cell ADRB2 enhances insulin secretion through intracellular cAMP-PKA/EPAC2 signaling.

Pathways (A) and (B) align with findings that β-agonist stimulation of isolated islets or β-cell lines enhances insulin secretion (Lacey et al., 1990, 1993; Malaisse et al., 1967; Meister et al., 2022; Zhang et al., 1989; Zielmann et al., 1985), and that these effects are inhibited by propranolol (Lacey et al., 1990, 1993). However, these findings cannot be solely attributed to pathway (B), as their β-cells do not express ADRB2.

(C) The activation of ADRB2 in adipocytes or intestinal K-cells triggers the release of FFA and GIP, respectively, into the bloodstream. This hypothesis aligns with the findings that isoproterenol infusion elevates circulating levels of FFA and GIP, along with insulin, effects that are counteracted by β_2-blockers (Flaten et al., 1982; Harms et al., 1978).

In summary, β-agonists enhance insulin secretion via ADRB2, although the precise site of action remains unidentified. Further experiments using tissue-specific ADRB2 knockout mice are necessary to clarify the underlying mechanisms.

Effects of β-blockers on insulin secretion

How do β-blockers influence insulin secretion in the absence of agonist administration? The experimental results are summarized in Table 29.3.

Propranolol infusion has been shown to reduce plasma insulin levels in both humans and dogs (Chap et al., 1986; Giugliano et al., 1979; Loubatières et al., 1971; Robertson & Porte, 1973). Similarly, the insulin response to a glucose challenge was suppressed following the oral administration of propranolol and metoprolol (Schluter et al., 1982). These findings suggest that β-blockers inhibit insulin secretion in vivo.

However, these effects were not replicated in vitro. It is important to note that the therapeutic range of β-blockers is typically <1 μM. For instance, plasma propranolol levels generally range from about 200 to 400 nM (0.05–0.10 μg/mL), with toxicity occurring above 7 μM (Kerns II, 2007; Sharifpour et al., 2022). However, β-blockers at concentrations below 10 μM have no effect on insulin secretion

TABLE 29.2 Effects of β-agonists on insulin secretion.

β-Agonist	Selectivity	Method	Species		Effect on insulin secretion	Drug interactions	References
In vitro							
Isoproterenol	$\beta_{1/2}$	Static	Rat	Pancreas fragment	Increased	Co-administered with Pbz	Malaisse et al. (1967)
Isoproterenol	$\beta_{1/2}$	Peri	Mouse	Islet	Increased at low conc. (0.1–1 μM); decreased at high conc. (10–50 μM)	Decrease at high conc was counteracted by Rau	Zielmann et al. (1985)
Isoproterenol	$\beta_{1/2}$	Static	Hamster	HIT cell line	Increased (insulin secretion and cAMP)	–	Zhang et al. (1989)
Clenbuterol, salbutamol, albuterol	β_2	Static and Peri	Human	Islet	Increased at 20 mM Glc, unaffected at 4 mM Glc.	Effect counteracted by Pro	Lacey et al. (1990)
Clenbuterol	β_2	Static	Human	Islet	Increased	Effect counteracted by Pro	Lacey et al. (1993)
Clenbuterol	β_2	Static	Mouse	Islet	Increased at 16 mM Glc, unaffected at 3 mM Glc.	–	Meister et al. (2022)
In vivo					**Effect on plasma insulin**		
Isoproterenol	$\beta_{1/2}$	IV infusion	Human		Increased	Effect counteracted by Pro	Harms et al. (1978)
Isoproterenol	$\beta_{1/2}$	IV infusion	Human		Increased	–	Imura et al. (1971)
Isoproterenol and salbutamol	Iso ($\beta_{1/2}$), Sal (β_2)	IV infusion	Dog		Increased	Effect counteracted by Pro but not Pra	Loubatières et al. (1971)
Isoproterenol	$\beta_{1/2}$	IV infusion	Human		Increased (insulin and glucagon)	–	Palmer et al. (1979)
Terbutaline	β_2	IV infusion	Human (pregnancy)		Increased	–	Cotton et al. (1981)
Terbutaline	β_2	IV infusion	Rat		Increased	Effect counteracted by Pro and ICI	Ahrén et al. (1981)
Terbutaline	β_2	IV bolus	Rat		Increased	–	Jansson et al. (1989)
Clenbuterol	β_2	IP injection	Mouse		Increased	–	Meister et al. (2022)

Method: Static, static incubation; Peri, perifusion; IV, intravenous; IP, intraperitoneal. Effect: Glc: glucose. Drug interactions: Pbz, phenoxybenzamine ($\alpha_{1/2}$ -blocker); Rau, rauwolscine (α_2 -blocker); Pro, propranolol; Pra, practolol (β_1 -blocker); ICI, ICI-118,551 (β_2 -blocker).

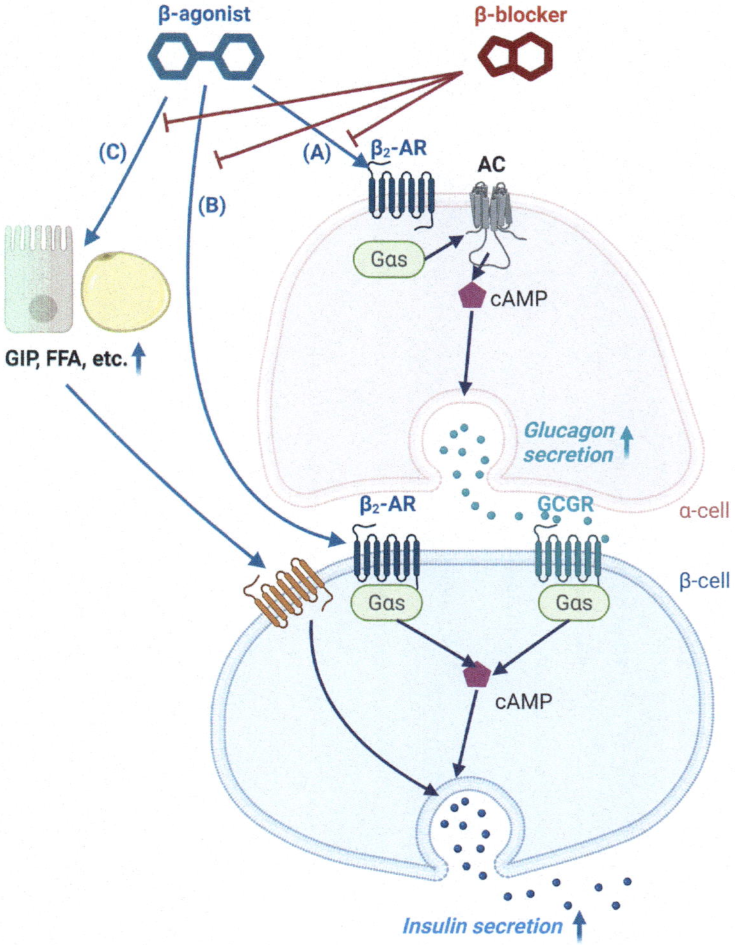

FIGURE 29.3 Potential mechanisms of potentiation of insulin secretion by β-agonists. (A) β-Agonists stimulating glucagon secretion by binding β_2-AR (β_2-adrenoceptor, ADRB2) on α-cells. (B) β-Agonists directly stimulating insulin secretion by binding β_2-AR (β_2-adrenoceptor, ADRB2) on β-cells. (C) β-Agonists acting on other tissues increasing circulating insulinotropic hormones and nutrients, which indirectly potentiates insulin secretion from β-cells. Refer to the main text for a detailed explanation of each pathway. Abbreviations are the same as those in Figure 1. Figure drawn using BioRender.com.

from islets or cell lines in vitro (Garrino & Henquin, 1990; Murao et al., 2025).

One potential explanation for the discrepancy between the in vivo and in vitro results is that β-blockers may suppress insulin secretion by counteracting endogenous ADRB2 stimulation, either on β-cells or in distant organs, thereby allowing β-cell ADRA2A to prevail in vivo. Vascular effects, such as reduced portal blood flow, may also contribute to this phenomenon (Chap et al., 1986).

Interestingly, β-blockers at supratherapeutic concentrations (>10–100 μM) significantly enhanced insulin secretion (Garrino & Henquin, 1990; Murao et al., 2025; Northrop et al., 1973). This paradoxical effect remained unexplained for a long time until the authors recently clarified the mechanism using propranolol as a model (Murao et al., 2025). Remarkably, we found that high levels of β-blockers increased intracellular cAMP levels and amplified the Ca^{2+} response in mouse β-cells, thereby enhancing insulin secretion. This mechanism is illustrated in Fig. 29.4.

(A) β-Blockers may function as inverse agonists on ADRA2A at supratherapeutic concentrations, thereby reducing Gαi/o activity and alleviating the tonic inhibition of AC. It is important to note that the binding of β-blockers to ADRA2A has not been experimentally proven. However, a similar interaction involving a β-specific ligand with ADRA2A was suggested in a previous study, which demonstrated that the suppression of insulin secretion at high isoproterenol levels was prevented by an α2-blocker (Table 29.2, Zielmann et al., 1985).

(B) Elevated cAMP levels activate protein kinase A (PKA), which in turn phosphorylates ryanodine receptor 2 (RYR2) and facilitates Ca^{2+}-induced Ca^{2+} release (CICR) from the endoplasmic reticulum (ER). It is important to note that while RYR2 phosphorylation increases its sensitivity to CICR, it cannot independently mobilize Ca^{2+} from the ER when cytosolic $[Ca^{2+}]$ is low (Hain et al., 1995; Islam, 2002; Islam et al., 1998). Therefore, it is plausible to hypothesize that propranolol-stimulated cAMP also initiates Ca^{2+} influx via unidentified mechanisms.

(C) CICR subsequently activates transient receptor potential cation channel subfamily M member 5 (TRPM5), enhancing membrane depolarization and promoting Ca^{2+} influx through VDCCs.

TABLE 29.3 Effects of β-blockers on insulin secretion.

β-Blocker	Selectivity	Test condition	Method	Species		Effect on insulin secretion	References
In vitro							
Propranolol	$\beta_{1/2}$	169 μM at 16.7 mM Glc	Static	Rat	Islet	Increased	Northrop et al. (1973)
Propranolol	$\beta_{1/2}$	100 μM at 10 mM Glc	Static	Mouse	Islet	Increased: Pro >100 μM No effect: Pro <10 μM	Garrino and Henquin (1990).
Metoprolol	β_1	100 μM at 10 mM Glc	Static	Mouse	Islet	Increased: Met >100 μM No effect: Met <10 μM	Garrino and Henquin (1990).
ICI-118,551	β_2	100 μM at 10 mM Glc	Static	Mouse	Islet	Increased: ICI >10 μM No effect: ICI <1 μM	Garrino and Henquin (1990).
Propranolol	$\beta_{1/2}$	0.01–100 μM	Static	Mouse	Islet, MIN6-K8 cell line	Increased: Pro >10 μM under 11 mM Glc; No effect: Pro <1 μM or 2.8 mM Glc.	Murao et al. (2025)
Bisoprolol	β_1	0.01–1000 μM	Static	Mouse	Islet, MIN6-K8 cell line	Increased: Bis >100 μM under 11 mM Glc; No effect: Bis <10 μM or 2.8 mM Glc.	Murao et al. (2025)
In vivo						**Effect on plasma insulin levels**	
Propranolol	$\beta_{1/2}$	0.3 mg/kg	IV infusion	Dog		Decreased	Loubatières et al. (1971)
Propranolol	$\beta_{1/2}$	10 mg	IV infusion	Human		Unchanged	Imura et al. (1971)
Propranolol	$\beta_{1/2}$	0.08 mg/min, 60 min	IV infusion	Human		Decreased	Robertson and Porte (1973)
Propranolol	$\beta_{1/2}$	0.08 mg/min	IV infusion	Human		Decreased	Giugliano et al. (1979)
Propranolol	$\beta_{1/2}$	160 mg	Oral	Human		Decreased insulin response to oral Glc (100 g)	Schluter et al. (1982)
Metoprolol	β_1	200 mg	Oral	Human		Decreased insulin response to oral Glc (100 g)	Schluter et al. (1982)
Propranolol	$\beta_{1/2}$	7 μg/kg	IV infusion	Dog		Decreased (portal vein)	Chap et al. (1986)

Test condition: concentrations of β-blockers used in the experiments. Glc, glucose. Method: Static, static incubation; Peri, perifusion; IV, intravenous. Effect: Observed effects on insulin secretion (in vitro) along with the respective β-blocker and glucose (Glc) concentrations are indicated. Pro, propranolol; Met, metoprolol; Bis, bisoprolol; ICI, ICI-118,551.

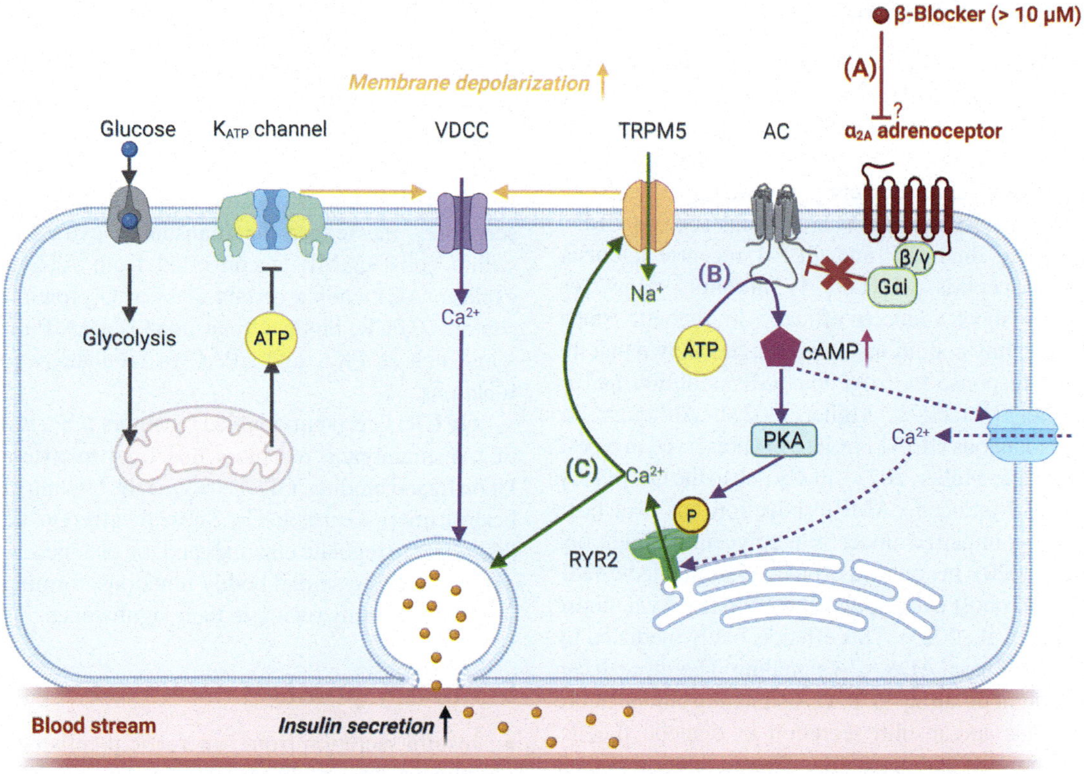

FIGURE 29.4 Mechanisms of potentiation of insulin secretion by high-concentration β-blockers. (A) β-Blockers at high concentration (>10 μM) may act on α_{2A}-adrenoceptor (ADRA2A) as an inverse agonist. (B) cAMP activates PKA, which phosphorylates ryanodine receptor 2 (RYR2). cAMP also triggers Ca^{2+} influx via an unknown mechanism (dashed arrows). (C) Activation of transient receptor potential cation channel subfamily M member 5 (TRPM5) augments membrane depolarization. Refer to the main text for a detailed explanation of each pathway. Abbreviations are the same as those in Figure 1. P, phosphorylation. Figure drawn using BioRender.com.

The observed effects do not align with the traditional understanding of β-blocker pharmacology, underscoring the importance of caution when employing high concentrations of β-blockers in experiments. In summary, although β-blockers generally reduce insulin secretion in vivo, they do not have the same effect in vitro, illustrating the variability in their actions based on the experimental context.

Other glycemic effects of β-blockers

While this chapter focused on β-blockers and insulin secretion, these drugs are also clinically linked to other glycemic factors. Compared to other classes of antihypertensive drugs, β-blockers are associated with a higher incidence of new-onset diabetes (Bangalore et al., 2007; Nazarzadeh et al., 2021). This may be partly due to the suppression of insulin secretion, as discussed earlier. Additionally, a reduction in insulin sensitivity is another adverse metabolic outcome of β-blockers (Pollare et al., 1989) and is potentially linked to the suppression of endothelial vasodilation (Kveiborg et al., 2010). Notably, newer β-blockers with vasodilating effects, such as carvedilol and nebivolol, outperform older β-blockers in terms of glycemic control (Bakris et al., 2004; Poirier et al., 2001).

Another significant adverse glycemic effect associated with β-blockers is hypoglycaemia. The mechanism behind β-blocker-induced hypoglycaemia likely involves the direct inhibition of hepatic glucose production and glucagon secretion from α-cells. Additionally, β-blockers suppress the counter-regulatory effect of adrenaline, which would otherwise stimulate glycogenolysis and gluconeogenesis, and inhibit insulin-dependent glucose disposal during hypoglycaemia. One might hypothesize that hypoglycaemia is also linked to the stimulation of insulin secretion by high concentrations of β-blockers observed in vitro (Garrino & Henquin, 1990; Murao et al., 2025; Northrop et al., 1973). However, this is unlikely because (1) the required concentrations are extremely high (>10–100 μM); and (2) nadolol, which completely lacks an insulinotropic effect in vitro (Murao et al., 2025), is associated with a higher hypoglycaemic risk than other β-blockers (Carnovale et al., 2021). Carvedilol has been shown to be safer than selective β-blockers (metoprolol/atenolol) (Dungan et al., 2019).

Applications of the material

As discussed earlier, the direct repositioning of β-blockers as insulinotropic agents in patients presents significant challenges. Nonetheless, the ability of high-concentration β-blockers to elevate cAMP levels remains noteworthy from a therapeutic perspective. Currently, the strategy to enhance β-cell cAMP involves incretin-related drugs such as GLP-1 receptor agonists, dual GIP and GLP-1 receptor agonists, and dipeptidyl peptidase-4 (DPP-4) inhibitors. However, these methods show reduced efficacy in diabetic states compared to healthy conditions. Research has shown that the insulinotropic response to exogenous GIP is diminished in type-2 diabetes (Nauck & Müller, 2023). Although the response to exogenous GLP-1 is relatively preserved in type-2 diabetes (Nauck & Müller, 2023), likely due to the Gαq/PLCβ pathway, GLP-1-induced cAMP production has also been suggested to be impaired under hyperglycemic conditions (Oduori et al., 2020). In contrast, β-blocker-stimulated cAMP production was robust even in diabetic NSY.B6-Ay/a mouse β-cells (Murao et al., 2025). This effect is likely mediated by the suppression of basal ADRA2A signaling. Therefore, it can be inferred that modulating ADRA2A signaling could restore cAMP signaling and insulin secretion in diabetic β-cells. Understanding β-cell adrenergic signaling in relation to their selective modulators may offer further insights into novel therapeutic strategies for T2DM.

Mini-dictionary of terms

Adrenoceptor: An adrenoceptor is a type of GPCR that binds to adrenaline and noradrenaline, transmitting signals into the cell. There are two primary families, α and β, comprising nine subtypes: α_1 (α_{1A}, α_{1B}, and α_{1D}), α_2 (α_{2A}, α_{2B}, and α_{2C}), and β (β_1, β_2, and β_3). They play a crucial role in the regulation of various physiological functions, including heart rate, vascular and airway tone, and metabolism.

β-Agonist: β-Agonists are a class of compounds that bind to β-adrenoceptors, activating them and mimicking the effects of adrenaline and noradrenaline. This activation increases intracellular cAMP levels, leading to bronchodilation, as well as increased heart rate and contractility. Clinically, they are primarily used for bronchodilation in asthma, chronic obstructive pulmonary disease (COPD), acute cardiac support, and tocolysis.

β-Blocker: β-Blockers are a class of compounds that bind to β-adrenoceptors and prevent adrenaline, noradrenaline, and β-agonists from activating them. These β-blockers differ in their capacity to block β_1- and β_2-adrenoceptors. Clinically, they are used to treat conditions such as hypertension, angina, post-myocardial infarction care, heart failure with reduced ejection fraction, arrhythmias, hyperthyroidism, migraine prophylaxis, essential tremors, portal hypertension, and glaucoma.

β-Cell: β-Cells are a type of endocrine cell found in the pancreas. They detect increases in blood glucose levels and release insulin into the bloodstream, facilitating the uptake and storage of glucose by tissues. Their activity is influenced by the cellular energy state (ATP/ADP ratio) and is further modulated by other nutrients and hormones.

cAMP: Cyclic adenosine monophosphate (cAMP) is a secondary messenger that transduces extracellular signals within cells. cAMP is converted from ATP by adenylyl cyclase (AC) when a certain class of G protein-coupled receptors (GPCR) binds to their ligands. cAMP activates proteins such as PKA and EPAC to regulate various cellular functions.

GPCR: G protein-coupled receptors (GPCRs) are a class of transmembrane receptors that bind to external ligands. Upon ligand binding, GPCRs trigger the binding of GTP to the heterotrimeric G protein Gα. Subsequently, Gα-GTP and Gβγ dissociate to regulate enzymes and ion channels. GPCRs play a role in many essential bodily functions, forming the largest drug-target family owing to their significance.

Summary points

- Insulin secretion from pancreatic β-cells is triggered by glucose and modulated by various nutrients and hormones through GPCRs.
- Catecholamines suppress insulin secretion by activating β-cell ADRA2A.
- β-Agonists enhance insulin secretion via ADRB2, though the site of action remains unclear.
- Therapeutic concentrations of β-blockers decrease insulin secretion in vivo but do not affect insulin secretion in vitro.
- Supratherapeutic concentrations of β-blockers lead to an increase in cAMP and insulin secretion in vitro.
- Gaining a deeper understanding of β-cell adrenergic signaling in connection with their selective modulators could provide valuable insights into new therapeutic strategies for T2DM.

Use of artificial intelligence (AI)

During the preparation of this work, the authors used ChatGPT 5.0 (OpenAI, CA, USA) and Paperpal (Cactus Communications Services, Singapore), solely to improve readability and language usage. Under no circumstances has it been used to draw insights from data, conclude, or produce recommendations. After using this tool/service, the authors reviewed and edited the content as needed and took full responsibility for the content of the publication.

References

Ahrén, B., Järhult, J., & Lundquist, I. (1981). Insulin secretion induced by glucose and by stimulation of β2-adrenoceptors in the rat. Different sensitivity to somatostatin. *Acta Physiologica Scandinavica, 112*(4), 421–426.

Bakris, G. L., Fonseca, V., Katholi, R. E., McGill, J. B., Messerli, F. H., Phillips, R. A., et al. (2004). Metabolic effects of carvedilol vs metoprolol in patients with type 2 diabetes mellitus and hypertension: A randomized controlled trial. *JAMA, 292*(18), 2227–2236.

Bangalore, S., Parkar, S., Grossman, E., & Messerli, F. H. (2007). A meta-analysis of 94,492 patients with hypertension treated with beta blockers to determine the risk of new-onset diabetes mellitus. *The American Journal of Cardiology, 100*(8), 1254–1262.

Baron, M., Veres, A., Wolock, S. L., Faust, A. L., Gaujoux, R., Vetere, A., et al. (2016). A single-cell transcriptomic map of the human and mouse pancreas reveals inter-and intra-cell population structure. *Cell Systems, 3*(4), 346–360. e344.

Campbell, J. E., & Newgard, C. B. (2021). Mechanisms controlling pancreatic islet cell function in insulin secretion. *Nature Reviews Molecular Cell Biology, 22*(2), 142–158.

Carnovale, C., Gringeri, M., Battini, V., Mosini, G., Invernizzi, E., Mazhar, F., et al. (2021). Beta-blocker-associated hypoglycaemia: New insights from a real-world pharmacovigilance study. *British Journal of Clinical Pharmacology, 87*(8), 3320–3331.

Ceasrine, A. M., Lin, E. E., Lumelsky, D. N., Iyer, R., & Kuruvilla, R. (2018). Adrb2 controls glucose homeostasis by developmental regulation of pancreatic islet vasculature. *eLife, 7*, e39689.

Chap, Z., Ishida, T., Chou, J., Michael, L., Hartley, C., Entman, M., et al. (1986). Effects of alpha and beta adrenergic blockade on hepatic glucose balance before and after oral glucose. Role of insulin and glucagon. *Journal of Clinical Investigation, 77*(4), 1357–1369.

Cotton, D. B., Strassner, H. T., Lipson, L. G., & Goldstein, D. A. (1981). The effects of terbutaline on acid base, serum electrolytes, and glucose homeostasis during the management of pre term labor. *American Journal of Obstetrics and Gynecology, 141*(5), 617–624.

Dickerson, M. T., Dadi, P. K., Zaborska, K. E., Nakhe, A. Y., Schaub, C. M., Dobson, J. R., et al. (2022). Gi/o protein-coupled receptor inhibition of beta-cell electrical excitability and insulin secretion depends on Na+/K+ ATPase activation. *Nature Communications, 13*(1), 6461.

Dungan, K., Merrill, J., Long, C., & Binkley, P. (2019). Effect of beta blocker use and type on hypoglycemia risk among hospitalized insulin requiring patients. *Cardiovascular Diabetology, 18*(1), 163.

Farzam, K., & Jan, A. (2023). *Beta blockers*. StatPearls: Treasure Island (FL): National Library of Medicine.

Flaten, O., Sand, T., & Myren, J. (1982). Beta-adrenergic stimulation and blockade of the release of gastric inhibitory polypeptide and insulin in man. *Scandinavian Journal of Gastroenterology, 17*(2), 283–288.

Garrino, M., & Henquin, J.-C. (1990). B cell adrenoceptors and sulphonylurea-induced insulin release in mouse islets. *Diabetologia, 33*(3), 145–147.

Giugliano, D., Torella, R., & Sgambato, S. (1979). A comparative study of metabolic and hormonal responses to somatostatin and propranolol in man. *Acta Diabetologia Latina, 16*(1), 77–84.

Hain, J., Onoue, H., Mayrleitner, M., Fleischer, S., & Schindler, H. (1995). Phosphorylation modulates the function of the calcium release channel of sarcoplasmic reticulum from cardiac muscle. *Journal of Biological Chemistry, 270*(5), 2074–2081.

Harms, H., Gooren, L., Spoelstra, A., Hesse, C., & Verschoor, L. (1978). Blockade of isoprenaline-induced changes in plasma free fatty acids, immunoreactive insulin levels and plasma renin activity in healthy human subjects, by propranolol, pindolol, practolol, atenolol, metoprolol and acebutolol. *British Journal of Clinical Pharmacology, 5*(1), 19.

Henquin, J.-C. (2000). Triggering and amplifying pathways of regulation of insulin secretion by glucose. *Diabetes, 49*(11), 1751–1760.

Henquin, J.-C. (2009). Regulation of insulin secretion: A matter of phase control and amplitude modulation. *Diabetologia, 52*(5), 739–751.

Henquin, J.-C. (2021). Non-glucose modulators of insulin secretion in healthy humans:(dis) similarities between islet and in vivo studies. *Metabolism, 122*, 154821.

Imura, H., Kato, Y., Ikeda, M., Morimoto, M., & Yawata, M. (1971). Effect of adrenergic-blocking or-stimulating agents on plasma growth hormone, immunoreactive insulin, and blood free fatty acid levels in man. *The Journal of Clinical Investigation, 50*(5), 1069–1079.

International Diabetes Federation. (2025). *Diabetes around the world in 2024: Global factsheet (IDF diabetes Atlas)* (11th ed.). Brussels: International Diabetes Federation.

Islam, M. S. (2002). The ryanodine receptor calcium channel of β-cells: Molecular regulation and physiological significance. *Diabetes, 51*(5), 1299–1309.

Islam, M. S., Leibiger, I., Leibiger, B., Rossi, D., Sorrentino, V., Ekström, T. J., et al. (1998). In situ activation of the type 2 ryanodine receptor in pancreatic beta cells requires cAMP-dependent phosphorylation. *Proceedings of the National Academy of Sciences, 95*(11), 6145–6150.

Ito, K., Dezaki, K., Yoshida, M., Yamada, H., Miura, R., Rita, R. S., et al. (2017). Endogenous α_{2A}-adrenoceptor–operated sympathoadrenergic tones attenuate insulin secretion via cAMP/TRPM2 signalling. *Diabetes, 66*(3), 699–709.

Jansson, L., EiZirik, D. L., & Sandler, S. (1989). Terbutaline decreases the blood flow of the pancreatic islets but does not reduce the diabetogenic action of streptozotocin in the rat. *European Journal of Pharmacology, 161*(1), 79–83.

Kerns II, W. (2007). Management of β-adrenergic blocker and calcium channel antagonist toxicity. *Emergency Medicine Clinics of North America, 25*(2), 309–331.

Kveiborg, B., Hermann, T. S., Major-Pedersen, A., Christiansen, B., Rask-Madsen, C., Raunsø, J., et al. (2010). Metoprolol compared to carvedilol deteriorates insulin-stimulated endothelial function in patients with type 2 diabetes-a randomized study. *Cardiovascular Diabetology, 9*(1), 21.

Lacey, R., Berrow, N., London, N., Lake, S., James, R., Scarpello, J., et al. (1990). Differential effects of β-adrenergic agonists on insulin secretion from pancreatic islets isolated from rat and man. *Journal of Molecular Endocrinology, 5*(1), 49–54.

Lacey, R., Cable, H., James, R., London, N., Scarpello, J., & Morgan, N. (1993). Concentration-dependent effects of adrenaline on the profile of insulin secretion from isolated human islets of langerhans. *Journal of Endocrinology, 138*(3), 555–563.

Lacey, R., Chan, S., Cable, H., James, R., Perret, C., Scarpello, J., et al. (1996). Expression of α2-and β-adrenoceptor subtypes in human islets of langerhans. *Journal of Endocrinology, 148*(3), 531–543.

Lerner, R. L., & Porte, D. (1971). Epinephrine: Selective inhibition of the acute insulin response to glucose. *The Journal of Clinical Investigation, 50*(11), 2453–2457.

Lorza-Gil, E., Gerst, F., Oquendo, M. B., Deschl, U., Häring, H.-U., Beilmann, M., et al. (2019). Glucose, adrenaline and palmitate antagonistically regulate insulin and glucagon secretion in human pseudoislets. *Scientific Reports, 9*(1), 10261.

Loubatières, A., Mariani, M., Sorel, G., & Savi, L. (1971). The action of β-adrenergic blocking and stimulating agents on insulin secretion. Characterization of the type of β receptor. *Diabetologia, 7*(3), 127–132.

Malaisse, W., Malaisse-Lagae, F., & Mayhew, D. (1967). A possible role for the adenylcyclase system in insulin secretion. *Journal of Clinical Investigation, 46*(11), 1724–1734.

Meguid, M. M., Moore, M. C., Fitzpatrick, G., & Moore, F. D. (1975). Norepinephrine-induced insulin and substrate changes in normal man: Incomplete reversal by phentolamine. *Journal of Surgical Research, 18*(4), 365–369.

Meister, J., Bone, D. B., Knudsen, J. R., Barella, L. F., Liu, L., Lee, R., et al. (2022). In vivo metabolic effects after acute activation of skeletal muscle Gs signalling. *Molecular Metabolism, 55*, 101415.

Murao, N., Morikawa, R., Seino, Y., Shimomura, K., Maejima, Y., Yamada, Y., et al. (2025). β-Adrenergic blockers increase cAMP and stimulate insulin secretion through a PKA/RYR2/TRPM5 pathway in pancreatic β-Cells in vitro. *Pharmacology Research & Perspectives, 13*(2), e70092.

Nauck, M. A., & Müller, T. D. (2023). Incretin hormones and type 2 diabetes. *Diabetologia, 66*(10), 1780–1795.

Nazarzadeh, M., Bidel, Z., Canoy, D., Copland, E., Wamil, M., Majert, J., et al. (2021). Blood pressure lowering and risk of new-onset type 2 diabetes: An individual participant data meta-analysis. *The Lancet, 398*(10313), 1803–1810.

Northrop, G., Ryan, W. G., & Schwartz, T. B. (1973). Propranolol-induced insulin release in isolated rat islets of langerhans. *Diabetes, 22*(2), 91–93.

Oduori, O. S., Murao, N., Shimomura, K., Takahashi, H., Zhang, Q., Dou, H., et al. (2020). Gs/Gq signalling switch in β cells defines incretin effectiveness in diabetes. *Journal of Clinical Investigation, 130*(12), 6639–6655.

Palmer, J. P., Halter, J., & Werner, P. L. (1979). Differential effect of isoproterenol on acute glucagon and insulin release in man. *Metabolism, 28*(3), 237–240.

Pan, M., Yang, G., Cui, X., & Yang, S.-N. (2011). Subthreshold α2-Adrenergic activation counteracts glucagon-like Peptide-1 potentiation of glucose-stimulated insulin secretion. *Journal of Diabetes Research, 2011*(1), 604989.

Poirier, L., Cléroux, J., Nadeau, A., & Lacourcière, Y. (2001). Effects of nebivolol and atenolol on insulin sensitivity and haemodynamics in hypertensive patients. *Journal of Hypertension, 19*(8), 1429–1435.

Pollare, T., Lithell, H., Selinus, I., & Berne, C. (1989). Sensitivity to insulin during treatment with atenolol and metoprolol: A randomised, double blind study of effects on carbohydrate and lipoprotein metabolism in hypertensive patients. *British Medical Journal, 298*(6681), 1152–1157.

Porte, D. (1967). A receptor mechanism for the inhibition of insulin release by epinephrine in man. *Journal of Clinical Investigation, 46*(1), 86–94.

Porte, J.,D., & Williams, R. H. (1966). Inhibition of insulin release by norepinephrine in man. *Science, 152*(3726), 1248–1250.

Robertson, R. P., & Porte, J.,D. (1973). Adrenergic modulation of basal insulin secretion in man. *Diabetes, 22*(1), 1–8.

Rorsman, P., & Ashcroft, F. M. (2018). Pancreatic β-cell electrical activity and insulin secretion: Of mice and men. *Physiological Reviews, 98*(1), 117–214.

Rosengren, A. H., Jokubka, R., Tojjar, D., Granhall, C., Hansson, O., Li, D.-Q., et al. (2010). Overexpression of alpha2A-adrenergic receptors contributes to type 2 diabetes. *Science, 327*(5962), 217–220.

Schluter, K., Aellig, W., Petersen, K., Rieband, H.-C., Wehrli, A., & Kerp, L. (1982). The influence of beta-adrenoceptor blocking drugs with and without intrinsic sympathomimetic activity on the hormonal responses to hypo-and hyperglycaemia. *British Journal of Clinical Pharmacology, 13*(S2), 407S–417S.

Schuit, F., & Pipeleers, D. (1986). Differences in adrenergic recognition by pancreatic A and B cells. *Science, 232*(4752), 875–877.

Segerstolpe, Å., Palasantza, A., Eliasson, P., Andersson, E.-M., Andréasson, A.-C., Sun, X., et al. (2016). Single-cell transcriptome profiling of human pancreatic islets in health and type 2 diabetes. *Cell Metabolism, 24*(4), 593–607.

Seino, S., & Shibasaki, T. (2005). PKA-dependent and PKA-independent pathways for cAMP-regulated exocytosis. *Physiological Reviews, 85*(4), 1303–1342.

Sharifpour, A., Sadeghi, M., Zakariaei, Z., & Soleymani, M. (2022). Seizures and irreversible cardiogenic shock following propranolol poisoning: Report of 2 cases and literature review. *Clinical Medicine Insights: Case Reports, 15*, 11795476221126981.

Sluga, N., Križančić Bombek, L., Kerčmar, J., Sarikas, S., Postić, S., Pfabe, J., et al. (2022). Physiological levels of adrenaline fail to stop pancreatic beta cell activity at unphysiologically high glucose levels. *Frontiers in Endocrinology, 13*, 1013697.

Straub, S. G., & Sharp, G. W. (2012). Evolving insights regarding mechanisms for the inhibition of insulin release by norepinephrine and heterotrimeric G proteins. *American Journal of Physiology: Cell Physiology, 302*(12), C1687–C1698.

Ullrich, S., & Wollheim, C. (1984). Islet cyclic AMP levels are not lowered during alpha 2-adrenergic inhibition of insulin release. *Journal of Biological Chemistry, 259*(7), 4111–4115.

Varney, M. J., & Benovic, J. L. (2024). The role of G protein–coupled receptors and receptor kinases in pancreatic β-Cell function and diabetes. *Pharmacological Reviews, 76*(2), 267–299.

Zhang, H.-J., Walseth, T. F., & Robertson, R. P. (1989). Insulin secretion and cAMP metabolism in HIT cells: Reciprocal and serial passage-dependent relationships. *Diabetes, 38*(1), 44–48.

Zielmann, S., Schütte, G., Lenzen, S., & Panten, U. (1985). Effects of isoprenaline and glucagon on insulin secretion from pancreatic islets. *Naunyn-Schmiedeberg's Archives of Pharmacology, 329*(3), 299–304.

Chapter 30

Focused review: Clozapine safety and effects on pancreatitis

De las Cuevas C.

Department of Internal Medicine, Dermatology, and Psychiatry and Instituto Universitario de Neurociencia (IUNE), University of La Laguna, San Cristobal de La Laguna, Spain

List of abbreviations

ADR Adverse Drug Reaction
AhR Aryl Hydrocarbon Receptor
CBC Complete Blood Count
CIP Clozapine-Induced Pancreatitis
CRP C-Reactive Protein
CT Computed Tomography
DAMPs Danger-Associated Molecular Patterns
DIP Drug-Induced Pancreatitis
EMR Electronic Medical Record
FAERS FDA Adverse Event Reporting System
HbA1c Glycated hemoglobin (HbA1c)
HIF1 Hypoxia-Inducible Factor 1
HOMA-IR Homeostatic Model Assessment of Insulin Resistance
HPA axis Hypothalamic–Pituitary–Adrenal Axis
IC Information Component
IC025 Lower Bound of the 95% Confidence Interval of the Information Component
ICSR Individual Case Safety Report
IL-1β Interleukin-1 Beta
IL-6 Interleukin-6
MRI Magnetic Resonance Imaging
NLR Neutrophil-to-Lymphocyte Ratio
OR Odds Ratio
PRR Proportional Reporting Ratio
ROR Reporting Odds Ratio
SDM Shared Decision-Making
SIRS Systemic Inflammatory Response Syndrome
TNF-α Tumor Necrosis Factor-Alpha
TRS Treatment-Resistant Schizophrenia
WHO World Health Organization

Introduction

Clozapine remains the most efficacious antipsychotic for individuals with treatment-resistant schizophrenia, offering unmatched benefits in symptom reduction, functional recovery, and reduction in suicidal behavior (Kane et al., 1988; Meltzer et al., 2003). Despite its therapeutic superiority, clozapine is underprescribed worldwide, largely due to concerns about its complex safety profile, which includes life-threatening but rare adverse drug reactions (ADRs) such as agranulocytosis, myocarditis, pneumonia, and seizures (De Las Cuevas et al., 2024a; Gurrera et al., 2022). Among these, acute pancreatitis represents a lesser-known but increasingly recognized adverse event, with potentially severe consequences including systemic inflammatory response syndrome (SIRS), organ failure, and death (de Filippis et al., 2024; Lally et al., 2018).

Although clozapine-induced pancreatitis (CIP) is considered rare, growing evidence from pharmacovigilance sources, including the WHO global database VigiBase, suggests a rare but clinically relevant occurrence, often associated with severe or fatal outcomes (de Filippis et al., 2024; De Las Cuevas et al., 2024b). The condition typically presents as an idiosyncratic reaction, frequently occurring within the first weeks to months of treatment initiation, though delayed presentations have also been documented (Lally et al., 2018). The underrecognition of CIP may be due in part to diagnostic overshadowing in psychiatric settings and a lack of routine pancreatic monitoring in psychopharmacological protocols (Qubad & Bittner, 2023). This phenomenon, well-documented in the psychiatric literature, often leads to the misattribution of somatic symptoms to the underlying mental illness, delaying appropriate investigation and management.

Here we frame CIP within a broader neuroendocrine framework, emphasizing how endocrine axes and immune pathways converge to affect pancreatic resilience. Clozapine's impact on pancreatic function is multifactorial: it modulates endocrine axes—including the hypothalamic–pituitary–adrenal (HPA) axis and the

Hormones in Neuroscience. https://doi.org/10.1016/B978-0-443-30263-3.00013-0

insulin–glucagon regulatory system—while simultaneously influencing immunological and inflammatory pathways implicated in the disruption of pancreatic endocrine and exocrine homeostasis (Alvarez-Herrera et al., 2020; Smith et al., 2014). The neuroendocrine model thus provides a valuable lens through which to understand the pathophysiological plausibility of CIP, beyond traditional toxicological paradigms.

To provide a comprehensive and structured approach to this multifaceted issue, the chapter is organized as follows: First, we explore the neuroendocrine and immunological mechanisms that may underlie pancreatic vulnerability during clozapine treatment. Next, we analyze clinical and epidemiological characteristics of CIP, leveraging data from international pharmacovigilance databases. This is followed by a discussion of the psychological and behavioral consequences of pancreatitis in patients with severe mental illness, and the clinical dilemmas associated with clozapine withdrawal or rechallenge. The final sections provide translational insights, including interdisciplinary management strategies and proposals for integrating pharmacovigilance findings into safer prescribing practices. The chapter concludes with an integrative summary, a mini-dictionary of key terms, and practical points for clinicians, researchers, and trainees. Building on our recent pharmacovigilance work in VigiBase (De las Cuevas et al., 2024a, 2024b), we highlight practice-oriented implications that we have repeatedly encountered in clinical and monitoring settings.

Neuroendocrine and pathophysiological mechanisms

Clozapine may compromise pancreatic function through converging mechanisms. These include neuroendocrine disruption, immune activation, mitochondrial dysfunction, and metabolic imbalance. Rather than a purely dose-dependent toxicological effect, CIP likely emerges from idiosyncratic biological interactions affecting stress-response systems and pancreatic homeostasis. Fig. 30.1 summarizes the converging neuroendocrine, immuno-inflammatory, and metabolic pathways that may underpin clozapine-induced pancreatic stress.

HPA axis activation and inflammatory cytokines

Clozapine modulates the hypothalamic–pituitary–adrenal (HPA) axis. Preclinical studies indicate that clozapine can normalize glucocorticoid receptor sensitivity and modulate cortisol levels, potentially impacting stress responses and glucose metabolism (Chan et al., 2010; Reynolds & McGowan, 2017). Clinical studies indicate that atypical

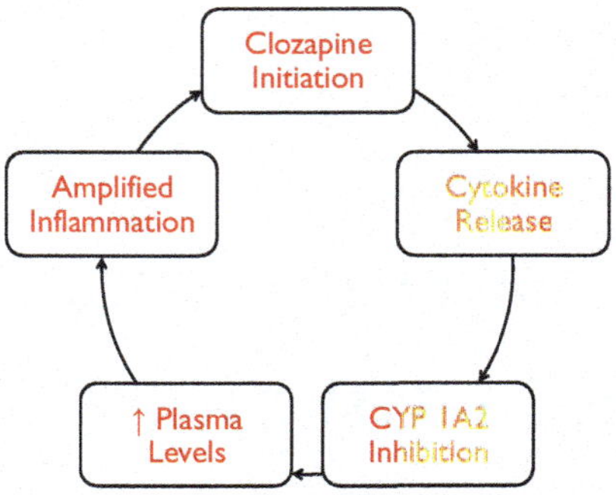

FIGURE 30.1 Pathophysiological mechanisms underlying clozapine-induced pancreatitis (CIP). This figure visually synthesizes the converging neuroendocrine, immunological, and metabolic pathways implicated in CIP.

antipsychotics, including clozapine, can influence adrenal hormone secretion (Nikolić et al., 2024). Simultaneously, clozapine induces a proinflammatory cytokine profile, elevating interleukin-6 (IL-6), C-reactive protein (CRP), tumor necrosis factor-alpha (TNF-α), and interleukin-1β (IL-1β), which are known contributors to pancreatic inflammation (Leung et al., 2023; Røge et al., 2012). Leung et al. (2023) demonstrated through a systematic review that clozapine-associated inflammation—especially during the early titration phase—is characterized by transient elevations of inflammatory markers including CRP, IL-6, and TNF-α, even in the absence of infection. These inflammatory states have been shown to downregulate CYP1A2 activity, impairing clozapine metabolism and thereby increasing plasma concentrations, a mechanism that may exacerbate systemic toxicity.

Disruption of pancreatic hormone secretion and metabolic regulation

Studies using isolated human islets and in vivo metabolic assessments have shown that clozapine directly increases insulin and glucagon secretion, disturbing the insulin–glucagon axis (Alvarez-Herrera et al., 2020; Smith et al., 2014). Hyperglucagonemia, whether centrally or peripherally mediated, may drive gluconeogenesis and alter pancreatic endocrine load, increasing vulnerability to injury (Habegger et al., 2010). Clozapine has also been linked to insulin resistance and impaired glucose tolerance in clamp studies (Yuen et al., 2021).

Mitochondrial toxicity and innate immune activation

Clozapine-induced mitochondrial dysfunction has been reported in hepatocytes, monocytes, and adipocytes, characterized by reduced mitochondrial membrane potential, ATP production, and increased oxidative stress

(Coccurello & Moles, 2010; Contreras-Shannon et al., 2013). These effects may extend to pancreatic cells, contributing to cell death and local inflammation. Mitochondrial injury may also promote the release of danger-associated molecular patterns (DAMPs), triggering innate immune responses and sustaining cytokine production (Faas & de Vos, 2020; Nakahira et al., 2015).

A fragile neuroendocrine–metabolic equilibrium

The convergence of HPA axis activation, systemic inflammation, insulin–glucagon imbalance, and mitochondrial distress undermines pancreatic resilience. In genetically or physiologically susceptible individuals, this milieu may suffice to precipitate CIP, particularly during the early treatment phase when compensatory systems are destabilized (Zhang et al., 2005).

This framework aligns with de Leon's three-phase model of clozapine-induced inflammation (de Leon, 2022), wherein rapid titration triggers cytokine release, metabolic inhibition of CYP1A2, and rising clozapine levels that may amplify inflammation. This cycle may progress toward localized autoimmune manifestations, including pancreatitis. This feedback loop, based on the model proposed by de Leon, is illustrated in Fig. 30.2.

From a translational standpoint, these insights suggest the need to monitor inflammatory markers (e.g., CRP, IL-6), metabolic indices (e.g., fasting glucose, insulin levels), and potentially mitochondrial function to detect early signs of pancreatic stress (Leung et al., 2023; Onodera et al., 2025).

Finally, although no direct immunopathogenic mechanism has been confirmed for clozapine, it is worth noting that drug-induced pancreatitis (DIP) may, in some cases, result from delayed-type hypersensitivity reactions. According to Badalov et al. (2007), immune-mediated mechanisms—including T-cell activation—have been proposed in non–dose-dependent DIP presentations involving other drugs. While speculative in the context of clozapine, this possibility aligns with its idiosyncratic profile and with clinical patterns such as late-onset reactions and recurrence upon rechallenge.

Clinical and epidemiological features of clozapine-induced pancreatitis

CIP is an infrequent but potentially life-threatening adverse drug reaction (ADR) that may present with a wide clinical spectrum, ranging from mild epigastric discomfort and transient enzyme elevation to fulminant pancreatitis with systemic inflammatory response syndrome (SIRS), multiorgan failure, or death. The condition is typically considered idiosyncratic, non–dose dependent, and unpredictable. Nonetheless, accumulating clinical and pharmacovigilance evidence highlights certain recurrent features that merit clinical awareness.

Clinical presentation and outcomes

According to the systematic review by Lally et al. (2018), most reported cases of CIP occur within the first 6 months of clozapine initiation, with a notable cluster during the first 2–8 weeks. Presenting symptoms include abdominal pain (especially in the epigastric region), nausea, vomiting, fever, and leukocytosis. Serum lipase and/or amylase are frequently elevated, though not always to diagnostic thresholds. Computed tomography (CT) or magnetic resonance imaging (MRI) may reveal pancreatic edema, necrosis, or peripancreatic fluid collections.

Outcomes vary widely. While many cases resolve with drug discontinuation and supportive treatment, others require intensive care management, with mortality reported in 5%–10% of severe presentations. Re-exposure to clozapine has led to recurrence in over 50% of rechallenge cases, highlighting the risk of immune sensitization or persistent vulnerability (Lally et al., 2018).

Pharmacovigilance data

Data derived from international pharmacovigilance systems provide a broader perspective on the epidemiology

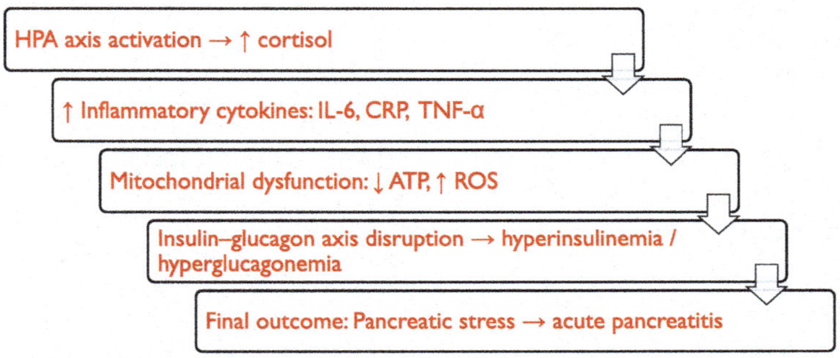

FIGURE 30.2 Inflammatory feedback cycle during clozapine titration: the de Leon Model. Illustrates the self-reinforcing inflammatory loop that can escalate to systemic or organ-specific toxicity, highlighting the temporal vulnerability window during early titration.

and seriousness of CIP. A recent VigiBase analysis focusing on clozapine-associated inflammatory organ disorders in pediatric patients identified pancreatitis as one of the top inflammatory ADRs, often co-reported with fever and elevated CRP (De Filippis et al., 2024). In the adult population, a study based on VigiBase records from over 30 countries found that reports of CIP, although rare, were disproportionately associated with fatal outcomes compared to pancreatitis linked to other antipsychotics (De las Cuevas et al., 2024a, 2024b). In line with those patterns, our VigiBase analyses also revealed disproportionate reporting of severe inflammatory outcomes, which informed the monitoring approach developed later in this chapter (De las Cuevas et al., 2024a, 2024b).

Complementary data from the FDA Adverse Event Reporting System (FAERS) indicate that antipsychotics as a drug class are among the top pharmacological groups associated with drug-induced pancreatitis (Zhang et al., 2023).

Briefly, VigiBase is the WHO global repository of individual case safety reports (ICSRs) curated by the Uppsala Monitoring Center. Disproportionality metrics such as Proportional Reporting Ratio (PRR), Reporting Odds Ratio (ROR), and the Bayesian Information Component (IC; IC025 > 0 indicates a signal) are used to flag potential safety concerns (Bate et al., 1998).

As of July 24, 2025, the public VigiAccess interface of the database reported 230,289 ADRs associated with clozapine, including 21,386 gastrointestinal disorders. Of these, 446 cases of pancreatitis and 157 cases of acute pancreatitis were documented, underscoring the rarity but persistence of this adverse reaction.

Furthermore, pediatric-specific data from a prior VigiBase extract (June 2022) (de Filippis et al., 2024) revealed 16 reports of pancreatitis in clozapine-treated individuals under 18 years of age. Although limited in absolute number, disproportionality analysis showed a robust signal: OR = 5.32 (95% CI: 3.25–8.71); IC025 = 1.43, confirming statistical significance; PRR = 5.30 (95% CI: 3.24–8.65)

These findings highlight a markedly elevated reporting probability of pancreatitis in this vulnerable age group. Although causality cannot be established from spontaneous reports, the statistical signal, combined with clinical plausibility and temporal association, supports the need for increased awareness and systematic monitoring—especially in off-label pediatric use.

Incidence, timing, and recurrence

Estimating the true incidence of CIP remains inherently difficult due to underreporting, diagnostic overshadowing in psychiatric populations, and the absence of standardized diagnostic thresholds across studies. Although no large-scale epidemiological studies have reported precise incidence rates for CIP, some reviews cite a tentative annual incidence of 0.05%–0.3% among clozapine users, extrapolated from clinical case series and pharmacovigilance reports. Case definitions, ascertainment practices, and thresholds for enzyme elevation vary across reports, which likely widens the confidence intervals around any crude incidence estimate.

Data from a systematic review by Lally et al. (2018) of published case series and reports revealed that CIP most often occurs within the first 2–6 weeks of treatment initiation, with a mean onset of 17.9 days (±11.2; range 4–35 days). Importantly, recurrence upon rechallenge was observed in five of seven cases (71.4%), often with more rapid and severe presentations—on average 3.5 days after re-initiation. These findings highlight the idiosyncratic nature of the reaction and the considerable risk associated with re-exposure.

In summary, CIP is a rare but potentially serious complication of clozapine treatment. While precise incidence remains undefined, the clinical and pharmacovigilance literature emphasizes the need for early recognition, high clinical suspicion during the initial treatment phase, and immediate discontinuation upon symptom onset. Given the substantial risk of recurrence, rechallenge should be considered only in exceptional cases, with interdisciplinary consensus and close monitoring. Finally, data from international pharmacovigilance databases such as VigiBase reinforce the importance of systematically documenting inflammatory adverse drug reactions associated with clozapine to guide safer prescribing practices.

Psychological and behavioral consequences of pancreatitis in psychiatric patients

The diagnosis of pancreatitis in individuals with severe mental illness, particularly those undergoing clozapine treatment for schizophrenia, carries significant psychological and behavioral implications. These consequences affect not only the trajectory of the primary psychiatric disorder but also patient attitudes toward medication adherence, illness perception, and engagement with healthcare.

Impact on schizophrenia course

Clozapine is often the last-line antipsychotic option for patients with treatment-resistant schizophrenia (TRS), offering substantial benefits in symptom reduction and suicidality (Meltzer et al., 2003). An acute episode of pancreatitis that necessitates the abrupt discontinuation of clozapine can precipitate psychiatric relapse, exacerbation of psychotic symptoms, hospitalization, or increased risk of self-harm (de Leon, Ruan, Schoretsanitis, & De Las Cuevas, 2020). The risk is heightened by the limited efficacy of alternative antipsychotic agents in TRS populations.

Furthermore, the physiological stress associated with pancreatitis—including pain, metabolic imbalance, and systemic inflammation—may aggravate cognitive and affective symptoms in schizophrenia, increasing disorganization, paranoia, or affective dysregulation. In support of this, a meta-analysis involving 2398 schizophrenia patients demonstrated a significant inverse relationship between systemic inflammatory markers (IL-6, IL-1β, TNF-α, CRP) and performance across key cognitive domains (e.g., processing speed, working memory, executive function), suggesting that inflammation exacerbates core psychopathological features (Patlola et al., 2023).

Effects on adherence, risk perception, and somatic distress

Experiencing a life-threatening adverse reaction like pancreatitis may lead patients to develop heightened health anxiety, mistrust of medications, or refusal to resume clozapine even when clinically indicated. Studies have shown that negative drug experiences can significantly impair attitudes toward pharmacotherapy, increase perceived vulnerability, and result in partial or full nonadherence (Byerly et al., 2005; De las Cuevas et al., 2021; Sendt et al., 2015). In the case of CIP, patients may also conflate abdominal discomfort or unrelated gastrointestinal symptoms with prodromal signs of relapse, further reinforcing anxiety and avoidance behaviors.

In addition, somatic distress may be magnified in patients with schizophrenia who exhibit impaired interoceptive accuracy and higher levels of somatization, contributing to diagnostic overshadowing or misinterpretation of early symptoms (Cohen et al., 2006; Yee et al., 2023). This can delay the recognition of true recurrence or lead to inappropriate medical consultations, compounding the emotional burden and undermining continuity of care.

Risks of abrupt clozapine withdrawal

Abrupt discontinuation of clozapine—whether prompted by pancreatitis or other medical emergencies—represents a major clinical challenge. Multiple case reports and systematic reviews have highlighted the rapid resurgence of psychotic symptoms, affective instability, and behavioral disorganization that can emerge within days of cessation. In more severe cases, catatonia and cholinergic rebound syndromes have also been described (Blackman & Oloyede, 2021; Shiovitz et al., 1996). The probability of psychiatric decompensation is particularly elevated when clozapine is stopped suddenly rather than tapered (Blackman et al., 2022).

From a broader perspective, clozapine withdrawal has been associated with inferior long-term psychiatric outcomes compared to discontinuation of other antipsychotics (Miura et al., 2022). These clinical risks are compounded by findings linking sudden cessation to elevated risk of suicide and mortality—issues further developed in Section 5.1.

Given these considerations, any unplanned withdrawal of clozapine should be treated as a high-risk event requiring careful multidisciplinary management, patient education, and, where feasible, contingency planning for potential rechallenge under strict monitoring.

Clinical dilemmas: Clozapine discontinuation and rechallenge

Managing clozapine treatment following an episode of pancreatitis involves critical clinical decisions, often in the absence of robust evidence. The risks associated with abrupt discontinuation must be carefully weighed against the potential for recurrence upon rechallenge. In this context, shared decision-making and interdisciplinary coordination are essential.

Risks of clozapine discontinuation

Epidemiological consequences of clozapine discontinuation

Beyond the immediate clinical deterioration described above, the long-term psychiatric and survival risks of discontinuing clozapine have also been extensively documented. Patchan et al. (2015) documented three suicides shortly after clozapine cessation in patients with schizophrenia and a history of suicidality. Similarly, a population-based study from Finland showed significantly higher all-cause mortality in patients who discontinued clozapine compared to those who continued treatment (Tiihonen et al., 2009).

These findings emphasize that discontinuation decisions must weigh not only the immediate medical risks, but also the broader psychiatric and survival implications. In the context of CIP, where discontinuation may be medically mandated, these risks underscore the need for careful monitoring, structured support, and consideration of rechallenge only under strict conditions.

Rechallenge after pancreatitis: Evidence and clinical guidance

Published evidence on clozapine rechallenge after pancreatitis is limited to case reports and small case series. Several reports describe recurrence of pancreatitis upon re-exposure to clozapine, while others document successful rechallenge under close monitoring conditions (Huang et al., 2009; Ratcliff et al., 2022; Rodríguez et al., 2020). In the largest systematic review of clozapine rechallenge after major adverse effects, Manu et al. (2018) noted that rechallenge may be considered in selected cases, though CIP was not specifically analyzed.

Consensus recommendations suggest that rechallenge might be cautiously pursued when (1) the initial episode was mild, non-necrotizing, and fully resolved; (2) no viable antipsychotic alternatives exist; and (3) intensive monitoring can be ensured.

Consistent with pharmacovigilance signals on inflammatory ADRs, a conservative re-exposure strategy is warranted if rechallenge is considered at all, with explicit multidisciplinary agreement and predefined stop criteria. Slow titration is increasingly recognized as a cornerstone of safe clozapine initiation, particularly during the vulnerable first weeks of treatment, when inflammatory or metabolic responses may emerge. In patients at increased risk of systemic inflammation or metabolic dysregulation, a conservative escalation schedule may reduce the likelihood of triggering pancreatic stress or inflammatory cascades (De Las Cuevas et al., 2024a; de Leon, 2024).

Suggested protocols include slow titration, regular laboratory monitoring (amylase, lipase, CRP), and early imaging in case of abdominal symptoms (Rodríguez et al., 2020).

Shared decision-making and clinical strategy

Given the complexity and uncertainty involved in CIP management, shared decision-making (SDM) is critical. Patients and caregivers should be engaged in detailed discussions regarding the psychiatric risks of discontinuation, the potential for recurrence, and the rationale for proposed management strategies. Multidisciplinary collaboration—including psychiatry, gastroenterology, and internal medicine—is essential to formulate a personalized and ethically sound approach (De las Cuevas and Peñate, 2014; De las Cuevas et al., 2014).

When discontinuation is unavoidable, tapering strategies should be employed whenever possible to minimize withdrawal-related phenomena such as cholinergic rebound, catatonia, or affective dysregulation (Blackman et al., 2022). Psychological support and relapse prevention planning are also key components of safe discontinuation (Horowitz et al., 2021).

Translational insights and future directions

Integrating mechanistic insights from immuno-neuroendocrine disruption into clinical practice provides opportunities to improve safety and efficacy in clozapine use. The following translational recommendations address proactive monitoring, interdisciplinary care, and pharmacovigilance-informed protocols.

Proactive monitoring: Inflammatory and metabolic markers

- **C-reactive protein (CRP)** and **troponin** are already recommended during the initial 4–8 weeks of clozapine titration to screen for myocarditis; similar protocols could be extended to pancreatitis risk profiling (Correll et al., 2022; de Leon et al., 2022).
- **Metabolic monitoring** should include fasting glucose, HbA1c, insulin levels, lipid panel, weight, and waist circumference, with an initial assessment at baseline and repeat testing at ~12 weeks and annually (Fehsel, 2024; Van Alphen et al., 2022)
- A quality-improvement initiative demonstrated that mental health and primary care collaboration—via shared EMRs, regular care conferences, and educational outreach—improves adherence to metabolic monitoring protocols (HbA1c, lipids) (Schneiderhan et al., 2023).

These measures can serve as early biomarkers not only for metabolic side effects but also for detecting early warning signs of pancreatic stress, including subclinical inflammation, insulin–glucagon imbalance, or mitochondrial strain. Additional indices such as the neutrophil-to-lymphocyte ratio (NLR), a validated marker of systemic inflammatory response, may provide a cost-effective tool for early detection of subclinical inflammation in clozapine-treated patients, especially during the titration phase or in those with comorbid risk (Zahorec, 2021).

Integration of endocrinology into psychiatric practice

- As clozapine can disturb glucose and lipid metabolism, a collaborative care model involving endocrinologists and psychiatrists ensures timely intervention for prediabetes, insulin resistance, and metabolic syndrome (Correll et al., 2022).
- National guidelines propose built-in screenings (e.g., glucose, insulin, lipids) within clozapine clinics, ideally with support from primary care, dietitians, or endocrinology services (Cohn & Sernyak, 2006).
- Treating metabolic and systemic inflammation early may reduce not only cardiovascular risk but also mitigate potential drivers of pancreatic vulnerability. In support of this, vascular inflammation—a key component of metabolic syndrome—has been identified as a critical mediator of cardiometabolic disease, suggesting that early anti-inflammatory and metabolic interventions could confer multi-organ protection, including to the pancreas (Koh et al., 2005).

Pharmacovigilance-based clinical protocols

- Drawing on structured monitoring protocols established for clozapine-induced myocarditis and supported by case reports of asymptomatic CIP, we propose a preventive protocol to detect early signs of CIP during the titration phase and beyond. Fig. 30.3 provides a practical outline for clinical monitoring during the first 8 weeks of clozapine treatment:
- In the context of a slow titration schedule, baseline and weekly (weeks 0–4): CRP, amylase, lipase, fasting glucose, insulin, and weight. This schedule is modeled after validated myocarditis protocols, where weekly CRP and troponin screening during the first month of clozapine use successfully identified inflammatory toxicity early in treatment (McNutt et al., 2021; Ronaldson et al., 2015).
- Weeks 5–8: Continue biweekly metabolic and pancreatic enzyme monitoring, especially in patients with comorbid metabolic risk or prior signs of inflammatory activation.
- When both fasting glucose and insulin are collected, calculating the Homeostatic Model Assessment of Insulin Resistance (HOMA-IR) may offer added clinical value. This index provides an estimate of insulin resistance and has been validated in psychiatric populations, where it correlates with both metabolic risk and systemic inflammatory load (Vancampfort et al., 2013).
- In the presence of abdominal symptoms: Immediate testing of CRP, amylase, and lipase should be undertaken, along with abdominal ultrasound or CT imaging within 24–48 hours. This is supported by case-based recommendations where asymptomatic or mild presentations progressed to clinically significant pancreatitis (Bergemann et al., 1999).

In line with best practices in pharmacovigilance, **all clinically relevant findings**—including elevated biomarkers, imaging results, or rechallenge outcomes—should be recorded in international databases such as **VigiBase**. This will enhance future signal detection and enable pattern recognition of high-risk biological profiles for CIP, especially when paired with structured data collection on inflammatory and metabolic parameters.

Future research directions

1. Prospective biomarker validation studies: Monitor CRP, IL-6, glucose, insulin, amylase, lipase, and HOMA-IR prospectively in clozapine initiators to establish predictive profiles for CIP.
2. Integration of mitochondrial health markers: Given preclinical evidence of mitochondrial toxicity, explore ATP, oxidative stress markers, or iron metabolism (e.g., ferritin, HIF-1 expression) as early indicators.
3. Randomized trials of protocol implementation: Compare standard versus enhanced monitoring on CIP detection rate, health outcomes, and cost-effectiveness.
4. Telehealth and shared-care interventions: Build on successful models (e.g., Schneiderhan et al., 2023) that link psychiatry, primary care, and endocrinology through technology-supported, interdisciplinary care forums.

Baseline & Weeks 0–4
- Baseline plus weekly monitoring: CRP, amylase, lipase, fasting glucose, insulin, and weight
- Modeled after validated clozapine myocarditis protocols (weekly CRP/troponin)
- Applied within a slow titration schedule

Weeks 5–8
- Continue biweekly metabolic and pancreatic enzyme monitoring
- Increased vigilance in patients with metabolic risk or prior inflammatory signs

Optional HOMA-IR
- If fasting glucose and insulin are collected → calculate HOMA-IR
- Validated in psychiatric populations for both metabolic and inflammatory risk

If Abdominal Symptoms
- Immediate CRP, amylase, and lipase testing
- Abdominal ultrasound or CT within 24–48 hours
- Based on case reports where mild/asymptomatic cases progressed to pancreatitis

Pharmacovigilance Reporting
- Record elevated biomarkers, imaging results, and rechallenge outcomes in VigiBase
- Supports early detection of high-risk biological profiles for CIP

FIGURE 30.3 Proposed pharmacovigilance-based clinical monitoring protocol for early detection of clozapine-induced pancreatitis (CIP).

Practical recommendations

For clinicians

- Monitor abdominal symptoms and biochemical changes during clozapine titration.
- Include amylase, lipase, and CRP in baseline labs with CBC and metabolic panels.
- Avoid rechallenge after severe, necrotizing, or recurrent pancreatitis; only consider after full recovery from mild cases and with strict monitoring.
- Use slow, individualized titration in the first 4–6 weeks, especially with metabolic comorbidities, obesity, or high baseline inflammation.

For researchers

- Conduct prospective studies to identify CIP predictive markers.
- Investigate genetic, immunologic, and mitochondrial factors increasing pancreatic risk.
- Report CIP cases to pharmacovigilance systems with detailed biomarker data.

For psychiatry residents/trainees

- Recognize CIP as multifactorial, not just "drug toxicity."
- Differentiate CIP from common GI side effects.
- Collaborate with endocrinology, gastroenterology, and primary care when adverse events occur.

Mini-dictionary of terms

Term	Definition
C-reactive protein (CRP)	An acute-phase protein synthesized by the liver in response to inflammation. It serves as an early and nonspecific marker of systemic inflammatory activity.
Amylase	A digestive enzyme primarily produced by the pancreas to break down carbohydrates. Elevated serum amylase is a common, though nonspecific, indicator of pancreatic irritation.
Lipase	A pancreatic enzyme essential for lipid digestion. Serum lipase is more specific than amylase for diagnosing acute pancreatitis and remains elevated longer.
Fasting glucose	The concentration of glucose in the blood after an overnight fast. It is a basic screening tool for detecting dysglycemia and insulin resistance.
Insulin	A hormone secreted by pancreatic β-cells that regulates glucose uptake and metabolism. Alterations in fasting insulin levels can indicate β-cell dysfunction or insulin resistance.
IL-6 (Interleukin-6)	A pro-inflammatory cytokine involved in acute-phase responses and immune regulation. It plays a central role in clozapine-induced inflammatory syndromes and may contribute to pancreatic stress.
HOMA-IR	Homeostatic model assessment of insulin resistance. A surrogate index calculated from fasting glucose and insulin to estimate the degree of insulin resistance.

Pathophysiological processes

Term	Definition
Clozapine-induced pancreatitis (CIP)	A rare but potentially severe inflammatory reaction of the pancreas triggered by clozapine treatment, typically presenting within weeks of initiation.
Neuroendocrine disruption	Interference with physiological axes—such as the HPA axis and insulin–glucagon system—that regulate metabolic and immune homeostasis.
Mitochondrial distress	A pathological state of impaired mitochondrial function, leading to reduced cellular energy production and increased oxidative stress, which may exacerbate inflammatory processes.
Inflammatory cascade	A coordinated sequence of immune responses involving cytokines, acute-phase proteins, and cellular mediators activated by infection, tissue damage, or drug reactions.
Diagnostic overshadowing	The misattribution of somatic complaints to psychiatric symptoms, which delays recognition and treatment of true medical pathology in psychiatric patients.

Clinical practice concepts

Term	Definition
Rechallenge	The cautious reintroduction of a medication after an adverse drug reaction. In the context of CIP, it requires strict monitoring and is generally discouraged unless no alternatives exist.
Pharmacovigilance	The systematic process of detecting, evaluating, understanding, and preventing adverse effects of pharmaceuticals, often through global reporting systems such as VigiBase and FAERS.

Slow titration	A gradual, individualized increase in medication dose to minimize the risk of adverse effects, particularly relevant during clozapine initiation.
Shared decision-making	A collaborative process through which clinicians and patients jointly make healthcare decisions, integrating clinical evidence with patient values and preferences.

Summary points

- Clozapine remains the gold standard for treatment-resistant schizophrenia. Its efficacy in symptom control and suicide prevention is unmatched.
- Clozapine-induced pancreatitis (CIP), although rare, is a severe, recurrent, and potentially fatal complication that remains underrecognized in psychiatric practice.
- CIP should be conceptualized as a neuroendocrine-–inflammatory phenomenon, involving HPA axis modulation, insulin–glucagon imbalance, and systemic inflammatory activation.
- Most cases occur within the first 4–8 weeks of treatment; recurrence rates upon rechallenge are high, necessitating cautious, individualized decision-making.
- Diagnostic overshadowing and the absence of routine pancreatic monitoring in psychiatry contribute to delayed diagnosis and worse outcomes.
- Proactive monitoring—especially during titration—should include CRP, amylase, lipase, fasting glucose, and insulin, with rapid evaluation of abdominal symptoms.
- Integration of psychiatry with endocrinology and internal medicine can optimize early detection, management, and prevention of CIP.
- Pharmacovigilance data from sources such as VigiBase should inform clinical protocols and contribute to building robust risk profiles.
- Addressing the psychological impact of CIP, including anxiety, reduced adherence, and relapse risk after abrupt discontinuation, is essential for holistic care.

References

Alvarez-Herrera, S., Escamilla, R., Medina-Contreras, O., Saracco, R., Flores, Y., Hurtado-Alvarado, G., Maldonado-García, J. L., Becerril-Villanueva, E., Pérez-Sánchez, G., & Pavón, L. (2020). Immunoendocrine peripheral effects induced by atypical antipsychotics. *Frontiers in Endocrinology, 11*, 195.

Badalov, N., Baradarian, R., Iswara, K., Li, J., Steinberg, W., & Tenner, S. (2007). Drug-induced acute pancreatitis: An evidence-based review. *Clinical Gastroenterology and Hepatology, 5*(6), 648–661.

Bate, A., Lindquist, M., Edwards, I. R., Olsson, S., Orre, R., Lansner, A., & De Freitas, R. M. (1998). A Bayesian neural network method for adverse drug reaction signal generation. *European Journal of Clinical Pharmacology, 54*(4), 315–321.

Bergemann, N., Ehrig, C., Diebold, K., Mundt, C., & von Einsiedel, R. (1999). Asymptomatic pancreatitis associated with clozapine. *Pharmacopsychiatry, 32*(2), 78–80.

Blackman, G., & Oloyede, E. (2021). Clozapine discontinuation withdrawal symptoms in schizophrenia. *Therapeutic Advances in Psychopharmacology, 11*, 20451253211032053.

Blackman, G., Oloyede, E., Horowitz, M., Harland, R., Taylor, D., MacCabe, J., & McGuire, P. (2022). Reducing the risk of withdrawal symptoms and relapse following clozapine discontinuation—Is it feasible to develop evidence-based guidelines? *Schizophrenia Bulletin, 48*(1), 176–189.

Byerly, M., Fisher, R., Whatley, K., Holland, R., Varghese, F., Carmody, T., Magouirk, B., & Rush, A. J. (2005). A comparison of electronic monitoring vs. clinician rating of antipsychotic adherence in outpatients with schizophrenia. *Psychiatry Research, 133*(2–3), 129–133.

Chan, H. Y., Chang, C. J., Chiang, S. C., Chen, J. J., Chen, C. H., Sun, H. J., Hwu, H. G., & Lai, M. S. (2010). A randomised controlled study of risperidone and olanzapine for schizophrenic patients with neuroleptic-induced acute dystonia or parkinsonism. *Journal of Psychopharmacology, 24*(1), 91–98.

Coccurello, R., & Moles, A. (2010). Potential mechanisms of atypical antipsychotic-induced metabolic derangement: Clues for understanding obesity and novel drug design. *Pharmacology & Therapeutics, 127*(3), 210–251.

Cohen, A. S., Forbes, C. B., Mann, M. C., & Blanchard, J. J. (2006). Specific cognitive deficits and differential domains of social functioning impairment in schizophrenia. *Schizophrenia Research, 81*(2–3), 227–238.

Cohn, T. A., & Sernyak, M. J. (2006). Metabolic monitoring for patients treated with antipsychotic medications. *Canadian Journal of Psychiatry, 51*(8), 492–501.

Contreras-Shannon, V., Heart, D. L., Paredes, R. M., Navaira, E., Catano, G., Maffi, S. K., & Walss-Bass, C. (2013). Clozapine-induced mitochondria alterations and inflammation in brain and insulin-responsive cells. *PLoS One, 8*(3), e59012.

Correll, C. U., Agid, O., Crespo-Facorro, B., de Bartolomeis, A., Fagiolini, A., Seppälä, N., & Howes, O. D. (2022). A guideline and checklist for initiating and managing clozapine treatment in patients with treatment-resistant schizophrenia. *CNS Drugs, 36*(7), 659–679. Erratum in *CNS Drugs*, *36*(9), 1015.

de Filippis, R., De Las Cuevas, C., Sanz, E. J., Schoretsanitis, G., Correll, C. U., & de Leon, J. (2024). Clozapine-associated pericarditis and pancreatitis in children and adolescents: A systematic literature review and pharmacovigilance study using the VigiBase database. *Schizophrenia Research, 268*, 118–130.

De Las Cuevas, C., Baptista, T., Motuca, M., Villasante-Tezanos, G. A., Lazary, J., Pogany, L., & De Leon, J. (2021). Poor adherence to oral psychiatric medication in adults with schizophrenia may be influenced by pharmacophobia, high internal health locus of control and treatment duration. *Neuropsychopharmacologia Hungarica, 23*(4), 388–404.

De Las Cuevas, C., & Peñate, W. (2014). To what extent psychiatric patients feel involved in decision making about their mental health care? Relationships with socio-demographic, clinical, and psychological variables. *Acta Neuropsychiatrica, 26*(6), 372–381.

De Las Cuevas, C., Peñate, W., & de Rivera, L. (2014). To what extent is treatment adherence of psychiatric patients influenced by their participation in shared decision making? *Patient Preference and Adherence, 8*, 1547–1553.

De Las Cuevas, C., Sanz, E. J., & de Leon, J. (2024a). Adverse drug reactions and their fatal outcomes in clozapine patients in VigiBase: Comparing the top four reporting countries (US, UK, Canada and Australia). *Schizophrenia Research, 268*, 165–174.

De Las Cuevas, C., Sanz, E. J., & de Leon, J. (2024b). Pharmacovigilance in action: Utilizing VigiBase data to improve clozapine safety. *Patient Preference and Adherence, 18*, 2261–2280.

de Leon, J. (2022). Reflections on the complex history of the concept of clozapine-induced inflammation during titration. *Psychiatria Danubina, 34*(3), 411–421.

de Leon, J. (2024). Can slow personalized titration using C-reactive protein monitoring decrease the high rates and mortality of clozapine-associated myocarditis seen in some countries? A call for research. *Journal of Clinical Psychopharmacology, 44*(3), 212–219.

de Leon, J., Ruan, C. J., Schoretsanitis, G., & De Las Cuevas, C. (2020). A rational use of clozapine based on adverse drug reactions, pharmacokinetics, and clinical pharmacopsychology. *Psychotherapy and Psychosomatics, 89*(4), 200–214.

de Leon, J., Ruan, C. J., Schoretsanitis, G., Rohde, C., Yağcıoğlu, A. E., Baptista, T., Kirilochev, O. O., De Las Cuevas, C., & Correll, C. U. (2022). An international guideline with six personalised titration schedules for preventing myocarditis and pneumonia associated with clozapine. *General Psychiatry, 35*(3), e100773.

Faas, M. M., & de Vos, P. (2020). Mitochondrial function in immune cells in health and disease. *Biochimica et Biophysica Acta - Molecular Basis of Disease, 1866*(10), 165845.

Fehsel, K. (2024). Metabolic side effects from antipsychotic treatment with clozapine linked to aryl hydrocarbon receptor (AhR) activation. *Biomedicines, 12*(10), 2294.

Gurrera, R. J., Gearin, P. F., Love, J., Li, K. J., Xu, A., Donaghey, F. H., & Gerace, M. R. (2022). Recognition and management of clozapine adverse effects: A systematic review and qualitative synthesis. *Acta Psychiatrica Scandinavica, 145*(5), 423–441.

Habegger, K. M., Heppner, K. M., Geary, N., Bartness, T. J., DiMarchi, R., & Tschöp, M. H. (2010). The metabolic actions of glucagon revisited. *Nature Reviews Endocrinology, 6*(12), 689–697.

Horowitz, M. A., Jauhar, S., Natesan, S., Murray, R. M., & Taylor, D. (2021). A method for tapering antipsychotic treatment that may minimize the risk of relapse. *Schizophrenia Bulletin, 47*(4), 1116–1129. Erratum in: *Schizophrenia Bulletin, 49*(2), 533.

Huang, Y. J., Lane, H. Y., Liao, C. H., & Huang, C. C. (2009). Recurrent pancreatitis without eosinophilia on clozapine rechallenge. *Progress in Neuro-Psychopharmacology & Biological Psychiatry, 33*(8), 1561–1562.

Kane, J., Honigfeld, G., Singer, J., & Meltzer, H. (1988). Clozapine for the treatment-resistant schizophrenic. A double-blind comparison with chlorpromazine. *Archives of General Psychiatry, 45*(9), 789–796.

Koh, K. K., Han, S. H., & Quon, M. J. (2005). Inflammatory markers and the metabolic syndrome: Insights from therapeutic interventions. *Journal of the American College of Cardiology, 46*(11), 1978–1985.

Lally, J., Al Kalbani, H., Krivoy, A., Murphy, K. C., Gaughran, F., & MacCabe, J. H. (2018). Hepatitis, interstitial nephritis, and pancreatitis in association with clozapine treatment: A systematic review of case series and reports. *Journal of Clinical Psychopharmacology, 38*(5), 520–527.

Leung, J. G., Zhang, L., Markota, M., Ellingrod, V. L., Gerberi, D. J., & Bishop, J. R. (2023). A systematic review of clozapine-associated inflammation and related monitoring. *Pharmacotherapy: The Journal of Human Pharmacology and Drug Therapy, 43*(12), 1364–1396.

Manu, P., Lapitskaya, Y., Shaikh, A., & Nielsen, J. (2018). Clozapine rechallenge after major adverse effects: Clinical guidelines based on 259 cases. *American Journal of Therapeutics, 25*(2), e218–e223.

McNutt, S. E., Morrow, G., Waters, K., Cooper, C., & Crish, E. (2021). Health system pharmacist protocol for myocarditis monitoring in clozapine-naïve inpatients: Two years' experience. *Journal of Pharmacy Practice, 34*(4), 542–546.

Meltzer, H. Y., Alphs, L., Green, A. I., Altamura, A. C., Anand, R., Bertoldi, A., Bourgeois, M., Chouinard, G., Islam, M. Z., Kane, J., Krishnan, R., Lindenmayer, J. P., Potkin, S., & International Suicide Prevention Trial Study Group. (2003). Clozapine treatment for suicidality in schizophrenia: International Suicide Prevention Trial (InterSePT). *Archives of General Psychiatry, 60*(1), 82–91.

Miura, G., Tanaka, K., Kemuriyama, T., Misawa, F., Uchida, H., Mimura, M., & Takeuchi, H. (2022). Clinical outcomes after clozapine discontinuation in patients with schizophrenia: A systematic review. *Pharmacopsychiatry, 55*(4), 181–192.

Nakahira, K., Hisata, S., & Choi, A. M. (2015). The roles of mitochondrial damage-associated molecular patterns in diseases. *Antioxidants & Redox Signaling, 23*(17), 1329–1350.

Nikolić, T., Bogosavljević, M. V., Stojković, T., Kanazir, S., Lončarević-Vasiljković, N., Radonjić, N. V., Popić, J., & Petronijević, N. (2024). Effects of antipsychotics on the hypothalamus–pituitary–adrenal axis in a phencyclidine animal model of schizophrenia. *Cells, 13*(17), 1425.

Onodera, B., Sakata, M., Ikawa, K., Kume, D., Horikawa, N., Komatsu, H., Yamaguchi, T., Tomita, H., & Kikuchi, Y. (2025). Elevated neutrophil-to-lymphocyte ratios correlate with increased clozapine concentration-to-dose ratios during titration. *Schizophrenia (Heidelberg), 11*(1), 96.

Patchan, K. M., Richardson, C., Vyas, G., & Kelly, D. L. (2015). The risk of suicide after clozapine discontinuation: Cause for concern. *Annals of Clinical Psychiatry, 27*(4), 253–256.

Patlola, S. R., Donohoe, G., & McKernan, D. P. (2023). The relationship between inflammatory biomarkers and cognitive dysfunction in patients with schizophrenia: A systematic review and meta-analysis. *Progress in Neuro-Psychopharmacology & Biological Psychiatry, 121*, 110668.

Qubad, M., & Bittner, R. A. (2023). Second to none: Rationale, timing, and clinical management of clozapine use in schizophrenia. *Therapeutic Advances in Psychopharmacology, 13*, 20451253231158152.

Røge, R., Møller, B. K., Andersen, C. R., Correll, C. U., & Nielsen, J. (2012). Immunomodulatory effects of clozapine and their clinical implications: What have we learned so far? *Schizophrenia Research, 140*(1–3), 204–213.

Ratcliff, K., Juneja, V., Mbbs, A. G., & Bastiampillai, T. (2022). Clozapine rechallenge with recurrent pancreatitis. *Primary Care Companion for CNS Disorders, 24*(6), 22cr03237.

Reynolds, G. P., & McGowan, O. O. (2017). Mechanisms underlying metabolic disturbances associated with psychosis and antipsychotic drug treatment. *Journal of Psychopharmacology, 31*(11), 1430–1436.

Rodriguez, V., Hanley, K., Arias, A. J., Quattrone, D., Kuforiji, J., Whiskey, E., & Shergill, S. S. (2020). Successful clozapine rechallenge following recurrent clozapine-associated pancreatitis: A case report. *BMC Pharmacology and Toxicology, 21*(1), 35.

Ronaldson, K. J., Fitzgerald, P. B., & McNeil, J. J. (2015). Clozapine-induced myocarditis, a widely overlooked adverse reaction. *Acta Psychiatrica Scandinavica, 132*(4), 231–240.

Schneiderhan, M. E., O'Donnell, C., Jackson, S. K., MacDonald, D. A., Yapel, A. M., Renier, C. M., Albee, J. N., & Hager, K. D. (2023). Bridging community mental health and primary care to improve medication monitoring and outcomes for patients with mental illness taking second-generation antipsychotics—Phase 2: Quality improvement initiative over 15 months. *Primary Care Companion for CNS Disorders, 25*(3), 22m03432.

Sendt, K. V., Tracy, D. K., & Bhattacharyya, S. (2015). A systematic review of factors influencing adherence to antipsychotic medication in schizophrenia-spectrum disorders. *Psychiatry Research, 225*(1–2), 14–30.

Shiovitz, T. M., Welke, T. L., Tigel, P. D., Anand, R., Hartman, R. D., Sramek, J. J., Kurtz, N. M., & Cutler, N. R. (1996). Cholinergic rebound and rapid onset psychosis following abrupt clozapine withdrawal. *Schizophrenia Bulletin, 22*(4), 591–595.

Smith, G. C., Zhang, Z. Y., Mulvey, T., Petersen, N., Lach, S., Xiu, P., Phillips, A., Han, W., Wang, M. W., & Shepherd, P. R. (2014). Clozapine directly increases insulin and glucagon secretion from islets: Implications for impairment of glucose tolerance. *Schizophrenia Research, 157*(1–3), 128–133.

Tiihonen, J., Lönnqvist, J., Wahlbeck, K., Klaukka, T., Niskanen, L., Tanskanen, A., & Haukka, J. (2009). 11-year follow-up of mortality in patients with schizophrenia: A population-based cohort study (FIN11 study). *Lancet, 374*(9690), 620–627.

Van Alphen, M. U., Vyas, C. M., Paudel, S., Makhlouf, A. T., Arjyal, S., & Stern, T. A. (2022). Managing clozapine in primary care settings. *Primary Care Companion for CNS Disorders, 24*(1), 21f02948.

Vancampfort, D., Probst, M., Scheewe, T., De Herdt, A., Sweers, K., Knapen, J., van Winkel, R., & De Hert, M. (2013). Relationships between physical fitness, physical activity, smoking and metabolic and mental health parameters in people with schizophrenia. *Psychiatry Research, 207*(1–2), 25–32.

Yee, B., Looi, J. C. L., Agaciak, M., Allison, S., Chan, S. K. W., & Bastiampillai, T. (2023). Non-psychosis symptoms of clozapine withdrawal: A systematic review. *East Asian Archives of Psychiatry, 33*(2), 44–64.

Yuen, J. W. Y., Kim, D. D., Procyshyn, R. M., Panenka, W. J., Honer, W. G., & Barr, A. M. (2021). A focused review of the metabolic side-effects of clozapine. *Frontiers in Endocrinology, 12*, 609240.

Zahorec, R. (2021). Neutrophil-to-lymphocyte ratio, past, present and future perspectives. *Bratislavske Lekarske Listy, 122*(7), 474–488.

Zhang, L., Mao, W., Liu, D., Hu, B., Lin, X., Ran, J., Li, X., & Hu, J. (2023). Risk factors for drug-related acute pancreatitis: An analysis of the FDA adverse event reporting system (FAERS). *Frontiers in Pharmacology, 14*, 1231320.

Zhang, X. Y., Zhou, D. F., Cao, L. Y., Wu, G. Y., & Shen, Y. C. (2005). Cortisol and cytokines in chronic and treatment-resistant patients with schizophrenia: Association with psychopathology and response to antipsychotics. *Neuropsychopharmacology, 30*(8), 1532–1538.

Section 6

Diseases of the pancreas or pancreas hormones status and impact on the brain and behavior

Chapter 31

Pancreatic cancer—An overview

Stanisław Surma[1], Dawid Szkudłapski[2] and Kazimierz Ciechanowski[3]

[1]*Department of Internal Medicine and Clinical Pharmacology, Medical University of Silesia, Katowice, Poland;* [2]*Collegium Medicum, WSB University, Dąbrowa Górnicza, Poland;* [3]*Department of Nephrology, Transplantology and Internal Medicine, Pomeranian Medical University, Szczecin, Poland*

List of all abbreviations

95% CI 95% confidence interval
BRPC Borderline resectable pancreatic cancer
CA19-9 Carbohydrate antigen 19–9
CEA Carcinoembryonic antigen
CT Computed tomography
dMMR Deficient mismatch repair
EUS Endoscopic ultrasound
EUS-FNA Endoscopic ultrasound-guided fine-needle aspiration
FDA Food and Drug Administration
FOLFIRINOX 5-FU, leucovorin, irinotecan, oxaliplatin
GemNabP Gemcitabine plus nab-paclitaxel
HR hazard ratio
HRD Homologous recombination deficiency
IPMN Intraductal papillary mucinous neoplasm
LAPC Locally advanced pancreatic cancer
MRCP Magnetic resonance cholangiopancreatography
MRI Magnetic resonance imaging
MSI-H Microsatellite instability-high
NCCN National Comprehensive Cancer Network
PanIN Pancreatic intraepithelial neoplasia
PARP Poly(ADP-ribose) polymerase
PDAC Pancreatic ductal adenocarcinoma
PET Positron emission tomography
RPC Resectable pancreatic cancer
RR relative risk
SIR standardized incidence ratio
TGF-β - transforming growth factor β
TNM Tumor–Node–Metastasis

Introduction

Pancreatic cancer, most notably pancreatic ductal adenocarcinoma, is among the malignancies characterized by an exceptionally aggressive course and very high mortality, and it frequently presents with vague, non-specific symptoms that contribute to delayed diagnosis. The median overall survival, regardless of disease stage at the time of presentation, is approximately 4 months, while the 5-year survival rate reaches 13%, having increased from 4% over the past 3 decades primarily due to advances in oncologic therapies, refinements in surgical techniques, and the centralization of care in specialized centers (Hayat et al., 2025; Stoop et al., 2025). The poor prognosis reflects both the intrinsic biological aggressiveness of the tumor and its asymptomatic or minimally symptomatic early course, resulting in more than half of patients having metastatic disease at the time of diagnosis. Early systemic dissemination and the limited effectiveness of available systemic treatments make disease control challenging even in individuals with radiologically localized tumors (Hayat et al., 2025; Stoop et al., 2025). The cornerstone of management remains the combination of chemotherapy and surgical resection; however, only a small proportion of patients are eligible for operative treatment. The introduction of multi-agent chemotherapy regimens has improved survival outcomes in the palliative setting for patients with metastatic disease, as well as in the (neo) adjuvant and induction settings for those with borderline resectable and locally advanced tumors. Given projections that pancreatic cancer will become the second leading cause of cancer-related death by 2030, the development of more effective therapeutic strategies is urgently needed, alongside surveillance of individuals with familial pancreatic cancer, specific genetic mutations, and a high risk of developing intraductal papillary mucinous neoplasms, all of whom represent populations at increased risk for this malignancy (Hayat et al., 2025; Stoop et al., 2025).

Pancreatic cancer: Epidemiology and risk factors

Among all malignancies, pancreatic cancer ranks 12th in incidence and sixth in overall mortality. It is projected that

Hormones in Neuroscience. https://doi.org/10.1016/B978-0-443-30263-3.00039-7

by 2030 it will become the second leading cause of cancer-related death in the United States (Tonini & Zanni, 2025). The estimated global lifetime risk of developing pancreatic cancer is 0.89% (95% CI 0.88–0.89), with marked geographic variation—from 0.15% (0.13–0.18) in Central Africa to 2.06% (2.04–0.08) in Western Europe. The corresponding lifetime risk of death worldwide is 0.85% (95% CI 0.85–0.85) (Tonini & Zanni, 2025). Between 2010 and 2019, the age-standardized incidence increased from 6.3 to 6.6 per 100,000 individuals, a trend largely driven by countries with low or middle sociodemographic index levels. A rising incidence has also been observed among young adults; in the 15–39-year age group, the global age-standardized incidence is 0.2%, partly attributable to the growing prevalence of obesity (Tonini & Zanni, 2025).

This worsening epidemiological situation suggests that, beyond global population aging, exposure to risk factors is becoming more widespread. Both modifiable lifestyle factors and inherited predispositions have been associated with pancreatic cancer risk (Grigorescu et al., 2024) (Fig. 31.1 and Table 31.1).

Tobacco smoking remains the most significant modifiable determinant, conferring a relative risk (RR) of 1.8 (95% CI 1.7–1.9) compared with never-smokers. Approximately 21% (95% CI 19%–24%) of age-standardized pancreatic cancer mortality is attributable to smoking, followed by elevated fasting plasma glucose (9%; 95% CI 2%–19%) and high body mass index (6%; 95% CI 3%–11%). Smoking cessation reduces risk, as reflected by the lower relative risk among former smokers compared with those who have never smoked (RR 1.2; 95% CI 1.1–1.2), with longer durations of abstinence conferring additional benefit. Heavy alcohol consumption is likewise associated with increased risk, which may decline after approximately 10 years of abstinence (Grigorescu et al., 2024; Stoop et al., 2025).

Diabetes mellitus represents another established risk factor (RR 1.5; 95% CI 1.4–1.6), although it may also constitute an early manifestation of pancreatic cancer, referred to as type 3c (pancreatogenic) diabetes. Its dual role is supported by the stronger association observed with recently diagnosed diabetes (≤three to four years) compared with long-standing diabetes (>three to four years). Higher HbA1c levels in individuals with newly diagnosed diabetes correlate with an elevated risk of pancreatic cancer. Moreover, even in individuals without diagnosed diabetes, increased fasting glucose levels are associated with greater risk. Metabolic syndrome is similarly linked to an increased likelihood of pancreatic cancer (RR 1.3; 95% CI 1.2–1.5), with risk rising in proportion to the number of its components; resolution of metabolic syndrome is associated with a reduction in this excess risk (Grigorescu et al., 2024; Sharma et al., 2025; Stoop et al., 2025).

Pancreatitis is a strong and well-established risk factor, particularly in its chronic form (standardized incidence ratio [SIR] 22.6; 95% CI 14.4–35.4), with cumulative incidence increasing over the course of the disease. In contrast, among patients with acute pancreatitis, the risk of pancreatic cancer is highest within the first 2 months after symptom onset, as acute pancreatitis may represent the initial manifestation of an underlying malignancy (HR 172.8; 95% CI 54.9–544.7). The risk gradually declines over time and is no longer elevated 10 years after diagnosis (Grigorescu et al., 2024; Sharma et al., 2025; Stoop et al., 2025).

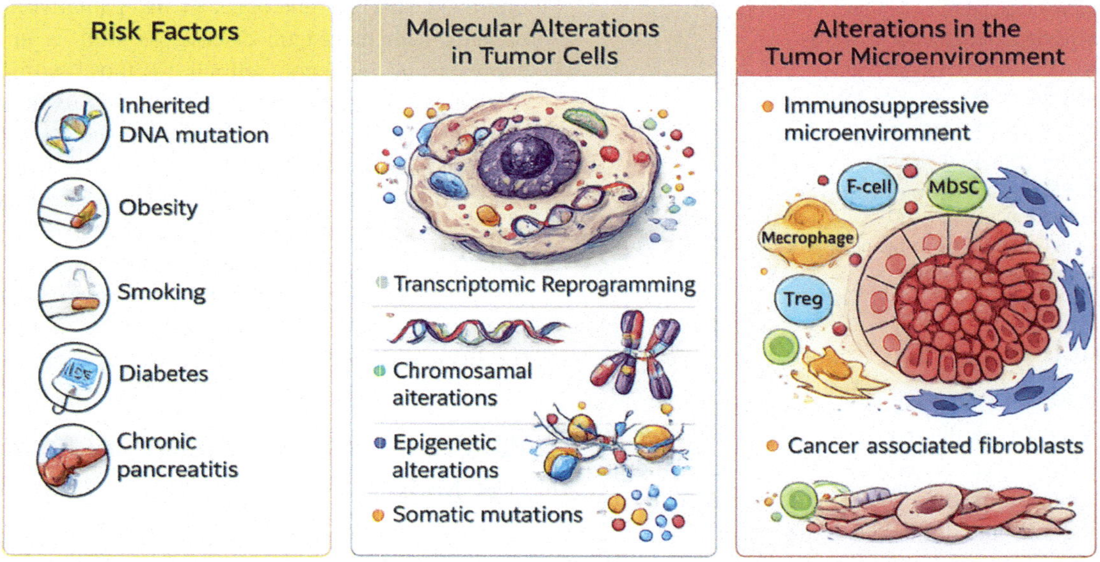

FIGURE 31.1 Mechanisms of pancreatic cancer pathogenesis. Based on information from (Rozengurt & Eibl, 2026).

TABLE 31.1 Risk factors (A – modifiable; B – inherited/genetic) of pancreatic cancer. Based on information from (Gu & Minko, 2024; Stoop et al., 2025; Wood et al., 2022).

A) Modifiable risk factors		
Risk factor	**Risk measure**	**95% CI**
Acute pancreatitis	SIR	172.8 (54.9–544.7)
Chronic pancreatitis	SIR	22.6 (14.4–35.4)
Cigarette smoking	RR	1.8 (1.7–1.9)
Heavy alcohol intake ≥60 g/day (vs. none) ≥9 drinks/day (vs. < 1/day)	RR OR	1.6 (1.0–2.5) 1.6 (1.2–2.2)
Diabetes	RR	1.5 (1.4–1.6)
Metabolic syndrome	RR	1.3 (1.2–1.5)
BMI (per 5-unit increase)	RR	1.1 (1.1–1.1)
B) Inherited/genetic risk factors		
Risk factor	**Risk measure**	**95% CI**
Familial pancreatic cancer	SIR	4.9 (4.0–5.9)
BRCA1	RR	2.4 (1.5–3.7)
BRCA2	RR	3.3 (2.2–5.1)
PALB2	RR	2.4 (1.2–4.5)
ATM	RR	6.5 (4.5–9.5)
MLH1/MSH2/MSH6 (Lynch syndrome)	HR	8.6 (4.7–15.7)
MLH1	OR	6.7 (2.5–15.0)
MSH2	OR	1.6 (0.1–7.5)
MSH6	OR	2.0 (0.8–4.1)
TP53	RR	6.7 (2.5–15.0)
CDKN2A (FAMMM)	OR	12.3 (5.4–25.6)
STK11/LKB1 (Peutz–Jeghers syndrome)	RR	132 (CI not reported)
PRSS1 (hereditary pancreatitis); other genes: SPINK1, CTRC, CFTR, and CPA1	SIR	63.4 (45.4–88.5)

A hereditary component is suspected in approximately 21%–36% of cases, although the exact proportion remains uncertain. Familial pancreatic cancer—defined as the presence of more than one affected first-degree relative—substantially increases the risk of developing the disease (SIR 4.9; 95% CI 4.0–5.9). The magnitude of risk rises with the number of affected first-degree relatives; it is higher in individuals with three or more such relatives (SIR 10.8; 95% CI 4.7–10.1) than in those with one (SIR 3.46; 95% CI 2.52–4.76) or two (SIR 5.44; 95% CI 4.07–7.26). Additionally, risk is greater in families in which cancer occurred at a young age (<50 years). Nevertheless, germline mutations are identified in only about 10% (95% CI 8%–12%) of individuals with familial pancreatic cancer. Carriers of germline mutations face a higher risk than individuals with a positive family history alone. Multiple germline mutations and hereditary syndromes have been implicated in susceptibility to pancreatic cancer (Grigorescu et al., 2024; Sharma et al., 2025; Stoop et al., 2025).

Pancreatic cancer: Pathogenesis

Pancreatic cancers are classified according to their cellular origin into two principal categories: exocrine (ductal) tumors and neuroendocrine neoplasms (Fig. 31.2).

The exocrine component of the pancreas, functionally integrated with the digestive system, is responsible for the production and secretion of digestive enzymes. Malignancies arising from this compartment account for approximately 95% of all pancreatic cancer cases and comprise several histological subtypes. The most prevalent form is pancreatic ductal adenocarcinoma (PDAC), which represents more than 90% of exocrine pancreatic malignancies. Other variants—such as squamous cell carcinoma,

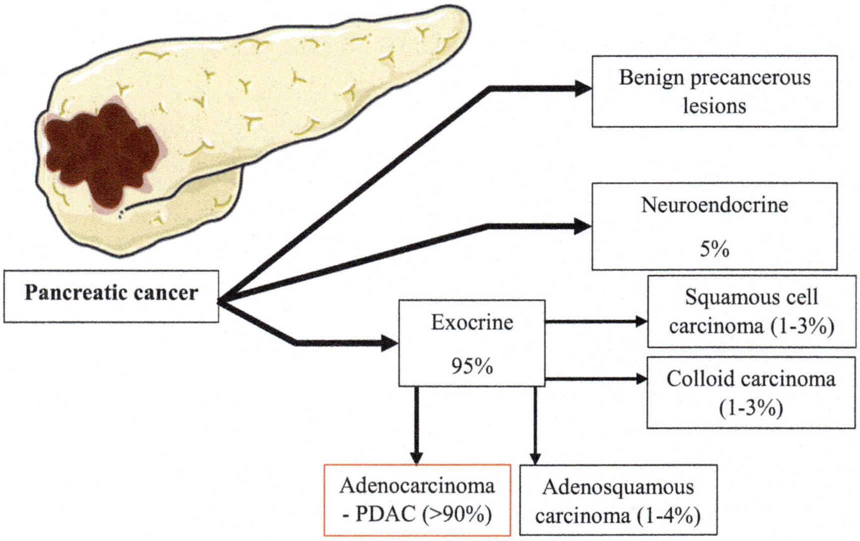

FIGURE 31.2 Classification of pancreatic cancer. Based on information from (McGuigan et al., 2018). PDAC - pancreatic ductal adenocarcinoma.

adenosquamous carcinoma, and colloid carcinoma—are encountered far less frequently. It is also important to note that pancreatic cysts and certain benign neoplastic lesions may serve as precursor conditions for the development of pancreatic cancer (Hayat et al., 2025; McGuigan et al., 2018; Stoop et al., 2025).

Pancreatic cancer is characterized by a pronounced propensity for early dissemination and metastatic spread, as well as by the difficulty of detecting the disease at an initial stage, factors that collectively contribute to its poor prognosis. In contrast, neuroendocrine tumors generally exhibit lower mortality rates and are associated with substantially more favorable clinical outcomes. These neoplasms originate from the islet cells of the pancreas, which secrete hormones such as insulin and glucagon that regulate glucose homeostasis. Neuroendocrine pancreatic cancers are relatively uncommon and constitute less than 5% of all pancreatic malignancies. The present work focuses on pancreatic ductal adenocarcinoma, the most lethal form of pancreatic cancer (Hayat et al., 2025; McGuigan et al., 2018; Stoop et al., 2025).

Pancreatic cancer represents the predominant form of solid tumors of this organ and accounts for more than 95% of all pancreatic malignancies. The vast majority of these tumors—approximately 95%—arise from precursor lesions known as pancreatic intraepithelial neoplasia (PanIN). Additional precursor conditions include cystic lesions of the pancreas, such as intraductal papillary mucinous neoplasms (IPMN) and mucinous cystic neoplasms. Notably, in nearly half of patients with pancreatic cancer coexisting with IPMN, the invasive carcinoma does not directly originate from the cystic lesion itself (Hayat et al., 2025; McGuigan et al., 2018; Stoop et al., 2025).

Pancreatic cysts are detected in roughly 16% of the general population (95% CI 13–18), with IPMNs being both the most prevalent subtype and the one associated with the highest risk of malignant transformation. In low-risk IPMN, the cumulative probability of progression to malignancy over 10 years is approximately 8% (95% CI 4–12), whereas in individuals exhibiting high-risk features, this probability may reach 25% (95% CI 15–36) (Stoop et al., 2025).

The pathogenesis of pancreatic cancer involves numerous molecular alterations, including oncogene activation, loss of tumor suppressor gene function, and defects in DNA damage repair mechanisms, particularly in homologous recombination pathways. In approximately 91% of cases, these alterations are exclusively somatic, while germline mutations are identified in about 9% of patients. The most frequently observed somatic mutations in PDAC affect oncogene *KRAS* (90%–95%) and tumor suppressor genes such as *TP53* (61%–74%), *CDKN2A* (16%–44%), and *SMAD4* (20%–22%). A pivotal genetic event in the development of pancreatic ductal adenocarcinoma is somatic mutation of the *KRAS* oncogene, detected in over 90% of patients already at the stage of low-grade PanIN. Mutant *KRAS* remains active throughout the progression from preinvasive epithelial lesions to invasive carcinoma, promoting cellular proliferation, survival, migration, and invasive capacity. Through its influence on the tumor stroma and microenvironment, *KRAS* also plays a critical role in metastatic spread and in the development of resistance to cytotoxic therapy (Hayat et al., 2025; Rozengurt & Eibl, 2026; Stoop et al., 2025).

The tumorigenesis process leading to PDAC is shown in Fig. 31.3.

Transcriptomic analyses have identified several distinct molecular subtypes of pancreatic cancer, among which the classification into classical and basal-like variants appears

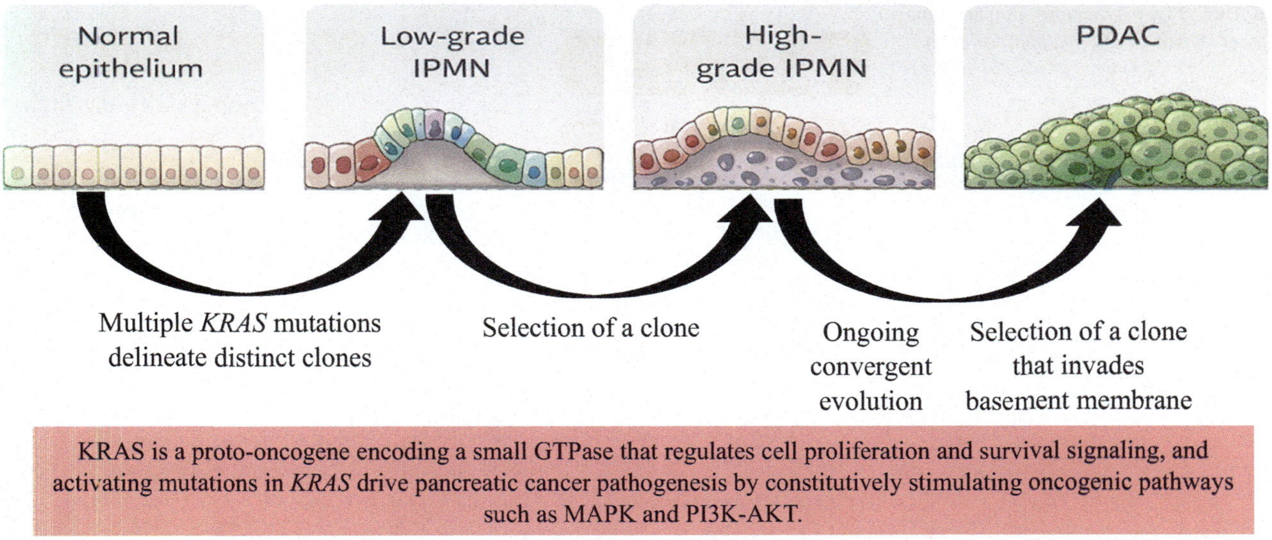

FIGURE 31.3 Tumorigenesis process leading to pancreatic ductal adenocarcinoma (PDAC). IPMNs - intraductal papillary mucinous neoplasms.

to have the greatest clinical relevance in terms of prognosis and responsiveness to chemotherapy. Basal-like tumors are enriched in features associated with epithelial–mesenchymal transition (EMT), enhanced cell cycle progression, and activation of TGF-β signaling, and they are linked to less favorable clinical outcomes. EMT represents a key mechanism underlying the reprogramming of malignant cells (Hayat et al., 2025; Rozengurt & Eibl, 2026; Stoop et al., 2025).

A hallmark of pancreatic cancer is the presence of a pronounced desmoplastic stroma, composed primarily of extracellular matrix components, vascular elements, and cancer-associated fibroblasts. This dense stromal architecture increases intratumoral pressure, thereby limiting the effective delivery of cytotoxic agents. Tumor cells disseminate along blood vessels, neural structures, and collagen fibers, which explains the high incidence of perineural invasion (approximately 62%) and lymphatic invasion (approximately 54%), both of which are associated with reduced overall survival (Hayat et al., 2025; Rozengurt & Eibl, 2026; Stoop et al., 2025).

Pancreatic cancer: Screening

Early detection of pancreatic cancer remains challenging due to the absence of specific clinical symptoms, reliable biomarkers, and effective screening strategies. Statistical modeling suggests that approximately 10 years may elapse between the emergence of the first mutations and the development of a non-metastatic tumor confined to the pancreas. The retroperitoneal location of the pancreas further hampers early diagnosis. Consequently, symptoms such as jaundice, abdominal pain, nausea and vomiting, or anorexia typically indicate advanced-stage disease (Fig. 31.4) (Hayat et al., 2025; McGuigan et al., 2018; Stoop et al., 2025).

At present, screening for pancreatic cancer in asymptomatic adults is not recommended, primarily because of the relatively low incidence in the general population and the lack of sufficiently sensitive and specific screening tools. Newly diagnosed diabetes has been considered a potential indication for targeted screening; however, given the standardized incidence ratio of 1.5 (95% CI 1.4–1.6), the limited performance of available predictive models, and the absence of clear survival benefit, routine screening in this subgroup is not currently advised (Stoop et al., 2025; Waleleng et al., 2022).

Genetic testing is recommended for individuals at elevated risk, including those with Li-Fraumeni syndrome, Lynch syndrome, familial pancreatic cancer, or relatives of carriers of confirmed pathogenic mutations. The National Comprehensive Cancer Network (NCCN) guidelines emphasize the critical importance of thorough patient counseling prior to initiating surveillance, addressing potential drawbacks such as uncertainty regarding benefit, false-positive findings, and financial costs (NCCN, 2025).

The goal of surveillance programs is to detect high-grade precursor lesions or early-stage pancreatic cancer. Surveillance is generally initiated at age 50 or 10 years earlier than the youngest age at diagnosis within the family. Recommended modalities include annual endoscopic ultrasound (EUS) and magnetic resonance imaging (MRI) with magnetic resonance cholangiopancreatography (MRCP). The diagnostic performance of EUS and MRI in detecting high-grade dysplasia or early pancreatic ductal

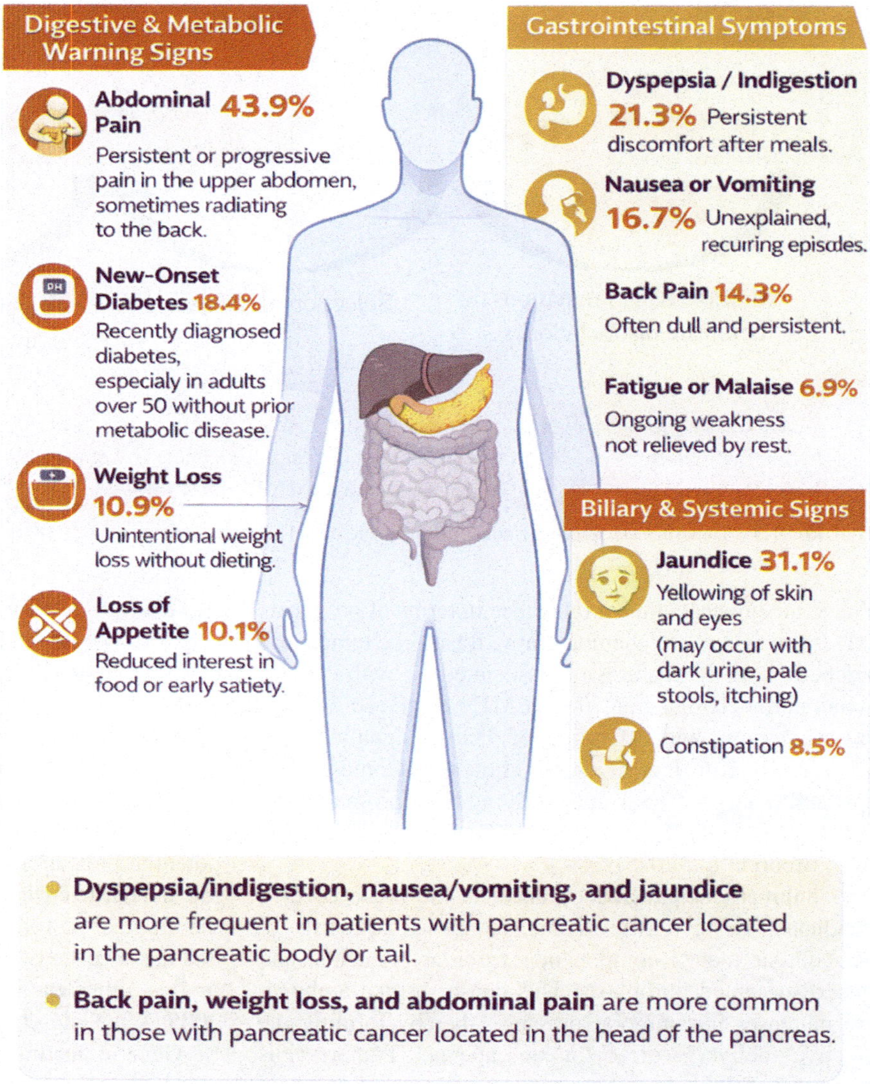

FIGURE 31.4 Symptoms of pancreatic cancer. *Based on information from (Hayat et al., 2025; McGuigan et al., 2018; Stoop et al., 2025).*

adenocarcinoma appears comparable; however, the added value of MRI/MRCP as a complement to EUS remains debated. At present, accurate liquid biomarkers are not available (Hayat et al., 2025; Lee & Lee, 2014).

In high-risk populations, the number needed to screen (NNS) to identify one high-risk lesion using EUS or MRI is 135 (95% CI 88–303). This number is lower in individuals with specific cancer-predisposing mutations, particularly those with hereditary pancreatitis (NNS 130), Peutz–Jeghers syndrome (NNS 71), and germline *CDKN2A* mutations (NNS 51) (Stoop et al., 2025).

Guidelines also recommend surveillance for individuals with radiologically diagnosed IPMN in the absence of absolute indications for surgical intervention. Such indications include specific imaging features (e.g., dilation of the main pancreatic duct, enlarged lymph nodes, enhancing mural nodules, or cyst growth) and clinical features (e.g., acute pancreatitis or new-onset diabetes) (Conroy et al., 2023).

Pancreatic cancer detected through surveillance is more frequently diagnosed at an early stage compared with cases identified in the general patient population, a difference that likely translates into improved overall survival. Nevertheless, annual imaging does not eliminate the possibility of surveillance failure, as nearly half of newly identified lesions emerge at a median of 11 months following the most recent EUS or MRI/MRCP examination (Conroy et al., 2023; Stoop et al., 2025).

Pancreatic cancer: Metastases

At present, most patients are diagnosed at the metastatic stage (mPDAC). The most common sites of distant spread include the liver (76%), lungs (20%), lymph nodes (9.4%), and bones (6.8%) (Gouton et al., 2024).

Brain metastases (BM) represent the most common intracranial tumors and occur most frequently in lung cancer (19.9%), melanoma (6.9%), renal cell carcinoma (6.5%), and breast cancer (5.1%). Their presence is associated with an unfavorable prognosis and mortality rates ranging from 18% to 50% (Gouton et al., 2024). Among gastrointestinal (GI) malignancies, the incidence of brain metastases largely depends on the type of primary tumor. Barnholtz-Sloan et al. reported that colorectal cancer was the most common GI malignancy leading to brain metastases, with an incidence of approximately 1.8% (Barnholtz-Sloan et al., 2004). Distinct biological characteristics have also been identified in GI-derived brain metastases, including HER2 overexpression and alterations in DNA repair mechanisms, such as homologous recombination deficiency (HRD) and mismatch repair deficiency (MMRd) (Gouton et al., 2024).

In PDAC, the incidence of brain metastases is even lower, ranging from only 0.25%–0.6% of cases. This low reported frequency may be attributable to the rarity of neurological symptoms in these patients or to the short overall survival associated with PDAC. Furthermore, brain imaging is not routinely recommended in the diagnostic workup of PDAC, which may result in underdiagnosis of asymptomatic brain metastases. This hypothesis is supported by autopsy studies reporting a brain metastasis rate of 7.9% among patients with PDAC (Gouton et al., 2024).

Pancreatic cancer: Diagnostics

In contrast to many other malignancies, circulating biomarkers currently have a limited role in the early detection of pancreatic cancer. Carbohydrate antigen 19–9 (CA19-9) remains the most extensively investigated and routinely applied serum marker in PDAC, serving both diagnostic and prognostic purposes (Conroy et al., 2023; Gu & Minko, 2024; Stoop et al., 2025).

Among symptomatic patients, CA19-9 demonstrates a sensitivity of approximately 79%–81% and a specificity ranging from 82% to 90%. In asymptomatic individuals, reported sensitivity may reach 100%, with specificity around 98.5%. Despite these seemingly favorable parameters, the positive predictive value remains modest—about 72% in symptomatic populations and as low as 0.9% in those without symptoms—substantially limiting its effectiveness as a population-based screening tool. An additional constraint is that approximately 5%–10% of individuals of Caucasian descent exhibit the Lewis-negative phenotype and therefore are unable to synthesize detectable levels of CA19-9. Elevated CA19-9 concentrations are also observed in other malignancies, including colorectal and biliary tract cancers, as well as in non-malignant conditions such as obstructive jaundice and liver cirrhosis. Moreover, nearly half of PDAC tumors measuring less than 3 cm do not produce increased CA19-9 levels, further complicating clinical interpretation. Nevertheless, CA19-9 remains the only FDA-recognized biomarker for PDAC and is widely used to differentiate pancreatic cancer from chronic pancreatitis, with reported sensitivity between 70% and 90% and specificity of 68%–91% (Conroy et al., 2023; Gu & Minko, 2024; Stoop et al., 2025).

Carcinoembryonic antigen (CEA) may also be elevated in a range of cancers, including pancreatic, colorectal, lung, and breast malignancies. It is estimated that 30%–60% of patients with PDAC exhibit increased CEA levels. However, its diagnostic utility is limited due to relatively low sensitivity (25%–54%) and variable specificity (75%–91%), making it inadequate as a standalone diagnostic marker. To improve early detection accuracy, ongoing research is exploring the integration of CA19-9 and/or CEA with additional biomarkers to establish multiparametric diagnostic panels (Conroy et al., 2023; Gu & Minko, 2024; Stoop et al., 2025).

Beyond serum biomarkers, imaging plays a central role in the diagnosis of PDAC. Standard imaging modalities include computed tomography (CT), MRI, EUS, positron emission tomography (PET), and conventional angiography. Each modality has inherent limitations. CT remains the primary imaging technique for both diagnosis and staging. A meta-analysis encompassing 52 studies and 3567 patients reported that CT-based assessment achieved a sensitivity of approximately 90%, specificity of 87%, and overall diagnostic accuracy of 89%. However, its performance decreases for tumors smaller than 20 mm; in such cases, limited contrast resolution reduces sensitivity and accuracy to roughly 69% and 73%, respectively (Toft et al., 2017).

MRI generally provides higher sensitivity and specificity compared to CT, with comparable overall diagnostic accuracy, particularly in detecting small primary lesions and metastatic spread. Nonetheless, broader implementation is restricted by issues of cost and availability. Endoscopic ultrasound is especially valuable for evaluating lesions under 20 mm in diameter. When combined with fine-needle aspiration (EUS-FNA), it enables minimally invasive tissue sampling and significantly enhances diagnostic performance, increasing sensitivity to approximately 92% and specificity to 96% (Conroy et al., 2023; Gu & Minko, 2024; Stoop et al., 2025).

Both imaging interpretation and biomarker analysis require substantial clinical expertise, and variability in operator experience may contribute to diagnostic discrepancies. Overall, the advancement of standardized, high-performance diagnostic approaches with improved sensitivity and specificity remains essential to ensure greater consistency and precision in the diagnosis of PDAC (Conroy et al., 2023; Gu & Minko, 2024; Stoop et al., 2025).

Pancreatic cancer: Staging

At initial presentation, nearly 57% of patients with pancreatic cancer are found to have metastatic disease. The liver is the predominant site of dissemination (75%–80%), followed by the peritoneum (13%–30%), lungs (15%–18%), and distant lymph nodes (12%). In patients without detectable distant metastases, tumors are categorized anatomically as resectable (RPC), borderline resectable (BRPC), or locally advanced (LAPC). This classification is determined by the extent of tumor contact with or invasion of critical peripancreatic vascular structures, including the superior mesenteric artery, the celiac trunk, hepatic arterial branches, and the portal–mesenteric venous axis. In routine practice, the resectability framework most commonly applied is that outlined in the NCCN guidelines (NCCN 2025; Stoop et al., 2025).

While this system primarily reflects technical operability, prognostic stratification additionally relies on the Tumor–Node–Metastasis (TNM) classification established by the American Joint Committee on Cancer. TNM stage is assigned either on the basis of imaging studies before treatment or, in patients undergoing resection, according to histopathological findings from the surgical specimen (McGuigan et al., 2018; Stoop et al., 2025; Zhao & Liu, 2020).

Precise evaluation of vascular involvement necessitates high-quality, contrast-enhanced, multiphasic computed tomography performed using a dedicated pancreatic protocol. Because interpretation may vary considerably among radiologists, structured reporting templates are recommended to improve consistency and reduce interobserver variability. Importantly, staging investigations should be completed within 4 weeks prior to treatment initiation and before any biliary drainage procedure is undertaken (McGuigan et al., 2018; Stoop et al., 2025; Zhao & Liu, 2020).

Beyond anatomical staging alone, refinement of clinical assessment can be achieved by integrating tumor biology and patient performance status into the decision-making process, as proposed in the A-B-C classification model. When serum CA19-9 levels are not elevated, alternative markers such as CEA and CA-125 may be evaluated. In patients with obstructive cholestasis, CA19-9 measurement should be deferred until bilirubin levels have normalized and ideally performed close to the start of definitive oncologic therapy, ensuring both accurate baseline staging and reliable post-chemotherapy response assessment (McGuigan et al., 2018; Stoop et al., 2025; Zhao & Liu, 2020).

Evidence supporting the prognostic relevance of biological factors comes from a nationwide observational study of 688 patients with anatomically resectable pancreatic cancer who underwent primary surgery. In this cohort, patients meeting criteria for biologically borderline resectable disease—defined by CA19-9 concentrations ≥500 U/mL—demonstrated significantly inferior overall survival. These findings were further validated in a binational observational study of 1835 patients with locally advanced pancreatic cancer treated with first-line (modified) FOLFIRINOX (5-fluorouracil, leucovorin, irinotecan, and oxaliplatin). This analysis identified three independent adverse prognostic domains: anatomical extent (BRPC or LAPC), elevated CA19-9 (≥500 U/mL), and compromised functional status (WHO performance status ≥1). Five-year overall survival varied markedly according to combined A-B-C risk profiles, ranging from 5% in patients with the least favorable constellation of factors to 47% in those with the most favorable combination (McGuigan et al., 2018; Stoop et al., 2025; Zhao & Liu, 2020).

Collectively, these findings demonstrate that optimal therapeutic planning in pancreatic cancer requires a multidimensional assessment strategy that incorporates anatomical resectability, tumor biological behavior, and patient-related clinical parameters, thereby enabling more accurate prognostic stratification and more individualized treatment selection (McGuigan et al., 2018; Stoop et al., 2025; Zhao & Liu, 2020).

Pancreatic cancer: Treatment

Treatment options for pancreatic cancer are summarized in Fig. 31.5.

Localized pancreatic cancer

Pancreatic cancer ranks among the foremost causes of cancer-related mortality in the United States, and the majority of patients are diagnosed when the disease is already metastatic or technically unresectable. Only approximately 10%–15% of individuals present with disease confined to the pancreas. In this subset, coordinated multidisciplinary care has been shown to be essential in order to maximize therapeutic benefit and improve clinical outcomes (Abe et al., 2026).

Surgical approach

The NCCN guidelines stratify localized PDAC according to surgical resectability, thereby providing a framework for

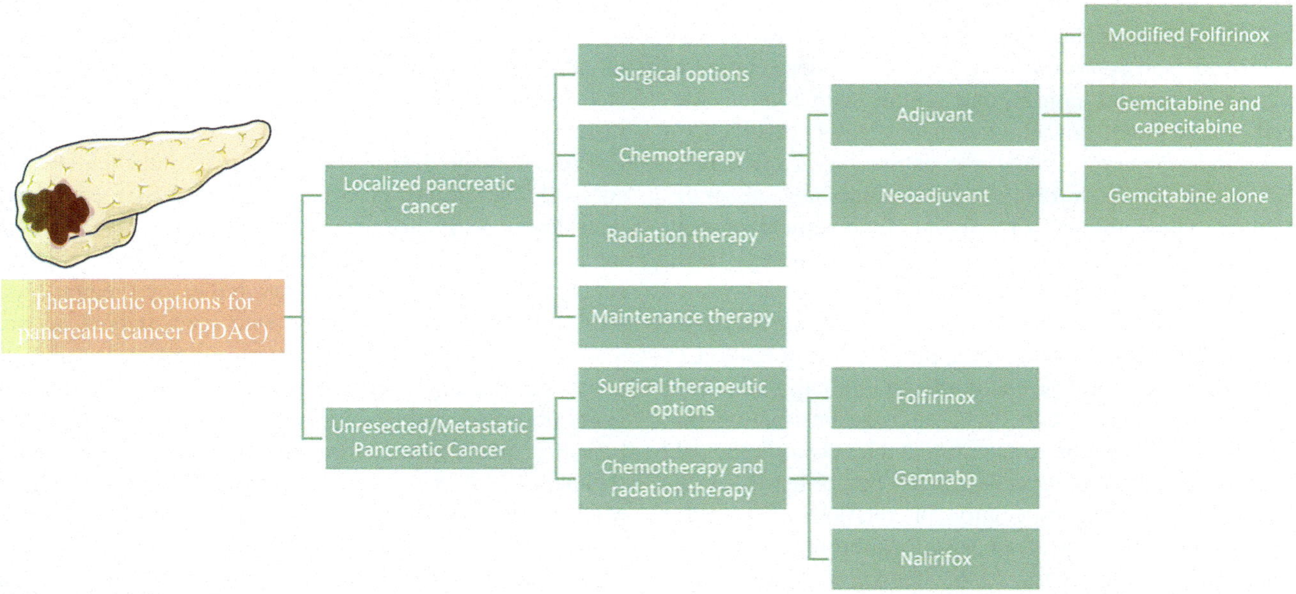

FIGURE 31.5 Treatment options for pancreatic cancer (PDAC). *Based on information from (Abe et al., 2026; Conroy et al., 2023; Hayat et al., 2025; Stoop et al., 2025).*

treatment planning. A tumor is considered resectable when there is no meaningful involvement of major arterial structures—specifically the celiac axis, superior mesenteric artery, or common hepatic artery—and either no venous invasion or ≤180 degrees contact with the superior mesenteric vein or portal vein without distortion of the vessel contour. Borderline resectable PDAC is defined by tumor abutment of ≤180 degrees with the superior mesenteric vein/portal vein, provided that an adequate segment of uninvolved vein remains to permit safe vascular reconstruction. In contrast, locally advanced or unresectable PDAC includes tumors demonstrating >180 degrees arterial encasement of the superior mesenteric vein or celiac axis, extension to the aorta, or venous involvement of the superior mesenteric vein/portal vein that cannot be surgically reconstructed because of tumor infiltration or vessel occlusion (Abe et al., 2026; Conroy et al., 2023; Hayat et al., 2025; Stoop et al., 2025).

Surgical management is largely determined by tumor location within the pancreas. Lesions arising in the body or tail are typically managed with distal pancreatectomy. Tumors involving the pancreatic head or uncinate process require pancreaticoduodenectomy (PD) to achieve oncologic clearance. The origins of PD date back to 1889, when the procedure was first attempted by the Italian surgeon Codivilla; it was subsequently standardized in 1935 by Whipple through his two-stage operative description. Over recent decades, improvements in surgical expertise and the incorporation of more aggressive techniques—including vascular resection and reconstruction—have led to better operative outcomes (Abe et al., 2026; Conroy et al., 2023; Hayat et al., 2025; Stoop et al., 2025).

Minimally invasive techniques have also been introduced. In 1994, Gagner and Pomp described the first laparoscopic pancreaticoduodenectomy (LPD), performed for chronic pancreatitis affecting the pancreatic head in the setting of pancreas divisum. Since that time, the laparoscopic approach has gained increasing acceptance. Contemporary data indicate that LPD yields complication rates, overall survival, and 5-year survival comparable to those of conventional open PD, while offering the advantage of reduced hospital stay. Robotic pancreaticoduodenectomy (RPD) has likewise emerged as an alternative approach, although the supporting evidence is derived primarily from observational studies rather than randomized trials (Abe et al., 2026; Conroy et al., 2023; Hayat et al., 2025; Stoop et al., 2025).

Despite technical advancements, PD remains associated with considerable perioperative risk. The most frequently encountered complications include postoperative pancreatic fistula, with an incidence of approximately 5%–15%, and delayed gastric emptying, occurring in roughly 20% of cases. Nevertheless, the long-term results of surgery for PDAC have improved in parallel with growing surgical experience and the routine performance of vascular resections when indicated. In carefully selected patients, overall survival rates of 20%–30% have been reported. Even so, recurrence after surgical resection remains common, likely reflecting the presence of occult micrometastatic disease at the time of surgery (Abe et al., 2026; Conroy et al., 2023; Hayat et al., 2025; Stoop et al., 2025).

The surgical resection criteria proposed by NCCN and those used at MD Anderson Cancer Center are broadly aligned, particularly in defining resectable tumors as lesions confined to the pancreas without major central vascular involvement. However, notable differences exist. MD Anderson applies a more permissive interpretation of vascular invasion, considering tumors with limited arterial encasement or short-segment vessel involvement potentially operable when reconstruction is technically achievable. In contrast, NCCN guidelines generally require either absence of vascular involvement or only minimal contact without contour abnormality. Both frameworks classify tumors with distant metastases or nodal spread beyond the regional field as unresectable. Importantly, MD Anderson adopts a wider definition of borderline resectability, including certain cases requiring adjacent organ resection or vascular reconstruction that NCCN would categorize as unresectable. Overall, the MD Anderson approach reflects greater flexibility in evaluating resectability, especially in relation to vascular encasement and invasion of neighboring structures (NCCN 2025; Abe et al., 2026; Stoop et al., 2025; Hayat et al., 2025; Conroy et al., 2023).

Table 31.2 outlines the operative criteria applied in pancreatic cancer resection.

Adjuvant chemotherapy

Even when PDAC is detected at a localized and technically resectable stage, long-term outcomes remain unsatisfactory. Surgical treatment alone is associated with an approximate 20% 5-year survival rate; the addition of postoperative (adjuvant) chemotherapy increases this figure to roughly 30%–40%. Over the last 2 decades, multiple prospective studies have firmly established the value of adjuvant systemic therapy following curative-intent resection, demonstrating that combined modality treatment provides the only realistic opportunity for durable survival or potential cure in patients with nonmetastatic disease (Abe et al., 2026; Conroy et al., 2023; Gu & Minko, 2024; Hayat et al., 2025; McGuigan et al., 2018; Stoop et al., 2025; Zhao & Liu, 2020).

Choice of regimen is guided primarily by the patient's performance status. Individuals with good functional capacity are generally offered modified FOLFIRINOX (fluorouracil, oxaliplatin, irinotecan, and leucovorin), whereas those with reduced tolerance may receive gemcitabine in combination with capecitabine or gemcitabine monotherapy. In the CONKO-001 trial, adjuvant gemcitabine resulted in a modest prolongation of median overall survival (22.8 months) compared with observation alone (20.2 months) within 6 months after surgical resection. The addition of erlotinib to gemcitabine failed to produce further benefit in the postoperative setting (Abe et al., 2026; Conroy et al., 2023; Gu & Minko, 2024; Hayat et al., 2025; McGuigan et al., 2018; Stoop et al., 2025; Zhao & Liu, 2020).

A phase II study evaluating gemcitabine plus cisplatin in 22 patients reported a median overall survival of 35.5 months. Although administrable, the regimen was associated with substantial toxicity, with 59% of patients experiencing grade 3 or four adverse events. The pivotal ESPAC-1 study demonstrated improved median overall survival with adjuvant 5-fluorouracil (5-FU) combined with leucovorin, increasing survival from 14.0 to 19.7 months. Subsequently, the phase III ESPAC-4 trial showed superior outcomes with combination therapy (gemcitabine plus capecitabine) compared with gemcitabine alone. Collectively, these trials indicated a consistent survival advantage for adjuvant chemotherapy, with incremental gains observed when additional agents such as capecitabine or cisplatin were incorporated into gemcitabine-based regimens (Abe et al., 2026; Conroy et al., 2023; Gu & Minko, 2024; Hayat et al., 2025; McGuigan et al., 2018; Stoop et al., 2025; Zhao & Liu, 2020).

Gemcitabine combined with nab-paclitaxel has also been investigated in the adjuvant context in the APACT study. Initial findings did not reveal a statistically significant difference; however, a 2021 updated analysis with extended follow-up demonstrated a significant improvement in median overall survival (41.8 versus 37.7 months; $P = .0091$) (Hayat et al., 2025).

The PRODIGE-24 trial, published in 2018 and enrolling patients with excellent performance status, reported a marked survival advantage for modified FOLFIRINOX, achieving a median overall survival of 54.4 months compared with 35 months in those treated with gemcitabine alone. The comparatively prolonged survival observed in the control arm likely reflected the favorable baseline characteristics of the study population. Consequently, adjuvant FOLFIRINOX is generally reserved for patients with preserved functional capacity, specifically those with an ECOG performance status of 0–1 (Abe et al., 2026; Conroy et al., 2023; Gu & Minko, 2024; Hayat et al., 2025; McGuigan et al., 2018; Stoop et al., 2025; Zhao & Liu, 2020).

Neoadjuvant chemotherapy

Neoadjuvant chemotherapy has become increasingly integrated into the treatment paradigm over the past decade, supported by multiple studies demonstrating its efficacy in patients with borderline resectable pancreatic cancer (BRPC) and locally advanced pancreatic cancer (LAPC). When this preoperative strategy is employed, patients should undergo repeat staging with imaging; in the absence of radiographic progression, they may then proceed to surgical exploration. Importantly, in pancreatic cancer, radiologic response does not consistently correspond to pathological response. Therefore, the absence of significant tumor shrinkage on imaging alone should not

TABLE 31.2 Outlines the operative criteria applied in pancreatic cancer resection

Criteria	Resectable (NCCN)	Resectable (MD Anderson Cancer Center)	Borderline Resectable (NCCN)	Borderline Resectable (MD Anderson Cancer Center)	Unresectable (NCCN)	Unresectable (MD Anderson Cancer Center)
Tumor size	Tumor confined to the pancreas, no involvement of major vessels	Tumor confined to the pancreas, no involvement of major vessels	Tumor involves major vessels but is amenable to resection with venous reconstruction	Tumor may involve local structures but with the potential for resection	Involvement of major vessels or distant metastasis, not amenable to resection	Extensive local invasion, distant metastasis, or unreconstructable vascular involvement
Vascular involvement	No involvement of major blood vessels (e.g., SMA, portal vein, hepatic artery), or ≤180 degrees contact without vein contour irregularity	No involvement of major blood vessels (e.g., SMA, portal vein, hepatic artery), patent SMV-PV confluence	Vascular encasement of veins (portal/superior mesenteric) of >180 degrees or ≤ 180 degrees with contour irregularity or arterial involvement that can be resected or reconstructed	Tumor abutment of ≤180 degrees of the circumference of SMA, short-segment encasement of the CHA or gastroduodenal artery, SMV, or SMV-PV, and patent vessel above and below	Encases or obstructs major blood vessels such as the SMA, celiac axis with >180 degrees contact and unreconstructable SMV/PV	Encasement of major vessels (SMA, portal vein, celiac axis) > 180 degrees, interface with the aorta, or unresectable venous occlusion
Lymph node involvement	No regional lymph node involvement (negative regional nodes)	No regional lymph node involvement (negative regional nodes)	Possible lymph node involvement, but resection may still be considered	Lymph node involvement that may be resectable	Extensive regional or distant lymph node involvement	Distant lymph node metastasis
Distant metastasis	No distant metastasis	No distant metastasis	No distant metastasis	No distant metastasis	Distant metastasis (e.g., liver, lungs)	Distant metastasis (e.g., liver, lungs)
Adjacent organ involvement	No invasion into adjacent organs like the duodenum, bile ducts, or stomach	No invasion into adjacent organs like the duodenum, bile ducts, or stomach	It may involve adjacent organs but is amenable to surgical resection	Involvement of adjacent organs but still potentially resectable	Invasion of adjacent organs such as the liver, stomach, or duodenum	Invasion of adjacent organs making resection not possible
Overall prognosis	Likely to benefit from curative surgery	Likely to benefit from curative surgery	High risk, but resection can still be performed with some degree of certainty	High risk, borderline, but resection may offer palliative outcomes	Poor prognosis, resection not an option	Poor prognosis, resection not an option

automatically exclude a patient from surgical consideration (Hayat et al., 2025).

The NUPAT-01 study reported promising results with neoadjuvant regimens such as FOLFIRINOX or gemcitabine plus nab-paclitaxel (GemNabP) in BRPC, achieving R0 resection rates of 67.4% and a median overall survival (mOS) of 39.4 months. A multicenter study similarly demonstrated higher rates of margin-negative resections among patients with BRPC treated with neoadjuvant therapy. Additionally, an analysis derived from the National Cancer Database found that, in stage I/II pancreatic cancer, patients receiving neoadjuvant therapy followed by resection experienced superior overall survival compared with propensity score–matched individuals undergoing upfront surgery (median OS 26 months *vs.* 21 months; HR = 0.72, $P < 0.01$) (Abe et al., 2026; Conroy et al., 2023; Gu & Minko, 2024; Hayat et al., 2025; Kwaśniewska et al., 2025; McGuigan et al., 2018; Stoop et al., 2025; Zhao & Liu, 2020).

Given that a considerable proportion of patients are unable to receive adjuvant systemic therapy because of postoperative complications or functional decline, neoadjuvant treatment may increase the number of individuals ultimately exposed to systemic therapy. It may also contribute to the eradication of occult micrometastatic disease at an earlier stage. Nevertheless, the precise benefit of neoadjuvant therapy in clearly resectable PDAC remains under active investigation. One potential limitation of this approach is the risk of disease progression in tumors that do not respond adequately to preoperative treatment (Abe et al., 2026; Conroy et al., 2023; Gu & Minko, 2024; Hayat et al., 2025; Kwaśniewska et al., 2025; McGuigan et al., 2018; Stoop et al., 2025; Zhao & Liu, 2020).

Radiation therapy

The contribution of radiotherapy to the management of PDAC remains debated, and the most appropriate treatment protocol has yet to be clearly defined. In the PREOPANC study, neoadjuvant chemoradiation based on gemcitabine was associated with improved overall survival compared with immediate surgery followed by adjuvant gemcitabine in patients with resectable and borderline resectable disease. Similarly, the phase II ESPAC-5F trial reported higher 1-year survival rates among patients treated with preoperative chemoradiotherapy. Additional investigations have also suggested potential benefits of neoadjuvant chemoradiation in selected populations (Abe et al., 2026; Conroy et al., 2023; Gu & Minko, 2024; Hayat et al., 2025; Kwaśniewska et al., 2025; McGuigan et al., 2018; Stoop et al., 2025; Zhao & Liu, 2020).

However, findings have not been uniformly positive. In the ALLIANCE trial, patients receiving neoadjuvant modified FOLFIRINOX (mFOLFIRINOX) were randomized to receive hypofractionated radiation therapy or not; no improvement in median overall survival was demonstrated, leaving the true impact of radiation in this setting uncertain. Furthermore, results from ESPAC-1 indicated worse outcomes when radiotherapy was added to 5-fluorouracil–based adjuvant regimens (Abe et al., 2026; Conroy et al., 2023; Gu & Minko, 2024; Hayat et al., 2025; Kwaśniewska et al., 2025; McGuigan et al., 2018; Stoop et al., 2025; Zhao & Liu, 2020).

The LAP-07 study assessed capecitabine-based chemoradiation in patients with unresectable pancreatic cancer whose disease had not progressed after 4 months of gemcitabine therapy. Compared with continued gemcitabine for an additional 2 months, chemoradiation did not significantly prolong overall survival. A modest improvement in progression-free survival was observed, suggesting that this strategy may be reasonable for patients seeking a temporary break from systemic chemotherapy (Abe et al., 2026; Conroy et al., 2023; Gu & Minko, 2024; Hayat et al., 2025; Kwaśniewska et al., 2025; McGuigan et al., 2018; Stoop et al., 2025; Zhao & Liu, 2020).

Overall, the role of chemoradiation in locally advanced pancreatic cancer remains incompletely defined. It may be considered in selected cases following R1 resection in the adjuvant setting or preoperatively after multidisciplinary evaluation. While small prospective studies and retrospective analyses have reported higher rates of R0 resection with neoadjuvant chemoradiation, confirmation from larger randomized trials is still required. Notably, combining radiation with FOLFIRINOX in the neoadjuvant setting is not recommended based on unfavorable findings from the SWOG study (Abe et al., 2026; Conroy et al., 2023; Gu & Minko, 2024; Hayat et al., 2025; Kwaśniewska et al., 2025; McGuigan et al., 2018; Stoop et al., 2025; Zhao & Liu, 2020).

Maintenance therapy

An estimated 5%–9% of patients with pancreatic cancer harbor either germline or somatic mutations in *BRCA1* or *BRCA2*. Evidence suggests that tumors with these alterations, particularly those demonstrating platinum sensitivity, may respond to poly(ADP-ribose) polymerase (PARP) inhibitors. The phase III POLO trial evaluated maintenance therapy with olaparib in this population and demonstrated a significant improvement in median progression-free survival, although no statistically significant difference in overall survival was observed at interim analysis. In December 2019, olaparib received approval in the United States for patients with BRCA-mutated pancreatic cancer who had stable disease following first-line chemotherapy, marking the first biomarker-driven targeted therapy authorized for this malignancy. Despite this advance, maintenance treatment strategies in pancreatic cancer continue to evolve, and their definitive role has yet to be fully established (Abe et al., 2026; Conroy et al., 2023; Gu & Minko, 2024; Hayat et al., 2025; Kwaśniewska et al., 2025; McGuigan et al., 2018; Stoop et al., 2025; Zhao & Liu, 2020).

Advanced cancer or metastasis present

Management of advanced pancreatic cancer requires a carefully planned therapeutic strategy. Patients face competing risks of locoregional progression and distant metastatic spread, both of which significantly influence overall survival. Notably, approximately one-third of patients die from uncontrolled local disease in the absence of distant metastases. Systemic treatment, administered alone or in conjunction with locoregional interventions, represents the mainstay of care in this setting (Abe et al., 2026; Conroy et al., 2023; Gu & Minko, 2024; Hayat et al., 2025; Kwaśniewska et al., 2025; McGuigan et al., 2018; Stoop et al., 2025; Zhao & Liu, 2020).

In unresectable disease, surgical management is primarily palliative and aims to alleviate biliary or duodenal obstruction and to control pain. For malignant biliary obstruction, endoscopic stent placement is generally favored over surgical bypass due to its minimally invasive nature and shorter recovery time. A review of 29 randomized studies confirmed the superiority of stenting for palliation. The type of stent selected depends largely on life expectancy: plastic stents (e.g., polyethylene) are typically chosen for patients expected to survive less than 6 months because of their lower cost, whereas self-expanding metal stents (e.g., Wallstents) are preferred in patients with longer anticipated survival due to a reduced risk of recurrent biliary obstruction (Abe et al., 2026; Conroy et al., 2023; Gu & Minko, 2024; Hayat et al., 2025; Kwaśniewska et al., 2025; McGuigan et al., 2018; Stoop et al., 2025; Zhao & Liu, 2020).

In cases of duodenal obstruction, endoscopic stenting is considered appropriate for patients with limited prognosis, while surgical gastrojejunostomy may be offered to those with better functional status or combined obstructions. Pain related to pancreatic ductal adenocarcinoma (PDAC) can be managed with celiac plexus block or thoracoscopic splanchnicectomy, particularly when standard analgesic regimens are insufficient. Although endoscopic approaches are often preferred because of their tolerability and effectiveness, the choice of intervention should be individualized according to expected survival and overall patient condition (Abe et al., 2026; Conroy et al., 2023; Gu & Minko, 2024; Hayat et al., 2025; Kwaśniewska et al., 2025; McGuigan et al., 2018; Stoop et al., 2025; Zhao & Liu, 2020).

Combination chemotherapy constitutes the cornerstone of treatment for metastatic PDAC. While single-agent regimens have been explored, multi-agent protocols have demonstrated superior efficacy. In a phase II randomized study, the FIRGEM regimen—which incorporates irinotecan administered before and after 5-fluorouracil/leucovorin (FOLFIRI.3)—was compared with fixed-dose gemcitabine monotherapy. Median progression-free survival (PFS) was longer with FOLFIRI.3 (6 months) than with gemcitabine (3.4 months; HR 0.59; 95% CI 0.38–0.90). However, higher rates of hematologic toxicity were observed in the combination arm (Abe et al., 2026; Conroy et al., 2023; Gu & Minko, 2024; Hayat et al., 2025; Kwaśniewska et al., 2025; McGuigan et al., 2018; Stoop et al., 2025; Zhao & Liu, 2020).

FOLFIRINOX has demonstrated substantial benefit over gemcitabine, producing marked improvements in both median PFS (6.4 *vs.* 3.3 months, $P < 0.001$) and median overall survival (11.1 *vs.* 6.8 months, $P < 0.001$). Various gemcitabine-based combinations have also been evaluated. The MPACT trial reported improved survival outcomes in patients receiving gemcitabine plus albumin-bound paclitaxel (nab-paclitaxel). In contrast, combinations of gemcitabine with 5-fluorouracil, erlotinib, cisplatin, or capecitabine have generally demonstrated limited or no additional benefit in metastatic PDAC (Abe et al., 2026; Conroy et al., 2023; Gu & Minko, 2024; Hayat et al., 2025; Kwaśniewska et al., 2025; McGuigan et al., 2018; Stoop et al., 2025; Zhao & Liu, 2020).

The NAPOLI-3 trial, initiated during the COVID-19 pandemic and conducted across 18 countries, introduced nalirifox—a regimen consisting of liposomal irinotecan, fluorouracil, leucovorin, and oxaliplatin—as a promising first-line therapy for metastatic pancreatic adenocarcinoma. A randomized controlled trial demonstrated superior efficacy of nalirifox compared with gemcitabine plus nab-paclitaxel as initial treatment. Median overall survival was 11.1 months with nalirifox versus 9.2 months with gemnab (HR 0.83; $P = .04$). Although the overall incidence of adverse events was similar (87% *vs.* 86%), toxicity profiles differed: nalirifox was associated more frequently with gastrointestinal events, whereas gemnab produced more hematologic toxicity. Treatment-related mortality occurred in 2% of patients in each group (six in the nalirifox arm and eight in the gemnab arm) (Abe et al., 2026; Conroy et al., 2023; Gu & Minko, 2024; Hayat et al., 2025; Kwaśniewska et al., 2025; McGuigan et al., 2018; Stoop et al., 2025; Zhao & Liu, 2020).

A systematic review and meta-analysis by Nichetti et al. compared three principal first-line regimens—nalirifox, FOLFIRINOX, and gemnab—across seven phase III trials including 2581 patients (383 treated with nalirifox, 433 with FOLFIRINOX, and 1756 with gemnab). Median PFS was longer with nalirifox (7.4 months; 95% CI 6.1–7.7) and FOLFIRINOX (7.3 months; 95% CI 6.5–7.9) compared with gemnab (5.7 months; 95% CI 5.6–6.1).

Gemnab was associated with shorter median overall survival (10.4 months; 95% CI 9.8–10.8) compared with nalirifox (HR 1.18; 95% CI 1.00–1.39; $P = 0.05$). No statistically significant difference in overall survival was observed between FOLFIRINOX (11.7 months; 95% CI 10.4–13.0) and nalirifox (11.1 months; 95% CI 10.1–12.3; HR 1.06; 95% CI 0.81–1.39; $P = 0.65$) (Abe et al., 2026; Conroy et al., 2023; Gu & Minko, 2024; Hayat et al., 2025; Kwaśniewska et al., 2025; McGuigan et al., 2018; Stoop et al., 2025; Zhao & Liu, 2020).

In patients with metastatic pancreatic adenocarcinoma harboring germline *BRCA* mutations, maintenance therapy with olaparib following platinum-based chemotherapy has been shown to improve progression-free survival compared with placebo (7.4 *vs.* 3.8 months; HR for progression or death 0.53; 95% CI 0.35–0.82; $P = 0.004$) (Abe et al., 2026; Conroy et al., 2023; Gu & Minko, 2024; Hayat et al., 2025; Kwaśniewska et al., 2025; McGuigan et al., 2018; Stoop et al., 2025; Zhao & Liu, 2020).

In 2023, the U.S. Food and Drug Administration approved pembrolizumab, an immune checkpoint inhibitor, for patients with unresectable or metastatic solid tumors characterized by high microsatellite instability (MSI-H) or deficient mismatch repair (dMMR) who have progressed after prior therapy and lack satisfactory alternative options. However, only approximately 1%–2% of pancreatic cancers exhibit MSI-H or dMMR status, limiting the proportion of patients eligible for this treatment(Abe et al., 2026; Conroy et al., 2023; Gu & Minko, 2024; Hayat et al., 2025; Kwaśniewska et al., 2025; McGuigan et al., 2018; Stoop et al., 2025; Zhao & Liu, 2020).

Conclusions

PDAC remains a highly lethal malignancy due to its aggressive biology, delayed clinical presentation, and resistance to conventional therapies. Although advances in surgical techniques, systemic chemotherapy, and molecularly targeted treatment have modestly improved survival, long-term outcomes remain poor. The integration of anatomical staging with biological markers such as CA19-9 and patient performance status has enhanced prognostic stratification and therapeutic decision-making. Neoadjuvant approaches increasingly play a central role, particularly in borderline resectable and locally advanced disease, while maintenance and targeted therapies offer benefit in selected molecular subgroups. Early detection strategies remain limited to high-risk populations. Future progress depends on improved understanding of tumor biology, stromal interactions, molecular heterogeneity, and resistance mechanisms. Precision oncology, biomarker development, and novel therapeutic combinations are critical for meaningful improvements in survival and quality of life for patients with pancreatic cancer.

Summary points

PDAC accounts for over 90% of pancreatic malignancies and has extremely poor prognosis.

Smoking, diabetes, obesity, chronic pancreatitis, and genetic mutations significantly increase risk.

KRAS mutation is the key molecular driver in most cases.

Surgical resection with systemic chemotherapy remains the only curative approach.

Targeted and maintenance therapies benefit selected molecular subgroups.

Mini-dictionary of terms

Desmoplastic stroma – a dense fibrotic tumor microenvironment that impairs drug delivery and promotes tumor progression.

BRPC – tumors with limited vascular involvement that may become operable after neoadjuvant therapy.

HRD – impaired DNA repair mechanism that increases sensitivity to platinum agents and PARP inhibitors.

MSI-H – a condition of defective DNA mismatch repair associated with responsiveness to immunotherapy.

PanIN – a microscopic precursor lesion from which most PDACs arise.

Other hormones, endocrine tissues, and pathways

Although the chapter focuses primarily on exocrine pancreatic ductal adenocarcinoma, the pancreas also serves as a major endocrine organ. Islet cells produce insulin, glucagon, somatostatin, and pancreatic polypeptide, which regulate glucose metabolism and systemic energy homeostasis. Disruption of endocrine function is clinically relevant in PDAC, particularly in the context of new-onset diabetes (type 3 diabetes), which may represent both a risk factor and an early manifestation of malignancy. Tumor-induced alterations in insulin signaling pathways, inflammatory mediators, and cytokine networks contribute to systemic metabolic dysregulation.

Beyond pancreatic hormones, broader endocrine interactions are implicated in cancer biology. Obesity-related hyperinsulinemia, insulin resistance, and elevated insulin-like growth factor (IGF) signaling promote mitogenic and anti-apoptotic pathways that may facilitate tumorigenesis. Additionally, stress hormones and glucocorticoids can influence immune surveillance and tumor microenvironment dynamics. The interaction between KRAS-driven oncogenic signaling and metabolic pathways further highlights the integration of endocrine and molecular mechanisms. Understanding these systemic hormonal influences expands the conceptual framework of PDAC beyond local tumor growth to include metabolic and endocrine cross-talk that may offer future therapeutic targets.

Applications of the material

The knowledge presented in this chapter has direct implications for clinical practice, research, and public health. Clinically, understanding risk factors enables identification of high-risk populations eligible for genetic counseling and surveillance programs. Recognition of new-onset diabetes as a potential early indicator of PDAC may improve diagnostic vigilance. Integration of imaging modalities with biomarker assessment enhances diagnostic accuracy and staging precision.

From a therapeutic perspective, multidimensional staging that incorporates anatomical, biological, and functional parameters allows for individualized treatment planning. Selection between upfront surgery, neoadjuvant chemotherapy, or systemic therapy depends on vascular involvement, CA19-9 levels, and patient performance status. Molecular profiling facilitates targeted treatment decisions, including PARP inhibitors for BRCA-mutated tumors and immunotherapy in MSI-H/dMMR cases.

In research, insights into KRAS signaling, stromal biology, and tumor microenvironment interactions guide the development of novel agents aimed at overcoming chemoresistance. Public health strategies focusing on smoking cessation, obesity reduction, and metabolic disease control may contribute to primary prevention. Ultimately, application of this material supports a precision medicine approach, combining molecular diagnostics, tailored systemic therapy, optimized surgical techniques, and supportive care to improve survival and quality of life.

References

Abe, T., Nakamura, M., Morizane, C., Yoshida, M., Kitano, M., Ohtsuka, T., Nakata, K., Someya, M., Mizuno, N., Nakai, Y., Ozaka, M., Hanada, K., & Masamune, A. (2026). Committee for the revision of clinical guidelines for pancreatic cancer of the Japan pancreas society. Clinical practice guidelines for pancreatic cancer 2025 from the Japan pancreas society: A synopsis. *Pancreas, 55*(1), e120–e133. https://doi.org/10.1097/MPA.0000000000002600. PMID: 41342569.

Barnholtz-Sloan, J. S., Sloan, A. E., Davis, F. G., Vigneau, F. D., Lai, P., & Sawaya, R. E. (2004). Incidence proportions of brain metastases in patients diagnosed (1973 to 2001) in the metropolitan Detroit cancer surveillance system. *Journal of Clinical Oncology, 22*(14), 2865–2872. https://doi.org/10.1200/JCO.2004.12.149. PMID: 15254054.

Conroy, T., Pfeiffer, P., Vilgrain, V., Lamarca, A., Seufferlein, T., O'Reilly, E. M., Hackert, T., Golan, T., Prager, G., Haustermans, K., Vogel, A., Ducreux, M., & ESMO Guidelines Committee. Electronic address: clinicalguidelines@esmoorg. (2023). Pancreatic cancer: ESMO clinical practice guideline for diagnosis, treatment and follow-up. *Annals of Oncology, 34*(11), 987–1002. https://doi.org/10.1016/j.annonc.2023.08.009. Epub 2023 Sep 9. PMID: 37678671.

Gouton, E., Gilabert, M., Launay, S., Loir, E., Tyran, M., Rochigneux, P., Turrini, O., Garnier, J., Mitry, E., & Chanez, B. (2024). Management and outcomes of brain metastases from pancreatic adenocarcinoma: A pooled analysis and literature review. *Frontiers in Oncology, 13*, 1326676. https://doi.org/10.3389/fonc.2023.1326676. PMID: 38260832; PMCID: PMC10800932.

Grigorescu, R. R., Husar-Sburlan, I. A., & Gheorghe, C. (2024). Pancreatic cancer: A review of risk factors. *Life, 14*(8), 980. https://doi.org/10.3390/life14080980

Gu, X., & Minko, T. (2024). Targeted nanoparticle-based diagnostic and treatment options for pancreatic cancer. *Cancers (Basel), 16*(8), 1589. https://doi.org/10.3390/cancers16081589. PMID: 38672671; PMCID: PMC11048786.

Hayat, U., Croce, P. S., Saadeh, A., Desai, K., Appiah, J., Khan, S., Khan, Y. I., Kumar, K., & Hanif, A. (2025). Current and emerging treatment options for pancreatic cancer: A comprehensive review. *Journal of Clinical Medicine, 14*(4), 1129. https://doi.org/10.3390/jcm14041129. PMID: 40004658; PMCID: PMC11856716 https://www.nccn.org/guidelines/guidelines-detail?id=1455.

Kwaśniewska, D., Fudalej, M., Badowska-Kozakiewicz, A. M., Czerw, A., & Deptała, A. (August 6, 2025). Neoadjuvant therapy or upfront surgery for pancreatic Cancer-To whom, when, and how? *Cancers (Basel), 17*(15), 2584. https://doi.org/10.3390/cancers17152584. PMID: 40805279; PMCID: PMC12346865.

Lee, E. S., & Lee, J. M. (2014). Imaging diagnosis of pancreatic cancer: A state-of-the-art review. *World Journal of Gastroenterology, 20*(24), 7864–7877. https://doi.org/10.3748/wjg.v20.i24.7864. PMID: 24976723; PMCID: PMC4069314.

McGuigan, A., Kelly, P., Turkington, R. C., Jones, C., Coleman, H. G., & McCain, R. S. (2018). Pancreatic cancer: A review of clinical diagnosis, epidemiology, treatment and outcomes. *World Journal of Gastroenterology, 24*(43), 4846–4861. https://doi.org/10.3748/wjg.v24.i43.4846. PMID: 30487695; PMCID: PMC6250924.

NCCN, 2025. NCCN Guidelines for patients: Pancreatic cancer. https://www.nccn.org/patients/guidelines/content/PDF/pancreatic-patient.pdf.

Rozengurt, E., & Eibl, G. (January 3, 2026). Pancreatic cancer: Molecular pathogenesis and emerging therapeutic strategies. *Signal Transduction and Targeted Therapy, 11*(1), 6. https://doi.org/10.1038/s41392-025-02499-y. PMID: 41484057; PMCID: PMC12764932.

Sharma, R., Ghosh, R., Kumar, S., Komal, K., & Kumar, M. (2025). Decoding pancreatic cancer: Key risk factors across genetics and lifestyle. *Expert Review of Molecular Diagnostics, 25*(4), 95–99. https://doi.org/10.1080/14737159.2025.2464563. Epub 2025 Feb 8. PMID: 39918844.

Stoop, T. F., Javed, A. A., Oba, A., Koerkamp, B. G., Seufferlein, T., Wilmink, J. W., & Besselink, M. G. (2025). Pancreatic cancer. Erratum in: Lancet. 2025 May 17;405(10491):1742 *Lancet, 405*(10485), 1182–1202. https://doi.org/10.1016/S0140-6736(25)00992-4. PMID: 40187844.

Toft, J., Hadden, W. J., Laurence, J. M., Lam, V., Yuen, L., Janssen, A., & Pleass, H. (July 2017). Imaging modalities in the diagnosis of pancreatic adenocarcinoma: A systematic review and meta-analysis of sensitivity, specificity and diagnostic accuracy. *European Journal of Radiology, 92*, 17–23. https://doi.org/10.1016/j.ejrad.2017.04.009. Epub 2017 Apr 24. PMID: 28624015.

Tonini, V., & Zanni, M. (2025). Pancreatic cancer in 2025: Have we found a solution? *World Journal of Gastroenterology, 31*(43), 111433. https://doi.org/10.3748/wjg.v31.i43.111433. PMID: 41358172; PMCID: PMC12678954.

Waleleng, B. J., Adiwinata, R., Wenas, N. T., Haroen, H., Rotty, L., Gosal, F., Rotty, L., Winarta, J., Waleleng, A., & Simadibrata, M. (2022). Screening of pancreatic cancer: Target population, optimal timing and how? *Annals of Medicine and Surgery, 84*, 104814. https://doi.org/10.1016/j.amsu.2022.104814. PMID: 36582884; PMCID: PMC9793126.

Wood, L. D., Canto, M. I., Jaffee, E. M., & Simeone, D. M. (2022). Pancreatic cancer: Pathogenesis, screening, diagnosis, and treatment. *Gastroenterology, 163*(2), 386–402.e1. https://doi.org/10.1053/j.gastro.2022.03.056. Epub 2022 Apr 7. PMID: 35398344; PMCID: PMC9516440.

Zhao, Z., & Liu, W. (2020). Pancreatic cancer: A review of risk factors, diagnosis, and treatment, 1533033820962117 *Technology in Cancer Research & Treatment, 19*. https://doi.org/10.1177/1533033820962117. PMID: 33357065; PMCID: PMC7768873.

Chapter 32

Acute pancreatitis and its impact on mental health

Mikaela Hoard[1] and Zachary L. Smith[2]
[1]West Virginia School of Osteopathic Medicine, Lewisburg, WV, United States; [2]Division of Gastroenterology and Hepatology, Wisconsin Institute for Research in Endoscopy, Medical College of Wisconsin, Milwaukee, WI, United States

Abbreviations

ANP Acute necrotizing pancreatitis
AP Acute pancreatitis
BMD Bone mineral density
CP Chronic pancreatitis
CRP C-reactive protein
ERCP Endoscopic retrograde cholangiopancreatography
GAD-7 Generalized Anxiety Disorder 7-item scale
GIP Glucose-dependent insulinotropic polypeptide
GLP-1 Glucagon-like peptide-1
GLP-2 Glucagon-like peptide-2
HRQOL Health-related quality of life
IGF-1 Insulin-like growth factor-1
IL-6 Interleukin-6
PCA Patient-controlled analgesia
PHQ-9 Patient Health Questionnaire-9
PRSS1 Protease, serine 1 (cationic trypsinogen)
QOL Quality of life
RAP Recurrent acute pancreatitis
SAP Severe acute pancreatitis
SPINK1 Serine peptidase inhibitor, Kazal type 1
WON Walled-off necrosis

Introduction

Acute pancreatitis (AP) is an acute inflammatory condition of the pancreas with a broad clinical spectrum. Most patients develop interstitial pancreatitis, which is self-limited and resolves with supportive care. A smaller but clinically significant subset progresses to necrotizing pancreatitis, characterized by pancreatic or peripancreatic necrosis, a systemic inflammatory response, and the potential for organ failure (Banks et al., 2013). Common presenting features include acute upper abdominal pain, nausea, and vomiting. While most cases follow a benign course, severe disease may result in prolonged hospitalization, local complications such as pseudocysts or walled-off necrosis, and high short-term morbidity and mortality.

The burden of acute pancreatitis

AP is one of the most common gastrointestinal causes of hospital admission worldwide. In the United States, it accounts for hundreds of thousands of emergency visits and hospitalizations annually, with admission rates of 30–40 cases per 100,000 persons per year (Peery et al., 2022). The cost burden is considerable, with median hospitalization charges exceeding $25,000 per admission and nearly one in six patients requiring readmission within 30 days (Peery et al., 2022). Although most patients experience only a single lifetime episode, a substantial minority develop recurrent acute pancreatitis (RAP) (Li et al., 2023; Yu et al., 2020). Observational studies consistently report recurrence rates between 15% and 40% after an index episode, with alcohol, smoking, and necrotizing disease as major risk factors. RAP drives much of the long-term healthcare utilization, as it predisposes patients to chronic pancreatitis (CP), pancreatic insufficiency, and an elevated lifetime risk of pancreatic cancer (de Rijk et al., 2024; Li et al., 2023; Munigala et al., 2023).

Causes of acute pancreatitis

The etiology of AP varies by region, but gallstones and alcohol are the most common causes worldwide (Mederos et al., 2021). Gallstone-related AP typically occurs in older adults and carries a lower risk of recurrence once the biliary tract is cleared (Li et al., 2023). Alcohol-associated AP is more likely to recur, particularly with ongoing use, and is frequently associated with smoking as a synergistic risk factor. Hypertriglyceridemia, certain medications,

Hormones in Neuroscience. https://doi.org/10.1016/B978-0-443-30263-3.00004-X

post–endoscopic retrograde cholangiopancreatography (ERCP), and genetic variants such as PRSS1 or SPINK1 mutations account for additional cases (Mederos et al., 2021). Patients with necrotizing AP, alcohol use disorder, or tobacco exposure are at greatest risk for RAP, with recurrence rates approaching 30%–40% in some series (Ahmed Ali et al., 2016; Yadav et al., 2012). Idiopathic AP also carries a higher risk of recurrence, although occult microlithiasis, early CP, or genetic predisposition may explain many of these cases (Cote et al., 2012; Hajibandeh et al., 2023; Sirtl et al., 2024).

Acute to chronic pancreatitis: the continuum

Progression from AP to CP reflects the cumulative impact—either clinical or subclinical—of recurrent pancreatic injury. CP is defined by irreversible fibrosis and glandular dysfunction, often accompanied by calcifications and ductal changes (Beyer et al., 2020). Clinically, it is characterized by chronic abdominal pain, recurrent flares of inflammation, and progressive loss of both exocrine and endocrine function (Angelopoulos et al., 2005; Aslam et al., 2021; de la Iglesia-Garcia et al., 2017; de Rijk et al., 2022). Exocrine insufficiency leads to steatorrhea, weight loss, and malnutrition, while endocrine insufficiency manifests as pancreatogenic diabetes. Patients with CP often experience a markedly reduced quality of life, with disability, opioid dependence, and psychiatric comorbidity contributing substantially to the disease burden (Blachnio et al., 2023; Kjellesvig et al., 2022; Machicado et al., 2017; Mokrowiecka et al., 2011; Ratnayake et al., 2020). The natural history underscores the importance of RAP as the key transitional state, linking an initially reversible condition to a chronic, debilitating disease with systemic consequences.

Other sequelae of acute pancreatitis

Additionally, AP can have sequelae such as pancreatic fluid collections (PFCs)—including pseudocysts (PCs) and walled-off necrosis (WON) (Banks et al., 2013). These collections are characterized by the accumulation of fluid and necrotic debris in and around the pancreas. Some resolve spontaneously, but others persist and require intervention for symptoms, infection, or obstruction. Endoscopic transmural drainage is now the preferred approach, but complete resolution often necessitates multiple procedures. The need for repeated interventions, coupled with the risk of exocrine or endocrine insufficiency, has a measurable adverse effect on health-related quality of life (Smith et al., 2019). Patients undergoing drainage frequently report impaired physical and mental well-being, and a notable proportion develop chronic opioid use, adding further complexity to long-term management.

The psychosocial consequences of pancreatitis

RAP and CP are also the forms most strongly associated with mental health disorders. The recurrent pain, physical disability, and loss of pancreatic function that define these conditions contribute directly to depression, anxiety, and substance misuse (Blachnio et al., 2023; Machicado et al., 2017; Mokrowiecka et al., 2011; Ratnayake et al., 2020). The management of chronic pain frequently leads to opioid exposure, which carries its own psychiatric risks, including dependence and overdose (Issa et al., 2020; Maydeo et al., 2023). In addition, the systemic inflammation and hormonal disturbances that accompany recurrent and chronic disease may influence neuropsychiatric outcomes through biological mechanisms. As such, RAP and CP represent not only the clinical bridge between acute and chronic pancreatic disease but also the point at which mental health complications begin to emerge most prominently, a relationship that will be further explored in the next section.

Recurrent disease and CP are firmly established determinants of psychiatric morbidity; however, evidence increasingly illustrates a clinical predisposition for new onset mental health disorders following an episode of AP As the incidence of acute pancreatitis steadily rises, the impact of pancreatic pain on patient quality of life and psychiatric health has become an increasingly recognized facet of the disease burden.

Pain as a driver of psychosocial wellbeing

Among the vastly differing clinical manifestations of pancreatitis, pain remains the strongest single determinant in patient mental health and quality of life. Classically, pain in AP is often described as epigastric abdominal pain that often radiates to the back, driven by pancreatic inflammation and systemic illness. Patients' discomfort may be compounded by visceral pain in the setting of systemic effects such as hypovolemia and organ dysfunction, furthering the symptom burden (Dunbar et al., 2021). While this is a classical presentation, pain on an individual level is complex and poorly understood.

The psychological effects of pain in AP are an increasingly recognized aspect of holistic patient care. Notably, pain in AP patients is a major contributing factor in long-term distress and impairment in quality of life. Among those with acute pancreatitis, the reported prevalence of anxiety is 29%, while depression is reported at 35.7% (Ma et al., 2022). Patients who experience greater pain intensity may develop new or exacerbate mental health comorbidities, illustrating the dynamic relationship between the physical and psychological burden of AP.

Incident mental health disorders after acute pancreatitis

Recent cohort studies have provided compelling evidence of the association between AP and new onset mental health disorders. In a large U.S. propensity matched study, patients following a single episode of AP were two times more likely to develop a new onset mental health disorder within 1 year in comparison to matched controls. Among these new onset mental health disorders, 1 year incident rates of anxiety (7.3%) and mood disorders (6.3%) were most common (Khoja et al., 2023). New onset mental health disorders among AP patients diverge most prominently within the first 120 days in comparison to controls, suggesting the need for prompt mental health screening and early intervention for pancreatitis patients. Importantly, some data suggest that a single episode of AP may trigger the emergence of new onset mental health disorders well before the onset of RAP or CP (Khoja et al., 2023). While RAP and CP are often classically associated with mental health disorders, these findings demonstrate psychological sequelae can precede these stages, indicating the mental health burden may begin earlier than traditionally recognized.

Impact of disease severity on mental health disorders

While many patients experience a mild course of AP, 20%–30% will progress into severe acute pancreatitis, compounding the long-term impact of disease (Wang et al., 2023). Severe acute pancreatitis (SAP) is defined by the presence of persistent organ failure lasting more than 48 hours, with accompanied systemic inflammatory response syndrome (Banks et al., 2013; Khoja et al., 2023). SAP is associated with frequent complications, prolonged hospitalizations, and intensive interventions. Alongside the physical burden, patients frequently face financial hardship, reduced employment opportunities, and a diminished quality of life, intensifying the effect of the disease and showcasing its long-term consequences. Among SAP patients, the prevalence of anxiety and depression was 33.6% and 34.3%, respectively (Wang et al., 2023). Most notably, anxiety had the largest confounding effect on health-related quality of life, worsening outcomes by increasing depression, highlighting the bidirectional relationship between anxiety and depression. The recognition of psychiatric burden among varying disease stages is a pivotal factor in shaping long-term healthcare outcomes.

As disease severity progresses, 5%–10% of patients develop necrosis of the pancreatic parenchyma, peripancreatic tissue or both, a condition known as acute necrotizing pancreatitis (ANP). The severity of the disease presents a critical determinant in the overall patient burden. The risk of developing mental health disorders is increasingly elevated in those who suffer from ANP, particularly mood and anxiety disorders, compared to those who experienced non-necrotizing disease (Khoja et al., 2023). All-cause mortality for ANP is significantly higher when compared to acute pancreatitis without necrosis. Psychiatric comorbidity was significantly higher in ANP than in healthy controls and those with non-pancreatic inflammatory disorders. ANP patients experience a higher rate of invasive procedures, hospitalizations, critical care, and exocrine insufficiency, all of which are likely to compound psychological stress. These findings are reinforced by a greater use of antidepressants and anxiolytics among ANP patients, illustrating a clear parallel between disease severity and the likelihood of developing psychiatric conditions (Khoja et al., 2023).

Psychiatric comorbidities in chronic pancreatitis

AP is an isolated event; however, a subset of patients experience recurring attacks or transition to CP. Following an acute attack, psychiatric comorbidities often develop within this period, suggesting that many patients are anxiously awaiting another acute disease attack (Pezzilli et al., 2005). This uncertainty of the future is common among those who experience acute pancreatitis. The anxiety patients experience during the course of the disease is amplified due to the fear of aggravation or relapse. In a study among patients with AP, 46% live with ongoing, measurable anxiety of disease recurrence. Their concerns centered on recurring abdominal pain, infections such as ascites, new-onset diabetes, dietary restrictions, and the possibility of malignancy. This fear extends beyond physical health, reshaping the way patients view themselves, their interpersonal connections, and their expectations for the future (Ma et al., 2022).

The effects of CP on mental health and patient quality of life have been widely studied (Blachnio et al., 2023; Ma et al., 2022; Machicado et al., 2017; Mokrowiecka et al., 2011; Ratnayake et al., 2020; Smith et al., 2019). Studies in chronic pain highlight the complexity of chronic pancreatitis, noting the shared neural networks between pain perception and emotional regulation. Recurrent or prolonged pain inputs alter neural processing, recruiting corticolimbic pathways and shifting the pain experience from purely sensory processing toward emotional processing. This neuronal mechanism is rational for the emergence of anxiety and depression following episodes of pancreatitis. Other contributing factors include central sensitization, genetic susceptibility, and systemic inflammation, all of which amplify pain and ultimately, psychiatric burden. The persistence of pain in CP is the dominant factor in patient mental health and quality of life. In CP patients, anxiety and/or depression is highly prevalent. In a cohort study of 171 CP patients, nearly half of patients experienced clinically significant anxiety and over one third experienced depression, with considerable overlap between the two. Of these patients who experience anxiety and depression, many also report considerable interference with daily activities. Prospective studies demonstrate the impact of CP on quality of life is similar to heart, liver, kidney, and lung disease, and comparatively worse than diabetes mellitus, rheumatoid arthritis and non-skin cancers, underlying the comprehensive dysfunction of chronic pancreatitis (Dunbar et al., 2021). The psychosocial sequela of CP highlights the importance of early intervention and psychosocial support in those who experience AP.

The role of opioid analgesics

The burden of pain in AP is reinforced by its pharmacologic approach of treatment. Clinical reliance on opioids is an important facet of treatment in AP and reflects the lack of alternatives available to patients, introducing nuances to long-term psychiatric health and quality of life. AP is historically reliant on opioids, due to limited alternatives for nociceptive and neuropathic pain. Many patients who experience a form of pancreatitis are predisposed to chronic opioid therapy. Current consensus recommends opioids for pain relief in the United States. Oral opioids are used for mild symptom control, with nonsteroidal anti-inflammatory drugs given as adjunct therapy. In severe cases, pain control is managed with parenteral opioids such as hydromorphone or fentanyl. In patients who experience refractory pain, opioids can be administered with a patient-controlled analgesia pump (PCA) (Basurto Ona et al., 2013). Under the current guidelines of pain management therapy in AP, many patients receive parenteral opioids as initial therapy. Following discharge, approximately one in 10 patients remain on opioids after 6 months (Nelson et al., 2022). Incidence of mood and anxiety disorders has been observed to parallel the rate of new onset opioid prescriptions, suggesting an association between the two comorbidities in AP (Khoja et al., 2023).

Other hormones, endocrine tissues, and pathways

Having established the clinical course of acute and chronic pancreatitis and their strong associations with impaired quality of life and psychiatric comorbidity, it is important to examine the underlying biological mechanisms that may mediate these outcomes. A growing body of literature highlights the role of gut and pancreatic hormones in modulating both metabolic and neuropsychiatric states (Al Qassab et al., 2025; Bharmal et al., 2021; Ganeshalingam et al., 2025; Hiroyoshi et al., 1999; Hornum et al., 2010; Knop et al., 2003, 2007b, 2007c; Lach et al., 2018; Llach et al., 2025). AP, RAP, and CP are characterized not only by inflammation and parenchymal injury but also by disruption of normal endocrine and exocrine signaling (Angelopoulos et al., 2005; Aslam et al., 2021; Beyer et al., 2020; de la Iglesia-Garcia et al., 2017; de Rijk et al., 2022). These disturbances extend beyond glucose metabolism to involve incretin hormones, stress axes, and sex steroids, all of which may influence mood regulation and cognitive function.

In this section, we will first review alterations in pancreatic and gut hormone activity observed in AP and CP, including insulin resistance, glucagon-like peptide-1 (GLP-1) and glucose-dependent insulinotropic polypeptide (GIP) and infer potential mechanisms linking pancreatitis-related endocrine disruption to psychiatric outcomes. Table 32.1 summarizes the best-characterized hormonal disturbances in CP—centered on the incretin axis—and highlights clinical modifiers and why these signals plausibly intersect with mental health.

Incretin hormone alterations in pancreatitis

Among the best characterized hormonal disturbances in CP are changes in the incretin system, comprising GLP-1 and GIP (Bharmal et al., 2021; Hiroyoshi et al., 1999; Hornum et al., 2010; Knop et al., 2007c). These peptides are secreted from intestinal L- and K-cells, respectively, in response to nutrient ingestion. They potentiate glucose-dependent insulin secretion, suppress glucagon release, slow gastric emptying, and promote satiety. Collectively, this "incretin effect" accounts for up to 70% of the insulin response after oral glucose compared with isoglycemic intravenous glucose administration (Baggio & Drucker, 2007; Samms et al., 2020).

In patients with CP and preserved glucose tolerance, the incretin effect remains relatively intact, with both GLP-1 and GIP eliciting insulinotropic responses. However, as glucose tolerance deteriorates and secondary diabetes develops, a selective defect emerges: the insulinotropic action of GIP becomes markedly impaired, whereas GLP-1 retains activity (Hiroyoshi et al., 1999; Knop et al., 2003, 2007a, 2007c). This mirrors findings in type 2 diabetes, where GIP's late-phase insulin effect is virtually absent (Knop et al., 2007a). Studies employing hyperglycemic clamp protocols in CP demonstrate that GIP can augment insulin secretion in those with normal glucose tolerance but loses this capacity in patients with CP-related diabetes (Knop et al., 2007a). In contrast, exogenous GLP-1 infusion continues to stimulate insulin and suppress glucagon secretion, even in insulin-treated individuals (Knop et al., 2003).

Interestingly, secretion patterns of GLP-1 may also be altered in CP (Knop et al., 2007c). Some studies have reported elevated plasma GLP-1 responses to oral glucose in

TABLE 32.1 Incretin and related gut–pancreas hormones in chronic pancreatitis (CP): alterations, clinical modifiers, and mental-health relevance.

Hormone	What changes in CP	Key clinical modifiers	Metabolic correlate	Mental health relevance
GLP-1	Insulinotropic action generally preserved even as glucose tolerance worsens; GLP-1 still suppresses glucagon and stimulates residual insulin secretion.	Exocrine insufficiency and maldigestion can alter post-prandial GLP-1 patterns; enzyme replacement can normalize/augment responses.	Lowers glucose without hypoglycemia in CP-related diabetes.	Receptors in mood/stress circuits; links to appetite, energy homeostasis, and CNS signaling relevant to affect.
GIP	Selective loss of insulinotropic effect as CP progresses to pancreatogenic (type 3c) diabetes; relatively intact in CP with normal glucose tolerance.	Glycemic status (normal vs. CP-DM) is the main determinant; enzyme therapy may raise secretion but does not restore insulinotropic effect in CP-DM.	Deteriorating GIP action tracks with worsening insulin resistance/glycemic variability.	Potential impact on reward/energy balance; emerging CNS roles make it a candidate pathway for mood links.
GLP-2	Exaggerated post-prandial GLP-2 in CP with exocrine insufficiency (adaptive L-cell response to maldigestion).	Exocrine insufficiency; nutrient delivery to distal gut; enzyme replacement.	Intestinal trophic effects; may influence absorption and nutrition status.	Nutrition and weight trajectories intersect with mood, fatigue, and QoL.
Insulin (type 3c DM)	Progressive insulin deficiency + insulin resistance in CP; frequent glycemic volatility. (Summarized in draft.)	Pain, inflammation, malnutrition, and enzyme replacement; presence of diabetes.	Drives hyperglycemia and variability; GLP-1 remains actionable.	Insulin resistance and glycemic swings are linked to depressive symptoms and lower mental HRQoL in chronic disease contexts.

patients with pancreatic diabetes, while others observed preserved postprandial GLP-1 and GIP secretion that increased further after pancreatic enzyme supplementation in those with exocrine insufficiency (Hiroyoshi et al., 1999; Knop et al., 2003). These findings suggest that nutrient maldigestion and delivery to the distal gut modulate incretin release, and that exocrine–endocrine crosstalk is critical in CP. Additional work has shown exaggerated postprandial GLP-2 responses in CP with exocrine insufficiency, further underscoring the adaptive role of L-cell hormones in this context (Hornum et al., 2010).

Therapeutically, GLP-1 administration in CP-related diabetes lowers glucose, suppresses glucagon, and enhances residual insulin secretion without provoking hypoglycemia (Knop et al., 2003). These properties differentiate GLP-1 from GIP and raise the possibility that selective or dual incretin receptor agonism may have unique utility in CP. The relevance of these pathways extends beyond glycemic control: incretins exert effects on satiety, adipose tissue metabolism, and central nervous system signaling, domains increasingly recognized as pertinent to mental health outcomes (Baggio & Drucker, 2007).

The alteration of incretin biology in CP may have implications that extend beyond metabolic regulation. GLP-1 receptors are expressed in brain regions involved in mood, cognition, and stress response, while GIP signaling may intersect with pathways of neuroplasticity (Al Qassab et al., 2025; Baggio & Drucker, 2007; Ganeshalingam et al., 2025). Disruption of incretin function in CP could therefore contribute to changes in appetite regulation, weight fluctuation, and energy homeostasis, each of which are closely tied to psychiatric vulnerability (Bremner et al., 2020). Moreover, the persistence of GLP-1 activity in the face of lost GIP responsiveness suggests that incretin-based therapies might influence not only metabolic outcomes but also neuropsychiatric health. These observations provide a rationale for exploring incretin hormones as mediators of the link between pancreatic disease and mental health.

Sex hormone dysregulation and sexual health in chronic pancreatitis

Beyond metabolic hormones, disturbances in sex steroid pathways are increasingly recognized in patients with chronic pancreatitis (CP). Several mechanisms converge to disrupt gonadal function, including chronic systemic inflammation, malnutrition, diabetes, and long-term opioid therapy for pain control.

Hypogonadism and bone health

Prospective data demonstrate that hypogonadism is common in CP, with a prevalence of approximately 27% among nonmenopausal patients (Gupta et al., 2019). In men, testosterone levels inversely correlate with opioid exposure, reflecting opioid-induced suppression of the hypothalamic–pituitary–gonadal axis (Shah et al., 2024; Umar et al., 2020). Hypogonadism in CP has been linked to reduced lumbar spine bone mineral density (BMD), highlighting its role in the high burden of osteoporosis and fracture risk in this population (Gupta et al., 2019). These findings underscore that the endocrine sequelae of CP extend beyond glucose metabolism to skeletal integrity, mediated in part by sex hormone deficiency.

Sexual dysfunction in CP

Sexual dysfunction (SD) is another underappreciated consequence of CP. Large-scale population analyses reveal that patients with CP have significantly higher odds of developing SD compared with individuals without CP, affecting both men and women (Umar et al., 2020). In men, prospective evaluation shows that as many as 60% report erectile or orgasmic dysfunction, ranging from mild to severe. Pain severity, daily opioid use, and systemic inflammation (elevated IL-6 and CRP) appear to be major contributors. SD in CP is strongly associated with distress, impaired psychosocial well-being, and lower health-related quality of life (HRQOL), even if this is not always fully captured by generic QOL instruments (Shah et al., 2024; Umar et al., 2020).

Sarcopenia and hormonal crosstalk

A related dimension is sarcopenia, increasingly recognized in CP as a manifestation of chronic inflammation, malabsorption, and hormonal imbalance. Experimental models demonstrate reductions in anabolic hormones such as testosterone and oxytocin, accompanied by increases in catabolic mediators such as myostatin (Tomaszewska et al., 2024). These changes tilt the anabolic–catabolic balance toward muscle wasting, exacerbating frailty and further impairing QOL. In human studies, sarcopenia prevalence in CP ranges from 17% to 62% and is tightly linked with sex hormone deficiencies.

The convergence of hypogonadism, sexual dysfunction, and sarcopenia highlights the broad influence of sex hormones on both physical and psychological health in CP (Amiaz & Seidman, 2008). Testosterone and estradiol modulate neurotransmitter systems involved in mood and cognition, while oxytocin has recognized roles in social bonding and stress regulation (Froemke & Young, 2021; Leng et al., 2022). Deficiency states may therefore heighten vulnerability to depression, fatigue, and anxiety in patients already burdened by chronic pain and disability. The interaction of opioid therapy, systemic inflammation, and gonadal suppression further reinforces this cycle, positioning sex hormone dysregulation as a plausible biological mediator of the mental health complications observed in chronic pancreatitis.

Future directions

Endocrine–psychiatric targets

Among the most promising areas of future investigation is the role of incretin-based therapies in mitigating both metabolic and psychiatric sequelae of pancreatitis. GLP-1 receptor agonists and dual GLP-1/GIP agonists, such as tirzepatide, have demonstrated robust efficacy in improving glycemic control and weight regulation in diabetes and obesity (Aronne et al., 2024; Frias et al., 2021; Garvey et al., 2023; Hendershot et al., 2025; Jastreboff et al., 2022, 2025; Malhotra et al., 2024; Siskind et al., 2025). Beyond these metabolic effects, emerging evidence suggests GLP-1 receptor agonists exert antidepressant and anxiolytic properties, potentially through direct central nervous system signaling, modulation of inflammatory pathways, and effects on satiety and energy homeostasis (Chen et al., 2024; de Paiva et al., 2024; Pozzi et al., 2019).

Patients with CP and post-pancreatitis diabetes represent a population in which these therapies may have dual benefit. The persistence of GLP-1 activity in CP, despite impaired GIP responsiveness, provides a mechanistic rationale for their use. Preliminary clinical observations indicate that GLP-1 analogs can enhance residual insulin secretion, suppress glucagon, and reduce glycemic variability in CP-related diabetes without provoking hypoglycemia (Knop et al., 2003). Importantly, these metabolic benefits may coincide with improvements in mood and quality of life, though this remains to be formally tested.

Future clinical trials should prospectively evaluate incretin-based therapies in pancreatitis cohorts with standardized psychiatric endpoints. Such studies could clarify whether GLP-1 and dual GLP-1/GIP agonists not only improve metabolic outcomes but also alleviate depression and anxiety that frequently complicate the disease course. This line of inquiry is particularly relevant given the rising prevalence of pancreatitis, the lack of effective psychiatric interventions specific to this population, and the growing recognition of shared endocrine and neuropsychiatric pathways.

Sex hormones and related pathways

Sex hormone dysregulation represents another important avenue for future study. Hypogonadism is prevalent in CP and is linked to reduced bone mineral density, sarcopenia, and impaired quality of life. These endocrine changes extend beyond musculoskeletal health, as testosterone and estradiol play recognized roles in mood regulation, energy, and cognitive function. Observational data in CP suggest that hypogonadism and sexual dysfunction are associated with higher rates of depression and anxiety, yet systematic evaluation of sex hormone replacement or modulation in this population is lacking (Gupta et al., 2019; Umar et al., 2020).

Future research should address whether targeted hormone replacement — including testosterone therapy in men and estrogen-based strategies in women — can improve not only skeletal health but also psychiatric outcomes. Clinical trials incorporating validated measures of depression, fatigue, and quality of life are warranted to clarify the extent of benefit. Beyond gonadal hormones, oxytocin and anabolic-catabolic regulators such as myostatin represent additional candidates for investigation, given their established links with social bonding, stress reactivity, and muscle wasting.

Importantly, the interaction between opioid therapy, systemic inflammation, and sex hormone suppression highlights a potentially modifiable cycle. Studies examining whether opioid-sparing pain regimens or adjunctive endocrine therapy can interrupt this cycle would have major implications for both physical and psychiatric health in CP. Ultimately, integrating sex hormone assessment into the routine care of pancreatitis patients may offer a pathway to reduce psychiatric morbidity and enhance long-term quality of life.

Study design and research priorities

To establish a causal link between endocrine dysregulation and psychiatric morbidity in pancreatitis, future studies must pair detailed endocrine profiling with validated mental health assessments. Prospective cohort studies should include serial measurements of incretin hormones, insulin sensitivity, glucagon, and sex steroid levels in patients following AP, RAP, and CP. Parallel use of standardized psychiatric instruments—such as the PHQ-9 for depression, GAD-7 for anxiety, and validated quality-of-life measures—would allow investigators to map temporal associations between hormonal changes and mental health outcomes.

Interventional studies are also needed. Randomized controlled trials evaluating GLP-1 receptor agonists, dual GLP-1/GIP agonists, or testosterone replacement in CP should incorporate psychiatric endpoints alongside metabolic and clinical outcomes. Study designs might include stratification by baseline psychiatric risk factors (e.g., prior depression, high pain burden, opioid use) or by hormonal phenotype (e.g., hypogonadism, impaired incretin response). Such stratification would enable precision approaches and maximize the likelihood of detecting treatment effects.

Clinical implementation and multidisciplinary care

Beyond mechanistic studies and interventional trials, there is a pressing need to translate emerging knowledge into routine clinical practice. Psychiatric morbidity after pancreatitis is common, underrecognized, and frequently under-treated. Systematic mental health screening should be incorporated into standard follow-up for patients with. Simple, validated screening tools can be integrated into outpatient visits with minimal time burden, allowing for early detection of depression and anxiety.

Optimal management of pancreatitis requires a multidisciplinary framework that includes gastroenterology, endocrinology, psychiatry, nutrition, and pain medicine. This integrated approach can address overlapping drivers of psychiatric morbidity—such as poor glycemic control, hypogonadism, malnutrition, sarcopenia, and opioid dependence—while also providing access to evidence-based psychiatric care. Particular attention should be given to opioid stewardship, as reliance on opioids for pain management has been consistently associated with higher rates of new-onset mood and anxiety disorders (Dunbar et al., 2021; Kjellesvig et al., 2022; Maydeo et al., 2023). Implementing opioid-sparing strategies, alongside endocrine and psychiatric interventions, may reduce long-term psychiatric burden.

In addition, patient education and psychosocial support should be embedded into pancreatitis care pathways. Counseling patients regarding recurrence risk, dietary modifications, and lifestyle interventions may reduce uncertainty and alleviate health-related anxiety. Support groups and behavioral interventions can further improve coping and resilience, particularly in those with chronic pain or recurrent disease. By embedding psychiatric screening and multidisciplinary care into pancreatitis management, clinicians may not only improve mental health outcomes but also enhance adherence to medical therapies, reduce hospital readmissions, and improve overall quality of life.

Translational research and biomarkers

A comprehensive understanding of the gut–pancreas–brain axis in pancreatitis will require translational approaches that extend beyond clinical observation. Biomarker discovery represents a key priority. Panels incorporating incretin hormones, glucagon, sex steroids, inflammatory mediators, and markers of muscle metabolism (e.g., myostatin, IGF-1) could help stratify patients at highest risk for psychiatric morbidity. Such biomarkers may also identify subgroups most likely to benefit from targeted interventions, including incretin-based therapies or sex hormone replacement.

Advanced imaging techniques offer another path forward. Functional MRI and PET studies could evaluate whether endocrine therapies modulate activity in corticolimbic networks involved in mood and pain perception. These approaches would allow investigators to connect biochemical disturbances with neural correlates of depression, anxiety, and altered pain processing in pancreatitis. Coupling imaging endpoints with clinical trials of hormonal therapies would provide critical mechanistic insights.

Finally, preclinical models of pancreatitis can be leveraged to explore molecular mechanisms linking endocrine disruption, systemic inflammation, and behavior. Animal studies using selective GLP-1 or GIP agonists, sex steroid supplementation, or oxytocin modulation may uncover specific pathways that can then be validated in human cohorts. Together, biomarker-driven patient stratification, neuroimaging, and preclinical experimentation form a translational framework that can accelerate progress toward targeted, mechanism-based strategies for reducing psychiatric morbidity in pancreatitis.

Conclusion

Pancreatitis imposes a substantial burden that extends beyond gastrointestinal morbidity to include persistent psychiatric sequelae. While RAP and CP are most strongly associated with depression, anxiety, and impaired quality of life, accumulating evidence demonstrates that even a single episode of AP can precipitate new-onset mental health disorders. Pain, endocrine disruption, and uncertainty regarding recurrence all contribute to this vulnerability. Increasing recognition of hormonal perturbations—including alterations in incretins, insulin, glucagon, and sex steroids—offers a mechanistic framework linking pancreatic disease with psychiatric morbidity. These insights underscore the need for integrative research strategies, targeted endocrine-based interventions, and multidisciplinary models of care. Future work bridging the gut–pancreas–brain axis holds promise not only for improving physical outcomes but also for addressing the underappreciated psychiatric dimensions of pancreatitis.

Applications of the material

The material presented here can be translated into several concrete practices that align with routine pancreatitis care while addressing the mental health burden that accompanies acute and chronic disease. First, standardized mental health screening can be incorporated into existing clinical pathways without adding substantial complexity. Patients with severe or necrotizing acute pancreatitis, recurrent acute pancreatitis, or chronic pancreatitis—and those reporting constant or severe pain—represent high-risk groups in whom brief, validated instruments (e.g., PHQ-9 for depression, GAD-7 for anxiety) can be administered at the index specialty visit, repeated during early recovery, and reassessed at follow-up. Positive screens should prompt documentation of a plan, safety assessment when indicated, and timely referral to behavioral health. Embedding this cadence within pancreatitis follow-up acknowledges the natural history of the disease and the periods when psychiatric symptoms most commonly emerge.

Second, psychiatric screening is best paired with systematic attention to exocrine and endocrine sequelae. In chronic pancreatitis, evaluation for exocrine pancreatic insufficiency (dietary history, weight trajectory, fat-soluble vitamin status) and disturbed glucose homeostasis (fasting glucose, A1c, oral glucose tolerance testing as indicated) can be performed alongside symptom scales. Optimizing enzyme replacement, nutrition, and glycemic stability is expected to improve energy, sleep, and concentration—domains patients experience as "quality of life" and that are frequently captured in HRQOL measures. In this way, core gastroenterology interventions are linked to outcomes that matter to patients and that intersect with the psychological dimensions of illness.

Third, pain management can be oriented around function and quality of life rather than absolute pain intensity. Multimodal, non-opioid regimens should be prioritized, with judicious use of opioids when necessary and clear plans for tapering as the clinical course allows. For inpatients requiring parenteral agents or patient-controlled analgesia, transition plans at discharge can help reduce long-term exposure. Tracking opioid use in parallel with PHQ-9 and GAD-7 supports more deliberate adjustments and may mitigate the downstream psychiatric risk that accompanies prolonged opioid therapy.

Fourth, the endocrine context outlined in this chapter supports a pragmatic research agenda within clinical practice. Programs with the capacity to do so may collect a lean set of hormonal and metabolic data—standardized fasting and post-prandial glucose and insulin, with incretin measures where feasible—in selected patients, and relate these to serial PHQ-9, GAD-7, and HRQOL results. These observations can inform pilot trials that evaluate whether improving endocrine disturbances translates into measurable mental health benefits in chronic pancreatitis. Importantly, such work should stabilize exocrine and nutritional therapy prior to enrollment, prespecify psychiatric endpoints, and include pancreatitis-specific safety monitoring.

Finally, a multidisciplinary framework is central to implementation. Gastroenterology, endocrinology, pain medicine, nutrition, and behavioral health each address elements that converge on the patient's lived experience: pain, maldigestion, weight loss, glycemic variability, and uncertainty about recurrence. Clear role definition—who screens, who follows positive results, how enzyme therapy is optimized, when to escalate to specialty psychiatry—allows teams to deliver consistent, reproducible care. Patient education about recurrence risk, diet, and activity, together with accessible psychosocial support, can reduce illness-related anxiety and improve engagement. By integrating these steps, clinics can move from recognizing the psychiatric burden of pancreatitis to measuring it, addressing it, and, ultimately, improving long-term outcomes.

Summary points

- Psychiatric morbidity is common across the pancreatitis spectrum and is highest with severe or necrotizing acute pancreatitis and with constant or severe pain.
- Compared with non-necrotizing acute pancreatitis, acute necrotizing pancreatitis has higher all-cause mortality and a higher prevalence of psychiatric comorbidity, alongside greater use of antidepressants and anxiolytics.
- Patients with chronic pancreatitis demonstrate clinically meaningful decrements in physical and mental quality-of-life scores on validated instruments.
- Opioid exposure is frequent in pancreatitis care and is associated with worse mental-health indices and greater pain interference.
- Chronic pancreatitis often results in endocrine disruption, including post-pancreatitis (type 3c) diabetes and glycemic variability, which are associated with increased risk of mental health disorders.
- Exocrine insufficiency and maldigestion in chronic pancreatitis plausibly alter gut–brain signaling via changes in incretin responses.
- A research agenda is proposed prioritizing hormonal phenotyping and exploratory incretin-targeted trials (including GLP-1/GIP agonists) with standardized mental-health outcomes.

Mini-dictionary of terms

- **Acute necrotizing pancreatitis (ANP).** A severe form of acute pancreatitis characterized by pancreatic and/or peripancreatic tissue death, often requiring intensive care and invasive interventions.
- **Chronic pancreatitis (CP).** A long-standing fibroinflammatory disorder of the pancreas marked by progressive structural damage, persistent or recurrent pain, maldigestion, and disturbances in glucose regulation.
- **Endoscopic retrograde cholangiopancreatography (ERCP).** An endoscopic procedure that combines endoscopy and fluoroscopy to diagnose or treat pancreatic and biliary duct disorders (e.g., stone extraction, stenting).
- **Exocrine pancreatic insufficiency (EPI).** Inadequate secretion or function of pancreatic digestive enzymes leading to maldigestion, steatorrhea, weight loss, and fat-soluble vitamin deficiency.
- **GIP.** An incretin hormone released from intestinal K-cells after meals that helps regulate post-prandial blood-sugar control and energy balance.

- **GLP-1.** An incretin hormone released from intestinal L-cells after meals that supports blood-sugar regulation, appetite control, and gut motility; also active in brain regions linked to mood and cognition.
- **Gut–brain axis.** Bidirectional communication network linking the gastrointestinal tract and the central nervous system through neural, immune, and hormonal pathways.
- **Health-related quality of life (HRQOL).** Patient-reported assessment of health across physical, mental, and social domains, often captured with validated surveys.
- **Incretin.** A class of meal-stimulated gut hormones that enhance post-prandial blood-sugar control and influence appetite and satiety.
- **Walled-off necrosis (WON).** A mature, encapsulated collection of pancreatic or peripancreatic necrotic tissue that typically develops more than 4 weeks after acute pancreatitis.

References

Ahmed Ali, U., Issa, Y., Hagenaars, J. C., Bakker, O. J., van Goor, H., Nieuwenhuijs, V. B., Bollen, T. L., van Ramshorst, B., Witteman, B. J., Brink, M. A., Schaapherder, A. F., Dejong, C. H., Spanier, B. W., Heisterkamp, J., van der Harst, E., van Eijck, C. H., Besselink, M. G., Gooszen, H. G., van Santvoort, H. C., & Dutch Pancreatitis Study, G. (2016). Risk of recurrent pancreatitis and progression to chronic pancreatitis after a first episode of acute pancreatitis. *Clinical Gastroenterology and Hepatology, 14*(5), 738–746. https://doi.org/10.1016/j.cgh.2015.12.040

Al Qassab, M., Mneimneh, M., Jradi, A., Derbas, B., Dabboussi, D., Khoury Baini, J., Katrib, N., Chaarani, N., Attieh, P., Kanaan, A., Harb, F., Azar, S., & Ghadieh, H. E. (2025). The expanding role of GLP-1 receptor agonists: Advancing clinical outcomes in metabolic and mental health. *Current Issues in Molecular Biology, 47*(4). https://doi.org/10.3390/cimb47040285

Amiaz, R., & Seidman, S. N. (2008). Testosterone and depression in men. *Current Opinion in Endocrinology Diabetes and Obesity, 15*(3), 278–283. https://doi.org/10.1097/MED.0b013e3282fc27eb

Angelopoulos, N., Dervenis, C., Goula, A., Rombopoulos, G., Livadas, S., Kaltsas, D., Kaltzidou, V., & Tolis, G. (2005). Endocrine pancreatic insufficiency in chronic pancreatitis. *Pancreatology, 5* (2–3), 122–131. https://doi.org/10.1159/000085264

Aronne, L. J., Sattar, N., Horn, D. B., Bays, H. E., Wharton, S., Lin, W. Y., Ahmad, N. N., Zhang, S., Liao, R., Bunck, M. C., Jouravskaya, I., Murphy, M. A., & Investigators, S.-. (2024). Continued treatment with tirzepatide for maintenance of weight reduction in adults with obesity: The SURMOUNT-4 randomized clinical trial. *JAMA, 331*(1), 38–48. https://doi.org/10.1001/jama.2023.24945

Aslam, M., Jagtap, N., Karyampudi, A., Talukdar, R., & Reddy, D. N. (2021). Risk factors for development of endocrine insufficiency in chronic pancreatitis. *Pancreatology, 21*(1), 15–20. https://doi.org/10.1016/j.pan.2020.11.011

Baggio, L. L., & Drucker, D. J. (2007). Biology of incretins: GLP-1 and GIP. *Gastroenterology, 132*(6), 2131–2157. https://doi.org/10.1053/j.gastro.2007.03.054

Banks, P. A., Bollen, T. L., Dervenis, C., Gooszen, H. G., Johnson, C. D., Sarr, M. G., Tsiotos, G. G., Vege, S. S., & Acute Pancreatitis Classification Working, G. (2013). Classification of acute pancreatitis–2012: Revision of the Atlanta classification and definitions by international consensus. *Gut, 62*(1), 102–111. https://doi.org/10.1136/gutjnl-2012-302779

Basurto Ona, X., Rigau Comas, D., & Urrútia, G. (2013). Opioids for acute pancreatitis pain. *Cochrane Database of Systematic Reviews, 7*. https://doi.org/10.1002/14651858.CD009179.pub2

Beyer, G., Habtezion, A., Werner, J., Lerch, M. M., & Mayerle, J. (2020). Chronic pancreatitis. *Lancet, 396*(10249), 499–512. https://doi.org/10.1016/S0140-6736(20)31318-0

Bharmal, S. H., Kimita, W., Ko, J., & Petrov, M. S. (2021). Pancreatic and gut hormones as predictors of new-onset prediabetes after non-necrotising acute pancreatitis: A prospective longitudinal cohort study. *Endocr Connect, 10*(7), 715–724. https://doi.org/10.1530/EC-21-0229

Blachnio, K., Szymocha, A., Kovalsky, M., Pawlak, A., Juzwiszyn, J., & Chabowski, M. (2023). Quality of life and pain in patients with chronic pancreatitis. *Pancreas, 52*(6), e321–e327. https://doi.org/10.1097/MPA.0000000000002262

Bremner, J. D., Moazzami, K., Wittbrodt, M. T., Nye, J. A., Lima, B. B., Gillespie, C. F., Rapaport, M. H., Pearce, B. D., Shah, A. J., & Vaccarino, V. (2020). Diet, stress and mental health. *Nutrients, 12*(8). https://doi.org/10.3390/nu12082428

Chen, X., Zhao, P., Wang, W., Guo, L., & Pan, Q. (2024). The antidepressant effects of GLP-1 receptor agonists: A systematic review and meta-analysis. *American Journal of Geriatric Psychiatry, 32*(1), 117–127. https://doi.org/10.1016/j.jagp.2023.08.010

Cote, G. A., Imperiale, T. F., Schmidt, S. E., Fogel, E., Lehman, G., McHenry, L., Watkins, J., & Sherman, S. (2012). Similar efficacies of biliary, with or without pancreatic, sphincterotomy in treatment of idiopathic recurrent acute pancreatitis. *Gastroenterology, 143*(6), 1502–1509. https://doi.org/10.1053/j.gastro.2012.09.006

de la Iglesia-Garcia, D., Huang, W., Szatmary, P., Baston-Rey, I., Gonzalez-Lopez, J., Prada-Ramallal, G., Mukherjee, R., Nunes, Q. M., Dominguez-Munoz, J. E., Sutton, R., & Group, N. P. B. R. U. P. A. (2017). Efficacy of pancreatic enzyme replacement therapy in chronic pancreatitis: Systematic review and meta-analysis. *Gut, 66*(8), 1354–1355. https://doi.org/10.1136/gutjnl-2016-312529

de Paiva, I. H. R., da Silva, R. S., Mendonca, I. P., de Souza, J. R. B., & Peixoto, C. A. (2024). Semaglutide attenuates anxious and depressive-like behaviors and reverses the cognitive impairment in a type 2 diabetes mellitus mouse model via the microbiota-gut-brain axis. *Journal of Neuroimmune Pharmacology, 19*(1), 36. https://doi.org/10.1007/s11481-024-10142-w

de Rijk, F., Sissingh, N. J., Boel, T. T., Timmerhuis, H. C., de Jong, M., Pauw, H. S., van Veldhuisen, C. L., Hallensleben, N. D., Anten, M., Brink, M. A., Curvers, W. L., van Duijvendijk, P., Hazen, W. L., Kuiken, S. D., Poen, A. C., Quispel, R., Romkens, T., Spanier, B., Tan, A., & Dutch Pancreatitis Study, G. (2024). Development of pancreatic diseases during long-term follow-up after acute pancreatitis: A post-hoc analysis of a prospective multicenter cohort. *Journal of Gastroenterology and Hepatology, 39*(4), 674–684. https://doi.org/10.1111/jgh.16453

de Rijk, F. E. M., van Veldhuisen, C. L., Besselink, M. G., van Hooft, J. E., van Santvoort, H. C., van Geenen, E. J. M., Hegyi, P., Lohr, J. M., Dominguez-Munoz, J. E., de Jonge, P. J. F., Bruno, M. J., Verdonk, R. C., & Dutch Pancreatitis Study, G. (2022). Diagnosis and treatment of exocrine pancreatic insufficiency in chronic pancreatitis: An international expert survey and case vignette study. *Pancreatology, 22*(4), 457–465. https://doi.org/10.1016/j.pan.2022.03.013

Dunbar, E. K., Saloman, J. L., Phillips, A. E., & Whitcomb, D. C. (2021). Severe pain in chronic pancreatitis patients: Considering mental health and associated genetic factors. *Journal of Pain Research, 14*, 773–784. https://doi.org/10.2147/JPR.S274276

Frias, J. P., Davies, M. J., Rosenstock, J., Perez Manghi, F. C., Fernandez Lando, L., Bergman, B. K., Liu, B., Cui, X., Brown, K., & Investigators, S. (2021). Tirzepatide versus semaglutide once weekly in patients with type 2 diabetes. *New England Journal of Medicine, 385*(6), 503–515. https://doi.org/10.1056/NEJMoa2107519

Froemke, R. C., & Young, L. J. (2021). Oxytocin, neural plasticity, and social behavior. *Annual Review of Neuroscience, 44*, 359–381. https://doi.org/10.1146/annurev-neuro-102320-102847

Ganeshalingam, A. A., Uhrenholt, N., Arnfred, S., Gaede, P., During, S., Stenager, E. N., Bunger, N., Pedersen, A. K., Bilenberg, N., & Frystyk, J. (2025). Semaglutide treatment of antipsychotic-treated patients with schizophrenia, prediabetes, and obesity: The HISTORI randomized clinical trial. *JAMA Psychiatry, 82*, 1065–1074. https://doi.org/10.1001/jamapsychiatry.2025.2332

Garvey, W. T., Frias, J. P., Jastreboff, A. M., le Roux, C. W., Sattar, N., Aizenberg, D., Mao, H., Zhang, S., Ahmad, N. N., Bunck, M. C., Benabbad, I., Zhang, X. M., & investigators, S. (2023). Tirzepatide once weekly for the treatment of obesity in people with type 2 diabetes (SURMOUNT-2): A double-blind, randomised, multicentre, placebo-controlled, phase 3 trial. *Lancet, 402*(10402), 613–626. https://doi.org/10.1016/S0140-6736(23)01200-X

Gupta, N., Singh, S., Vargas, L., Moore, T. E., Shostrom, V. K., & Boerner, B. P. (2019). Prevalence of low bone density and comorbid hypogonadism in patients with chronic pancreatitis. *Pancreas, 48*(3), 387–395. https://doi.org/10.1097/MPA.0000000000001257

Hajibandeh, S., Jurdon, R., Heaton, E., Hajibandeh, S., & O'Reilly, D. (2023). The risk of recurrent pancreatitis after first episode of acute pancreatitis in relation to etiology and severity of disease: A systematic review, meta-analysis and meta-regression analysis. *Journal of Gastroenterology and Hepatology, 38*(10), 1718–1733. https://doi.org/10.1111/jgh.16264

Hendershot, C. S., Bremmer, M. P., Paladino, M. B., Kostantinis, G., Gilmore, T. A., Sullivan, N. R., Tow, A. C., Dermody, S. S., Prince, M. A., Jordan, R., McKee, S. A., Fletcher, P. J., Claus, E. D., & Klein, K. R. (2025). Once-weekly semaglutide in adults with alcohol use disorder: A randomized clinical trial. *JAMA Psychiatry, 82*(4), 395–405. https://doi.org/10.1001/jamapsychiatry.2024.4789

Hiroyoshi, M., Tateishi, K., Yasunami, Y., Maeshiro, K., Ono, J., Matsuoka, Y., & Ikeda, S. (1999). Elevated plasma levels of glucagon-like peptide-1 after oral glucose ingestion in patients with pancreatic diabetes. *American Journal of Gastroenterology, 94*(4), 976–981. https://doi.org/10.1111/j.1572-0241.1999.01043.x

Hornum, M., Pedersen, J. F., Larsen, S., Olsen, O., Holst, J. J., & Knop, F. K. (2010). Increased postprandial response of glucagon-like peptide-2 in patients with chronic pancreatitis and pancreatic exocrine insufficiency. *Pancreatology, 10*(2–3), 201–207. https://doi.org/10.1159/000243729

Issa, Y., Kempeneers, M. A., Bruno, M. J., Fockens, P., Poley, J. W., Ahmed Ali, U., Bollen, T. L., Busch, O. R., Dejong, C. H., van Duijvendijk, P., van Dullemen, H. M., van Eijck, C. H., van Goor, H., Hadithi, M., Haveman, J. W., Keulemans, Y., Nieuwenhuijs, V. B., Poen, A. C., Rauws, E. A., & Dutch Pancreatitis Study, G. (2020). Effect of early surgery vs endoscopy-first approach on pain in patients with chronic pancreatitis: The escape randomized clinical trial. *JAMA, 323*(3), 237–247. https://doi.org/10.1001/jama.2019.20967

Jastreboff, A. M., Aronne, L. J., Ahmad, N. N., Wharton, S., Connery, L., Alves, B., Kiyosue, A., Zhang, S., Liu, B., Bunck, M. C., Stefanski, A., & Investigators, S.-. (2022). Tirzepatide once weekly for the treatment of obesity. *New England Journal of Medicine, 387*(3), 205–216. https://doi.org/10.1056/NEJMoa2206038

Jastreboff, A. M., le Roux, C. W., Stefanski, A., Aronne, L. J., Halpern, B., Wharton, S., Wilding, J. P. H., Perreault, L., Zhang, S., Battula, R., Bunck, M. C., Ahmad, N. N., Jouravskaya, I., & Investigators, S. (2025). Tirzepatide for obesity treatment and diabetes prevention. *New England Journal of Medicine, 392*(10), 958–971. https://doi.org/10.1056/NEJMoa2410819

Khoja, K., Sadiq, O., Chisholm, P. R., Dua, K. S., Madhavan, S., & Smith, Z. L. (2023). The incidence of new mental health disorders after acute pancreatitis: A large, propensity-matched, observational study. *Pancreatology, 23*(2), 163–170. https://doi.org/10.1016/j.pan.2023.01.008

Kjellesvig, S., Madhavan, S., & Smith, Z. L. (2022). S57 opioid use patterns in patients with chronic pancreatitis and concomitant mental illness: A propensity-matched cohort analysis. *American Journal of Gastroenterology, 117*(10S), e42–e43.

Knop, F. K., Vilsboll, T., Hojberg, P. V., Larsen, S., Madsbad, S., Holst, J. J., & Krarup, T. (2007a). The insulinotropic effect of GIP is impaired in patients with chronic pancreatitis and secondary diabetes mellitus as compared to patients with chronic pancreatitis and normal glucose tolerance. *Regulatory Peptides, 144*(1–3), 123–130. https://doi.org/10.1016/j.regpep.2007.07.002

Knop, F. K., Vilsboll, T., Hojberg, P. V., Larsen, S., Madsbad, S., Volund, A., Holst, J. J., & Krarup, T. (2007b). Reduced incretin effect in type 2 diabetes: Cause or consequence of the diabetic state? *Diabetes, 56*(8), 1951–1959. https://doi.org/10.2337/db07-0100

Knop, F. K., Vilsboll, T., Larsen, S., Hojberg, P. V., Volund, A., Madsbad, S., Holst, J. J., & Krarup, T. (2007c). Increased postprandial responses of GLP-1 and GIP in patients with chronic pancreatitis and steatorrhea following pancreatic enzyme substitution. *American Journal of Physiology. Endocrinology and Metabolism, 292*(1), E324–E330. https://doi.org/10.1152/ajpendo.00059.2006

Knop, F. K., Vilsboll, T., Larsen, S., Madsbad, S., Holst, J. J., & Krarup, T. (2003). No hypoglycemia after subcutaneous administration of glucagon-like peptide-1 in lean type 2 diabetic patients and in patients with diabetes secondary to chronic pancreatitis. *Diabetes Care, 26*(9), 2581–2587. https://doi.org/10.2337/diacare.26.9.2581

Lach, G., Schellekens, H., Dinan, T. G., & Cryan, J. F. (2018). Anxiety, depression, and the microbiome: A role for gut peptides. *Neurotherapeutics, 15*(1), 36–59. https://doi.org/10.1007/s13311-017-0585-0

Leng, G., Leng, R. I., & Ludwig, M. (2022). Oxytocin-a social peptide? Deconstructing the evidence. *Philosophical Transactions of the Royal Society of London B Biological Sciences, 377*(1858), 20210055. https://doi.org/10.1098/rstb.2021.0055

Li, S., Gao, L., Gong, H., Cao, L., Zhou, J., Ke, L., Liu, Y., Tong, Z., & Li, W. (2023). Recurrence rates and risk factors for recurrence after first episode of acute pancreatitis: A systematic review and meta-analysis. *European Journal of Internal Medicine, 116*, 72–81. https://doi.org/10.1016/j.ejim.2023.06.006

Llach, C. D., Badulescu, S., Tabassum, A., Shah, H., Gill, H., Le, G. H., Vieta, E., McIntyre, R. S., Rosenblat, J. D., & Mansur, R. B. (2025). Glucagon-like peptide-1 receptor agonists as emerging therapeutics in bipolar disorder: A narrative review of preclinical and clinical evidence. *Molecular Psychiatry, 31*, 456–479. https://doi.org/10.1038/s41380-025-03261-0

Ma, S., Yang, X., He, H., Gao, Y., Chen, Y., Qin, J., Zhang, C., Lu, G., Gong, W., Chen, W., & Ren, Y. (2022). Psychological experience of inpatients with acute pancreatitis: A qualitative study. *BMJ Open, 12*(6), e060107. https://doi.org/10.1136/bmjopen-2021-060107

Machicado, J. D., Amann, S. T., Anderson, M. A., Abberbock, J., Sherman, S., Conwell, D. L., Cote, G. A., Singh, V. K., Lewis, M. D., Alkaade, S., Sandhu, B. S., Guda, N. M., Muniraj, T., Tang, G., Baillie, J., Brand, R. E., Gardner, T. B., Gelrud, A., Forsmark, C. E., & Yadav, D. (2017). Quality of life in chronic pancreatitis is determined by constant pain, disability/unemployment, current smoking, and associated co-morbidities. *American Journal of Gastroenterology, 112*(4), 633–642. https://doi.org/10.1038/ajg.2017.42

Malhotra, A., Grunstein, R. R., Fietze, I., Weaver, T. E., Redline, S., Azarbarzin, A., Sands, S. A., Schwab, R. J., Dunn, J. P., Chakladar, S., Bunck, M. C., Bednarik, J., & Investigators, S.-O. (2024). Tirzepatide for the treatment of obstructive sleep apnea and obesity. *New England Journal of Medicine, 391*(13), 1193–1205. https://doi.org/10.1056/NEJMoa2404881

Maydeo, A., Kamat, N., Dalal, A., & Patil, G. (2023). Advances in the management of pain in chronic pancreatitis. *Current Gastroenterology Reports, 25*(10), 260–266. https://doi.org/10.1007/s11894-023-00898-1

Mederos, M. A., Reber, H. A., & Girgis, M. D. (2021). Acute pancreatitis: A review. *JAMA, 325*(4), 382–390. https://doi.org/10.1001/jama.2020.20317

Mokrowiecka, A., Pinkowski, D., & Malecka-Panas, E. (2011). Assessment of quality of life in patients with chronic pancreatitis. *Medical Science Monitor, 17*(10), CR583–CR588. https://doi.org/10.12659/msm.881985

Munigala, S., Almaskeen, S., Subramaniam, D. S., Bandi, S., Bowe, B., Xian, H., Sheth, S. G., Burroughs, T. E., & Agarwal, B. (2023). Acute pancreatitis recurrences augment long-term pancreatic cancer risk. *American Journal of Gastroenterology, 118*(4), 727–737. https://doi.org/10.14309/ajg.0000000000002081

Nelson, A. D., Lugo-Fagundo, N. S., Mahapatra, S. J., Cheungpastiporn, W., Thongprayoon, C., Wijarnpreecha, K., Muktesh, G., Kochhar, R., Garg, P., Wallace, M., & Bi, Y. (2022). A systematic review and meta-analysis of opioids vs nonopioids in acute pancreatitis. *Gastro Hep Advances, 1*(1), 83–92. https://doi.org/10.1016/j.gastha.2021.09.006

Peery, A. F., Crockett, S. D., Murphy, C. C., Jensen, E. T., Kim, H. P., Egberg, M. D., Lund, J. L., Moon, A. M., Pate, V., Barnes, E. L., Schlusser, C. L., Baron, T. H., Shaheen, N. J., & Sandler, R. S. (2022). Burden and cost of gastrointestinal, liver, and pancreatic diseases in the United States: Update 2021. *Gastroenterology, 162*(2), 621–644. https://doi.org/10.1053/j.gastro.2021.10.017

Pezzilli, R., Morselli Labate, A. M., Ceciliato, R., Frulloni, L., Cavestro, G. M., Comparato, G., Ferri, B., Corinaldesi, R., & Gullo, L. (2005). Quality of life in patients with chronic pancreatitis. *Digestive and Liver Disease, 37*(3), 181–189. https://doi.org/10.1016/j.dld.2004.10.007

Pozzi, M., Mazhar, F., Peeters, G., Vantaggiato, C., Nobile, M., Clementi, E., Radice, S., & Carnovale, C. (2019). A systematic review of the antidepressant effects of glucagon-like peptide 1 (GLP-1) functional agonists: Further link between metabolism and psychopathology: Special section on "Translational and Neuroscience Studies in Affective Disorders". Section Editor, Maria Nobile MD, PhD. This section of JAD focuses on the relevance of translational and neuroscience studies in providing a better understanding of the neural basis of affective disorders. The main aim is to briefly summaries relevant research findings in clinical neuroscience with particular regards to specific innovative topics in mood and anxiety disorders. *Journal of Affective Disorders, 257*. https://doi.org/10.1016/j.jad.2019.05.044

Ratnayake, C. B. B., Kamarajah, S. K., Loveday, B. P. T., Nayar, M., Oppong, K., White, S., French, J. J., Windsor, J. A., & Pandanaboyana, S. (2020). A network meta-analysis of surgery for chronic pancreatitis: Impact on pain and quality of life. *Journal of Gastrointestinal Surgery, 24*(12), 2865–2873. https://doi.org/10.1007/s11605-020-04718-z

Samms, R. J., Coghlan, M. P., & Sloop, K. W. (2020). How may GIP enhance the therapeutic efficacy of GLP-1? *Trends in Endocrinology and Metabolism, 31*(6), 410–421. https://doi.org/10.1016/j.tem.2020.02.006

Shah, I., Anderson, K., Bocchino, R., Freedman, S. D., Carrasquillo, R., & Sheth, S. G. (2024). Prospective evaluation of sexual dysfunction in men with chronic pancreatitis: A pilot study. *Pancreas, 53*(2), e187–e192. https://doi.org/10.1097/MPA.0000000000002286

Sirtl, S., Hohmann, E., Ahmad, M., Bretthauer, K., Junge, M., Vornhulz, M., Goni, E., Saka, D., Knoblauch, M., Aghamaliyev, U., Schulz, C., Zorniak, M., Mahajan, U. M., Mayerle, J., & Beyer, G. (2024). Severity and outcome of a first episode of idiopathic acute pancreatitis is not more severe than pancreatitis of other etiologies. *Pancreatology, 24*(6), 840–846. https://doi.org/10.1016/j.pan.2024.08.004

Siskind, D., Baker, A., Arnautovska, U., Warren, N., Russell, A., DeMonte, V., Halstead, S., Iyer, R., Korman, N., McKeon, G., Medland, S., Parker, S., Stedman, T., & Trott, M. (2025). Efficacy and safety of semaglutide versus placebo for people with schizophrenia on clozapine with obesity (COaST): A phase 2, multi-centre, participant and investigator- blinded, randomised controlled trial in Australia. *The Lancet Psychiatry, 12*(7), 493–503. https://doi.org/10.1016/S2215-0366(25)00129-4

Smith, Z. L., Gregory, M. H., Elsner, J., Alajlan, B. A., Kodali, D., Hollander, T., Sayuk, G. S., Lang, G. D., Das, K. K., Mullady, D. K., Early, D. S., & Kushnir, V. M. (2019). Health-related quality of life and long-term outcomes after endoscopic therapy for walled-off pancreatic necrosis. *Digestive Endoscopy, 31*(1), 77–85. https://doi.org/10.1111/den.13264

Tomaszewska, E., Wojtysiak, D., Grzegorzewska, A., Swiatkiewicz, M., Donaldson, J., Arciszewski, M. B., Dresler, S., Puzio, I., Szymanczyk, S., Dobrowolski, P., Bonior, J., Mielnik-Blaszczak, M., Kuc, D., & Muszynski, S. (2024). Understanding secondary sarcopenia development in young adults using pig model with chronic pancreatitis. *International Journal of Molecular Sciences, 25*(16). https://doi.org/10.3390/ijms25168735

Umar, S., Syed, A., Seoud, T., & Thakkar, S. (2020). S0100 chronic pancreatitis increases risk for sexual dysfunction: A national population-based analysis. *American Journal of Gastroenterology, 115*, S48.

Wang, X., Zhan, W., Huang, L., Guo, Y., Wang, Y., Tan, H., & Wang, L. (2023). The effect of anxiety and depression on the health-related quality of life of severe acute pancreatitis survivors: Structural equation modeling approach. *Frontiers in Psychiatry, 14*, 1160807. https://doi.org/10.3389/fpsyt.2023.1160807

Yadav, D., O'Connell, M., & Papachristou, G. I. (2012). Natural history following the first attack of acute pancreatitis. *American Journal of Gastroenterology, 107*(7), 1096–1103. https://doi.org/10.1038/ajg.2012.126

Yu, B., Li, J., Li, N., Zhu, Y., Chen, Y., He, W., & Lu, N. (2020). Progression to recurrent acute pancreatitis after a first attack of acute pancreatitis in adults. *Pancreatology, 20*(7), 1340–1346. https://doi.org/10.1016/j.pan.2020.09.006

Section 7

Diet and nutrition and impact on the pancreas and neuroscientific domains or behavior

Chapter 33

Diet, diabetes and gene expression in the brain

Jung Han Kim and Lawrence Grover
Department of Biomedical Sciences, Joan C. Edwards School of Medicine, Marshall University, Huntington, WV, United States

List of abbreviations

BMI body mass index
HbA1c glycated hemoglobin A1C
miRNA micro ribonucleic acid
MRI magnetic resonance imaging
ROS reactive oxygen species

Introduction

Diabetes is one of the most prevalent diseases in humans. According to the latest report from the International Diabetes Federation, 10.5% (536.6 million) of adults worldwide have diabetes and this number is expected to be 12.2% (783.2 million) in 2045 (Sun et al., 2022). Type 2 diabetes is the most common form of human diabetes, accounting for 90%–95% of diagnosed diabetes patients (Ali et al., 2022). The development of type 2 diabetes is attributed to the complex interplay between environmental factors and a genetic predisposition to the disease (Ortega et al., 2017). Even among populations who are genetically predisposed to type 2 diabetes the prevalence of type 2 diabetes is modulated by changes in environmental factors (Franks & Merino, 2018). Diets are a major environmental factor contributing the development of type 2 diabetes (Forouhi, 2023).

Type 2 diabetes is known to be associated with neurobehavioral alterations including cognitive impairment (Damanik & Yunir, 2021). Further, poorly controlled diabetic patients had greater decline in cognitive function than those whose diabetes was controlled (Rawlings et al., 2014). Studies provided evidence that dietary intake can impact on cognitive and neuronal functions (Gomez-Pinilla & Tyagi, 2013). For example, intake of diets rich in fat is associated with impaired cognitive performance both in animals and humans (Greenwood & Winocur, 2005).

A major mechanism by which dietary components modify cellular and metabolic pathways is through modulating gene expression levels (Berná et al., 2014). On the other hand, gene expression may be regulated by causal genetic polymorphisms underlying diseases including type 2 diabetes (Shah et al., 2025) and cognitive impairments (Cai et al., 2022). In this chapter we examine the contribution of diet to type 2 diabetes, and the role of type 2 diabetes in altering gene expression in areas of the brain that are involved in cognitive functions.

Diets in type 2 diabetes

A systemic analysis of the risk factors associated with type 2 diabetes for the Global Burden of Disease Study 2021 revealed that high body mass index (BMI) is the primary risk factor for type 2 diabetes worldwide, followed by dietary risks (GBD 2021 Diabetes Collaborators, 2023).

Dietary fat consumption is considered a large risk factor contributing to type 2 diabetes (Hu, 2011). Total and saturated fat intake is associated with an increased risk of type 2 diabetes, but this association is not independent of BMI (van Dam et al., 2002). A systemic review and meta-analysis of observational prospective studies demonstrated that the association between dietary fat and risk of type 2 diabetes is from specific fats and fatty acids, rather than total fat intake; a high intake of vegetable fat is inversely associated with type 2 diabetes incidence (Neuenschwander et al., 2020).

Considering the dietary fat quality, research evidence on randomized controlled trials signified that a high intake of saturated fatty acids increases the risk of type 2 diabetes, while a high intake of polyunsaturated fatty acids reduces the risk (Telle-Hansen et al., 2019). The sources of polyunsaturated fatty acids are both marine and vegetable oils, and it remains unclear whether polyunsaturated fatty acids

Hormones in Neuroscience. https://doi.org/10.1016/B978-0-443-30263-3.00035-X

from marine or vegetable sources differently affect glycemic regulation (Telle-Hansen et al., 2019).

A high intake of saturated fat is significantly associated with cognitive decline in humans (Cao et al., 2019). In contrast, high intake of polyunsaturated fatty acids was associated with better cognitive health (Eskelinen et al., 2008). Potential mechanisms rendering neuronal dysfunction and cognitive deficits with increased dietary fat intake include oxidative stress, neuro-inflammation, insulin resistance, transcriptional dysregulation, and plasticity dysfunction (Sharma, 2021).

Diet is a complex entity with interaction between foods, and therefore it is difficult to evaluate the effects of individual dietary components in humans. In this context, dietary patterns analysis may provide more information about the linkage between diets and diseases (Alhazmi et al., 2014). Recommended dietary patterns for the management of type 2 diabetes highlights (1) increased intake of fruits, vegetables, and legumes; (2) minimizing added sugars and refined grains; (3) choosing natural and minimally processed foods rather than ultra-processed foods (Minari et al., 2023).

Many types of dietary patterns have been studied for nutritional management of type 2 diabetes, and adherence to the Mediterranean diet has the highest level of scientific evidence (Pavlidou et al., 2023). The Mediterranean diet consists of fruits, vegetable, legumes, seeds, whole grains, fish, lean meats, skimmed dairy products, and olive oil (Minari et al., 2023). The intake of dietary fiber, unsaturated fatty acids, vitamins, and polyphenols might explain the beneficial effects of Mediterranean diet on type 2 diabetes. Antioxidant and anti-inflammatory activities of polyphenols and the effects of unsaturated fatty acids on lipid metabolism are considered as part of the underlying mechanism (Farias-Pereira et al., 2023). Adhering to the Mediterranean diet also supported maintaining optimal cognitive functions in type 2 diabetic patients, which further sustained by glycemic control (Mattei et al., 2019). The Mediterranean diet intervention was as effective as the diabetic drug metformin in preventing cognitive decline in patients with type 2 diabetes (Soldevila-Domenech et al., 2021).

What is cognition?

Before we examine effects of type 2 diabetes on gene expression in brain areas involved in cognitive function, we first need to define cognition and describe the specific brain structurers supporting cognitive functions. The term cognition can be defined in a number of different ways (Bayne et al., 2019), with a wide range of functions potentially included within the definition. Despite the potential variability, cognition can be reduced to two essential functions, executive control and memory. Executive control in turn has three core functions (Diamond, 2013): inhibitory control, working memory (memory retained under attentional control while performing a task, Baddeley, 2010), and cognitive flexibility. Long-term memory is widely considered to involve multiple systems within the brain that differ in key brain areas and type of information being stored (Squire & Zola, 1996). According to one widely held view (Squire, 2004), these memory systems can be divided into a declarative memory system (memory for facts and events), and nondeclarative memory systems (for retention of skills and habits, emotional associations, and conditioned responses).

Key brain areas in cognition

Prefrontal cortex contains key brain areas for executive control (Levy, 2023). Ventrolateral prefrontal cortex, in particular the inferior frontal gyrus, is a key brain area in inhibitory control (Tabibnia et al., 2011, 2014). Dorsolateral prefrontal cortex is considered a key area for working memory (Funahashi, 2006). Cognitive flexibility involves these same lateral prefrontal areas, but also inferior parietal lobule, posterior inferior temporal lobe, and midcingulate gyrus (Uddin, 2021). The executive functions of the prefrontal cortex are supported by the striatum, which contributes to anatomically and functionally defined cortio-striatal-thalmo-cortical loops (Bornstein & Daw, 2011; Foerde & Shohamy, 2011; Grahn et al., 2008; Uddin, 2021).

The declarative memory system includes structures of the medial temporal lobes (hippocampus, parahippocampal gyrus), medial nuclei of the thalamus, and multiple association cortex areas (Eichenbaum, 2006; Squire, 2004). The hippocampus and adjacent medial temporal lobe structures seem to be most important in early stages of long-term retention, with information eventually consolidated for permanent storage in a distributed manner across association areas of neocortex (Wiltgen et al., 2004).

Non-declarative memory systems involve different structures than the declarative memory system. The basal ganglia are a key structure in nondeclarative memory for skills and habits (Foerde & Shohamy, 2011; Squire, 2004). The cerebellum has long been known for its importance in motor learning, including conditioned responses and skilled movements (Manto et al., 2012). Emotional memory depends especially on the amygdala and related medial frontal cortex (LeDoux, 1993; Tovote et al., 2015).

Effects of type 2 diabetes on cognition

Type 2 diabetes has adverse effects on memory, attention, information processing speed, and executive function. Several longitudinal aging studies have demonstrated greater declines in memory, reasoning, and executive function in those with known diabetes compared with normoglycemic individuals (Rawlings et al., 2014; Spauwen et al., 2013; Tuligenga et al., 2014). MRI studies reveal effects on both gray and white matter. Wang et al. (2014) reported white matter loss (type 2 diabetes compared to control) in right superior and inferior temporal gyri, and left inferior occipital gyrus, and right inferior temporal gyrus, as well as the right posterior lobe of the cerebellum. Kamiyama et al. (2010) reported loss of hippocampal as well as whole brain volume in aged (59–79 years compared to 35–70 years in Wang et al., 2014) diabetics. Similarly, Moran et al. (2013) found lower total gray, white, and hippocampal volumes in an older (>55 years) population.

A meta-analysis examining effects of type 1 diabetes and type 2 diabetes on gray matter volume (Yu et al., 2021) found overall a greater loss of volume in type 2 diabetes compared to type 1 diabetes, despite a longer disease duration in type 1 diabetes in the population included in the meta-analysis. Both type 1 and type 2 diabetes were associated with reduced gray matter in the left caudate, right middle temporal lobe, and left cuneus, whereas only type 2 diabetes led to gray matter loss in left posterior cingulate gyrus, right inferior temporal lobe, right caudate, and left occipital lobe.

Yoon et al. (2017) asked whether overweight or obesity might alter type 2 diabetes effects on brain structure and cognitive function. They found global effects of type 2 diabetes on gray matter thickness and fractional anisotropy (a measure of integrity), and greater loss of thickness in type 2 diabetes for overweight or obesity compared to normal weight. Memory and psychomotor speed, but not executive function, were significantly affected by type 2 diabetes.

Zhang et al. (2014) examined the relationship between white matter integrity and cognitive function in type 2 diabetes. Patients and controls received a battery of tests for episodic memory, attention, visuospatial ability, language, executive function, working memory, and logical reasoning. Compared to controls, type 2 diabetes patients showed significantly reduced performance in executive function, spatial processing, attention, and working memory. Type 2 diabetes showed significantly reduced fractional anisotropy in whole corpus callosum, bilateral corona radiata, internal capsule, posterior thalamic radiation, cingulum, superior longitudinal fasciculus, inferior longitudinal fasciculus, superior fronto-occipital fasciculus, external capsule, fornix, stria terminalis, uncinate fasciculus, tapetum, and left cingulum. Type 2 diabetes patients showed significantly increased mean diffusivity (along with fractional anisotropy, a second index of white matter integrity) in whole corpus callosum, bilateral corona radiata, internal capsule (except right posterior limb), external capsule, posterior thalamic radiation, cingulum, superior longitudinal fasciculus, and tapetum.

Transcriptomic analysis of brain gene expression in type 2 diabetes

To avoid bias that might arise from studies that chose to examine a limited number of genes, we focused our review on studies employing transcriptomic methods to examine human brain tissue.

Zhou et al. (2019) analyzed the transcriptome of 13 brain regions from type 2 diabetes and control patient profiles provided by the Genotype-Tissue Expression project (Lonsdale et al., 2013). The number of type 2 diabetes-associated genes varied greatly by brain region, with caudate nucleus and hippocampus showing the greatest number (Table 33.1). Gene Ontology analysis showed that downregulated genes in both caudate and hippocampus were primarily associated with synaptic functions, whereas upregulated genes differed in function between caudate (cilium movement and organization) and hippocampus (muscle function and muscle tissue morphogenesis). Kyoto Encyclopedia of Genes and Genomes pathway enrichment analysis showed that type 2 diabetes-associated genes in caudate are associated with neurodegenerative diseases. The large observed transcriptomic changes in hippocampus and caudate and the roles of hippocampus in memory and caudate in cognitive functions help explain the functional and structural losses seen in type 2 diabetes. The Zhou et al. study did not investigate potential consequences of disease severity (e.g., HbA1c) or duration. Also, the approach used by Zhou et al. might not detect changes in gene expression or pathway enrichment in small but important cell population if these differ from changes occurring in large cell populations, for example, cortical inhibitory neurons versus principal neurons, or capillary endothelial cells versus astrocytes.

Based on the known role of type 2 diabetes in small and large vessel disease, Bury et al. (2021) investigated possible type 2 diabetes effects on the brain neurovascular unit, which is composed of endothelial cells, pericytes, astrocytes, and neurons. Specimens of lateral temporal cortex (Brodmann areas 21 and 22, superior and middle temporal gyri) from type 2 diabetes and age and sex matched controls were used for laser capture microdissection to obtain neuron-, astrocyte-, and endothelial cell-enriched samples for transcriptomic analysis. Over 4000 differentially expressed genes were identified. Only five of these were commonly altered in all three cell types, 47 in both neurons and astrocytes, 34 in neurons and endothelial cells, and 101 in astrocytes and endothelial cells. Pathway analysis was used to identify functional groups, summarized in Table 33.2.

TABLE 33.1 Total number of differentially expressed genes per brain region at false-discovery rate of 5% in type 2 diabetes (summarized from Zhou et al., 2019).

Brain region	Number of differentially expressed genes
Caudate	2939
Hippocampus	486
Amygdala	6
Cerebellar hemisphere	2
Anterior cingulate cortex, spinal cord	1
Cerebellum, cortex, frontal cortex, hypothalamus, nucleus accumbens, putamen, substantia nigra	0

Key findings from the Bury et al. (2021) study are that although there was little overlap in terms of individual genes, changes in expression of several functional groups were shared among cell types, most notably in insulin and other signaling pathways. These findings are consistent with the idea of brain, in addition to peripheral, insulin resistance as a component of type 2 diabetes. Conclusions from the Bury et al. study regarding cell type-specific

TABLE 33.2 Pathway analysis for altered gene expression in neurons, astrocytes and endothelial cells in type 2 diabetes (summarized from Bury et al., 2021).

Neurons	Astrocytes	Endothelial cells
Signaling pathways: PI3K-Akt, MAPK, Rap1, TNF, retrograde endocannabinoid	Signaling pathways: PI3K-Akt, MAPK, sphingolipid, oxytocin, Rap1, neurotrophin, ErbB, phosphatidylinositol, chemokine, T cell receptor, GnRH, Ras, AMPK, mTOR, estrogen, Wnt, p53	Signaling pathways: PI3K-Akt, MAPK, Rap1, oxytocin, phospholipase D, GnRH, thyroid hormone, cGMP-PKG, Wnt
Insulin signaling pathway	Insulin signaling	Insulin signaling pathway
Infection	Infection	Infection
Cancer	Cancer	Cancer
Focal adhesion	Focal adhesion	Adherens junction
Cell cycle	Cell cycle	Lysosome
Regulation of actin cytoskeleton	Actin cytoskeleton	AGE-RAGE signaling pathway
Cellular senescence	RNA transport	RNA transport
Alzheimer's disease	Ribosome	Ribosome
HIF-1 signaling pathway	Fluid shear stress and atherosclerosis	Fluid shear stress and atherosclerosis
Oxidative phosphorylation	Membrane trafficking	Membrane trafficking
Hematopoietic cell lineage	Spliceosome	Spliceosome
Cytokine-cytokine receptor interaction	Axon guidance	Axon guidance
	Apoptosis	Innate immunity
	Protein processing in ER	Pancreatic secretion
	Longevity regulating pathway	

Continued

TABLE 33.2 Pathway analysis for altered gene expression in neurons, astrocytes and endothelial cells in type 2 diabetes (summarized from Bury et al., 2021).—cont'd

Neurons	Astrocytes	Endothelial cells
	RNA degradation	
	Autophagy	
	Metabolism	
	Steroid hormone biosynthesis	
	Renin secretion	
	Vacular smooth muscle contraction	
	Synaptic transmission	
	Non-alcoholic fatty liver disease	
	Tight junction	
	Necroptosis	

changes are limited by the method used which resulted in cell type enrichment, but not purification, in the samples analyzed. Also, as in the Zhou et al. (2019) study, there is no information on type 2 diabetes duration or severity.

Type 2 diabetes has been identified as a risk factor for Alzheimer's disease. To determine if the two conditions might share common molecular mechanisms, Shu et al. (2022) searched for differentially expressed genes common to both diseases using datasets from the National Center for Biotechnology Information Gene Expression Omnibus database. The datasets analyzed by Shu et al. were from temporal cortex and included type 2 diabetes and Alzheimer's disease (and matched controls). Differentially expressed genes were subjected to a range of analyses, including gene ontology and pathway, protein-protein interaction, transcription factor interaction, miRNA interaction, and subcellular localization. The authors identified a relatively short list of common differentially expressed genes in type 2 diabetes and Alzheimer's disease (see Table 33.3). Further analyses suggested involvement of astrocytes and microglia, as well as neurons, and the processes of apoptosis, autophagy, inflammation and innate

TABLE 33.3 Differentially expressed genes common to type 2 diabetes and Alzheimer's disease (summarized from Shu et al., 2022).

Regulation	Gene	Protein product
Upregulated in both	SERPINA3	Serpin Family A Member 3
	MERTK	Tyrosine-protein kinase Mer
	C1QB	Complement C1q B Chain
	GFAP	Glial fibrillary acidic protein
	CD44	CD44 molecule
	ITGB2	Integrin Subunit Beta 2
	S100A8	S100 Calcium Binding Protein A8
	GEM	GTP binding protein overexpressed in skeletal muscle
	SFN	Stratifin

Continued

TABLE 33.3 Differentially expressed genes common to type 2 diabetes and Alzheimer's disease (summarized from Shu et al., 2022).—cont'd

Regulation	Gene	Protein product
Downregulated in both	SLC22A9	Solute Carrier Family 22 Member 9/Organic anion transporter 7
Downregulated in Alzheimer's disease and upregulated in T2DM	ETNK1	Ethanolamine kinase 1
	SST	Somatostatin
	OLA1	Obg-Like ATPase 1
	PPEF1	Serine/threonine-protein phosphatase with EF-hands
	PCSK1	Proprotein Convertase Subtilisin/Kexin Type 1
	C2orf80	Chromosome 2 Open Reading Frame 80

immunity, hemostasis, and infection, as well as insulin resistance. Limitations noted above for Zhou et al. (2019) and Bury et al. (2021) studies also apply to the Shu et al. study.

Conclusions

High BMI and a diet rich in saturated fats are major factors leading to type 2 diabetes (Fig. 33.1). Type 2 diabetes in turn alters brain structure in areas that are of critical importance in cognition, causing impairment in cognitive functions. Altered gene expression, particularly for insulin signaling and inflammatory pathways may be key factors in brain structural and functional disruption in type 2 diabetes. Moreover, the effects of type 2 diabetes on brain gene expression overlap with effects of Alzheimer's disease, suggesting shared mechanisms and explaining the increased risk of Alzheimer's disease in type 2 diabetes (Lemche et al., 2024; Mattson, 2004).

Although these basic conclusions are well-supported by existing experimental literature, several key questions about diet, diabetes, and brain gene expression remain unanswered. First, does altered brain gene expression and its resulting dysfunction develop in series or in parallel with peripheral type 2 diabetes? In other words, do BMI and diet drive peripheral type 2 diabetes, which drives effects on brain gene expression, structure and function (series), or do BMI and diet independently drive peripheral

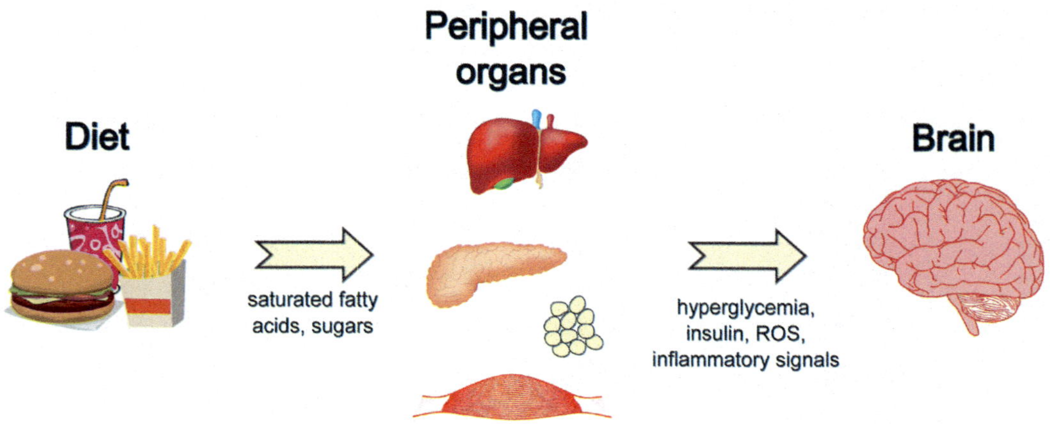

FIGURE 33.1 Dietary fat and sugar drive type 2 diabetes and brain changes leading to impaired cognition. Dietary saturated fats and sugars are key factors leading to the development of type 2 diabetes with associated dysfunctions of hepatic, pancreatic, adipose, and muscular systems. Altered function of peripheral organs in type 2 diabetes leads to hyperglycemia, increased insulin secretion, production of reactive oxygen species (ROS), and other inflammatory signals which induce changes in brain gene expression, leading to altered brain structure and function. Brain areas required for normal cognitive function are among those that are adversely affected in type 2 diabetes, resulting in the cognitive impairments seen in type 2 diabetes (Artwork based on illustration used with permission from https://openclipart.org/ and https://pixabay.com/).

type 2 diabetes and brain changes (parallel)? These issues could be addressed in future brain imaging and transcriptomic studies by including disease stage and severity (e.g., pre-diabetes, diabetes, HbA1c, insulin dependence) and duration as factors. Second, do altered brain gene expression and structure recover with treatment for type 2 diabetes? This could be addressed in future studies by including treatment type, duration, and effectiveness as factors. Last, what biological mechanisms link diet and type 2 diabetes to dysregulated brain gene expression, altered structure, and impaired cognitive function? Many possibilities exist (Lemche et al., 2024; Ma et al., 2020), including: vascular injury and disruption of the blood-brain barrier, altered calcium signaling and calcium homeostasis, neuroinflammation, brain insulin resistance, advanced glycation end products. Answers to these questions should lead to more effective prevention and reversal of cognitive loss in type 2 diabetes.

Summary points

- This chapter focuses on dietary factors which promote type 2 diabetes, and the effects of type 2 diabetes on brain leading to loss of cognitive functions.
- Saturated fats and sugars are key dietary components in type 2 diabetes.
- Brain regions affected in type 2 diabetes include areas critical for cognitive functions: lateral prefrontal cortex areas involved in executive functions and medial temporal lobe areas involved in declarative memory.
- Transcriptomic analysis suggests altered insulin signaling and neuroinflammation as potential mechanisms linking type 2 diabetes to brain injury with effects on neurons, glia, and vascular cells.

References

Alhazmi, A., Stojanovski, E., McEvoy, M., & Garg, M. L. (2014). The association between dietary patterns and type 2 diabetes: A systematic review and meta-analysis of cohort studies. *Journal of Human Nutrition and Dietetics, 27*(3), 251–260.

Ali, M. K., Pearson-Stuttard, J., Selvin, E., & Gregg, E. W. (2022). Interpreting global trends in type 2 diabetes complications and mortality. *Diabetologia, 65*(1), 3–13.

Baddeley, A. (2010). Working memory. *Current Biology, 20*(4), R136–R140.

Bayne, T., Brainard, D., Byrne, R. W., Chittka, L., Clayton, N., Heyes, C., Mather, J., Ölveczky, B., Shadlen, M., Suddendorf, T., & Webb, B. (2019). What is cognition? *Current Biology, 29*(13), R608–R615.

Berná, G., Oliveras-López, M. J., Jurado-Ruíz, E., Tejedo, J., Bedoya, F., Soria, B., & Martín, F. (2014). Nutrigenetics and nutrigenomics insights into diabetes etiopathogenesis. *Nutrients, 6*(11), 5338–5369.

Bornstein, A. M., & Daw, N. D. (2011). Multiplicity of control in the basal ganglia: Computational roles of striatal subregions. *Current Opinion in Neurobiology, 21*(3), 374–380.

Bury, J. J., Chambers, A., Heath, P. R., Ince, P. G., Shaw, P. J., Matthews, F. E., Brayne, C., Simpson, J. E., Wharton, S. B., & the Cognitive Function and Ageing Study. (2021). Type 2 diabetes mellitus-associated transcriptome alterations in cortical neurones and associated neurovascular unit cells in the ageing brain. *Acta Neuropathologica Communications, 9*(1), 5.

Cai, S., Yang, F., Wang, X., Wu, S., Alzheimers Disease Neuroimaging Initiative, Huang, L. (2022). Structural brain characteristics and gene co-expression analysis: A study with outcome label from normal cognition to mild cognitive impairment. *Neurobiology of Learning and Memory, 191*, 107620.

Cao, G.-Y., Li, M., Han, L., Tayie, F., Yao, S.-S., Huang, Z., Ai, P., Liu, Y.-Z., Hu, Y.-H., & Xu, B. (2019). Dietary fat intake and cognitive function among older populations: A systematic review and meta-analysis. *The Journal of Prevention of Alzheimer's Disease, 6* (3), 204–211.

Damanik, J., & Yunir, E. (2021). Type 2 diabetes mellitus and cognitive impairment. *Acta Medica Indonesiana, 53*(2), 2.

Diamond, A. (2013). Executive functions. *Annual Review of Psychology, 64*, 135–168.

Eichenbaum, H. (2006). Remembering: Functional organization of the declarative memory system. *Current Biology, 16*(16), R643–R645.

Eskelinen, M. H., Ngandu, T., Helkala, E.-L., Tuomilehto, J., Nissinen, A., Soininen, H., & Kivipelto, M. (2008). Fat intake at midlife and cognitive impairment later in life: A population-based CAIDE study. *International Journal of Geriatric Psychiatry, 23*(7), 741–747.

Farias-Pereira, R., Zuk, J. B., & Khavaran, H. (2023). Plant bioactive compounds from Mediterranean diet improve risk factors for metabolic syndrome. *International Journal of Food Sciences & Nutrition, 74*(4), 403–423.

Foerde, K., & Shohamy, D. (2011). The role of the basal ganglia in learning and memory: Insight from Parkinson's disease. *Neurobiology of Learning and Memory, 96*(4), 624–636.

Forouhi, N. G. (2023). Embracing complexity: Making sense of diet, nutrition, obesity and type 2 diabetes. *Diabetologia, 66*(5), 786–799.

Franks, P. W., & Merino, J. (2018). Gene-lifestyle interplay in type 2 diabetes. *Current Opinion in Genetics & Development, 50*, 35–40.

Funahashi, S. (2006). Prefrontal cortex and working memory processes. *Neuroscience, 139*(1), 251–261.

GBD 2021 Diabetes Collaborators. (2023). Global, regional, and national burden of diabetes from 1990 to 2021, with projections of prevalence to 2050: A systematic analysis for the Global Burden of Disease Study 2021. *Lancet (London, England), 402*(10397), 203–234.

Gomez-Pinilla, F., & Tyagi, E. (2013). Diet and cognition: Interplay between cell metabolism and neuronal plasticity. *Current Opinion in Clinical Nutrition and Metabolic Care, 16*(6), 726–733.

Grahn, J. A., Parkinson, J. A., & Owen, A. M. (2008). The cognitive functions of the caudate nucleus. *Progress in Neurobiology, 86*(3), 141–155.

Greenwood, C. E., & Winocur, G. (2005). High-fat diets, insulin resistance and declining cognitive function. *Neurobiology of Aging, 26* (Suppl. 1), 42–45.

Hu, F. B. (2011). Globalization of diabetes: The role of diet, lifestyle, and genes. *Diabetes Care, 34*(6), 1249–1257.

Kamiyama, K., Wada, A., Sugihara, M., Kurioka, S., Hayashi, K., Hayashi, T., Yoshisako, T., Yamamoto, N., Tsuchie, Y., Yamaguchi, S., Sugimoto, T., & Kitagaki, H. (2010). Potential hippocampal region atrophy in diabetes mellitus type 2: A voxel-based morphometry VSRAD study. *Japanese Journal of Radiology, 28*(4), 266–272.

LeDoux, J. E. (1993). Emotional memory systems in the brain. *Behavioural Brain Research, 58*(1–2), 69–79.

Lemche, E., Killick, R., Mitchell, J., Caton, P. W., Choudhary, P., & Howard, J. K. (2024). Molecular mechanisms linking type 2 diabetes mellitus and late-onset Alzheimer's disease: A systematic review and qualitative meta-analysis. *Neurobiology of Disease, 196*, 106485.

Levy, R. (2023). The prefrontal cortex: From monkey to man. *Brain, 147* (3), 794–815.

Lonsdale, J., Thomas, J., Salvatore, M., Phillips, R., Lo, E., Shad, S., Hasz, R., Walters, G., Garcia, F., Young, N., Foster, B., Moser, M., Karasik, E., Gillard, B., Ramsey, K., Sullivan, S., Bridge, J., Magazine, H., Syron, J., … Moore, H. F. (2013). The genotype-tissue expression (GTEx) project. *Nature Genetics, 45*(6), 580–585.

Ma, W.-X., Tang, J., Lei, Z.-W., Li, C.-Y., Zhao, L.-Q., Lin, C., Sun, T., Li, Z.-Y., Jiang, Y.-H., Jia, J.-T., Liang, C.-Z., Liu, J.-H., & Yan, L.-J. (2020). Potential biochemical mechanisms of brain injury in diabetes mellitus. *Aging and Disease, 11*(4), 978–987.

Manto, M., Bower, J. M., Conforto, A. B., Delgado-García, J. M., da Guarda, S. N. F., Gerwig, M., Habas, C., Hagura, N., Ivry, R. B., Mariën, P., Molinari, M., Naito, E., Nowak, D. A., Ben Taib, N. O., Pelisson, D., Tesche, C. D., Tilikete, C., & Timmann, D. (2012). Consensus paper: Roles of the cerebellum in motor control—The diversity of ideas on cerebellar involvement in movement. *The Cerebellum, 11*(2), 457–487.

Mattei, J., Bigornia, S. J., Sotos-Prieto, M., Scott, T., Gao, X., & Tucker, K. L. (2019). The Mediterranean diet and 2-year change in cognitive function by status of type 2 diabetes and glycemic control. *Diabetes Care, 42*(8), 1372–1379.

Mattson, M. P. (2004). Pathways towards and away from Alzheimer's disease. *Nature, 430*(7000), 631–639.

Minari, T. P., Tácito, L. H. B., Yugar, L. B. T., Ferreira-Melo, S. E., Manzano, C. F., Pires, A. C., Moreno, H., Vilela-Martin, J. F., Cosenso-Martin, L. N., & Yugar-Toledo, J. C. (2023). Nutritional strategies for the management of type 2 diabetes mellitus: A narrative review. *Nutrients, 15*(24), 5096.

Moran, C., Phan, T. G., Chen, J., Blizzard, L., Beare, R., Venn, A., Münch, G., Wood, A. G., Forbes, J., Greenaway, T. M., Pearson, S., & Srikanth, V. (2013). Brain atrophy in type 2 diabetes. *Diabetes Care, 36*(12), 4036–4042.

Neuenschwander, M., Barbaresko, J., Pischke, C. R., Iser, N., Beckhaus, J., Schwingshackl, L., & Schlesinger, S. (2020). Intake of dietary fats and fatty acids and the incidence of type 2 diabetes: A systematic review and dose-response meta-analysis of prospective observational studies. *PLoS Medicine, 17*(12), e1003347.

Ortega, Á., Berná, G., Rojas, A., Martín, F., & Soria, B. (2017). Gene-diet interactions in type 2 diabetes: The chicken and egg debate. *International Journal of Molecular Sciences, 18*(6), 1188.

Pavlidou, E., Papadopoulou, S. K., Fasoulas, A., Papaliagkas, V., Alexatou, O., Chatzidimitriou, M., Mentzelou, M., & Giaginis, C. (2023). Diabesity and dietary interventions: Evaluating the impact of Mediterranean diet and other types of diets on obesity and type 2 diabetes management. *Nutrients, 16*(1), 34.

Rawlings, A. M., Sharrett, A. R., Schneider, A. L. C., Coresh, J., Albert, M., Couper, D., Griswold, M., Gottesman, R. F., Wagenknecht, L. E., Windham, B. G., & Selvin, E. (2014). Diabetes in midlife and cognitive change over 20 years: A cohort study. *Annals of Internal Medicine, 161*(11), 785–793.

Shah, A., Bush, C. O., & Perry, R. J. (2025). Genetic underpinnnings of type 2 diabetes. *Advances in Genetics, 113*, 54–75.

Sharma, S. (2021). High fat diet and its effects on cognitive health: Alterations of neuronal and vascular components of brain. *Physiology & Behavior, 240*, 113528.

Shu, J., Li, N., Wei, W., & Zhang, L. (2022). Detection of molecular signatures and pathways shared by Alzheimer's disease and type 2 diabetes. *Gene, 810*, 146070.

Soldevila-Domenech, N., Cuenca-Royo, A., Babio, N., Forcano, L., Nishi, S., Vintró-Alcaraz, C., Gómez-Martínez, C., Jiménez-Murcia, S., Fernández-Carrión, R., Gomis-González, M., Alvarez-Sala, A., Carlos, S., Pintó, X., Corella, D., Díez-Espino, J., Castañer, O., Fernández-Aranda, F., Salas-Salvadó, J., & de la Torre, R. (2021). Metformin use and cognitive function in older adults with type 2 diabetes following a Mediterranean diet intervention. *Frontiers in Nutrition, 8*, 742586.

Spauwen, P. J. J., Köhler, S., Verhey, F. R. J., Stehouwer, C. D. A., & van Boxtel, M. P. J. (2013). Effects of type 2 diabetes on 12-year cognitive change: Results from the Maastricht aging study. *Diabetes Care, 36*(6), 1554–1561.

Squire, L. R. (2004). Memory systems of the brain: A brief history and current perspective. *Neurobiology of Learning and Memory, 82*(3), 171–177.

Squire, L. R., & Zola, S. M. (1996). Structure and function of declarative and nondeclarative memory systems. *Proceedings of the National Academy of Sciences of the United States of America, 93*(24), 13515.

Sun, H., Saeedi, P., Karuranga, S., Pinkepank, M., Ogurtsova, K., Duncan, B. B., Stein, C., Basit, A., Chan, J. C. N., Mbanya, J. C., Pavkov, M. E., Ramachandaran, A., Wild, S. H., James, S., Herman, W. H., Zhang, P., Bommer, C., Kuo, S., Boyko, E. J., & Magliano, D. J. (2022). IDF Diabetes Atlas: Global, regional and country-level diabetes prevalence estimates for 2021 and projections for 2045. *Diabetes Research and Clinical Practice, 183*, 109119.

Tabibnia, G., Creswell, J. D., Kraynak, T., Westbrook, C., Julson, E., & Tindle, H. A. (2014). Common prefrontal regions activate during self-control of craving, emotion, and motor impulses in smokers. *Clinical Psychological Science, 2*(5), 611–619.

Tabibnia, G., Monterosso, J. R., Baicy, K., Aron, A. R., Poldrack, R. A., Chakrapani, S., Lee, B., & London, E. D. (2011). Different forms of self-control share a neurocognitive substrate. *Journal of Neuroscience, 31*(13), 4805–4810.

Telle-Hansen, V. H., Gaundal, L., & Myhrstad, M. C. W. (2019). Polyunsaturated fatty acids and glycemic control in type 2 diabetes. *Nutrients, 11*(5), 1067.

Tovote, P., Fadok, J. P., & Lüthi, A. (2015). Neuronal circuits for fear and anxiety. *Nature Reviews Neuroscience, 16*(6), 317–331.

Tuligenga, R. H., Dugravot, A., Tabák, A. G., Elbaz, A., Brunner, E. J., Kivimäki, M., & Singh-Manoux, A. (2014). Midlife type 2 diabetes and poor glycaemic control as risk factors for cognitive decline in early old age: A post-hoc analysis of the Whitehall II cohort study. *Lancet Diabetes & Endocrinology, 2*(3), 228–235.

Uddin, L. Q. (2021). Cognitive and behavioural flexibility: Neural mechanisms and clinical considerations. *Nature Reviews Neuroscience, 22*(3), 167–179.

van Dam, R. M., Willett, W. C., Rimm, E. B., Stampfer, M. J., & Hu, F. B. (2002). Dietary fat and meat intake in relation to risk of type 2 diabetes in men. *Diabetes Care, 25*(3), 417–424.

Wang, C., Fu, K., Liu, H., Xing, F., & Zhang, S. (2014). Brain structural changes and their correlation with vascular disease in type 2 diabetes mellitus patients: A voxel-based morphometric study. *Neural Regeneration Research, 9*(16), 1548–1556.

Wiltgen, B. J., Brown, R. A. M., Talton, L. E., & Silva, A. J. (2004). New circuits for old memories: The role of the neocortex in consolidation. *Neuron, 44*(1), 101–108.

Yoon, S., Cho, H., Kim, J., Lee, D.-W., Kim, G. H., Hong, Y. S., Moon, S., Park, S., Lee, S., Lee, S., Lee, S., Bae, S., Simonson, D. C., & Lyoo, I. K. (2017). Brain changes in overweight/obese and normal-weight adults with type 2 diabetes mellitus. *Diabetologia, 60*(7), 1207–1217.

Yu, K. K. K., Cheing, G. L. Y., Cheung, C., Kranz, G. S., & Cheung, A. K.-K. (2021). Gray matter abnormalities in type 1 and type 2 diabetes: A dual disorder ALE quantification. *Frontiers in Neuroscience, 15*, 638861.

Zhang, J., Wang, Y., Wang, J., Zhou, X., Shu, N., Wang, Y., & Zhang, Z. (2014). White matter integrity disruptions associated with cognitive impairments in type 2 diabetic patients. *Diabetes, 63*(11), 3596–3605.

Zhou, Z., Zhu, Y., Liu, Y., & Yin, Y. (2019). Comprehensive transcriptomic analysis indicates brain regional specific alterations in type 2 diabetes. *Aging, 11*(16), 6398–6421.

Chapter 34

Diet, behavior and diabetes—an overview

Stanisław Surma[1], Karol Momot[2, 3], Artur Mamcarz[3], Kazimierz Ciechanowski[4] and Marcin Barylski[5]

[1]Department of Internal Medicine and Clinical Pharmacology, Medical University of Silesia, Katowice, Poland; [2]Department of Experimental and Clinical Physiology, Medical University of Warsaw, Warsaw, Poland; [3]3rd Department of Internal Medicine and Cardiology, Medical University of Warsaw, Warsaw, Poland; [4]Department of Nephrology, Transplantology and Internal Medicine, Pomeranian Medical University, Szczecin, Poland; [5]Department of Internal Medicine and Cardiac Rehabilitation, Medical University of Lodz, Łódź, Poland

List of all abbreviations

AHEI Alternate Healthy Eating Index
BCAAs Branched-Chain Amino Acids
BMI Body Mass Index
CQI Carbohydrate Quality Index
DALYs Disability-Adjusted Life Years
DASH Dietary Approaches to Stop Hypertension
DPPOS Diabetes Prevention Program Outcomes Study
GBD Global Burden of Disease
GI Glycemic Index
GL Glycemic Load
GLP-1RA Glucagon-Like Peptide-1 Receptor Agonist
GLUT-4 Glucose Transporter Type 4
HPA axis Hypothalamic-Pituitary-Adrenal axis
i-IFG Isolated Impaired Fasting Glucose
IDF International Diabetes Federation
IGF Impaired Fasting Glucose
IGT Impaired Glucose Tolerance
MET Metabolic Equivalent of Task
MUFAs Monounsaturated Fatty Acids
PDI Plant-Based Diet Index
PET Positron Emission Tomography
PUFAs Polyunsaturated Fatty Acids
RCT Randomized Controlled Trial
SFA Saturated Fatty Acids
SGLT2 Sodium-Glucose Cotransporter-2
T2D Type 2 Diabetes
TRE Time-Restricted Eating
UPF Ultra-Processed Foods
YFAS Yale Food Addiction Scale

Introduction

Diabetes, particularly type 2 diabetes (T2D), represents one of the most serious public health challenges of the 21st century due to its increasing prevalence, treatment costs, and the burden of cardiovascular, renal, and neurological complications. The latest estimates based on the IDF Diabetes Atlas (11th edition) indicate that in 2024, diabetes will affect approximately 589 million adults worldwide, and this increase is projected to continue until mid-century (Genitsaridi et al., 2026). In turn, the Global Burden of Disease (GBD) analysis shows that by 2021, the number of people living with diabetes had already reached approximately 529 million, and the assessment of the burden (DALYs) and projections to 2050 emphasize the dominant role of T2D and the growing role of modifiable risk factors, including high BMI, unfavorable dietary patterns, and low physical activity (American Diabetes Association Professional Practice Committee, 2024, GBD 2021 Diabetes Collaborators, 2023).

Fig. 34.1 summarizes the risk factors (modifiable and non-modifiable) and complications of type 2 diabetes.

In this context, the chapter focuses on two groups of determinants: dietary factors and behavioral factors, which contribute to the risk of developing T2D, influence the course of the disease, and determine the effectiveness of metabolic control. Contemporary data increasingly emphasize that diabetogenic risk does not result solely from the total energy or macronutrient intake, but also from the quality of the diet (e.g., degree of food processing, carbohydrate quality, fat profile) and established behaviors (e.g., eating patterns, sleep, activity, stress response). For example, a meta-analysis of longitudinal studies encompassing nearly 1.1 million people demonstrated a dose-response relationship between the consumption of ultra-processed foods and the risk of T2D (moderate consumption: RR ~1.12; high consumption: RR ~1.31) (Delpino et al., 2022). These findings are supported by

Hormones in Neuroscience. https://doi.org/10.1016/B978-0-443-30263-3.00023-3

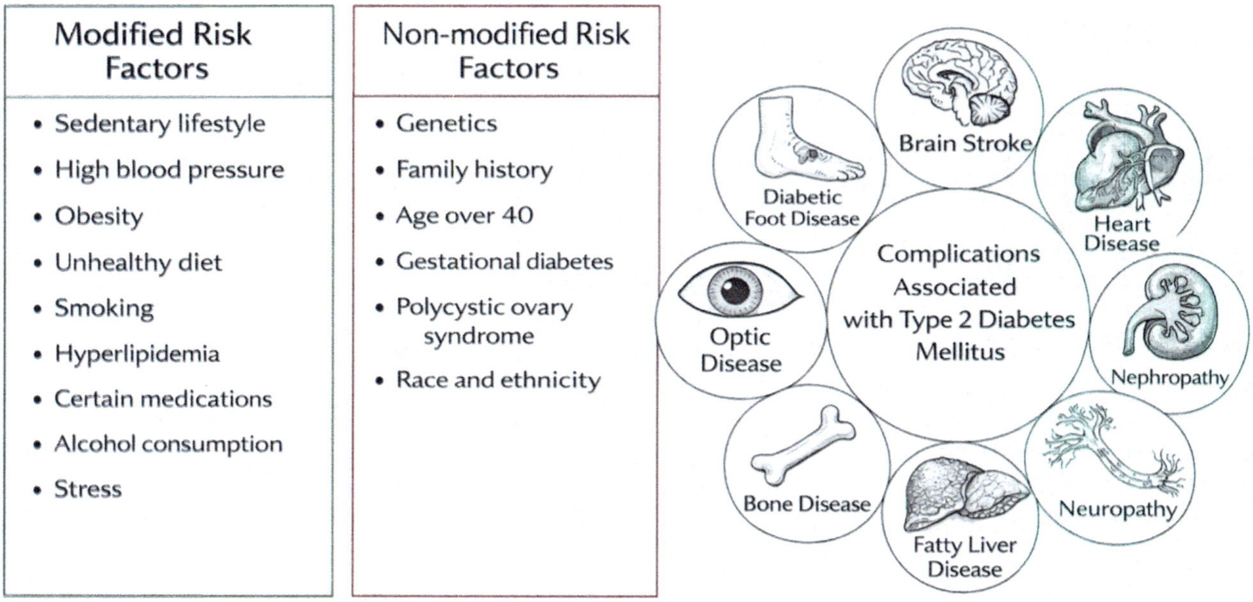

FIGURE 34.1 Risk factors (modifiable and non-modifiable) and complications of type 2 diabetes.

large prospective US cohorts, in which higher consumption of ultra-processed foods was associated with an increased risk of T2D, although the association varied within food subgroups (Chen et al., 2023). At the same time, there is growing evidence that lifestyle interventions are effective, but their effectiveness depends on the glycemic phenotype and the quality and durability of the behavioral change. An individual-participant meta-analysis of RCTs showed that "conventional" lifestyle interventions reduce the incidence of T2D, especially in people with impaired glucose tolerance (IGT), while the effect may be weaker in people with isolated impaired fasting glucose (i-IFG) (Sathish et al., 2023). In turn, the latest meta-analysis of RCTs confirms that long-term, structured, intensive lifestyle interventions reduce the risk of progression from prediabetes to T2D, although the certainty of the evidence is limited by the risk of bias in some studies (Portal Teixeira et al., 2024).

As a result, the key challenge becomes not only "what to recommend" (dietarily), but also "how to effectively implement and sustain" behavioral change. Therefore, in the following sections of this chapter, we summarize the current evidence on specific dietary and behavioral components and mechanisms (including postprandial glucose, insulin resistance, low-grade inflammation, and the gut-brain axis), with an emphasis on practical implications for prevention and treatment.

Part I: dietary factors

Carbohydrates and their quality

In the context of diabetes (T2D), it is increasingly clear that "how many carbohydrates" is less informative than "what the quality of carbohydrates is." This quality includes, among other things, glycemic response (GI/GL), degree of processing, fiber content, proportion of whole grains, and the form of sugar consumption (e.g., liquid *vs.* solid). Modeling the global burden of disease showed that a significant proportion of new T2D cases were associated with a suboptimal diet, with key factors including low whole grain consumption and high consumption of refined grains (rice and wheat) (O'Hearn et al., 2023).

Glycemic index (GI) and glycemic load (GL)

GI and GL are practical indicators of how quickly and how strongly a diet raises postprandial glucose levels. A meta-analysis of large cohorts ("mega cohorts" of $\geq$100,000 individuals) found that a high GI is associated with a higher incidence of T2D (RR of approximately 1.27 for extreme quantile comparisons), and a high GL also correlates with an increased risk (RR of approximately 1.15) (Jenkins et al., 2024). These results support the approach of reducing GI/GL as a tool comparable (in terms of the direction of effect) to increasing fiber and whole grain intake.

Whole grains versus refined grains

Whole grains represent a model of "high-quality carbohydrates": they provide fiber, magnesium, polyphenols, and resistant starch, while typically having a lower energy density and a more favorable postprandial profile. A recent meta-analysis (cohorts + RCTs) found that whole grain consumption was associated with a lower risk of T2D, and the RCTs observed reductions in fasting glucose and a dose-response relationship for several glycemic markers (Ying et al., 2024). Complementarily, a meta-analysis of

RCTs comparing whole grains to refined grains showed improved postprandial glycemia and insulinemia and a modest improvement in HbA1c, providing a mechanistic rationale for the effects observed in prospective studies (Sanders et al., 2023).

Added sugars and sweetened beverages

Sugar-sweetened beverages are a particularly unfavorable carbohydrate carrier, providing rapidly absorbed energy without the satiety typical of solid foods. A meta-analysis of prospective studies demonstrated a significant increased risk of T2D in individuals with higher consumption of sweetened beverages (Della Corte et al., 2025). Clinically, an interaction with physical activity was also assessed in large US cohorts: high SSB consumption remained unfavorable, and activity does not necessarily completely eliminate the risk associated with SSBs (Pacheco et al., 2025). In practice, this means that limiting SSBs is one of the most "high-yielding" dietary interventions for T2D prevention.

Carbohydrate quality indicators (CQIs)—attempts at operationalization

Because single indicators (e.g., GI) do not describe the entire nutritional context, an approach based on "carbohydrate quality indices" (e.g., carbohydrate:fiber:free sugar ratios) is being developed. However, cohort results are inconsistent: in the Malmö Diet and Cancer Study, favorable associations were revealed mainly after sensitivity analyses (e.g., after excluding potential misreporting), suggesting that carbohydrate quality may be important, but its measurement in population-based studies is methodologically challenging (Ramstedt et al., 2023). In turn, in a prospective study in the Chinese population, non-linear (U-shaped) relationships were observed for selected quality indicators, which highlights the role of cultural context and carbohydrate sources (Cui et al., 2023).

Fats

Dietary fats have long attracted interest in the context of T2D, as their composition and source influence metabolic pathways—from insulin sensitivity and cell membrane function to inflammation and lipid profile. It's not just about energy intake from fat, but also about the type of fatty acids and their metabolic consequences. Unsaturated fatty acids (MUFAs and PUFAs) may have beneficial effects, while excess saturated or trans fats have been associated with adverse metabolic effects.

Fat quality and the risk of T2D

In a cohort analysis of the US National Handbook of Evidence (NHANES), higher intakes of monounsaturated (MUFAs) and polyunsaturated fatty acids (PUFAs) were associated with a lower risk of both prediabetes and T2D, but this effect was attenuated with age (Jiang et al., 2023). This suggests that consumption of unsaturated fatty acids—especially PUFA-18:2 (linoleic acid) and MUFA-18:1 (oleic acid)—may promote better glycemic regulation and insulin sensitivity by influencing cell membranes and lipid metabolism.

Biological mechanisms of fat action

Recent mechanistic reviews have emphasized that fatty acid composition modifies glucose metabolism by influencing insulin resistance, inflammation, and cell membrane composition. Unsaturated fatty acids alter membrane fluidity and insulin receptor signaling, which may improve glucose transport to peripheral tissues. Furthermore, PUFAs are precursors of eicosanoids with anti-inflammatory effects, which may reduce the chronic inflammation associated with the pathogenesis of T2D (Sivri & Akdevelioğlu, 2025).

Omega-3 and omega-6 fatty acids

In the context of PUFAs, omega-3 (n-3) fatty acids—such as EPA and DHA—are particularly studied for their effects on metabolism and diabetes risk. Although clinical findings regarding the risk of T2D itself are mixed, longitudinal observations and metabolic analyses increasingly suggest that higher n-3 PUFA intake is associated with favorable changes in lipid profiles and a reduction in inflammatory markers, which are often elevated in people with T2D (Wang et al., 2024). In a large US population-based study, fish oil supplementation and higher plasma n-3 PUFA concentrations were associated with a lower risk of microvascular and macrovascular complications in people already with T2D (Tian et al., 2025). Some reports also point to potential benefits of omega-6 PUFAs—for example, linoleic acid, which correlated with lower inflammatory markers and better glycemic control in a population.

Saturated and trans fats

Regarding saturated fats (SFAs), results from a prospective analysis suggest that overall SFA intake does not show a strong, linear relationship with T2D risk, and the effects may depend on the specific fatty acid chain and food source (Neuenschwander et al., 2020). However, health organizations still recommend limiting SFAs in the diet due to their impact on lipid profile and cardiovascular risk, which is a significant prognostic factor in T2D (Neuenschwander et al., 2020).

Trans fats, especially industrial ones, are clearly associated with negative metabolic effects and an increased risk of insulin resistance and T2D, which is confirmed by epidemiological observations and WHO positions recommending their elimination from food (de Souza et al., 2015).

Protein

Dietary protein is one of the three main macronutrients and plays a key role in tissue structure, transport, and metabolic signaling. In recent years, a growing number of prospective and meta-analytical studies have examined the association between protein intake and its sources and the risk of developing T2D. Although the mechanisms are complex and depend on factors such as amino acid profile, inflammatory markers, and dietary composition, there is consensus that protein quality—particularly the difference between plant and animal protein—plays an important role.

Total protein intake and the risk of T2D

Recent cohort studies indicate an association between total protein intake and the risk of T2D, but this relationship is not linear. Results from the large Shanghai Suburban Adult Cohort and Biobank (SSACB) cohort show a U-shaped relationship between total dietary protein (% of energy) and the risk of T2D—the lowest risk was observed with moderate protein intake (approximately 12%–17% of energy), while both low and very high intakes were associated with higher risk (Xu et al., 2025). Furthermore, in this cohort, plant protein intake was associated with a lower risk of T2D, independent of other risk factors (Xu et al., 2025). A meta-analytic approach also indicates that the association between total protein intake and T2D risk is complex and dependent on other dietary components and lifestyle. Some prospective observations suggest that higher intakes of total and especially animal protein are associated with a higher risk of T2D, particularly in obese women (e.g., in the large EPIC-InterAct and Nurses' Health Study cohorts), although the biological details of this relationship are still being explored (Mensink, 2024).

Animal versus plant protein—differences in T2D risk

One of the most frequently highlighted aspects is the difference between animal and plant protein. A 2024 meta-analysis combining data from multiple population-based studies found that animal protein consumption was associated with a significant and dose-dependent increase in the risk of T2D, whereas plant protein did not show this association and, in substitution analyses, could lead to a reduced risk of T2D (Fotouhi Ardakani et al., 2024). These results support the hypothesis that replacing some animal protein intake with plant protein may be beneficial for glucose metabolism.

Mechanistically, these differences may be related, among other things, to the amino acid profile. Branched-chain amino acids (BCAAs), which are more abundant in animal protein, have been associated with insulin resistance and adverse metabolic changes in numerous metabolomic analyses, although their role in the etiology of T2D is still under active investigation (Mensink, 2024). Other meat components—such as saturated fats and processed products—may also contribute to the observed increased risk through inflammatory and oxidative mechanisms.

Not all animal protein sources are equally associated with risk. In prospective analyses, red and processed meat consumption showed the strongest associations with T2D, while protein consumption from low-fat milk, yogurt, and legumes was associated with a lower risk or a neutral effect (Mensink, 2024). This indicates that the effects of protein on glucose metabolism are partially dependent on the food matrix and other co-occurring components (e.g., fat, fiber).

Biological mechanisms

Different protein sources influence metabolic pathways through multiple mechanisms: modulation of insulin signaling and insulin sensitivity, impact on the production of satiety hormones, alteration of gut microbiota (particularly in the case of plant proteins), amino acid profiles related to BCAAs and other glucogenic amino acids (Mensink, 2024; Xu et al., 2025).

Dietary patterns

Unlike analyses focused on individual nutrients, dietary pattern analysis provides a broader picture of how cumulative dietary habits influence the risk of T2D. Such patterns combine various foods and dietary components, allowing for the identification of their synergistic or antagonistic metabolic effects. The most commonly studied patterns include the Mediterranean diet, plant-based diets, and unhealthy patterns such as the Western diet and patterns based on ultra-processed foods.

Mediterranean diet

Current cohort studies and meta-analyses confirm that high adherence to the Mediterranean diet is associated with a lower risk of developing T2D. A recent, large meta-analysis of prospective studies ($n = 22$ cohorts) demonstrated a significant dose-response relationship between adherence to a Mediterranean diet and a lower risk of T2D, both in analyses from several countries and overall (Wallerer et al., 2025).

Biological mechanisms of this relationship include higher consumption of vegetables, fruits, legumes, olive oil, and nuts (rich in antioxidants, fiber, and unsaturated fatty acids), along with reduced red meat and highly processed foods—which promotes improved insulin sensitivity and reduced inflammation (Wallerer et al., 2025).

Additional observational studies suggest that higher adherence to this pattern may be associated with a lower risk of prediabetes in high-risk individuals, as well as a better metabolic profile in the general population (Wallerer et al., 2025).

Plant-based and plant-based diets

Another widely studied pattern is the plant-based or plant-based diet, encompassing vegetarian and vegan eating styles and plant-based indices (Plant-Based Diet Indexes, PDIs). A meta-analysis including data from multiple populations showed that high adherence to plant-based diets is associated with a significantly lower risk of T2D (particularly for vegan and lacto-ovo-vegetarian diets), while an "unhealthy" PDI—based primarily on processed plant foods—correlated with a higher risk of the disease (Murciano et al., 2025).

It is important to distinguish between a "healthy" plant-based diet (rich in vegetables, fruits, legumes, whole grains, and nuts) and a plant-based diet with a high proportion of processed foods—the former is associated with a reduced risk of T2D through mechanisms including, among others, higher dietary fiber content, lower glycemic index of meals, and beneficial modulation of gut microbiota (Murciano et al., 2025).

Other healthy eating patterns

Besides classic patterns like the MedDiet and plant-based diets, research also indicates benefits from other healthy eating patterns, such as DASH (Dietary Approaches to Stop Hypertension), the Alternate Healthy Eating Index (AHEI), and patterns based on dietary recommendations (Healthy Eating Patterns). Systematic analyses suggest that these patterns are also associated with a lower risk of T2D compared to ultra-processed diets or standard ("control") diets, although the differences between them can be difficult to disentangle in comparative analyses.

Western diet and ultra-processed foods

In contrast to healthy patterns, the Western diet—characterized by a high proportion of red meat, sweetened beverages, ultra-processed foods, and refined carbohydrates—has been consistently associated with an increased risk of T2D in prospective studies. High consumption of ultra-processed foods is a significant risk factor for insulin resistance, increased body weight and metabolic dysfunction, which mechanistically promotes the development of T2D (Souza et al., 2025).

Part II: behavioral factors

Eating patterns

Besides the quality and quantity of food consumed, eating patterns are a significant determinant of T2D risk, including meal frequency, regularity of consumption, time of day, and episodes of excessive energy consumption. Accumulating evidence indicates that not only "what" but also "when" and "how" we eat has metabolic significance.

One of the most frequently studied phenomena is irregular meal intake. A meta-analysis demonstrated that skipping breakfast is associated with an increased risk of T2D, regardless of body mass index and other lifestyle factors (Bi et al., 2015). Mechanistically, this may be due to impaired insulin sensitivity in the morning and compensatory overeating later in the day (Bi et al., 2015).

Time-restricted eating (TRE) is also considered important. A randomized controlled trial was shown that limiting the eating window (usually 8–10 hours) leads to improved fasting glucose levels and indicators of insulin resistance, although this effect may be partially dependent on weight loss (Zhang et al., 2022). Synchronizing energy intake with the circadian rhythm may improve glucose metabolism by influencing the expression of biological clock genes. Another factor is emotional eating and episodes of loss of control over eating, which correlate with poorer glycemic control and a higher risk of insulin resistance (Burnatowska et al., 2022). Chronic stress and HPA axis hyperactivity may lead to excessive consumption of high-calorie foods, exacerbating metabolic risk (Burnatowska et al., 2022).

Physical activity and sedentary behavior

Physical activity is one of the best-documented protective factors in the context of T2D. Regular exercise increases skeletal muscle glucose uptake through the translocation of GLUT-4 transporters independently of insulin, improves insulin sensitivity, and reduces visceral adipose tissue.

A meta-analysis of over one million participants demonstrated a significant inverse association between moderate and vigorous physical activity levels and the risk of T2D (Aune et al., 2015). Each increase of 10 MET-hours per week was associated with a reduced risk of the disease.

Moderate-intensity activity, such as brisk walking, is also of particular importance. The UK Biobank analysis showed that walking pace was a strong predictor of T2D risk—individuals who walked briskly had a significantly lower risk of developing the disease (Boonpor et al., 2023). At the same time, evidence for the negative impact of sedentary behavior is growing. Prolonged sitting (>8 h per day) is associated with a higher risk of T2D, even after accounting for physical activity levels (Åsvold et al., 2017). Mechanisms include reduced lipoprotein lipase activity, impaired insulin sensitivity, and impaired lipid metabolism.

Interventions that interrupt sitting with short bouts of movement (2–3 minutes every 30 minutes) have been shown to improve postprandial glycemia (Quan et al., 2021). This suggests that reducing sedentary time may be as important as increasing planned physical activity.

Sleep and circadian rhythm

Sleep disorders and circadian rhythm dysregulation are increasingly recognized as independent risk factors for T2D. Both short (<6 hours) and long (>9 hours) sleep duration are associated with an increased risk of developing the disease (Liu et al., 2025).

Mechanistically, sleep deprivation leads to increased cortisol, sympathetic nervous system activation, and impaired insulin sensitivity. Experimental sleep restriction has demonstrated decreased insulin sensitivity after just a few days (Liu et al., 2025; Surma et al., 2023).

Shift work, which leads to circadian rhythm dysregulation, also increases the risk of T2D. A studies including shift workers showed a significantly higher risk of developing the disease compared to day workers (Zhu & Mi, 2025). Sleep optimization—regular bedtimes, adequate sleep duration, and limiting light exposure at night—can support metabolic control and be part of a lifestyle intervention (Surma et al., 2023).

Psychosocial stress

Chronic psychosocial stress is a significant risk factor for T2D through activation of the hypothalamic-pituitary-adrenal (HPA) axis and increased cortisol secretion. Long-term hypercortisolemia promotes insulin resistance, visceral fat accumulation, and metabolic disturbances.

A meta-analysis of prospective studies found that high levels of occupational and life stress are associated with an increased risk of T2D (especially in women) (Sui et al., 2016). Depressive and anxiety symptoms also correlate with an increased risk of carbohydrate metabolism disorders (Sui et al., 2016).

Stress also has an indirect effect through behaviors—increased consumption of high-calorie foods, reduced physical activity, and sleep disturbances. Stress-reducing interventions, including mindfulness techniques and cognitive-behavioral therapy, have shown improvements in metabolic parameters in pilot studies (Burnatowska et al., 2022).

Part III: diet-behavior interactions

Reward system and overconsumption

Interactions between diet and behavior largely occur through the reward system, which integrates taste stimuli, metabolic signals, and habit learning. Products high in refined sugars and/or fat and high "hyperpalatability" promote the reinforcement of consumption behaviors, which can lead to chronic energy surplus and the progression of insulin resistance. In clinical terms, the construct of "food addiction," measured by tools such as the YFAS, is gaining importance; behavioral similarities to substance use disorders (e.g., loss of control, craving, continuing despite harm) are emphasized (Gearhardt et al., 2023). Importantly, a population-based study found that higher consumption of ultra-processed foods (UPF) and specific eating patterns were associated with a higher likelihood of meeting criteria for food addiction (Silva Júnior et al., 2023), providing an important link between dietary exposure and behavior.

Mechanistically, the hypothesis of the "addictive" potential of UPF often invokes dopaminergic reward coding. At the same time, recent neurobiological data indicate that the dopamine response to food stimuli is variable across individuals and does not always take the form of a simple "dopamine surge after ingestion." A PET study found no significant increase in striatal dopamine after ingestion of an ultra-processed high-sugar and high-fat cocktail, with high interindividual variability (Darcey et al., 2025). In practice, this means that overconsumption may result not only from post-consumption dopamine itself, but also from cue-reactivity, habit learning, stress, and food availability—mechanisms that are strongly dependent on the environmental and behavioral context (Gearhardt et al., 2023).

Food microenvironment

The food microenvironment (food environment) encompasses factors such as the availability and density of retail outlets, prices, marketing exposure, convenience of purchase, and social norms. It is a key moderator of the diet-behavior relationship, as it determines which choices are "default" and how easily they are maintained over time. A longitudinal ecological study found that changes in the food environment (including the increase in the number of fast-food/pizza outlets and the decline in full-service restaurants) were associated with a higher incidence of T2D at the neighborhood level (Zick et al., 2023). This supports the idea that the environment may "push" populations toward choices with higher energy density and a higher proportion of UPFs.

Simultaneously, the association between socioeconomic conditions and the risk of T2D is being analyzed. A study involving three large samples (including a cohort of veterans) found that poorer socioeconomic conditions at the neighborhood level were associated with a higher risk of T2D, while mediation by simple indicators of the food environment (share of fast food *vs.* supermarkets) was limited or variable across regions (Thorpe et al., 2022). This suggests that the "microenvironment" encompasses not only the number of stores but also the quality of the offerings, prices, transportation, neighborhood safety, and chronic stress.

A meta-analysis of longitudinal studies on the "built environment" provides a synthetic approach: less favorable environments (including unhealthy food environments) were found to be associated with a higher incidence of T2D, while greater exposure to green spaces and improved walkability were associated with a slightly lower incidence (Feyissa et al., 2024). From an intervention perspective, this means that interventions targeted at the individual may be less effective without parallel improvements in environmental conditions.

An integrated model

An integrated model of diet-behavior interactions in diabetes can be conceptualized as a feedback system: (1) environment → (2) behavior → (3) diet → (4) metabolism → (5) biological signals influencing behavior. At the dietary level, ultra-processed foods are an example of an exposure with strong population-level impact; a meta-analysis of longitudinal studies has shown that higher UPF intake is associated with an increased risk of T2D (Delpino et al., 2022). At the behavioral level, key "nodes" include physical activity, sleep, and stress, which modify insulin sensitivity and eating patterns while also being shaped by the environment.

An important element of the model is the fact that the effectiveness of lifestyle interventions may depend on the metabolic risk phenotype. An individual-participant meta-analysis of RCTs showed that classic lifestyle interventions reduced the risk of progression to T2D, particularly in individuals with IGT, while the effect was weaker in individuals with isolated impaired fasting glucose (Sathish et al., 2023). In practice, this supports a "precision prevention" approach: combining dietary and behavioral modifications with the diagnosis of the dominant mechanism (e.g., hepatic vs. muscular insulin resistance). Overall, diabetes remains a multifactorial disease in which risk is accumulated through the coexistence of biological, dietary, behavioral, and environmental determinants (Ismail et al., 2021). From a public health perspective, it is important to simultaneously address: (a) the availability of healthy food and limiting exposure to UPFs, (b) support for behavior change (self-regulation, habits, stress), and (c) identification of the highest-risk groups in which interventions are most effective.

Part IV: clinical implications

Type 2 diabetes prevention

T2D prevention relies primarily on lifestyle modification, with particular emphasis on diet, physical activity, weight loss, and sleep and stress regulation. The strongest clinical evidence comes from long-term randomized trials.

The 21-year follow-up results of the Diabetes Prevention Program Outcomes Study (DPPOS) demonstrated a sustained reduction in the incidence of T2D in individuals receiving an intensive lifestyle intervention compared with placebo; the effect was greater than that of metformin (Goldberg et al., 2022). The intervention included weight loss of ≥7% and ≥150 minutes of moderate activity per week. These data confirm that lifestyle modification remains the gold standard for primary prevention. An individual-participant meta-analysis of RCTs showed that the effectiveness of lifestyle interventions depends on the prediabetes phenotype—the greatest risk reduction was observed in individuals with IGT, while the effect was weaker in those with isolated IGF (Sathish et al., 2023). This suggests the need to personalize preventive strategies. Diet quality also plays a role. A meta-analysis of prospective studies found that higher consumption of ultra-processed foods (UPF) is associated with an increased risk of T2D (Delpino et al., 2022). In clinical practice, this means limiting UPF, added sugars, and red and processed meats while increasing the intake of vegetables, legumes, whole grains, and unsaturated fats. Secondary prevention (in individuals with prediabetes) also includes pharmacotherapy in selected cases—metformin is recommended for those at high risk of progression, particularly those with a BMI ≥35 kg/m^2 and in younger patients (Goldberg et al., 2022).

Treatment of T2D

Treatment of T2D is based on an integrated model encompassing lifestyle modification, pharmacotherapy, and—in selected cases—bariatric surgery. Current clinical approaches emphasize the need for individualized therapy based on age, body weight, cardiovascular complications, and patient preferences.

Dietary interventions remain the cornerstone of treatment. The randomized Look AHEAD trial demonstrated that intensive lifestyle intervention led to sustained weight loss and improved glycemic control, although no significant reduction in the primary cardiovascular endpoint was demonstrated (Look AHEAD Research Group, 2022). Nevertheless, improved quality of life and metabolic control confirm the importance of behavioral therapy.

In recent years, incretin-based medications (GLP-1RAs) and SGLT2 inhibitors have gained significant traction, demonstrating beneficial cardiovascular and nephroprotective effects in addition to lowering glycemia (ADA Standards, 2024). Integrating pharmacotherapy with dietary interventions can increase treatment effectiveness.

A growing body of evidence also indicates the effectiveness of low-calorie diets in achieving remission in early-stage T2D. The DiRECT study demonstrated that intensive weight loss can lead to remission in a significant percentage of patients (Lean et al., 2024). The mechanisms include a reduction in visceral and liver fat and improved β-cell function (Lean et al., 2024).

Behavioral factors—sleep quality, stress reduction, and psychosocial support—that influence adherence and metabolic control should also be considered in the context of treatment (ADA Standards 2024).

Conclusions

This chapter presents a multidimensional model of the relationship between diet, behavior, and the risk T2D. Table 34.1 summarizes key information on diet, behavior and diabetes.

The quality of the food consumed is crucial—especially limiting ultra-processed foods, added sugars, and red meat, while simultaneously increasing the consumption of whole grains, legumes, vegetables, and unsaturated fatty acids. Dietary patterns such as the Mediterranean diet and healthy plant-based models demonstrate protective effects. Behavioral factors are equally important: physical activity, reduced sedentary behavior, adequate sleep, and reduced stress. The effectiveness of interventions depends on the metabolic phenotype and the sustainability of behavioral changes. Prevention and treatment of T2D require an integrated approach, combining individual interventions with modification of the nutritional and social environments.

Summary points

1. Diet quality is more important than the quantity of macronutrients.
2. Ultra-processed foods significantly increase the risk of T2D.
3. Mediterranean diets and healthy plant-based diets reduce the risk of diabetes.
4. Physical activity, sleep, and stress reduction modulate insulin resistance.
5. The effectiveness of prevention depends on the metabolic phenotype and environment.

Mini-dictionary of terms

Insulin resistance (IR)—a state of reduced tissue sensitivity to insulin, leading to impaired glucose metabolism.

Glycemic index (GI)—a measure of the rate of increase in blood glucose after consuming a given product.

Ultra-processed foods (UPF)—industrial products containing numerous technological additives and highly processed.

Gut-brain axis—a bidirectional communication system between the gut microbiota and the central nervous system.

Low-grade inflammation—chronic, subclinical activation of the immune system, promoting the development of metabolic diseases.

Other hormones, endocrine tissues, and pathways

The pathogenesis of T2D extends beyond insulin and involves a complex hormonal network. Adipose tissue, as an endocrine organ, secretes adipokines (leptin, adiponectin, resistin), which modulate insulin sensitivity and lipid metabolism. Decreased adiponectin and increased leptin resistance promote atherogenic dyslipidemia (Wełnicki et al., 2025). The hypothalamic-pituitary-adrenal axis (HPA), through cortisol, enhances gluconeogenesis, lipolysis, and redistribution of visceral adipose tissue. Incretins (GLP-1, GIP) regulate postprandial insulin secretion and influence appetite. Disturbances in the gut-liver axis promote fatty liver disease and overproduction of VLDL. Thyroid hormones modulate basal metabolic rate and lipid profile. Interactions between these systems link lipid abnormalities, inflammation, and insulin resistance, forming a common pathophysiological basis for T2D and cardiovascular disease.

Applications of the material

The data presented have direct clinical and population-level applications. In medical practice, they should support a shift from a reductionist approach based solely on dietary caloric intake to an assessment of product quality and dietary patterns. Patient education should focus on limiting ultra-processed foods, sweetened beverages, and processed meats, as well as increasing the consumption of vegetables, legumes, whole grains, and unsaturated fatty acids. Interventions should also include sleep hygiene, stress reduction, and regular physical activity.

In a precision prevention model, it is recommended to differentiate strategies based on metabolic phenotype (e.g., IGT *vs.* isolated fasting hyperglycemia). In the treatment of T2D, integrating pharmacotherapy (e.g., incretin-based medications, SGLT2 inhibitors) with intensive lifestyle interventions increases the effectiveness of therapy and improves cardiovascular prognosis. At the public health level, environmental interventions are necessary, including improved access to healthy foods, regulations regarding UPF, and supporting physical activity through the design of spaces that encourage movement.

TABLE 34.1 Dietary and behavioral determinants of type 2 diabetes: direction of effect, mechanisms, and strength of evidence.

Domain	Factor	Direction of association with T2D risk	Key biological mechanisms	Clinical relevance
Carbohydrates	High glycemic index (GI)/Glycemic load (GL)	Increased risk	Postprandial hyperglycemia; β-cell stress; hyperinsulinemia	Prefer low-GI carbohydrate sources
	Whole grains	Reduced risk	Fiber; magnesium; improved insulin sensitivity; lower postprandial glucose	≥3 servings/day recommended
	Sugar-sweetened beverages (SSBs)	Increased risk	Liquid calories; hepatic lipogenesis; low satiety	Strong recommendation to limit
Fats	MUFAs/PUFAs	Reduced risk	Improved membrane fluidity; enhanced insulin signaling; anti-inflammatory effects	Prefer plant oils, nuts, fish
	Trans fats	Increased risk	Inflammation; endothelial dysfunction; adverse lipid profile	Eliminate from diet
Protein	Animal protein (red/processed meat)	Increased risk	BCAAs; heme iron; inflammatory pathways	Replace partially with plant protein
	Plant protein	Reduced risk	Fiber co-intake; improved gut microbiota; lower inflammatory burden	Promote legumes and nuts
Dietary patterns	Mediterranean diet	Reduced risk	Anti-inflammatory; antioxidant; improved lipid profile	Strong recommendation
	Western diet	Increased risk	High UPF; refined carbohydrates; red meat; inflammation	Avoid pattern
Physical activity	Moderate–vigorous activity	Reduced risk	↑ GLUT-4 translocation; ↓ visceral adiposity; improved insulin sensitivity	≥150 minutes/week recommended
Sedentary behavior	Prolonged sitting (>8 h/day)	Increased risk	↓ Lipoprotein lipase activity; impaired glucose uptake	Interrupt sitting regularly
Sleep and stress	Short/long sleep duration	Increased risk	↑ Cortisol; sympathetic activation; insulin resistance	7–8 hours/night recommended
	Chronic psychosocial stress	Increased risk	HPA axis activation; visceral adiposity; inflammatory signaling	Stress-reduction interventions recommended

References

American Diabetes Association Professional Practice Committee. (2024). Introduction and methodology: Standards of care in diabetes-2024. *Diabetes Care, 47*(Suppl. ment_1), S1–S4. https://doi.org/10.2337/dc24-SINT

Åsvold, B. O., Midthjell, K., Krokstad, S., Rangul, V., & Bauman, A. (2017). Prolonged sitting may increase diabetes risk in physically inactive individuals: An 11 year follow-up of the HUNT Study, Norway. *Diabetologia, 60*(5), 830–835. https://doi.org/10.1007/s00125-016-4193-z

Aune, D., Norat, T., Leitzmann, M., Tonstad, S., & Vatten, L. J. (2015). Physical activity and the risk of type 2 diabetes: A systematic review and dose-response meta-analysis. *European Journal of Epidemiology, 30*(7), 529–542. https://doi.org/10.1007/s10654-015-0056-z

Bi, H., Gan, Y., Yang, C., Chen, Y., Tong, X., & Lu, Z. (2015). Breakfast skipping and the risk of type 2 diabetes: A meta-analysis of observational studies. *Public Health Nutrition, 18*(16), 3013–3019. https://doi.org/10.1017/S1368980015000257

Boonpor, J., Parra-Soto, S., Gore, J., Talebi, A., Lynskey, N., Raisi, A., Welsh, P., Sattar, N., Pell, J. P., Gill, J. M. R., Gray, S. R., Ho, F. K., & Celis-Morales, C. A. (2023). Association between walking pace and incident type 2 diabetes by adiposity level: A prospective cohort study from the UK Biobank. *Diabetes, Obesity and Metabolism, 25*(7), 1900–1910. https://doi.org/10.1111/dom.15053

Burnatowska, E., Surma, S., & Olszanecka-Glinianowicz, M. (2022). Relationship between mental health and emotional eating during the COVID-19 pandemic: A systematic review. *Nutrients, 14*(19), 3989. https://doi.org/10.3390/nu14193989

Chen, Z., Khandpur, N., Desjardins, C., Wang, L., Monteiro, C. A., Rossato, S. L., Fung, T. T., Manson, J. E., Willett, W. C., Rimm, E. B., Hu, F. B., Sun, Q., & Drouin-Chartier, J. P. (2023). Ultra-processed food consumption and risk of type 2 diabetes: Three large prospective U.S. cohort studies. *Diabetes Care, 46*(7), 1335–1344. https://doi.org/10.2337/dc22-1993

Cui, Z., Wu, M., Liu, K., Wang, Y., Kang, T., Meng, S., & Meng, H. (2023). Associations between conventional and emerging indicators of dietary carbohydrate quality and new-onset type 2 diabetes mellitus in Chinese adults. *Nutrients, 15*(3), 647. https://doi.org/10.3390/nu15030647

Darcey, V. L., Guo, J., Chi, M., Chung, S. T., Courville, A. B., Gallagher, I., Herscovitch, P., Joseph, P. V., Howard, R., La Noire, M., Milley, L., Schick, A., Stagliano, M., Turner, S., Urbanski, N., Yang, S., Zhai, N., Zhou, M. S., & Hall, K. D. (2025). Brain dopamine responses to ultra-processed milkshakes are highly variable and not significantly related to adiposity in humans. *Cell Metabolism, 37*(3), 616–628.e5. https://doi.org/10.1016/j.cmet.2025.02.002

de Souza, R. J., Mente, A., Maroleanu, A., Cozma, A. I., Ha, V., Kishibe, T., Uleryk, E., Budylowski, P., Schünemann, H., Beyene, J., & Anand, S. S. (2015). Intake of saturated and trans unsaturated fatty acids and risk of all cause mortality, cardiovascular disease, and type 2 diabetes: Systematic review and meta-analysis of observational studies. *BMJ, 351*, h3978. https://doi.org/10.1136/bmj.h3978

Della Corte, K. A., Bosler, T., McClure, C., Buyken, A. E., LeCheminant, J. D., Schwingshackl, L., & Della Corte, D. (2025). Dietary sugar intake and incident type 2 diabetes risk: A systematic review and dose-response meta-analysis of prospective cohort studies. *Advances in Nutrition, 16*(5), 100413. https://doi.org/10.1016/j.advnut.2025.100413

Delpino, F. M., Figueiredo, L. M., Bielemann, R. M., da Silva, B. G. C., Dos Santos, F. S., Mintem, G. C., Flores, T. R., Arcêncio, R. A., & Nunes, B. P. (2022). Ultra-processed food and risk of type 2 diabetes: A systematic review and meta-analysis of longitudinal studies. *International Journal of Epidemiology, 51*(4), 1120–1141. https://doi.org/10.1093/ije/dyab247

Feyissa, T. R., Wood, S. M., Vakil, K., Mc Namara, K., Coffee, N. T., Alsharrah, S., Daniel, M., & Versace, V. L. (2024). The built environment and its association with type 2 diabetes mellitus incidence: A systematic review and meta-analysis of longitudinal studies. *Social Science & Medicine, 361*, 117372. https://doi.org/10.1016/j.socscimed.2024.117372

Fotouhi Ardakani, A., Anjom-Shoae, J., Sadeghi, O., Marathe, C. S., Feinle-Bisset, C., & Horowitz, M. (2024). Association between total, animal, and plant protein intake and type 2 diabetes risk in adults: A systematic review and dose-response meta-analysis of prospective cohort studies. *Clinical Nutrition, 43*(8), 1941–1955. https://doi.org/10.1016/j.clnu.2024.07.001

GBD 2021 Diabetes Collaborators. (2023). Global, regional, and national burden of diabetes from 1990 to 2021, with projections of prevalence to 2050: A systematic analysis for the Global Burden of Disease Study 2021. *Lancet, 402*(10397), 203–234. https://doi.org/10.1016/S0140-6736(23)01301-6

Gearhardt, A. N., Bueno, N. B., DiFeliceantonio, A. G., Roberto, C. A., Jiménez-Murcia, S., & Fernandez-Aranda, F. (2023). Social, clinical, and policy implications of ultra-processed food addiction. *BMJ, 383*, e075354. https://doi.org/10.1136/bmj-2023-075354

Genitsaridi, I., Salpea, P., Salim, A., Sajjadi, S. F., Tomic, D., James, S., Thirunavukkarasu, S., Issaka, A., Chen, L., Basit, A., Luk, A. O. Y., Ma, R. C. W., Mbanya, J. C., Ramachandran, A., Wild, S. H., Duncan, B. B., Boyko, E. J., & Magliano, D. J. (2026). 11th edition of the IDF Diabetes Atlas: Global, regional, and national diabetes prevalence estimates for 2024 and projections for 2050. *Lancet Diabetes & Endocrinology, 14*(2), 149–156. https://doi.org/10.1016/S2213-8587(25)00299-2

Goldberg, R. B., Orchard, T. J., Crandall, J. P., Boyko, E. J., Budoff, M., Dabelea, D., Gadde, K. M., Knowler, W. C., Lee, C. G., Nathan, D. M., Watson, K., Temprosa, M., & Diabetes Prevention Program Research Group. (2022). Effects of long-term metformin and lifestyle interventions on cardiovascular events in the diabetes prevention program and its outcome study. *Circulation, 145*(22), 1632–1641. https://doi.org/10.1161/CIRCULATIONAHA.121.056756

Ismail, L., Materwala, H., & Al Kaabi, J. (2021). Association of risk factors with type 2 diabetes: A systematic review. *Computational and Structural Biotechnology Journal, 19*, 1759–1785. https://doi.org/10.1016/j.csbj.2021.03.003

Jenkins, D. J. A., Willett, W. C., Yusuf, S., Hu, F. B., Glenn, A. J., Liu, S., Mente, A., Miller, V., Bangdiwala, S. I., Gerstein, H. C., Sieri, S., Ferrari, P., Patel, A. V., McCullough, M. L., Le Marchand, L., Freedman, N. D., Loftfield, E., Sinha, R., Shu, X. O., , … Sievenpiper, J. L., & Clinical Nutrition & Risk Factor Modification Centre Collaborators. (2024). Association of glycaemic index and glycaemic load with type 2 diabetes, cardiovascular disease, cancer, and all-cause mortality: A meta-analysis of mega cohorts of more than 100 000 participants. *Lancet Diabetes & Endocrinology, 12*(2), 107–118. https://doi.org/10.1016/S2213-8587(23)00344-3

Jiang, S., Yang, W., Li, Y., Feng, J., Miao, J., Shi, H., & Xue, H. (2023). Monounsaturated and polyunsaturated fatty acids concerning prediabetes and type 2 diabetes mellitus risk among participants in the

National Health and Nutrition Examination Surveys (NHANES) from 2005 to March 2020. *Frontiers in Nutrition, 10*, 1284800. https://doi.org/10.3389/fnut.2023.1284800

Lean, M. E., Leslie, W. S., Barnes, A. C., Brosnahan, N., Thom, G., McCombie, L., Kelly, T., Irvine, K., Peters, C., Zhyzhneuskaya, S., Hollingsworth, K. G., Adamson, A. J., Sniehotta, F. F., Mathers, J. C., McIlvenna, Y., Welsh, P., McConnachie, A., McIntosh, A., Sattar, N., & Taylor, R. (2024). 5-year follow-up of the randomised Diabetes Remission Clinical Trial (DiRECT) of continued support for weight loss maintenance in the UK: An extension study. *Lancet Diabetes & Endocrinology, 12*(4), 233–246. https://doi.org/10.1016/S2213-8587(23)00385-6

Liu, H., Zhu, H., Lu, Q., Ye, W., Huang, T., Li, Y., Li, B., Wu, Y., Wang, P., Chen, T., Xu, J., & Ji, L. (2025). Sleep features and the risk of type 2 diabetes mellitus: A systematic review and meta-analysis. *Annals of Medicine, 57*(1), 2447422. https://doi.org/10.1080/07853890.2024.2447422

Look AHEAD Research Group. (2022). Effects of intensive lifestyle intervention on all-cause mortality in older adults with type 2 diabetes and overweight/obesity: Results from the look AHEAD Study. *Diabetes Care, 45*(5), 1252–1259. https://doi.org/10.2337/dc21-1805

Mensink, M. (2024). Dietary protein, amino acids and type 2 diabetes mellitus: A short review. *Frontiers in Nutrition, 11*, 1445981. https://doi.org/10.3389/fnut.2024.1445981

Murciano, A., Malavolti, M., Fairweather-Tait, S., Paduano, S., Vinceti, M., & Filippini, T. (2025). Plant-based diets and risk of type 2 diabetes: Systematic review and dose-response meta-analysis. *British Journal of Nutrition, 134*(4), 277–296. https://doi.org/10.1017/S0007114525104157

Neuenschwander, M., Barbaresko, J., Pischke, C. R., Iser, N., Beckhaus, J., Schwingshackl, L., & Schlesinger, S. (2020). Intake of dietary fats and fatty acids and the incidence of type 2 diabetes: A systematic review and dose-response meta-analysis of prospective observational studies. *PLoS Medicine, 17*(12), e1003347. https://doi.org/10.1371/journal.pmed.1003347

O'Hearn, M., Lara-Castor, L., Cudhea, F., Miller, V., Reedy, J., Shi, P., Zhang, J., Wong, J. B., Economos, C. D., Micha, R., Mozaffarian, D., & Global Dietary Database. (2023). Incident type 2 diabetes attributable to suboptimal diet in 184 countries. *Nature Medicine, 29*(4), 982–995. https://doi.org/10.1038/s41591-023-02278-8

Pacheco, L. S., Tobias, D. K., Haslam, D. E., Drouin-Chartier, J. P., Li, Y., Bhupathiraju, S. N., Willett, W. C., Ludwig, D. S., Ebbeling, C. B., Hu, F. B., & Guasch-Ferré, M. (2025). Sugar-sweetened or artificially sweetened beverage consumption, physical activity and risk of type 2 diabetes in US adults. *Diabetologia, 68*(4), 792–800. https://doi.org/10.1007/s00125-024-06351-w

Portal Teixeira, P., Pozzer Zucatti, K., Strassburger Matzenbacher, L., Fink Wayerbacher, L., Zhang, M., Colpani, V., & Gerchman, F. (2024). Long-term lifestyle intervention can reduce the development of type 2 diabetes mellitus in subjects with prediabetes: A systematic review and meta-analysis. *Diabetes Research and Clinical Practice, 210*, 111637. https://doi.org/10.1016/j.diabres.2024.111637

Quan, M., Xun, P., Wu, H., Wang, J., Cheng, W., Cao, M., Zhou, T., Huang, T., Gao, Z., & Chen, P. (2021). Effects of interrupting prolonged sitting on postprandial glycemia and insulin responses: A network meta-analysis. *Journal of Sport and Health Science, 10*(4), 419–429. https://doi.org/10.1016/j.jshs.2020.12.006

Ramstedt, M., Janzi, S., Olsson, K., González-Padilla, E., Ramne, S., Borné, Y., Ericson, U., & Sonestedt, E. (2023). Comparisons of different carbohydrate quality indices for risk of type 2 diabetes in the Malmö Diet and Cancer Study. *Nutrients, 15*(18), 3870. https://doi.org/10.3390/nu15183870

Sanders, L. M., Zhu, Y., Wilcox, M. L., Koecher, K., & Maki, K. C. (2023). Whole grain intake, compared to refined grain, improves postprandial glycemia and insulinemia: A systematic review and meta-analysis of randomized controlled trials. *Critical Reviews in Food Science and Nutrition, 63*(21), 5339–5357. https://doi.org/10.1080/10408398.2021.2017838

Sathish, T., Khunti, K., Narayan, K. M. V., Mohan, V., Davies, M. J., Yates, T., Oldenburg, B., Thankappan, K. R., Tapp, R. J., Bajpai, R., Anjana, R. M., Weber, M. B., Ali, M. K., & Shaw, J. E. (2023). Effect of conventional lifestyle interventions on type 2 diabetes incidence by glucose-defined prediabetes phenotype: An individual participant data meta-analysis of randomized controlled trials. *Diabetes Care, 46*(11), 1903–1907. https://doi.org/10.2337/dc23-0696

Silva Júnior, A. E. D., Gearhardt, A. N., & Bueno, N. B. (2023). Association between food addiction with ultra-processed food consumption and eating patterns in a Brazilian sample. *Appetite, 186*, 106572. https://doi.org/10.1016/j.appet.2023.106572

Sivri, D., & Akdevelioğlu, Y. (2025). Effect of fatty acids on glucose metabolism and type 2 diabetes. *Nutrition Reviews, 83*(5), 897–907. https://doi.org/10.1093/nutrit/nuae165

Souza, M., Moura, F. S., Lima, L. C. V., & Amaral, M. J. M. (2025). Association between higher consumption of ultra-processed foods and risk of diabetes and its complications: A systematic review & updated meta-analysis. *Metabolism, 165*, 156134. https://doi.org/10.1016/j.metabol.2025.156134

Sui, H., Sun, N., Zhan, L., Lu, X., Chen, T., & Mao, X. (2016). Association between work-related stress and risk for type 2 diabetes: A systematic review and meta-analysis of prospective cohort studies. *PLoS One, 11*(8), e0159978. https://doi.org/10.1371/journal.pone.0159978

Surma, S., Szyndler, A., & Narkiewicz, K. (2023). Deprywacja snu a ryzyko rozwoju i skuteczność leczenia nadciśnienia tętniczego. *Choroby Serca i Naczyń, 20*(1), 1–13. https://doi.org/10.5603/ChSiN.2023.0007

Thorpe, L. E., Adhikari, S., Lopez, P., Kanchi, R., McClure, L. A., Hirsch, A. G., Howell, C. R., Zhu, A., Alemi, F., Rummo, P., Ogburn, E. L., Algur, Y., Nordberg, C. M., Poulsen, M. N., Long, L., Carson, A. P., DeSilva, S. A., Meeker, M., Schwartz, B. S., … Elbel, B. (2022). Neighborhood socioeconomic environment and risk of type 2 diabetes: Associations and mediation through food environment pathways in three independent study samples. *Diabetes Care, 45*(4), 798–810. https://doi.org/10.2337/dc21-1693

Tian, S., Guo, T., Qian, F., Qiu, Z., Lu, Q., Li, R., Zhu, K., Li, L., Yu, H., Li, R., Ou, Y., Pan, A., & Liu, G. (2025). Fish oil, plasma n-3 PUFAs, and risk of macro- and microvascular complications among individuals with type 2 diabetes. *Journal of Clinical Endocrinology & Metabolism, 110*(5), e1687–e1696. https://doi.org/10.1210/clinem/dgae482

Wallerer, S., Stadelmaier, J., Floegel, E., Kiesswetter, E., Bantle, G., Hoffmann, G., & Schwingshackl, L. (2025). Adherence to Mediterranean diet and risk of type 2 diabetes: An updated systematic review and dose-response meta-analysis. *Advances in Nutrition, 16*(12), 100562. https://doi.org/10.1016/j.advnut.2025.100562

Wang, L., Huang, X., Sun, M., Zheng, T., Zheng, L., Lin, X., Ruan, J., & Lin, F. (2024). New light on ω-3 polyunsaturated fatty acids and diabetes debate: A population pharmacokinetic-pharmacodynamic modelling and intake threshold study. *Nutrition & Diabetes, 14*(1), 8. https://doi.org/10.1038/s41387-024-00262-w

Wełnicki, M., Gorczyca-Głowacka, I., Kapłon-Cieślicka, A., Janiszewski, M., Kasprzak, J. D., Leszek, P., Tomaszuk-Kazberuk, A., Łukasiewicz, M., Tomaniak, M., Surma, S., Mickiewicz, A., Narkiewicz, K., Jaguszewski, M., Krasiński, Z., Śliż, D., Mamcarz, A., Szymański, F. M., Barylski, M., Wożakowska-Kapłon, B., & Filipiak, K. J. (2025). Treatment of dyslipidemia in Poland - Common diagnostics, early combined therapy. Expert position statement endorsed by the Polish Cardiac Society Working Group on Cardiovascular Pharmacotherapy. *Cardiology Journal, 32*(2), 107–119. https://doi.org/10.5603/cj.103680

Xu, M., Zheng, J., Ying, T., Zhu, Y., Du, J., Li, F., Chen, B., Liu, Y., & He, G. (2025). Dietary protein and risk of type 2 diabetes: Findings from a registry-based cohort study and a meta-analysis of prospective cohort studies. *Nutrition & Diabetes, 15*(1), 25. https://doi.org/10.1038/s41387-025-00380-z

Ying, T., Zheng, J., Kan, J., Li, W., Xue, K., Du, J., Liu, Y., & He, G. (2024). Effects of whole grains on glycemic control: A systematic review and dose-response meta-analysis of prospective cohort studies and randomized controlled trials. *Nutrition Journal, 23*(1), 47. https://doi.org/10.1186/s12937-024-00952-2

Zhang, L. M., Liu, Z., Wang, J. Q., Li, R. Q., Ren, J. Y., Gao, X., Lv, S. S., Liang, L. Y., Zhang, F., Yin, B. W., Sun, Y., Tian, H., Zhu, H. C., Zhou, Y. T., & Ma, Y. X. (2022). Randomized controlled trial for time-restricted eating in overweight and obese young adults. *iScience, 25*(9), 104870. https://doi.org/10.1016/j.isci.2022.104870

Zhu, Y., & Mi, J. (2025). Unraveling the complex relationship between night shift work and diabetes: Exploring mechanisms and potential interventions. *Frontiers in Public Health, 13*, 1539679. https://doi.org/10.3389/fpubh.2025.1539679

Zick, C. D., Curtis, D. S., Meeks, H., Smith, K. R., Brown, B. B., Kole, K., & Kowaleski-Jones, L. (2023). The changing food environment and neighborhood prevalence of type 2 diabetes. *SSM, Population Health, 21*, 101338. https://doi.org/10.1016/j.ssmph.2023.101338

Chapter 35

Brain insulin signaling and nutritional modulation: the role of microRNAs and the gut–brain axis

Javier Mateu-Fabregat, Nil Novau-Ferré, Laura Panisello-Melo, Jaume Folch and Mònica Bulló
Department of Biochemistry and Biotechnology, School of Medicine and Life Sciences, University Rovira I Virgili, Reus, Spain

List of abbreviations

ADF alternate-day fasting
AhR indole–aryl hydrocarbon receptor
AMPK AMP-activated protein kinase
BA bile acid
BBB blood–brain barrier
CNS central nervous system
DGCR8 DiGeorge syndrome critical region 8
fMRI functional magnetic resonance imaging
FMT fecal microbiota transplantation
FXR farnesoid X receptor
GABA γ-aminobutyric acid
GLP-1 glucagon-like peptide-1
HFD high-fat diet
IF intermittent fasting
IGFR insulin-like growth factor receptor
INSR insulin receptor
IPA indole-3-propionic acid
IRS insulin receptor substrate
KP kynurenine pathway
miRNA microRNA
pre-miRNA precursor microRNA
pri-miRNA primary microRNA
PYY peptide YY
RISC RNA-induced silencing complex
SCFA short-chain fatty acid
T2D type 2 diabetes
TGR5 Takeda G protein-coupled receptor 5
TMAO trimethylamine N-oxide
TRE time-restricted eating
WD Western diet

Introduction

Although the brain was once considered an insulin-insensitive, it is now regarded as an insulin-sensitive organ, with widespread, yet selective, expression of insulin receptors (INSR) in regions such as the olfactory bulb, hypothalamus, hippocampus, cerebellum, amygdala and cerebral cortex. Insulin receptors, insulin-like growth factor receptors, and signaling partners are distributed throughout the brain. Insulin can cross the blood-brain barrier (BBB) and can also be secreted locally within the brain, regulating important functions in the central nervous system (Gray et al., 2014). This is particularly relevant to diet and metabolic health, as nutrient intake, circulating glucose, and peripheral insulin levels influence central insulin signaling. Dysregulation of this system, through poor diet, obesity, or type 2 diabetes (T2D), impairs brain insulin sensitivity, linking metabolic disorders with cognitive decline and neurodegenerative conditions.

Insulin action in the brain

Insulin receptor signaling in the brain is critical for neuronal development, feeding behavior, body weight regulation, and cognitive processes including learning, memory, attention, and executive function (Banks et al., 2012). In the hippocampus and cerebral cortex, INSR regulates neurogenesis, synaptic maintenance, dendritic outgrowth, and synaptic plasticity, supporting memory and

Hormones in Neuroscience. https://doi.org/10.1016/B978-0-443-30263-3.00037-3

neuronal survival (Derakhshan & Toth, 1966; Folch et al., 2018). Reduced brain insulin signaling impairs neurotransmitter availability, long-term potentiation, and is associated with mood disorders, depression, cognitive decline, and neurodegenerative diseases such as Alzheimer's (Steen et al., 2005). INSR in the hypothalamus controls feeding behavior and body weight by modulating neuropeptides such as neuropeptide Y, agouti-related peptide, and POMC (Dodd & Tiganis, 2017). Brain insulin also influences systemic metabolism, including hepatic glucose production, lipolysis, lipogenesis, and glucose sensing via multiple GLUT transporters, with insulin resistance impairing their function and contributing to metabolic and neurodegenerative pathology (Akhtar & Sah, 2020).

Mechanism of insulin receptor signaling in brain

The INSR is a glycoprotein composed of an extracellular α subunit and a β-transmembrane subunit. The binding of insulin to the α subunit induces receptor dimerization and autophosphorylation of β-subunit tyrosines, initiating signaling through IRS-1/2 and regulating GLUT-4 translocation in hippocampal neurons. INSR activation triggers the PI3K/AKT pathway: PI3K phosphorylates phosphatidylinositol, leading to AKT activation, which in turn inactivates GSK-3β. Dysfunction in this pathway promotes GSK-3β activity and tau hyperphosphorylation, contributing to neurofibrillary tangles and cognitive decline. INSR also signals through the Ras/MAPK pathway, including ERK1/2, which phosphorylate CREB to enhance synaptic plasticity, long-term potentiation, and memory formation. Stress-activated JNKs (JNK1-3) are modulated by MAP kinases (MKK4/7) and participate in stress responses, learning, and memory. Chronic JNK activation, commonly seen in obesity, T2D, and other diet-related metabolic disorders, disrupts insulin signaling, impairs glucose metabolism, and contributes to neuroinflammation and cognitive dysfunction. These mechanisms illustrate how peripheral metabolic status, shaped by diet and systemic insulin sensitivity, directly influences brain insulin signaling, neuronal health, cognition, and vulnerability to neurodegenerative disease (Fig. 35.1).

Emerging evidence indicates that diet, microRNAs and gut microbes modulate brain insulin signaling pathways. These factors influence central insulin sensitivity by regulating INSR, IRS, and downstream signaling components, thereby linking nutritional and microbial cues to neuronal insulin responsiveness. Together, these interactions demonstrate how nutrition, systemic metabolism, and the gut–brain axis converge to shape brain insulin function, impacting cognition, neuronal health, and vulnerability to neurodegenerative disease.

Diet-induced modulation of regional brain insulin signaling action

Much is known about the role of diet in regulating peripheral insulin sensitivity, particularly in muscle, adipose tissue, and liver. In contrast, evidence directly linking nutritional factors to central insulin action remains still relatively limited. Nonetheless, emerging findings suggest that dietary strategies and specific dietary components can modulate not only peripheral insulin sensitivity but also central insulin function (Table 35.1).

Energy intake

Accumulating evidence indicates that total energy intake is one of the most direct modulators of brain insulin sensitivity. Within this context, caloric restriction has emerged as a promising dietary intervention, offering broad health benefits such as improved cardiometabolic function, enhanced cognitive performance, and potentially increased lifespan. In rats, ~40% reduction in caloric intake improves peripheral insulin sensitivity and increases hippocampal insulin content compared with ad libitum feeding (Todorovic et al., 2024). Interestingly, both the duration of dietary restriction (3–18 months) and the age at intervention onset differently modulate hippocampal insulin signaling proteins (e.g., INSRβ, IRS-1, AKT). Specifically, short-term and middle-term energy restriction are associated with reduced pAKT levels, whereas long-term restriction exerts a more pronounced effect by lowering INSRβ activation and the pINSRβ/INSRβ ratio. Dietary restriction extends beyond total caloric intake, with strategies like intermittent fasting, which includes approaches such astime-restricted eating (TRE) and alternate-day fasting (ADF). TRE limits food consumption to a specific daily window (e.g., 16 hours fasting followed by 8 hours eating), while ADF alternates fasting periods with days of ad libitum intake (e.g., 5 ad libitum days combined with two non-consecutive fasting days). In male C57BL/6 mice, a specific ADF intervention consisting of 24 hours of fasting followed by 24 hours of ad libitum feeding during 6 weeks modulated metabolic pathways, leading to a downregulation of brain insulin signaling (Lu et al., 2011). Specifically, AKT phosphorylation at Ser473 and its downstream target GSK3β at Ser9 were reduced, suggesting an overall attenuation of insulin pathway activity in the brain compared with mice fed ad libitum. Importantly, a reduction of insulin pathway proteins does not directly imply impaired insulin sensitivity. Rather, the net effect depends on insulin transport, availability and receptor activity. Extending an ADF regimen of alternating feeding and fasting days for up to 28 days in diabetic mouse models (homozygous $Lepr^{db/db}$ mice) has been proven to modulate hippocampal insulin signaling (Liu et al., 2020). This was

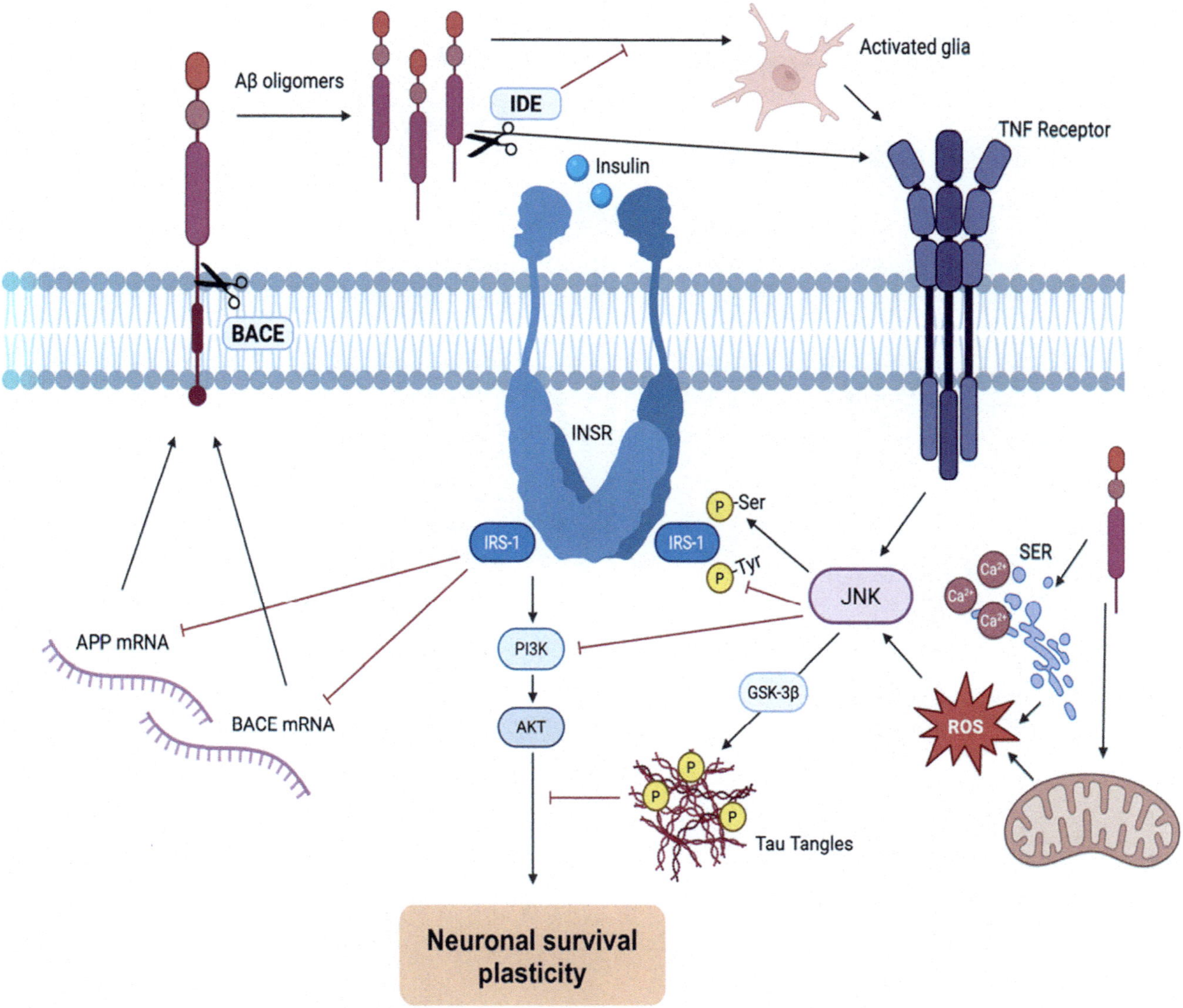

FIGURE 35.1 **Insulin signaling dysregulation and its interaction with amyloid-β (Aβ) and tau pathology in neuronal cells.** Insulin action in hippocampus promotes neuronal survival and it also regulates both central and peripheral functions in the brain. Consequences of this include reduced glucose metabolism and increased oxidative stress which modulates APP and JNK activity. Moreover, reduced GSK3β phosphorylation leads to upregulation in TAU phosphorylation and Aβ formation. Likewise, Aβ promotes the activation of microglia increasing the levels of cytokines, mainly TNFα that activates JNK that subsequently inhibits the brain IR.

evidenced by increased IRS-1 phosphorylation at Tyr896 and enhanced AKT activation. These molecular changes were accompanied by upregulation of brain-derived neurotrophic factor (BDNF), suppression of NFκB activity, and downregulation of the JNK/p38 pathway, compared with diabetic mice with no ADF regimen. Notably, energy intake did not differ between groups. However, when ADF-treated diabetic mice were compared with non-diabetic controls (heterozygous Lepr$^{db/m}$ mice), no significant differences were observed, suggesting that the beneficial effects of ADF on hippocampal insulin signaling may be specific to diabetic conditions and not generalized to healthy animals. Importantly, the beneficial effects of caloric restriction could diminish with shorter intervention durations and could even become detrimental if initiated later in life for only a brief period.

Preclinical and clinical evidence indicate that chronic overnutrition and sustained positive energy balance impair brain insulin signalling. Functional magnetic resonance imaging (fMRI) with intranasal insulin administration shows that even short-term overeating can reduce insulin responsiveness. For instance, a 5-day hypercaloric diet

TABLE 35.1 Summary of the studies on diet-induced modulation of regional brain insulin signaling action.

Type of intervention		Population	Description of the intervention	Comparator	Duration	Effect on brain insulin signaling	Other effects	References
Energy intake	Hypocaloric intake	Male Wistar rats ($n = 8$ per group)	Dietary restriction of 40% the average daily food intake.	Age-matched controls fed ad libitum with standard laboratory chow pellets	LTDR (12 and 18 months), MTDR (6 months) and STDR (3 months)	STDR and MTDR are associated with reduced pAKT levels, whereas LTDR exerts an effect by lowering IRβ activation and the pIRβ/IRβ ratio.	Dietary restriction increases hippocampal insulin content and improves peripheral insulin sensitivity (as assessed by HOMA-IR and QUICKI)	Todorovic et al. (2024)
	Hypercaloric intake	Men ($n = 18$ in the intervention group and $n = 11$ as controls)	High-caloric diet (1500 kcal from ultra-processed snacks in addition to the regular diet)	Regular diet	5 days of high-caloric diet, followed by 7 days of regular diet	Administration of intranasal insulin combined with fMRI directly after the intervention showed higher insulin activity in the right insular cortex, left rolandic operculum and right midbrain/pons, while lower brain insulin activity in the right hippocampus and bilateral fusiform gyrus 7 days after resuming regular diet compared with the control group.	The group with high-caloric diet displayed an increase in liver fat after the 5-day hypercaloric period.	Kullmann et al. (2025)
Intermittent fasting (IF)	Alternate day fasting (ADF)	Male C57BL/6 mice ($n = 5$ per group)	ADF (24 hours of fasting and 24 hours of food access) and ADF with antioxidant-supplemented chow	Ad libitum diet	Six weeks	AKT phosphorylation at Ser473 and its downstream target GSK3β at Ser9 were reduced in brain in the ADF groups.	Pathways that mediate mitochondrial biogenesis were not activated; AMPK phosphorylation, SIRT1 phosphorylation, PGC1a levels, and COX4 levels did not change.	Lu et al. (2011)
	ADF	Male homozygous $Lepr^{db/db}$ mice ($n = 10$ per group)	ADF (24 hours of fasting and 24 hours of food access)	Homozygous $Lepr^{db/db}$ or heterozygous $Lepr^{db/m}$ mice fed ad libitum	28 days	Increased IRS1 phosphorylation at Tyr896 and enhanced AKT activation in hippocampus	Upregulation of BDNF, suppression of NFκB activity, and downregulation of the JNK/p38 pathway, compared with diabetic mice with no ADF regimen. No significant differences were found when comparing ADF-treated diabetic mice with controls.	Liu et al. (2020)

Dietary patterns	Western diet (WD)	Male C57BL/6J ($n = 5$ per group)	WD (30% fat, 16.3% sugar)	Control diet	Starting from 13 weeks of age until they reached 20 months	Increase in p-IRS-1(Ser616) starting from 8 months of age, which indicates the development of insulin resistance in the entorhinal cortex but not in the hippocampus after 5 months of WD	Increased glucose levels in 4-month-old mice fed a WD, and higher insulin levels in the WD group observed in the 12- and 16-month-old groups of animals. The concomitant increase in HOMA-IR in the 12-month-old mice relative to the control group indicates peripheral insulin resistance.	Mietelska-Porowska et al. (2022)
Macronutrients	High-fat diet (HFD)	Male rats ($n = 12$ per group)	HFD (40% fat)	Low-fat diet (4% fat)	Chronic experiment (70 days) and acute experiment (3 days)	Insulin administration into the 3rd cerebral ventriclei nduced increases in phosphorylation of AKT in the animals maintained on the low-fat diet, but not in the animals maintained on the HFD diet, suggesting that in the presence of a HFD, insulin is less effective to activate its downstream signaling cascade.	Insulin administered into the 3rd-cerebral ventricle decreased food intake and body weight of low-fat diet rats but had no effect on HFD rats in either the chronic or the acute experiment.	Clegg et al. (2011)
	High-fat and high-fructose diet	Male rats ($n = 20–22$ per group)	HFD (10% lard) and 20% of high-fructose corn syrup in water	Standard chow	7 days	Reduced hippocampal IRS-1 and AKT phosphorylation, indicating that 7 days of high-fat and high-fructose diet are enough to modify hippocampal insulin signaling	Intraperitoneal insulin injection significantly reduced blood glucose levels, increased leptin levels in animals fed a high-fat and high-fructose diet as compared with controls.	Calvo-Ochoa et al. (2014)
	High-fat, high-fructose and sucrose diet	Male C57BL/6NHsd mice ($n = 8$ per group)	HFD (40% fat) and sugary water (42 g/L) made of 55% of fructose and 45% of sucrose	Normal chow diet (12% lipid)	14 weeks	Decreased levels of IR phosphorylation at Tyr1150/1151, increased phosphorylation in IRS-1 at Ser^{307} and $Ser^{336/339}$ and hyperactivation of AKT at Ser^{473} not at Thr^{308} in whole brain of mice fed a high-fat, high-fructose and sucrose diet compared to controls.	Increase in circulating HOMA-IR index, decreased expression of plasma membrane GLUT1 and GLUT3 and activation of the major components of the inflammatory pathways (increased phosphorylation of MAPK, p38, and JNK and higher expression of NFκB) in whole brain compared to controls.	Kothari et al. (2017)

Continued

TABLE 35.1 Summary of the studies on diet-induced modulation of regional brain insulin signaling action.—cont'd

Type of intervention		Population	Description of the intervention	Comparator	Duration	Effect on brain insulin signaling	Other effects	References
	High-fructose diet	Female Sprague–Dawley rats ($n = 12$ per group)	Standard diet and fructose supplementation (10% w/v in water)	Control group with no supplementation or isocaloric-glucose supplementation	7 months	IRS-2 and AKT protein expression in the frontal cortex was reduced in fructose-supplemented rats. A trend toward reduced AKT phosphorylation in the hippocampus was also observed. These effects were not found in isocaloric glucose-supplemented rats, suggesting a fructose-specific impairment of brain insulin signaling independent of caloric intake	Fructose-supplemented rats showed significant hyperinsulinemia and an increase in body weight at the end of treatment period and displayed postprandial and fasting hypertriglyceridemia	Sangüesa et al. (2018)

disrupts insulin signalling in key regions like the hippocampus, as indicated by blunted insulin-induced changes in cerebral blood flow compared to controls, and these impairments persisted even after the intervention ended (Kullmann et al., 2025).

Overeating appears to induce an initial hyperactivation of brain insulin pathways, which over time progresses to central insulin resistance. Consequently, neuronal responsiveness to insulin is impaired, contributing to metabolic disturbances and an elevated risk of cognitive decline. Importantly, these central alterations emerge even when peripheral insulin sensitivity remains unaffected, suggesting that brain insulin dysfunction may precede detectable changes in body weight, systemic insulin sensitivity, and broader metabolic function. Of note, the impact of overeating on brain insulin signalling appears to depend not only on total energy intake but also on macronutrient composition. Yet, isolating the contribution of individual nutrients driving these effects remains challenging. In this context, dietary patterns, which integrate both nutrient quality and overall eating behaviours, offer a more comprehensive framework for understanding how diet influences brain insulin function.

Dietary patterns and macronutrients

The Western Diet (WD) is characterized by high intake of sugar, saturated fats, and salt. In male mice, a WD (30% fat, 16% sugar) led to increased phosphorylation of IRS-1 at Ser616 after ~5 months compared with standard-fed controls. This alteration, indicative of brain insulin resistance, was observed in the entorhinal cortex but not consistently in the hippocampus (Mietelska-Porowska et al., 2022). Complementary evidence comes from studies combining high-sugar and high-fat diets (HFD). In rats, just 7 days of a high-fat-and-fructose diet reduced hippocampal IRS-1 and AKT phosphorylation (Calvo-Ochoa et al., 2014). Similarly, 7 months of fructose supplementation significantly reduced IRS-2 protein expression and AKT phosphorylation in the frontal cortex, with a similar trend in the hippocampus. Notably, these effects were not found in isocaloric glucose-supplemented rats, suggesting a fructose-specific impairment of brain insulin signaling independent of caloric intake (Sangüesa et al., 2018). Other models showed that 12 weeks of a HFD with sugary water decreased IRS-1 phosphorylation at Tyr1150/1151, indicating inactivation and downstream insulin signaling disruption in brain. Importantly, HFD also activates neuroinflammatory pathways in the brain, with NFKB, MAPK, p38 and JNK mediators differently upregulated, indicating neuroinflammation as a mechanism linking dietary fat to disrupted brain insulin signaling (Kothari et al., 2017). Consistently, in HFD-fed rats, insulin administration into the third cerebral ventricle failed to induce AKT phosphorylation or reduce food intake and body weight, demonstrating that HFD diminishes central insulin effectiveness by impairing downstream signaling (Clegg et al., 2011).

While overeating and high-fat diets are well-stablished contributors to obesity, as well as systemic and central insulin resistance, tailored nutritional strategies such as dietary restriction, optimized macronutrients distribution, and intermittent fasting require further investigation to determine their efficacy in preventing or reversing diet-induced impairments in insulin signaling and to clarify their long-term impact on brain metabolism.

miRNA as mechanisms of nutritional regulation of brain insulin signaling

The biological and molecular mechanisms through which diet modulates insulin sensitivity remain an active area of research. In particular, dietary intake can leave a molecular imprint with notably effects on gene regulation by influencing noncoding RNAs, particularly microRNAs (miRNAs). As master regulators of gene expression, miRNAs orchestrate critical processes such as proliferation, cell cycle progression, and apoptosis. Accordingly, dietary-driven changes in miRNA expression can affect genes involved in insulin-related protein synthesis, secretion, and signaling (Capetini et al., 2024).

miRNA and their regulatory functions

miRNAs are small non-coding RNAs, 18–25 nucleotides in length, that regulate gene expression. Most miRNAs are transcribed into primary miRNAs (pri-miRNAs) by RNA polymerase II and processed into precursor miRNAs (pre-miRNAs) by the microprocessor complex, which includes the nuclear RNase III Drosha and the DiGeorge Syndrome Critical Region 8 (DGCR8). Pre-miRNAs are cleaved by Dicer to form mature miRNAs, which guide the RNA-induced silencing complex (RISC) to target mRNAs. Binding occurs via the miRNA seed region (nucleotides 2–8) and can silence gene expression through translation inhibition, mRNA destabilization, or degradation. Perfect complementarity leads to degradation of both miRNA and mRNA, whereas imperfect pairing blocks translation without mRNA decay. Each miRNA can target multiple mRNAs, and conversely, a single mRNA may be regulated by multiple miRNAs, highlighting their extensive regulatory potential (O'Brien et al., 2018). Growing evidence indicates that miRNAs interact with insulin signaling-related protein expression, supporting their role as both regulators and potential biomarkers in the early stages of insulin resistance and related complications (Ağagündüz et al., 2025) (Fig. 35.2).

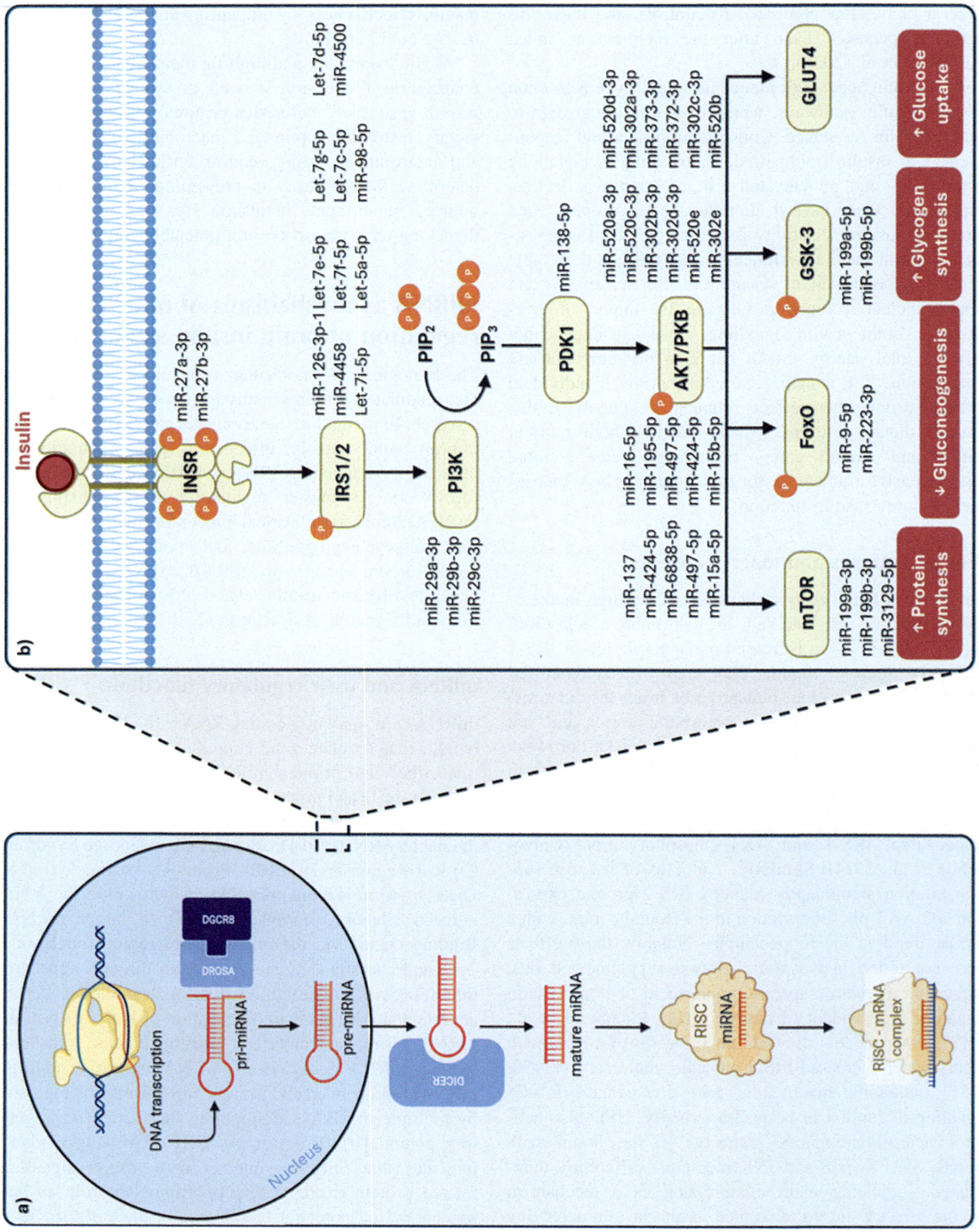

FIGURE 35.2 miRNA biogenesis (a) and miRNAs targeting insulin signaling-related protein expression (b), according to TargetScan.

Cross-tissue regulation of insulin signaling by miRNAs

miRNAs are key modulators of systemic insulin function, as they directly regulate insulin-related protein expression. Their influence is further amplified by their activity across multiple tissues and their stable transport through body fluids such as blood, urine, milk, saliva, and cerebrospinal fluid, enabling coordinated regulation of insulin signaling throughout the body. Unlike other RNA species, miRNAs are extracellularly stable, circulating between tissues and organs. This stability is facilitated by their short length and protection from ribonucleases by carriers such as membrane-derived vesicles, proteins, lipoproteins, and ribonucleoprotein complexes. Delivery to recipient cells occurs through direct fusion, internalization, or receptor-mediated interactions, allowing them to modify cell function and acting as extracellular signaling molecules (D'Amico et al., 2024).

Disruption of glucose and insulin metabolism affects not only the pancreas and the adipose tissue, but also brain function. As discussed earlier, proper insulin signaling is essential for brain health. miRNAs can be transported along the brain-fat axis, a bidirectional network connecting metabolic tissues with the central nervous system. In this context, they may play a central role in mediating cross-talk, regulating pathways involved in appetite, lipid homeostasis, inflammation and BBB integrity (Wu et al., 2025).

Potential therapeutics targeting miRNAs

The key role of miRNAs in regulating the expression of proteins within the insulin signaling cascade, affecting both insulin resistance and brain function, makes them attractive therapeutic targets. Two main strategies have been developed depending on the miRNA's role: restoring the expression of downregulated miRNAs using synthetic miRNA mimics or viral expression vectors and inhibiting overexpressed miRNAs via antisense oligonucleotides (anti-miRNA) or competitive miRNA inhibitors (miRNA sponges) (Palihaderu et al., 2022).

miRNAs also have several characteristics that make them ideal biomarkers, including sensitivity, predictive value, specificity, robustness, and non-invasiveness. However, some challenges remain for clinical applications, such as experimental variability due to differences in sample collection, storage, RNA isolation and accurate quantification. In conclusion, miRNAs are promising targets for the prevention, diagnosis and treatment of cognitive dysfunction associated with disrupted insulin signaling (Chakraborty et al., 2021).

Dietary interventions and miRNA modulation

As discussed in a previous section on this chapter, nutrition plays a significant role in modulating insulin signaling, with important implications for brain function. Certain nutrients and phytochemicals can alter miRNA expression and activity, offering a natural mechanism for dietary interventions to influence miRNA-mediated regulation. By targeting miRNAs involved in insulin signaling, dietary strategies may help prevent or manage cognitive dysfunction associated with insulin resistance (Yu et al., 2021).

Accumulating evidence indicates that diet-responsive miRNAs play a pivotal role in the post-transcriptional regulation of insulin signaling (Table 35.2). Saturated fatty acids, such as palmitic acid, as well as high-fat diets, have been shown to modulate miR-195, miR-155-5p, miR-141-3p, and miR-19a-3p, thereby altering key nodes of the insulin signaling cascade, including INSR, PI3K/AKT, and GSK3β, and contributing to impaired glucose tolerance and inflammatory responses (Prévost et al., 2025; Rojas-Criollo et al., 2024; Wang et al., 2025; Yang et al., 2014). In contrast, diets rich in simple carbohydrates reduce the expression of miR-200c-3p and miR-126a-3p, favoring the development of insulin resistance (Barbosa de Queiroz et al., 2022). Conversely, polyphenol supplementation, such as chlorogenic acid, influences miR-146a and miR-363-3p expression, which is associated with improved insulin sensitivity through attenuation of pro-inflammatory cytokines and activation of the PI3K/AKT pathway (Alenezi et al., 2025; Shu et al., 2020). Collectively, these observations highlight the potential of dietary modulation to shape insulin-related processes via the regulatory activity of specific miRNAs.

The role of the gut-brain axis and microbiota in modulating central insulin response

Overview of the gut-brain axis

Since the mid-nineteenth century, a bidirectional influence of the gastrointestinal tract on brain function has been recognized, with the gut–brain axis now understood as a homeostatic communication pathway involving neural, hormonal, and immunological mechanisms. The gut microbiota, mainly composed of bacteria but also viruses, protozoa, and fungi, reflects a long coevolution with the host. Research on gut microbes and their metabolites has accelerated due to the accessibility of stool and blood samples.

Diet, gut microbiota, and brain interactions are central to regulating host metabolism, particularly glucose and

TABLE 35.2 Dietary modulation of miRNAs targeting insulin signaling.

Dietary intervention	miRNA	Effects on insulin sensitivity	References
Saturated fatty acids			
HepG2 cells treated with palmitic acid (0.5 mM) for 18h. C57BL/6J mice fed a HFD (60% calories from fat) for 14 weeks.	↑ miR-195	↓ INSR, pINSR, pIRS-1, pAKT2, pGSK3β ↑ Glycogen synthesis	Yang et al. (2014)
C57BL/6J mice fed a HFD (45% calories from fat) for 3 days, 8 days or 8 weeks.	↑ miR-155-5p	↑ Neuroinflammation ↑ Glucose and insulin intolerance ↑ IL1β, TNFα, NFκb, and TLR4 in hypothalamus	Prévost et al. (2025)
C57BL/6 mice fed a HFD (60% calories from fat) for 8 weeks. Mouse hepatocytes (AML12 cells) transfected with extracellular vesicles containing miR-141-3p inhibitor.	↓ miR-141-3p	↓ Glucose and insulin tolerance ↓ Insulin stimulated glucose uptake ↓ AKT phosphorylation	Wang et al. (2025)
C57BL/6J mice fed a HFD (60% calories from fat) for 6 months.	↑ miR-19a-3p, miR-34a-5p, miR-146a-5p, miR-29c-3p, miR-338-3p, miR-125b-5p	↓ Glucose and insulin tolerance	Rojas-Criollo et al. (2024)
Simple carbohydrates			
Wistar rats fed a high sugar diet (68% carbohydrates) for 4-, 8, or 12-week.	↓ miR-200c-3p, miR-126a-3p	Insulin resistance	Barbosa de Queiroz et al. (2022)
Polyphenols			
Chlorogenic acid administered at 10 and 100 mg/kg/day for the last 4 weeks of a 12-week HFD protocol in Sprague Dawley rats.	↑ miR-146a	↓ Fasting glucose and insulin ↓ Pro-inflammatory cytokines	Alenezi et al. (2025)
C57BL/6J mice fed a HFD for 8 weeks, followed by 6 weeks of resveratrol treatment	↑ miR-363-3p	Improved PI3K/AKT signaling pathway	Shu et al. (2020)

energy homeostasis. However, it was not until 2010 when compositional changes in the intestinal microbiota were linked to T2D (Larsen et al., 2010). Maintaining microbial balance is essential for gut barrier integrity and for preserving systemic and brain insulin sensitivity through the regulation of intestinal and BBB permeability.

Microbiota as modulator of insulin metabolism

Gut microbial composition

Dysbiosis refers to an imbalance in the gut microbiota composition, characterized by reduced microbial diversity, loss of beneficial species, or overgrowth of potentially harmful microorganisms. The composition of the gut microbiota is regulated by factors such as diet, host genetics, immune responses, and microbial interactions. Dietary changes, particularly in macronutrient composition, can rapidly alter bacterial and fungal populations, leading to physiological consequences, including gastrointestinal, and metabolic disorders.

Studies by Craciun and collaborators have shown that individuals with T2D exhibit significantly lower levels of Firmicutes and higher levels of Bacteroidetes (Craciun et al., 2022). When microbial homeostasis is disrupted, dysbiosis can reduce the expression of gut tight junction proteins, induce intestinal inflammation, and increase permeability, often referred to as a "leaky gut". These changes facilitate the translocation of lipopolysaccharides and other harmful metabolites into circulation, triggering metabolic endotoxemia. Subsequent activation of inflammatory pathways elevates cytokine production, interfering with insulin signaling, particularly by promoting serine phosphorylation of IRS-1, thereby impairing its function (Amabebe et al., 2020).

Microbiota-derived metabolites

Gut microbiota produces a wide range of metabolites that act as key signaling molecules influencing host physiology (Table 35.3). Among these, short-chain fatty acids (SCFAs) such as acetate, propionate, and butyrate, produced through fermentation of indigestible carbohydrates, contribute up to 10% of the host's daily energy supply and act as signaling molecules. SCFAs stimulate the secretion of satiety hormones including glucagon-like peptide-1 (GLP-1) and peptide YY (PYY), thereby regulating appetite and glucose homeostasis. They also attenuate proinflammatory cytokine release in adipose tissue, enhance glucose uptake, and promote glycogen storage in skeletal muscle through AMP-activated protein kinase (AMPK) activation and GPCR-mediated signaling, supporting improved metabolic homeostasis and peripheral insulin responsiveness (Yamashita et al., 2009).

In addition to SCFAs, bile acids (BAs), traditionally recognized for their role in digestion, also act as signaling molecules that influence glucose homeostasis, lipid metabolism, and brain function. Synthesized in hepatocytes from cholesterol and modified by gut bacteria, BAs activate the nuclear receptor farnesoid X receptor (FXR) and the membrane-bound Takeda G protein-coupled receptor 5 (TGR5), resulting in downregulation of gluconeogenic genes, promotion of glycogen synthesis, and stimulation of enteroendocrine hormone secretion, including GLP-1. Both FXR and TGR5 are expressed in the brain, where their activation modulates neuroinflammation and synaptic plasticity (De Aguiar Vallim et al., 2013).

Additionally, microbiota-derived neurotransmitters, such as γ-aminobutyric acid (GABA) and serotonin, along with tryptophan derivatives, further regulate neuroinflammation, serotonergic signaling, and central nervous system activity. GABA, enhances insulin sensitivity by upregulating IRS-1 and AKT, facilitating GLUT4 translocation (Rezazadeh et al., 2021). Islet-derived serotonin supports β-cell proliferation, stimulates insulin secretion, and reduces oxidative stress, thereby reinforcing metabolic control in diabetic states. On the other hand, tryptophan metabolism, a key component of host–microbiota interactions, proceeds through three main pathways: the indole–aryl hydrocarbon receptor (AhR) axis, the kynurenine pathway (KP), and the serotonin (5-HT) pathway (Agus et al., 2018). Microbial conversion of tryptophan into indoles stimulates GLP-1 secretion, enhancing

TABLE 35.3 Microbial-derived main metabolites and impact in insulin signaling pathways.

Metabolite	Representative substances	Bacteria and genera	Metabolic and physiological effects	References
Inversely associated with insulin resistance				
SCFAs	Acetate, propionate, butyrate	Predominantly *Bacillota*, Lachnospiraceae, Ruminococcaceae, Oscillospiraceae	↑ GLP-1 & PYY secretion ↑ glucose uptake & glycogen storage ↓ proinflammatory cytokines	Mukhopadhya & Louis (2025); Yamashita et al. (2009)
Bile acids	Deoxycholic acid, lithocholic acid	Predominantly *Bacteroides, Clostridium, Escherichia, Eggerthella*	↑ FXR & TGR5; modulation of neuroinflammation & synaptic plasticity ↓ gluconeogenesis ↑ glycogen synthesis ↑ GLP-1 secretion	De Aguiar Vallim et al. (2013); Doden and Ridlon (2021)
Tryptophan metabolites	Indoles, indole-3-propionic acid (IPA), kynurenine, serotonin (5-HT)	Predominantly *Bacteroides, Clostridium, Bifidobacterium, Lactobacillus*	↑ GLP-1 & ↑ β-cell insulin release ↑ Neurotransmitter synthesis (serotonin)	Agus et al. (2018); Hou et al. (2023); Liu et al. (2020)
Positively associated with insulin resistance				
TMAO	TMAO	Predominantly Firmicutes, Proteobacteria	↑ Neuroinflammation (microglia, astrocyte activation) ↓ insulin secretion ↑ β-cell dysfunction & disrupt glucose metabolism	Govindarajulu et al. (2020); Qiao et al. (2023); Romano et al. (2015); Zhou et al. (2023)

pancreatic β-cell insulin release. In contrast, chronic low-grade inflammation, as observed in obesity, overactivates the KP, impairing insulin biosynthesis and disrupting peripheral signaling. Although gut-derived 5-HT cannot cross the BBB, it contributes to peripheral metabolic regulation by modulating satiety and energy balance and, along with indole-3-propionic acid (IPA), responds to interventions such as intermittent fasting in db/db mice (Liu et al., 2020). Collectively, these microbiota-derived metabolites serve as critical messengers linking the gut to systemic and brain insulin signaling, energy balance, and overall host health, supporting metabolic homeostasis and peripheral insulin responsiveness.

Trimethylamine N-oxide (TMAO) provides another example of microbiota-derived signaling with systemic effects. Elevated TMAO levels in diabetes were first reported in 1998, when observed higher urinary concentrations in diabetic patients (Liu et al., 2020). TMAO is primarily produced by gut microbiota metabolism from nutrients such as choline, L-carnitine, phosphatidylcholine, and betaine, which are abundant in red meat. Age-related dysbiosis often favors TMAO-producing taxa, leading to elevated systemic levels. Elevated TMAO levels have been associated with neuroinflammation, activating microglia and astrocytes (Qiao et al., 2023), and contributing to neuronal loss and synaptic dysfunction via mTOR pathway signaling (Zhou et al., 2023). In peripheral tissues, TMAO impairs insulin secretion, promotes β-cell dysfunction, and disrupts glucose metabolism through PERK–FoxO1 signaling, whereas reducing TMAO levels preserves β-cell function and improves glucose tolerance in preclinical models (Govindarajulu et al., 2020).

Overall, microbial metabolites serve as a critical link between diet, gut microbiota, and systemic metabolism. While most evidence is in peripheral tissues, certain pathways also affect insulin signaling in specific brain regions, suggesting a potential role for gut-derived signals in region-specific regulation of brain insulin function.

Targeting the gut microbiota through nutrition and microbiome interventions

Modulating the intestinal microbiota is an established therapeutic strategy for recurrent *Clostridium difficile* infection in patients with Crohn's disease. By contrast, its potential therapeutic roles in brain and mental conditions (Novau-Ferré et al., 2025) and in metabolic disorders such as insulin resistance remain under active investigation. Diet is a major determinant of gut microbial composition and plays a critical role in shaping the gut–brain axis. High-fiber diets, particularly those rich in whole grains, vegetables, and fruits, foster the growth of beneficial gut bacteria and increase SCFA production, thereby promoting metabolic health. In contrast, diets rich in saturated fats and refined sugars can induce dysbiosis and contribute to insulin resistance.

Beyond general dietary patterns, microbiome-targeted strategies such as biotics and related supplements show promise for managing metabolic disorders and improving insulin sensitivity. Probiotics, administered through fermented foods or supplements, have been associated with reductions in glycated hemoglobin and improvements in pancreatic β-cell function (Govindarajulu et al., 2020). Prebiotics have also demonstrated efficacy in improving glycemic control and reducing insulin resistance. For example, phloretin, an apple-derived phytonutrient, has been shown in mouse models to exert prebiotic effects with metabolic benefits (Shelke et al., 2024).

In addition to pre- and probiotics, more experimental approaches such as fecal microbiota transplantation (FMT) are being investigated for their metabolic benefits. FMT has shown potential to enhance insulin sensitivity in patients with metabolic syndrome, and early evidence suggests that combining it with daily low-fermentable fiber supplementation may further amplify these effects by reshaping microbial composition and modulating the enteroendocrine axis (Vrieze et al., 2012). Nonetheless, the long-term efficacy, safety, and standardized protocols for FMT in metabolic disease remain to be determined.

Translational and clinical perspectives: future directions

The growing understanding of how insulin works within the brain, particularly regulating both metabolic and cognitive processes, opens promising avenues for translational applications in clinical settings. One such approach is intranasal insulin delivery, which bypasses the BBB and directly targets central insulin pathways through the olfactory and trigeminal nerve pathways. Early clinical trials have shown potential benefits for cognition and metabolic regulation, although long-term efficacy, safety, and individual variability require further investigation. Beyond direct insulin delivery, targeting specific miRNAs may enhance insulin sensitivity, regulate energy balance, and support cognitive function. Clinically, miRNA-based approaches could serve as biomarkers for early detection, therapeutic targets to restore insulin signaling, or adjuncts to personalized nutrition. Interventions targeting the gut microbiota, including prebiotics, probiotics and symbiotic, offer scalable non-invasive ways to influence miRNA expression and central insulin pathways. By shaping gut–brain signaling, these microbiota-targeted approaches

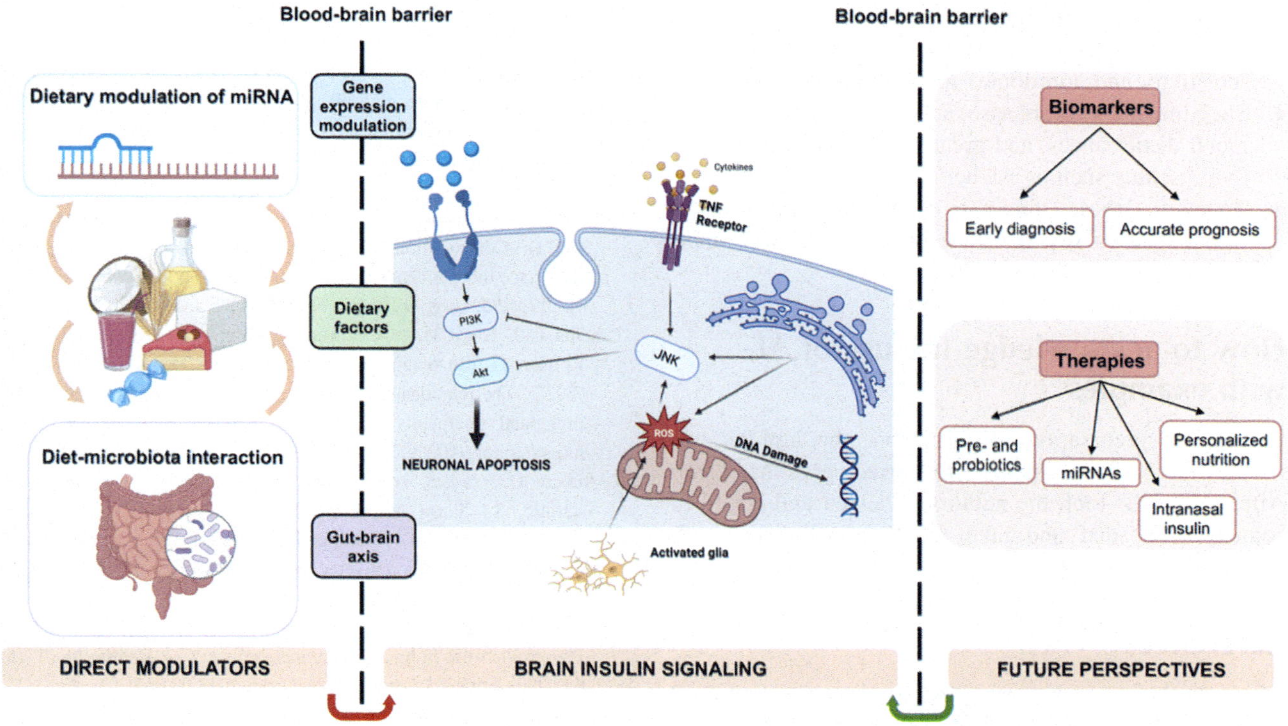

FIGURE 35.3 Interactions between diet, miRNA regulation and gut microbiota in brain insulin signaling: Current insights and future directions.

may enhance insulin sensitivity, metabolic regulation and cognitive function. Translating these strategies into practice will require rigorous validation in humans, optimization of delivery methods, and careful assessment of safety and long-term efficacy.

Despite these advances, implementing these interventions at the community level remains challenging compared with nutritional strategies. Personalized nutrition, guided by metabolic phenotyping and multi-omics approaches, including epi-transcriptomics and microbiome profiling, is transforming dietary interventions. Such tailored approaches move beyond one-size-fits-all solutions, offering potential for disease prevention and management both clinically and at the population level, where nutritional modulation could complement or even replace pharmacological treatments (Fig. 35.3).

Translating these insights into real-word applications requires addressing several critical gaps of knowledge. Region-specific human studies are required to validate preclinical findings and better understand the role of central insulin signaling in disease progression. Reliable biomarkers reflecting central insulin activity, ideally detectable through less-invasive sampling methods, such as blood, saliva or urine, are essential for precise diagnosis, monitoring and design of tailored interventions. Factors such as sex, age and ethnicity differences, which likely influence responses to dietary or insulin-based interventions, remain underexplored. Finally, longitudinal studies are essential to assess the sustained impact of dietary modulation on cognitive and metabolic health over time.

Moving forward, bridging clinical research with personalized, nutrition-based approaches holds significant potential, not only for managing disease, but also for promoting lifelong brain health and overall well-being.

Summary points

- The brain, once considered insulin-insensitive, is now recognized as an insulin-sensitive organ with insulin receptors distributed in regions such as the hippocampus, hypothalamus, cortex, cerebellum, and amygdala.
- Insulin receptors, insulin-like growth factor receptors, and signaling partners like IRS-1 and IRS-2 are widely expressed throughout the brain, enabling insulin regulation of central nervous system functions.
- Insulin can cross the blood–brain barrier and is also secreted locally in the brain, where it regulates cognition, metabolism, and neuronal survival.
- Central insulin signaling is shaped by diet, circulating glucose, and peripheral insulin levels, creating a link between systemic metabolism and brain function.

- Disruption of this signaling, often caused by obesity, type 2 diabetes, or poor diet, impairs brain insulin sensitivity and contributes to cognitive decline
- Such impairments are associated with reduced memory, mood disturbances, and greater risk of neurodegenerative diseases such as Alzheimer's.
- Together, these findings highlight brain insulin signaling as a key interface between nutrition, metabolism, and cognitive health.

How to acknowledge the use of AI, with examples

During the preparation of this work the authors used ChatGPT 3.5, from OpenAI, to correct grammar errors. After using this tool, the authors reviewed and edited the content as needed and take full responsibility for the publication's content.

References

Ağagündüz, D., Çelik, M. N., Deniz Güneş, B., Atabilen, B., Sarikaya, B., Icer, M. A., & Budán, F. (2025). Involvement of miRNAs in the cluster of metabolic factors of MetS: Nutrition-genome-MetS axis. *Journal of Clinical Medicine, 14*(12). https://doi.org/10.3390/JCM14124234

Agus, A., Planchais, J., & Sokol, H. (2018). Gut microbiota regulation of tryptophan metabolism in health and disease. *Cell Host & Microbe, 23*(6), 716–724. https://doi.org/10.1016/j.chom.2018.05.003

Akhtar, A., & Sah, S. P. (2020). Insulin signaling pathway and related molecules: Role in neurodegeneration and Alzheimer's disease. *Neurochemistry International, 135*. https://doi.org/10.1016/j.neuint.2020.104707

Alenezi, R. F., Abdelkhalek, A., El-Sayed, G., Pet, I., Ahmadi, M., Sherbini, E. S. El, Puşcaşiu, D., & Arisha, A. H. (2025). A natural polyphenol, chlorogenic acid, attenuates obesity-related metabolic disorders in male rats via miR-146a-IRAK1-TRAF6 and NRF2-mediated antioxidant pathways. *Biomolecules, 15*(8), 1086. https://doi.org/10.3390/BIOM15081086

Amabebe, E., Robert, F. O., Agbalalah, T., & Orubu, E. S. F. (2020). Microbial dysbiosis-induced obesity: Role of gut microbiota in homoeostasis of energy metabolism. *The British Journal of Nutrition, 123*(10), 1127–1137. https://doi.org/10.1017/S0007114520000380

Banks, W. A., Owen, J. B., & Erickson, M. A. (2012). Insulin in the brain: There and back again. *Pharmacology & Therapeutics, 136*(1), 82–93. https://doi.org/10.1016/j.pharmthera.2012.07.006

Barbosa de Queiroz, K., Evangelista, E. A., & Guerra-Sa, R. (2022). Metabolic imprinting induced by a high-sugar diet: Effects on microRNA expression and insulin resistance in young rats. *Molecular Biology Reports, 49*(8), 8173–8178. https://doi.org/10.1007/S11033-022-07473-3

Calvo-Ochoa, E., Hernández-Ortega, K., Ferrera, P., Morimoto, S., & Arias, C. (2014). Short-term high-fat-and-fructose feeding produces insulin signaling alterations accompanied by neurite and synaptic reduction and astroglial activation in the rat hippocampus. *Journal of Cerebral Blood Flow and Metabolism, 34*(6), 1001–1008. https://doi.org/10.1038/JCBFM.2014.48

Capetini, V. C., Quintanilha, B. J., Garcia, B. R. E. V., & Rogero, M. M. (2024). Dietary modulation of microRNAs in insulin resistance and type 2 diabetes. *Journal of Nutritional Biochemistry, 133*. https://doi.org/10.1016/j.jnutbio.2024.109714

Chakraborty, C., Sharma, A. R., Sharma, G., & Lee, S. S. (2021). Therapeutic advances of miRNAs: A preclinical and clinical update. *Journal of Advanced Research, 28*, 127–138. https://doi.org/10.1016/J.JARE.2020.08.012

Clegg, D. J., Gotoh, K., Kemp, C., Wortman, M. D., Benoit, S. C., Brown, L. M., D'Alessio, D., Tso, P., Seeley, R. J., & Woods, S. C. (2011). Consumption of a high-fat diet induces central insulin resistance independent of adiposity. *Physiology & Behavior, 103*(1), 10. https://doi.org/10.1016/J.PHYSBEH.2011.01.010

Craciun, C. I., Neag, M. A., Catinean, A., Mitre, A. O., Rusu, A., Bala, C., Roman, G., Buzoianu, A. D., Muntean, D. M., & Craciun, A. E. (2022). The relationships between gut microbiota and diabetes mellitus, and treatments for diabetes mellitus. *Biomedicines, 10*(2). https://doi.org/10.3390/BIOMEDICINES10020308

D'Amico, G., Carista, A., Manna, O. M., Paladino, L., Picone, D., Sarullo, S., Sausa, M., Cappello, F., Vitale, A. M., & Caruso Bavisotto, C. (2024). Brain–periphery axes: The potential role of extracellular vesicles-delivered miRNAs. *Biology, 13*, 1056. https://doi.org/10.3390/BIOLOGY13121056

De Aguiar Vallim, T. Q., Tarling, E. J., & Edwards, P. A. (2013). Pleiotropic roles of bile acids in metabolism. *Cell Metabolism, 17*(5), 657–669. https://doi.org/10.1016/j.cmet.2013.03.013

Derakhshan, F., & Toth, C. (1966). Insulin and the brain. *Current Diabetes Reviews, 9*(2), 102–116. https://doi.org/10.2174/1573399811309020002

Dodd, G. T., & Tiganis, T. (2017). Insulin action in the brain: Roles in energy and glucose homeostasis. *Journal of Neuroendocrinology, 29*(10). https://doi.org/10.1111/JNE.12513

Doden, H. L., & Ridlon, J. M. (2021). Microbial hydroxysteroid dehydrogenases: From alpha to omega. *Microorganisms, 9*(3), 1–24. https://doi.org/10.3390/MICROORGANISMS9030469

Folch, J., Ettcheto, M., Busquets, O., Sánchez-López, E., Castro-Torres, R. D., Verdaguer, E., Manzine, P. R., Poor, S. R., García, M. L., Olloquequi, J., Beas-Zarate, C., Auladell, C., & Camins, A. (2018). The implication of the brain insulin receptor in late onset Alzheimer's disease dementia. *Pharmaceuticals, 11*(1). https://doi.org/10.3390/PH11010011

Govindarajulu, M., Pinky, P. D., Steinke, I., Bloemer, J., Ramesh, S., Kariharan, T., Rella, R. T., Bhattacharya, S., Dhanasekaran, M., Suppiramaniam, V., & Amin, R. H. (2020). Gut metabolite TMAO induces synaptic plasticity deficits by promoting endoplasmic reticulum stress. *Frontiers in Molecular Neuroscience, 13*. https://doi.org/10.3389/FNMOL.2020.00138

Gray, S. M., Meijer, R. I., & Barrett, E. J. (2014). Insulin regulates brain function, but how does it get there? *Diabetes, 63*(12), 3992–3997. https://doi.org/10.2337/DB14-0340

Hou, Y., Li, J., & Ying, S. (2023). Tryptophan metabolism and gut microbiota: A novel regulatory axis integrating the microbiome, immunity, and cancer. *Metabolites, 13*(11). https://doi.org/10.3390/METABO13111166

Kothari, V., Luo, Y., Tornabene, T., O'Neill, A. M., Greene, M. W., Geetha, T., & Babu, J. R. (2017). High fat diet induces brain insulin resistance and cognitive impairment in mice. *Biochimica et Biophysica Acta (BBA) - Molecular Basis of Disease, 1863*(2), 499–508. https://doi.org/10.1016/J.BBADIS.2016.10.006

Kullmann, S., Wagner, L., Hauffe, R., Kühnel, A., Sandforth, L., Veit, R., Dannecker, C., Machann, J., Fritsche, A., Stefan, N., Preissl, H., Kroemer, N. B., Heni, M., Kleinridders, A., & Birkenfeld, A. L. (2025). A short-term, high-caloric diet has prolonged effects on brain insulin action in men. *Nature Metabolism, 7*(3), 469–477. https://doi.org/10.1038/S42255-025-01226-9

Larsen, N., Vogensen, F. K., Van Den Berg, F. W. J., Nielsen, D. S., Andreasen, A. S., Pedersen, B. K., Al-Soud, W. A., Sørensen, S. J., Hansen, L. H., & Jakobsen, M. (2010). Gut microbiota in human adults with type 2 diabetes differs from non-diabetic adults. *PLoS One, 5*(2). https://doi.org/10.1371/JOURNAL.PONE.0009085

Liu, Z., Dai, X., Zhang, H., Shi, R., Hui, Y., Jin, X., Zhang, W., Wang, L., Wang, Q., Wang, D., Wang, J., Tan, X., Ren, B., Liu, X., Zhao, T., Wang, J., Pan, J., Yuan, T., Chu, C., … Liu, X. (2020). Gut microbiota mediates intermittent-fasting alleviation of diabetes-induced cognitive impairment. *Nature Communications, 11*(1), 1–14. https://doi.org/10.1038/S41467-020-14676-4

Lu, J., Lezi, E., Wang, W. F., Frontera, J., Zhu, H., Wang, W. T., Lee, P., Choi, I. Y., Brooks, W. M., Burns, J. M., Aires, D., & Swerdlow, R. H. (2011). Alternate day fasting impacts the brain insulin signaling pathway of young adult male C57BL/6 mice. *Journal of Neurochemistry, 117*(1), 154. https://doi.org/10.1111/J.1471-4159.2011.07184.X

Mietelska-Porowska, A., Domańska, J., Want, A., Więckowska-Gacek, A., Chutorański, D., Koperski, M., & Wojda, U. (2022). Induction of brain insulin resistance and Alzheimer's molecular changes by Western diet. *International Journal of Molecular Sciences, 23*(9), 4744. https://doi.org/10.3390/IJMS23094744/S1

Mukhopadhya, I., & Louis, P. (2025). Gut microbiota-derived short-chain fatty acids and their role in human health and disease. *Nature Reviews Microbiology, 23*(10), 635–651. https://doi.org/10.1038/S41579-025-01183-W

Novau-Ferré, N., Papandreou, C., Rojo-Marticella, M., Canals-Sans, J., & Bulló, M. (2025). Gut microbiome differences in children with attention deficit hyperactivity disorder and autism spectrum disorder and effects of probiotic supplementation: A randomized controlled trial. *Research in Developmental Disabilities, 161*. https://doi.org/10.1016/j.ridd.2025.105003

O'Brien, J., Hayder, H., Zayed, Y., & Peng, C. (2018). Overview of microRNA biogenesis, mechanisms of actions, and circulation. *Frontiers in Endocrinology, 9*. https://doi.org/10.3389/FENDO.2018.00402

Palihaderu, P. A. D. S., Mendis, B. I. L. M., Premarathne, J. M. K. J. K., Dias, W. K. R. R., Yeap, S. K., Ho, W. Y., Dissanayake, A. S., Rajapakse, I. H., Karunanayake, P., Senarath, U., & Satharasinghe, D. A. (2022). Therapeutic potential of miRNAs for type 2 diabetes mellitus: An overview. *Epigenetics Insights, 15*. https://doi.org/10.1177/25168657221130041

Prévost, M., Crépin, D., Rifai, S. Al, Poizat, G., Gonçalves, M., van Barneveld, F., Shadpay, R., Taouis, K., Riffault, L., Benomar, Y., & Taouis, M. (2025). The Resistin/TLR4/miR-155-5p axis: A novel signaling pathway in the onset of hypothalamic neuroinflammation. *Journal of Neuroinflammation, 22*(1), 1–16. https://doi.org/10.1186/S12974-025-03522-3/FIGURES/4

Qiao, C. M., Quan, W., Zhou, Y., Niu, G. Y., Hong, H., Wu, J., Zhao, L. P., Li, T., Cui, C., Zhao, W. J., & Shen, Y. Q. (2023). Orally induced high serum level of trimethylamine N-oxide worsened glial reaction and neuroinflammation on MPTP-induced acute Parkinson's disease model mice. *Molecular Neurobiology, 60*(9), 5137–5154. https://doi.org/10.1007/S12035-023-03392-X

Rezazadeh, H., Sharifi, M. R., & Soltani, N. (2021). Gamma-aminobutyric acid attenuates insulin resistance in type 2 diabetic patients and reduces the risk of insulin resistance in their offspring. *Biomedicine & Pharmacotherapy, 138*(5), 111440. https://doi.org/10.1016/J.BIOPHA.2021.111440

Rojas-Criollo, M., Novau-Ferré, N., Gutierrez-Tordera, L., Ettcheto, M., Folch, J., Papandreou, C., Panisello, L., Cano, A., Mostafa, H., Mateu-Fabregat, J., Carrasco, M., Camins, A., & Bulló, M. (2024). Effects of a high-fat diet on insulin-related miRNAs in plasma and brain tissue in APPSwe/PS1dE9 and wild-type C57BL/6J mice. *Nutrients, 16*(7), 955. https://doi.org/10.3390/NU16070955/S1

Romano, K. A., Vivas, E. I., Amador-Noguez, D., & Rey, F. E. (2015). Intestinal microbiota composition modulates choline bioavailability from diet and accumulation of the proatherogenic metabolite trimethylamine-N-oxide. *mBio, 6*(2). https://doi.org/10.1128/MBIO.02481-14

Sangüesa, G., Cascales, M., Griñán, C., Sánchez, R. M., Roglans, N., Pallàs, M., Laguna, J. C., & Alegret, M. (2018). Impairment of novel object recognition memory and brain insulin signaling in fructose-but not glucose-drinking female rats. *Molecular Neurobiology, 55*(8), 6984–6999. https://doi.org/10.1007/S12035-017-0863-1/FIGURES/8

Shelke, V., Kale, A., Kulkarni, Y. A., & Gaikwad, A. B. (2024). Phloretin: A comprehensive review of its potential against diabetes and associated complications. *Journal of Pharmacy and Pharmacology, 76*(3), 201–212. https://doi.org/10.1093/JPP/RGAE010

Shu, L., Zhao, H., Huang, W., Hou, G., Song, G., & Ma, H. (2020). Resveratrol upregulates mmu-miR-363-3p via the PI3K-Akt pathway to improve insulin resistance induced by a high-fat diet in mice. *Diabetes, Metabolic Syndrome and Obesity: Targets and Therapy, 13*, 391–403. https://doi.org/10.2147/DMSO.S240956

Steen, E., Terry, B. M., Rivera, E. J., Cannon, J. L., Neely, T. R., Tavares, R., Xu, X. J., Wands, J. R., & De La Monte, S. M. (2005). Impaired insulin and insulin-like growth factor expression and signaling mechanisms in Alzheimer's disease–is this type 3 diabetes? *Journal of Alzheimer's Disease, 7*(1), 63–80. https://doi.org/10.3233/JAD-2005-7107

Todorovic, S., Simeunovic, V., Prvulovic, M., Dakic, T., Jevdjovic, T., Sokanovic, S., Kanazir, S., & Mladenovic, A. (2024). Dietary restriction alters insulin signaling pathway in the brain. *BioFactors, 50*(3), 450–466. https://doi.org/10.1002/BIOF.2018

Vrieze, A., Van Nood, E., Holleman, F., Salojärvi, J., Kootte, R. S., Bartelsman, J. F. W. M., Dallinga-Thie, G. M., Ackermans, M. T., Serlie, M. J., Oozeer, R., Derrien, M., Druesne, A., Van Hylckama Vlieg, J. E. T., Bloks, V. W., Groen, A. K., Heilig, H. G. H. J., Zoetendal, E. G., Stroes, E. S., De Vos, W. M., … Nieuwdorp, M. (2012). Transfer of intestinal microbiota from lean donors increases insulin sensitivity in individuals with metabolic syndrome. *Gastroenterology, 143*(4). https://doi.org/10.1053/j.gastro.2012.06.031

Wang, Z., Ma, T., Bai, G., Fang, Q., Ou, B., Chen, M., Xu, P., Tian, M., Xu, A., & Ma, Y. (2025). Adipose tissue-derived extracellular vesicles loaded with miR-141-3p regulate obesity-induced insulin resistance by targeting glycogen synthesis and gluconeogenesis. *International Journal of Nanomedicine, 20*, 5709–5726. https://doi.org/10.2147/IJN.S511842

Wu, Y., Yang, S., Zhang, Y., & Su, Z. (2025). Exosomes in metabolic diseases: From molecular mechanisms to clinical applications. *MedComm – Future Medicine, 4*(3), e70031. https://doi.org/10.1002/MEF2.70031

Yamashita, H., Maruta, Y. H., Jozuka, M., Kimura, R., Iwabuchi, H., Yamato, M., Saito, T., Fujisawa, K., Takahashi, Y., Kimoto, M.,

Hiemori, M., & Tsuji, H. (2009). Effects of acetate on lipid metabolism in muscles and adipose tissues of type 2 diabetic Otsuka Long-Evans Tokushima Fatty (OLETF) rats. *Bioscience, Biotechnology, and Biochemistry, 73*(3), 570–576. https://doi.org/10.1271/BBB.80634

Yang, W. M., Jeong, H. J., Park, S. Y., & Lee, W. (2014). Saturated fatty acid-induced miR-195 impairs insulin signaling and glycogen metabolism in HepG2 cells. *FEBS Letters, 588*(21), 3939–3946. https://doi.org/10.1016/J.FEBSLET.2014.09.006

Yu, Y., Zhang, J., Wang, J., & Sun, B. (2021). MicroRNAs: The novel mediators for nutrient-modulating biological functions. *Trends in Food Science & Technology, 114*, 167–175. https://doi.org/10.1016/J.TIFS.2021.05.028

Zhou, S., Liu, J., Sun, Y., Xu, P., Liu, J. ling, Sun, S., Zhu, B., & Wu, H. (2023). Dietary choline metabolite TMAO impairs cognitive function and induces hippocampal synaptic plasticity declining through the mTOR/P70S6K/4EBP1 pathway. *Food & Function, 14*(6), 2881–2895. https://doi.org/10.1039/D2FO03874A

Chapter 36

Vitamins, analog and effects on insulin signaling in the brain

Camila A.E.F. Cardinali, Yandara A. Martins and Andrea S. Torrão
Departamento de Fisiologia e Biofísica, Universidade de Sao Paulo, Sao Paulo, Brazil

Abbreviations

AD Alzheimer's disease
Akt Protein kinase B
ATP Adenosine triphosphate
Aβ Amyloid-β
Bcl-2 B-cell lymphoma 2
BDNF Brain-derived neurotrophic factor
BFT Benfotiamine
cAMP Cyclic Adenosine Monophosphate
CNS Central nervous system
CoA Coenzyme A
CREB cAMP response element-binding protein
DBT O,S-dibenzoyl thiamine
ERK Extracellular signal-regulated kinase
FAD Flavin adenine dinucleotide
FMN Flavin mononucleotide
FOXO Forkhead box O transcription factors
Gas6 Growth arrest–specific protein 6
GGCX γ-Glutamyl carboxylase
IGF Insulin-like growth factor
IL-1β Interleukin 6
IR Insulin receptor
IRS-1 Insulin receptor substrate 1
MAPK Mitogen-activated protein kinase
MAT Methionine adenosyltransferase
MEK MAPK/ERK kinase
mTOR Mammalian target of rapamycin
NAD Nicotinamide adenine dinucleotide
NADP Nicotinamide adenine dinucleotide phosphate
NF-κB Nuclear factor kappa-light-chain-enhancer of activated B cells
NMDA N-Methyl-D-Aspartate
Nrf2 Nuclear factor erythroid 2–related factor 2
p70S6K p70 ribosomal protein S6 kinase
PI3K Phosphoinositide 3-kinase
PIP2 Phosphatidylinositol 4,5-bisphosphate
PIP3 Phosphatidylinositol 3,4,5-trisphosphate
PLP Pyridoxal-5′-phosphate
PP2A Protein phosphatase 2A
PTEN Phosphatase and tensin homolog
PTP1B Protein tyrosine phosphatase 1B
RAF RAF kinase
RAR Retinoic acid receptor
RAS Monomeric G protein
ROS Reactive oxygen species
RXR Retinoid X receptor
SAH S-adenosylhomocysteine
SAHH S-adenosylhomocysteine hydrolase
SAM S-adenosylmethionine
SH Serine hydroxymethyltransferase
SHC Src homology protein
SOS Guanine nucleotide exchange factor
STZ Streptozotocin
ThDP Thiamine diphosphate
THF Tetrahydrofolate
TNF-α Tumor necrosis factor alpha
TRF Tocotrienol-rich fractions
TrkB Tropomyosin receptor kinase B
VDR Vitamin D receptor
VDRE Vitamin D response element
VKDP K-Dependent proteins

Introduction

The concept of the brain as an insulin-independent organ has changed over the years. Insulin signaling is crucial for brain health and dysregulation of this pathway has been linked to central insulin resistance, as well as metabolic and neurodegenerative diseases. Insulin acts as a neurotrophic, neuromodulatory, and neuroendocrine molecule (Milstein & Ferris, 2021), supporting glucose uptake by neurons, neuronal growth and proliferation, synaptic plasticity, and long-term potentiation (Blázquez et al., 2014; Nakabeppu, 2019). The high density of insulin receptors in cognition-related brain regions further underscores the important role of insulin (Milstein & Ferris, 2021).

Hormones in Neuroscience. https://doi.org/10.1016/B978-0-443-30263-3.00028-2

Recent evidence suggests that some vitamins directly modulate insulin signaling in the central nervous system (CNS) at multiple levels of the pathway, including the regulation of insulin receptor expression and activation, modulation of receptor substrates and enzymatic cofactors, and modulation of downstream intracellular cascades such as the phosphoinositide 3-kinase/protein kinase B (PI3K/Akt), glycogen synthase kinase-3 beta (GSK-3β), and mitogen-activated protein kinase (MAPK). Other vitamins indirectly modulate insulin signaling through antioxidative and anti-inflammatory effects (Hussein, 2025; Manning et al., 2004; Zaulkffali et al., 2019). Altogether, current evidence suggests that the mechanism of action of vitamins may provide effective therapeutic strategies targeting neuroprotection and preventing insulin-related neurological disorders. This chapter will review available literature on how vitamins and their analogs influence insulin signaling in the brain, highlighting their potential as modulators of the insulin signaling pathway.

Vitamins

Vitamins are organic compounds essential for normal physiological functioning, including those related to brain function. They are primarily acquired through dietary intake and, in some cases, from supplementation, because the human body cannot synthesize them endogenously. There is a total of 13 vitamins, four fat soluble vitamins (A, D, E, K) and nine water soluble vitamins (vitamin C and the complex of B vitamins) (Kennedy, 2016).

Not only do vitamins have a role as metabolic cofactors, but they also modulate cell function and directly affect neurological health through, for example, regulation of mitochondrial function, gene expression, neurotransmitter synthesis, and myelogenesis, and as anti-inflammatory and antioxidant agents (Alisi et al., 2019; Calderón-Ospina & Nava-Mesa, 2019; Moretti & Rodrigues, 2022; Sailike et al., 2024; Tafti & Ghyselinck, 2007; Ulatowski & Manor, 2015). Evidence suggests that vitamins may have therapeutic effects in chronic and neurodegenerative diseases (Orywal et al., 2025). This dual perspective, encompassing both preventive and therapeutic roles, underscores the growing interest in understanding how vitamins and their analogs influence key signaling pathways, such as the insulin signaling pathway.

Vitamins and brain insulin signaling

As previously discussed, insulin is an important player in brain health and neuronal function (Steen et al., 2005). Its effects are mediated by binding to the insulin receptor, which triggers receptor autophosphorylation and the recruitment of multiple associated proteins that initiate downstream signaling cascades. As illustrated in Fig. 36.1, insulin receptor activation triggers two major pathways, the PI3K/Akt pathway and the MAPK/ERK pathway (Saltiel, 2021).

It is important to emphasize, however, that the molecular components of the insulin signaling pathway are not unique to insulin. Many are shared across different signaling pathways, being modulated by a variety of extracellular cues (Duan et al., 2012; Liu et al., 2013; Pomytkin & Pinelis, 2021; Talbot et al., 2012; Tian, 2005; Valvezan & Klein, 2012; Wang et al., 2008). Akt, for example, functions as a central signaling hub in neurons. It can be activated not only by insulin and insulin-like growth factor (IGF) receptors, but also by tropomyosin receptor kinase B (TrkB) receptors in response to brain-derived neurotrophic factor (BDNF), as well as by N-methyl-D-aspartate (NMDA) receptors (Duan et al., 2012; Li et al., 2009; Talbot et al., 2012). Conversely, Akt activity is impaired by inflammatory mediators and oxidative stress (Pomytkin & Pinelis, 2021; Tian, 2005). Similarly, extracellular signal-regulated kinase (ERK) signaling is modulated by neurotrophins, neurotransmitters, and cytokines (Liu et al., 2013). Altogether, these examples highlight that brain insulin signaling is not an isolated pathway but part of an intricate network of multiple extracellular inputs to regulate neuronal survival, synaptic function, and metabolic homeostasis.

In this context, vitamins and vitamin analogs emerge as modulators of insulin signaling. Some studies report direct effects on the insulin receptor and its downstream cascades, while others demonstrate modulation of individual components, such as Akt or GSK3, which cannot be strictly interpreted as insulin-specific effects, given that these proteins participate in multiple pathways. Therefore, while vitamins may enhance or impair insulin signaling, their role in this pathway must be interpreted with caution.

Direct impact in the insulin signaling pathway

Vitamin K (VK), for example, contributes to neuroprotection by engaging intracellular cascades that converge with insulin signaling mechanisms (Shankar et al., 2006), although direct evidence of VK effects on brain insulin signaling remains limited. VK is a cofactor of γ-glutamyl carboxylase (GGCX), which activates vitamin K-dependent proteins (VKDPs) (Ferland, 2012). One of the most extensively studied VKDPs in the CNS is the growth arrest–specific protein 6 (Gas6) (Allen et al., 1999; Prieto et al., 2007; Shankar et al., 2003).

In neuronal models, Gas6 protected gonadotropin-releasing hormone neurons from apoptosis induced by serum deprivation through the activation of ERK and Akt (Allen et al., 1999). In hippocampal and cortical neurons,

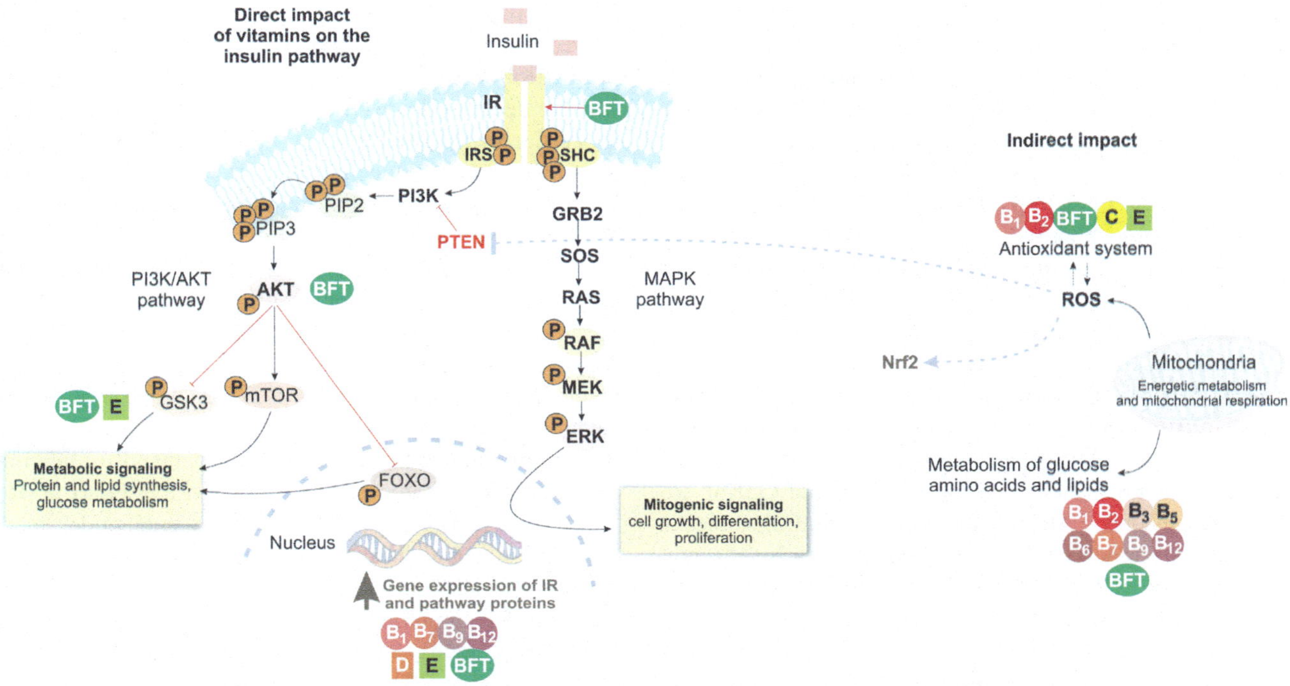

FIGURE 36.1 Direct and indirect impacts of vitamin in the insulin signaling pathway. Vitamins direct effects include regulation of insulin receptor gene expression and activation of the downstream pathways PI3K/Akt and MAPK. Indirect effects occur through regulation of mitochondrial metabolism and the antioxidant system, supporting insulin sensitivity, redox balance, and the metabolism of glucose, lipids, and amino acids. *ERK*, Extracellular signal-regulated kinase; *FOXO*, Forkhead box O transcription factors; *GRB2*, Growth factor receptor-bound protein 2; *GSK3*, Glycogen synthase kinase 3; *IR*, Insulin receptor; *IRS*, Insulin receptor substrate; *MAPK*, Mitogen-activated protein kinase; *MEK*, MAPK/ERK kinase; *mTOR*, Mechanistic target of rapamycin; *NRF2*, Nuclear factor erythroid 2–related factor 2; *PI3K*, Phosphoinositide 3-kinase; *PIP2*, Phosphatidylinositol 4,5-bisphosphate; *PIP3*, Phosphatidylinositol 3,4,5-trisphosphate; *PTEN*, Phosphatase and tensin homolog; *RAF*, RAF kinase; *RAS*, Monomeric G protein; *ROS*, Reactive oxygen species; *SHC*, Src homology protein; *SOS*, Guanine nucleotide exchange factor.

Gas6 promoted survival by activating both MAPK and PI3K cascades, recruiting ERK, cAMP response element-binding protein (CREB), mechanistic target of rapamycin (mTOR), and p70 ribosomal protein S6 kinase (p70S6K) (Prieto et al., 2007). Similar protective effects were observed in glial cells. In primary oligodendrocyte cultures derived from human fetal spinal cord, Gas6 supplementation enhanced cell survival and prevented apoptosis through the activation of the PI3K/Akt pathway, even under insulin deprivation, suggesting it may act as a compensatory factor to sustain cell viability (Shankar et al., 2003).

Epidemiological studies also reported that higher circulating VK levels are associated with improved insulin sensitivity and reduced risk of type 2 diabetes (Ho et al., 2020). Although direct studies on VK and brain insulin signaling remain scarce, the convergence of Gas6-mediated PI3K/Akt activation with canonical insulin pathways provides mechanistic support to consider VK a modulator of cerebral insulin sensitivity and neuronal survival (Shankar et al., 2006).

Pyridoxine (B6 vitamin), a vitamin involved in over 140 coenzymatic reactions (Parra et al., 2018; Rivlin & Pinto, 2007), also seems to have a role in brain insulin signaling. In *Drosophila* models with impaired PI3K function (PI3K92E RNAi), for example, supplementation with pyridoxal-5′-phosphate (PLP), the active form of B6, rescued diabetic-like phenotypes. PLP seems to enhance residual PI3K activity, thereby improving insulin sensitivity through increased Akt phosphorylation and upregulation of Akt downstream genes (Mascolo et al., 2022). Similarly, in insulin-resistant mice, B6 supplementation enhanced liver insulin sensitivity by potentiating Akt signaling (Liu et al., 2015). Together, these findings point to a potential role of B6 in modulating the PI3K/Akt pathway, a central cascade in brain insulin signaling, although direct evidence in mammalian CNS is still lacking.

Analogs of vitamin B1 (thiamine) also demonstrate effects in the insulin pathway. O,S-dibenzoylthiamine (DBT), a lipophilic thiamine precursor with antioxidant and anti-inflammatory properties (Sambon et al., 2020), normalized GSK-3β brain levels in a transgenic mouse model of amyotrophic lateral sclerosis (Probert et al., 2022). Suggesting that DBT influences insulin-related signaling pathways, particularly those involving PI3K/Akt and its downstream targets.

Benfotiamine (BFT), another lipophilic thiamine derivative, has been extensively studied in an insulin-related context. In intracerebroventricular streptozotocin (STZ)-induced models of sporadic Alzheimer's disease (AD), BFT restored insulin signaling in the hippocampus and entorhinal cortex by enhancing PI3K/Akt pathway activity and inhibiting GSK3α/β. These effects coincided with improved mitochondrial function, reduced neuronal apoptosis, and regulation of NMDA receptors and glucose transporters (GLUT1, GLUT3) (Cardinali et al., 2024; Moraes et al., 2020, 2022).

Evidence from transgenic AD and tauopathy models further supports the insulin-related actions of BFT (Pan et al., 2010; Tapias et al., 2018). These studies show decreased amyloid-β (Aβ) deposition and tau hyperphosphorylation, which correlate with reduced GSK3β activity and enhanced Akt phosphorylation. Clinical studies in patients with mild cognitive impairment or early AD indicate that oral BFT is safe and can improve cognitive function (Gibson et al., 2020; Pan et al., 2016). This suggests that its therapeutic effects may extend beyond thiamine replacement to include restoration of brain insulin sensitivity. Taken together, these findings position BFT as a promising therapeutic agent targeting insulin signaling–related mechanisms in neurodegeneration.

These studies illustrate that vitamins and their analogs can modulate elements of the insulin signaling network in the brain, either by directly acting the pathway or by influencing overlapping signaling cascades. However, due to the interconnected nature of these pathways, effects observed at the level of shared proteins such as Akt and GSK3 cannot be ascribed exclusively to insulin signaling.

Indirect impact in the insulin signaling pathway

Energy metabolism

B complex vitamins, including thiamine (B1), riboflavin (B2), niancin (B3), and pantothenic acid (B5), are key cofactors in cellular energy metabolism (Depeint et al., 2006; Rivlin & Pinto, 2007). They participate as coenzymes in mitochondrial aerobic respiration, the citric acid cycle (Krebs cycle), and the electron transport chain, sustaining ATP production (Fig. 36.2). Beyond their direct role in substrate oxidation, these vitamins also support the metabolism of glucose, fatty acids, and amino acids, contributing with substrates to the Krebs cycle (Kennedy, 2016). The deficiency of these vitamins impairs neuronal energy metabolism and may indirectly compromise the insulin signaling pathway, since both the insulin receptor and downstream molecules are ATP-dependent.

Thiamine (B1), in its biologically active form thiamine diphosphate (ThDP), acts as a coenzyme in the pentose phosphate pathway, glycolysis, and the Krebs cycle, where it is directly involved in cellular energy production. The symptoms of B1 deficiency mainly involve decreased activity of enzymes that require ThDP as a cofactor (Mrowicka et al., 2023). BFT, an analog of thiamine, is converted into ThDP, also exerting effects in mitochondrial energy metabolism (Sambon et al., 2021).

Riboflavin (B2) is the precursor of two flavoprotein coenzymes, flavin mononucleotide (FMN) and flavin adenine dinucleotide (FAD), both crucial rate-limiting factors in mitochondrial respiration and cellular energy metabolism (Rivlin & Pinto, 2007). FMN and FAD act as electron carriers in the respiratory chain, driving ATP production to meet the high metabolic demands of neurons and glial cells (Plantone et al., 2021; Rivlin & Pinto, 2007).

Niacin (B3) is the precursor of the nucleotides nicotinamide adenine dinucleotide (NAD) and nicotinamide adenine dinucleotide phosphate (NADP), indispensable cofactors for numerous brain enzymes involved in energy metabolism, redox reactions, and biosynthetic pathways (Miller & Dulay, 2008; Rivlin & Pinto, 2007). Vitamin B5 is the precursor of coenzyme A (CoA), essential for the synthesis of acetyl-CoA, the main substrate for the citric acid cycle (Rivlin & Pinto, 2007).

B vitamins regulate mitochondrial enzymes through their active metabolites and support aerobic respiration, ATP synthesis, and neuronal energy homeostasis. By ensuring an ATP supply, they indirectly support the integrity of insulin signaling in the brain.

Oxidative stress

Oxidative stress is a well-known disruptor of the insulin signaling at multiple levels of the pathway (Barone et al., 2021; Eriksson, 2007). It arises from an imbalance between reactive oxygen species (ROS) and antioxidant defenses. ROS are mostly toxic byproducts of the mitochondrial aerobic metabolism. Evidence indicates that ROS play a role in insulin signaling by targeting key regulatory proteins. For instance, ROS inhibit the protein tyrosine phosphatase 1B (PTP1B), a negative regulator of the insulin receptor and Insulin receptor substrate 1 (IRS-1), and affect the lipid phosphatase PTEN, which dephosphorylates phosphatidylinositol 3,4,5-trisphosphate (PIP3) to terminate PI3K signaling (Onoue et al., 2016). Also, hydrogen peroxide has been shown to increase Akt phosphorylation at serine 473, downregulating insulin receptor and contributing to insulin resistance (Pomytkin & Pinelis, 2021; Tian, 2005).

Vitamin E, in the form of tocopherol and tocotrienol, acts as a potent exogenous antioxidant (Fig. 36.3) (Bouayed & Bohn, 2010). Tocotrienol-rich fractions (TRF) from palm oil have been suggested to enhance insulin

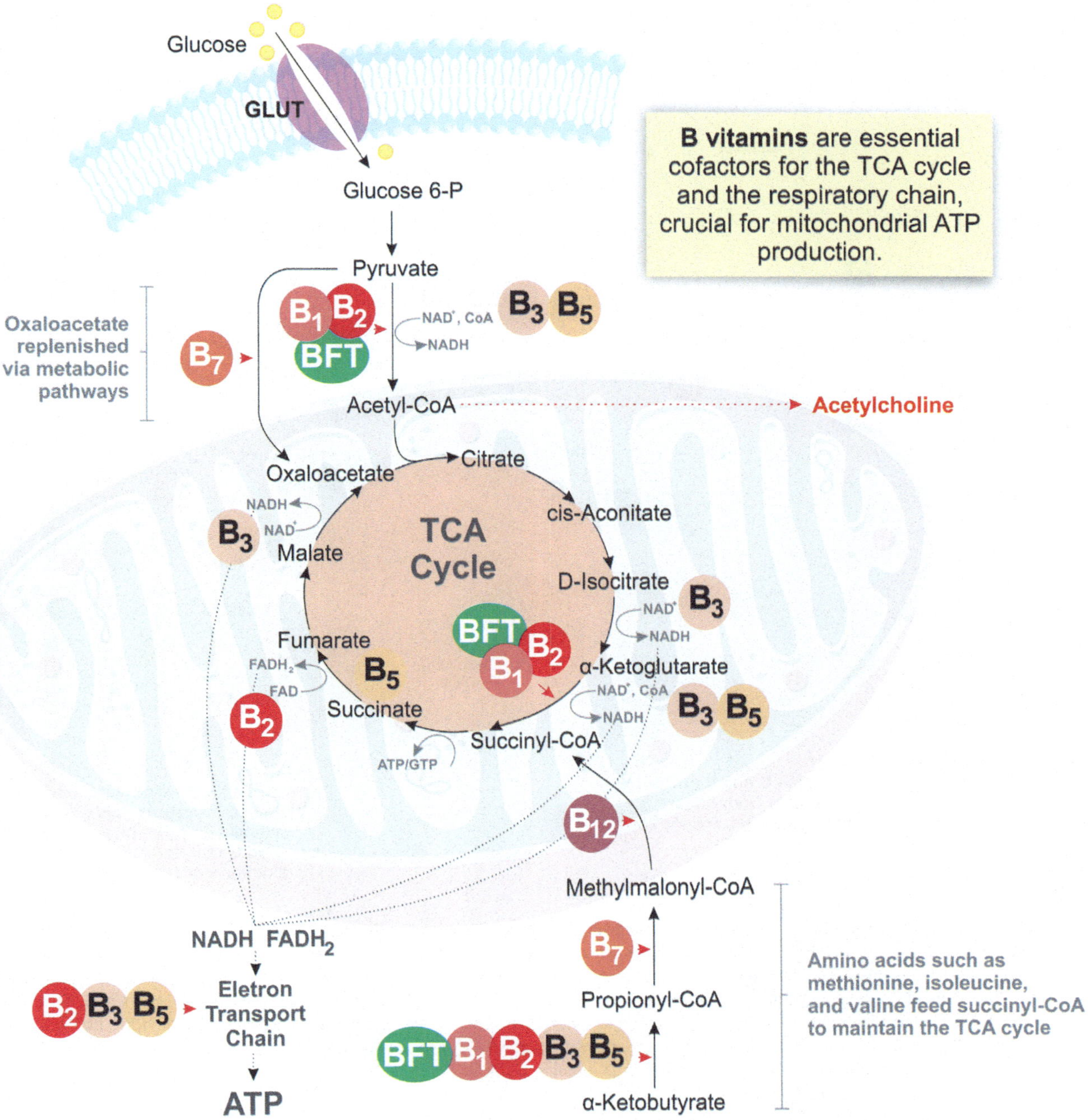

FIGURE 36.2 Impact of B complex vitamins and BFT (thiamine pathway) on mitochondrial energy production. B vitamins act as cofactors in the citric acid cycle, promoting the generation of NADH and $FADH_2$ and sustaining the electron transport chain for ATP synthesis. They also contribute to amino acid utilization and the replenishment of cycle intermediates, ensuring the continuity of energy metabolism. *BFT*, Benfotiamine; *CoA*, Coenzyme A; *FAD/FADH₂*, Flavin adenine dinucleotide (oxidized/reduced); *NAD*, Nicotinamide adenine dinucleotide (+/H = oxidized/reduced).

sensitivity and reduce the diabetes risk by mitigating oxidative stress (Manning et al., 2004; Zaulkffali et al., 2019). In models of chronic insulin resistance, vitamin E reduced ROS by quenching free radicals and preventing chain-breaking antioxidant properties, counteracting the oxidative stress that impairs insulin receptor substrate phosphorylation and inhibits downstream PI3K/Akt signaling (Ingold et al., 1987; Zaulkffali et al., 2019). Consistently, vitamin E normalized GSK-3β expression induced by insulin and oxidative (Zaulkffali et al., 2019).

At a molecular level, α-T pre-incubation protected cortical neurons from hydrogen peroxide-induced oxidative stress, not only through ROS scavenging but also by modulating signaling pathways. Under oxidative stress conditions, α-T enhanced the antiapoptotic effects of insulin by activating Akt, upregulating B-cell lymphoma 2

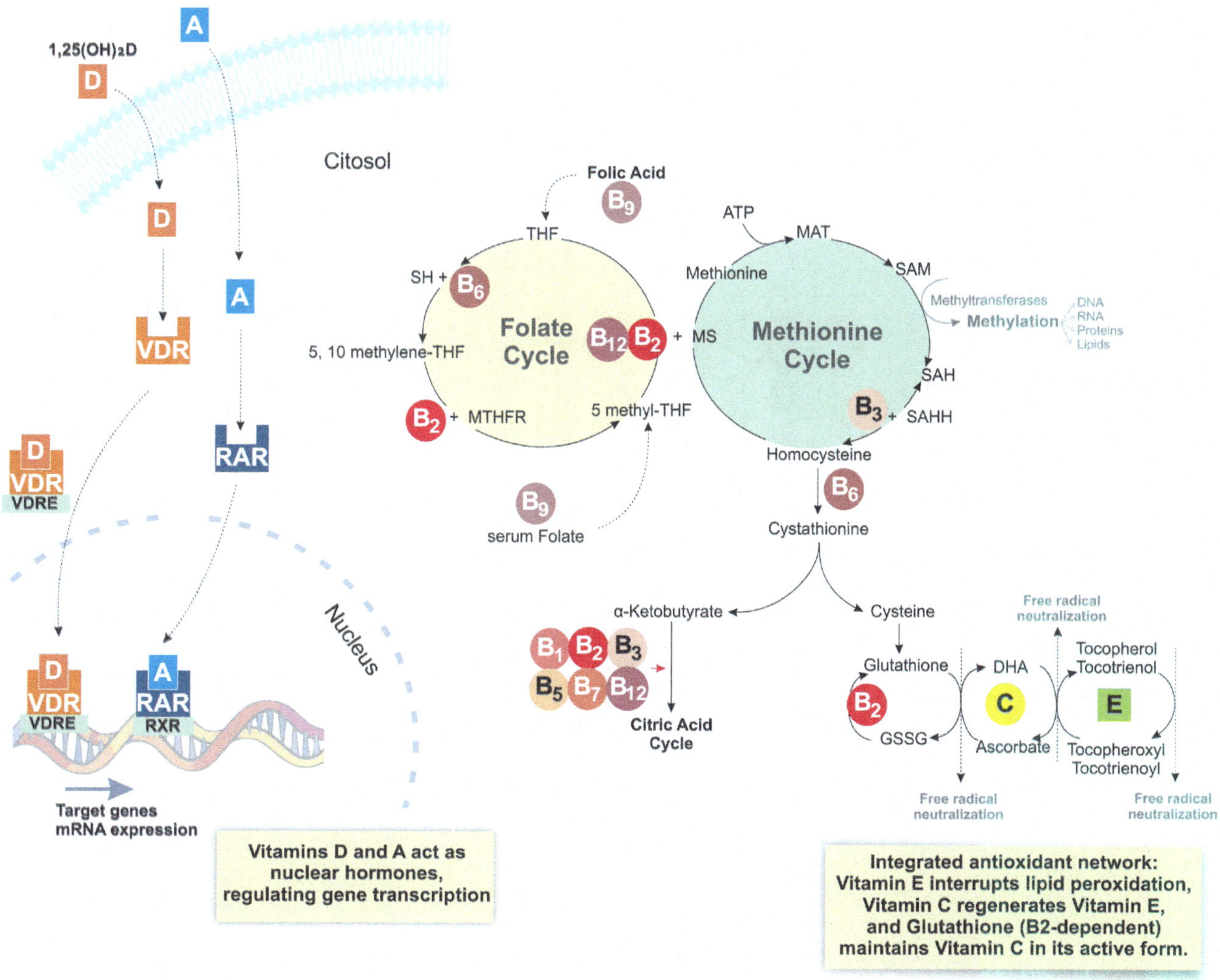

FIGURE 36.3 Impact of vitamins on amino acid metabolism, gene expression, and antioxidant defense. B-complex vitamins participate in the folate and methionine cycles, which are essential for methylation reactions, DNA/RNA synthesis, and regulation of gene expression. Additionally, vitamins A and D act as nuclear hormones modulating gene transcription, while vitamins C and E, together with glutathione (B2-dependent), are part of the antioxidant network, preventing cellular oxidative damage. *ATP*, Adenosine triphosphate; *DNA*, Deoxyribonucleic acid; *GSSG*, Oxidized glutathione.; *MAT*, Methionine adenosyltransferase; *mRNA*, Messenger RNA; *RAR*, Retinoic acid receptor; *RNA*, Ribonucleic acid; *RXR*, Retinoid X receptor; *SAH*, S-adenosylhomocysteine; *SAHH*, S-adenosylhomocysteine hydrolase; *SAM*, S-adenosylmethionine; *SH*, Serine hydroxymethyltransferase; *THF*, Tetrahydrofolate; *VDR*, Vitamin D receptor; *VDRE*, Vitamin D response element.

(Bcl-2) (anti-apoptotic), and inactivating GSK-3β via Ser9 phosphorylation. Furthermore, the combination of insulin and α-T intensified Akt activation, inhibited GSK-3β, and normalized ERK1/2 activity under oxidative stress (Luo et al., 2021; Vélez et al., 2020; Wang et al., 2021; Zakharova et al., 2017).

Vitamin C (ascorbic acid) also exerts antioxidant effects through its ability to donate electrons, neutralizing ROS and regenerating other antioxidants (Fig. 36.3) (Figueroa-Méndez & Rivas-Arancibia, 2015). Chronic supplementation with ascorbic acid attenuated brain oxidative stress in STZ-induced diabetic rats, enhancing the activity of antioxidant enzymes (Iwata et al., 2010, 2014). Also, vitamin C modulated the expression of genes involved in oxidative stress responses and antioxidant defenses (Hussein, 2025).

Oxidative stress disrupts brain insulin signaling, contributing to neuronal dysfunction and insulin resistance. Vitamins C and E, through their antioxidant properties, play crucial roles in keeping the oxidative stress under control, indirectly supporting insulin signaling.

Neuroinflammation

Neuroinflammation, mediated by molecules such as Tumor necrosis factor alpha (TNF-α) and nuclear factor kappa-light-chain-enhancer of activated B cells (NF-κB) has been shown to negatively impact brain insulin signaling, contributing to insulin resistance (Clemenzi et al., 2019; Darwish et al., 2024). Biotin (B7), a cofactor involved in key metabolic reactions, has also been linked to brain insulin signaling, likely because of its anti-inflammatory properties.

In $AlCl_3$-induced models of AD, B7 supplementation attenuated neuroinflammation and restored insulin signaling. This was evidenced by reduced inhibitory serine phosphorylation of IRS-1 and enhanced tyrosine phosphorylation, improving downstream Akt activation. B7 additionally promoted Akt phosphorylation and inhibited caspase-3 activation, highlighting its potential to enhance neuronal survival and protect against apoptosis (Attia et al., 2020). Mechanistically, these effects are thought to involve the suppression of NF-κB activity and consequent downregulation of pro-inflammatory cytokines such as TNF-α, Interleukin 1 beta (IL-1β) and interleukin 6 (IL-6), preventing their stimulatory effects on serine kinases that impair insulin signaling (Attia et al., 2020). Consistently, B7 has also demonstrated antidiabetic effects in peripheral tissues, modulating IRS-1 and NF-κB expression in type 2 diabetic models (Sahin et al., 2013).

In summary, neuroinflammation impairs brain insulin signaling and vitamins, like biotin can mitigate these effects, restoring insulin receptor function and Akt activation, highlighting the potential of anti-inflammatory micronutrients to support neuronal insulin sensitivity.

Gene expression

Insulin sensitivity in the brain depends not only on ATP availability but also on the transcriptional regulation of the insulin receptor and downstream signaling proteins. Emerging evidence indicates that some vitamins can influence insulin signaling at transcriptional and epigenetic levels, thereby affecting insulin responsiveness.

At a molecular level, folate (B9 vitamin) and cobalamin (B12 vitamin) deficiencies have been shown to impair brain insulin signaling though epigenetic mechanisms. In SK-N-BE human neuroblastoma cells and TgCRND8 mouse model of AD, combined B9 and B12 deficiency increased GSK-3β and PP2A (protein phosphatase 2A) gene activity by hindering methylation (Nicolia et al., 2010). In diet-induced obese mice, supplementation with folic acid, a synthetic form of B9, decreased insulin resistance, followed by changes in methylation and expression patterns of the insulin receptor gene in the hypothalamus, suggesting that B9 modulates insulin signaling through epigenetic regulation (Nicolia et al., 2010). Similarly, in STZ-induced diabetic mice, folic acid supplementation decreased hippocampal tau hyperphosphorylation by modulating Protein phosphatase 2A (PP2A) methylation, although it had no impact on GSK-3β phosphorylation or expression (Zheng et al., 2017). In SH-SY5Y human neuroblastoma cells, folic acid downregulated tau protein phosphorylation by inhibiting the PP2A demethylation (Li et al., 2015). Together, these results further support the role of B9 in protecting neuronal insulin signaling.

Vitamin A has also been implicated in transcriptional regulation of insulin signaling (Fig. 36.3). In both wild-type C57 and APP/PS1 transgenic mice, vitamin A deficiency decreased insulin receptor and IRS1 expression, while increasing insulin-degrading enzyme (IDE) expression in APP/PS1 animals. These findings suggest that vitamin A deficiency may worsen cerebral insulin resistance and impair glucose uptake (Li et al., 2025). Retinoic acid, the most active vitamin A metabolite, acts as a ligand for nuclear retinoid receptors and regulates gene transcription (Chambon, 1996). In an ischemic brain injury model (middle cerebral artery occlusion), retinoic acid activated the PI3K/Akt-Bad pathway, modulated Bcl-2 family protein interactions, and reduced apoptosis by lowering caspase-3 activation (Kang & Koh, 2024). Furthermore, it protected neurons against proteasome inhibition–induced cell death through sustained Akt activation and Bcl-2 regulation (Cheng et al., 2013), further supporting its neuroprotective role via insulin-related pathways.

Vitamin D likewise exerts transcriptional effects on insulin signaling (Fig. 36.3) (Muneeb et al., 2022). In type 2 diabetic mice, vitamin D supplementation improved hippocampal insulin signaling by upregulating insulin receptor expression and increasing the Akt and GSK-3β phosphorylation, which in turn reduced tau hyperphosphorylation and Aβ deposition (Muneeb et al., 2022). These protective effects may involve a cross-talk with Wnt/β-catenin signaling, as vitamin D increased hippocampal Wnt5a expression, RhoA and Rac1 activity, Akt phosphorylation, GSK-3β inhibition, Tau dephosphorylation, and Aβ clearance, ultimately restoring noncanonical Wnt signaling and improving learning and memory (Al-Harbi et al., 2017; Mehrabadi & Sadr, 2020; Muneeb et al., 2022). Similarly, in hypothalamic neurons (both in vivo and in vitro), vitamin D enhanced insulin signaling through PI3K-dependent mechanisms by increasing insulin-mediated Akt phosphorylation, regulating transcription of PI3K-related genes (Irs2, Pik3r1, Pik3r3), and modulating neuronal excitability (da Silva Teixeira et al., 2020). In SK-N-SH human neuroblastoma cells, vitamin D upregulated insulin receptor expression, restored PI3K function, and enhanced GLUT3 and GLUT4 activity, improving glucose homeostasis (Zaulkffali et al., 2019). This transcriptional effect is partially explained by the proximity of the insulin receptor promoter region to the vitamin D response element (Leung, 2016; Zaulkffali et al., 2019).

Collectively, these studies demonstrate that vitamins such as B9, B12, A, and D can regulate brain insulin signaling not only through energy metabolism but also by transcriptional modulation.

Concluding remarks

Brain insulin signaling plays a fundamental role in brain health, and its dysregulation is increasingly recognized as a key contributor to both metabolic and neurodegenerative disorders. The evidence reviewed in this chapter highlights vitamins as modulators of the insulin signaling pathway through a wide range of mechanisms, including direct regulation of insulin receptor expression and downstream signaling, support of mitochondrial energy metabolism and ATP availability, antioxidant defense, and anti-inflammatory effects.

From a translational perspective, these findings suggest that nutritional interventions or vitamin-based supplementation strategies could represent promising approaches to restore brain insulin sensitivity and protect against insulin-related cognitive decline. Nevertheless, mechanistic studies associating vitamins with brain insulin signaling are scarce. Future research should combine mechanistic studies with interventional studies, integrating molecular, neuroimaging, and cognitive outcomes to clarify the therapeutic potential of vitamins in preventing or mitigating central insulin resistance.

Mini-dictionary of terms

Benfotiamine (BFT). Lipophilic thiamine derivative that acts directly on PI3K/Akt signaling, enhances mitochondrial function, and reduces neuronal apoptosis.

Vitamin B1 (Thiamine). Coenzyme involved in energy metabolism, supporting ATP production and indirectly influencing insulin signaling in the brain.

Vitamin K. Cofactor that activates Gas6, promoting neuronal survival via PI3K/Akt and MAPK.

Vitamin D. Regulates the insulin receptor and Akt/GSK-3β phosphorylation, protecting neurons.

Vitamin B7. Metabolic cofactor that reduces neuroinflammation and enhances Akt, supporting insulin signaling.

Summary points

- Vitamins and their analogs modulate insulin signaling in the brain through direct and indirect mechanisms.
- Metabolism of glucose, mitochondrial activity, oxidative stress, inflammation, and the antioxidant system indirectly regulate insulin signaling in neurons.
- B-complex vitamins act as cofactors in energy metabolism, sustaining ATP production and supporting the integrity of insulin pathways.
- Vitamins A and D modulate gene expression, regulating insulin receptor levels, downstream signaling, and neuronal responsiveness.
- BFT enhances insulin signaling through both direct activation of PI3K/Akt and indirect support of mitochondrial and antioxidant functions.

During the preparation of this work the authors used ChatGPT 3.5, from OpenAI, strictly to improve readability and correct grammar errors. Under no circumstances has it been used to draw insights from data, conclude or produce recommendations. After using this tool/service, the authors reviewed and edited the content as needed and take full responsibility for the content of the publication.

References

Al-Harbi, A. N., Khan, K. M., & Rahman, A. (2017). Developmental vitamin D deficiency affects spatial learning in Wistar rats. *Journal of Nutrition, 147*(9), 1795–1805.

Alisi, L., Cao, R., De Angelis, C., Cafolla, A., Caramia, F., Cartocci, G., Librando, A., & Fiorelli, M. (2019). The relationships between vitamin k and cognition: A review of current evidence. *Frontiers in Neurology, 10*, 416803.

Allen, M. P., Zeng, C., Schneider, K., Xiong, X., Meintzer, M. K., Bellosta, P., Basilico, C., Varnum, B., Heidenreich, K. A., & Wierman, M. E. (1999). Growth arrest-specific gene 6 (gas6)/ adhesion related kinase (Ark) signaling promotes gonadotropin-releasing hormone neuronal survival via extracellular signal-regulated kinase (ERK) and AKT. *Molecular Endocrinology, 13* (2), 191–201.

Attia, H., Albuhayri, S., Alaraidh, S., Alotaibi, A., Yacoub, H., Mohamad, R., & Al-Amin, M. (2020). Biotin, coenzyme Q10, and their combination ameliorate aluminium chloride-induced Alzheimer's disease via attenuating neuroinflammation and improving brain insulin signaling. *Journal of Biochemical and Molecular Toxicology, 34*(9), 173–187.

Barone, E., Di Domenico, F., Perluigi, M., & Butterfield, D. A. (2021). The interplay among oxidative stress, brain insulin resistance and AMPK dysfunction contribute to neurodegeneration in type 2 diabetes and Alzheimer disease. *Free Radical Biology & Medicine, 176*, 16.

Blázquez, E., Velázquez, E., Hurtado-Carneiro, V., & Ruiz-Albusac, J. M. (2014). Insulin in the brain: Its pathophysiological implications for states related with central insulin resistance, type 2 diabetes and Alzheimer's disease. *Frontiers in Endocrinology, 5*, 161.

Bouayed, J., & Bohn, T. (2010). Exogenous antioxidants - Double-edged swords in cellular redox state: Health beneficial effects at physiologic doses versus deleterious effects at high doses. *Oxidative Medicine and Cellular Longevity, 3*(4), 228–237.

Calderón-Ospina, C. A., & Nava-Mesa, M. O. (2019). B Vitamins in the nervous system: Current knowledge of the biochemical modes of action and synergies of thiamine, pyridoxine, and cobalamin. *CNS Neuroscience and Therapeutics, 26*(1), 5.

Cardinali, C. A. E. F., Martins, Y. A., Moraes, R. C. M., Costa, A. P., Alencar, M. B., Silber, A. M., & Torrão, A. S. (2024). Exploring the therapeutic potential of benfotiamine in a sporadic Alzheimer's-like disease rat model: Insights into insulin signaling and cognitive function. *ACS Chemical Neuroscience, 15*(16), 2982–2994.

Chambon, P. (1996). A decade of molecular biology of retinoic acid receptors. *The FASEB Journal, 10*(9), 940–954.

Cheng, B., Martinez, A. A., Morado, J., Scofield, V., Roberts, J. L., & Maffi, S. K. (2013). Retinoic acid protects against proteasome inhibition associated cell death in SH-SY5Y cells via the AKT pathway. *Neurochemistry International, 62*(1), 31–42.

Clemenzi, M. N., Wellhauser, L., Aljghami, M. E., & Belsham, D. D. (2019). Tumour necrosis factor α induces neuroinflammation and insulin resistance in immortalised hypothalamic neurones through independent pathways. *Journal of Neuroendocrinology, 31*(1), e12678.

da Silva Teixeira, S., Harrison, K., Uzodike, M., Rajapakshe, K., Coarfa, C., He, Y., Xu, Y., & Sisley, S. (2020). Vitamin D actions in neurons require the PI3K pathway for both enhancing insulin signaling and rapid depolarizing effects. *The Journal of Steroid Biochemistry and Molecular Biology, 200*, 105690.

Darwish, R., Alcibahy, Y., Bucheeri, S., Albishtawi, A., Tama, M., Shetty, J., & Butler, A. E. (2024). The role of hypothalamic microglia in the onset of insulin resistance and type 2 diabetes: A neuro-immune perspective. *International Journal of Molecular Sciences, 25*(23), 13169.

Depeint, F., Bruce, W. R., Shangari, N., Mehta, R., & O'Brien, P. J. (2006). Mitochondrial function and toxicity: Role of the B vitamin family on mitochondrial energy metabolism. *Chemico-Biological Interactions, 163*(1–2), 94–112.

Duan, B., Liu, D. S., Huang, Y., Zeng, W. Z., Wang, X., Yu, H., Zhu, M. X., Chen, Z. Y., & Xu, T. Le (2012). PI3-kinase/Akt pathway-regulated membrane insertion of acid-sensing ion channel 1a underlies BDNF-induced pain hypersensitivity. *Journal of Neuroscience, 32*(18), 6351–6363.

Eriksson, J. W. (2007). Metabolic stress in insulin's target cells leads to ROS accumulation - A hypothetical common pathway causing insulin resistance. *FEBS Letters, 581*(19), 3734–3742.

Ferland, G. (2012). Vitamin K and the nervous system: An overview of its actions. *Advances in Nutrition, 3*(2), 204.

Figueroa-Méndez, R., & Rivas-Arancibia, S. (2015). Vitamin C in health and disease: Its role in the metabolism of cells and redox state in the brain. *Frontiers in Physiology, 6*, 163187.

Gibson, G. E., Luchsinger, J. A., Cirio, R., Chen, H., Franchino-Elder, J., Hirsch, J. A., Bettendorff, L., Chen, Z., Flowers, S. A., Gerber, L. M., Grandville, T., Schupf, N., Xu, H., Stern, Y., Habeck, C., Jordan, B., & Fonzetti, P. (2020). Benfotiamine and cognitive decline in Alzheimer's disease: Results of a randomized placebo-controlled phase IIA clinical trial. *Journal of Alzheimer's Disease, 78*(3), 989.

Ho, H. J., Komai, M., & Shirakawa, H. (2020). Beneficial effects of vitamin K status on glycemic regulation and diabetes mellitus: A mini-review. *Nutrients, 12*(8), 2485.

Hussein, W. M. (2025). Vitamin C as adjuvant therapy in diabetes management. *Islamic University Journal of Applied Sciences, 7*(1), 266–274.

Ingold, K. U., Webb, A. C., Witter, D., Burton, G. W., Metcalfe, T. A., & Muller, D. P. R. (1987). Vitamin E remains the major lipid-soluble, chain-breaking antioxidant in human plasma even in individuals suffering severe vitamin E deficiency. *Archives of Biochemistry and Biophysics, 259*(1), 224–225.

Iwata, N., Okazaki, M., Kamiuchi, S., & Hibino, Y. (2010). Protective effects of oral administrated ascorbic acid against oxidative stress and neuronal damage after cerebral ischemia/reperfusion in diabetic rats. *Journal of Health Science, 56*(1), 20–30.

Iwata, N., Okazaki, M., Xuan, M., Kamiuchi, S., Matsuzaki, H., & Hibino, Y. (2014). Orally administrated ascorbic acid suppresses neuronal damage and modifies expression of SVCT2 and GLUT1 in the brain of diabetic rats with cerebral ischemia-reperfusion. *Nutrients, 6*(4), 1554–1577.

Kang, J. B., & Koh, P. O. (2024). Retinoic acid alleviates the reduction of Akt and Bad phosphorylation and regulates Bcl-2 family protein interactions in animal models of ischemic stroke. *PLoS One, 19*(5), e0303213.

Kennedy, D. O. (2016). B vitamins and the brain: Mechanisms, dose and efficacy—A review. *Nutrients, 8*(2), 68.

Leung, P. S. (2016). The potential protective action of vitamin D in hepatic insulin resistance and pancreatic islet dysfunction in type 2 diabetes mellitus. *Nutrients, 8*(3), 147.

Li, W., Jiang, M., Xiao, Y., Zhang, X., Cui, S., & Huang, G. (2015). Folic acid inhibits tau phosphorylation through regulation of PP2A methylation in SH-SY5Y cells. *The Journal of Nutrition, Health & Aging, 19*(2), 123–129.

Li, J., Wang, H., & Rosenberg, P. A. (2009). Vitamin K prevents oxidative cell death by inhibiting activation of 12-lipoxygenase in developing oligodendrocytes. *Journal of Neuroscience Research, 87*(9), 1997.

Li, P., Xu, J., Guo, Y., Ma, X., Wang, X., Liu, L., Liu, Y., Ren, X., Li, J., Wang, Y., Meng, L., Zhou, S., & Yuan, L. (2025). Impact of vitamin A on aged people's cognition and Alzheimer's disease progression in an animal model. *NPJ Science of Food, 9*(1), 67.

Liu, Z., Li, P., Zhao, Z. H., Zhang, Y., Ma, Z. M., & Wang, S. X. (2015). Vitamin B6 prevents endothelial dysfunction, insulin resistance, and hepatic lipid accumulation in Apoe −/− mice fed with high-fat diet. *Journal of Diabetes Research, 2016*, 1748065.

Liu, Y., Lijian, T., Fu, X., Zhao, Y., & Xu, X. (2013). BDNF protects retinal neurons from hyperglycemia through the TrkB/ERK/MAPK pathway. *Molecular Medicine Reports, 7*(6), 1773–1778. https://doi.org/10.3892/MMR.2013.1433

Luo, P., He, W. X., Li, C., & Chang, M. J. (2021). Enteric glial cells exert neuroprotection from hyperglycemia-induced damage via Akt/GSK3β pathway. *NeuroReport, 32*(10), 875–881.

Manning, P. J., Sutherland, W. H. F., Walker, R. J., Williams, S. M., De Jong, S. A., Ryalls, A. R., & Berry, E. A. (2004). Effect of high-dose vitamin E on insulin resistance and associated parameters in overweight subjects. *Diabetes Care, 27*(9), 2166–2171.

Mascolo, E., Liguori, F., Merigliano, C., Schiano, L., Gnocchini, E., Pilesi, E., Volonté, C., Di Salvo, M. L., Contestabile, R., Tramonti, A., & Vernì, F. (2022). Vitamin B6 rescues insulin resistance and glucose-induced DNA damage caused by reduced activity of Drosophila PI3K. *Journal of Cellular Physiology, 237*(9), 3578.

Mehrabadi, S., & Sadr, S. S. (2020). Administration of vitamin D3 and E supplements reduces neuronal loss and oxidative stress in a model of rats with Alzheimer's disease. *Neurological Research, 42*(10), 862–868.

Miller, C. L., & Dulay, J. R. (2008). The high-affinity niacin receptor HM74A is decreased in the anterior cingulate cortex of individuals with schizophrenia. *Brain Research Bulletin, 77*(1), 33–41.

Milstein, J. L., & Ferris, H. A. (2021). The brain as an insulin-sensitive metabolic organ. *Molecular Metabolism, 52*, 101234.

Moraes, R. C. M. de, Lima, G. C. A., Cardinali, C. A. E. F., Gonçalves, A. C., Portari, G. V., Guerra-Shinohara, E. M., Leboucher, A., Donato, J., Kleinridders, A., & Torrão, A. da S. (2022). Benfotiamine protects against hypothalamic dysfunction in a STZ-induced model of neurodegeneration in rats. *Life Sciences, 306*, 120841.

Moraes, R. C. M., Singulani, M. P., Gonçalves, C., Portari, G. V., & Torrão, A. S. (2020). Oral benfotiamine reverts cognitive deficit and increase thiamine diphosphate levels in the brain of a rat model of neurodegeneration. *Experimental Gerontology, 141*, 111097.

Moretti, M., & Rodrigues, A. L. S. (2022). Functional role of ascorbic acid in the central nervous system: A focus on neurogenic and synaptogenic processes. *Nutritional Neuroscience, 25*(11), 2431–2441.

Mrowicka, M., Mrowicki, J., Dragan, G., & Majsterek, I. (2023). The importance of thiamine (vitamin B1) in humans. *Bioscience Reports, 43*(10).

Muneeb, M., Mansou, S. M., Saleh, S., & Mohammed, R. A. (2022). Vitamin D and rosuvastatin alleviate type-II diabetes-induced cognitive dysfunction by modulating neuroinflammation and canonical/noncanonical Wnt/β-catenin signaling. *PLoS One, 17*(11), e0277457.

Nakabeppu, Y. (2019). Origins of brain insulin and its function. *Advances in Experimental Medicine and Biology, 1128*, 1–11.

Nicolia, V., Fuso, A., Cavallaro, R. A., Di Luzio, A., & Scarpa, S. (2010). B vitamin deficiency promotes tau phosphorylation through regulation of GSK3β and PP2A. *Journal of Alzheimer's Disease, 19*(3), 895–907.

Onoue, T., Goto, M., Tominaga, T., Sugiyama, M., Tsunekawa, T., Hagiwara, D., Banno, R., Suga, H., Sugimura, Y., & Arima, H. (2016). Reactive oxygen species mediate insulin signal transduction in mouse hypothalamus. *Neuroscience Letters, 619*, 1–7.

Orywal, K., Socha, K., Iwaniuk, P., Kaczyński, P., Farhan, J. A., Zoń, W., Łozowicka, B., Perkowski, M., & Mroczko, B. (2025). Vitamins in the prevention and support therapy of neurodegenerative diseases. *International Journal of Molecular Sciences, 26*(3), 1333.

Pan, X., Chen, Z., Fei, G., Pan, S., Bao, W., Ren, S., Guan, Y., & Zhong, C. (2016). Long-term cognitive improvement after benfotiamine administration in patients with Alzheimer's disease. *Neuroscience Bulletin, 32*(6), 591. https://doi.org/10.1007/S12264-016-0067-0

Pan, X., Gong, N., Zhao, J., Yu, Z., Gu, F., Chen, J., Sun, X., Zhao, L., Yu, M., Xu, Z., Dong, W., Qin, Y., Fei, G., Zhong, C., & Xu, T. Le (2010). Powerful beneficial effects of benfotiamine on cognitive impairment and β-amyloid deposition in amyloid precursor protein/presenilin-1 transgenic mice. *Brain, 133*(5), 1342–1351. https://doi.org/10.1093/BRAIN/AWQ069

Parra, M., Stahl, S., & Hellmann, H. (2018). Vitamin B6 and its role in cell metabolism and physiology. *Cells, 7*(7), 84.

Plantone, D., Pardini, M., & Rinaldi, G. (2021). Riboflavin in neurological diseases: A narrative review. *Clinical Drug Investigation, 41*(6), 513–527.

Pomytkin, I., & Pinelis, V. (2021). Brain insulin resistance: Focus on insulin receptor-mitochondria interactions. *Life, 11*(3), 262.

Prieto, A. L., O'Dell, S., Varnum, B., & Lai, C. (2007). Localization and signaling of the receptor protein tyrosine kinase Tyro3 in cortical and hippocampal neurons. *Neuroscience, 150*(2), 319.

Probert, F., Gorlova, A., Deikin, A., Bettendorff, L., Veniaminova, E., Nedorubov, A., Chaprov, K. D., Ivanova, T. A., Anthony, D. C., & Strekalova, T. (2022). In FUS[1−359]-tg mice O,S-dibenzoyl thiamine reduces muscle atrophy, decreases glycogen synthase kinase 3 beta, and normalizes the metabolome. *Biomedicine & Pharmacotherapy, 156*.

Rivlin, R. S., & Pinto, J. T. (2007). In J. Zempleni, R. Rucker, D. McCormick, & J. Suttie (Eds.), *Handbook of vitamins*.

Sahin, K., Tuzcu, M., Orhan, C., Sahin, N., Kucuk, O., Ozercan, I. H., Juturu, V., & Komorowski, J. R. (2013). Anti-diabetic activity of chromium picolinate and biotin in rats with type 2 diabetes induced by high-fat diet and streptozotocin. *British Journal of Nutrition, 110*(2), 197–205.

Sailike, B., Onzhanova, Z., Akbay, B., Tokay, T., & Molnár, F. (2024). Vitamin D in central nervous system: Implications for neurological disorders. *International Journal of Molecular Sciences, 25*(14), 7809.

Saltiel, A. R. (2021). Insulin signaling in health and disease. *The Journal of Clinical Investigation, 131*(1), e142241.

Sambon, M., Gorlova, A., Demelenne, A., Alhama-Riba, J., Coumans, B., Lakaye, B., Wins, P., Fillet, M., Anthony, D. C., Strekalova, T., & Bettendorff, L. (2020). Dibenzoylthiamine has powerful antioxidant and anti-inflammatory properties in cultured cells and in mouse models of stress and neurodegeneration. *Biomedicines, 8*(9), 361.

Sambon, M., Wins, P., & Bettendorff, L. (2021). Neuroprotective effects of thiamine and precursors with higher bioavailability: Focus on benfotiamine and dibenzoylthiamine. *International Journal of Molecular Sciences, 22*(11), 5418. https://doi.org/10.3390/IJMS22115418

Shankar, S. L., O'Guin, K., Cammer, M., McMorris, F. A., Stitt, T. N., Basch, R. S., Varnum, B., & Shafit-Zagardo, B. (2003). The growth arrest-specific gene product Gas6 promotes the survival of human oligodendrocytes via a phosphatidylinositol 3-kinase-dependent pathway. *Journal of Neuroscience, 23*(10), 4208.

Shankar, S. L., O'Guin, K., Kim, M., Varnum, B., Lemke, G., Brosnan, C. F., & Shafit-Zagardo, B. (2006). Gas6/Axl signaling activates the phosphatidylinositol 3-kinase/Akt1 survival pathway to protect oligodendrocytes from tumor necrosis factorα-induced apoptosis. *Journal of Neuroscience, 26*(21), 5638.

Steen, E., Terry, B. M., Rivera, E. J., Cannon, J. L., Neely, T. R., Tavares, R., Xu, X. J., Wands, J. R., & de la Monte, S. M. (2005). Impaired insulin and insulin-like growth factor expression and signaling mechanisms in Alzheimer's disease–is this type 3 diabetes? *Journal of Alzheimer's Disease, 7*(1), 63–80.

Tafti, M., & Ghyselinck, N. B. (2007). Functional implication of the vitamin A signaling pathway in the brain. *Archives of Neurology, 64*(12), 1706–1711.

Talbot, K., Wang, H. Y., Kazi, H., Han, L. Y., Bakshi, K. P., Stucky, A., Fuino, R. L., Kawaguchi, K. R., Samoyedny, A. J., Wilson, R. S., Arvanitakis, Z., Schneider, J. A., Wolf, B. A., Bennett, D. A., Trojanowski, J. Q., & Arnold, S. E. (2012). Demonstrated brain insulin resistance in Alzheimer's disease patients is associated with IGF-1 resistance, IRS-1 dysregulation, and cognitive decline. *The Journal of Clinical Investigation, 122*(4), 1316–1338.

Tapias, V., Jainuddin, S., Ahuja, M., Stack, C., Elipenahli, C., Vignisse, J., Gerges, M., Starkova, N., Xu, H., Starkov, A. A., Bettendorff, L.,

Hushpulian, D. M., Smirnova, N. A., Gazaryan, I. G., Kaidery, N. A., Wakade, S., Calingasan, N. Y., Thomas, B., Gibson, G. E., ... Beal, M. F. (2018). Benfotiamine treatment activates the Nrf2/ARE pathway and is neuroprotective in a transgenic mouse model of tauopathy. *Human Molecular Genetics, 27*(16), 2874–2892.

Tian, R. (2005). Another role for the celebrity: Akt and insulin resistance. *Circulation Research, 96*(2), 139–140. https://doi.org/10.1161/01.RES.0000156076.17807.1F/ASSET/777D1908-FD5A-46A1-8AD4-67F41528BFE1/ASSETS/GRAPHIC/1FF1.JPEG

Ulatowski, L. M., & Manor, D. (2015). Vitamin E and neurodegeneration. *Neurobiology of Disease, 84*, 78–83.

Valvezan, A. J., & Klein, P. S. (2012). GSK-3 and Wnt signaling in neurogenesis and bipolar disorder. *Frontiers in Molecular Neuroscience, 5*, 17824.

Vélez, D. E., Mestre-Cordero, V. E., Hermann, R., Perego, J., Harriet, S., Fernandez-Pazos, M. de las M., Mourglia, J., & Marina-Prendes, M. G. (2020). Rosuvastatin protects isolated hearts against ischemia-reperfusion injury: Role of Akt-GSK-3β, metabolic environment, and mitochondrial permeability transition pore. *Journal of Physiology & Biochemistry, 76*(1), 85–98.

Wang, L., Chen, Y., Sternberg, P., & Cai, J. (2008). Essential roles of the PI3 kinase/Akt pathway in regulating Nrf2-dependent antioxidant functions in the RPE. *Investigative Ophthalmology & Visual Science, 49*(4), 1671.

Wang, X., Sun, Q., Jiang, Q., Jiang, Y., Zhang, Y., Cao, J., Lu, L., Li, C., Wei, P., Wang, Q., & Wang, Y. (2021). Cryptotanshinone ameliorates doxorubicin-induced cardiotoxicity by targeting akt-GSK-3β-mPTP pathway in vitro. *Molecules, 26*(5), 1460.

Zakharova, I. O., Sokolova, T. V., Vlasova, Y. A., Bayunova, L. V., Rychkova, M. P., & Avrova, N. F. (2017). α-Tocopherol at nanomolar concentration protects cortical neurons against oxidative stress. *International Journal of Molecular Sciences, 18*(1), 216.

Zaulkffali, A. S., Razip, N. N. M., Alwi, S. S. S., Jalil, A. A., Mutalib, M. S. A., Gopalsamy, B., Chang, S. K., Zainal, Z., Ibrahim, N. N., Zakaria, Z. A., & Khaza'ai, H. (2019). Vitamins D and E stimulate the PI3K-AKT signalling pathway in insulin-resistant SK-N-SH neuronal cells. *Nutrients, 11*, 2525.

Zheng, M., Zou, C., Li, M., Huang, G., Gao, Y., & Liu, H. (2017). Folic acid reduces tau phosphorylation by regulating PP2A methylation in streptozotocin-induced diabetic mice. *International Journal of Molecular Sciences, 18*(4), 861.

Chapter 37

Recommended resources for the study of hormones in neuroscience: The Pancreas and Pancreatic Hormones

Rajkumar Rajendram[1, 2, 3], Vinood B. Patel[4] and Victor R. Preedy[5, 6, 7]

[1]Research Unit, Department of Medical Education, College of Medicine, King Saud bin Abdulaziz University for Health Sciences, Riyadh, Saudi Arabia; [2]Medical Protocol Department, King Abdulaziz Medical City, Ministry of National Guard Health Affairs, Riyadh, Saudi Arabia; [3]Research Offices, King Abdullah International Medical Research Center, Riyadh, Saudi Arabia; [4]School of Life Sciences, University of Westminster, London, United Kingdom; [5]Department of Clinical Biochemistry, King's College Hospital, London, United Kingdom; [6]Faculty of Life Sciences and Medicine, King's College London, London, United Kingdom; [7]University of Suffolk, Ipswich, United Kingdom

Introduction

Early physiologists noted that the brain influences organ secretions. In the middle of the 20th century, work on the hypothalamic-pituitary axis revealed that neuronal populations project to peripheral glands, establishing two-way communication (Choi & Kim, 2022). Subsequently, mapping hormone receptors in the brainstem and cerebral cortex demonstrated that pancreatic hormones act on central targets (Shaughness et al., 2020). These discoveries highlighted the neuromodulatory role of pancreatic hormones.

Hormonal signals from the pancreas shape brain activity as much as they regulate glucose. Examining the effects of insulin, glucagon and amylin on neural pathways may link nutrition and metabolism to cognition, mood and neuro-degeneration. Insulin released from β-cells in the pancreatic islets of Langerhans lowers blood glucose and simultaneously binds neuronal insulin receptors, influencing synaptic plasticity (Shaughness et al., 2020). Glucagon and amylin influence appetite and energy expenditure via effects on the hypothalamus (Yan et al., 2025). These pathways explain why pancreatic islet dysfunction affects the central nervous system.

Pancreatic hormones modulate neurotransmitter release and inflammatory cascades. Insulin enhances long-term potentiation, while glucagon influences dopaminergic signaling in reward circuits (Aslanoglou et al., 2021). Recent exosome analyses reveal that vesicles from the islets of Langerhans carry microRNAs that can direct astrocytic gene expression (Brown & Tzanakakis, 2023). These observations illustrate how endocrine cues fine-tune neural networks.

Modern investigations rely on integrated platforms that preserve tissue architecture while analyzing molecular depth. Spatial transcriptomics maps hormone receptor distribution across brain regions, uncovering previously hidden cell types (Wang et al., 2025). Co-culture systems that pair isolated islets with cortical neurons allow real-time monitoring of insulin-driven calcium fluxes (Brown & Tzanakakis, 2023). Computational models now synthesize electrophysiological and metabolic datasets, predicting system-level outcomes of pharmacologic interventions.

It is important for those active in the field to keep up to date and abreast of current research into the endocrine aspects of neuroscience. To address this, we have compiled a list of resources as recommended by researchers, physicians, practitioners, and others who actively work in this field. By providing a curated collection of impactful web based links, we enhance the access of current knowledge in this vital field. Information on over 100 organizations, scientific societies, regulatory bodies and other web-based resources are included.

This list of material in these tables is to provide general information only. It does not constitute any recommendation or endorsement of the activities of these sites, facilities, or other resources listed in this chapter, by the authors or editors of this book. Readers should visit the listed websites sites at their own discretion and ensure they have

Hormones in Neuroscience. https://doi.org/10.1016/B978-0-443-30263-3.00011-7

appropriate security measures, such as up-to-date antivirus and anti-malware protection, in place.

Tables 37.1–37.4 list the most up-to-date information on the regulatory bodies (Table 37.1), professional societies (Table 37.2), emerging technologies and platforms (Table 37.3), and other resources of interest (Table 37.4) that are relevant to an evidence-based approach to the investigation of the endocrine aspects of neuroscience, the pancreas, related hormones and clinical condition. Some organizations are listed in more than one table as they occasionally fulfill multiple roles.

TABLE 37.1 Websites, regulatory bodies or organizations associated with the study of hormones in neuroscience, the pancreas, related hormones and associated clinical conditions.

Regulatory Body or Organisation	Web address
American Diabetes Association (ADA)	https://diabetes.org/
American Pancreatic Association	https://www.american-pancreatic-association.org/
Australian Diabetes Society (ADS)	https://diabetessociety.com.au/
Canadian Diabetes Association (Diabetes Canada)	https://www.diabetes.ca/
Centers for Disease Control and Prevention (CDC)—Division of Diabetes Translation	https://www.cdc.gov/diabetes/
Endocrine Society	https://www.endocrine.org/
European Association for the Study of Diabetes (EASD)	https://easd.org/
European Association of Neurology	https://www.ean.org/
European Commission—Health and Food Safety	https://health.ec.europa.eu/
European Medicines Agency (EMA)	https://www.ema.europa.eu/en/homepage
European Pancreatic Club (EPC)	https://www.europeanpancreaticclub.org/
Human Pancreas Analysis Consortium (HIRN-HPAC)	https://hirnetwork.org/consortium/hpac
International Association of Pancreatology (IAP)	https://internationalpancreatology.org/
International Diabetes Federation (IDF)	https://idf.org/
International Society for Pediatric and Adolescent Diabetes (ISPAD)	https://www.ispad.org/
Ministry of Health of Ukraine	https://moz.gov.ua/
National Institute for Health and Care Excellence (NICE)	https://www.nice.org.uk/
National Institute of Diabetes and Digestive and Kidney Diseases	https://www.niddk.nih.gov/
National Institutes of Health (NIH)	https://www.nih.gov
National Pancreas Foundation (NPF)	https://pancreasfoundation.org
Pancreatic Cancer Action Network (PanCAN)	https://pancan.org
The Human Pancreas Analysis Program (HPAP)	https://hpap.pmacs.upenn.edu/
The National Pancreas Foundation (NPF)	https://pancreasfoundation.org/
United States (U.S.) Food and Drug Administration (FDA)	https://www.fda.gov/
United States Food and Drug Administration (FDA) Adverse Event Reporting System (FAERS) Public Dashboard [section 3 or 4]	https://www.fda.gov/drugs/questions-and-answers-fdas-adverse-event-reporting-system-faers/fda-adverse-event-reporting-system-faers-public-dashboard
World Health Organization (WHO)	https://www.who.int

This table lists the regulatory bodies and organizations relevant to the study of hormones in neuroscience, the pancreas, related hormones and associated clinical conditions. The links were accurate at the time of going to press but may move or alter. In these cases, the use of the "Search" tabs should be explored at the parent address or site. In some cases, links direct the reader to relevant pages within parent sites. Some societies and organizations prefer shortened terms, such as acronyms and abbreviations. Some organizations provide a dual role of imparting information on specific conditions and acting as professional or regulatory bodies. Where possible we provided translations of the name of the site if not already provided by the invited authors. See also Table 37.2.

TABLE 37.2 Professional societies that are associated with the study of hormones in neuroscience, the pancreas, related hormones and associated clinical conditions.

Society name	Web address
American Association of Clinical Endocrinology (AACE)	https://www.aace.com/
American Diabetes Association (ADA)	https://diabetes.org/
American Heart Association (AHA)	https://www.heart.org/
American Medical Association	https://www.ama-assn.org/
American Pancreatic Association (APA)	https://www.american-pancreatic-association.org/
American Physiological Society	https://www.physiology.org/
Asian Association for the Study of Diabetes	https://www.aa-sd.org/
Asociación Española de Pancreatología (AESPANC) [Spanish Association of Pancreatology]	https://aespanc.es
Breakthrough T1D	https://www.breakthrought1d.org
Diabetes Technology Society (DTS)	https://www.diabetestechnology.org/
Diabetes UK	https://www.diabetes.org.uk/
Endocrine Society	https://www.endocrine.org/
European Association for the Study of Diabetes (EASD)	https://www.easd.org/home/
European Foundation for the Study of Diabetes	https://www.europeandiabetesfoundation.org/home/
European Pancreatic Club	https://www.europeanpancreaticclub.org/
European Society of Endocrinology (ESE)	https://www.ese-hormones.org/
International Association of Pancreatology	https://internationalpancreatology.org/
International Hepato-Pancreato-Biliary Association (IHPBA)	https://www.ihpba.org/
International Pancreas and Islet Transplant Association (IPITA)	https://tts.org/ipita-about/ipita-presidents-message
International Society for Pediatric and Adolescent Diabetes (ISPAD)	https://www.ispad.org/
International Society of Endocrinology (ISE)	https://www.isendo.org/
Japan Association for Diabetes Education and Care (JADEC)	https://www.nittokyo.or.jp/
Japan Diabetes Society	https://www.jds.or.jp/modules/en/index.php?content_id=1
Japan Pancreas Society	https://www.suizou.org/english/
Pancreas Club	https://www.pancreasclub.com
Pancreatic Society of Great Britain and Ireland (PSGBI)	https://www.psgbi.org/
Royal Society of Medicine	https://www.rsm.ac.uk/
Sociedad Española de Endocrinología y Nutrición (SEEN) [Spanish Society of Endocrinology and Nutrition]	https://www.seen.es/
Society for Endocrinology (UK)	https://www.endocrinology.org/
Ukrainian Society for Neuroscience (USN)	http://usn.org.ua/

This table lists the professional societies relevant to the study of hormones in neuroscience, the pancreas, related hormones and associated clinical conditions. See also Table 37.1.

Other resources

The Wellcome Collection (https://wellcomecollection.org/collections) and The British Library (www.bl.uk/) also hold material on topics related to Hormones in Neuroscience: The Pancreas and Pancreatic Hormones.

Other chapters on resources (recommended by authors and practitioners) may also be relevant to the study of the pancreas, related hormones and associated clinical conditions in neuroscience (Rajendram and Preedy, 2017, 2019; Rajendram et al., 2013a, 2013b,

TABLE 37.3 Emerging techniques, instruments, tools and platforms associated with hormones in neuroscience, the pancreas, related hormones and associated clinical conditions.

Organisation or company name	Technique/instrument/platform	Web address
10x Genomics	Single-cell RNA sequencing for pancreatic islets	https://www.10xgenomics.com/
Abbott (Freestyle Libre)	Continuous glucose monitoring (CGM) platforms	https://www.freestyle.abbott/
Addgene	Optogenetics tools	https://www.addgene.org/
Bio-Plex (Bio-Rad)	Multiplex hormone detection	https://www.bio-rad.com/en-us/category/bio-plex-multiplex-immunoassay-system?ID=85824182-db2a-4c6e-ac82-4e07dd9ab904
Biorep Technologies	Slet microperifusion system for hormone secretion studies	https://biorep.com/
Bogomolets National Medical University	Medical teaching and research	https://nmuofficial.com/
Bruker	NMR spectroscopy and associated instrumentation	https://www.bruker.com/en.html
Bruker	Mass spectrometry imaging (MSI)	https://www.bruker.com/
Carl Zeiss	Confocal microscope	https://www.zeiss.com/microscopy/en/products/light-microscopes/confocal-microscopes/airyscan.html
CMA Microdialysis AB	Microdialysis (continuous extracellular fluid sampling)	https://www.microdialysis.se/
CRISPR Therapeutics	CRISPR-Cas9 genome editing tools for beta-cell research	https://crisprtx.com/
Dexcom	Continuous glucose monitoring device	https://www.dexcom.com
EMBL-EBI Biomodels Repository	Model repository	https://www.ebi.ac.uk/biomodels/
Emulate Inc.	Pancreas-on-a-chip	https://emulatebio.com/
Glytec	A commercial system that analyses patient data (insulin, glucose, etc.) to provide personalized insulin dosing recommendations.	https://glytec.com
Hubrecht Organoid Technology (HUB)	Organoid and stem-cell derived pancreatic models	https://huborganoids.nl/
Illumina NovaSeq	Omics platforms	https://emea.illumina.com/systems/sequencing-platforms/novaseq.html
Kansai Electric Power Medical Research Institute	Institue associated with research	http://kepmri.org/
Leica Microsystems	Coherent Raman imaging (BCARS)	https://www.leica-microsystems.com/
McGill University, The Douglas Research Centre, Lundmer Centre	Expression-based polygenic score (ePRS)	https://github.com/SilveiraLab
Medtronic	Closed-loop/artificial pancreas systems	https://www.medtronic.com/
Mettler Toledo	Analytical platforms and various measuring devices	https://www.mt.com/
Nanomath	A commercial compartmental modeling and systems-analysis software for biological systems	https://www.nanomath.us/saam2

Nuclear Medicine Departments/Hospitals [move to other are as this is Wiki	Octreotide scan (somatostatin receptor scintigraphy)	https://en.wikipedia.org/wiki/Octreotide_scan
Olympus Corporation	Endoscopic ultrasound (EUS) with elastography	https://www.olympus-global.com/
Open source artificial pancreas system (OpenAPS)	Open and transparent effort to make artificial pancreas system technology widely available	https://openaps.org
Physiome Project Model Repository	Model repository	http://models.physiomeproject.org/
ProteinSimple (Bio-Techne)	Single-cell Western blot (scWestern)	https://www.proteinsimple.com/
STEMCELL Technologies	Pancreatic organoid culture system for functional studies	https://www.stemcell.com/
Thermo Fisher	Liquid Chromatographyâ€"Mass Spectrometry (LCâ€"MS)	https://www.thermofisher.com/
Thermo Fisher	Antibodies for immunohistochemistry and Northern blotting	https://www.thermofisher.com/
Thermo Orbitrap Fusion,	Mass spectrometry	https://tools.thermofisher.com/
Tidepool	Open-source and interoperable software tools to help collect, unify, visualize, and share diabetes device data	https://www.tidepool.org
University of Alberta IsletCore	Islet cell isolation and transplantation platforms	https://www.ualberta.ca/en/alberta-diabetes/core-services/isletcore.html

This table lists some emerging technologies and online resources relevant to hormones in neuroscience, the pancreas, related hormones and associated clinical conditions. Some companies are mentioned more than once and in some cases, we have merged suggestions into a single entry. The links were accurate at the time of going to press but may move or alter. In these cases, the use of the "Search" tabs should be explored at the parent address or site. In some cases, links direct the reader to relevant pages within parent sites. Some material suggested by authors is no longer available at specific sites, so we have inserted the parent URL.

TABLE 37.4 Other resources of interest or relevance for health care professionals or patients associated with the study of hormones in neuroscience, the pancreas, related hormones and clinical conditions.

Organisation or company name	Technique/instrument/platform	Web address
Alzheimer's Disease Neuroimaging Initiative (ADNI)	Biomarker and imaging database	http://adni.loni.usc.edu/
American Association of Clinical Endocrinology	Clinical guidance: diabetes	https://pro.aace.com/clinical-guidance/diabetes
American Diabetes Association (ADA)	Diabetes and pancreatic hormone research	https://diabetes.org/
American Diabetes Association (ADA)	Insulin pump introduction and options.	https://consumerguide.diabetes.org/collections/pumps
American Diabetes Association (ADA)	Insulin pen introduction and options	https://consumerguide.diabetes.org/collections/insulin-pens
American Diabetes Association (ADA)	Standards of Care in Diabetes	https://professional.diabetes.org/standards-of-care
American Diabetes Association (ADA)	Guidelines on diabetes management, pancreatic hormone therapies, and translational research	https://diabetesjournals.org/care
American Pancreatic Association	Clinical guidelines and research on pancreatic diseases	https://www.american-pancreatic-association.org/
American Pancreatic Association	Peer-reviewed publication dedicated to pancreatic research	https://journals.lww.com/pancreasjournal
American Pancreatic Association	Pancreapedia—exocrine pancreas knowledge base	https://www.pancreapedia.org/
Beta Cell Biology Consortium	Data for pancreatic islet research, tools, and protocols	https://bcbcnetdb.org
ClinicalTrials.gov	National Library of Medicine (NLM) material on clinical studies around the world	https://clinicaltrials.gov/
Diabetes UK	Information on HbA1c for the general public	https://www.diabetes.org.uk/about-diabetes/looking-after-diabetes/hba1c
Endocrine Society	Global endocrine research and clinical resources	https://www.endocrine.org/
EUROPAC	Rare pancreatic disorders and genetic research	https://www.europac-research.org/
European Medicine Agency EudraVigilance.	European Union database of adverse drug reaction (ADR) reports	https://www.adrreports.eu/
European Society of Endocrinology	European endocrine research and conferences	https://www.ese-hormones.org/
Google Scholar	Portal to access medical literature and their authors	https://scholar.google.com/
Guidelines for Working with Human Tissue and Patient Information (University College London; UCL)	Working with clinical samples. Printable PDF with guidance notes	chrome-extension://efaidnbmnnnibpcajpcglclefindmkaj/https://www.ucl.ac.uk/human-tissue/sites/human_tissue/files/1i_working_with_human_tissue_and_patient_information_-_ucl_guidelines_20231107v.4-a.pdf
Human Protein Atlas	Human Protein Atlas—pancreatic tissue expression profiles	https://www.proteinatlas.org/
Human Protein Atlas—Pancreas	GTEx Portal—tissue expression	https://www.proteinatlas.org/humanproteome/tissue/pancreas

Medicines and Healthcare products Regulatory Agency (MHRA) Yellow Card (UK pharmacovigilance)	Yellow Card (UK pharmacovigilance) for reporting adverse reactions.	https://yellowcard.mhra.gov.uk/
MitoTox	Comprehensive mitochondrial toxicity electronic database	https://www.mitotox.org/
National Institute for Health and Clinical Excellence (NICE)	Guidelines on health care (various areas such as diabetes and UK based)	https://www.nice.org.uk
National Institutes of Health (NIH) & Centers for Disease Control and Prevention (CDC)	National Diabetes Education Program (NDEP)—patient and professional resources	https://www.cdc.gov/diabetes-prevention/index.html
National Institutes of Health (NIH) & Chronic pancreatitis, diabetes and cancer (CPDPC) Consortium	Chronic pancreatitis, diabetes and cancer studies	https://cpdpc.mdanderson.org/
National Pancreas Foundation	Educational material, clinical resources, pancreatic disease research funding and patient support	https://pancreasfoundation.org/
National Pancreas Foundation	Clinical Trial Resource Center. Details on clinical studies such as eligibility, consent etc	https://pancreasfoundation.org/patient-resources/clinical-trial-resource-center/
Pancreatic Cancer Action Network (PanCAN)	Pancreatic Cancer Action Network (PanCAN)—research and advocacy	https://pancan.org/
Pancreatic Cancer Action Network (PanCAN)	Pancreatic cancer early detection and research	https://pancan.org/
T1D Exchange	T1D exchange—clinical research network and data-sharing platform	https://t1dexchange.org/
United States (U.S.) Centers for Disease Control and Prevention	Clinical guidance and resources for diabetes	https://www.cdc.gov/diabetes/hcp/clinical-guidance/index.html
World Health Organization	Quality of life questionnaire	https://www.who.int/tools/whoqolh
World Health Organization (WHO) Uppsala Monitoring Centre—VigiAccess	Global side effects of medical products	https://vigiaccess.org/

This table lists some other resources relevant to hormones in neuroscience, the pancreas, related hormones and associated clinical conditions. The links were accurate at the time of going to press but may move or alter. In these cases, the use of the "Search" tabs should be explored at the parent address or site. In some cases, links direct the reader to relevant pages within parent sites.

2017, 2019a, 2019b, 2022a, 2022b, 2023, 2020, 2025a, 2025b, 2025c, 2025d, 2025e, 2025f, 2026a, 2026b).

Acknowledgments

We thank the following authors for their contributions to the development of this resource. We apologise if some of the suggested material was not included in this chapter (for example for being expired sites, incomplete or flagged as potential security threats by university cybersecurity software) or has been moved to different tables. We are also grateful for the anonymous colleagues who provided useful information.

Aghoram R
Ajele KW
Alizadeh A
Carosi Arcangeli I
Chang X
De las Cuevas C
Deravi N
Djabir YY
Félix-Martínez GJ
Forouzanmehr B
Godugu C
Grover L
Idemudia ES
Khan A
Kim JH
Lukman HY
Macri S
Madide T
Murao N
Natrus L
Peter C
Rahmanian M
Sabiu S
Silveira P
Xue J
Yaribeygi H
Ziouzenkova O

References

Aslanoglou, D., Bertera, S., Sánchez-Soto, M., Free, R. B., Lee, J., Zong, W., Xue, X., Shrestha, S., Brissova, M., Logan, R., Wollheim, C., Trucco, M., Yechoor, V., Sibley, D., Bottino, R., & Freyberg, Z. (2021). Dopamine regulates pancreatic glucagon and insulin secretion via adrenergic and dopaminergic receptors. *Translational Psychiatry, 11*, 1171. https://doi.org/10.1038/s41398-020-01171-z

Brown, A., & Tzanakakis, E. (2023). Mathematical modeling clarifies the paracrine roles of insulin and glucagon on the glucose-stimulated hormonal secretion of pancreatic alpha- and beta-cells. *Frontiers in Endocrinology, 14*, 1212749. https://doi.org/10.3389/fendo.2023.1212749

Choi, J. H., & Kim, M.-S. (2022). Homeostatic regulation of glucose metabolism by the central nervous system. *Endocrinology and Metabolism, 37*(1), 13–24. https://doi.org/10.3803/EnM.20211364

Michailidis, M., Moraitou, D., Tata, D. A., Kalinderi, K., Papamitsou, T., & Papaliagkas, V. (2022). Alzheimer's disease as type 3 diabetes: Common pathophysiological mechanisms between Alzheimer's disease and type 2 diabetes. *International Journal of Molecular Sciences, 23*(5), 2687. https://doi.org/10.3390/ijms23052687

Rajendram, R., Gyamfi, D., Patel, V. B., & Preedy, V. R. (2022a). Recommended resources for biomarkers of nutrition. In V. R. Preedy, & V. B. Patel (Eds.), *Biomarkers of nutrition*. New York, USA: Elsevier.

Rajendram, R., Gyamfi, D., Patel, V. B., & Preedy, V. R. (2022b). Recommended resources for biomarkers in diabetes: Methods, discoveries, and applications. In V. B. Patel, & V. R. Preedy (Eds.), *Biomarkers in diabetes: Biomarkers in disease: Methods, discoveries and applications*. Berlin, Germany: Springer.

Rajendram, R., Gyamfi, D., Patel, V. B., & Preedy, V. R. (2023). Recommended resources on diabetes in relation to foods, plants, herbs and spices in human health. In R. Rajendram, V. B. Patel, & V. R. Preedy (Eds.), *Ancient and traditional foods, plants, herbs and spices used in diabetes*. Boca Raton, USA: CRC Press.

Rajendram, R., Patel, V. B., & Preedy, V. R. (2017). Recommended resources on maternal nutrition. In R. Rajendram, V. B. Patel, & V. R. Preedy (Eds.), *Nutrition and diet in maternal diabetes* (pp. 495–500). New York, USA: Springer.

Rajendram, R., Patel, V. B., & Preedy, V. R. (2019a). Resources in famine, starvation, and nutrient deprivation. In V. B. Patel, & V. R. Preedy (Eds.), *Famine, starvation, and nutrient deprivation* (pp. 2399–2406). New York, USA: Springer.

Rajendram, R., Patel, V. B., & Preedy, V. R. (2019b). Resources in diet, nutrition and epigenetics. In V. B. Patel, & V. R. Preedy (Eds.), *Nutrition and epigenetics* (pp. 2309–2314). New York, USA: Springer.

Rajendram, R., Patel, V. B., & Preedy, V. R. (2020). Recommended resources for nutrition, oxidative stress, and dietary antioxidants. In C. R. Martin, & V. R. Preedy (Eds.), *Nutrition, oxidative stress, and dietary antioxidants* (pp. 393–397). New York, USA: Elsevier.

Rajendram, R., Patel, V. B., & Preedy, V. R. (2025a). Recommended resources on protein intake in health and disease. In V. R. Preedy (Ed.), *Protein intake in health and disease*. Boca Raton: Taylor and Francis.

Rajendram, R., Patel, V. B., & Preedy, V. R. (2025b). Recommended resources on plant and animal proteins in health and disease prevention. In V. R. Preedy (Ed.), *Plant and animal proteins in health and disease prevention*. Boca Raton: Taylor and Francis.

Rajendram, R., Patel, V. B., & Preedy, V. R. (2025c). Recommended resources for the scientific basis of fatigue. In C. R. Martin, V. R. Preedy, V. B. Patel, & R. Rajendram (Eds.), *The scientific basis of fatigue* (pp. 585–594). Academic Press.

Rajendram, R., Patel, V. B., & Preedy, V. R. (2025d). Recommended resources for the behavior and psychology of disease. In V. R. Preedy (Ed.), *Handbook of the behavior and psychology of disease* (pp. 2927–2950). Springer.

Rajendram, R., Patel, V. B., & Preedy, V. R. (2025e). Recommended resources for the neuroscience of disease. In V. R. Preedy (Ed.), *Handbook of the neuroscience of disease* (pp. 1745–1760). Springer.

Rajendram, R., Patel, V. B., & Preedy, V. R. (2025f). Recommended resources for the biology and pathology of mental disorders. In C. R. Martin, V. R. Preedy, V. B. Patel, & R. Rajendram (Eds.),

Handbook of the biology and pathology of mental disorders (pp. 1–15). Springer Nature.

Rajendram, R., & Preedy, V. R. (2017). Recommended resources on biomarkers in alcoholic liver disease. In V. B. Patel, & V. R. Preedy (Eds.), *Biomarkers in liver disease* (pp. 821–836). Berlin, Germany: Springer.

Rajendram, R., & Preedy, V. R. (2019). Resources for the neuroscience of alcohol. In V. R. Preedy (Ed.), *Neuroscience of alcohol: Mechanisms and treatment* (pp. 853–862). Elsevier.

Rajendram, R., Rajendram, R., Patel, V. B., & Preedy, V. R. (2013a). Interlinking diet, nutrition, the menopause and recommended resources. In C. J. Hollins-Martin, R. R. Watson, & V. R. Preedy (Eds.), *Nutrition and diet in menopause*. Berlin, Germany: Springer.

Rajendram, R., Rajendram, R., Patel, V. B., & Preedy, V. R. (2013b). Diet quality: What more is there to know? In V. R. Preedy, L.-A. Hunter, & V. B. Patel (Eds.), *Diet quality: An evidence-based approach* (pp. 397–401). Berlin, Germany: Springer.

Rajendram, R., Siraj, M. A., Patel, V. B., & Preedy, V. R. (2026a). Recommended resources for the biological measures of well-being. In C. R. Martin, V. R. Preedy, V. B. Patel, & R. Rajendram (Eds.), *Biological measures of well-being*. New York, USA: Elsevier.

Rajendram, R., Siraj, M. A., Patel, V. B., & Preedy, V. R. (2026b). Recommended resources for hormones in neuroscience: The thyroid and thyroid hormones. In V. R. Preedy, V. B. Patel, & R. Rajendram (Eds.), *Hormones in neuroscience: The thyroid and thyroid hormones*. New York, USA: Elsevier.

Shaughness, M. C., Acs, D., Brabazon, F., Hockenbury, N., & Byrnes, K. (2020). Role of insulin in neurotrauma and neurodegeneration: A review. *Frontiers in Neuroscience, 14*, 547175. https://doi.org/10.3389/fnins.2020.547175

Wang, Q., Zhu, H., Deng, L., Xu, S., Xie, W., Li, M., Wang, R., Tie, L., Zhan, L., & Yu, G. (2025). Spatial transcriptomics: Biotechnologies, computational tools, and neuroscience applications. *Small Methods, 9*(4), 1107. https://doi.org/10.1002/smtd.202401107

Yan, C., Burley, G., Gao, H., & Shi, Y.-C. (2025). Emerging insights into brown adipose tissue crosstalk with pancreatic β-cells in metabolic regulation. *Endocrinology, 166*(3), bqaf118. https://doi.org/10.1210/endocr/bqaf118

Index

Note: 'Page numbers followed by *f* indicate figures and *t* indicate tables.'

B

C

D

E

H

K

L

M

N

Q

R

S

T

U

V

W

Z

Printed and bound by CPI Group (UK) Ltd, Croydon, CR0 4YY
14/09/2026
02216388-0009